Environment

Investment

Science

Sports

Statistics/Demographics

ELEMENTARY ALGEBRA

SECOND EDITION

Elementary Algebra

MARK DUGOPOLSKI

Southeastern Louisiana University

SECOND EDITION

ADDISON-WESLEY PUBLISHING COMPANY

Reading, Massachusetts • Menlo Park, California • New York • Don Mills, Ontario • Wokingham, England
Amsterdam • Bonn • Sydney • Singapore • Tokyo • Madrid • San Juan • Milan • Paris

Sponsoring Editor *Jason Jordan*
Managing Editor *Karen Guardino*
Art and Design Director *Karen Rappaport*
Development Editor *Elka Block*
Sr. Production Supervisor *Karen Wernholm*
Prepress Buying Manager *Sarah McCracken*
Technical Art Buyer *Joseph Vetere*
Marketing Manager *Michelle Babinec*
Manufacturing Manager *Roy Logan*
Text Designer *Margaret Ong Tsao*
Art Editor *Alena Konecny*
Technical Illustrator *Burmar Technical Corporation*
Situational Illustrator *Scientific Illustrators*
Photo Researcher *Susan Van Etten*
Compositor/Production Coordination *Burmar Technical Corporation*

Photo Credits
Page 1 Peter Menzel/Stock Boston; pages 8, 65, 125, 130, 196, 206, 268, 274, 324, 405, 409, © Susan Van Etten; page 165, Wide World Photos; page 359, courtesy, Alcan Aluminum Corporation; page 373, Bobby Lanier, Martin Marietta

Library of Congress Cataloging-in-Publication Data

Dugopolski, Mark
Elementary algebra/Mark Dugopolski.—2nd ed.
p. cm.
Includes index.
ISBN 0-201-59563-X
1. Algebra. I. Title.
QA152.2.D83 1996
512.9—dc20 95-13813
CIP

2 3 4 5 6 7 8 9 10-DOW-9897

To my wife and daughters,
Cheryl, Sarah, and Alisha

Contents

Chapter 3 POLYNOMIALS AND EXPONENTS 125

Chapter 4 FACTORING 165

Chapter 5 RATIONAL EXPRESSIONS 206

Chapter 9 QUADRATIC EQUATIONS AND FUNCTIONS 405

APPENDIXES A-1

ANSWERS TO SELECTED EXERCISES A-5

INDEX I-1

Preface

Elementary Algebra is designed to provide students with the algebra background needed for further algebra and college-level mathematics courses. The unifying theme of this text is the development of the skills necessary for solving equations and inequalities, followed by the application of those skills to solving applied problems. My primary goal in writing the second edition of *Elementary Algebra* has been to retain the features that made the first edition so successful, while incorporating more of the new standards in mathematics education and the comments and suggestions of first-edition users. I have endeavored to write a text that students can read, understand, and enjoy while they gain confidence in their ability to use mathematics.

Content Changes

This edition has increased topic coverage and depth (as described below), an increased emphasis on applications and graphing (throughout), increased coordination with the intermediate-level companion text, and increased attention to the gradation of difficulty of the examples and exercises.

- Within Chapter 1, Arithmetic and Algebra, there is new instruction on reading a graph, as explicit preparation for graph work. This enables students to work with real-data applications from the beginning of the course. In Chapter 2, Linear Equations and Inequalities, the previous single section on verbal problems has been expanded into two sections, 2.5 and 2.6 (Applications and More Applications), for more emphasis in this traditionally difficult area. The strategy for solving linear equations now is framed in language consistent with that of the intermediate level.
- Chapter 3, Polynomials and Exponents, now contains the topic of positive integral exponents, which has been moved to 3.6 from Chapter 6 of the first edition. (Negative integral exponents are discussed later.) In Chapter 4, Factoring, the *ac* method and the trial-and-error method are now explained more fully. Factoring the difference or sum of two cubes is newly added as a natural extension of the chapter topics.

- Rational Expressions, Chapter 5, has substantially expanded examples. For instance, Section 5.2 has 14 illustrative parts in 6 examples compared with 8 parts in 3 examples in the first edition.
- Two chapters (Linear Equations and Their Graphs and Systems of Equations and Inequalities) now appear as Chapters 6 and 7, moved from Chapters 8 and 9 in the first edition, in order to give students earlier experiences with equations and inequalities. Within the revised Chapter 6 are the following changes: a new discussion on graphing a line using intercepts, a new introduction of the vertical line test, an expanded discussion on slope (Section 6.2), and more graph reading. Chapter 7 contains expanded discussions on inconsistent and dependent systems and more applications of solving systems.
- In Chapter 8, Powers and Roots, scientific notation and the preparatory work on negative integral exponents have been brought forward to lead off the chapter, in answer to instructors' requests to teach scientific notation at the beginning of the chapter instead of at the end. Chapter 9, Quadratic Equations and Functions, now includes the topic of quadratic functions, their graphs, and applications, as a culmination of the chapter and book.
- More geometric applications and examples have been added throughout the book. This material not only demonstrates the relevance of the geometry and encourages skill development and practice, but also reviews basic geometric facts and provides visual applications of algebra, helping students make connections through spatial representations. A reference summary of geometric facts is found in Appendix A: Geometry Review.

In addition to these changes, the exercise sets have been carefully revised throughout. Particular care has been given to achieving an appropriate balance of problems that progressively increase in difficulty from routine, practice exercises in the beginning of the set to more challenging exercises at the end of the set. Also, more simple fractions and decimals are used in the exercises and throughout the text discussions to help reinforce the basic arithmetic skills that are necessary for success in algebra.

Features

- Each chapter now begins with a **Chapter Opener** that discusses a real application of algebra. The discussion is accompanied by a photograph and, in most cases, by a real-data application graph that helps students visualize algebra and more fully understand the underlying math. In addition, each chapter contains a **Math at Work** feature, which profiles a real person and the mathematics he or she uses on the job. These two features have corresponding real data exercises.
- An increased emphasis on real-data applications is a focus of the second edition. Applications have been added throughout the text to help demonstrate concepts, motivate students, and give students practice using new skills. The exercise sets include exercises that involve real data and real situations. Many of these new, applied exercises also require students to read and interpret graphs containing real data in order to solve the problem. An **Index of Applications** listing applications by subject matter is included at the front of the text.

- Every section of the second edition begins with **In This Section,** a list of topics that shows the student what will be covered. Since the topics correspond to the headings within each section, students will find it easy to locate and study specific concepts.
- Important ideas, such as definitions, rules, summaries, and strategies, are set apart in boxes for quick reference. Color is used to highlight these boxes as well as other important points in the text.
- A new **Calculator Close-up** feature, which is integrated throughout the text, gives instructions on the use of scientific and graphing calculators. Exercises requiring a scientific calculator are designated by a logo ( ). There are also groups of exercises specifically requiring graphing calculators ( ).
- Each section ends with **Warm-up** exercises, a set of ten simple statements that are to be answered true or false. These exercises are designed to provide a smooth transition between the ideas and the exercise sets, and they can also be used to stimulate discussion of the concepts presented in the section. Many of the false statements point out examples of common student errors, which can help students identify areas they may need to work on. The answers to all Warm-ups are given in the answer section at the end of the book.
- The end-of-section **Exercises** follow the same order as the textual material and contain exercises that are keyed to the examples, as well as numerous exercises that are not keyed to examples. This organization allows an instructor to cover only part of a section and easily determine which exercises are appropriate to assign. The keyed exercises give the student a place to start practicing and building confidence, whereas the nonkeyed exercises are designed to bring all of the chapter ideas together. New **Getting More Involved** exercises have been added to the end-of-section exercise sets to encourage critical thinking, writing, group discussion, exploration, and cooperative learning.
- At the end of each chapter there is a **Collaborative Activities** feature, which is designed to encourage interaction and learning in groups. Research has shown that students who study together do better than those who try to make it on their own. Students should read the discussion below for helpful information concerning the Collaborative Activities. Instructions and suggestions for using these activities and answers to all problems can be found in the Instructor's Solutions Manual.
- Each chapter ends with a four-part **Wrap-up,** which includes the following:

  The **Chapter Summary** lists important concepts along with brief illustrative examples.

  The **Review Exercises** contain problems that are keyed to the sections of the chapter as well as numerous miscellaneous exercises.

  The **Chapter Test** is designed to help the student assess his or her readiness for a test. To aid the student in working independently of the chapter examples and exercises, the Chapter Test has no keyed exercises.

  The **Tying It All Together** exercises are designed to help students "tie together," or synthesize, the new material with ideas from previous chapters and, in some cases, review material necessary for success in the upcoming chapter. Each Tying It All Together section includes at least one applied exercise that requires ideas from one or more of the previous chapters.

These features are highlighted in the pages immediately following this preface.

Collaborative Activities—Information to Students

One of the skills employers look for when hiring new employees is the ability to work with others. These collaborative activities are designed to give you opportunities to work with your fellow students. Take the time to read over these activities and, if your instructor does not assign them, work through them with a tutor or a group of fellow students. Some of these activities will be *open-ended*—there will be more than one correct answer; some will stretch your mathematical abilities and thinking; some will require different learning patterns; and all should help you think about mathematics differently.

After you have joined a group, it is important to work together effectively. Collaborative learning does not occur just because you are in a group. You will need to develop collaborative learning skills. One suggested way to work together effectively is to have assigned roles. Suggested roles are:

Moderator This person keeps the group on task, asks appropriate questions, and encourages everyone to participate.

Quality Manager This person makes sure the work and the finished activity is the best the group can produce.

Recorder This person keeps track of ideas and solutions during the group interaction.

Messenger This person interfaces with the group and the instructor by asking questions that arise in working through the activities.

You may find yourselves switching or combining roles during the time you work together. Eventually, you will fall into a natural rhythm of working together with the four roles suggested or with other appropriate roles.

When getting started on an activity, make sure each member of your group understands the activity before you begin working. The following list gives suggestions for getting started.

1. Brainstorm together ways to solve the problem. If you are stuck, don't be afraid to ask your instructor for help.
2. Try out the ideas you talked about while brainstorming. Find the one that best solves the problem.
3. Review and reread the activity as a group. Make sure you have actually solved the problem posed in the activity and decide whether you are finished.

You may find yourselves doing these steps all at the same time. The point is to manage your work in a productive and fair manner so you get the work done on time and no one member of the group is generating more effort than the others. Incorporating the assignment of roles and using the tips on how to work together in groups will help to make collaborative learning a reality. It may seem awkward at first, as new skills many times are, but with continued practice you will discover the benefits and rewards of working together.

Supplements for the Instructor

Instructor's Solutions Manual

Prepared by Mark Dugopolski, this supplement contains detailed worked solutions to all the exercises in the text. The solution processes match those shown in the text. Instructions and suggestions for using the Collaborative Activities feature in the text are also included in the Instructor's Solutions Manual.

Answer Book

The Answer Book contains answers to all exercises in the text. Instructors may make quick reference to all answers, or have quantities of these booklets made available for sale if they want students to have all the answers. Contact your local Addison-Wesley representative for details or ordering information.

Printed Test Bank

Prepared by Rebecca Muller, the Printed Test Bank is an extensive collection of alternate chapter test forms, including the following:

- Two alternate test forms for each chapter, with free-response questions modeled after the Chapter Tests in the text.
- One test form for each chapter, with multiple-choice questions modeled after the Chapter Tests in the text.
- Four alternate forms of the final examination—two in free-response form, and two in multiple-choice form.
- Answer keys for all alternate test forms and final examinations.

Computerized Testing: OmniTest³

Addison-Wesley's algorithm-driven computerized testing system for Macintosh and DOS computers features a brand new graphical user interface for the DOS version and a substantial increase in the number of test items available for each chapter of the text.

The new graphical user interface for DOS is a Windows look-alike. It allows users to choose items by test item number or by reviewing all the test items available for a specific text topic. Users can choose the exact iteration of the test item they wish to have on their test or allow the computer to generate iterations for them. Users can also preview all the items for a test on screen and make changes to them during the preview process. They can control the format of the test, including the appearance of the test header, the spacing between items, and the layout of the test and the answer sheet. In addition, users can now save the exact form of the test they have created so that they can modify it for later use. Users can also create their own items using OmniTest³'s WYSIWYG editor, and have access to a library of preloaded graphics.

Course Management and Testing System

InterAct Math Plus for Windows or Macintosh (available from Addison-Wesley) combines a powerful and flexible system for course management and on-line testing with the features of the basic tutorial software (see "Supplements for the Student") to create an invaluable teaching resource. Consult your Addison-Wesley representative for details.

Supplements for the Student

Student's Solutions Manual

Prepared by Mark Dugopolski, the Student's Solutions Manual contains complete worked-out solutions to all the odd-numbered exercises in the text. It may be purchased by your students from Addison-Wesley.

Math at Work Videotapes

The *Math at Work* videos are new to this edition of *Elementary Algebra.* They feature an easy-to-follow *MathCad* narration of worked examples, many of which are taken from the exposition of the text, along with motivational math applications that are tied to the text. Question-and-answer narratives between a student and an instructor are used to highlight important points and to help make the videotapes interactive and engaging. A complete set of *Math at Work* videotapes is free to qualifying adopters.

InterAct Math Tutorial Software

InterAct Math Tutorial Software, new to this edition of *Elementary Algebra,* has been developed and designed by professional software engineers working closely with a team of experienced developmental math teachers.

InterAct Math Tutorial Software includes exercises that are linked with the textbook and require the same computational and problem-solving skills as their companion exercises in the text. Each exercise has an example and an interactive guided solution that are designed to involve students in the solution process and to help them identify precisely where they are having trouble. In addition, the software recognizes common student errors and provides students with appropriate customized feedback.

Both free-response and multiple-choice items are contained in the tutorial. It also tracks student activity and scores for each section, which can then be printed out.

Available for both DOS-based and Macintosh computers, the software is free to qualifying adopters.

Acknowledgments

First of all I thank all of the students and professors who used the first edition of this text, for without their support there would not be a second edition. I sincerely appreciate the efforts of the reviewers who made many helpful suggestions for improving the first edition:

Barbara Armenta, *Pima Community College*

Carol Atnip, *University of Louisville*

Joan Bookbinder, *Johnson & Wales College*

Martin Bonsangue
Richard Butterworth, *Massasoit Community College*
Linda Crabtree, *Longview Community College*
Ginny Crisonino, *Union County College*
David Dudley, *Phoenix College*
Terry Fung, *Kean College of New Jersey*
Dale Hughes, *Johnson County Community College*
Joyce Huntington, *Walla Walla Community College*
Judy Kasabian, *El Camino Community College*
Joanne Kennedy, *Hunter College*
Mark Khebiel
Giles Wilson Maloof, *Boise State University*
Carolyn Meitler
Ronald Mesa, *Xavier University of Louisiana*
Julia Monte, *Daytona Beach Community College*
Doris Nice, *University of Wisconsin–Parkside*
Michelle Olsen, *College of the Redwoods*
Linda Padilla, *Joliet Junior College*
C. Phillips, *Pace University*
Ken Seydel, *Skyline College*
Brenda Wood, *Florida Community College*
Deborah Woods, *University of Cincinnati*
Caryn Yetter, *Slippery Rock University*

I also thank Rebecca Muller, Southeastern Louisiana University, for error checking and writing the printed test bank. I thank Robert Brown, Southeastern Louisiana University, for error checking the printed test bank. I thank Elaine W. Cohen, New Mexico State University and Visioneering Research Laboratory, and Vickie J. Aldrich, Dona Aña Branch Community College, for writing the Collaborative Activities. I thank Paula Maida and Bob Martin for error checking the manuscript.

I thank the staff at Addison-Wesley for all of their help and encouragement throughout the revision process, especially Jason Jordan, Greg Tobin, Elka Block, Bobbie Lewis, Helen Curtis, Karen Wernholm, Michelle Babinec, Kate Derrick, Barbara Willette, and Alena Konecny.

I also want to express my sincere appreciation to my wife, Cheryl, and my daughters, Sarah and Alisha, for their invaluable patience and support.

Hammond, Louisiana M.D.

Chapter Openers

Chapter openers use real-life examples of mathematical concepts discussed in the chapters. Several exercises within the chapter relate to the chapter opening example.

For a closer look at this feature see page 206.

CHAPTER 5

Rational Expressions

Advanced technical developments have made sports equipment faster, lighter, and more responsive to the human body. Behind the more flexible skis, lighter bats, and comfortable athletic shoes lies the science of biomechanics, which is the study of human movement and the factors that influence it.

Designing and testing an athletic shoe go hand in hand. While a shoe is being designed, it is tested in a multitude of ways, including long-term wear, rear foot stability, and strength of materials. Testing basketball shoes usually includes an evaluation of the force applied to the ground by the foot during running, jumping, and landing. Many biomechanics laboratories have a special platform that can measure the force exerted when a player cuts from side to side as well as the force against the bottom of the shoe. Force exerted in landing from a lay-up shot can be as high as 14 times the weight of the body. Side-to-side force is usually about 1 to 2 body weights in a cutting movement.

In Exercises 47 and 48 of Section 5.7 you will see how designers of athletic shoes use proportions to find the amount of force on the foot and soles of shoes for activities such as running and jumping.

...shoe, a wall stud, two ...l floor, before starting ...he total height of the ...

...dition:

$= 107\frac{19}{8}$

$= 109\frac{3}{8}$

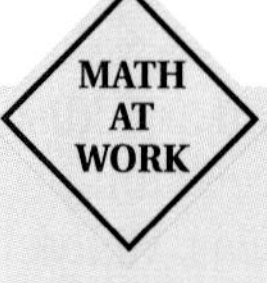

Figure 1.3

The total height of the framing shown is $109\frac{3}{8}$ in. ■

Math at Work

A motivational **Math at Work** feature in each chapter illustrates how people in various careers use the mathematics discussed in the chapter. Exercises relating to each Math at Work feature appear in exercise sets.

For a closer look at this feature see page 8.

MATH AT WORK

Building Contractor

Building a new house can be a complicated and daunting task. Shirley Zaborowski, project manager for Court Construction, is responsible for estimating, pricing, negotiating, subcontracting, and scheduling all portions of new house construction.

Ms. Zaborowski works from drawings and first does a "take off" or estimate for the quantity of material needed. The quantity of concrete is measured in cubic yards and the amount of wood is measured in board feet. If masonry is being used, it is measured in bricks or blocks per square foot.

Scheduling is another important part of the project manager's responsibility and it is based on the take off. Certain industry standards help Ms. Zaborowski estimate how many carpenters are needed and how much time it takes to frame the house and how many electricians and plumbers are needed to wire the house, install the heating systems, and put in the bathrooms. Of course, common sense says that the foundation is done before the framing and the roof. However, some rough plumbing and electrical work can be done simultaneously with the framing. Ideally the estimates of time and cost are accurate and the homeowner can move in on schedule.

In Exercise 96 of this section you will use operations with fractions to find the volume of concrete needed to construct a rectangular patio.

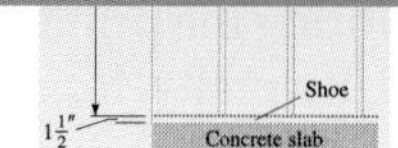

Calculator Close-up

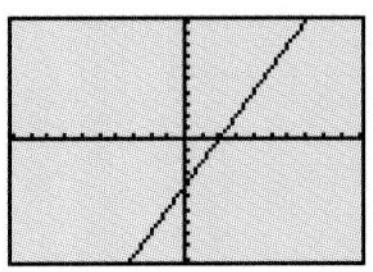

Figure 6.28

A graphing calculator (or a computer graphing utility) can find many ordered pairs that satisfy an equation and then plot them on its screen. Because a graphing calculator screen is made up of a finite number of pixels, a graphing calculator can plot only a finite number of ordered pairs, even if there are infinitely many ordered pairs that satisfy the equation. However, the graph shown on the graphing calculator can usually give you a good idea what the graph of the equation looks like.

To plot some ordered pairs that satisfy $y = 2x - 4$, enter the equation into the calculator using the Y= key and then press the GRAPH key. (Because there are many differences in how different brands of graphing calculators graph equations, you should consult your manual at this point.) Your calculator might show ordered pairs as in Fig. 6.28. This output is consistent with the fact that $y = 2x - 4$ is a linear equation and its graph is a straight line.

The viewing window used in Fig. 6.28 has x-values ranging from a minimum of −10 to a maximum of 10 and y-values ranging from a minimum of −10 to a maximum of 10, the **standard viewing window.** If the points satisfying $y = 2x - 4$ do not appear on your calculator, your viewing window might be set so that it does not contain any of the points. You can change the viewing window by pressing the RANGE key. Consult your manual. Exercises for graphing calculators are included in this section and in others for the remainder of the text.

Warm-ups

True or false? Explain your answer.

1. The formula $y = m(x - x_1)$ is the point-slope form for a line.
2. It is impossible to find the equation of a line through (2, 5) and (−3, 1).
3. The point-slope form will not work for the line through (3, 4) and (3, 6).
4. The equation of the line through the origin with slope 1 is $y = x$.
5. The slope of the line $5x + y = 4$ is 5.
6. The slope of any line perpendicular to the line $y = 4x - 3$ is $-\frac{1}{4}$.
7. The slope of any line parallel to the line $x + y = 1$ is −1.
8. The line $2x - y = -1$ goes through the point (−2, −3).
9. The lines $2x + y = 4$ and $y = -2x + 7$ are parallel.
10. The equation of the line through (0, 0) perpendicular to $y = x$ is $y = -x$.

6.4 EXERCISES

Write each equation in slope-intercept form. See Example 1.

1. $y - 1 = 5(x + 2)$
2. $y + 3 = -3(x - 6)$
3. $3x - 4y = 80$
4. $2x + 3y = 90$
5. $y - \frac{1}{2} = \frac{2}{3}\left(x - \frac{1}{4}\right)$
6. $y + \frac{2}{3} = -\frac{1}{2}\left(x - \frac{2}{5}\right)$

Find the equation of each line. Write each answer in slope-intercept form. See Example 1.

7. The line through (2, 3) with slope $\frac{1}{3}$
8. The line through (1, 4) with slope $\frac{1}{4}$

ny of
ever,
learn
blem

r dis-
prob-
mber

, and
that
t the

sum of the integers is 48.

$$x + (x + 1) + (x + 2) = 48$$

$3x + 3 = 48$ Combine like terms.
$3x = 45$ Subtract 3 from each side.
$x = 15$ Divide each side by 3.
$x + 1 = 16$ If x is 15, then $x + 1$ is 16 and $x + 2$ is 17.
$x + 2 = 17$

Because $15 + 16 + 17 = 48$, the three consecutive integers that have a sum of 48 are 15, 16, and 17.

General Strategy for Solving Verbal Problems

You should use the following steps as a guide for solving problems.

Strategy for Solving Problems

1. Read the problem as many times as necessary.
2. If possible, draw a diagram to illustrate the problem.
3. Choose a variable and *write* what it represents.
4. Write algebraic expressions for any other unknowns in terms of that variable.
5. Write an equation that describes the situation.
6. Solve the equation.
7. Check your possible answer in the original problem (not the equation).
8. Answer the original question.

Calculator Close-ups

Calculator Close-ups—optional boxes occurring throughout the book—give students guidance on using either a scientific or graphing calculator.

Warm-ups

Warm-ups are sets of ten true/false statements that precede exercise sets in each section and are designed to check student understanding of concepts.

For a closer look at these features see page 296.

Strategy Boxes

Strategy boxes teach students how to approach new problems and techniques with a step-by-step method; some boxes are particularly helpful in learning how to solve word problems.

For a closer look at this feature see page 95.

Geometry

Geometry is integrated in the presentation of algebra throughout the book. A **Geometry Review** appendix is in the back of the book.

Real Data

Many of the exercises in the book reference **real data** to help motivate and interest students. An **Index of Applications** is also included.

Application Graphs

Four-color **application graphs** based on real data appear throughout the text. These help students visualize and understand the mathematics.

For a closer look at these features see page 147.

Getting More Involved

Getting More Involved exercise sets present exploratory, writing, discussion, and collaborative learning exercises to promote critical thinking and conceptualization.

Calculator Exercises

Numerous **calculator exercises** in the exercise sets are identified with an icon.

For a closer look at these features see page 73.

3.4 Special Products **147**

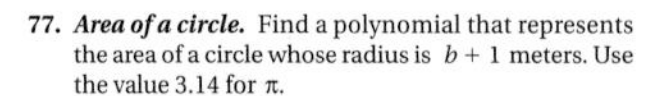

77. ***Area of a circle.*** Find a polynomial that represents the area of a circle whose radius is $b + 1$ meters. Use the value 3.14 for π.

78. ***Comparing dart boards.*** A toy store sells two sizes of circular dartboards. The larger of the two has a radius that is 3 inches greater than that of the other. The radius of the smaller dartboard is t inches. Find a polynomial that represents the difference in area between the two dartboards?

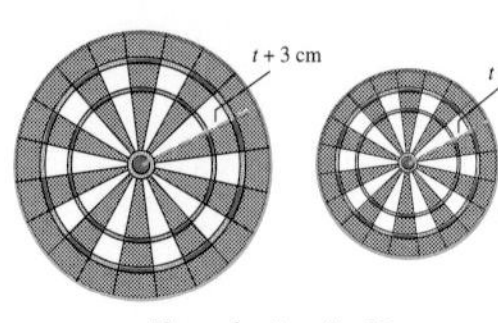

Figure for Exercise 78

79. ***Poiseuille's law.*** According to the nineteenth-century physician Poiseuille, the velocity (in centimeters per second) of blood r centimeters from the center of an artery of radius R centimeters is given by

$$v = k(R - r)(R + r),$$

where k is a constant. Rewrite the formula using a special product rule.

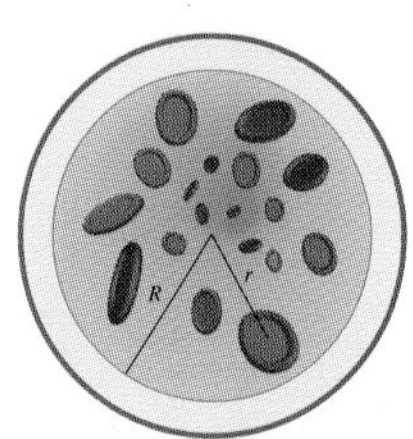

Figure for Exercise 79

80. ***Going in circles.*** A promoter is planning a circular race track with an inside radius of r feet and a width of w feet. The cost in dollars for paving the track is given by the formula

$$C = 1.2\pi[(r + w)^2 - r^2].$$

Use a special product rule to simplify this formula. What is the cost of paving the track if the inside radius is 1000 feet and the width of the track is 40 feet?

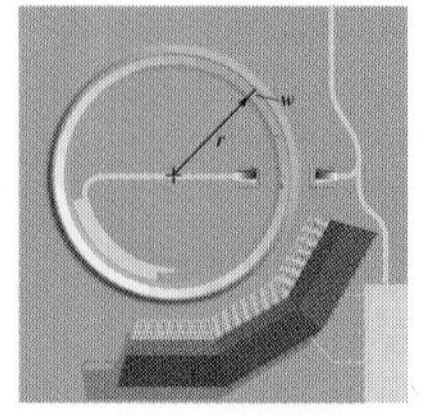

Figure for Exercise 80

81. ***Compounded annually.*** P dollars is invested at annual interest rate r for 2 years. If the interest is compounded annually, then the polynomial $P(1 + r)^2$ represents the value of the investment after 2 years. Rewrite this expression without parentheses. Evaluate the polynomial if $P = \$200$ and $r = 10\%$.

82. ***Compounded semiannually.*** P dollars is invested at annual interest rate r for 1 year. If the interest is compounded semiannually, then the polynomial $P\left(1 + \frac{r}{2}\right)^2$ represents the value of the investment after 1 year. Rewrite this expression without parentheses. Evaluate the polynomial if $P = \$200$ and $r = 10\%$.

83. ***Investing in treasury bills.*** An investment advisor uses the polynomial $P(1 + r)^{10}$ to predict the value in 10 years of a client's investment of P dollars with average annual return r. The accompanying graph shows historic average annual returns for various investments *(Stocks, Bonds, Bills, and Inflation: 1995 Yearbook)*. What is the predicted value in 10 years of an investment of \$10,000 in U.S. treasury bills?

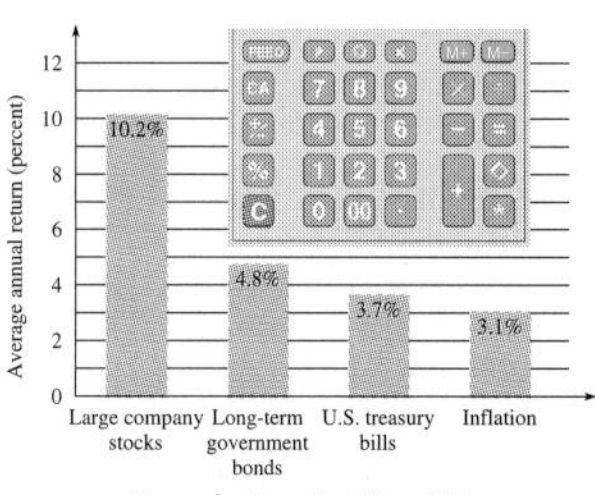

Figure for Exercises 83 and 84

ar Equations in One Variable **73**

ure for Exercise 89

triangle. If a triangle has sides of and $x + 2$ meters and a perimeter en the value of x can be found by $1) + (x + 2) = 12$. Find the values of 2.

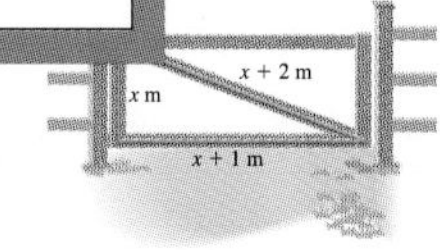

Figure for Exercise 90

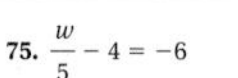

75. $\frac{w}{5} - 4 = -6$

76. $\frac{q}{2} + 13 = -22$

77. $\frac{2}{3}y - 5 = 7$

78. $\frac{3}{4}u - 9 = -6$

79. $4 - \frac{2n}{5} = 12$

80. $9 - \frac{2m}{7} = 19$

81. $-\frac{1}{3}p - \frac{1}{2} = \frac{1}{2}$

82. $-\frac{3}{4}z - \frac{2}{3} = \frac{1}{3}$

83. $3x - 2(x - 4) = 4 - (x - 5)$

84. $5 - 5(5 - x) = 6(x + 7) - 3$

85. $3.5x - 23.7 = -38.75$

86. $3(x - 0.87) - 2x = 4.98$

Solve each problem.

87. ***Celsius temperature.*** If the air temperature is 68° Fahrenheit, then the solution to the equation $\frac{9}{5}C + 32 = 68$ gives the Celsius temperature of the air. Find the Celsius temperature.

88. ***Fahrenheit temperature.*** If the temperature of hot tap water is 70°C, then the solution to the equation $70 = \frac{5}{9}(F - 32)$ gives the Fahrenheit temperature of the water. Find the Fahrenheit temperature of the water.

89. ***Rectangular patio.*** If a rectangular patio has a length that is 3 feet longer than its width and a perimeter of 42 feet, then the width can be found by solving the equation $2x + 2(x + 3) = 42$. What is the width?

Getting More Involved

91. ***Writing.*** If we multiply or divide each side of an equation by zero, do we get an equivalent equation?

92. ***Writing.*** If we add zero to each side of an equation, do we get an equivalent equation?

93. ***Discussion.*** Examine the following solution:

$$2(x - 3) = 8$$
$$2(x - 3) - 8 = 0$$
$$2x - 6 - 8 = 0$$
$$2x - 14 = 0$$
$$2x - 4 = 10$$
$$2x = 14$$
$$x = 7$$

Are all of the equations equivalent? How could you improve on the method used here?

94. ***Discussion.*** Examine the following solution:

$$3x - 4 = 5x - 6$$
$$2x - 4 = 6$$
$$2x = 2$$
$$x = 1$$

Does 1 satisfy the original equation. Are all of the equations equivalent?

200 Chapter 4 Factoring

COLLABORATIVE ACTIVITIES

The Puzzle Box

Grouping: 4 students
Topic: Multiplying and factoring polynomials

After graduating from college, you get a job working for a small business that makes jigsaw puzzles. You are on the team to design the cover of a puzzle box. Your design will use 525 square centimeters. The production manager has found a great deal on the price of cardboard. The cardboard is precut to 29 cm by 33 cm. The production manager tells your team that the depth of the box can vary to fit the needs of your design. He asks your team not only to come up with a design, but also to determine what the depth of the box should be so that production can fold it correctly to allow for your design.

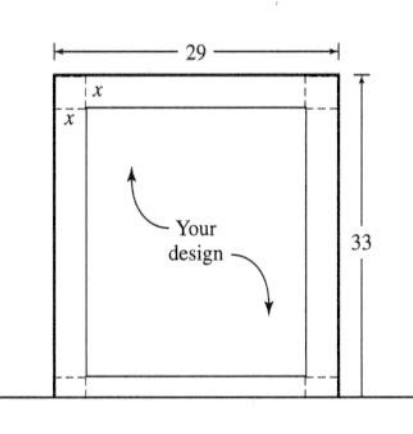

Part I: Finding the depth and dimensions

1. Write an equation using the given information and the diagram.
2. Solve the equation to find the depth of the box.
3. Use the depth of the box to find the dimensions for your design.

Part II: Presenting your final design. For shipping purposes, management has decided to make the boxes only 3 cm deep. This will change the dimensions of the box top. There isn't time for you to come up with a new design. Your team must find a way to incorporate your present design with its original dimensions onto the new box top. You will need to make an exact drawing in centimeters of your solution, using the new dimensions for the box top.

1. Using the same precut cardboard with the new depth, what will be the dimensions of the top of the box?
2. Make an *exact* drawing of your puzzle box top using the new dimensions.
3. Include the puzzle name in your drawing.

160 Chapter 3 Polynomials and Exponents

Wrap-up CHAPTER 3

SUMMARY

Polynomials		Examples
Term	A number or the product of a number and one or more variables raised to powers	$5x^3, -4x, 7$
Polynomial	A single term or a finite sum of terms	$2x^5 - 9x^2 + 11$
Degree of a polynomial	The highest degree of any of the terms	Degree of $2x - 9$ is 1. Degree of $5x^3 - x^2$ is 3.
Adding, Subtracting, and Multiplying Polynomials		**Examples**
Add or subtract polynomials	Add or subtract the like terms.	$(x + 1) + (x - 4) = 2x - 3$ $(x^2 - 3x) - (4x^2 - x)$ $= -3x^2 - 2x$
Multiply monomials	Use the product rule for exponents	$-2x^5 \cdot 6x^8 = -12x^{13}$
Multiply polynomials	Multiply each term of one polynomial by every term of the other polynomial, then combine like terms.	$x^2 + 2x + 5$ $x - 1$ $-x^2 - 2x - 5$ $x^3 + 2x^2 + 5x$ $x^3 + x^2 + 3x - 5$
Binomials		**Examples**
FOIL	A method for multiplying two binomials quickly	$(x - 2)(x + 3) = x^2 + x - 6$
Square of a sum	$(a + b)^2 = a^2 + 2ab + b^2$	$(x + 3)^2 = x^2 + 6x + 9$
Square of a difference	$(a - b)^2 = a^2 - 2ab + b^2$	$(m - 5)^2 = m^2 - 10m + 25$
Product of a sum and a difference	$(a - b)(a + b) = a^2 - b^2$	$(x + 2)(x - 2) = x^2 - 4$
Dividing Polynomials		**Examples**
Dividing monomials	Use the quotient rule for exponents	$8x^5 \div (2x^2) = 4x^3$
Divide a polynomial by a monomial	Divide each term of the polynomial by the monomial.	$\frac{3x^5 + 9x}{3x} = x^4 + 3$
Divide a polynomial by a binomial	If the divisor is a binomial, use long division. (divisor)(quotient) + (remainder) = dividend	$x - 7$ ← Quotient Divisor → $x + 2\overline{)x^2 - 5x - 4}$ ← Dividend $x^2 + 2x$ $-7x - 4$ $-7x - 14$ 10 ← Remainder

Collaborative Activities

Collaborative Activities provide opportunity for group work in each chapter and stimulate critical thinking, student collaboration, and team problem-solving.

For a closer look at this feature see page 200.

Wrap-ups

Each chapter in the book has a four-part **Wrap-up.**

- Chapter Summary
- Chapter Review
- Chapter Test
- Tying It All Together

Chapter Summary

The comprehensive **Chapter Summary** lists key chapter concepts and formulas, along with examples, for convenient review.

For a closer look at these features see page 160.

Review Exercises

Chapter **Review Exercises** are grouped and keyed to each section of the chapter to assist students with directed review and practice.

For a closer look at this feature see page 120.

Miscellaneous Exercises

Miscellaneous exercises in the Review Exercises are designed to test the student's ability to synthesize various concepts.

For a closer look at this feature see page 122.

REVIEW EXERCISES

2.1 *Solve each equation.*

1. $2x - 5 = 9$
2. $5x - 8 = 38$
3. $3r - 7 = 0$
4. $3t + 5 = 0$
5. $3 - 4y = 11$
6. $4 - 3k = -8$
7. $2(h - 7) = -14$
8. $2(t - 7) = 0$
9. $3(w - 5) = 6(w + 2) - 3$
10. $2(a - 4) + 4 = 5(9 - a)$
11. $-\frac{2}{3}b = 20$
12. $\frac{3}{4}b = -6$
13. $\frac{1}{3}x = \frac{1}{7}$
14. $\frac{x}{2} = -\frac{2}{5}$
15. $0.24c + 1 = 97$
16. $1.05q = 420$
17. $p - 0.1p = 90$
18. $m + 0.05m = 2.1$

2.2 *Solve each equation. Identify each equation as a conditional equation, an inconsistent equation, or an identity.*

19. $2(x - 7) - 5 = 5 - (3 - 2x)$
20. $2(x - 7) + 5 = -(9 - 2x)$
21. $2(w - w) = 0$
22. $2y - y = 0$
23. $\frac{3r}{3r} = 1$
24. $\frac{3t}{3} = 1$
25. $\frac{1}{\ }a - 5 = \frac{1}{\ }a - 1$
26. $\frac{1}{2}b - \frac{1}{2} = \frac{1}{4}b$
27. $0.06q + 14 = 0.3q$ …
28. $0.05(z + $ …
30. $0.06x + 0.08(x + 1)$ …

Solve each equation.

31. $2x + \frac{1}{2} = 3x + \frac{1}{4}$
33. $\frac{x}{2} - \frac{3}{4} = \frac{x}{6} + \frac{1}{8}$
35. $\frac{5}{6}x = -\frac{2}{3}$
37. $-\frac{1}{2}(x - 10) = \frac{3}{4}x$
39. $3 - 4(x - 1) + 6 =$ …
40. $6 - 5(1 - 2x) + 3 = -3(1 - 2x) - 1$
41. $5 - 0.1(x - 30) = 18 + 0.05(x + 100)$
42. $0.6(x - 50) = 18 - 0.3(40 - 10x)$

2.3 *Solve each equation for x.*

43. $ax + b = 0$
44. $mx + e = t$
45. $ax - 2 = b$
46. $b = 5 - x$
47. $LWx = V$
48. $3xy = 6$
49. $2x - b = 5x$
50. $t - 5x = 4x$

Solve each equation for y. Write the answer in the form $y = mx + b$, where m and b are real numbers.

51. $5x + 2y = 6$
52. $5x - 3y + 9 = 0$
53. $y - 1 = -\frac{1}{2}(x - 6)$
54. $y + 6 = \frac{1}{2}(x + 8)$
55. $\frac{1}{2}x + \frac{1}{4}y = 4$
56. $-\frac{x}{3} + \frac{y}{2} = 1$

Find the value of y in each formula if $x = -3$.

57. $y = 3x - 4$
58. $2x - 3y = -7$
59. $5xy = 6$
60. $3xy - 2x = -12$
61. $y - 3 = -2(x - 4)$
62. $y + 1 = 2(x - 5)$

2.4 *Translate each verbal expression into an algebraic expression.*

63. The sum of a number and 9

Miscellaneous

Use an equation, inequality, or formula to solve each problem.

111. ***Flat yield curve.*** The accompanying graph shows that the *yield curve* for U.S. Treasury bonds was relatively flat from 2 years out to 30 years as of February 16, 1995 (*Fidelity Investments Special Report*, Volume 1, Number 6). In this situation there is not much benefit to be obtained from long-term investing. Use the interest rate in the graph to find the amount of interest earned in the first year on a 2-year bond of \$10,000. How much more interest would you earn in the first year on a 30-year bond of \$10,000?

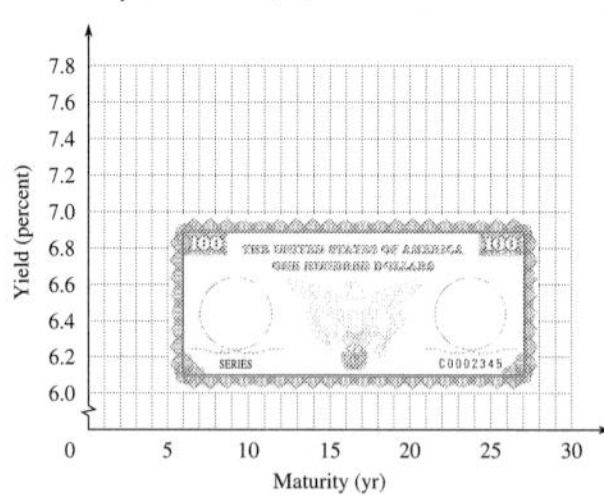

Figure for Exercises 111 and 112

112. ***Reading the curve.*** Use the accompanying graph to find the maturity of a U.S. Treasury bond that had a yield of 7.4% on February 16, 1995.

113. ***Combined videos.*** The owners of ABC Video discovered that they had no movies in common with XYZ Video and bought XYZ's entire stock. Although XYZ had 200 titles, they had no children's movies, while 60% of ABC's titles were children's movies. If 40% of the movies in the combined stock are children's movies, then how many movies did ABC have before the merger?

114. ***Living comfortably.*** Gary has figured that he needs to take home \$30,400 a year to live comfortably. If the government takes 24% of Gary's income, then what must his income be for him to live comfortably?

115. ***Bracing a gate.*** The diagonal brace on a rectangular gate forms an angle with the horizontal side with degree measure x and an angle with the vertical side with degree measure $2x - 3$. Find x.

116. ***Digging up the street.*** A contractor wants to install a pipeline connecting point *A* with point *C* on opposite sides of a road as shown in the figure. To save money, the contractor has decided to lay the pipe to point *B* and then under the road to point *C*. Find the measure of the angle marked *x* in the figure.

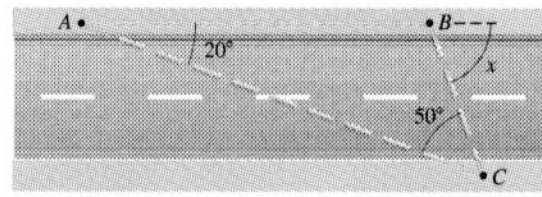

Figure for Exercise 116

117. ***Perimeter of a triangle.*** One side of a triangle is 1 foot longer than the shortest side, and the third side is twice as long as the shortest side. If the perimeter is less than 25 feet, then what is the range of the length of the shortest side?

118. ***Restricted hours.*** Alana makes \$5.80 per hour working in the library. To keep her job, she must make at least \$116 per week; but to keep her scholarship, she must not earn more than \$145 per week. What is the range of the number of hours per week that she may work?

CHAPTER 2 TEST

Solve each equation.

1. $-10x - 6 + 4x = -4x + 8$
2. $5(2x - 3) = x + 3$
3. $-\frac{2}{3}x + 1 = 7$
4. $x + 0.06x = 742$

Solve for the indicated variable.

5. $2x - 3y = 9$ for y
6. $m = aP - w$ for a

Write an inequality that describes the graph.

7. −5 −4 −3 −2 −1 0 1 2 3 4 5
8. −2 −1 0 1 2 3 4 5 6 7 8

Solve and graph each inequality.

9. $4 - 3(w - 5) < -2w$
10. $1 < \frac{1 - 2x}{3} < 5$
11. $1 < 3x - 2 < 7$
12. $-\frac{2}{3}y < 4$

Solve each equation.

13. $2(x + 6) = 2x - 5$
14. $x + 7x = 8x$
15. $x - 0.03x = 0.97$
16. $6x - 7 = 0$

Write a complete solution to each problem.

17. The perimeter of a rectangle is 72 meters. If the width is 8 meters less than the length, then what is the width of the rectangle?
18. If the area of a triangle is 54 square inches and the base is 12 inches, then what is the height?
19. How many liters of a 20% alcohol solution should Maria mix with 50 liters of a 60% alcohol solution to obtain a 30% solution?
20. Brandon gets a 40% discount on loose diamonds where he works. The cost of the setting is \$250. If he plans to spend at most \$1450, then what is the price range (list price) of the diamonds that he can afford?
21. If the degree measure of the smallest angle of a triangle is one-half of the degree measure of the second largest angle and one-third of the degree measure of the largest angle, then what is the degree measure of each angle?

Chapter Tests

The **Chapter Tests** are carefully developed to help students check their readiness for tests.

For a closer look at this feature see page 123.

Tying It All Together CHAPTERS 1–6

Simplify each expression.

1. $3^2 - 2^3$
2. $3^2 \cdot 2^3$
3. $10^4 \cdot 10^9$
4. $2^{12} \div 2^{10}$
5. $(34 \cdot 258)^0$
6. $(8^0 - 3^2)^3$
7. $\left(\frac{1}{2}\right)^3 + \left(\frac{2}{3}\right)^2$
8. $\left(-\frac{3}{2}\right)^3 - \left(-\frac{3}{4}\right)^2$
9. $\left(\frac{3}{5}\right)^3 \cdot \left(\frac{5}{6}\right)^2$
10. $\left(-\frac{3}{5}\right)^3 \div \left(-\frac{6}{5}\right)^4$
11. $\dfrac{\frac{1}{4} - \frac{1}{8}}{\frac{3}{4} + \frac{1}{2}}$
12. $\dfrac{\frac{1}{3} - \frac{1}{5}}{\frac{3}{10} + \frac{1}{20}}$

Perform the indicated operations.

13. $-3(2x - 7)$
14. $x - 3(2x - 7)$
15. $(x - 3)(2x - 7)$
16. $(2x - 1)^2$
17. $(z + 5)^2$
18. $(w - 7)(w + 7)$
19. $3x^2y^3 \cdot 12xy^4$
20. $(2x^2y)^3 \cdot 5x^2y^6$

Sketch a graph of each equation.

21. $y = \frac{1}{3}x$
22. $y = 3x$
23. $y = -3x$
24. $y = -\frac{1}{3}x$
25. $y = 3x + 1$
26. $y = 3x - 2$
27. $y = 3$
28. $x = 3$

Solve each equation for y.

29. $3\pi y + 2 = t$
30. $x = \frac{y - b}{m}$
31. $3x - 3y - 12 = 0$
32. $2y - 3 = 9$
33. $y^2 - 3y - 40 = 0$
34. $\frac{y}{2} - \frac{y}{4} = \frac{1}{5}$

Solve each equation.

35. $5 = 4x - 7$
36. $5 = 4x^2 - 11$
37. $(3x - 4)(x + 9) = 0$
38. $\frac{2}{3} - \frac{x}{6} = \frac{1}{2} + \frac{x}{4}$
39. $2x^2 - 7x = 0$
40. $\frac{3}{x} = \frac{x - 1}{2}$

Solve.

41. ***Financial planning.*** Financial advisors at Fidelity Investments use the information in the accompanying graph as a guide for retirement investing.
    a) What is the slope of the line segment for ages 35 through 50?
    b) What is the slope of the line segment for ages 50 through 65?
    c) If a 38-year-old man is making \$40,000 per year, then what percent of his income should he be saving?
    d) If a 58-year-old woman has an annual salary of \$60,000, then how much should she have saved and how much should she be saving per year?

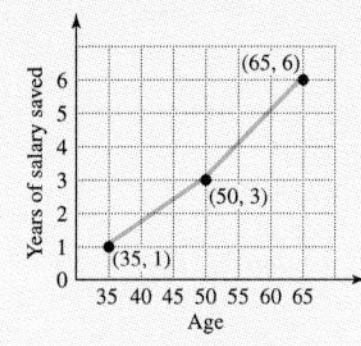

Figure for Exercise 41

Tying It All Together

The **Tying It All Together** feature at the end of each chapter serves as a cumulative review for all of the previous chapters of the text. At least one problem is a word problem to give students important and continual practice with problem solving.

For a closer look at this feature see page 323.

CHAPTER 1

Arithmetic and Algebra

It has been said that baseball is the "great American pastime." All of us who have played the game or who have only been spectators believe we understand the game. But do we realize that a pitcher must aim for an invisible three-dimensional target that is about 20 inches wide by 23 inches high by 17 inches deep and that a pitcher must throw so that the batter has difficulty hitting the ball? A curve ball may deflect 14 inches to skim over the outside corner of the plate, or a knuckle ball can break 11 inches off center when it is 20 feet from the plate and then curve back over the center of the plate.

The batter is trying to hit a rotating ball that can travel up to 120 miles per hour and must make split-second decisions about shifting his weight, changing his stride, and swinging the bat. The size of the bat each batter uses depends on his strengths, and pitchers in turn try to capitalize on a batter's weaknesses.

Millions of baseball fans enjoy watching this game of strategy and numbers. Many watch their favorite teams at the local ball parks, while others cheer for the home team on television. Of course, baseball fans are always interested in which team is leading the division and the number of games that their favorite team is behind the leader. Finding the number of games behind for each team in the division involves both arithmetic and algebra. Algebra provides the formula for finding games behind, and arithmetic is used to do the computations. In Exercise 87 of Section 1.6 we will find the number of games behind for each team in the American League West.

1.1 Fractions

In this section:

- Equivalent Fractions
- Multiplying Fractions
- Dividing Fractions
- Adding and Subtracting Fractions
- Fractions, Decimals, and Percents
- Applications

Algebra is an extension of arithmetic. Many of the ideas we encounter in algebra are based on the basic operations with fractions. In this section we review the operations with fractions that we will use in algebra.

Equivalent Fractions

In a fraction the number above the fraction bar is called the **numerator,** and the number below the fraction bar is called the **denominator:**

$$\frac{3}{7} \quad \begin{matrix} \leftarrow \text{Numerator} \\ \leftarrow \text{Denominator} \end{matrix}$$

Two fractions are **equivalent** if they represent the same value. If we multiply the numerator and denominator of a fraction by the same nonzero number, then we obtain an equivalent fraction. Every fraction can be written in infinitely many equivalent forms. Consider the following equivalent forms of the fraction $\frac{2}{3}$:

$$\frac{2}{3} = \frac{4}{6} = \frac{6}{9} = \frac{8}{12} = \frac{10}{15} = \underbrace{\cdots}$$

The three dots mean "and so on."

To add fractions, it may be necessary to convert each fraction to an equivalent fraction with a larger denominator. This is called **building up** the fraction. For example, we can build up $\frac{2}{3}$ to $\frac{10}{15}$ by multiplying the numerator and denominator of $\frac{2}{3}$ by 5:

$$\frac{2}{3} = \frac{2 \cdot 5}{3 \cdot 5} = \frac{10}{15} \quad \text{The raised dot indicates multiplication.}$$

We can state a general rule for building up fractions. When we state rules, we use letters to represent numbers. A letter used to represent some numbers is called a **variable.** Variables are used extensively in algebra.

Building Up Fractions

If $b \neq 0$ and $c \neq 0$, then

$$\frac{a}{b} = \frac{a \cdot c}{b \cdot c}.$$

EXAMPLE 1 **Building up fractions**

Build up each fraction so that it is equivalent to the fraction with the indicated denominator.

a) $\frac{3}{4} = \frac{?}{28}$ **b)** $\frac{5}{3} = \frac{?}{30}$

Solution

a) Because $4 \cdot 7 = 28$, we multiply both the numerator and denominator by 7:

$$\frac{3}{4} = \frac{3 \cdot 7}{4 \cdot 7} = \frac{21}{28}$$

b) Because $3 \cdot 10 = 30$, we multiply both the numerator and denominator by 10:

$$\frac{5}{3} = \frac{5 \cdot 10}{3 \cdot 10} = \frac{50}{30}$$

Converting a fraction to an equivalent fraction with a smaller denominator is called **reducing** the fraction. For example, to reduce $\frac{10}{15}$, we *factor* 10 as $2 \cdot 5$ and 15 as $3 \cdot 5$, and then divide out the *common factor* 5:

$$\frac{10}{15} = \frac{2 \cdot \cancel{5}}{3 \cdot \cancel{5}} = \frac{2}{3}$$

The fraction $\frac{2}{3}$ cannot be reduced further because the numerator 2 and the denominator 3 have no factors (other than 1) in common. So we say that $\frac{2}{3}$ is in **lowest terms.**

Reducing Fractions

If $b \neq 0$ and $c \neq 0$, then

$$\frac{a \cdot c}{b \cdot c} = \frac{a}{b}.$$

EXAMPLE 2 **Reducing fractions**

Reduce each fraction to lowest terms.

a) $\frac{15}{24}$

b) $\frac{42}{30}$

Solution

For each fraction, factor the numerator and denominator and then divide by the common factor:

a) $\frac{15}{24} = \frac{\cancel{3} \cdot 5}{\cancel{3} \cdot 8} = \frac{5}{8}$

b) $\frac{42}{30} = \frac{7 \cdot \cancel{6}}{5 \cdot \cancel{6}} = \frac{7}{5}$

Strategy for Obtaining Equivalent Fractions

Equivalent fractions can be obtained by multiplying or dividing the numerator and denominator by the same nonzero number.

Multiplying Fractions

To multiply two fractions, we multiply their numerators and multiply their denominators.

EXAMPLE 3 **Multiplying fractions**

Find the product, $\frac{2}{3} \cdot \frac{5}{8}$.

Solution

Multiply the numerators and the denominators:

$$\frac{2}{3} \cdot \frac{5}{8} = \frac{10}{24}$$

$$= \frac{\not{2} \cdot 5}{\not{2} \cdot 12} \quad \text{Factor the numerator and denominator.}$$

$$= \frac{5}{12} \quad \text{Divide out the common factor 2.}$$

In general, we have the following definition.

Multiplication of Fractions

If $b \neq 0$ and $d \neq 0$, then

$$\frac{a}{b} \cdot \frac{c}{d} = \frac{a \cdot c}{b \cdot d}.$$

It is usually easier to reduce before multiplying, as shown in the next example.

EXAMPLE 4 **Reducing before multiplying**

Find the indicated products.

a) $\frac{1}{3} \cdot \frac{3}{4}$ **b)** $\frac{4}{5} \cdot \frac{15}{22}$ **c)** $3\frac{1}{4} \cdot \frac{8}{5}$

Solution

a) $\frac{1}{3} \cdot \frac{3}{4} = \frac{1}{\not{3}} \cdot \frac{\not{3}}{4} = \frac{1}{4}$

b) Factor the numerators and denominators, and then divide out the common factors before multiplying:

$$\frac{4}{5} \cdot \frac{15}{22} = \frac{2 \cdot \not{2}}{\not{5}} \cdot \frac{3 \cdot \not{5}}{\not{2} \cdot 11} = \frac{6}{11}$$

c) $3\frac{1}{4} \cdot \frac{8}{5} = \frac{13}{4} \cdot \frac{8}{5}$ Write $3\frac{1}{4}$ as an improper fraction.

$$= \frac{13}{\not{4}} \cdot \frac{2 \cdot \not{4}}{5} \quad \text{Factor.}$$

$$= \frac{26}{5} \quad \text{Divide out the common factor 4.}$$

Dividing Fractions

If $a \div b = c$, then b is called the **divisor** and c is called the **quotient** of a and b. We also refer to both $a \div b$ and $\frac{a}{b}$ as the quotient of a and b. To find the quotient for two fractions, we invert the divisor and multiply.

EXAMPLE 5 **Dividing fractions**

Find the indicated quotients.

a) $\frac{1}{3} \div \frac{7}{6}$ **b)** $\frac{2}{3} \div 5$

Solution

In each case we invert the divisor (the number on the right) and multiply.

a) $\frac{1}{3} \div \frac{7}{6} = \frac{1}{3} \cdot \frac{6}{7}$ Invert the divisor.

$= \frac{1}{\cancel{3}} \cdot \frac{2 \cdot \cancel{3}}{7}$ Reduce.

$= \frac{2}{7}$ Multiply.

b) $\frac{2}{3} \div 5 = \frac{2}{3} \div \frac{5}{1} = \frac{2}{3} \cdot \frac{1}{5} = \frac{2}{15}$ ■

We can state the definition for dividing fractions as follows.

Division of Fractions

If $b \neq 0$, $c \neq 0$, and $d \neq 0$, then

$$\frac{a}{b} \div \frac{c}{d} = \frac{a}{b} \cdot \frac{d}{c}.$$

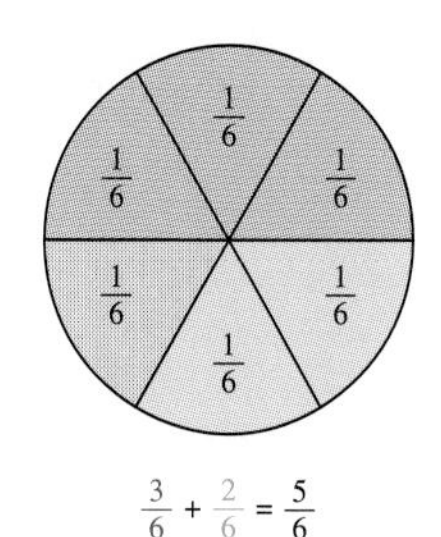

Figure 1.1

Adding and Subtracting Fractions

To add or subtract fractions with identical denominators, we add or subtract their numerators and write the result over the common denominator. See Fig. 1.1.

EXAMPLE 6 **Adding and subtracting fractions**

Perform the indicated operations.

a) $\frac{1}{7} + \frac{2}{7}$ **b)** $\frac{7}{10} - \frac{3}{10}$

Solution

a) $\frac{1}{7} + \frac{2}{7} = \frac{3}{7}$

b) $\frac{7}{10} - \frac{3}{10} = \frac{4}{10} = \frac{\cancel{2} \cdot 2}{\cancel{2} \cdot 5} = \frac{2}{5}$ ■

The definition for adding and subtracting fractions is stated as follows.

Addition and Subtraction of Fractions

If $b \neq 0$, then

$$\frac{a}{b} + \frac{c}{b} = \frac{a+c}{b} \quad \text{and} \quad \frac{a}{b} - \frac{c}{b} = \frac{a-c}{b}.$$

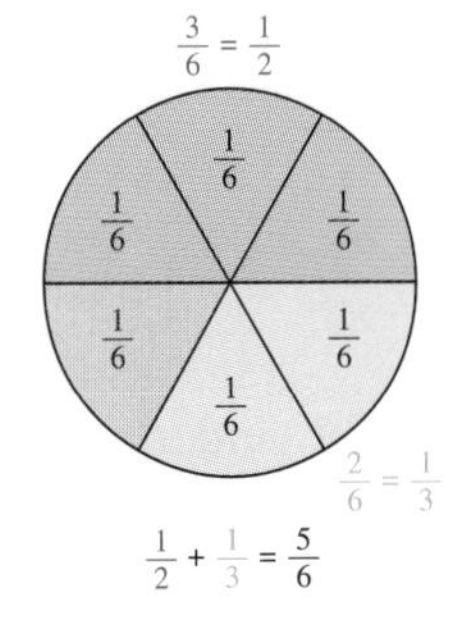

Figure 1.2

If the fractions have different denominators, we must convert them to equivalent fractions with the same denominator and then add or subtract. For example, to add the fractions $\frac{1}{2}$ and $\frac{1}{3}$, we build up each fraction to get a denominator of 6. See Fig. 1.2. The denominator 6 is the smallest number that is a multiple of both 2 and 3. For this reason, 6 is called the **least common denominator (LCD).**

EXAMPLE 7 **Adding fractions**

Perform the indicated operations.

a) $\frac{1}{2} + \frac{1}{3}$ **b)** $\frac{1}{3} - \frac{1}{12}$ **c)** $\frac{3}{4} - \frac{1}{6}$ **d)** $2\frac{1}{3} + \frac{5}{9}$

Solution

a) $\frac{1}{2} + \frac{1}{3} = \frac{1 \cdot 3}{2 \cdot 3} + \frac{1 \cdot 2}{3 \cdot 2}$ The LCD is 6.

$= \frac{3}{6} + \frac{2}{6}$ Build each denominator to a denominator of 6.

$= \frac{5}{6}$ Then add.

b) The LCD for the denominators 3 and 12 is 12. In this case we change the denominator of only one of the fractions:

$\frac{1}{3} - \frac{1}{12} = \frac{1 \cdot 4}{3 \cdot 4} - \frac{1}{12}$ The LCD is 12.

$= \frac{4}{12} - \frac{1}{12}$ Build up $\frac{1}{3}$ to get a denominator of 12.

$= \frac{3}{12}$ Subtract.

$= \frac{1}{4}$ Reduce to lowest terms.

c) The smallest number that is a multiple of both 4 and 6 is 12.

$\frac{3}{4} - \frac{1}{6} = \frac{3 \cdot 3}{4 \cdot 3} - \frac{1 \cdot 2}{6 \cdot 2}$ The LCD is 12.

$= \frac{9}{12} - \frac{2}{12}$ Build up each fraction to a denominator of 12.

$= \frac{7}{12}$

d) $2\frac{1}{3} + \frac{5}{9} = \frac{7}{3} + \frac{5}{9}$ Write $2\frac{1}{3}$ as an improper fraction.

$= \frac{7 \cdot 3}{3 \cdot 3} + \frac{5}{9}$ The LCD is 9.

$= \frac{21}{9} + \frac{5}{9} = \frac{26}{9}$

Fractions, Decimals, and Percents

In the decimal number system, fractions with a denominator of 10, 100, 1000, and so on are written as decimal numbers. For example,

$$\frac{3}{10} = 0.3, \quad \frac{25}{100} = 0.25, \quad \text{and} \quad \frac{5}{1000} = 0.005.$$

Fractions with a denominator of 100 are often written as percents. Think of the percent symbol (%) as representing the denominator of 100. For example,

$$\frac{25}{100} = 25\%, \quad \frac{5}{100} = 5\%, \quad \text{and} \quad \frac{300}{100} = 300\%.$$

The next example illustrates further how to convert from any one of the forms (fraction, decimal, percent) to the others.

EXAMPLE 8 **Changing forms**

Convert each given fraction, decimal, or percent into its other two forms.

a) $\frac{1}{5}$ **b)** 6% **c)** 0.1

Solution

a) $\frac{1}{5} = \frac{1 \cdot 20}{5 \cdot 20} = \frac{20}{100} = 20\%$ and $\frac{1}{5} = \frac{1 \cdot 2}{5 \cdot 2} = \frac{2}{10} = 0.2$

So $\frac{1}{5} = 0.2 = 20\%$.

b) $6\% = \frac{6}{100} = 0.06$ and $\frac{6}{100} = \frac{\cancel{2} \cdot 3}{\cancel{2} \cdot 50} = \frac{3}{50}$

So $6\% = 0.06 = \frac{3}{50}$.

c) $0.1 = \frac{1}{10} = \frac{1 \cdot 10}{10 \cdot 10} = \frac{10}{100} = 10\%$

So $0.1 = \frac{1}{10} = 10\%$.

Applications

The dimensions for lumber used in construction are usually given in fractions. For example, a 2×4 stud used for framing a wall is actually $1\frac{1}{2}$ in. by $3\frac{1}{2}$ in. by $92\frac{5}{8}$ in. A 2×12 floor joist is actually $1\frac{1}{2}$ in. by $11\frac{1}{2}$ in.

EXAMPLE 9 **Framing a two-story house**

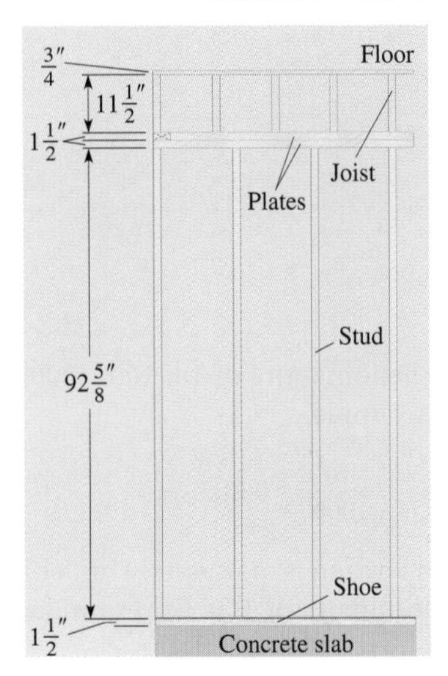

Figure 1.3

In framing a two-story house, a carpenter uses a 2×4 shoe, a wall stud, two 2×4 plates, then 2×12 floor joists, and a $\frac{3}{4}$-in. plywood floor, before starting the second level. Use the dimensions in Fig. 1.3 to find the total height of the framing shown.

Solution

We can find the total height using multiplication and addition:

$$3 \cdot 1\frac{1}{2} + 92\frac{5}{8} + 11\frac{1}{2} + \frac{3}{4} = 4\frac{1}{2} + 92\frac{5}{8} + 11\frac{1}{2} + \frac{3}{4}$$
$$= 4\frac{4}{8} + 92\frac{5}{8} + 11\frac{4}{8} + \frac{6}{8}$$
$$= 107\frac{19}{8}$$
$$= 109\frac{3}{8}$$

The total height of the framing shown is $109\frac{3}{8}$ in. ■

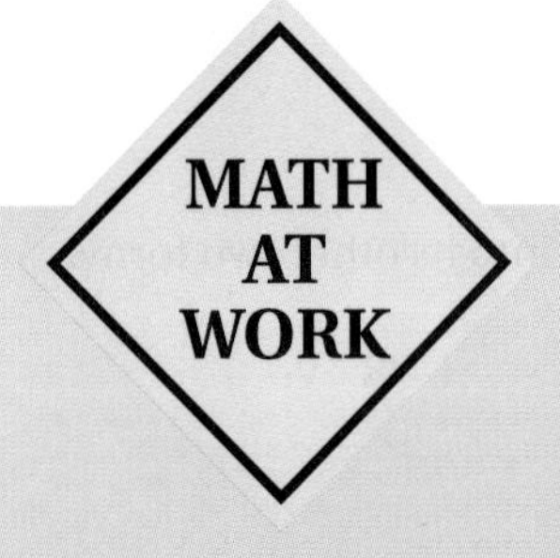

Building Contractor

Building a new house can be a complicated and daunting task. Shirley Zaborowski, project manager for Court Construction, is responsible for estimating, pricing, negotiating, subcontracting, and scheduling all portions of new house construction.

Ms. Zaborowski works from drawings and first does a "take off" or estimate for the quantity of material needed. The quantity of concrete is measured in cubic yards and the amount of wood is measured in board feet. If masonry is being used, it is measured in bricks or blocks per square foot.

Scheduling is another important part of the project manager's responsibility and it is based on the take off. Certain industry standards help Ms. Zaborowski estimate how many carpenters are needed and how much time it takes to frame the house and how many electricians and plumbers are needed to wire the house, install the heating systems, and put in the bathrooms. Of course, common sense says that the foundation is done before the framing and the roof. However, some rough plumbing and electrical work can be done simultaneously with the framing. Ideally the estimates of time and cost are accurate and the homeowner can move in on schedule.

In Exercise 96 of this section you will use operations with fractions to find the volume of concrete needed to construct a rectangular patio.

Warm-ups

True or false? Explain your answer.

1. Every fraction is equal to infinitely many equivalent fractions.
2. The fraction $\frac{8}{12}$ is equivalent to the fraction $\frac{4}{6}$.
3. The fraction $\frac{8}{12}$ reduced to lowest terms is $\frac{4}{6}$.
4. $\frac{1}{2} \cdot \frac{2}{3} = \frac{1}{3}$
5. $\frac{1}{2} \cdot \frac{3}{5} = \frac{3}{10}$
6. $\frac{1}{2} \cdot \frac{6}{5} = \frac{6}{10}$
7. $\frac{1}{2} \div 3 = \frac{1}{6}$
8. $5 \div \frac{1}{2} = 10$
9. $\frac{1}{2} + \frac{1}{4} = \frac{2}{6}$
10. $2 - \frac{1}{2} = \frac{3}{2}$

1.1 EXERCISES

Build up each fraction or whole number so that it is equivalent to the fraction with the indicated denominator. See Example 1.

1. $\frac{3}{4} = \frac{?}{8}$ **2.** $\frac{5}{7} = \frac{?}{21}$ **3.** $\frac{8}{3} = \frac{?}{12}$

4. $\frac{7}{2} = \frac{?}{8}$ **5.** $5 = \frac{?}{2}$ **6.** $9 = \frac{?}{3}$

7. $\frac{3}{4} = \frac{?}{100}$ **8.** $\frac{1}{2} = \frac{?}{100}$ **9.** $\frac{3}{10} = \frac{?}{100}$

10. $\frac{2}{5} = \frac{?}{100}$ **11.** $\frac{5}{3} = \frac{?}{42}$ **12.** $\frac{5}{7} = \frac{?}{98}$

Reduce each fraction to lowest terms. See Example 2.

13. $\frac{3}{6}$ **14.** $\frac{2}{10}$ **15.** $\frac{12}{18}$ **16.** $\frac{30}{40}$

17. $\frac{15}{5}$ **18.** $\frac{39}{13}$ **19.** $\frac{50}{100}$ **20.** $\frac{5}{1000}$

21. $\frac{200}{100}$ **22.** $\frac{125}{100}$ **23.** $\frac{18}{48}$ **24.** $\frac{34}{102}$

25. $\frac{26}{42}$ **26.** $\frac{70}{112}$ **27.** $\frac{84}{91}$ **28.** $\frac{121}{132}$

Find each product. See Examples 3 and 4.

29. $\frac{2}{3} \cdot \frac{5}{9}$ **30.** $\frac{1}{8} \cdot \frac{1}{8}$ **31.** $\frac{1}{3} \cdot 15$

32. $\frac{1}{4} \cdot 16$ **33.** $\frac{3}{4} \cdot \frac{14}{15}$ **34.** $\frac{5}{8} \cdot \frac{12}{35}$

35. $\frac{2}{5} \cdot \frac{35}{26}$ **36.** $\frac{3}{10} \cdot \frac{20}{21}$ **37.** $3\frac{1}{2} \cdot \frac{6}{5}$

38. $6\frac{1}{2} \cdot \frac{3}{5}$ **39.** $4\frac{1}{2} \cdot 3\frac{1}{3}$ **40.** $2\frac{3}{16} \cdot 3\frac{1}{7}$

Find each quotient. See Example 5.

41. $\frac{3}{4} \div \frac{1}{4}$ **42.** $\frac{2}{3} \div \frac{1}{2}$ **43.** $\frac{1}{3} \div 5$

44. $\frac{3}{5} \div 3$ **45.** $5 \div \frac{5}{4}$ **46.** $8 \div \frac{2}{3}$

47. $\frac{6}{10} \div \frac{3}{4}$ **48.** $\frac{2}{3} \div \frac{10}{21}$ **49.** $2\frac{3}{16} \div \frac{5}{2}$

50. $3\frac{1}{8} \div \frac{5}{16}$

Find each sum or difference. See Examples 6 and 7.

51. $\frac{1}{4} + \frac{1}{4}$ **52.** $\frac{1}{10} + \frac{1}{10}$ **53.** $\frac{5}{12} - \frac{1}{12}$

54. $\frac{17}{14} - \frac{5}{14}$ **55.** $\frac{1}{2} - \frac{1}{4}$ **56.** $\frac{1}{3} + \frac{1}{6}$

57. $\frac{1}{3} + \frac{1}{4}$ **58.** $\frac{1}{2} + \frac{3}{5}$ **59.** $\frac{3}{4} - \frac{2}{3}$

60. $\frac{4}{5} - \frac{3}{4}$ **61.** $\frac{1}{6} + \frac{5}{8}$ **62.** $\frac{3}{4} + \frac{1}{6}$

63. $\frac{5}{24} - \frac{1}{18}$ **64.** $\frac{3}{16} - \frac{1}{20}$ **65.** $3\frac{5}{6} + \frac{5}{16}$

66. $5\frac{3}{8} - \frac{15}{16}$

Convert each given fraction, decimal, or percent into its other two forms. See Example 8.

67. $\frac{3}{5}$ **68.** $\frac{19}{20}$ **69.** 9%

70. 60% **71.** 0.08 **72.** 0.4

73. $\frac{3}{4}$ **74.** $\frac{5}{8}$ **75.** 2%

76. 120% **77.** 0.01 **78.** 0.005

Perform the indicated operations.

79. $\frac{3}{8} \div \frac{1}{8}$

80. $\frac{7}{8} \div \frac{3}{14}$

81. $\frac{3}{4} \cdot \frac{28}{21}$

82. $\frac{5}{16} \cdot \frac{3}{10}$

83. $\frac{7}{12} + \frac{5}{32}$

84. $\frac{2}{15} + \frac{8}{21}$

85. $\frac{5}{24} - \frac{1}{15}$

86. $\frac{9}{16} - \frac{1}{12}$

87. $3\frac{1}{8} + \frac{15}{16}$

88. $5\frac{1}{4} - \frac{9}{16}$

89. $7\frac{2}{3} \cdot 2\frac{1}{4}$

90. $6\frac{1}{2} \div \frac{7}{2}$

91. $\frac{1}{2} + \frac{1}{3} + \frac{1}{4}$

92. $\frac{1}{2} + \frac{1}{3} - \frac{1}{6}$

93. $\frac{1}{2} \cdot \frac{1}{2} \cdot \frac{1}{2}$

94. $\frac{2}{3} \cdot \frac{2}{3} \cdot \frac{2}{3}$

Solve each problem. See Example 9.

95. ***Bundle of studs.*** A lumber yard receives 2×4 studs in a bundle that contains 25 rows (or layers) of studs with 20 studs in each row. Find the cross-sectional area of a bundle in square inches. Find the volume of a bundle in cubic feet. (The formula $V = LWH$ gives the volume of a rectangular solid.)

96. ***Concrete patio.*** A contractor plans to pour a concrete rectangular patio that is $12\frac{1}{2}$ feet long, $8\frac{3}{4}$ feet wide, and 4 inches thick. Find the volume of concrete in cubic feet and in cubic yards.

97. ***Stock prices.*** On Monday, GM stock opened at $54\frac{3}{4}$ per share and closed up $\frac{3}{16}$. On Tuesday it closed down $\frac{1}{8}$. On Wednesday it gained $\frac{5}{16}$. On Thursday it fell $\frac{1}{4}$. On Friday there was no change. What was the closing price on Friday? What was the percent change for the week?

98. ***Diversification.*** Helen has $\frac{1}{5}$ of her portfolio in U.S. stocks, $\frac{1}{8}$ of her portfolio in European stocks, and $\frac{1}{10}$ of her portfolio in Japanese stocks. The remainder is invested in municipal bonds. What fraction of her portfolio is invested in municipal bonds? What percent is invested in municipal bonds?

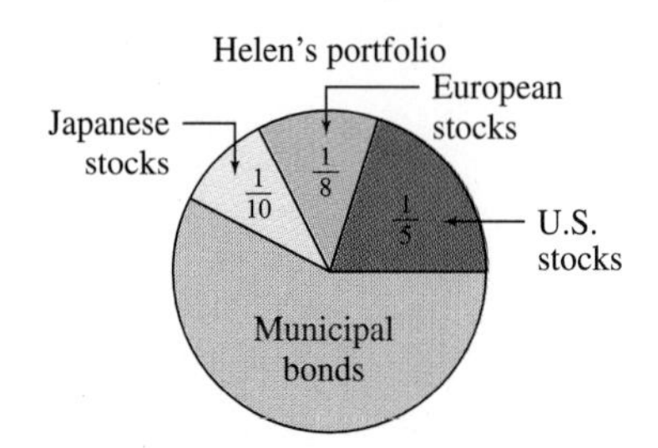

Figure for Exercise 98

Getting More Involved

99. ***Writing.*** Explain why we must use a common denominator when adding or subtracting fractions with different denominators.

100. ***Writing.*** Find an example of a real-life situation in which it is necessary to add two fractions.

101. ***Cooperative learning.*** Write a step-by-step procedure for adding two fractions with different denominators. Give your procedure to a classmate to try out on some addition problems. Refine your procedure as necessary.

1.2 The Real Numbers

In this section:

- The Integers
- The Rational Numbers
- The Number Line
- The Real Numbers
- Absolute Value

In arithmetic we use only positive numbers, but in algebra we use negative numbers also. The numbers that we use in algebra are called the real numbers. We start the discussion of the real numbers with some simpler sets of numbers.

The Integers

The most fundamental collection or **set** of numbers is the set of **counting numbers** or **natural numbers.** Of course, these are the numbers that we use for counting. The set of natural numbers is written in symbols as follows.

The Natural Numbers

$$\{1, 2, 3, \ldots\}$$

Figure 1.4

Braces, { }, are used to indicate a set of numbers. The three dots after 1, 2, and 3, which are read "and so on," mean that the pattern continues without end. There are infinitely many natural numbers.

The natural numbers, together with the number 0, are called the **whole numbers.** The set of whole numbers is written as follows.

The Whole Numbers

$$\{0, 1, 2, 3, \ldots\}$$

Although the whole numbers have many uses, they are not adequate for indicating losses or debts. A debt of $20 can be expressed by the negative number -20 (negative twenty). See Fig. 1.4. When a thermometer reads 10 degrees below zero on a Fahrenheit scale, we say that the temperature is -10°F. See Fig. 1.5. The whole numbers together with the negatives of the counting numbers form the set of **integers.**

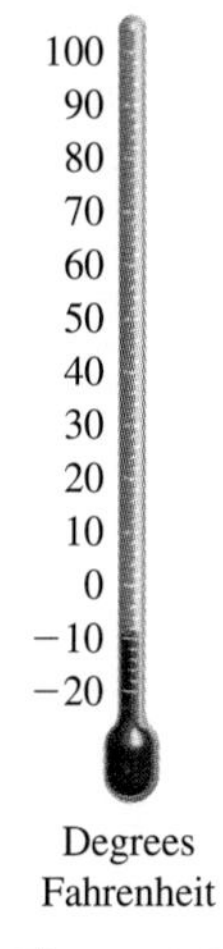

Figure 1.5

The Integers

$$\{\ldots, -3, -2, -1, 0, 1, 2, 3, \ldots\}$$

The Rational Numbers

The fractions that you studied in Section 1.1 were all positive numbers. The set of rational numbers includes both positive and negative fractions. A **rational number** is any number that can be expressed as a ratio (or quotient) of two integers. We cannot list the rational numbers as easily as we listed the numbers in the other sets we have been discussing. So we write the set of rational numbers in symbols using **set-builder notation** as follows.

The Rational Numbers

$$\left\{\frac{a}{b} \,\middle|\, a \text{ and } b \text{ are integers, with } b \neq 0\right\}$$

↑ The set of ↑ such that ↑ conditions

We read this notation as "the set of numbers of the form $\frac{a}{b}$ such that a and b are integers, with $b \neq 0$."

Examples of rational numbers are

$$\frac{3}{1}, \quad \frac{5}{4}, \quad -\frac{7}{10}, \quad \frac{0}{6}, \quad \frac{5}{1}, \quad -\frac{77}{3}, \quad \text{and} \quad \frac{-3}{-6}.$$

Note that we usually use simpler forms for some of these rational numbers. For instance, $\frac{3}{1} = 3$ and $\frac{0}{6} = 0$. The integers are rational numbers because any integer can be written with a denominator of 1.

If you divide the denominator into the numerator, then you can convert a rational number to decimal form. As a decimal, every rational number either repeats indefinitely (for example, $\frac{1}{3} = 0.333\ldots$) or terminates (for example, $\frac{1}{8} = 0.125$).

The Number Line

The number line is a diagram that helps us to visualize numbers and their relationships to each other. A number line is like the scale on the thermometer in Fig. 1.5. To construct a number line, we draw a straight line and label any convenient point with the number 0. Now we choose any convenient length and use it to locate other points. Points to the right of 0 correspond to the positive integers, and points to the left of 0 correspond to the negative integers. The number line is shown in Fig. 1.6.

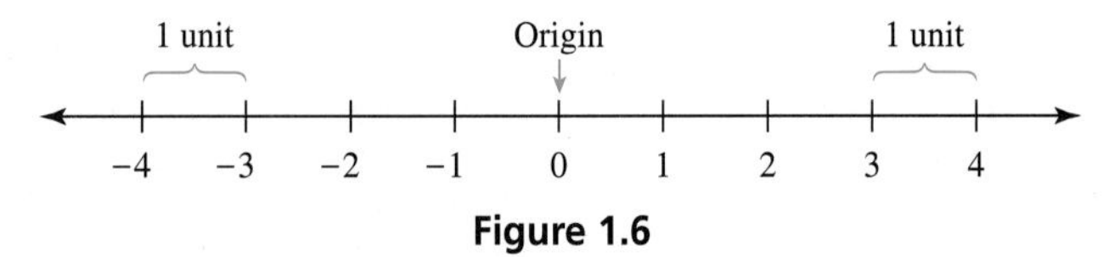

Figure 1.6

The numbers corresponding to the points on the line are called the **coordinates** of the points. The distance between two consecutive integers is called a **unit** and is the same for any two consecutive integers. The point with coordinate 0 is called the **origin.** The numbers on the number line increase in size from left to right. *When we compare the size of any two numbers, the larger number lies to the right of the smaller on the number line.*

EXAMPLE 1 **Comparing numbers on a number line**

Determine which number is the larger in each given pair of numbers.

a) $-3, 2$ **b)** $0, -4$ **c)** $-2, -1$

Solution

a) The larger number is 2 because 2 lies to the right of -3 on the number line. In fact, any positive number is larger than any negative number.

b) The larger number is 0, because 0 lies to the right of -4 on the number line.

c) The larger number is -1, because -1 lies to the right of -2 on the number line. ■

The set of integers is illustrated or **graphed** in Figure 1.7 by drawing a point for each integer. The three dots to the right and left below the number line indicate that the numbers go on indefinitely in both directions.

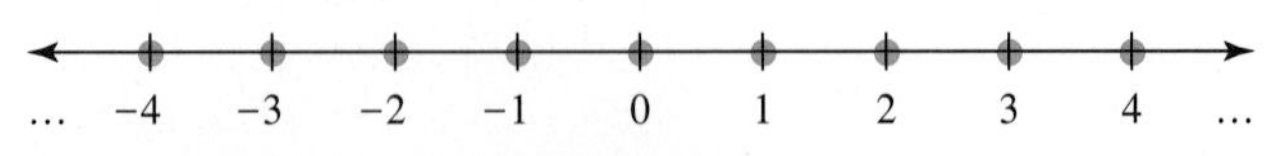

Figure 1.7

EXAMPLE 2 **Graphing numbers on a number line**

List the numbers described, and graph the numbers on a number line.

a) The whole numbers less than 4

b) The integers between 3 and 9

c) The integers greater than -3

Solution

a) The whole numbers less than 4 are 0, 1, 2, and 3. These numbers are shown in Fig. 1.8.

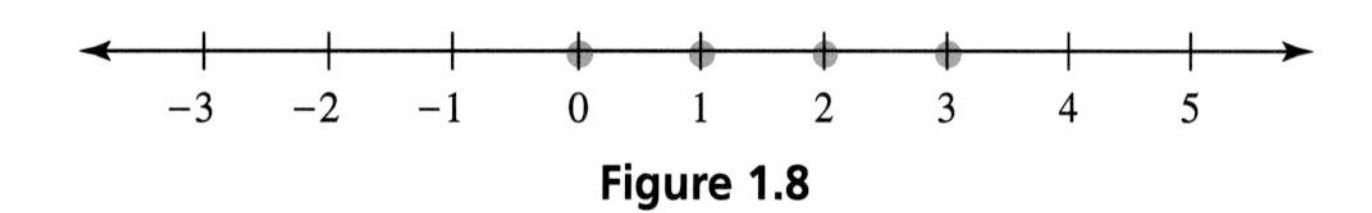

Figure 1.8

b) The integers between 3 and 9 are 4, 5, 6, 7, and 8. Note that 3 and 9 are not considered to be *between* 3 and 9. The graph is shown in Fig. 1.9.

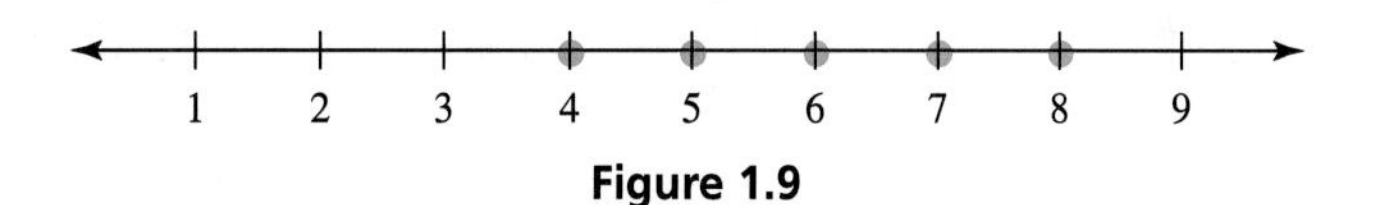

Figure 1.9

c) The integers greater than -3 are -2, -1, 0, 1, and so on. To indicate the continuing pattern, we use three dots on the graph shown in Fig. 1.10

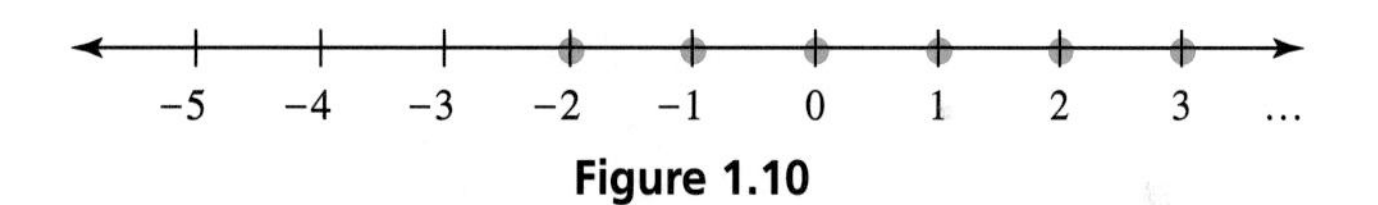

Figure 1.10

■

The Real Numbers

For every rational number there is a point on the number line. For example, the number $\frac{1}{2}$ corresponds to a point halfway between 0 and 1 on the number line, and $-\frac{5}{4}$ corresponds to a point one and one-quarter units to the left of 0, as shown in Fig. 1.11. Since there is a correspondence between numbers and points on the number line, the points are often referred to as numbers.

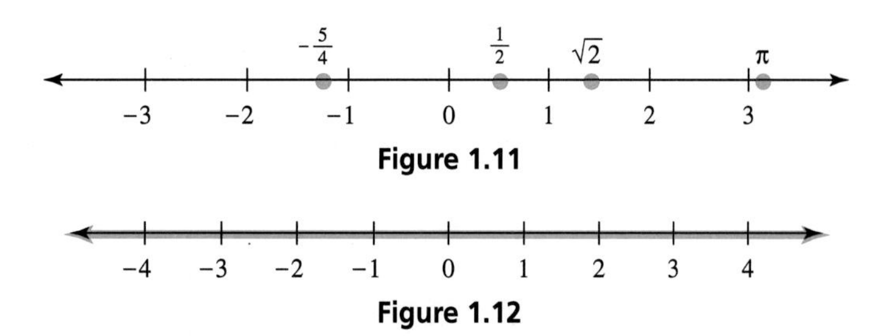

Figure 1.11

Figure 1.12

The set of numbers that corresponds to *all* points on a number line is called the set of **real numbers.** A graph of the real numbers is shown on a number line by shading all points as in Fig. 1.12. All rational numbers are real numbers, but there are points on the number line that do not correspond to rational numbers. Those real numbers that are not rational are called **irrational.** An irrational number cannot be written as a ratio of integers. It can be shown that numbers such as $\sqrt{2}$ (the square root of 2) and π (Greek letter pi) are irrational. The number $\sqrt{2}$ is a number that can be multiplied by itself to obtain 2 $(\sqrt{2} \cdot \sqrt{2} = 2)$. The number π is the ratio of the circumference and diameter of any circle. Irrational numbers are not as easy to represent as rational numbers. That is why we use symbols such as $\sqrt{2}$, $\sqrt{3}$, and π for irrational numbers. When we perform computations with irrational numbers, we use rational approximations for them. For example, $\sqrt{2} \approx 1.414$ and $\pi \approx 3.14$. The symbol $\approx$ means "is approximately equal to." Note that not all square roots are irrational. For example, $\sqrt{9} = 3$, because $3 \cdot 3 = 9$. We will deal with irrational numbers in greater depth when we discuss roots in Chapter 8.

Figure 1.13 summarizes the sets of numbers that make up the real numbers, and shows the relationships between them.

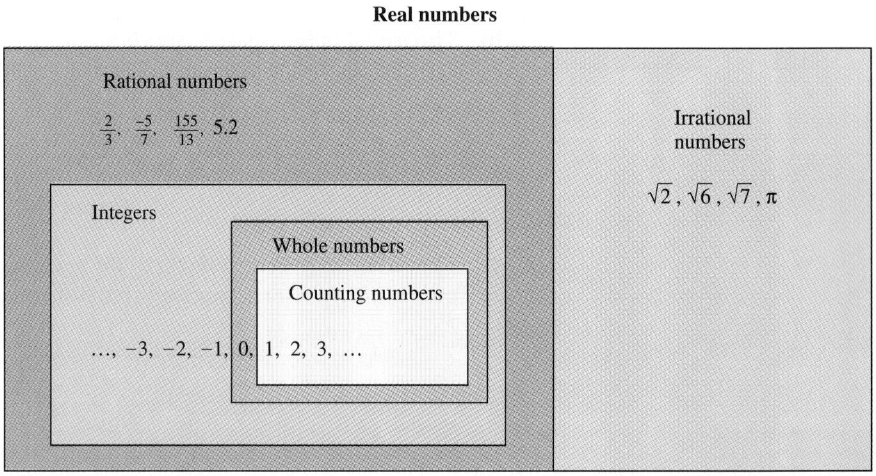

Real numbers

Rational numbers

$\frac{2}{3}$, $\frac{-5}{7}$, $\frac{155}{13}$, 5.2

Integers

Whole numbers

Counting numbers

..., −3, −2, −1, 0, 1, 2, 3, ...

Irrational numbers

$\sqrt{2}$, $\sqrt{6}$, $\sqrt{7}$, π

Figure 1.13

EXAMPLE 3 **Types of numbers**

Determine whether each statement is true or false.

a) Every rational number is an integer.

b) Every counting number is an integer.

c) Every irrational number is a real number.

Solution

a) False. For example, $\frac{1}{2}$ is a rational number that is not an integer.

b) True, because the integers consist of the counting numbers, the negatives of the counting numbers, and zero.

c) True, because the rational numbers together with the irrational numbers form the real numbers. ■

Absolute Value

The concept of absolute value will be used to define the basic operations with real numbers in Section 1.3. The **absolute value** of a number is the number's distance from 0 on the number line. For example, the numbers 5 and -5 are both five units away from 0 on the number line. So the absolute value of each of these numbers is 5. See Fig. 1.14. We write $|a|$ for "the absolute value of a." So

$$|5| = 5 \quad \text{and} \quad |-5| = 5.$$

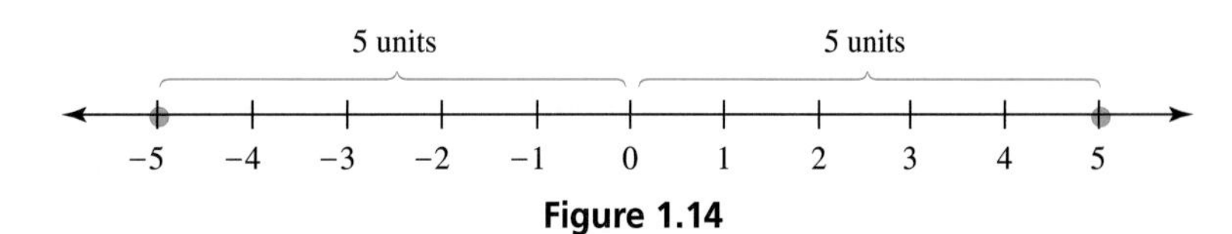

Figure 1.14

The notation $|a|$ represents distance, and distance is never negative. So $|a|$ is greater than or equal to zero for any real number a.

EXAMPLE 4 **Finding absolute value**

Evaluate.

a) $|3|$ **b)** $|-3|$ **c)** $|0|$ **d)** $\left|\frac{2}{3}\right|$ **e)** $|-0.39|$

Solution

a) $|3| = 3$ because 3 is three units away from 0.

b) $|-3| = 3$ because -3 is three units away from 0.

c) $|0| = 0$ because 0 is zero units away from 0.

d) $\left|\frac{2}{3}\right| = \frac{2}{3}$ **e)** $|-0.39| = 0.39$

Two numbers that are located on opposite sides of zero and have the same absolute value are called **opposites** of each other. The numbers 5 and -5 are opposites of each other. We say that the opposite of 5 is -5 and the opposite of -5 is 5. The symbol "$-$" is used to indicate "opposite" as well as "negative." When the negative sign is used before a number, it should be read as "negative." When it is used in front of parentheses or a variable, it should be read as "opposite." For example, $-(5) = -5$ means "the opposite of 5 is negative 5," and $-(-5) = 5$ means "the opposite of negative 5 is 5." Zero does not have an opposite in the same sense as nonzero numbers. Zero is its own opposite. We read $-(0) = 0$ as the "the opposite of zero is zero."

In general, $-a$ means "the opposite of a." If a is positive, $-a$ is negative. If a is negative, $-a$ is positive. Opposites have the following property.

Opposite of an Opposite

For any real number a,

$$-(-a) = a.$$

Remember that we have defined $|a|$ to be the distance between 0 and a on the number line. Using opposites, we can give a symbolic definition of absolute value.

Absolute Value

$$|a| = \begin{cases} a & \text{if } a \text{ is positive or zero} \\ -a & \text{if } a \text{ is negative} \end{cases}$$

According to this definition, the absolute value of a nonnegative number is that number. Using this definition, we write

$$|8| = 8$$

because 8 is positive. The second line of the definition says that the absolute value of a negative number is the opposite of that number. For example, the absolute value of -8 is the opposite of -8:

$$|-8| = -(-8) = 8$$

Warm-ups

True or false? Explain your answer.

1. The natural numbers and the counting numbers are the same.
2. The number 8,134,562,877,565 is a counting number.
3. Zero is a counting number.
4. Zero is not a rational number.
5. The opposite of negative 3 is positive 3.
6. The absolute value of 4 is -4.
7. $-(-9) = 9$
8. $-(-b) = b$ for any number b.
9. Negative six is greater than negative three.
10. Negative five is between four and six.

1.2 EXERCISES

Determine which number is the larger in each given pair of numbers. See Example 1.

1. $-3, 6$ **2.** $7, -10$ **3.** $0, -6$

4. $-8, 0$ **5.** $-3, -2$ **6.** $-5, -8$

7. $-12, -15$ **8.** $-13, -7$

List the numbers described and graph them on a number line. See Example 2.

9. The counting numbers smaller than 6

10. The natural numbers larger than 4

11. The whole numbers smaller than 5

12. The integers between -3 and 3

13. The whole numbers between -5 and 5

14. The integers smaller than -1

15. The counting numbers larger than -4

16. The natural numbers between -5 and 7

17. The integers larger than $\frac{1}{2}$

18. The whole numbers smaller than $\frac{7}{4}$

Determine whether each statement is true or false. Explain your answer. See Example 3.

19. Every integer is a rational number.

20. Every counting number is a whole number.

21. Zero is a counting number.

22. Every whole number is a counting number.

23. The ratio of the circumference and diameter of a circle is an irrational number.

24. Every rational number can be expressed as a ratio of integers.

25. Every whole number can be expressed as a ratio of integers.

26. Some of the rational numbers are integers.

27. Some of the integers are natural numbers.

28. There are infinitely many rational numbers.

29. Zero is an irrational number.

30. Every irrational number is a real number.

Determine the values of the following. See Example 4.

31. $|-6|$ **32.** $|4|$ **33.** $|0|$

34. $|2|$ **35.** $|7|$ **36.** $|-7|$

37. $|-9|$ **38.** $|-2|$ **39.** $|-45|$

40. $|-30|$ **41.** $\left|\frac{3}{4}\right|$ **42.** $\left|-\frac{1}{2}\right|$

43. $|-5.09|$ **44.** $|0.00987|$

Select the smaller number in each given pair of numbers.

45. $-16, 9$ **46.** $-12, -7$ **47.** $-\frac{5}{2}, -\frac{9}{4}$

48. $\frac{5}{8}, \frac{6}{7}$ **49.** $|-3|, 2$ **50.** $|-6|, 0$

51. $|-4|, 3$ **52.** $|5|, -4$

Which number in each given pair has the larger absolute value?

53. $-5, -9$ **54.** $-12, -8$

55. $16, -9$ **56.** $-12, 7$

True or false? Explain your answer.

57. If we add the absolute values of -3 and -5, we get 8.

58. If we multiply the absolute values of -2 and 5, we get 10.

59. The absolute value of any negative number is greater than 0.

60. The absolute value of any positive number is less than 0.

61. The absolute value of -9 is larger than the absolute value of 6.

62. The absolute value of 12 is larger than the absolute value of -11.

Getting More Involved

63. ***Writing.*** What is the difference between a rational number and an irrational number?

64. ***Writing.*** Find a real-life question for which the answer is a rational number that is not an integer.

65. ***Exploration.*** **a)** Find a rational number between $\frac{1}{3}$ and $\frac{1}{4}$. **b)** Find a rational number between -3.205 and -3.114. **c)** Find a rational number between $\frac{2}{3}$ and 0.6667. **d)** Explain how to find a rational number between any two given rational numbers.

66. ***Discussion.*** Suppose that a is a negative real number. Determine whether each of the following is positive or negative, and explain your answer.

a) $-a$ **b)** $|-a|$ **c)** $-|a|$

d) $-(-a)$ **e)** $-|-a|$

1.3 Addition and Subtraction of Real Numbers

In this section:

- Addition of Two Negative Numbers
- Addition of Numbers with Unlike Signs
- Subtraction of Signed Numbers

In arithmetic we add and subtract only positive numbers. In Section 1.2 we introduced the concept of absolute value of a number. Now we will use absolute value to extend the operations of addition and subtraction to the real numbers. We will work only with rational numbers in this chapter. You will learn to perform operations with irrational numbers in Chapter 8.

Addition of Two Negative Numbers

A good way to understand positive and negative numbers is to *think of the positive numbers as assets and the negative numbers as debts.* For this illustration we can think of assets simply as cash. For example, if you have \$3 and \$5 in cash, then your total cash is \$8. You get the total by adding two positive numbers.

Think of debts as unpaid bills such as the electric bill or the phone bill. If you have debts of \$7 and \$8, then your total debt is \$15. You can get the total debt by adding negative numbers:

$$\underset{\$7\text{ debt}}{\underset{\uparrow}{(-7)}} \quad \underset{\text{plus}}{\underset{\uparrow}{+}} \quad \underset{\$8\text{ debt}}{\underset{\uparrow}{(-8)}} \quad = \quad \underset{\$15\text{ debt}}{\underset{\uparrow}{-15}}$$

We think of this addition as adding the absolute values of -7 and -8 $(7 + 8 = 15)$, and then putting a negative sign on that result to get -15. These examples illustrate the following rule.

> **Sum of Two Numbers with Like Signs**
>
> To find the sum of two numbers with the same sign, add their absolute values. The sum has the same sign as the given numbers.

EXAMPLE 1 **Adding numbers with like signs**

Perform the indicated operations.

a) $23 + 56$ **b)** $(-12) + (-9)$

c) $(-3.5) + (-6.28)$ **d)** $\left(-\frac{1}{2}\right) + \left(-\frac{1}{4}\right)$

Solution

a) The sum of two positive numbers is a positive number: $23 + 56 = 79$.

b) The absolute values of -12 and -9 are 12 and 9, and $12 + 9 = 21$. So

$$(-12) + (-9) = -21.$$

c) Add the absolute values of -3.5 and -6.28, and put a negative sign on the sum. Remember to line up the decimal points when adding decimal numbers:

$$(-3.5) + (-6.28) = -9.78$$

d) $\left(-\frac{1}{2}\right) + \left(-\frac{1}{4}\right) = \left(-\frac{2}{4}\right) + \left(-\frac{1}{4}\right) = -\frac{3}{4}$ ■

Addition of Numbers with Unlike Signs

If you have a debt of \$5 and have only \$5 in cash, then your debts equal your assets (in absolute value), and your net worth is \$0. **Net worth** is the total of debts and assets. Symbolically,

$$\underset{\text{\$5 debt}}{\underset{\uparrow}{-5}} \quad + \quad \underset{\text{\$5 cash}}{\underset{\uparrow}{5}} \quad = \quad \underset{\text{Net worth}}{\underset{\uparrow}{0.}}$$

For any number a, a and its opposite, $-a$, have a sum of zero. For this reason, a and $-a$ are called **additive inverses** of each other. Note that the words "negative," "opposite," and "additive inverse" are often used interchangeably.

Additive Inverse Property

For any number a,

$$a + (-a) = 0 \quad \text{and} \quad (-a) + a = 0.$$

EXAMPLE 2 **Finding the sum of additive inverses**

Evaluate.

a) $34 + (-34)$ **b)** $-\frac{1}{4} + \frac{1}{4}$ **c)** $2.97 + (-2.97)$

Solution

a) $34 + (-34) = 0$

b) $-\frac{1}{4} + \frac{1}{4} = 0$

c) $2.97 + (-2.97) = 0$

To understand the sum of a positive and a negative number that are not additive inverses of each other, consider the following situation. If you have a debt of \$6 and \$10 in cash, you may have \$10 in hand, but your net worth is only \$4. Your assets exceed your debts (in absolute value), and you have a positive net worth. In symbols,

$$-6 + 10 = 4.$$

Note that to get 4, we actually subtract 6 from 10.

If you have a debt of \$7 but have only \$5 in cash, then your debts exceed your assets (in absolute value). You have a negative net worth of $-\$2$. In symbols,

$$-7 + 5 = -2.$$

Note that to get the 2 in the answer, we subtract 5 from 7.

As you can see from these examples, the sum of a positive number and a negative number (with different absolute values) may be either positive or negative. These examples help us to understand the rule for adding numbers with unlike signs and different absolute values.

Sum of Two Numbers with Unlike Signs (and Different Absolute Values)

To find the sum of two numbers with unlike signs (and different absolute values), subtract their absolute values.

- The answer is positive if the number with the larger absolute value is positive.
- The answer is negative if the number with the larger absolute value is negative.

EXAMPLE 3 **Adding numbers with unlike signs**

Evaluate.

a) $-5 + 13$

b) $6 + (-7)$

c) $-6.4 + 2.1$

d) $-5 + 0.09$

e) $\left(-\frac{1}{3}\right) + \left(\frac{1}{2}\right)$

f) $\frac{3}{8} + \left(-\frac{5}{6}\right)$

Solution

a) The absolute values of -5 and 13 are 5 and 13. Subtract them to get 8. Since the number with the larger absolute value is 13 and it is positive, the result is positive:

$$-5 + 13 = 8$$

b) The absolute values of 6 and -7 are 6 and 7. Subtract them to get 1. Since -7 has the larger absolute value, the result is negative:

$$6 + (-7) = -1$$

c) Line up the decimal points and subtract 2.1 from 6.4. Since 6.4 is larger than 2.1 and 6.4 has a negative sign, the sign of the answer is negative:

$$-6.4 + 2.1 = -4.3$$

d) Line up the decimal points and subtract 0.09 from 5.00. Since 5.00 is larger than 0.09, and 5.00 has the negative sign, the sign of the answer is negative:

$$-5 + 0.09 = -4.91$$

e) $$\left(-\frac{1}{3}\right) + \left(\frac{1}{2}\right) = \left(-\frac{2}{6}\right) + \left(\frac{3}{6}\right)$$
$$= \frac{1}{6}$$

f) $$\frac{3}{8} + \left(-\frac{5}{6}\right) = \frac{9}{24} + \left(-\frac{20}{24}\right)$$
$$= -\frac{11}{24}$$

■

Subtraction of Signed Numbers

Each subtraction problem with signed numbers is solved by doing an equivalent addition problem. So before attempting subtraction of signed numbers be sure that you understand addition of signed numbers.

Now think of subtraction as removing debts or assets, and think of addition as receiving debts or assets. If you have \$10 in cash and \$3 is taken from you, your resulting net worth is the same as if you have \$10 cash and a phone bill for \$3 arrives in the mail. In symbols,

$$\underset{}{10} \quad \underset{\text{Remove}}{\overset{\uparrow}{-}} \quad \underset{\text{Cash}}{\overset{\uparrow}{3}} \quad = \quad 10 \quad \underset{\text{Receive}}{\overset{\uparrow}{+}} \quad \underset{\text{Debt}}{\overset{\uparrow}{(-3)}}.$$

Removing cash is equivalent to receiving a debt.

Suppose you have \$15 but owe a friend \$5. Your net worth is only \$10. If the debt of \$5 is canceled or forgiven, your net worth will go up to \$15, the same as if you received \$5 in cash. In symbols,

$$10 \quad \underset{\text{Remove}}{\overset{\uparrow}{-}} \quad \underset{\text{Debt}}{\overset{\uparrow}{(-5)}} \quad = \quad 10 \quad \underset{\text{Receive}}{\overset{\uparrow}{+}} \quad \underset{\text{Cash}}{\overset{\uparrow}{5}}.$$

Removing a debt is equivalent to receiving cash.

Notice that each subtraction problem is equivalent to an addition problem in which we add the opposite of what we want to subtract. In other words, subtracting a number is the same as adding its opposite.

> **Subtraction of Real Numbers**
>
> For any real numbers a and b,
>
> $$a - b = a + (-b).$$

EXAMPLE 4 **Subtracting signed numbers**

Perform each subtraction.

a) $-5 - 3$ **b)** $5 - (-3)$ **c)** $-5 - (-3)$

d) $\frac{1}{2} - \left(-\frac{1}{4}\right)$ **e)** $-3.6 - (-5)$ **f)** $0.02 - 8$

Solution

To do *any* subtraction, we can change it to addition of the opposite.

a) $-5 - 3 = -5 + (-3) = -8$

b) $5 - (-3) = 5 + (3) = 8$

c) $-5 - (-3) = -5 + 3 = -2$

d) $\frac{1}{2} - \left(-\frac{1}{4}\right) = \frac{2}{4} + \frac{1}{4} = \frac{3}{4}$

e) $-3.6 - (-5) = -3.6 + 5 = 1.4$

f) $0.02 - 8 = 0.02 + (-8) = -7.98$ ■

Calculator Close-up

There are many exercises in this text that require a **scientific calculator.** If you do these exercises from the beginning, you will soon become proficient with your calculator. The most modern scientific calculators, called **graphing calculators,** can draw graphs and perform many other tasks that are not possible with ordinary scientific calculators. Ask your instructor about choosing a calculator for your course.

All scientific calculators can perform computations with signed numbers. A number entered into a scientific calculator is positive unless you change the sign to negative. The sign-change key [+/−] changes the sign of a displayed number. For example, to find 3.2 − (−1.5) with a scientific calculator, try the following keystrokes:

3.2 [−] 1.5 [+/−] [=]

The display will read

[4.7]

With a graphing calculator you enter negative numbers in the same way that you write them. Most graphing calculators have a negative-sign key [(−)] for entering a negative number. On a graphing calculator you can use parentheses to make the display look like the original problem. Then press [ENTER] or [EXE] to carry out the operations:

3.2−(⁻1.5) [ENTER]

[4.7]

On most graphing calculators you can also display 3.2 − ⁻1.5 and press [ENTER] to get 4.7. If you cannot perform this computation, see your calculator manual or ask your instructor for help.

Warm-ups

True or false? Explain your answer.

1. $-9 + 8 = -1$ **2.** $(-2) + (-4) = -6$ **3.** $0 - 7 = -7$

4. $5 - (-2) = 3$ **5.** $-5 - (-2) = -7$

6. The additive inverse of -3 is 0.

7. If b is a negative number, then $-b$ is a positive number.

8. The sum of a positive number and a negative number is a negative number.

9. The result of a subtracted from b is the same as b plus the opposite of a.

10. If a and b are negative numbers, then $a - b$ is a negative number.

1.3 EXERCISES

Perform the indicated operation. See Example 1.

1. $3 + 10$

2. $81 + 19$

3. $(-3) + (-10)$

4. $(-81) + (-19)$

5. $-0.25 + (-0.9)$

6. $-0.8 + (-2.35)$

7. $\left(-\frac{1}{3}\right) + \left(-\frac{1}{6}\right)$

8. $\frac{2}{3} + \frac{1}{12}$

Evaluate. See Examples 2 and 3.

9. $-8 + 8$

10. $20 + (-20)$

11 $-\frac{17}{50} + \frac{17}{50}$

12. $\frac{12}{13} + \left(-\frac{12}{13}\right)$

13. $-7 + 9$

14. $10 + (-30)$

15. $7 + (-13)$

16. $-8 + 20$

17. $8.6 + (-3)$

18. $-9.5 + 12$

19. $3.9 + (-6.8)$

20. $-5.24 + 8.19$

21. $\frac{1}{4} + \left(-\frac{1}{2}\right)$

22. $-\frac{2}{3} + 2$

Fill in the parentheses to make each statement correct. See Example 4.

23. $8 - 2 = 8 + (?)$

24. $3.5 - 1.2 = 3.5 + (?)$

25. $4 - 12 = 4 + (?)$

26. $\frac{1}{2} - \frac{5}{6} = \frac{1}{2} + (?)$

27. $-3 - (-8) = -3 + (?)$

28. $-9 - (-2.3) = -9 + (?)$

29. $8.3 - (-1.5) = 8.3 + (?)$

30. $10 - (-6) = 10 + (?)$

Perform the indicated operation. See Example 4.

31. $6 - 10$

32. $3 - 19$

33. $-3 - 7$

34. $-3 - 12$

35. $5 - (-6)$

36. $5 - (-9)$

37. $-6 - 5$

38. $-3 - 6$

39. $\frac{1}{4} - \frac{1}{2}$

40. $\frac{2}{5} - \frac{2}{3}$

41. $\frac{1}{2} - \left(-\frac{1}{4}\right)$

42. $\frac{2}{3} - \left(-\frac{1}{6}\right)$

43. $10 - 3$

44. $13 - 3$

45. $1 - 0.07$

46. $0.03 - 1$

47. $7.3 - (-2)$

48. $-5.1 - 0.15$

49. $-0.03 - 5$

50. $0.7 - (-0.3)$

Perform the indicated operations. Do not use a calculator.

51. $-5 + 8$

52. $-6 + 10$

53. $-6 + (-3)$

54. $(-13) + (-12)$

55. $-80 - 40$

56. $44 - (-15)$

57. $61 - (-17)$

58. $-19 - 13$

59. $(-12) + (-15)$

60. $-12 + 12$

61. $13 + (-20)$

62. $15 + (-39)$

63. $-102 - 99$

64. $-94 - (-77)$

65. $-161 - 161$

66. $-19 - 88$

67. $-16 + 0.03$

68. $0.59 + (-3.4)$

69. $0.08 - 3$

70. $1.8 - 9$

71. $-3.7 + (-0.03)$

72. $0.9 + (-1)$

73. $-2.3 - (-6)$

74. $-7.08 - (-9)$

75. $\frac{3}{4} + \left(-\frac{3}{5}\right)$

76. $-\frac{1}{3} + \frac{3}{5}$

77. $-\frac{1}{12} - \left(-\frac{3}{8}\right)$

78. $-\frac{1}{17} - \left(-\frac{1}{17}\right)$

Use a calculator to perform the indicated operations.

79. $45.87 + (-49.36)$

80. $-0.357 + (-3.465)$

81. $0.6578 + (-1)$

82. $-2.347 + (-3.5)$

83. $-3.45 - 45.39$

84. $9.8 - 9.974$

85. $-5.79 - 3.06$

86. $0 - (-4.537)$

Solve each problem.

87. ***Overdrawn.*** Willard opened his checking account with a deposit of \$97.86. He then wrote checks and had other charges as shown in his account register. Find his current balance.

Deposit		97.86
Wal-Mart	\$27.89	
Kmart	\$42.32	
ATM withdrawal	\$25.00	
Service charge	\$3.50	
Check printing	\$8.00	

Table for Exercise 87

88. ***Net worth.*** Melanie has \$45,000 equity in her house, \$21,236 in a savings account, and cars and other items worth a total of \$15,000. She owes \$78,422 on her mortgage, has \$9,477 in credit card debt, and owes \$6,131 to the credit union. What is Melanie's net worth?

89. ***Falling temperatures.*** At noon the temperature in Minneapolis was 14°F. By midnight the mercury had fallen 20°. What was the temperature at midnight?

90. ***Bitter cold.*** Monday night the overnight low temperature in Milwaukee was -13°F. The temperature went up 20 degrees during the day on Tuesday and then fell 15 degrees to reach Tuesday night's overnight low temperature. What was the overnight low Tuesday night?

Getting More Involved

91. ***Writing.*** What does absolute value have to do with adding signed numbers? Can you add signed numbers without using absolute value?

92. ***Discussion.*** Why do we learn addition of signed numbers before subtraction?

93. ***Discussion.*** Aimee and Joni are traveling south in separate cars on Interstate 5 near Stockton. While they are speaking to each other on cellular telephones, Aimee gives her location as mile marker x and Joni gives her location as mile marker y. Which of the following expressions gives the distance between them? Explain your answer.

a) $y - x$ **b)** $x - y$ **c)** $|x - y|$
d) $|y - x|$ **e)** $|x| + |y|$

1.4 Multiplication and Division of Real Numbers

In this section:

- Multiplication of Real Numbers
- Division of Real Numbers
- Division by Zero

In this section we will complete the study of the four basic operations with real numbers.

Multiplication of Real Numbers

The result of multiplying two numbers is referred to as the **product** of the numbers. The numbers multiplied are called **factors.** In algebra we use a raised dot between the factors to indicate multiplication, or we place symbols next to one another to indicate multiplication. Thus $a \cdot b$ or ab are both referred to as the product of a and b. When multiplying numbers, we may enclose them in parentheses to make the meaning clear. To write 5 times 3, we may write it as $5 \cdot 3$, $5(3)$, $(5)3$, or $(5)(3)$. In multiplying a number and a variable, no sign is used between them. Thus $5x$ is used to represent the product of 5 and x.

Multiplication is just a short way to do repeated additions. Adding together five 3's gives

$$3 + 3 + 3 + 3 + 3 = 15.$$

So we have the multiplication fact $5 \cdot 3 = 15$. Adding together five -3's gives

$$(-3) + (-3) + (-3) + (-3) + (-3) = -15.$$

So we should have $5(-3) = -15$. We can think of $5(-3) = -15$ as saying that taking on five debts of \$3 each is equivalent to a debt of \$15. Losing five debts of \$3 each is equivalent to gaining \$15, so we should have $(-5)(-3) = 15$.

These examples illustrate the rule for multiplying signed numbers.

Product of Signed Numbers

To find the product of two nonzero real numbers, multiply their absolute values.

- The product is *positive* if the numbers have *like* signs.
- The product is *negative* if the numbers have *unlike* signs.

EXAMPLE 1 **Multiplying signed numbers**

Evaluate each product.

a) $(-2)(-3)$ **b)** $3(-6)$ **c)** $-5 \cdot 10$

d) $\left(-\frac{1}{3}\right)\left(-\frac{1}{2}\right)$ **e)** $(-0.02)(0.08)$ **f)** $(-300)(-0.06)$

Solution

a) First find the product of the absolute values:

$$|-2| \cdot |-3| = 2 \cdot 3 = 6$$

Because -2 and -3 have the same sign, we get $(-2)(-3) = 6$.

b) First find the product of the absolute values:

$$|3| \cdot |-6| = 3 \cdot 6 = 18$$

Because 3 and -6 have unlike signs, we get $3(-6) = -18$.

c) $-5 \cdot 10 = -50$ Unlike signs, negative result

d) $\left(-\frac{1}{3}\right)\left(-\frac{1}{2}\right) = \frac{1}{6}$ Like signs, positive result

e) When multiplying decimals, we total the number of decimal places in the factors to get the number of decimal places in the product. Thus

$$(-0.02)(0.08) = -0.0016.$$

f) $(-300)(-0.06) = 18$ ■

Division of Real Numbers

We say that $10 \div 5 = 2$ because $2 \cdot 5 = 10$. This example illustrates how division is defined in terms of multiplication.

> **Division of Real Numbers**
>
> If a, b, and c are any real numbers with $b \neq 0$, then
>
> $$a \div b = c \quad \text{provided that} \quad c \cdot b = a.$$

Using the definition of division, we get

$$10 \div (-2) = -5$$

because $(-5)(-2) = 10$;

$$-10 \div 2 = -5$$

because $(-5)(2) = -10$; and

$$-10 \div (-2) = 5$$

because $(5)(-2) = -10$. From these examples we see that the rule for dividing signed numbers is similar to that for multiplying signed numbers.

Division of Signed Numbers

To find the quotient of nonzero real numbers, divide their absolute values.

- The quotient is *positive* if the numbers have *like* signs.
- The quotient is *negative* if the numbers have *unlike* signs.

Zero divided by any nonzero real number is zero.

EXAMPLE 2 **Dividing signed numbers**

Evaluate.

a) $(-8) \div (-4)$ **b)** $(-8) \div 8$ **c)** $8 \div (-4)$

d) $-4 \div \frac{1}{3}$ **e)** $-2.5 \div 0.05$ **f)** $0 \div (-6)$

Solution

a) $(-8) \div (-4) = 2$ Same sign, positive result

b) $(-8) \div 8 = -1$ Unlike signs, negative result

c) $8 \div (-4) = -2$

d) $-4 \div \frac{1}{3} = -4 \cdot \frac{3}{1}$ Invert and multiply.

$= -4 \cdot 3$

$= -12$

e) $-2.5 \div 0.05 = -50$

f) $0 \div (-6) = 0$ ■

Division can also be indicated by a fraction bar. For example,

$$24 \div 6 = \frac{24}{6} = 4.$$

If signed numbers occur in a fraction, we use the rules for dividing signed numbers. For example,

$$\frac{-9}{3} = -3, \quad \frac{9}{-3} = -3, \quad \frac{-1}{2} = \frac{1}{-2} = -\frac{1}{2}, \quad \text{and} \quad \frac{-4}{-2} = 2.$$

Note that if one negative sign appears in a fraction, the fraction has the same value whether the negative sign is in the numerator, in the denominator, or in front of the fraction. If the numerator and denominator of a fraction are both negative, then the fraction has a positive value.

Division by Zero

Why do we exclude division by zero from the definition of division? If we write $10 \div 0 = c$, we need to find a number c such that $c \cdot 0 = 10$. This is impossible. If we write $0 \div 0 = c$, we need to find a number c such that $c \cdot 0 = 0$. In fact, $c \cdot 0 = 0$ is true for any value of c. Having $0 \div 0$ equal to any number would be confusing in doing computations. Thus $a \div b$ is defined only for $b \neq 0$. Quotients such as

$$8 \div 0, \quad 0 \div 0, \quad \frac{8}{0}, \quad \text{and} \quad \frac{0}{0}$$

are said to be **undefined.**

Warm-ups

True or false? Explain your answer.

1. The product of 7 and y is written as $7y$.
2. The product of -2 and 5 is 10.
3. The quotient of x and 3 can be written as $x \div 3$ or $\frac{x}{3}$.
4. $0 \div 6$ is undefined.
5. $(-9) \div (-3) = 3$
6. $6 \div (-2) = -3$
7. $\left(-\frac{1}{2}\right)\left(-\frac{1}{2}\right) = \frac{1}{4}$
8. $(-0.2)(0.2) = -0.4$
9. $\left(-\frac{1}{2}\right) \div \left(-\frac{1}{2}\right) = 1$
10. $\frac{0}{0} = 0$

1.4 EXERCISES

Evaluate. See Example 1.

1. $-3 \cdot 9$
2. $6(-4)$
3. $(-12)(-11)$
4. $(-9)(-15)$
5. $-\frac{3}{4} \cdot \frac{4}{9}$
6. $\left(-\frac{2}{3}\right)\left(-\frac{6}{7}\right)$
7. $0.5(-0.6)$
8. $(-0.3)(0.3)$
9. $(-12)(-12)$
10. $(-11)(-11)$
11. $-3 \cdot 0$
12. $0(-7)$

Evaluate. See Example 2.

13. $8 \div (-8)$
14. $-6 \div 2$
15. $(-90) \div (-30)$
16. $(-20) \div (-40)$
17. $\frac{44}{-66}$
18. $\frac{-33}{-36}$
19. $\left(-\frac{2}{3}\right) \div \left(-\frac{4}{5}\right)$
20. $-\frac{1}{3} \div \frac{4}{9}$
21. $\frac{-125}{0}$
22. $-37 \div 0$
23. $0 \div \left(-\frac{1}{3}\right)$
24. $0 \div 43.568$
25. $40 \div (-0.5)$
26. $3 \div (-0.1)$
27. $-0.5 \div (-2)$
28. $-0.75 \div (-0.5)$

Perform the indicated operations.

29. $(25)(-4)$
30. $(5)(-4)$
31. $(-3)(-9)$
32. $(-51) \div (-3)$
33. $-9 \div 3$
34. $86 \div (-2)$
35. $20 \div (-5)$
36. $(-8)(-6)$
37. $(-6)(5)$
38. $(-18) \div 3$
39. $(-57) \div (-3)$
40. $(-30)(4)$
41. $(0.6)(-0.3)$
42. $(-0.2)(-0.5)$
43. $(-0.03)(-10)$
44. $(0.05)(-1.5)$
45. $(-0.6) \div (0.1)$
46. $8 \div (-0.5)$
47. $(-0.6) \div (-0.4)$
48. $(-63) \div (-0.9)$
49. $-\frac{12}{5}\left(-\frac{55}{6}\right)$
50. $-\frac{9}{10} \cdot \frac{4}{3}$
51. $-2\frac{3}{4} \div 8\frac{1}{4}$
52. $-9\frac{1}{2} \div \left(-3\frac{1}{6}\right)$
53. $(0.45)(-365)$
54. $8.5 \div (-0.15)$
55. $(-52) \div (-0.034)$
56. $(-4.8)(5.6)$

Perform the indicated operations.

57. $(-4)(-4)$

58. $-4 - 4$

59. $-4 + (-4)$

60. $-4 \div (-4)$

61. $-4 + 4$

62. $-4 \cdot 4$

63. $-4 - (-4)$

64. $0 \div (-4)$

65. $0.1 - 4$

66. $(0.1)(-4)$

67. $(-4) \div (0.1)$

68. $-0.1 - 4$

69. $(-0.1)(-4)$

70. $-0.1 + 4$

71. $|-0.4|$

72. $|0.4|$

73. $\frac{-0.06}{0.3}$

74. $\frac{2}{-0.04}$

75. $\frac{3}{-0.4}$

76. $\frac{-1.2}{-0.03}$

77. $-\frac{1}{5} + \frac{1}{6}$

78. $-\frac{3}{5} - \frac{1}{4}$

79. $\left(-\frac{3}{4}\right)\left(\frac{2}{15}\right)$

80. $-1 \div \left(-\frac{1}{4}\right)$

Use a calculator to perform the indicated operation. Round answers to three decimal places.

81. $\frac{45.37}{6}$

82. $(-345) \div (28)$

83. $(-4.3)(-4.5)$

84. $\frac{-12.34}{-3}$

85. $\frac{0}{6.345}$

86. $0 \div (34.51)$

87. $199.4 \div 0$

88. $\frac{23.44}{0}$

Getting More Involved

89. ***Discussion.*** If you divide $0 among five people, how much does each person get? If you divide $5 among zero people, how much does each person get? What do these questions illustrate?

90. ***Discussion.*** What is the difference between the nonnegative numbers and the positive numbers?

91. ***Writing.*** Why do we learn multiplication of signed numbers before division?

92. ***Writing.*** Try to rewrite the rules for multiplying and dividing signed numbers without using the idea of absolute value? Are your rewritten rules clearer than the original rules?

1.5 Arithmetic Expressions

In this section:

- Arithmetic Expressions
- Exponential Expressions
- The Order of Operations

In Sections 1.3 and 1.4 you learned how to perform operations with a pair of real numbers to obtain a third real number. In this section you will learn to evaluate expressions involving several numbers and operations.

Arithmetic Expressions

The result of writing numbers in a meaningful combination with the ordinary operations of arithmetic is called an **arithmetic expression** or simply an **expression.** Consider the expressions

$$(3 + 2) \cdot 5 \quad \text{and} \quad 3 + (2 \cdot 5).$$

The parentheses are used as **grouping symbols** and indicate which operation to perform first. Because of the parentheses, these expressions have different values:

$$(3 + 2) \cdot 5 = 5 \cdot 5 = 25$$
$$3 + (2 \cdot 5) = 3 + 10 = 13$$

Absolute value symbols and fraction bars are also used as grouping symbols. The numerator and denominator of a fraction are treated as if each is in parentheses.

EXAMPLE 1 **Using grouping symbols**

Evaluate each expression.

a) $(3 - 6)(3 + 6)$ **b)** $|3 - 4| - |5 - 9|$ **c)** $\dfrac{4 - (-8)}{5 - 9}$

Solution

a) $(3 - 6)(3 + 6) = (-3)(9)$ Evaluate within parentheses first.

$= -27$ Multiply.

b) $|3 - 4| - |5 - 9| = |-1| - |-4|$ Evaluate within absolute value symbols.

$= 1 - 4$ Find the absolute values.

$= -3$ Subtract.

c) $\dfrac{4 - (-8)}{5 - 9} = \dfrac{12}{-4}$ Evaluate the numerator and denominator.

$= -3$ Divide.

Exponential Expressions

An arithmetic expression with repeated multiplication can be written by using exponents. For example,

$$2 \cdot 2 \cdot 2 = 2^3 \quad \text{and} \quad 5 \cdot 5 = 5^2.$$

The 3 in 2^3 is the number of times that 2 occurs in the product $2 \cdot 2 \cdot 2$, while the 2 in 5^2 is the number of times that 5 occurs in $5 \cdot 5$. We read 2^3 as "2 cubed" or "2 to the third power." We read 5^2 as "5 squared" or "5 to the second power." In general, an expression of the form a^n is called an **exponential expression** and is defined as follows.

Exponential Expression

For any counting number n,

$$a^n = \underbrace{a \cdot a \cdot a \cdot \ldots \cdot a.}_{n \text{ factors}}$$

We call a the **base** and n the **exponent.**

The expression a^n is read "a to the nth power." If the exponent is 1, it is usually omitted. For example, $9^1 = 9$.

EXAMPLE 2 **Using exponential notation**

Write each product as an exponential expression.

a) $6 \cdot 6 \cdot 6 \cdot 6 \cdot 6$ **b)** $(-3)(-3)(-3)(-3)$ **c)** $\frac{3}{2} \cdot \frac{3}{2} \cdot \frac{3}{2}$

Solution

a) $6 \cdot 6 \cdot 6 \cdot 6 \cdot 6 = 6^5$

b) $(-3)(-3)(-3)(-3) = (-3)^4$

c) $\frac{3}{2} \cdot \frac{3}{2} \cdot \frac{3}{2} = \left(\frac{3}{2}\right)^3$ ■

EXAMPLE 3 **Writing an exponential expression as a product**

Write each exponential expression as a product without exponents.

a) y^6 **b)** $(-2)^4$ **c)** $\left(\frac{5}{4}\right)^3$ **d)** $(-0.1)^2$

Solution

a) $y^6 = y \cdot y \cdot y \cdot y \cdot y \cdot y$

b) $(-2)^4 = (-2)(-2)(-2)(-2)$

c) $\left(\frac{5}{4}\right)^3 = \frac{5}{4} \cdot \frac{5}{4} \cdot \frac{5}{4}$

d) $(-0.1)^2 = (-0.1)(-0.1)$ ■

To evaluate an exponential expression, write the base as many times as indicated by the exponent, then multiply the factors from left to right.

EXAMPLE 4 **Evaluating exponential expressions**

Evaluate.

a) 3^3 **b)** $(-2)^3$ **c)** $\left(\frac{2}{3}\right)^4$ **d)** $(0.4)^2$

Solution

a) $3^3 = 3 \cdot 3 \cdot 3 = 9 \cdot 3 = 27$

b)
$$\begin{aligned}(-2)^3 &= (-2)(-2)(-2)\\ &= 4(-2)\\ &= -8\end{aligned}$$

c)
$$\begin{aligned}\left(\frac{2}{3}\right)^4 &= \frac{2}{3} \cdot \frac{2}{3} \cdot \frac{2}{3} \cdot \frac{2}{3}\\ &= \frac{4}{9} \cdot \frac{2}{3} \cdot \frac{2}{3}\\ &= \frac{8}{27} \cdot \frac{2}{3}\\ &= \frac{16}{81}\end{aligned}$$

d) $(0.4)^2 = (0.4)(0.4) = 0.16$ ■

CAUTION Note that $3^3 \neq 9$. We do not multiply the exponent and the base when evaluating an exponential expression.

Be especially careful with exponential expressions involving negative numbers. An exponential expression with a negative base is written with parentheses around the base as in $(-2)^4$:

$$(-2)^4 = (-2)(-2)(-2)(-2) = 16$$

To evaluate $-(2^4)$, use the base 2 as a factor four times, then find the opposite:

$$-(2^4) = -(2 \cdot 2 \cdot 2 \cdot 2) = -(16) = -16$$

We often omit the parentheses in $-(2^4)$ and simply write -2^4. So

$$-2^4 = -(2^4) = -16.$$

To evaluate $-(-2)^4$, use the base -2 as a factor four times, then find the opposite:

$$-(-2)^4 = -(16) = -16$$

EXAMPLE 5 **Evaluating exponential expressions involving negative numbers**

Evaluate.

a) $(-10)^4$ **b)** -10^4 **c)** $-(-0.5)^2$ **d)** $-(5-8)^2$

Solution

a) $(-10)^4 = (-10)(-10)(-10)(-10)$ Use -10 as a factor four times.
$= 10{,}000$

b) $-10^4 = -(10^4)$ Rewrite using parentheses.
$= -(10{,}000)$ Find 10^4.
$= -10{,}000$ Then find the opposite of 10,000.

c) $-(-0.5)^2 = -(-0.5)(-0.5)$ Use -0.5 as a factor two times.
$= -(0.25)$
$= -0.25$

d) $-(5-8)^2 = -(-3)^2$ Evaluate within parentheses first.
$= -(9)$ Square -3 to get 9.
$= -9$ Take the opposite of 9 to get -9.

The Order of Operations

To simplify writing of expressions, parentheses are often omitted as in the expression $3 + 2 \cdot 5$. When no parentheses are present, we agree to perform multiplication before addition. So

$$3 + 2 \cdot 5 = 3 + 10 = 13.$$

To evaluate expressions consistently, we follow an accepted **order of operations.** When no grouping symbols are present, we perform operations in the following order:

Order of Operations

1. Evaluate each exponential expression (in order from left to right).
2. Perform multiplication and division (in order from left to right).
3. Perform addition and subtraction (in order from left to right).

EXAMPLE 6 **Using the order of operations**

Evaluate each expression.

a) $2^3 \cdot 3^2$ **b)** $2 \cdot 5 - 3 \cdot 4 + 4^2$ **c)** $2 \cdot 3 \cdot 4 - 3^3 + \frac{8}{2}$

Solution

a) $2^3 \cdot 3^2 = 8 \cdot 9$ Evaluate exponential expressions before multiplying.

$= 72$

b) $2 \cdot 5 - 3 \cdot 4 + 4^2 = 2 \cdot 5 - 3 \cdot 4 + 16$ Exponential expressions first

$= 10 - 12 + 16$ Multiplication second

$= 14$ Addition and subtraction from left to right

c) $2 \cdot 3 \cdot 4 - 3^3 + \frac{8}{2} = 2 \cdot 3 \cdot 4 - 27 + \frac{8}{2}$ Exponential expressions first

$= 24 - 27 + 4$ Multiplication and division second

$= 1$ Addition and subtraction from left to right ■

When grouping symbols are used, we perform operations within grouping symbols first. The order of operations is followed within the grouping symbols.

EXAMPLE 7 **Grouping symbols and the order of operations**

Evaluate.

a) $3 - 2(7 - 2^3)$ **b)** $3 - |7 - 3 \cdot 4|$ **c)** $\frac{9 - 5 + 8}{-5^2 - 3(-7)}$

Solution

a) $3 - 2(7 - 2^3) = 3 - 2(7 - 8)$ Evaluate within parentheses first.

$= 3 - 2(-1)$

$= 3 - (-2)$ Multiply.

$= 5$ Subtract.

b) $3 - |7 - 3 \cdot 4| = 3 - |7 - 12|$ Evaluate within the absolute value symbols first.

$= 3 - |-5|$

$= 3 - 5$ Evaluate the absolute value.

$= -2$ Subtract.

c) $\frac{9 - 5 + 8}{-5^2 - 3(-7)} = \frac{12}{-25 + 21} = \frac{12}{-4} = -3$ Numerator and denominator are treated as if in parentheses. ■

When grouping symbols occur within grouping symbols, we evaluate within the innermost grouping symbols first and then work outward. In this case, brackets [ ] can be used as grouping symbols along with parentheses to make the grouping clear.

EXAMPLE 8 **Grouping within grouping**

Evaluate each expression.

a) $6 - 4[5 - (7 - 9)]$

b) $-2|3 - (9 - 5)| - |-3|$

Solution

a)

$$\begin{aligned} 6 - 4[5 - (7 - 9)] &= 6 - 4[5 - (-2)] && \text{Innermost parentheses first} \\ &= 6 - 4[7] && \text{Next evaluate within the brackets.} \\ &= 6 - 28 && \text{Multiply.} \\ &= -22 && \text{Subtract.} \end{aligned}$$

b)

$$\begin{aligned} -2|3 - (9 - 5)| - |-3| &= -2|3 - 4| - |-3| && \text{Innermost grouping first} \\ &= -2|-1| - |-3| && \text{Evaluate within the first absolute value.} \\ &= -2 \cdot 1 - 3 && \text{Evaluate absolute values.} \\ &= -2 - 3 && \text{Multiply.} \\ &= -5 && \text{Subtract.} \end{aligned}$$

■

Calculator Close-up

You can use your calculator to evaluate exponential expressions. The x^y key is used to raise a number to a power. For example, to evaluate 2^5 using a scientific calculator, try the following keystrokes:

2 [x^y] 5 [=]

The display will read

32

If x^y is printed above another key, then it is the second function for that key and you must press [2nd] before pressing the key. Even though the x^y key can be used to find any power of a number, many calculators also have an x^2 key, which can be used for squaring a number.

On a graphing calculator the symbol ^ is used to indicate that a number is an exponent. So to evaluate 2^5, the display should read as follows:

2^5 [ENTER]

32

Most scientific and graphing calculators have parentheses that can be used for grouping, and they follow the order of operations when no parentheses are used. Use your calculator to verify the following equations:

$$4.1 - (3.2)(4.5) = -10.3$$

$$(4.8 + 2.7)6.2 = 46.5$$

$$4 \cdot 5.6^3 - 3.2^2 = 692.224$$

Warm-ups

True or false? Explain your answer.

1. $(-3)^2 = -6$
2. $5 - 3 \cdot 2 = 4$
3. $(5 - 3)2 = 4$
4. $|5 - 6| = |5| - |6|$
5. $5 + 6 \cdot 2 = (5 + 6) \cdot 2$
6. $(2 + 3)^2 = 2^2 + 3^2$
7. $5 - 3^3 = 8$
8. $(5 - 3)^3 = 8$
9. $6 - \frac{6}{2} = \frac{0}{2}$
10. $\frac{6 - 6}{2} = 0$

1.5 EXERCISES

Evaluate each expression. See Example 1.

1. $(4 - 3)(5 - 9)$
2. $(5 - 7)(-2 - 3)$
3. $|3 + 4| - |-2 - 4|$
4. $|-4 + 9| + |-3 - 5|$
5. $\frac{7 - (-9)}{3 - 5}$
6. $\frac{-8 + 2}{-1 - 1}$
7. $(-6 + 5)(7)$
8. $-6 + (5 \cdot 7)$
9. $(-3 - 7) - 6$
10. $-3 - (7 - 6)$
11. $-16 \div (8 \div 2)$
12. $(-16 \div 8) \div 2$

Write each product as an exponential expression. See Example 2.

13. $4 \cdot 4 \cdot 4 \cdot 4$
14. $1 \cdot 1 \cdot 1 \cdot 1 \cdot 1$
15. $(-5)(-5)(-5)(-5)$
16. $(-7)(-7)(-7)$
17. $(-y)(-y)(-y)$
18. $x \cdot x \cdot x \cdot x \cdot x$
19. $\frac{3}{7} \cdot \frac{3}{7} \cdot \frac{3}{7} \cdot \frac{3}{7} \cdot \frac{3}{7}$
20. $\frac{y}{2} \cdot \frac{y}{2} \cdot \frac{y}{2} \cdot \frac{y}{2}$

Write each exponential expression as a product without exponents. See Example 3.

21. 5^3
22. $(-8)^4$
23. b^2
24. $(-a)^5$
25. $\left(-\frac{1}{2}\right)^5$
26. $\left(-\frac{13}{12}\right)^3$
27. $(0.22)^4$
28. $(1.25)^6$

Evaluate each exponential expression. See Examples 4 and 5.

29. 3^4
30. 5^3
31. 0^9
32. 0^{12}
33. $(-5)^4$
34. $(-2)^5$
35. $(-6)^3$
36. $(-12)^2$
37. $(10)^5$
38. $(-10)^6$
39. $(-0.1)^3$
40. $(-0.2)^2$
41. $\left(\frac{1}{2}\right)^3$
42. $\left(\frac{2}{3}\right)^3$
43. $\left(-\frac{1}{2}\right)^2$
44. $\left(-\frac{2}{3}\right)^2$
45. -8^2
46. -7^2
47. $-(-8)^4$
48. $-(-7)^3$
49. $-(7 - 10)^3$
50. $-(6 - 9)^4$
51. $(-2^2) - (3^2)$
52. $(-3^4) - (-5^2)$

Evaluate each expression. See Example 6.

53. $3^2 \cdot 2^2$
54. $5 \cdot 10^2$
55. $-3 \cdot 2 + 4 \cdot 6$
56. $-5 \cdot 4 - 8 \cdot 3$
57. $(-3)^3 + 2^3$
58. $3^2 - 5(-1)^3$
59. $-21 + 36 \div 3^2$
60. $-18 - 9^2 \div 3^3$
61. $-3 \cdot 2^3 - 5 \cdot 2^2$
62. $2 \cdot 5 - 3^2 + 4 \cdot 0$
63. $\frac{-8}{2} + 2 \cdot 3 \cdot 5 - 2^3$
64. $-4 \cdot 2 \cdot 6 - \frac{12}{3} + 3^3$

Evaluate each expression. See Example 7.

65. $(-3 + 4^2)(-6)$
66. $-3 \cdot (2^3 + 4) \cdot 5$
67. $(-3 \cdot 2 + 6)^3$
68. $5 - 2(-3 + 2)^3$
69. $2 - 5(3 - 4 \cdot 2)$
70. $(3 - 7)(4 - 6 \cdot 2)$
71. $3 - 2 \cdot |5 - 6|$
72. $3 - |6 - 7 \cdot 3|$
73. $(3^2 - 5) \cdot |3 \cdot 2 - 8|$
74. $|4 - 6 \cdot 3| + |6 - 9|$
75. $\frac{3 - 4 \cdot 6}{7 - 10}$
76. $\frac{6 - (-8)^2}{-3 - (-1)}$
77. $\frac{7 - 9 - 3^2}{9 - 7 - 3}$
78. $\frac{3^2 - 2 \cdot 4}{-30 + 2 \cdot 4^2}$

Evaluate each expression. See Example 8.

79. $3 + 4[9 - 6(2 - 5)]$
80. $9 + 3[5 - (3 - 6)^2]$
81. $6^2 - [(2 + 3)^2 - 10]$
82. $3[(2 - 3)^2 + (6 - 4)^2]$
83. $4 - 5 \cdot |3 - (3^2 - 7)|$
84. $2 + 3 \cdot |4 - (7^2 - 6^2)|$
85. $-2|3 - (7 - 3)| - |-9|$
86. $[3 - (2 - 4)][3 + |2 - 4|]$

Evaluate each expression.

87. $1 + 2^3$

88. $(1 + 2)^3$

89. $(-2)^2 - 4(-1)(3)$

90. $(-2)^2 - 4(-2)(-3)$

91. $4^2 - 4(1)(-3)$

92. $3^2 - 4(-2)(3)$

93. $(-11)^2 - 4(5)(0)$

94. $(-12)^2 - 4(3)(0)$

95. $-5^2 - 3 \cdot 4^2$

96. $-6^2 - 5(-3)^2$

97. $[3 + 2(-4)]^2$

98. $[6 - 2(-3)]^2$

99. $|-1| - |-1|$

100. $4 - |1 - 7|$

101. $\dfrac{4 - (-4)}{-2 - 2}$

102. $\dfrac{3 - (-7)}{3 - 5}$

103. $3(-1)^2 - 5(-1) + 4$

104. $-2(1)^2 - 5(1) - 6$

105. $5 - 2^2 + 3^4$

106. $5 + (-2)^2 - 3^2$

107. $-2 \cdot |9 - 6^2|$

108. $8 - 3|5 - 4^2 + 1|$

109. $-3^2 - 5[4 - 2(4 - 9)]$

110. $-2[(3 - 4)^3 - 5] + 7$

111. $1 - 5|5 - (9 + 1)|$

112. $|6 - 3 \cdot 7| + |7 - (5 - 2)|$

Use a calculator to evaluate each expression.

113. $3.2^2 - 4(3.6)(-2.2)$

114. $(-4.5)^2 - 4(-2.8)(-4.6)$

115. $(5.63)^3 - [4.7 - (-3.3)^2]$

116. $9.8^3 - [1.2 - (4.4 - 9.6)^2]$

117. $\dfrac{3.44 - (-8.32)}{6.89 - 5.43}$

118. $\dfrac{-4.56 - 3.22}{3.44 - (-6.26)}$

Solve each problem.

119. ***Population of the United States.*** In 1989 the population of the United States was 246.3 million *(World Resources 1988–1989)*. If the population continues to grow at the annual rate of 0.86%, then the population in the year 2000 will be $246.3(1.0086)^{11}$ million. Find the predicted population in the year 2000 to the nearest tenth of a million.

120. ***Population of Mexico.*** In 1989 the population of Mexico was 87 million *(World Resources 1988–1989)*. If Mexico's population continues to grow at an annual rate of 2.39%, then the population in the year 2000 will be $87(1.0239)^{11}$ million. Find the predicted population in the year 2000 to the nearest tenth of a million. According to this exercise and Exercise 119, will the United States or Mexico have the greater increase in population between the years 1989 and 2000?

Getting More Involved

121. ***Writing.*** What is the purpose of the order of operations?

122. ***Discussion.*** Which operation symbols are used also as grouping symbols?

123. ***Discussion.*** What is wrong with each of the following expressions?

a) $-6(4 - (3 - (-5))$

b) $(4 - 6)(-5 - 3[-6 + 4) - 8]$

c) $\dfrac{-5 - (-7}{-3 - 2)}$

124. ***Discussion.*** How do the expressions $(-5)^3$, $-(5^3)$, -5^3, $-(-5)^3$, and $-1 \cdot 5^3$ differ?

125. ***Discussion.*** How do the expressions $(-4)^4$, $-(4^4)$, -4^4, $-(-4)^4$, and $-1 \cdot 4^4$ differ?

1.6 Algebraic Expressions

In this section:

- Identifying Algebraic Expressions
- Translating Algebraic Expressions
- Evaluating Algebraic Expressions
- Equations
- Applications

In Section 1.5 you studied arithmetic expressions. In this section you will study expressions that are more general—expressions that involve variables.

Identifying Algebraic Expressions

Since variables (or letters) are used to represent numbers, we can use variables in arithmetic expressions. The result of combining numbers and variables with the ordinary operations of arithmetic (in some meaningful way) is called an **algebraic expression** or simply an **expression.** For example,

$$x + 2, \quad \pi r^2, \quad b^2 - 4ac, \quad \text{and} \quad \frac{a - b}{c - d}$$

are algebraic expressions.

Expressions are often named by the last operation to be performed in the expression. For example, the expression $x + 2$ is a **sum** because the only operation in the expression is addition. The expression $a - bc$ is referred to as a **difference** because subtraction is the last operation to be performed. The expression $3(x - 4)$ is a **product,** while $\frac{3}{x-4}$ is a **quotient.** The expression $(a + b)^2$ is a **square** because the addition is performed before the square is found.

EXAMPLE 1 **Naming expressions**

Identify each expression as either a sum, difference, product, quotient, or square.

a) $3(x + 2)$

b) $b^2 - 4ac$

c) $\dfrac{a - b}{c - d}$

d) $(a - b)^2$

Solution

a) In $3(x + 2)$ we add before we multiply. So this expression is a product.

b) By the order of operations the last operation to perform in $b^2 - 4ac$ is subtraction. So this expression is a difference.

c) The last operation to perform in this expression is division. So this expression is a quotient.

d) In $(a - b)^2$ we subtract before we square. This expression is a square. ■

Translating Algebraic Expressions

Algebra is useful because it can be used to solve problems. Since problems are often communicated verbally, we must be able to translate verbal expressions into algebraic expressions and translate algebraic expressions into verbal expressions. Consider the following examples of verbal expressions and their corresponding algebraic expressions.

Verbal Expressions and Corresponding Algebraic Expressions

Verbal Expression	Algebraic Expression
The sum of $5x$ and 3	$5x + 3$
The product of 5 and $x + 3$	$5(x + 3)$
The sum of 8 and $\frac{x}{3}$	$8 + \dfrac{x}{3}$
The quotient of $8 + x$ and 3	$\dfrac{8 + x}{3}$, $(8 + x)/3$, or $(8 + x) \div 3$
The difference of 3 and x^2	$3 - x^2$
The square of $3 - x$	$(3 - x)^2$

Because of the order of operations, reading from left to right does not always describe an expression accurately. For example, $5x + 3$ and $5(x + 3)$ can both be read as "5 times x plus 3." The next example shows how the terms sum, difference, product, quotient, and square are used to describe expressions. (You will study verbal and algebraic expressions further in Section 2.4.)

EXAMPLE 2 **Algebraic expressions to verbal expressions**

Translate each algebraic expression into a verbal expression. Use the word sum, difference, product, quotient, or square.

a) $\frac{3}{x}$ **b)** $2y + 1$ **c)** $3x - 2$

d) $(a - b)(a + b)$ **e)** $(a + b)^2$

Solution

a) The quotient of 3 and x

b) The sum of $2y$ and 1

c) The difference of $3x$ and 2

d) The product of $a - b$ and $a + b$

e) The square of the sum $a + b$ ■

EXAMPLE 3 **Verbal expressions to algebraic expressions**

Translate each verbal expression into an algebraic expression.

a) The quotient of $a + b$ and 5 **b)** The difference of x^2 and y^2

c) The product of π and r^2 **d)** The square of the difference $x - y$

Solution

a) $\frac{a + b}{5}$, $(a + b) \div 5$, or $(a + b)/5$

b) $x^2 - y^2$

c) πr^2

d) $(x - y)^2$ ■

Evaluating Algebraic Expressions

The value of an algebraic expression depends on the values given to the variables. For example, the value of $x - 2y$ when $x = -2$ and $y = -3$ is found by replacing x and y by -2 and -3, respectively:

$$x - 2y = -2 - 2(-3) = -2 - (-6) = 4$$

If $x = 1$ and $y = 2$, the value of $x - 2y$ is found by replacing x by 1 and y by 2, respectively:

$$x - 2y = 1 - 2(2) = 1 - 4 = -3$$

Note that we use the order of operations when evaluating an algebraic expression.

EXAMPLE 4 **Evaluating algebraic expressions**

Evaluate each expression using $a = 3$, $b = -2$, and $c = -4$.

a) $a^2 + 2ab + b^2$

b) $(a - b)(a + b)$

c) $b^2 - 4ac$

d) $\dfrac{-a^2 - b^2}{c - b}$

Solution

a)
$$\begin{aligned} a^2 + 2ab + b^2 &= 3^2 + 2(3)(-2) + (-2)^2 && \text{Replace } a \text{ by 3 and } b \text{ by } -2. \\ &= 9 + (-12) + 4 && \text{Evaluate.} \\ &= 1 && \text{Add.} \end{aligned}$$

b)
$$\begin{aligned} (a - b)(a + b) &= [3 - (-2)][3 + (-2)] && \text{Replace.} \\ &= [5][1] && \text{Simplify within the brackets.} \\ &= 5 && \text{Multiply.} \end{aligned}$$

c)
$$\begin{aligned} b^2 - 4ac &= (-2)^2 - 4(3)(-4) && \text{Replace.} \\ &= 4 - (-48) && \text{Square } -2\text{, and then multiply before subtracting.} \\ &= 52 && \text{Subtract.} \end{aligned}$$

d) $$\frac{-a^2 - b^2}{c - b} = \frac{-3^2 - (-2)^2}{-4 - (-2)} = \frac{-9 - 4}{-2} = \frac{13}{2}$$ ■

Equations

An **equation** is a statement of equality of two expressions. For example,

$$11 - 5 = 6, \qquad x + 3 = 9, \qquad 2x + 5 = 13, \qquad \text{and} \qquad \frac{x}{2} - 4 = 1$$

are equations. In an equation involving a variable, any number that gives a true statement when we replace the variable by the number is said to **satisfy** the equation and is called a **solution** or **root** to the equation. For example, 6 is a solution to $x + 3 = 9$ because $6 + 3 = 9$ is true. Because $5 + 3 = 9$ is false, 5 is not a solution to the equation $x + 3 = 9$. We have **solved** an equation when we have found all solutions to the equation. You will learn how to solve certain equations in the next chapter.

EXAMPLE 5 **Satisfying an equation**

Determine whether the given number is a solution to the equation following it.

a) $6,\ 3x - 7 = 9$

b) $-3,\ \dfrac{2x - 4}{5} = -2$

c) $-5,\ -x - 2 = 3$

Solution

a) Replace x by 6 in the equation $3x - 7 = 9$:

$$3(6) - 7 = 9$$
$$18 - 7 = 9$$
$$11 = 9 \quad \text{False.}$$

The number 6 is not a solution to the equation $3x - 7 = 9$.

b) Replace x by -3 in the equation $\frac{2x - 4}{5} = -2$:

$$\frac{2(-3) - 4}{5} = -2$$
$$\frac{-10}{5} = -2$$
$$-2 = -2 \quad \text{True.}$$

The number -3 is a solution to the equation.

c) Replace x by -5 in $-x - 2 = 3$:

$$-(-5) - 2 = 3$$
$$5 - 2 = 3 \quad \text{True.}$$

The number -5 is a solution to the equation $-x - 2 = 3$. ■

Just as we translated verbal expressions into algebraic expressions, we can translate verbal sentences into algebraic equations.

EXAMPLE 6 **Writing equations**

Translate each sentence into an equation.

a) The sum of x and 7 is 12.

b) The product of 4 and x is the same as the sum of y and 5.

c) The quotient of $x + 3$ and 5 is equal to -1.

Solution

a) $x + 7 = 12$

b) $4x = y + 5$

c) $\frac{x + 3}{5} = -1$ ■

Applications

Algebraic expressions are used to describe or **model** real-life situations. We can evaluate an algebraic expression for many values of a variable to get a collection of data. A graph (picture) of this data can give us useful information. For example, a forensic scientist can use a graph to estimate the length of a person's femur from the person's height.

EXAMPLE 7 **Reading a graph**

A forensic scientist uses the expression $69.1 + 2.2F$ as an estimate of the height in centimeters of a male with a femur of length F centimeters (*American Journal of Physical Anthropology,* 1952).

a) If the femur of a male skeleton measures 50.6 cm, then what was the person's height?

b) Use the graph shown in Fig. 1.15 to estimate the length of a femur for a person who is 150 cm tall.

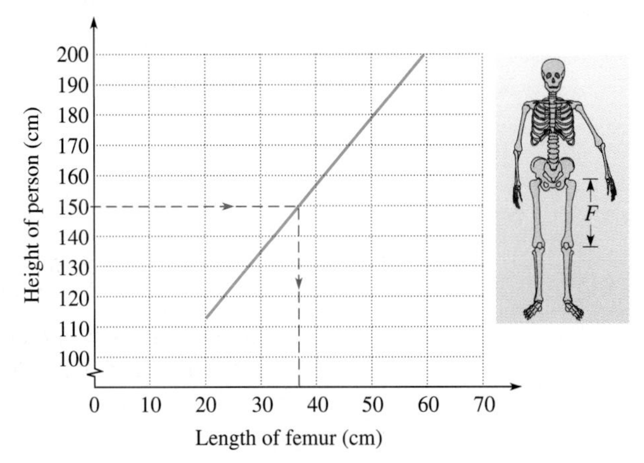

Figure 1.15

Solution

a) To find the height of the person, we use $F = 50.6$ in the expression $69.1 + 2.2F$:

$$69.1 + 2.2(50.6) \approx 180.4$$

So the person was approximately 180.4 cm tall.

b) To find the length of a femur for a person who is 150 cm tall, first locate 150 cm on the height scale of the graph in Fig. 1.15. Now draw a horizontal line to the graph and then a vertical line down to the length scale. So the length of femur for a person who is 150 cm tall is approximately 35 cm. ■

Warm-ups

True or false? Explain your answer.

1. The expression $2x + 3y$ is referred to as a sum.

2. The expression $5(y - 9)$ is a difference.

3. The expression $2(x + 3y)$ is a product.

4. The expression $\frac{x}{2} + \frac{y}{3}$ is a quotient.

5. The expression $(a - b)(a + b)$ is a product of a sum and a difference.

6. If x is -2, then the value of $2x + 4$ is 8.

7. If $a = -3$, then $a^3 - 5 = 22$.

8. The number 5 is a solution to the equation $2x - 3 = 13$.

9. The product of $x + 3$ and 5 is $(x + 3)5$.

10. The expression $2(x + 7)$ should be read as "the sum of 2 times x plus 7."

1.6 EXERCISES

Identify each expression as a sum, difference, product, quotient, square, or cube. See Example 1.

1. $a^3 - 1$

2. $b(b - 1)$

3. $(w - 1)^3$

4. $m^2 + n^2$

5. $3x + 5y$

6. $\dfrac{a - b}{b - a}$

7. $\dfrac{u}{v} - \dfrac{v}{u}$

8. $(s - t)^2$

9. $3(x + 5y)$

10. $a - \dfrac{a}{2}$

11. $\left(\dfrac{2}{z}\right)^2$

12. $(2q - p)^3$

Use the term sum, difference, product, quotient, square, or cube to translate each algebraic expression into a verbal expression. See Example 2.

13. $x^2 - a^2$

14. $a^3 + b^3$

15. $(x - a)^2$

16. $(a + b)^3$

17. $\dfrac{x - 4}{2}$

18. $2(x - 3)$

19. $\dfrac{x}{2} - 4$

20. $2x - 3$

21. $(ab)^3$

22. a^3b^3

Translate each verbal expression into an algebraic expression. See Example 3.

23. The sum of $2x$ and $3y$

24. The product of $5x$ and z

25. The difference of 8 and $7x$

26. The quotient of 6 and $x + 4$

27. The square of $a + b$

28. The difference of a^3 and b^3

29. The product of $x + 9$ and $x + 12$

30. The cube of x

31. The quotient of $x - 7$ and $7 - x$

32. The product of -3 and $x - 1$

Evaluate each expression using $a = -1$, $b = 2$, and $c = -3$. See Example 4.

33. $-(a - b)$

34. $b - a$

35. $-b^2 + 7$

36. $-c^2 - b^2$

37. $c^2 - 2c + 1$

38. $b^2 - 2b + 4$

39. $a^3 - b^3$

40. $b^3 - c^3$

41. $(a - b)(a + b)$

42. $(a - c)(a + c)$

43. $b^2 - 4ac$

44. $a^2 - 4bc$

45. $\dfrac{a - c}{a - b}$

46. $\dfrac{b - c}{b + a}$

47. $\dfrac{2}{a} + \dfrac{6}{b} - \dfrac{9}{c}$

48. $\dfrac{c}{a} + \dfrac{6}{c} - \dfrac{b}{a}$

49. $a \div |-a|$

50. $|a| \div a$

51. $|b| - |a|$

52. $|c| + |b|$

53. $-|-a - c|$

54. $-|-a - b|$

55. $(3 - |a - b|)^2$

56. $(|b + c| - 2)^3$

Determine whether the given number is a solution to the equation following it. See Example 5.

57. $2,\ 3x + 7 = 13$

58. $-1,\ -3x + 7 = 10$

59. $-2,\ \dfrac{3x - 4}{2} = 5$

60. $-3,\ \dfrac{-2x + 9}{3} = 5$

61. $-2,\ -x + 4 = 6$

62. $-9,\ -x + 3 = 12$

63 $4,\ 3x - 7 = x + 1$

64. $5,\ 3x - 7 = 2x + 1$

65. $3,\ -2(x - 1) = 2 - 2x$

66. $-8,\ x - 9 = -(9 - x)$

67. $1,\ x^2 + 3x - 4 = 0$

68. $-1,\ x^2 + 5x + 4 = 0$

69. $8,\ \dfrac{x}{x - 8} = 0$

70. $3,\ \dfrac{x - 3}{x + 3} = 0$

71. $-6,\ \dfrac{x + 6}{x + 6} = 1$

72. $9,\ \dfrac{9}{x - 9} = 0$

Translate each sentence into an equation. See Example 6.

73. The sum of $5x$ and $3x$ is $8x$.

74. The sum of $\frac{y}{2}$ and 3 is 7.

75. The product of 3 and $x + 2$ is equal to 12.

76. The product of -6 and $7y$ is equal to 13.

77. The quotient of x and 3 is the same as the product of x and 5.

78. The quotient of $x + 3$ and $5y$ is the same as the product of x and y.

79. The square of the sum of a and b is equal to 9.

80. The sum of the squares of a and b is equal to the square of c.

Use a calculator to find the value of $b^2 - 4ac$ for each of the following choices of a, b, and c.

81. $a = 4.2,\ b = 6.7,\ c = 1.8$

82. $a = -3.5,\ b = 9.1,\ c = 3.6$

83. $a = -1.2,\ b = 3.2,\ c = 5.6$

84. $a = 2.4,\ b = -8.5,\ c = -5.8$

Solve each problem. See Example 7.

85. ***Forensics.*** A forensic scientist uses the expression $81.7 + 2.4T$ to estimate the height in centimeters of a male with a tibia of length T centimeters. If a male skeleton has a tibia of length 36.5 cm, then what was the height of the person? Use the graph shown here to estimate the length of a tibia for a male with a height of 180 cm.

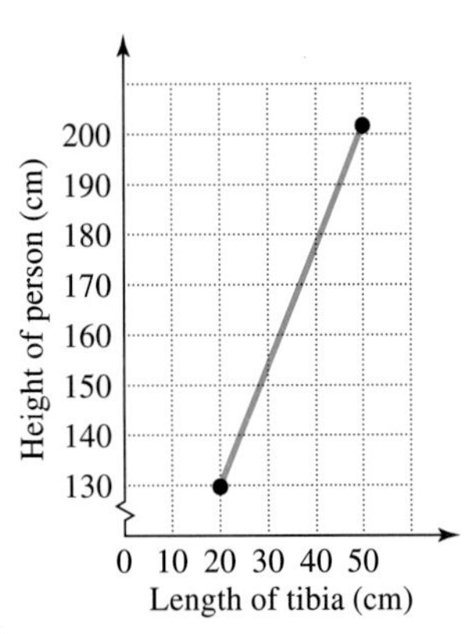

Figure for Exercise 85

86. ***Forensics.*** A forensic scientist uses the expression $72.6 + 2.5T$ to estimate the height in centimeters of a female with a tibia of length T centimeters. If a female skeleton has a tibia of length 32.4 cm, then what was the height of the person? Find the length of your tibia in centimeters, and use the expression from this exercise or the previous exercise to estimate your height.

87. ***Games behind.*** In baseball a team's standing is measured by its percentage of wins and by the number of games it is behind the leading team in its division. The expression

$$\frac{(X - x) + (y - Y)}{2}$$

gives the number of games behind for a team with x wins and y losses, where the division leader has X wins and Y losses. The table shown here gives the won-lost records for the American League West on June 17, 1994. On that date, Texas was the division leader. Fill in the column for the games behind (GB).

	W	L	Pct	GB
Texas	31	32	0.492	—
Seattle	27	36	0.429	?
California	27	39	0.409	?
Oakland	21	43	0.328	?

Table for Exercise 87

88. ***Fly ball.*** The approximate distance in feet that a baseball travels when hit at an angle of 45° is given by the expression

$$\frac{(v_0)^2}{32}$$

where v_0 is the initial velocity in feet per second. If Barry Bonds of the Giants hits a ball at a 45° angle with an initial velocity of 120 feet per second, then how far will the ball travel? Use the graph shown here to estimate the initial velocity for a ball that has traveled 370 feet.

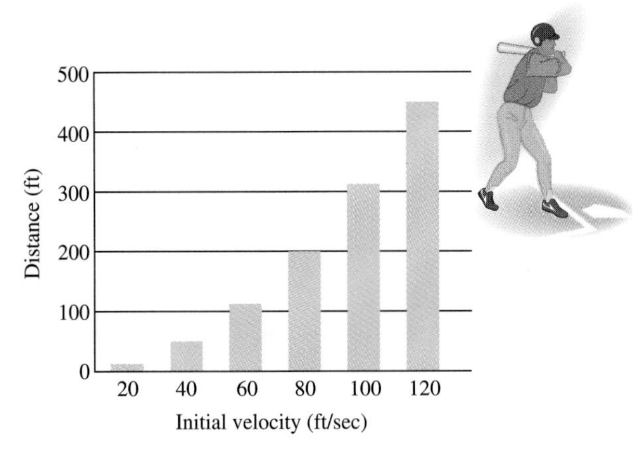

Figure for Exercise 88

89. ***Football field.*** The expression $2L + 2W$ gives the perimeter of a rectangle with length L and width W. What is the perimeter of a football field with length 100 yards and width 160 feet?

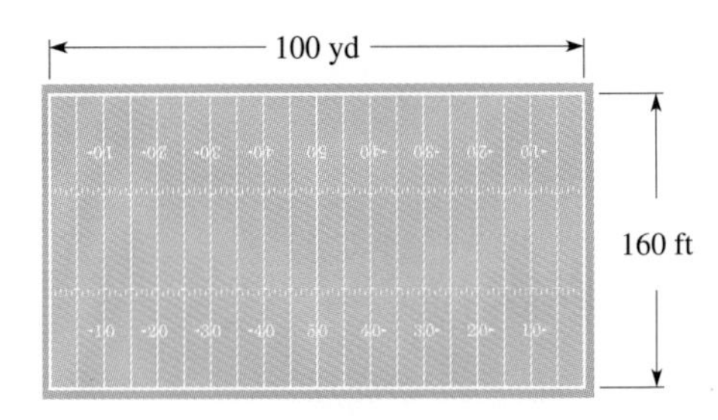

Figure for Exercise 89

90. ***Crop circles.*** The expression πr^2 gives the area of a circle with radius r. How many square meters of wheat were destroyed when an alien ship made a crop circle of diameter 25 meters in the wheat field at the Southwind Ranch?

Getting More Involved

91. ***Cooperative learning.*** Find some examples of algebraic expressions outside of this class, and explain to the class what they are used for?

92. ***Discussion.*** Why do we use letters to represent numbers? Wouldn't it be simpler to just use numbers?

93. ***Writing.*** Explain why the square of the sum of two numbers is different from the sum of the squares of two numbers?

1.7 Properties of the Real Numbers

In this section:

- The Commutative Properties
- The Associative Properties
- The Distributive Property
- The Identity Properties
- The Inverse Properties
- Multiplication Property of Zero
- Applications

Everyone knows that the price of a hamburger plus the price of a Coke is the same as the price of a Coke plus the price of a hamburger. But do you know that this example illustrates the commutative property of addition? The properties of the real numbers are commonly used by anyone who performs the operations of arithmetic. In algebra we must have a thorough understanding of these properties.

The Commutative Properties

We get the same result whether we evaluate $3 + 5$ or $5 + 3$. This example illustrates the commutative property of addition. The fact that $4 \cdot 6$ and $6 \cdot 4$ are equal illustrates the commutative property of multiplication.

Commutative Properties

For any real numbers a and b,

$$a + b = b + a \quad \text{and} \quad ab = ba.$$

EXAMPLE 1 **The commutative property of addition**

Use the commutative property of addition to rewrite each expression.

a) $2 + (-10)$ **b)** $8 + x^2$ **c)** $2y - 4x$

Solution

a) $2 + (-10) = -10 + 2$

b) $8 + x^2 = x^2 + 8$

c) $2y - 4x = 2y + (-4x) = -4x + 2y$ ■

EXAMPLE 2 **The commutative property of multiplication**

Use the commutative property of multiplication to rewrite each expression.

a) $n \cdot 3$ **b)** $(x + 2) \cdot 3$ **c)** $5 - yx$

Solution

a) $n \cdot 3 = 3 \cdot n = 3n$

b) $(x + 2) \cdot 3 = 3(x + 2)$

c) $5 - yx = 5 - xy$ ■

Addition and multiplication are commutative operations, but what about subtraction and division? Since $5 - 3 = 2$ and $3 - 5 = -2$, subtraction is not commutative. To see that division is not commutative, try dividing $8 among 4 people and $4 among 8 people.

The Associative Properties

Consider the computation of $2 + 3 + 6$. Using the order of operations, we add 2 and 3 to get 5 and then add 5 and 6 to get 11. If we add 3 and 6 first to get 9 and then add 2 and 9, we also get 11. So

$$(2 + 3) + 6 = 2 + (3 + 6).$$

We get the same result for either order of addition. This property is called the **associative property of addition.** The commutative and associative properties of addition are the reason that a hamburger, a Coke, and French fries cost the same as French fries, a hamburger, and a Coke.

We also have an **associative property of multiplication.** Consider the following two ways to find the product of 2, 3 and 4:

$$(2 \cdot 3)4 = 6 \cdot 4 = 24$$
$$2(3 \cdot 4) = 2 \cdot 12 = 24$$

We get the same result for either arrangement.

Associative Properties

For any real numbers a, b, and c,

$$(a + b) + c = a + (b + c) \qquad \text{and} \qquad (ab)c = a(bc).$$

EXAMPLE 3 **Using the properties of multiplication**

Use the commutative and associative properties of multiplication and exponential notation to rewrite each product.

a) $(3x)(x)$ **b)** $(xy)(5yx)$

Solution

a) $(3x)(x) = 3(x \cdot x) = 3x^2$

b) The commutative and associative properties of multiplication allow us to rearrange the multiplication in any order. We generally write numbers before variables, and we usually write variables in alphabetical order:

$$(xy)(5yx) = 5xxyy = 5x^2y^2$$

■

Consider the expression

$$3 - 9 + 7 - 5 - 8 + 4 - 13.$$

According to the accepted order of operations, we could evaluate this by computing from left to right. However, using the definition of subtraction, we can rewrite this expression as addition:

$$3 + (-9) + 7 + (-5) + (-8) + 4 + (-13)$$

The commutative and associative properties of addition allow us to add these numbers in any order we choose. It is usually faster to add the positive numbers, add the negative numbers, and then combine those two totals:

$$3 + 7 + 4 + (-9) + (-5) + (-8) + (-13) = 14 + (-35) = -21$$

Note that by performing the operations in this manner, we must subtract only once. There is no need to rewrite this expression as we have done here. We can sum the positive numbers and the negative numbers from the original expression and then combine their totals.

EXAMPLE 4 **Using the properties of addition**

Evaluate.

a) $3 - 7 + 9 - 5$ **b)** $4 - 5 - 9 + 6 - 2 + 4 - 8$

Solution

a) First add the positive numbers and the negative numbers:

$$\begin{aligned} 3 - 7 + 9 - 5 &= 12 + (-12) \\ &= 0 \end{aligned}$$

b) $$\begin{aligned} 4 - 5 - 9 + 6 - 2 + 4 - 8 &= 14 + (-24) \\ &= -10 \end{aligned}$$

■

It is certainly not essential that we evaluate the expressions of Example 4 as shown. We get the same answer by adding and subtracting from left to right. However, in algebra, just getting the answer is not always the most important point. Learning new methods often increases understanding.

Even though addition is associative, subtraction is not an associative operation. For example, $(8 - 4) - 3 = 1$ and $8 - (4 - 3) = 7$. So

$$(8 - 4) - 3 \neq 8 - (4 - 3).$$

We can also use a numerical example to show that division is not associative. For instance, $(16 \div 4) \div 2 = 2$ and $16 \div (4 \div 2) = 8$. So

$$(16 \div 4) \div 2 \neq 16 \div (4 \div 2).$$

The Distributive Property

If four men and five women pay \$3 each for a movie, there are two ways to find the total amount spent:

$$3(4 + 5) = 3 \cdot 9 = 27$$

$$3 \cdot 4 + 3 \cdot 5 = 12 + 15 = 27$$

Since we get \$27 either way, we can write

$$3(4 + 5) = 3 \cdot 4 + 3 \cdot 5.$$

We say that the multiplication by 3 is *distributed* over the addition. This example illustrates the **distributive property.**

Consider the following expressions involving multiplication and subtraction:

$$5(6 - 4) = 5 \cdot 2 = 10$$

$$5 \cdot 6 - 5 \cdot 4 = 30 - 20 = 10$$

Since both expressions have the same value, we can write

$$5(6 - 4) = 5 \cdot 6 - 5 \cdot 4.$$

Multiplication by 5 is distributed over each number in the parentheses. This example illustrates that multiplication distributes over subtraction.

Distributive Property

For any real numbers a, b, and c,

$$a(b + c) = ab + ac \quad \text{and} \quad a(b - c) = ab - ac.$$

The distributive property is used in two ways. If we start with $4(x + 3)$ and write

$$4(x + 3) = 4x + 4 \cdot 3 = 4x + 12,$$

we are using it to multiply 4 and $x + 3$ or to remove the parentheses. We wrote the product $4(x + 3)$ as the sum $4x + 12$. If we start with $3x + 15$ and write

$$3x + 15 = 3x + 3 \cdot 5 = 3(x + 5),$$

we are using it to write the sum $3x + 15$ as the product $3(x + 5)$. When we write a number or an expression as a product, we are **factoring.** In this case we factored out the common factor 3.

EXAMPLE 5 **Using the distributive property**

Use the distributive property to rewrite each product as a sum or difference and each sum or difference as a product.

a) $7x - 21$ **b)** $a(3 - b)$ **c)** $5a + 5$ **d)** $-3(x - 2)$

Solution

a) $7x - 21 = 7x - 7 \cdot 3$ Write 21 as $7 \cdot 3$.

$= 7(x - 3)$ Distributive property

b) $a(3 - b) = a3 - ab$ Distributive property

$= 3a - ab$ $a3 = 3a$

c) $5a + 5 = 5a + 5 \cdot 1$ Write 5 as $5 \cdot 1$.

$= 5(a + 1)$ Factor out the common factor 5.

d) $-3(x - 2) = -3x - (-3)(2)$ Distributive property

$= -3x - (-6)$ $(-3)(2) = -6$

$= -3x + 6$ Simplify. ■

The Identity Properties

The numbers 0 and 1 have special properties. Multiplication of a number by 1 does not change the number, and addition of 0 to a number does not change the number. That is why 1 is called the **multiplicative identity** and 0 is called the **additive identity.**

Identity Properties

For any real number a,

$$a \cdot 1 = 1 \cdot a = a \quad \text{and} \quad a + 0 = 0 + a = a.$$

The Inverse Properties

The idea of additive inverses was introduced in Section 1.3. Every real number a has an **additive inverse** or **opposite,** $-a$, such that $a + (-a) = 0$. Every nonzero real number a also has a **multiplicative inverse** or **reciprocal,** written $\frac{1}{a}$, such that $a \cdot \frac{1}{a} = 1$. Note that the sum of additive inverses is the additive identity and that the product of multiplicative inverses is the multiplicative identity.

Inverse Properties

For any real number a there is a number $-a$, such that

$$a + (-a) = 0.$$

For any nonzero real number a there is a number $\frac{1}{a}$ such that

$$a \cdot \frac{1}{a} = 1.$$

We are already familiar with multiplicative inverses for rational numbers. For example, the multiplicative inverse of $\frac{2}{3}$ is $\frac{3}{2}$ because

$$\frac{2}{3} \cdot \frac{3}{2} = \frac{6}{6} = 1.$$

EXAMPLE 6 **Multiplicative inverses**

Find the multiplicative inverse of each number.

a) 5 **b)** 0.3 **c)** $-\frac{3}{4}$ **d)** 1.7

Solution

a) The multiplicative inverse of 5 is $\frac{1}{5}$ because

$$5 \cdot \frac{1}{5} = 1.$$

b) To find the reciprocal of 0.3, we first write 0.3 as a ratio of integers:

$$0.3 = \frac{3}{10}$$

The multiplicative inverse of 0.3 is $\frac{10}{3}$ because

$$\frac{3}{10} \cdot \frac{10}{3} = 1.$$

c) The reciprocal of $-\frac{3}{4}$ is $-\frac{4}{3}$ because

$$\left(-\frac{3}{4}\right)\left(-\frac{4}{3}\right) = 1.$$

d) First convert 1.7 to a ratio of integers:

$$1.7 = 1\frac{7}{10} = \frac{17}{10}$$

The multiplicative inverse is $\frac{10}{17}$. ■

Multiplication Property of Zero

Zero has a property that no other number has. Multiplication involving zero always results in zero.

> **Multiplication Property of Zero**
>
> For any real number a,
>
> $$0 \cdot a = 0 \quad \text{and} \quad a \cdot 0 = 0.$$

EXAMPLE 7 **Identifying the properties**

Name the property that justifies each equation.

a) $5 \cdot 7 = 7 \cdot 5$

b) $4 \cdot \frac{1}{4} = 1$

c) $1 \cdot 864 = 864$

d) $6 + (5 + x) = (6 + 5) + x$

e) $3x + 5x = (3 + 5)x$

f) $6 + (x + 5) = 6 + (5 + x)$

g) $\pi x^2 + \pi y^2 = \pi(x^2 + y^2)$

h) $325 + 0 = 325$

i) $-3 + 3 = 0$

j) $455 \cdot 0 = 0$

Solution

a) Commutative

b) Multiplicative inverse

c) Multiplicative identity

d) Associative

e) Distributive

f) Commutative

g) Distributive

h) Additive identity

i) Additive inverse

j) Multiplication property of 0 ■

Calculator Close-up

Most scientific calculators have a reciprocal key [1/x] that gives the reciprocal of the number on the display. For example, you can evaluate

$$\frac{1}{6.25} + \frac{1}{12.5}$$

on your scientific calculator by using the following keystrokes:

6.25 [1/x] [+] 12.5 [1/x] [=]

The display will read

[0.24] Try it.

On a graphing calculator you can use the division key to write a reciprocal. The display should read as follows:

1/6.25+1/12.5 [ENTER]

[0.24]

Applications

Reciprocals are important in problems involving work. For example, if you wax one car in 3 hours, then your rate is $\frac{1}{3}$ of a car per hour. If you can wash one car in 12 minutes $\left(\frac{1}{5}\text{ of an hour}\right)$, then you are washing cars at the rate of 5 cars per hour. In general, if you can complete a task in x hours, then your rate is $\frac{1}{x}$ tasks per hour.

EXAMPLE 8 **Washing rates**

A car wash has two machines. The old machine washes one car in 0.1 hour, while the new machine washes one car in 0.08 hour. If both machines are operating, then at what rate (in cars per hour) are the cars being washed?

Solution

The old machine is working at the rate of $\frac{1}{0.1}$ cars per hour, and the new machine is working at the rate of $\frac{1}{0.08}$ cars per hour. Their rate working together is the sum of their individual rates:

$$\frac{1}{0.1} + \frac{1}{0.08} = 10 + 12.5 = 22.5$$

So working together, the machines are washing 22.5 cars per hour. ■

Warm-ups

True or false? Explain your answer.

1. $24 \div (4 \div 2) = (24 \div 4) \div 2$

2. $1 \div 2 = 2 \div 1$

3. $6 - 5 = -5 + 6$

4. $9 - (4 - 3) = (9 - 4) - 3$

5. Multiplication is a commutative operation.

6. $5x + 5 = 5(x + 1)$ for any value of x.

7. The multiplicative inverse of 0.02 is 50.

8. $-3(x - 2) = -3x + 6$ for any value of x.

9. $3x + 2x = (3 + 2)x$ for any value of x.

10. The additive inverse of 0 is 0.

1.7 EXERCISES

Use the commutative property of addition to rewrite each expression. See Example 1.

1. $9 + r$ **2.** $t + 6$ **3.** $3(2 + x)$

4. $P(1 + rt)$ **5.** $4 - 5x$ **6.** $b - 2a$

Use the commutative property of multiplication to rewrite each expression. See Example 2.

7. $x \cdot 6$ **8.** $y \cdot (-9)$ **9.** $(x - 4)(-2)$

10. $a(b + c)$ **11.** $4 - y \cdot 8$ **12.** $z \cdot 9 - 2$

Use the commutative and associative properties of multiplication and exponential notation to rewrite each product. See Example 3.

13. $(4w)(w)$ **14.** $(y)(2y)$ **15.** $3a(ba)$

16. $(x \cdot x)(7x)$ **17.** $(x)(9x)(xz)$ **18.** $y(y \cdot 5)(wy)$

Evaluate by finding first the sum of the positive numbers and then the sum of the negative numbers. See Example 4.

19. $8 - 4 + 3 - 10$

20. $-3 + 5 - 12 + 10$

21. $8 - 10 + 7 - 8 - 7$

22. $6 - 11 + 7 - 9 + 13 - 2$

23. $-4 - 11 + 7 - 8 + 15 - 20$

24. $-8 + 13 - 9 - 15 + 7 - 22 + 5$

25. $-3.2 + 2.4 - 2.8 + 5.8 - 1.6$

26. $5.4 - 5.1 + 6.6 - 2.3 + 9.1$

27. $3.26 - 13.41 + 5.1 - 12.35 - 5$

28. $5.89 - 6.1 + 8.58 - 6.06 - 2.34$

Use the distributive property to rewrite each product as a sum or difference and each sum or difference as a product. See Example 5.

29. $3(x - 5)$ **30.** $4(b - 1)$ **31.** $2m + 12$

32. $3y + 6$ **33.** $a(2 + t)$ **34.** $b(a + w)$

35. $-3(w - 6)$ **36.** $-3(m - 5)$ **37.** $-4(5 - y)$

38. $-3(6 - p)$ **39.** $4x - 4$ **40.** $6y + 6$

41. $-1(a - 7)$ **42.** $-1(c - 8)$ **43.** $-1(t + 4)$

44. $-1(x + 7)$ **45.** $4y - 16$ **46.** $5x + 15$

47. $4a + 8$ **48.** $7a - 35$

Find the multiplicative inverse (reciprocal) of each number. See Example 6.

49. $\frac{1}{2}$ **50.** $\frac{1}{3}$ **51.** -5

52. -6 **53.** 7 **54.** 8

55. 1 **56.** -1 **57.** -0.25

58. 0.75 **59.** 2.5 **60.** 3.5

Name the property that justifies each equation. See Example 7.

61. $3 \cdot x = x \cdot 3$ **62.** $x + 5 = 5 + x$

63. $2(x - 3) = 2x - 6$ **64.** $a(bc) = (ab)c$

65. $-3(xy) = (-3x)y$ **66.** $3(x + 1) = 3x + 3$

67. $4 + (-4) = 0$ **68.** $1.3 + 9 = 9 + 1.3$

69. $x^2 \cdot 5 = 5x^2$ **70.** $0 \cdot \pi = 0$

71. $1 \cdot 3y = 3y$ **72.** $(0.1)(10) = 1$

73. $2a + 5a = (2 + 5)a$ **74.** $3 + 0 = 3$

75. $-7 + 7 = 0$

76. $1 \cdot b = b$

77. $(2346)0 = 0$

78. $4x + 4 = 4(x + 1)$

79. $ay + y = y(a + 1)$

80. $ab + bc = b(a + c)$

Complete each equation, using the property named.

81. $a + y =$ _____, commutative

82. $6x + 6 =$ _____, distributive

83. $5(aw) =$ _____, associative

84. $x + 3 =$ _____, commutative

85. $\frac{1}{2}x + \frac{1}{2} =$ _____, distributive

86. $-3(x - 7) =$ _____, distributive

87. $6x + 15 =$ _____, distributive

88. $(x + 6) + 1 =$ _____, associative

89. $4(0.25) =$ _____, inverse property

90. $-1(5 - y) =$ _____, distributive

91. $0 = 96($_____$)$, multiplication property of zero

92. $3 \cdot ($_____$) = 3$, identity property

93. $0.33($_____$) = 1$, inverse property

94. $-8(1) =$ _____, identity property

Solve each problem. See Example 8.

95. ***Laying bricks.*** A bricklayer lays one brick in 0.04 hour, while his apprentice lays one brick in 0.05 hour. If both are working, then at what rate (in bricks per hour) are they laying bricks?

96. ***Recovering golf balls.*** Susan and Joan are diving for golf balls in a large water trap. Susan recovers a golf ball every 0.016 hour while Joan recovers a ball every 0.025 hour. If both are working, then at what rate (in golf balls per hour) are they recovering golf balls?

97. ***Population explosion.*** The population of the earth increases by one person every 0.34483 second (*World Population Data Sheet*, 1989). At what rate (in people per second) is the population of the earth increasing? At what rate in people per week is the population of the earth increasing?

98. ***Farmland conversion.*** The amount of farmland in the United States is decreasing by one hectare every 0.01111 hours as farmland is being converted to nonfarm use (Population Reference Bureau). At what rate in hectares per day is the farmland decreasing?

Getting More Involved

99. ***Writing.*** The perimeter of a rectangle is the sum of twice the length and twice the width. Write in words another way to find the perimeter that illustrates the distributive property.

100. ***Discussion.*** Eldrid bought a loaf of bread for \$1.69 and a gallon of milk for \$2.29. Using a tax rate of 5%, he correctly figured that the tax on the bread would be 8 cents and the tax on the milk would be 11 cents, for a total of \$4.17. However, at the cash register he was correctly charged \$4.18. How could this happen? Which property of the real numbers is in question in this case?

101. ***Exploration.*** Determine whether each of the following pairs of tasks are "commutative." That is, does the order in which they are performed produce the same result?

a) Put on your coat; put on your hat.
b) Put on your shirt; put on your coat.

Find another pair of "commutative" tasks and another pair of "noncommutative" tasks.

1.8 Using the Properties

In this section:

The properties of the real numbers can be helpful when we are doing computations. In this section we will see how the properties can be applied in arithmetic and algebra.

Using the Properties in Computation

The properties of the real numbers can often be used to simplify computations. For example, to find the product of 26 and 200, we can write

$$\begin{aligned}(26)(200) &= (26)(2 \cdot 100)\\ &= (26 \cdot 2)(100)\\ &= 52 \cdot 100\\ &= 5200\end{aligned}$$

It is the associative property that allows us to multiply 26 by 2 to get 52, then multiply 52 by 100 to get 5200.

EXAMPLE 1 **Using the properties**

Use the appropriate property to aid you in evaluating each expression.

a) $347 + 35 + 65$ **b)** $3 \cdot 435 \cdot \frac{1}{3}$ **c)** $6 \cdot 28 + 4 \cdot 28$

Solution

a) Notice that the sum of 35 and 65 is 100. So apply the associative property as follows:

$$\begin{aligned}347 + (35 + 65) &= 347 + 100\\ &= 447\end{aligned}$$

b) Use the commutative and associative properties to rearrange this product. We can then do the multiplication quickly:

$$\begin{aligned}3 \cdot 435 \cdot \frac{1}{3} &= 435\left(3 \cdot \frac{1}{3}\right) && \text{Commutative and associative properties}\\ &= 435 \cdot 1 && \text{Inverse property}\\ &= 435 && \text{Identity property}\end{aligned}$$

c) Use the distributive property to rewrite this expression.

$$\begin{aligned}6 \cdot 28 + 4 \cdot 28 &= (6 + 4)28\\ &= 10 \cdot 28\\ &= 280\end{aligned}$$

■

Like Terms

An expression containing a number or the product of a number and one or more variables raised to powers is called a **term.** For example,

$$-3, \qquad 5x, \qquad -3x^2y, \qquad a, \qquad \text{and} \qquad -abc$$

are terms. The number preceding the variables in a term is called the **coefficient.** In the term $5x$, the coefficient of x is 5. In the term $-3x^2y$ the coefficient of x^2y is -3. In the term a, the coefficient of a is 1 because $a = 1 \cdot a$. In the term $-abc$ the coefficient of abc is -1 because $abc = -1 \cdot abc$. If two terms contain the same variables with the same exponents, they are called **like terms.** For example, $3x^2$ and $-5x^2$ are like terms, but $3x^2$ and $-5x^3$ are not like terms.

Combining Like Terms

Using the distributive property on an expression involving the sum of like terms allows us to combine the like terms as shown in the next example.

EXAMPLE 2 **Combining like terms**

Use the distributive property to perform the indicated operations.

a) $3x + 5x$ **b)** $-5xy - (-4xy)$

Solution

a) $3x + 5x = (3 + 5)x$ Distributive property

$= 8x$ Add the coefficients.

Because the distributive property is valid for any real numbers, we have $3x + 5x = 8x$ no matter what number is used for x.

b) $-5xy - (-4xy) = [-5 - (-4)]xy$ Distributive property

$= -1xy$ $-5 - (-4) = -5 + 4 = -1$

$= -xy$ Multiplying by -1 is the same as taking the opposite. ■

Of course, we do not want to write out all of the steps shown in Example 2 every time we combine like terms. We can combine like terms as easily as we can add or subtract their coefficients.

EXAMPLE 3 **Combining like terms**

Perform the indicated operations.

a) $w + 2w$ **b)** $-3a + (-7a)$ **c)** $-9x + 5x$

d) $7xy - (-12xy)$ **e)** $2x^2 + 4x^2$

Solution

a) $w + 2w = 1w + 2w = 3w$

b) $-3a + (-7a) = -10a$

c) $-9x + 5x = -4x$

d) $7xy - (-12xy) = 19xy$

e) $2x^2 + 4x^2 = 6x^2$ ■

CAUTION There are no like terms in expressions such as

$$2 + 5x, \quad 3xy + 5y, \quad 3w + 5a, \quad \text{and} \quad 3z^2 + 5z$$

The terms in these expressions cannot be combined. ⊘

Products and Quotients

In the next example we use the associative property of multiplication to simplify the product of two expressions.

EXAMPLE 4 **Finding products**

Simplify.

a) $3(5x)$ **b)** $2\left(\frac{x}{2}\right)$ **c)** $(4x)(6x)$

Solution

a)
$$\begin{aligned} 3(5x) &= (3 \cdot 5)x && \text{Associative property} \\ &= (15)x && \text{Multiply.} \\ &= 15x && \text{Remove unnecessary parentheses.} \end{aligned}$$

b)
$$\begin{aligned} 2\left(\frac{x}{2}\right) &= 2\left(\frac{1}{2} \cdot x\right) && \text{Multiplying by } \tfrac{1}{2} \text{ is the same as dividing by 2.} \\ &= \left(2 \cdot \frac{1}{2}\right)x && \text{Associative property} \\ &= 1 \cdot x && \text{Multiplicative inverse} \\ &= x && \text{Multiplicative identity is 1.} \end{aligned}$$

c)
$$\begin{aligned} (4x)(6x) &= 4 \cdot 6 \cdot x \cdot x && \text{Commutative and associative properties} \\ &= 24x^2 && \text{Definition of exponent} \end{aligned}$$

CAUTION Be careful with expressions such as $3(5x)$ and $3(5 + x)$. In $3(5x)$ we multiply 5 by 3 to get $3(5x) = 15x$. In $3(5 + x)$, both 5 and x are multiplied by the 3 to get $3(5 + x) = 15 + 3x$.

In Example 4 we showed how the properties are used to simplify products. However, in practice we usually do not write out any steps for these problems—we can write just the answer.

EXAMPLE 5 **Finding products quickly**

Find each product.

a) $(-3)(4x)$ **b)** $(-4a)(-7a)$ **c)** $(-3a)\left(\frac{b}{3}\right)$ **d)** $6 \cdot \frac{x}{2}$

Solution

a) $-12x$ **b)** $28a^2$ **c)** $-ab$ **d)** $3x$

In Section 1.1 we found the quotient of two numbers by inverting the divisor and then multiplying. Since $a \div b = a \cdot \frac{1}{b}$, any quotient can be written as a product.

EXAMPLE 6 **Simplifying quotients**

Simplify.

a) $\frac{10x}{5}$ **b)** $\frac{4x + 8}{2}$

Solution

a) Since dividing by 5 is equivalent to multiplying by $\frac{1}{5}$, we have

$$\frac{10x}{5} = \frac{1}{5}(10x) = \left(\frac{1}{5} \cdot 10\right)x = (2)x = 2x.$$

Note that you can simply divide 10 by 5 to get 2.

b) Since dividing by 2 is equivalent to multiplying by $\frac{1}{2}$, we have

$$\frac{4x + 8}{2} = \frac{1}{2}(4x + 8) = 2x + 4.$$

Note that both 4 and 8 are divided by 2. ■

CAUTION It is not correct to divide only one term in the numerator by the denominator. For example,

$$\frac{4 + 7}{2} \neq 2 + 7$$

because $\frac{4+7}{2} = \frac{11}{2}$ and $2 + 7 = 9$. ⊘

Removing Parentheses

Multiplying a number by -1 merely changes the sign of the number. For example,

$$(-1)(7) = -7 \quad \text{and} \quad (-1)(-8) = 8.$$

So -1 times a number is the *opposite* of the number. Using variables, we write

$$(-1)x = -x \quad \text{or} \quad -1(y + 5) = -(y + 5).$$

When a minus sign appears in front of a sum, we can change the minus sign to -1 and use the distributive property. For example,

$$\begin{aligned} -(w + 4) &= -1(w + 4) \\ &= (-1)w + (-1)4 && \text{Distributive property} \\ &= -w + (-4) && \textit{Note: } -1 \cdot w = -w,\ -1 \cdot 4 = -4 \\ &= -w - 4 \end{aligned}$$

If a minus sign appears in front of a difference, we can rewrite the difference as a sum. For example,

$$\begin{aligned} -(x - 3) &= -1(x - 3) \\ &= (-1)x - (-1)3 \\ &= -x + 3 \end{aligned}$$

CAUTION A negative sign in front of a set of parentheses is distributed over *each* term in the parentheses, changing the sign of each term. ⊘

EXAMPLE 7 **Removing parentheses**

Simplify each expression.

a) $5 - (x + 3)$ **b)** $3x - 6 - (2x - 4)$ **c)** $-6x - (-x + 2)$

Solution

a) $5 - (x + 3) = 5 - x - 3$ Change the sign of each term in parentheses.

$= 5 - 3 - x$ Commutative property

$= 2 - x$ Combine like terms.

b) $3x - 6 - (2x - 4) = 3x - 6 - 2x + 4$ Remove parentheses.

$= 3x - 2x - 6 + 4$ Commutative property

$= x - 2$ Combine like terms.

c) $-6x - (-x + 2) = -6x + x - 2$ Remove parentheses.

$= -5x - 2$ Combine like terms. ■

The commutative and associative properties of addition allow us to rearrange the terms so that we may combine the like terms. However, it is not necessary to actually write down the rearrangement. We can identify the like terms and combine them without rearranging.

EXAMPLE 8 **Simplifying algebraic expressions**

Simplify.

a) $(-2x + 3) + (5x - 7)$ **b)** $-3x + 6x + 5(4 - 2x)$

c) $-2x(3x - 7) - (x - 6)$ **d)** $x - 0.02(x + 500)$

Solution

a) $(-2x + 3) + (5x - 7) = 3x - 4$ Combine like terms.

b) $-3x + 6x + 5(4 - 2x) = -3x + 6x + 20 - 10x$ Distributive property

$= -7x + 20$ Combine like terms.

c) $-2x(3x - 7) - (x - 6) = -6x^2 + 14x - x + 6$ Distributive property

$= -6x^2 + 13x + 6$ Combine like terms.

d) $x - 0.02(x + 500) = 1x - 0.02x - 10$ Distributive property

$= 0.98x - 10$ Combine like terms. ■

Warm-ups

True or false? Explain your answer.

A statement involving variables should be marked true only if it is true for all values of the variable.

1. $3(x + 6) = 3x + 18$

2. $-3x + 9 = -3(x + 9)$

3. $-1(x - 4) = -x + 4$

4. $3a + 4a = 7a$

5. $(3a)(4a) = 12a$

6. $3(5 \cdot 2) = 15 \cdot 6$

7. $x + x = x^2$

8. $x \cdot x = 2x$

9. $3 + 2x = 5x$

10. $-(5x - 2) = -5x + 2$

1.8 EXERCISES

Use the appropriate properties to evaluate the expressions. See Example 1.

1. $35(200)$

2. $15(300)$

3. $\frac{4}{3}(0.75)$

4. $5(0.2)$

5. $256 + 78 + 22$

6. $12 + 88 + 376$

7. $35 \cdot 3 + 35 \cdot 7$

8. $98 \cdot 478 + 2 \cdot 478$

9. $18 \cdot 4 \cdot 2 \cdot \frac{1}{4}$

10. $19 \cdot 3 \cdot 2 \cdot \frac{1}{3}$

11. $(120)(300)$

12. $150 \cdot 200$

13. $12 \cdot 375(-6 + 6)$

14. $354^2(-2 \cdot 4 + 8)$

15. $78 + 6 + 8 + 4 + 2$

16. $-47 + 12 - 6 - 12 + 6$

Combine like terms where possible. See Examples 2 and 3.

17. $5w + 6w$

18. $4a + 10a$

19. $4x - x$

20. $a - 6a$

21. $2x - (-3x)$

22. $2b - (-5b)$

23. $-3a - (-2a)$

24. $-10m - (-6m)$

25. $-a - a$

26. $a - a$

27. $10 - 6t$

28. $9 - 4w$

29. $3x^2 + 5x^2$

30. $3r^2 + 4r^2$

31. $-4x + 2x^2$

32. $6w^2 - w$

33. $5mw^2 - 12mw^2$

34. $4ab^2 - 19ab^2$

Simplify the following products or quotients. See Examples 4–6.

35. $3(4h)$

36. $2(5h)$

37. $6b(-3)$

38. $-3m(-1)$

39. $(-3m)(3m)$

40. $(2x)(-2x)$

41. $(-3d)(-4d)$

42. $(-5t)(-2t)$

43. $(-y)(-y)$

44. $y(-y)$

45. $-3a(5b)$

46. $-7w(3r)$

47. $-3a(2 + b)$

48. $-2x(3 + y)$

49. $-k(1 - k)$

50. $-t(t - 1)$

51. $\frac{3y}{3}$

52. $\frac{-9t}{9}$

53. $\frac{-15y}{5}$

54. $\frac{-12b}{2}$

55. $2\left(\frac{y}{2}\right)$

56. $6\left(\frac{m}{3}\right)$

57. $8y\left(\frac{y}{4}\right)$

58. $10\left(\frac{2a}{5}\right)$

59. $\frac{6a - 3}{3}$

60. $\frac{-8x + 6}{2}$

61. $\frac{-9x + 6}{-3}$

62. $\frac{10 - 5x}{-5}$

Simplify each expression. See Example 7.

63. $x - (3x - 1)$

64. $4x - (2x - 5)$

65. $5 - (y - 3)$

66. $8 - (m - 6)$

67. $2m + 3 - (m + 9)$

68. $7 - 8t - (2t + 6)$

69. $-3 - (-w + 2)$

70. $-5x - (-2x + 9)$

Simplify the following expressions by combining like terms. See Example 8.

71. $3x + 5x + 6 + 9$

72. $2x + 6x + 7 + 15$

73. $-2x + 3 + 7x - 4$

74. $-3x + 12 + 5x - 9$

75. $3a - 7 - (5a - 6)$

76. $4m - 5 - (m - 2)$

77. $2(a - 4) - 3(-2 - a)$

78. $2(w + 6) - 3(-w - 5)$

79. $-5m + 6(m - 3) + 2m$

80. $-3a + 2(a - 5) + 7a$

81. $5 - 3(x + 2) - 6$

82. $7 + 2(k - 3) - k + 6$

83. $x - 0.05(x + 10)$

84. $x - 0.02(x + 300)$

85. $4.5 - 3.2(x - 5.3) - 8.75$

86. $0.03(4.5x - 3.9) + 0.06(9.8x - 45)$

Simplify each expression.

87. $3x - (4 - x)$

88. $2 + 8x - 11x$

89. $y - 5 - (-y - 9)$

90. $a - (b - c - a)$

91. $7 - (8 - 2y - m)$

92. $x - 8 - (-3 - x)$

93. $\frac{1}{2}(10 - 2x) + \frac{1}{3}(3x - 6)$

94. $\frac{1}{2}(x - 20) - \frac{1}{5}(x + 15)$

95. $0.2(x + 3) - 0.05(x + 20)$

96. $0.08x + 0.12(x + 100)$

97. $2k + 1 - 3(5k - 6) - k + 4$

98. $2w - 3 + 3(w - 4) - 5(w - 6)$

99. $-3m - 3[2m - 3(m + 5)]$

100. $6h + 4[2h - 3(h - 9) - (h - 1)]$

Solve each problem.

101. ***Married filing jointly.*** Suppose that a married couple filing jointly has a taxable income of x dollars, where x is over \$38,000 but not over \$91,850. The expression $0.15(38,000) + 0.28(x - 38,000)$ gives the 1994 federal income tax for the couple (Internal Revenue Service). Simplify the expression. Find the amount of tax for a married couple with a taxable income of \$52,000. Use the graph shown here to estimate the 1994 federal income tax for a married couple with a taxable income of \$150,000.

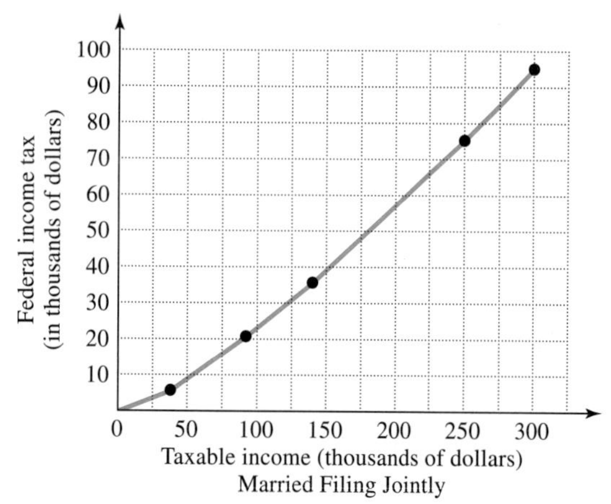

Figure for Exercise 101

102. ***Single taxpayer.*** Suppose that a single taxpayer has a taxable income of x dollars, where x is over \$22,750 but is not over \$55,100. The expression $0.15(22,750) + 0.28(x - 22,750)$ represents the 1994 federal income tax for the taxpayer. Simplify the expression. Find the amount of tax for a single taxpayer with a taxable income of \$52,000.

103. ***Perimeter of a corral.*** The perimeter of a rectangular corral that has width x feet and length $x + 40$ feet is $2(x) + 2(x + 40)$. Simplify the expression for the perimeter. Find the perimeter if $x = 30$ feet.

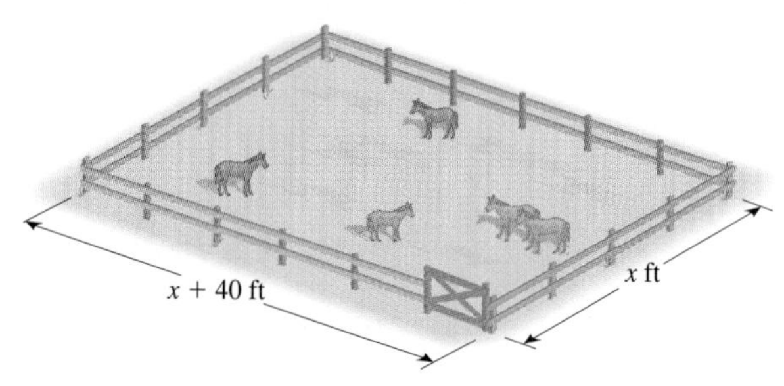

Figure for Exercise 103

Getting More Involved

104. ***Discussion.*** What is wrong with the way in which each of the following expressions is simplified?

a) $4(2 + x) = 8 + x$

b) $4(2x) = 8 \cdot 4x = 32x$

c) $\dfrac{4 + x}{2} = 2 + x$

d) $5 - (x - 3) = 5 - x - 3 = 2 - x$

105. ***Discussion.*** An instructor asked his class to evaluate the expression $1/2x$ for $x = 5$. Some students got 0.1; others got 2.5. Which answer is correct and why?

COLLABORATIVE ACTIVITIES

Remembering the Rules

Grouping: 4 students
Topic: Fractions, decimals, and signed numbers

This chapter reviews different types of numbers used in algebra. This activity will review the rules for the basic operations: addition, subtraction, multiplication, and division for fractions, decimals, and real numbers.

Part I: Remembering the rules. Have each member of your group choose an operation: addition, subtraction, multiplication, or division.

1. Fractions:
   a. Write the rules for working a fraction problem using the operation you have chosen. Use your book as a reference and consider the following sample problems:

$$\frac{1}{2}+\frac{2}{5},\quad \frac{1}{3}\cdot\frac{6}{7},\quad \frac{3}{5}-\frac{1}{3},\quad \frac{1}{3}\div\frac{2}{3}$$

   b. Starting with addition, each of you will share what he or she has written with the other members of the group. Make additions or corrections if needed.

Switch operations: Each member of the group now takes the operation of the person to his or her right.

2. Decimals: Repeat parts (a) and (b), in 1 above for the following sample problems:

$$0.012 + 3,\quad 2.1 - 0.25,\quad 3.2 \cdot 0.23,\quad 5.4 \div 1.2$$

Switch operations: Each member of the group now takes the operation of the person to his or her right.

3. Signed numbers: Repeat parts (a) and (b), in 1 above for the following sample problems:

$$-3 + 5,\quad 3 - (-2),\quad -2 \cdot 3,\quad -6 \div -2$$

Part II: Testing your rules. As a group, work through the following problems together, using the rules you have written from Part I. Add any new rules that come up while you work.

1. $\frac{1}{3}+\frac{2}{5}$

2. $0.076 + 7 + 2.005$

3. $-8 + 17$

4. $2\frac{9}{14} - 1\frac{5}{7}$

5. $8 - 3.024$

6. $-5 - (-19)$

7. $3\frac{1}{3} \cdot 2\frac{2}{5}$

8. $0.0723 \cdot 100$

9. $12 \cdot -3 \cdot -2$

10. $1\frac{2}{3} \div 5\frac{1}{2}$

11. $1.024 \div 3.2$

12. $-405 \div 15$

13. $-2\left(\frac{2}{3} - \frac{1}{4}\right) \div 2\frac{1}{2} + (3.052 - (-0.948))$

Extension: Before each exam, form a study group to review material. Write any new rules or definitions used in each chapter. Save these to study for your final exam.

Wrap-up CHAPTER 1

SUMMARY

Fractions		Examples
Reducing fractions	$\frac{a \cdot c}{b \cdot c} = \frac{a}{b}$	$\frac{4}{6} = \frac{2 \cdot 2}{2 \cdot 3} = \frac{2}{3}$
Building up fractions	$\frac{a}{b} = \frac{a \cdot c}{b \cdot c}$	$\frac{3}{8} = \frac{3 \cdot 5}{8 \cdot 5} = \frac{15}{40}$

Multiplying fractions	$\frac{a}{b} \cdot \frac{c}{d} = \frac{ac}{bd}$	$\frac{2}{3} \cdot \frac{4}{5} = \frac{8}{15}$
Dividing fractions	$\frac{a}{b} \div \frac{c}{d} = \frac{a}{b} \cdot \frac{d}{c}$	$\frac{2}{3} \div \frac{4}{5} = \frac{2}{3} \cdot \frac{5}{4} = \frac{10}{12} = \frac{5}{6}$
Adding or subtracting fractions	$\frac{a}{b} + \frac{c}{b} = \frac{a+c}{b}$	$\frac{1}{5} + \frac{2}{5} = \frac{3}{5}$
	$\frac{a}{b} - \frac{c}{b} = \frac{a-c}{b}$	$\frac{3}{5} - \frac{2}{5} = \frac{1}{5}$
Least common denominator	The smallest number that is a multiple of all denominators.	$\frac{1}{4} + \frac{1}{6} = \frac{3}{12} + \frac{2}{12} = \frac{5}{12}$

The Real Numbers

		Examples
Counting or natural numbers	$\{1, 2, 3, \ldots\}$	
Whole numbers	$\{0, 1, 2, 3, \ldots\}$	
Integers	$\{\ldots, -3, -2, -1, 0, 1, 2, 3, \ldots\}$	
Rational numbers	$\left\{\frac{a}{b} \middle\vert\, a \text{ and } b \text{ are integers with } b \neq 0\right\}$	$\frac{3}{2}, 5, -6, 0$
Irrational numbers	$\{x \mid x \text{ is a real number that is not rational}\}$	$\sqrt{2}, \sqrt{3}, \pi$
Real numbers	The set of real numbers consists of all rational numbers together with all irrational numbers.	

Operations with Real Numbers

		Examples
Absolute value	$\lvert a \rvert = \begin{cases} a & \text{if } a \text{ is positive or zero} \\ -a & \text{if } a \text{ is negative} \end{cases}$	$\lvert 3 \rvert = 3$, $\lvert 0 \rvert = 0$ $\lvert -3 \rvert = 3$
Sum of two numbers with like signs	Add their absolute values. The sum has the same sign as the given numbers.	$-3 + (-4) = -7$
Sum of two numbers with unlike signs (and different absolute values)	Subtract the absolute values of the numbers. The answer is positive if the number with the larger absolute value is positive. The answer is negative if the number with the larger absolute value is negative.	$-4 + 7 = 3$ $-7 + 4 = -3$
Sum of opposites	The sum of any number and its opposite is 0.	$-6 + 6 = 0$
Subtraction of signed numbers	$a - b = a + (-b)$ Subtract any number by adding its opposite.	$3 - 5 = 3 + (-5) = -2$ $4 - (-3) = 4 + 3 = 7$

Product or quotient	Like signs ↔ Positive result Unlike signs ↔ Negative result	$(-3)(-2) = 6$ $(-8) \div 2 = -4$
Definition of exponents	For any counting number n, $a^n = \underbrace{a \cdot a \cdot a \cdot \ldots \cdot a}_{n \text{ factors}}.$	$2^3 = 2 \cdot 2 \cdot 2 = 8$
Order of operations	No parentheses or absolute value present: 1. Exponential expressions 2. Multiplication and division 3. Addition and subtraction With parentheses or absolute value: First evaluate within each set of parentheses or absolute value, using the order of operations.	 $5 + 2^3 = 13$ $2 + 3 \cdot 5 = 17$ $4 + 5 \cdot 3^2 = 49$ $(2 + 3)(5 - 7) = -10$ $2 + 3\|2 - 5\| = 11$

Properties of the Real Numbers — **Examples**

Commutative properties	$a + b = b + a$ $a \cdot b = b \cdot a$	$5 + 7 = 7 + 5$ $6 \cdot 3 = 3 \cdot 6$
Associative properties	$a + (b + c) = (a + b) + c$ $a \cdot (b \cdot c) = (a \cdot b) \cdot c$	$1 + (2 + 3) = (1 + 2) + 3$ $2(3 \cdot 4) = (2 \cdot 3)4$
Distributive properties	$a(b + c) = ab + ac$ $a(b - c) = ab - ac$	$2(3 + x) = 6 + 2x$ $-2(x - 5) = -2x + 10$
Identity properties	$a + 0 = a$ and $0 + a = a$ Zero is the additive identity. $1 \cdot a = a$ and $a \cdot 1 = a$ One is the multiplicative identity.	$5 + 0 = 0 + 5 = 5$ $7 \cdot 1 = 1 \cdot 7 = 7$
Inverse properties	For any real number a, there is a number $-a$ (additive inverse or opposite) such that $a + (-a) = 0$ and $-a + a = 0$. For any nonzero real number a there is a number $\frac{1}{a}$ (multiplicative inverse or reciprocal) such that $a \cdot \frac{1}{a} = 1$ and $\frac{1}{a} \cdot a = 1.$	$3 + (-3) = 0$ $-3 + 3 = 0$ $3 \cdot \frac{1}{3} = 1$ $\frac{1}{3} \cdot 3 = 1$
Multiplication property of 0	$a \cdot 0 = 0$ and $0 \cdot a = 0$	$5 \cdot 0 = 0$ $0(-7) = 0$

REVIEW EXERCISES

1.1 *Perform the indicated operations.*

1. $\frac{1}{3}+\frac{3}{8}$

2. $\frac{2}{3}-\frac{1}{4}$

3. $\frac{3}{5}\cdot 10$

4. $\frac{3}{5}\div 10$

5. $\frac{2}{5}\cdot\frac{15}{14}$

6. $7\div\frac{1}{2}$

7. $4+\frac{2}{3}$

8. $\frac{7}{12}-\frac{1}{4}$

9. $\frac{1}{2}+\frac{1}{3}+\frac{1}{4}$

10. $\frac{3}{4}\div 9$

1.2 *Which of the numbers* $-\sqrt{5}, -2, 0, 1, 2, 3.14, \pi$, *and 10 are*

11. whole numbers?

12. natural numbers?

13. integers?

14. rational numbers?

15. irrational numbers?

16. real numbers?

True or false? Explain your answer.

17. Every whole number is a rational number.

18. Zero is not a rational number.

19. The counting numbers between -4 and 4 are $-3, -2, -1, 0, 1, 2$, and 3.

20. There are infinitely many integers.

21. The set of counting numbers smaller than the national debt is infinite.

22. The decimal number 0.25 is a rational number.

23. Every integer greater than -1 is a whole number.

24. Zero is the only number that is neither rational nor irrational.

1.3 *Evaluate.*

25. $-5+7$

26. $-9+(-4)$

27. $35-48$

28. $-3-9$

29. $-12+5$

30. $-12-5$

31. $-12-(-5)$

32. $-9-(-9)$

33. $-0.05+12$

34. $-0.03+(-2)$

35. $-0.1-(-0.05)$

36. $-0.3+0.3$

37. $\frac{1}{3}-\frac{1}{2}$

38. $-\frac{2}{3}+\frac{1}{4}$

39. $-\frac{1}{3}+\left(-\frac{2}{5}\right)$

40. $\frac{1}{3}-\left(-\frac{1}{4}\right)$

1.4 *Evaluate.*

41. $(-3)(5)$

42. $(-9)(-4)$

43. $(-8)\div(-2)$

44. $50\div(-5)$

45. $\frac{-20}{-4}$

46. $\frac{30}{-5}$

47. $\left(-\frac{1}{2}\right)\left(-\frac{1}{3}\right)$

48. $8\div\left(-\frac{1}{3}\right)$

49. $-0.09\div 0.3$

50. $4.2\div(-0.3)$

51. $(0.3)(-0.8)$

52. $0\div(-0.0538)$

53. $(-5)(-0.2)$

54. $\frac{1}{2}(-12)$

1.5 *Evaluate.*

55. $3+7(9)$

56. $(3+7)9$

57. $(3+4)^2$

58. $3+4^2$

59. $3+2\cdot|5-6\cdot 4|$

60. $3-(8-9)$

61. $(3-7)-(4-9)$

62. $3-7-4-9$

63. $-2-4(2-3\cdot 5)$

64. 3^2-7+5^2

65. $3^2-(7+5)^2$

66. $|4-6\cdot 3|-|7-9|$

67. $\frac{-3-5}{2-(-2)}$

68. $\frac{1-9}{4-6}$

69. $\frac{6+3}{3}-5\cdot 4+1$

70. $\frac{2\cdot 4+4}{3}-3(1-2)$

1.6 *Let* $a=-1$, $b=-2$, *and* $c=3$. *Find the value of each algebraic expression.*

71. b^2-4ac

72. a^2-4b

73. $(c-b)(c+b)$

74. $(a+b)(a-b)$

75. $a^2+2ab+b^2$

76. $a^2-2ab+b^2$

77. a^3-b^3

78. a^3+b^3

79. $\frac{b+c}{a+b}$

80. $\frac{b-c}{2b-a}$

81. $|a-b|$

82. $|b-a|$

83. $(a+b)c$

84. $ac+bc$

Determine whether the given number is a solution to the equation following it.

85. $4,\ 3x-2=10$

86. $1,\ 5(x+3)=20$

87. $-6,\ \frac{3x}{2}=9$

88. $-30,\ \frac{x}{3}-4=6$

89. $15, \frac{x+3}{2} = 9$

90. $1, \frac{12}{2x+1} = 4$

91. $4, -x - 3 = 1$

92. $7, -x + 1 = 6$

1.7 *Name the property that justifies each statement.*

93. $a(x + y) = ax + ay$

94. $3(4y) = (3 \cdot 4)y$

95. $(0.001)(1000) = 1$

96. $xy = yx$

97. $0 + y = y$

98. $325 \cdot 1 = 325$

99. $3 + (2 + x) = (3 + 2) + x$

100. $2x - 6 = 2(x - 3)$

101. $5 \cdot 200 = 200 \cdot 5$

102. $3 + (x + 2) = (x + 2) + 3$

103. $-50 + 50 = 0$

104. $43 \cdot 59 \cdot 82 \cdot 0 = 0$

105. $12 \cdot 1 = 12$

106. $3x + 1 = 1 + 3x$

1.8 *Simplify by combining like terms.*

107. $3a + 7 - (4a - 5)$

108. $2m + 6 - (m - 2)$

109. $2a(3a - 5) + 4a$

110. $3a(a - 5) + 5a(a + 2)$

111. $3(t - 2) - 5(3t - 9)$

112. $2(m + 3) - 3(3 - m)$

113. $0.1(a + 0.3) - (a + 0.6)$

114. $0.1(x + 0.3) - (x - 0.9)$

115. $0.05(x - 20) - 0.1(x + 30)$

116. $0.02(x - 100) + 0.2(x - 50)$

117. $5 - 3x(-5x - 2) + 12x^2$

118. $7 - 2x(3x - 7) - x^2$

119. $-(a - 2) - 2 - a$

120. $-(w - y) - 3(y - w)$

121. $x(x + 1) + 3(x - 1)$

122. $y(y - 2) + 3(y + 1)$

Miscellaneous

Evaluate each expression.

123. $752(-13) + 752(13)$

124. $75 - (-13)$

125. $|15 - 23|$

126. $4^2 - 6^2$

127. $-6^2 + 3(5)$

128. $(0.03)(-200)$

129. $\frac{2}{5} + \frac{1}{10}$

130. $\frac{2+1}{5+10}$

131. $(0.05) \div (-0.1)$

132. $(4 - 9)^2 + (2 \cdot 3 - 1)^2$

133. $2\left(-\frac{1}{2}\right)^2 + \left(-\frac{1}{2}\right) - 1$

134. $\left(-\frac{6}{7}\right)\left(\frac{21}{26}\right)$

Simplify each expression if possible.

135. $\frac{2x+4}{2}$

136. $4(2x)$

137. $4 + 2x$

138. $4(2 + x)$

139. $4 \cdot \frac{x}{2}$

140. $4 - (x - 2)$

141. $-4(x - 2)$

142. $(4x)(2x)$

143. $4x + 2x$

144. $2 + (x + 4)$

145. $4 \cdot \frac{x}{4}$

146. $4 \cdot \frac{3x}{2}$

147. $2 \cdot x \cdot 4$

148. $4 - 2(2 - x)$

Solve each problem.

149. ***Telemarketing.*** Brenda and Nicki sell memberships in an automobile club over the telephone. Brenda sells one membership every 0.125 hour, and Nicki sells one membership every 0.1 hour. At what rate (in memberships per hour) are the memberships being sold when both are working?

150. ***High-income bracket.*** Troy Aikman, quarterback for the Dallas Cowboys, had a taxable income of approximately \$2.5 million for 1993 (*Parade*, June 26, 1994). The expression $79{,}639.50 + 0.396(x - 250{,}000)$ represents the 1993 federal income tax in dollars for a taxpayer with x dollars of taxable income, where x is over \$250,000. Simplify the expression. Use the simplified expression to find the amount of federal income tax owed.

CHAPTER 1 TEST

Which of the numbers $-3, -\sqrt{3}, -\frac{1}{4}, 0, \sqrt{5}, \pi,$ *and* 8 *are*

1. whole numbers?

2. integers?

3. rational numbers?

4. irrational numbers?

Evaluate each expression.

5. $6 + 3(-9)$

6. $(-2)^2 - 4(-2)(-1)$

7. $\frac{-3^2 - 9}{3 - 5}$

8. $-5 + 6 - 12 + 4$

9. $0.05 - 1$

10. $(5 - 9)(5 + 9)$

11. $(878 + 89) + 11$

12. $6 + |3 - 5(2)|$

13. $8 - 3\,|7 - 10|$

14. $(839 + 974)[3(-4) + 12]$

15. $974(7) + 974(3)$

16. $-\frac{2}{3} + \frac{3}{8}$

17. $(-0.05)(400)$

18. $\left(-\frac{3}{4}\right)\left(\frac{2}{9}\right)$

19. $13 \div \left(-\frac{1}{3}\right)$

Identify the property that justifies each equation.

20. $2(x + 7) = 2x + 14$

21. $48 \cdot 1000 = 1000 \cdot 48$

22. $2 + (6 + x) = (2 + 6) + x$

23. $-348 + 348 = 0$

24. $1 \cdot (-6) = -6$

25. $0 \cdot 388 = 0$

Use the distributive property to write each sum or difference as a product.

26. $3x + 30$

27. $7w - 7$

Simplify each expression.

28. $6 + 4x + 2x$

29. $6 + 4(x - 2)$

30. $5x - (3 - 2x)$

31. $x + 10 - 0.1(x + 25)$

32. $2a(4a - 5) - 3a(-2a - 5)$

33. $\frac{6x + 12}{6}$

34. $8 \cdot \frac{t}{2}$

35. $(-9xy)(-6xy)$

Evaluate each expression if $a = -2$, $b = 3$, *and* $c = 4$.

36. $b^2 - 4ac$

37. $\frac{a - b}{b - c}$

38. $(a - c)(a + c)$

Determine whether the given number is a solution to the equation following it.

39. -2, $3x - 4 = 2$

40. 13, $\frac{x + 3}{8} = 2$

41. -3, $-x + 5 = 8$

Solve each problem.

42. Burke and Nora deliver pizzas for Godmother's Pizza. Burke averages one delivery every 0.25 hour, and Nora averages one delivery every 0.2 hour. At what rate (in deliveries per hour) are the deliveries made when both are working?

43. A forensic scientist uses the expression $80.405 + 3.660R - 0.06(A - 30)$ to estimate the height in centimeters for a male with a radius (bone in the forearm) of length R centimeters and age A in years, where A is over 30. Simplify the expression. Use the expression to estimate the height of an 80-year-old male with a radius of length 25 cm.

CHAPTER 2

Linear Equations and Inequalities

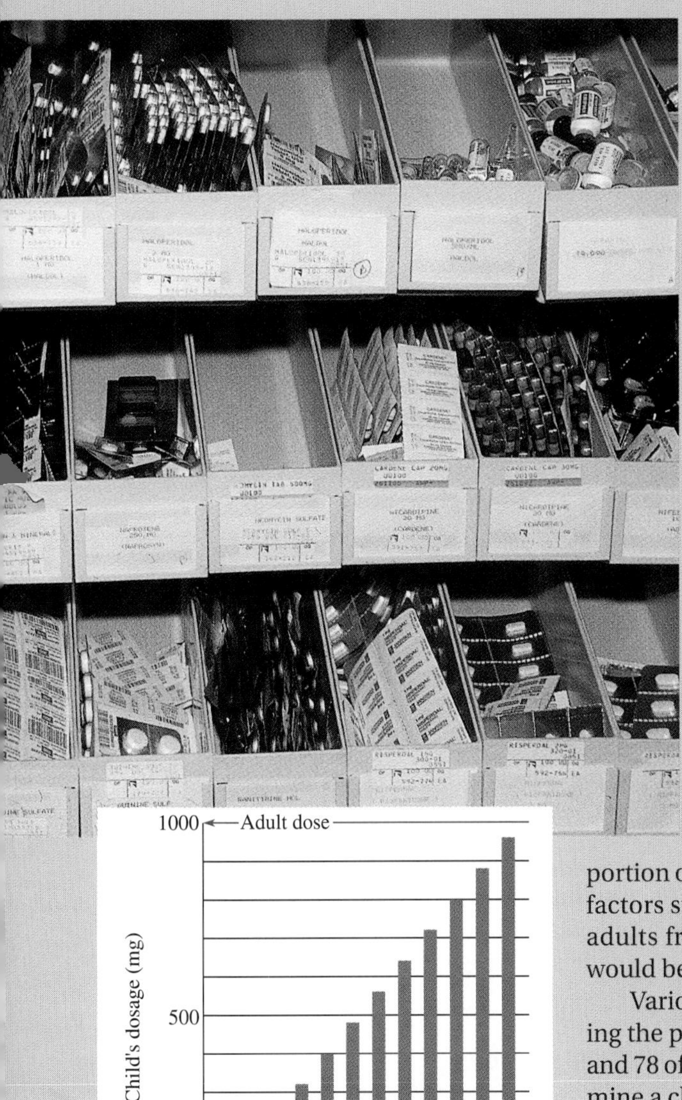

Some ancient peoples chewed on leaves to cure their headaches. Thousands of years ago, the Egyptians used honey, salt, cedar oil, and sycamore bark to cure illnesses. Currently, some of the indigenous people of the United States use black birch as a pain reliever.

Today, we are grateful for the miracles of modern medicine and the seemingly simple cures for illnesses. From our own experiences we know that just the right amount of a drug can work wonders but too much of a drug can do great harm. Even though physicians often prescribe the same drug for children and adults, the amount given must be tailored to the individual. The portion of a drug given to children is usually reduced on the basis of factors such as the weight and height of the child. Likewise, older adults frequently need a lower dosage of medication than what would be prescribed for a younger, more active person.

Various algebraic formulas have been developed for determining the proper dosage for a child and an older adult. In Exercises 77 and 78 of Section 2.3 you will see two formulas that are used to determine a child's dosage by using the adult dosage and the child's age.

2.1 Linear Equations in One Variable

In this section:

- Addition and Subtraction Properties of Equality
- Multiplication and Division Properties of Equality

In Section 1.6 you learned that an equation is a statement that two expressions are equal. You also learned how to determine whether a number is a solution to an equation. In this section you will learn systematic procedures for solving certain equations.

Addition and Subtraction Properties of Equality

Just as we use properties of the real numbers to simplify algebraic expressions, we use properties of equality to simplify and solve equations. The **addition property of equality** says that if we add the same real number to both sides of an equation, the resulting equation has the same solution as the original. Equations that have the same solution are called **equivalent equations.** In our first example we use the addition property to solve an equation.

EXAMPLE 1 **Using the addition property of equality**

Solve $x - 3 = -7$.

Solution

Note that the left side of this equation is a difference. We can remove the 3 from the left side of the equation by adding 3 to each side of the equation:

$$\begin{aligned} x - 3 &= -7 \\ x - 3 + 3 &= -7 + 3 && \text{Add 3 to each side.} \\ x + 0 &= -4 && \text{Simplify each side.} \\ x &= -4 && \text{Zero is the additive identity.} \end{aligned}$$

Only -4 satisfies the last equation. Check that -4 satisfies the original equation by replacing x by -4:

$$\begin{aligned} x - 3 &= -7 && \text{Original equation} \\ -4 - 3 &= -7 && \text{Replace } x \text{ by } -4. \end{aligned}$$

Since $-4 - 3 = -7$ is correct, -4 is the solution to the equation. ■

In Example 1 we used addition to isolate the variable on the left-hand side of the equation. We can also use subtraction to isolate the variable. The **subtraction property of equality** says that if we subtract the same real number from both sides of an equation, the resulting equation is equivalent to the original equation.

EXAMPLE 2 **Using the subtraction property of equality**

Solve $9 + x = -2$.

Solution

Note that the left side of this equation is a sum. We can remove the 9 from the left side by subtracting 9 from both sides of the equation:

$$\begin{aligned} 9 + x &= -2 \\ 9 + x - 9 &= -2 - 9 && \text{Subtract 9 from each side.} \\ x &= -11 && \text{Simplify each side.} \end{aligned}$$

Check that -11 satisfies the original equation by replacing x by -11:

$$9 + x = -2 \quad \text{Original equation}$$
$$9 + (-11) = -2 \quad \text{Replace } x \text{ by } -11.$$

Since $9 + (-11) = -2$ is correct, -11 is the solution to the equation. ■

Because algebraic expressions represent real numbers, we can use the properties of equality with algebraic expressions. In the next example we subtract an algebraic expression from each side of an equation to isolate the variable on the right side of the equation.

EXAMPLE 3 **Subtracting an algebraic expression from each side**

Solve $-9 + 6y = 7y$.

Solution

The expression $6y$ can be removed from the left side of the equation by subtracting $6y$ from both sides:

$$-9 + 6y = 7y$$
$$-9 + 6y - 6y = 7y - 6y \quad \text{Subtract } 6y \text{ from each side.}$$
$$-9 = y \quad \text{Simplify each side.}$$

Check by replacing y by -9 in the original equation:

$$-9 + 6(-9) = 7(-9)$$
$$-63 = -63$$

The solution to the equation is -9. ■

In Example 3 we would get the same result by adding $-6y$ to both sides of the equation. Since the subtraction property is so closely related to the addition property, we will summarize the two properties as follows.

Addition-Subtraction Property of Equality

Adding the same number to both sides of an equation or subtracting the same number from both sides of an equation gives an equation equivalent to the original equation. If $a = b$, then

$$a + c = b + c \quad \text{and} \quad a - c = b - c.$$

The idea in solving an equation is to isolate the variable on one side of the equation. In the process, we might get an equation such as $-x = 7$. Since every real number has a unique opposite, we get an equivalent equation by finding the opposite of each side of an equation. So if $-x = 7$, then $-(-x) = -7$, or $x = -7$.

Taking the Opposite of Each Side

The equation $-x = a$ is equivalent to $x = -a$.

In the next example we use the addition property and the subtraction property to isolate the variable.

EXAMPLE 4 **Using several properties**

Solve $2m - 4 = 3m - 9$.

Solution

To eliminate the 4 from the left side, we can add 4 to each side:

$$
\begin{aligned}
2m - 4 &= 3m - 9 \\
2m - 4 + 4 &= 3m - 9 + 4 && \text{Add 4 to each side.} \\
2m &= 3m - 5 && \text{Simplify each side.} \\
2m - 3m &= 3m - 5 - 3m && \text{Subtract } 3m \text{ from each side.} \\
-m &= -5 && \text{Simplify each side.} \\
-(-m) &= -(-5) && \text{Take the opposite of each side.} \\
m &= 5 && \text{Simplify.}
\end{aligned}
$$

To check, replace m by 5 in the original equation:

$$
\begin{aligned}
2m - 4 &= 3m - 9 && \text{Original equation} \\
2 \cdot 5 - 4 &= 3 \cdot 5 - 9 \\
6 &= 6
\end{aligned}
$$

Since $m = 5$ satisfies the original equation, the solution to the equation is 5. ■

Multiplication and Division Properties of Equality

Equations involving products or quotients cannot be solved with only the addition-subtraction property of equality. We need another property. The **multiplication property of equality** says that multiplying both sides of an equation by the same nonzero number produces an equation equivalent to the original.

EXAMPLE 5 **Using the multiplication property of equality**

Solve $\frac{z}{2} = 6$.

Solution

We isolate the variable z by multiplying each side of the equation by 2:

$$
\begin{aligned}
\frac{z}{2} &= 6 && \text{Original equation} \\
2 \cdot \frac{z}{2} &= 2 \cdot 6 && \text{Multiply each side by 2.} \\
z &= 12 && \text{Simplify.}
\end{aligned}
$$

Check 12 in the original equation to be sure that 12 is the solution. ■

Since z was divided by 2 in the original equation of Example 5, we multiplied each side by 2 to isolate z. If a multiple of a variable appears in the equation, then we can use division to isolate the variable. The **division property of equality** says that if both sides of an equation are divided by the same nonzero number, then the resulting equation is equivalent to the original equation.

EXAMPLE 6 **Using the division property of equality**

Solve $-5w = 30$.

Solution

Since w is multiplied by -5, we can isolate w by dividing each side by -5:

$$-5w = 30 \quad \text{Original equation}$$

$$\frac{-5w}{-5} = \frac{30}{-5} \quad \text{Divide each side by } -5.$$

$$w = -6 \quad \text{Simplify.}$$

Check -6 in the original equation to see that -6 is the solution. ■

We can state the multiplication and division properties together as follows.

Multiplication-Division Property of Equality

Multiplying or dividing both sides of an equation by the same nonzero number gives an equation equivalent to the original equation. If $a = b$ and $c \neq 0$, then

$$ac = bc \quad \text{and} \quad \frac{a}{c} = \frac{b}{c}.$$

In the next example the coefficient of the variable is a fraction. We could divide each side by the coefficient as we did in Example 6, but it is easier to multiply each side by the reciprocal of the coefficient.

EXAMPLE 7 **Multiplying by the reciprocal**

Solve $\frac{2}{3}p = 40$.

Solution

Multiply each side by $\frac{3}{2}$, the reciprocal of $\frac{2}{3}$, to isolate p on the left side:

$$\frac{2}{3}p = 40$$

$$\frac{3}{2} \cdot \frac{2}{3}p = \frac{3}{2} \cdot 40 \quad \text{Multiply each side by } \tfrac{3}{2}.$$

$$1 \cdot p = 60 \quad \text{Multiplicative inverses}$$

$$p = 60 \quad \text{Multiplicative identity}$$

Check 60 in the original equation. The solution is 60. ■

If the coefficient of the variable is an integer, we usually divide each side by that integer, as in Example 6. If the coefficient of the variable is a fraction, we usually multiply each side by the reciprocal of the fraction as in Example 7. However, to solve an equation such as $3x = \frac{1}{4}$, it might be simpler for you to multiply each side by $\frac{1}{3}$ than to divide each side by 3. Try it.

You will be learning how to solve many different types of equations. The simplest equations are the linear equations.

Linear Equation

A **linear equation in one variable** x is an equation of the form

$$ax + b = 0,$$

where a and b are real numbers and $a \neq 0$.

Many equations that are not exactly in the form $ax + b = 0$ are still called linear equations because they are equivalent to equations of that form. All of the equations in this section are called linear equations. To solve an equation of the form $ax + b = 0$, we use the addition-subtraction property and the multiplication-division property of equality.

EXAMPLE 8 **Using the addition and the division properties of equality**

Solve $3r - 5 = 0$.

Solution

To isolate r, first add 5 to each side, then divide each side by 3:

$3r - 5 = 0$	Original equation
$3r - 5 + 5 = 0 + 5$	Add 5 to each side.
$3r = 5$	Combine like terms.
$\frac{3r}{3} = \frac{5}{3}$	Divide each side by 3.
$r = \frac{5}{3}$	Simplify.

Checking $\frac{5}{3}$ in the original equation gives

$$3 \cdot \frac{5}{3} - 5 = 5 - 5 = 0.$$

So $\frac{5}{3}$ is the solution. ■

In the next example we simplify each side of the equation as much as possible before we use any properties of equality.

EXAMPLE 9 **Simplifying before using properties of equality**

Solve $2(q-3)+5q=8(q-1)$.

Solution

Remove parentheses and combine like terms on the left-hand side of the equation:

$2(q-3)+5q=8(q-1)$	Original equation
$2q-6+5q=8q-8$	Distributive property
$7q-6=8q-8$	Combine like terms.
$7q-6+6=8q-8+6$	Add 6 to each side.
$7q=8q-2$	Combine like terms.
$7q-8q=8q-2-8q$	Subtract $8q$ from each side.
$-q=-2$	
$-1(-q)=-1(-2)$	Multiply each side by -1.
$q=2$	Simplify.

To check, we replace q by 2 in the original equation and simplify:

$2(q-3)+5q=8(q-1)$	Original equation
$2(2-3)+5(2)=8(2-1)$	Replace q by 2.
$2(-1)+10=8(1)$	
$8=8$	

Because both sides have the same value, we can be sure that 2 is the solution. ■

In Example 9, we multiplied both sides of $-q=-2$ by -1 to get $q=2$. Finding the opposite of each side as in Example 4 would yield the same result because multiplying a number by -1 gives the opposite of the number.

It is not necessary to solve an equation exactly as done in the examples. Use the examples as guidelines. You can do the steps in a different order and still get the solution as long as you have equivalent equations at every step. When solving equations, you should keep in mind the following strategy for isolating the variable.

Strategy for Solving Equations

1. Remove parentheses and combine like terms to simplify each side as much as possible.
2. Use the addition-subtraction property of equality to get like terms from opposite sides onto the same side so that they may be combined.
3. The multiplication-division property of equality is generally used last.
4. The equation $-x=a$ is equivalent to $x=-a$.
5. Check that the solution satisfies the original equation.

Warm-ups

True or false? Explain your answer.

1. The solution to $2x - 5 = 5$ is 5.
2. The equation $\frac{x}{2} = 4$ is equivalent to the equation $x - 8 = 0$.
3. To solve $\frac{3}{4}y = 12$, we should multiply each side by $\frac{3}{4}$.
4. If $w + 5 = 0$, then $w = 5$.
5. Multiplying each side of an equation by the same real number will result in an equation that is equivalent to the original equation.
6. To isolate t in $3t - 7 = 9$, divide each side by 3 and then add 7 to each side.
7. To solve $\frac{2r}{3} = 30$, we should multiply each side by $\frac{3}{2}$.
8. The equation $2n + 3 = 0$ is equivalent to $2n = -3$.
9. The equation $x - (x - 3) = 5x$ is equivalent to $3 = 5x$.
10. The solution to $4 - x = -2x$ is -4.

2.1 EXERCISES

Use the properties of equality to solve each equation. Show your work and check your answer. See Examples 1 and 2.

1. $x - 6 = -5$
2. $-7 + x = -2$
3. $x + 13 = -4$
4. $x + 8 = -12$
5. $x - \frac{1}{2} = \frac{1}{2}$
6. $x - \frac{1}{4} = \frac{1}{2}$
7. $\frac{1}{3} + x = \frac{1}{3}$
8. $\frac{1}{3} + x = \frac{1}{2}$

Solve each equation. See Examples 3 and 4.

9. $5a = -2 + 4a$
10. $8y = 6 + 7y$
11. $-3y = 5 - 4y$
12. $-9y = 3 - 10y$
13. $-x + 6 = 5$
14. $-x - 2 = 9$
15. $-9 - a = -3$
16. $4 - r = 6$
17. $2q + 5 = q - 7$
18. $3z - 6 = 2z - 7$
19. $-3x + 1 = 5 - 2x$
20. $5 - 2x = 6 - x$
21. $-12 - 5x = -4x + 1$
22. $-3x - 4 = -2x + 8$
23. $3x + 0.3 = 2 + 2x$
24. $2y - 0.05 = y + 1$
25. $1.2k - 0.6 = 0.2k - 0.5$
26. $2.3h + 6 = 1.3h - 1$
27. $0.2x - 4 = 0.6 - 0.8x$
28. $0.3x = 1 - 0.7x$

Solve each linear equation. Show your work and check your answer. See Examples 5 and 6.

29. $\frac{x}{2} = -4$
30. $\frac{x}{3} = -6$
31. $0.03 = \frac{x}{60}$
32. $0.05 = \frac{x}{80}$
33. $-3x = 15$
34. $-5x = -20$
35. $20 = 4x$
36. $18 = -3x$
37. $\frac{a}{2} = \frac{1}{3}$
38. $\frac{b}{2} = \frac{1}{5}$
39. $\frac{c}{3} = \frac{1}{6}$
40. $\frac{d}{3} = \frac{1}{4}$
41. $0.5w = 10$
42. $0.4y = -8$
43. $-z = \frac{2}{3}$
44. $-t = -\frac{3}{5}$

Solve each linear equation. Show your work and check your answer. See Example 7.

45. $\frac{2}{3}x = 8$
46. $\frac{3}{4}x = 9$
47. $-\frac{3}{5}x = 6$
48. $-\frac{4}{5}x = 12$
49. $\frac{5a}{7} = -10$
50. $\frac{7r}{12} = -24$
51. $-\frac{10}{3} = -\frac{5t}{6}$
52. $-\frac{2}{15} = -\frac{3y}{5}$
53. $2m = \frac{1}{2}$
54. $3n = \frac{1}{4}$
55. $-2u = \frac{4}{5}$
56. $-3v = \frac{6}{7}$

Solve each linear equation. Show your work and check your answer. See Examples 8 and 9.

57. $2x - 3 = 0$

58. $5x - 7 = 0$

59. $-2x + 5 = 7$

60. $-3x + 4 = 13$

61. $-2x - 5 = 7$

62. $-3x - 7 = -1$

63. $-3(k - 6) = 2 - k$

64. $-2(h - 5) = 3 - h$

65. $2(p + 1) - p = 36$

66. $3(q + 1) - q = 23$

67. $7 - 3(5 - u) = 5(u - 4)$

68. $v - 4(4 - v) = -2(2v - 1)$

Solve each equation.

69. $4(x + 3) = 12$

70. $5(x - 3) = -15$

71. $-3x - 1 = 5 - 2x$

72. $-3x - 2 = -5 - 4x$

73. $0.3(x + 30) = 27$

74. $0.5(x - 12) = 6$

75. $\frac{w}{5} - 4 = -6$

76. $\frac{q}{2} + 13 = -22$

77. $\frac{2}{3}y - 5 = 7$

78. $\frac{3}{4}u - 9 = -6$

79. $4 - \frac{2n}{5} = 12$

80. $9 - \frac{2m}{7} = 19$

81. $-\frac{1}{3}p - \frac{1}{2} = \frac{1}{2}$

82. $-\frac{3}{4}z - \frac{2}{3} = \frac{1}{3}$

83. $3x - 2(x - 4) = 4 - (x - 5)$

84. $5 - 5(5 - x) = 6(x + 7) - 3$

85. $3.5x - 23.7 = -38.75$

86. $3(x - 0.87) - 2x = 4.98$

Solve each problem.

87. ***Celsius temperature.*** If the air temperature is 68° Fahrenheit, then the solution to the equation $\frac{9}{5}C + 32 = 68$ gives the Celsius temperature of the air. Find the Celsius temperature.

88. ***Fahrenheit temperature.*** If the temperature of hot tap water is 70°C, then the solution to the equation $70 = \frac{5}{9}(F - 32)$ gives the Fahrenheit temperature of the water. Find the Fahrenheit temperature of the water.

89. ***Rectangular patio.*** If a rectangular patio has a length that is 3 feet longer than its width and a perimeter of 42 feet, then the width can be found by solving the equation $2x + 2(x + 3) = 42$. What is the width?

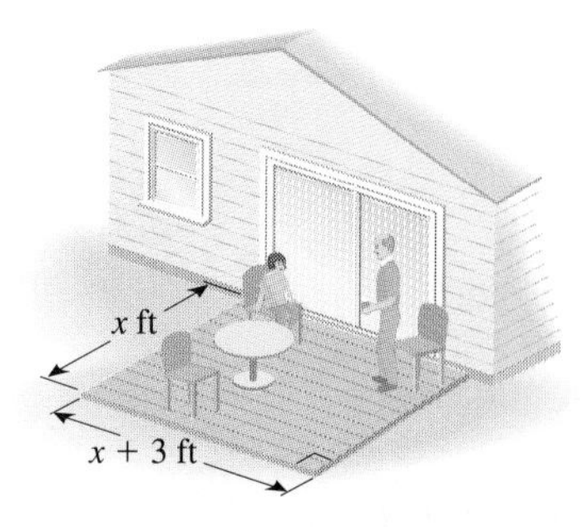

Figure for Exercise 89

90. ***Perimeter of a triangle.*** If a triangle has sides of length x, $x + 1$, and $x + 2$ meters and a perimeter of 12 meters, then the value of x can be found by solving $x + (x + 1) + (x + 2) = 12$. Find the values of x, $x + 1$, and $x + 2$.

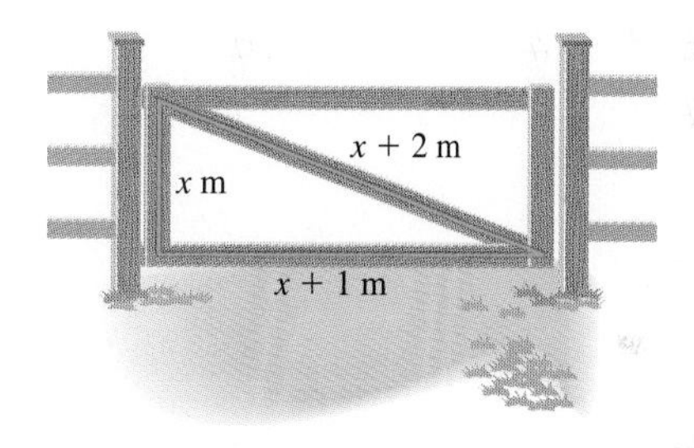

Figure for Exercise 90

Getting More Involved

91. ***Writing.*** If we multiply or divide each side of an equation by zero, do we get an equivalent equation?

92. ***Writing.*** If we add zero to each side of an equation, do we get an equivalent equation?

93. ***Discussion.*** Examine the following solution:

$$2(x - 3) = 8$$
$$2(x - 3) - 8 = 0$$
$$2x - 6 - 8 = 0$$
$$2x - 14 = 0$$
$$2x - 4 = 10$$
$$2x = 14$$
$$x = 7$$

Are all of the equations equivalent? How could you improve on the method used here?

94. ***Discussion.*** Examine the following solution:

$$3x - 4 = 5x - 6$$
$$2x - 4 = 6$$
$$2x = 2$$
$$x = 1$$

Does 1 satisfy the original equation. Are all of the equations equivalent?

2.2 More Linear Equations

In this section:

- Identities
- Conditional Equations
- Inconsistent Equations
- Equations Involving Fractions
- Equations Involving Decimals
- Solving Shortcuts

In this section we will solve more equations of the type that we solved in Section 2.1. However, some equations in this section have infinitely many solutions, and some have no solution.

Identities

It is easy to find equations that are satisfied by any real number that we choose as a replacement for the variable. For example, the equations

$$x \div 2 = \frac{1}{2}x, \qquad x + x = 2x, \qquad \text{and} \qquad x + 1 = x + 1$$

are satisfied by all real numbers. The equation

$$\frac{5}{x} = \frac{5}{x}$$

is satisfied by any real number except 0 because division by 0 is undefined.

> **Identity**
>
> An equation that is satisfied by every real number for which both sides are defined is called an **identity.**

We cannot recognize that the equation in the next example is an identity until we have simplified each side.

EXAMPLE 1 **Solving an identity**

Solve $7 - 5(x - 6) + 4 = 3 - 2(x - 5) - 3x + 28$.

Solution

We first use the distributive property to remove the parentheses:

$$7 - 5(x - 6) + 4 = 3 - 2(x - 5) - 3x + 28$$
$$7 - 5x + 30 + 4 = 3 - 2x + 10 - 3x + 28$$
$$41 - 5x = 41 - 5x \quad \text{Combine like terms.}$$

This last equation is true for any value of x because the two sides are identical. So all real numbers satisfy the original equation, and it is an identity. ■

CAUTION If you get an equation in which both sides are identical, as in Example 1, there is no need to continue to simplify the equation. If you do continue, you will eventually get $0 = 0$, from which you can still conclude that the equation is an identity. ⊘

Conditional Equations

The statement $2x + 4 = 10$ is true only on condition that we choose $x = 3$. The equation $x^2 = 4$ is satisfied only if we choose $x = 2$ or $x = -2$. These equations are called conditional equations.

Conditional Equation

A **conditional equation** is an equation that is satisfied by at least one real number but is not an identity.

Every equation that we solved in Section 2.1 is a conditional equation.

Inconsistent Equations

It is easy to find equations that are false no matter what number we use to replace the variable. Consider the equation

$$x = x + 1.$$

If we replace x by 3, we get $3 = 3 + 1$, which is false. If we replace x by 4, we get $4 = 4 + 1$, which is also false. Clearly, there is no number that will satisfy $x = x + 1$. Other examples of equations with no solutions include

$$x = x - 2, \qquad x - x = 5, \qquad \text{and} \qquad 0 \cdot x + 6 = 7.$$

Inconsistent Equation

An equation that has no solution is called an **inconsistent equation.**

EXAMPLE 2 **Solving an inconsistent equation**

Solve $2 - 3(x - 4) = 4(x - 7) - 7x$.

Solution

Use the distributive property to remove the parentheses:

$2 - 3(x - 4) = 4(x - 7) - 7x$	The original equation
$2 - 3x + 12 = 4x - 28 - 7x$	Distributive property
$14 - 3x = -28 - 3x$	Combine like terms on each side.
$14 - 3x + 3x = -28 - 3x + 3x$	Add $3x$ to each side.
$14 = -28$	Simplify.

This last equation is not true for any choice of x. So there is no solution to the original equation, and the equation is inconsistent. ■

Keep the following points in mind in solving equations.

Summary: Identities and Inconsistent Equations

1. An equation that is equivalent to an equation in which both sides are identical is an identity. The equation is satisfied by all real numbers for which both sides are defined.
2. An equation that is equivalent to an equation that is always false is inconsistent. The equation has no solution.

Equations Involving Fractions

We solved some equations involving fractions in Section 2.1. Here, we will solve equations with fractions by eliminating all fractions in the first step. All of the fractions will be eliminated if we multiply each side by the least common denominator.

EXAMPLE 3 **Multiplying by the least common denominator**

Solve $\frac{y}{2} - 1 = \frac{y}{3} + 1$

Solution

The least common denominator (LCD) for the denominators 2 and 3 is 6. Since both 2 and 3 divide into 6 evenly, multiplying each side by 6 will eliminate the fractions:

$$6\left(\frac{y}{2} - 1\right) = 6\left(\frac{y}{3} + 1\right) \quad \text{Multiply each side by 6.}$$

$$6 \cdot \frac{y}{2} - 6 \cdot 1 = 6 \cdot \frac{y}{3} + 6 \cdot 1 \quad \text{Distributive property}$$

$$3y - 6 = 2y + 6 \quad \text{Simplify: } 6 \cdot \frac{y}{2} = 3y$$

$$3y = 2y + 12 \quad \text{Add 6 to each side.}$$

$$y = 12 \quad \text{Subtract } 2y \text{ from each side.}$$

Check 12 in the original equation:

$$\frac{12}{2} - 1 = \frac{12}{3} + 1$$

$$5 = 5$$

Since 12 satisfies the original equation, the solution is 12. ■

Equations involving fractions are usually easier to solve if we first multiply each side by the LCD of the fractions.

Equations Involving Decimals

When an equation involves decimal numbers, we can work with the decimal numbers or we can eliminate all of the decimal numbers by multiplying both sides by 10, or 100, or 1000, and so on. Multiplying a decimal number by 10 moves the decimal point one place to the right. Multiplying by 100 moves the decimal point two places to the right, and so on.

EXAMPLE 4 **An equation involving decimals**

Solve $0.3p + 8.04 = 12.6$.

Solution

The largest number of decimal places appearing in the decimal numbers of the equation is two (in the number 8.04). Therefore we multiply each side of the equation by 100 because multiplying by 100 moves decimal points two places to the right:

$$
\begin{aligned}
0.3p + 8.04 &= 12.6 && \text{Original equation} \\
100(0.3p + 8.04) &= 100(12.6) && \text{Multiplication property of equality} \\
100(0.3p) + 100(8.04) &= 100(12.6) && \text{Distributive property} \\
30p + 804 &= 1260 \\
30p + 804 - 804 &= 1260 - 804 && \text{Subtract 804 from each side.} \\
30p &= 456 \\
\frac{30p}{30} &= \frac{456}{30} && \text{Divide each side by 30.} \\
p &= 15.2
\end{aligned}
$$

You can use a calculator to check that

$$0.3(15.2) + 8.04 = 12.6.$$

The solution is 15.2. ■

EXAMPLE 5 **Another equation with decimals**

Solve $0.5x + 0.4(x + 20) = 13.4$

Solution

First use the distributive property to remove the parentheses:

$$
\begin{aligned}
0.5x + 0.4(x + 20) &= 13.4 && \text{Original equation} \\
0.5x + 0.4x + 8 &= 13.4 && \text{Distributive property} \\
10(0.5x + 0.4x + 8) &= 10(13.4) && \text{Multiply each side by 10.} \\
5x + 4x + 80 &= 134 && \text{Simplify.} \\
9x + 80 &= 134 && \text{Combine like terms.} \\
9x + 80 - 80 &= 134 - 80 && \text{Subtract 80 from each side.} \\
9x &= 54 && \text{Simplify.} \\
x &= 6 && \text{Divide each side by 9.}
\end{aligned}
$$

Check 6 in the original equation:

$$
\begin{aligned}
0.5(6) + 0.4(6 + 20) &= 13.4 && \text{Replace } x \text{ by 6.} \\
3 + 0.4(26) &= 13.4 \\
3 + 10.4 &= 13.4
\end{aligned}
$$

Since both sides of the equation have the same value, 6 is the solution. ■

CAUTION If you multiply each side by 10 in Example 5 before using the distributive property, be careful how you handle the terms in parentheses:

$$
\begin{aligned}
10 \cdot 0.5x + 10 \cdot 0.4(x + 20) &= 10 \cdot 13.4 \\
5x + 4(x + 20) &= 134
\end{aligned}
$$

It is not correct to multiply 0.4 by 10 *and also* to multiply $x + 20$ by 10. ⊘

Solving Shortcuts

It is very important to develop the skill of solving equations in a systematic way, writing down every step as we have been doing. As you become more skilled at solving equations, you will probably want to speed up the process a bit. This can be done by writing only the result of performing an operation on each side. In some cases we can also save a step by isolating the variable on the right-hand side rather than the left-hand side. These two shortcuts are illustrated in the next example.

EXAMPLE 6 **Shortcuts in solving equations**

Solve each equation.

a) $2a - 3 = 0$ **b)** $2k + 5 = 3k + 1$

Solution

a) Add 3 to each side, then divide each side by 2:

$$2a - 3 = 0$$

$$2a = 3 \quad \text{Add 3 to each side.}$$

$$a = \frac{3}{2} \quad \text{Divide each side by 2.}$$

Check that $\frac{3}{2}$ satisfies the original equation. The solution is $\frac{3}{2}$.

b) For this equation we can get a single k on the right by subtracting $2k$ from each side. (If we subtract $3k$ from each side, we get $-k$, and then we need another step.)

$$2k + 5 = 3k + 1$$

$$5 = k + 1 \quad \text{Subtract } 2k \text{ from each side.}$$

$$4 = k \quad \text{Subtract 1 from each side.}$$

Check that 4 satisfies the original equation. The solution is 4. ■

Warm-ups

True or false? Explain your answer.

1. The equation $x - x = 99$ has no solution.
2. The equation $2n + 3n = 5n$ is an identity.
3. The equation $2y + 3y = 4y$ is inconsistent.
4. All real numbers satisfy the equation $1 \div x = \frac{1}{x}$.
5. The equation $5a + 3 = 0$ is an inconsistent equation.
6. The equation $2t = t$ is a conditional equation.
7. The equation $w - 0.1w = 0.9w$ is an identity.
8. The equation $0.2x + 0.03x = 8$ is equivalent to $20x + 3x = 8$.
9. The equation $\frac{x}{x} = 1$ is an identity.
10. The solution to $3h - 8 = 0$ is $\frac{8}{3}$.

2.2 EXERCISES

Solve each equation. Identify each as a conditional equation, an inconsistent equation, or an identity. See Examples 1 and 2.

1. $x + x = 2x$

2. $2x - x = x$

3. $a - 1 = a + 1$

4. $r + 7 = r$

5. $3y + 4y = 12y$

6. $9t - 8t = 7$

7. $-4 + 3(w - 1) = w + 2(w - 2) - 1$

8. $4 - 5(w + 2) = 2(w - 1) - 7w - 4$

9. $3(m + 1) = 3(m + 3)$

10. $5(m - 1) - 6(m + 3) = 4 - m$

11. $x + x = 2$

12. $3x - 5 = 0$

13. $2 - 3(5 - x) = 3x$

14. $3 - 3(5 - x) = 0$

15. $(3 - 3)(5 - z) = 0$

16. $(2 \cdot 4 - 8)p = 0$

17. $\frac{0}{x} = 0$

18. $\frac{2x}{2} = x$

19. $x \cdot x = x^2$

20. $\frac{2x}{2x} = 1$

Solve each equation by first eliminating the fractions. See Example 3.

21. $\frac{x}{2} + 3 = x - \frac{1}{2}$

22. $13 - \frac{x}{2} = x - \frac{1}{2}$

23. $\frac{x}{2} + \frac{x}{3} = 20$

24. $\frac{x}{2} - \frac{x}{3} = 5$

25. $\frac{w}{2} + \frac{w}{4} = 12$

26. $\frac{a}{4} - \frac{a}{2} = -5$

27. $\frac{3z}{2} - \frac{2z}{3} = -10$

28. $\frac{3m}{4} + \frac{m}{2} = -5$

29. $\frac{1}{3}p - 5 = \frac{1}{4}p$

30. $\frac{1}{2}q - 6 = \frac{1}{5}q$

31. $\frac{1}{6}v + 1 = \frac{1}{4}v - 1$

32. $\frac{1}{15}k + 5 = \frac{1}{6}k - 10$

Solve each equation by first eliminating the decimal numbers. See Examples 4 and 5.

33. $x - 0.2x = 72$

34. $x - 0.1x = 63$

35. $0.3x + 1.2 = 0.5x$

36. $0.4x - 1.6 = 0.6x$

37. $0.02x - 1.56 = 0.8x$

38. $0.6x + 10.4 = 0.08x$

39. $0.1a - 0.3 = 0.2a - 8.3$

40. $0.5b + 3.4 = 0.2b + 12.4$

41. $0.05r + 0.4r = 27$

42. $0.08t + 28.3 = 0.5t - 9.5$

43. $0.05y + 0.03(y + 50) = 17.5$

44. $0.07y + 0.08(y - 100) = 44.5$

45. $0.1x + 0.05(x - 300) = 105$

46. $0.2x - 0.05(x - 100) = 35$

Solve each equation. If you feel proficient enough, try using shortcuts, as described in Example 6.

47. $2x - 9 = 0$

48. $3x + 7 = 0$

49. $-2x + 6 = 0$

50. $-3x - 12 = 0$

51. $\frac{z}{5} + 1 = 6$

52. $\frac{s}{2} + 2 = 5$

53. $\frac{c}{2} - 3 = -4$

54. $\frac{b}{3} - 4 = -7$

55. $3 = t + 6$

56. $-5 = y - 9$

57. $5 + 2q = 3q$

58. $-4 - 5p = -4p$

59. $8x - 1 = 9 + 9x$

60. $4x - 2 = -8 + 5x$

61. $-3x + 1 = -1 - 2x$

62. $-6x + 3 = -7 - 5x$

Solve each equation.

63. $3x - 5 = 2x - 9$

64. $5x - 9 = x - 4$

65. $x + 2(x + 4) = 3(x + 3) - 1$

66. $u + 3(u - 4) = 4(u - 5)$

67. $23 - 5(3 - n) = -4(n - 2) + 9n$

68. $-3 - 4(t - 5) = -2(t + 3) + 11$

69. $0.05x + 30 = 0.4x - 5$

70. $x - 0.08x = 460$

71. $-\frac{2}{3}a + 1 = 2$

72. $-\frac{3}{4}t = \frac{1}{2}$

73. $\frac{y}{2} + \frac{y}{6} = 20$

74. $\frac{3w}{5} - 1 = \frac{w}{2} + 1$

75. $0.09x - 0.2(x + 4) = -1.46$

76. $0.08x + 0.5(x + 100) = 73.2$

77. $436x - 789 = -571$

78. $0.08x + 4533 = 10x + 69$

79. $\frac{x}{344} + 235 = 292$

80. $34(x - 98) = \frac{x}{2} + 475$

Solve each problem.

81. ***Eavesdropping.*** Reginald overheard his neighbor complaining to his wife that their federal income tax for 1994 was \$19,700. Reginald looked in his 1994 tax table and found that if he could solve the equation

$$0.15(38{,}000) + 0.28(x - 38{,}000) = 19{,}700,$$

he could find his neighbors' taxable income x. Find the taxable income of Reginald's neighbors.

82. ***Federal taxes.*** According to Bruce Harrell, CPA, the federal income tax for a class C corporation is found by solving a linear equation. The reason for the equation is that the amount x of federal tax is deducted before the state tax is figured, and the amount of state tax is deducted before the federal tax is figured. To find the amount of federal tax for a corporation with a taxable income of \$200,000, for which the federal tax rate is 25% and the state tax rate is 10%, Bruce must solve

$$x = 0.25[200{,}000 - 0.10(200{,}000 - x)].$$

Solve the equation for Bruce.

Getting More Involved

83. ***Writing.*** Explain in your own words the difference between a conditional equation, an identity, and an inconsistent equation.

84. ***Discussion.*** What are the two steps that you can use to solve $ax + b = 0$ with $a \neq 0$? How do you evaluate $ax + b$ for a specific value of x using the order of operations. What is the relationship between the steps used to evaluate $ax + b$ and the steps used to solve $ax + b = 0$?

2.3 Formulas

In this section:

- Solving for a Variable
- Finding the Value of a Variable

In this section you will learn to rewrite formulas using the same properties of equality that we used to solve equations. Formulas are often called **literal equations.** You will also learn how to find the value of one of the variables in a formula when we know the value of all of the others.

Solving for a Variable

A **formula** is an equation involving two or more variables. The formula

$$D = R \cdot T$$

expresses the relationship between distance, rate, and time of a moving object. The formula

$$C = \frac{5}{9}(F - 32)$$

expresses the relationship between the Fahrenheit and Celsius measurements of temperature. It is often necessary to rewrite a formula for one variable in terms of the others. We refer to this process as **solving for a certain variable.**

EXAMPLE 1 **Solving for a certain variable**

Solve the formula $D = RT$ for T:

Solution

$D = RT$ Original formula

$\frac{D}{R} = \frac{R \cdot T}{R}$ Divide each side by R.

$\frac{D}{R} = T$ Simplify.

$T = \frac{D}{R}$ It is customary to write the single variable on the left. ■

EXAMPLE 2 **Solving for a certain variable**

Solve the formula $C = \frac{5}{9}(F - 32)$ for F.

Solution

We could apply the distributive property to the right side of the equation, but it is simpler to proceed as follows:

$$C = \frac{5}{9}(F - 32)$$

$$\frac{9}{5}C = \frac{9}{5} \cdot \frac{5}{9}(F - 32) \quad \text{Multiply each side by } \tfrac{9}{5}\text{, the reciprocal of } \tfrac{5}{9}.$$

$$\frac{9}{5}C = F - 32 \quad \text{Simplify.}$$

$$\frac{9}{5}C + 32 = F - 32 + 32 \quad \text{Add 32 to each side.}$$

$$\frac{9}{5}C + 32 = F \quad \text{Simplify.}$$

The formula is usually written as $F = \frac{9}{5}C + 32$. ■

When solving for a variable that appears more than once in the equation, we must combine the terms to obtain a single occurrence of the variable. *When a formula has been solved for a certain variable, that variable will not occur on both sides of the equation.*

EXAMPLE 3 **Solving for a variable that appears on both sides**

Solve $5x - b = 3x + d$ for x.

Solution

First get all terms involving x onto one side and all other terms onto the other side:

$$5x - b = 3x + d \quad \text{Original formula}$$

$$5x - 3x - b = d \quad \text{Subtract } 3x \text{ from each side.}$$

$$5x - 3x = b + d \quad \text{Add } b \text{ to each side.}$$

$$2x = b + d \quad \text{Combine like terms.}$$

$$x = \frac{b + d}{2} \quad \text{Divide each side by 2.}$$

The formula solved for x is $x = \frac{b + d}{2}$. ■

In Chapter 7 it will be necessary to solve an equation involving x and y for y.

EXAMPLE 4 **Solving for y**

Solve $x + 2y = 6$ for y. Write the answer in the form $y = mx + b$, where m and b are real numbers.

Solution

$$x + 2y = 6 \quad \text{Original equation}$$

$$2y = 6 - x \quad \text{Subtract } x \text{ from each side.}$$

$$\frac{1}{2} \cdot 2y = \frac{1}{2}(6 - x) \quad \text{Multiply each side by } \tfrac{1}{2}.$$

$$y = 3 - \frac{1}{2}x \quad \text{Distributive property}$$

$$y = -\frac{1}{2}x + 3 \quad \text{Rearrange to get } y = mx + b \text{ form.}$$

Notice that in Example 4 we multiplied each side of the equation by $\frac{1}{2}$, and so we multiplied each term on the right-hand side by $\frac{1}{2}$. Instead of multiplying by $\frac{1}{2}$, we could have divided each side of the equation by 2. We would then divide each term on the right side by 2. This idea is illustrated in the next example.

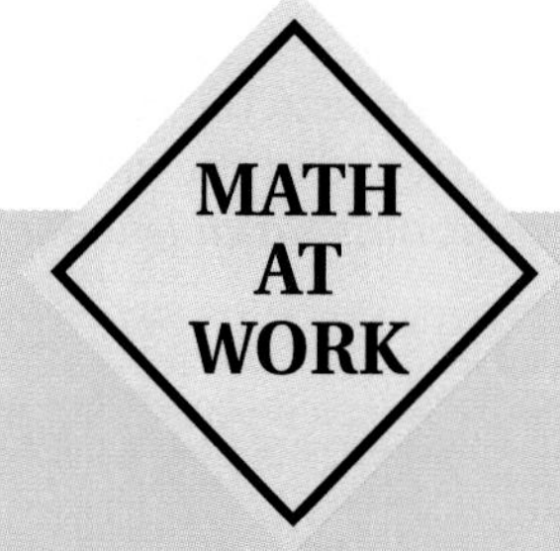

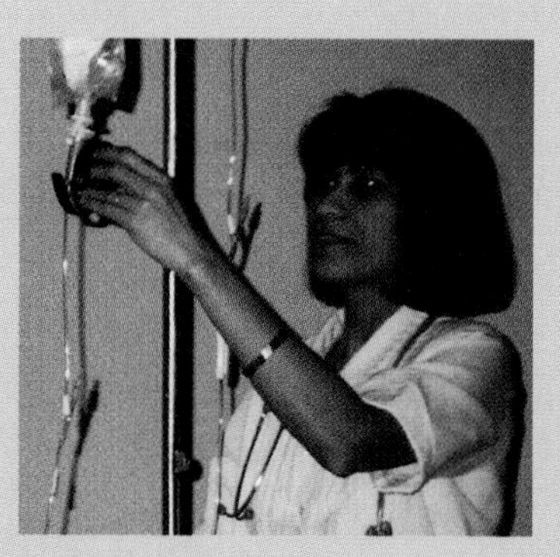

Nurse

Even before the days of Florence Nightingale, nurses around the world were giving comfort and aid to the sick and injured. Continuing in this tradition, Asenet Craffey, staff nurse at the Massachusetts Eye and Ear Infirmary, works in the intensive care unit. During her 12-hour shifts, Ms. Craffey is responsible for the full nursing care of four to eight patients. In the intensive care unit the nurse-to-patient ratio is usually one to one. When Ms. Craffey is assigned to this unit, she is responsible for overall care of a patient as well as being prepared for crisis care. Staff scheduling is an additional duty that Ms. Craffey performs, making sure that there is adequate nursing coverage for the day's planned surgeries and quality patient care. Full care means being directly involved in all of the patient's care: monitoring vital signs, changing dressings, helping to feed, following the prescribed orders left by the physicians, and administering drugs.

Many drugs come directly from the pharmacy in the exact dosage for a particular patient. Intravenous (IV) drugs, however, must be monitored so that the correct amount of drops per minute are administered. IV medications can be glucose solutions, antibiotics, or pain killers. Often the prescribed dosage is 1 gram per 100, 200, 500, or 1000 cubic centimeters of liquid. In Exercise 79 of this section you will calculate a drug dosage, just as Ms. Craffey would on the job.

EXAMPLE 5 **Solving for y**

Solve $2x - 3y = 9$ for y. Write the answer in the form $y = mx + b$, where m and b are real numbers.

Solution

$2x - 3y = 9$	Original equation
$-3y = -2x + 9$	Subtract $2x$ from each side.
$\dfrac{-3y}{-3} = \dfrac{-2x}{-3} + \dfrac{9}{-3}$	Divide each side by -3.
$y = \dfrac{2}{3}x - 3$	Simplify.

Finding the Value of a Variable

In many situations we know the values of all variables in a formula except one. We use the formula to determine the unknown value.

EXAMPLE 6 **Finding the value of a variable in a formula**

If $2x - 3y = 9$, find y when $x = 6$.

Solution

Method 1: First solve the equation for y. Because we have already solved this equation for y in Example 5, we will not repeat that process in this example. We have

$$y = \frac{2}{3}x - 3.$$

Now replace x by 6 in this equation:

$$\begin{aligned} y &= \frac{2}{3}(6) - 3 \\ &= 4 - 3 = 1 \end{aligned}$$

So, when $x = 6$, we have $y = 1$.

Method 2: First replace x by 6 in the original equation, then solve for y:

$2x - 3y = 9$	Original equation
$2 \cdot 6 - 3y = 9$	Replace x by 6.
$12 - 3y = 9$	Simplify.
$-3y = -3$	Subtract 12 from each side.
$y = 1$	Divide each side by -3.

So when $x = 6$, we have $y = 1$.

If we had to find the value of y for many different values of x, it would be best to solve the equation for y, then insert the various values of x. Method 1 of Example 6 would be the better method. If we must find only one value of y, it does not matter which method we use. When doing the exercises corresponding to this example, you should try both methods.

EXAMPLE 7 **Using the simple interest formula**

If the simple interest is \$120, the principal is \$400, and the time is 2 years, find the rate.

Solution

Simple interest is calculated by using the formula $I = Prt$, where I is the interest, P is the principal, r is the rate, and t is the time in years. First, we solve the formula for r, then insert the values of P, I, and t:

$Prt = I$	Simple interest formula
$\frac{Prt}{Pt} = \frac{I}{Pt}$	Divide each side by Pt.
$r = \frac{I}{Pt}$	Simplify.
$r = \frac{120}{400 \cdot 2}$	Substitute the values of I, P, and t.
$r = 0.15$	Simplify.
$r = 15\%$	Move the decimal point two places to the right.

In solving a geometric problem, it is always helpful to draw a diagram, as we do in the next example.

EXAMPLE 8 **Using a geometric formula**

The perimeter of a rectangle is 36 feet. If the width is 6 feet, then what is the length?

Solution

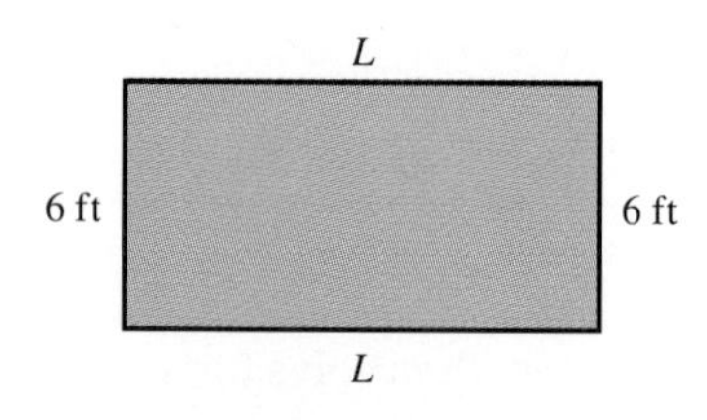

Figure 2.1

First, put the given information on a diagram as shown in Fig. 2.1. Substitute the given values into the formula for the perimeter of a rectangle found at the back of the book, and then solve for L. (We could solve for L first and then insert the given values.)

$P = 2L + 2W$	Perimeter of a rectangle
$36 = 2L + 2 \cdot 6$	Substitute 36 for P and 6 for W.
$36 = 2L + 12$	Simplify.
$24 = 2L$	Subtract 12 from each side.
$12 = L$	Divide each side by 2.

Check: If $L = 12$ and $W = 6$, then $P = 2(12) + 2(6) = 36$ feet. So we can be certain that the length is 12 feet.

EXAMPLE 9 **Finding the original price**

What was the original price of a stereo that sold for \$560 after a 20% discount.

Solution

We can use the formula $S = L - rL$, where S is the selling price, r is the rate of discount, and L is the list price:

$$\text{Selling price} = \text{list price} - \text{amount of discount}$$
$$560 = L - 0.2L$$
$$10(560) = 10(L - 0.2L) \quad \text{Multiply each side by 10.}$$
$$5600 = 10L - 2L \quad \text{Remove the parentheses.}$$
$$5600 = 8L \quad \text{Combine like terms.}$$
$$\frac{5600}{8} = \frac{8L}{8} \quad \text{Divide each side by 8.}$$
$$700 = L$$

Check: We find that 20% of \$700 is \$140 and \$700 − \$140 = \$560, the selling price. So we are certain that the original price was \$700. ■

Warm-ups

True or false? Explain your answer.

1. If we solve $D = R \cdot T$ for T, we get $T \cdot R = D$.
2. If we solve $a - b = 3a - m$ for a, we get $a = 3a - m + b$.
3. Solving $A = LW$ for L, we get $L = \frac{W}{A}$.
4. Solving $D = RT$ for R, we get $R = \frac{d}{t}$.
5. The perimeter of a rectangle is the product of its length and width.
6. The volume of a shoe box is the product of its length, width, and height.
7. The sum of the length and width of a rectangle is one-half of its perimeter.
8. Solving $y - x = 5$ for y gives us $y = x + 5$.
9. If $x = -1$ and $y = -3x + 6$, then $y = 3$.
10. The circumference of a circle is the product of its diameter and the number π.

2.3 EXERCISES

Solve each formula for the specified variable. See Examples 1 and 2.

1. $D = RT$ for R
2. $A = LW$ for W
3. $C = \pi D$ for π
4. $F = ma$ for a
5. $I = Prt$ for P
6. $I = Prt$ for t
7. $F = \frac{9}{5}C + 32$ for C
8. $y = \frac{3}{4}x - 7$ for x
9. $A = \frac{1}{2}bh$ for h
10. $A = \frac{1}{2}bh$ for b
11. $P = 2L + 2W$ for L
12. $P = 2L + 2W$ for W
13. $A = \frac{1}{2}(a + b)$ for a
14. $A = \frac{1}{2}(a + b)$ for b
15. $S = P + Prt$ for r
16. $S = P + Prt$ for t
17. $A = \frac{1}{2}h(a + b)$ for a
18. $A = \frac{1}{2}h(a + b)$ for b

Solve each equation for x. See Example 3.

19. $5x + a = 3x + b$

20. $2c - x = 4x + c - 5b$

21. $4(a + x) - 3(x - a) = 0$

22. $-2(x - b) - (5a - x) = a + b$

23. $3x - 2(a - 3) = 4x - 6 - a$

24. $2(x - 3w) = -3(x + w)$

25. $3x + 2ab = 4x - 5ab$

26. $x - a = -x + a + 4b$

Solve each equation for y. See Examples 4 and 5.

27. $x + y = -9$

28. $3x + y = -5$

29. $x + y - 6 = 0$

30. $4x + y - 2 = 0$

31. $2x - y = 2$

32. $x - y = -3$

33. $3x - y + 4 = 0$

34. $-2x - y + 5 = 0$

35. $x + 2y = 4$

36. $3x + 2y = 6$

37. $2x - 2y = 1$

38. $3x - 2y = -6$

39. $y + 2 = 3(x - 4)$

40. $y - 3 = -3(x - 1)$

41. $y - 1 = \frac{1}{2}(x - 2)$

42. $y - 4 = -\frac{2}{3}(x - 9)$

43. $\frac{1}{2}x - \frac{1}{3}y = -2$

44. $\frac{x}{2} + \frac{y}{4} = \frac{1}{2}$

For each equation that follows, find y given that x = 2. See Example 6.

45. $y = 3x - 4$

46. $y = -2x + 5$

47. $3x - 2y = -8$

48. $4x + 6y = 8$

49. $\frac{3x}{2} - \frac{5y}{3} = 6$

50. $\frac{2y}{5} - \frac{3x}{4} = \frac{1}{2}$

51. $y - 3 = \frac{1}{2}(x - 6)$

52. $y - 6 = -\frac{3}{4}(x - 2)$

53. $y - 4.3 = 0.45(x - 8.6)$

54. $y + 33.7 = 0.78(x - 45.6)$

Solve each of the following problems. Appendix A contains some geometric formulas that may be helpful. See Examples 7–9.

55. *Finding the rate.* If the simple interest on \$5000 for 3 years is \$600, then what is the rate?

56. *Finding the rate.* Wayne paid \$420 in simple interest on a loan of \$1000 for 7 years. What was the rate?

57. *Finding the time.* Kathy paid \$500 in simple interest on a loan of \$2500. If the rate was 5%, then what was the time?

58. *Finding the time.* Robert paid \$240 in simple interest on a loan of \$1000. If the rate was 8%, then what was the time?

59. *Finding the length.* The area of a rectangle is 28 square yards. The width is 4 yards. Find the length.

60. *Finding the width.* The area of a rectangle is 60 square feet. The length is 4 feet. Find the width.

61. *Finding the length.* If it takes 600 feet of wire fencing to fence a rectangular feed lot that has a width of 75 feet, then what is the length of the lot?

62. *Finding the depth.* If it takes 500 feet of fencing to enclose a rectangular lot that is 104 feet wide, then how deep is the lot?

63. *Finding the original price.* Find the original price if there is a 15% discount and the sale price is \$255.

64. *Finding the list price.* Find the list price if there is a 12% discount and the sale price is \$4400.

65. *Rate of discount.* Find the rate of discount if the discount is \$40 and the original price is \$200.

66. *Rate of discount.* Find the rate of discount if the discount is \$20 and the original price is \$250.

67. *Width of a football field.* The perimeter of a football field in the NFL, excluding the end zones, is 920 feet. How wide is the field?

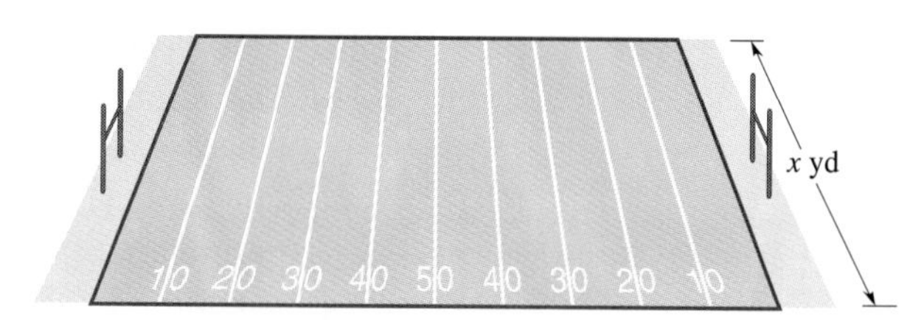

Figure for Exercise 67

68. *Perimeter of a frame.* If a picture frame is 16 inches by 20 inches, then what is its perimeter?

69. *Volume of a box.* A rectangular box measures 2 feet wide, 3 feet long, and 4 feet deep. What is its volume?

70. *Volume of a refrigerator.* The volume of a rectangular refrigerator is 20 cubic feet. If the top measures 2 feet by 2.5 feet, then what is the height?

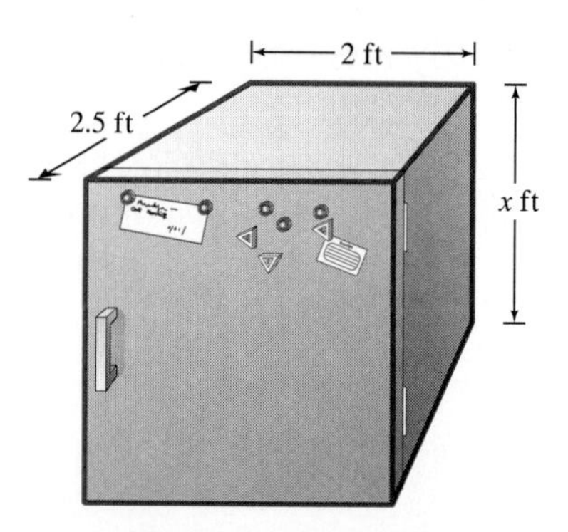

Figure for Exercise 70

71. ***Radius of a pizza.*** If the circumference of a pizza is 8π inches, then what is the radius?

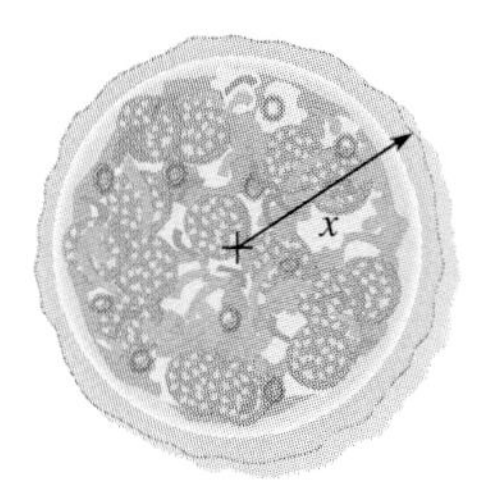

Figure for Exercise 71

72. ***Diameter of a circle.*** If the circumference of a circle is 4π meters, then what is the diameter?

73. ***Height of a banner.*** If a banner in the shape of a triangle has an area of 16 square feet with a base of 4 feet, then what is the height of the banner?

Figure for Exercise 73

74. ***Length of a leg.*** If a right triangle has an area of 14 square meters and one leg is 4 meters in length, then what is the length of the other leg?

75. ***Length of the base.*** A trapezoid with height 20 inches and lower base 8 inches has an area of 200 square inches. What is the length of its upper base?

76. ***Height of a trapezoid.*** The end of a flower box forms the shape of a trapezoid. The area of the trapezoid is 300 square centimeters. The bases are 16 centimeters and 24 centimeters in length. Find the height.

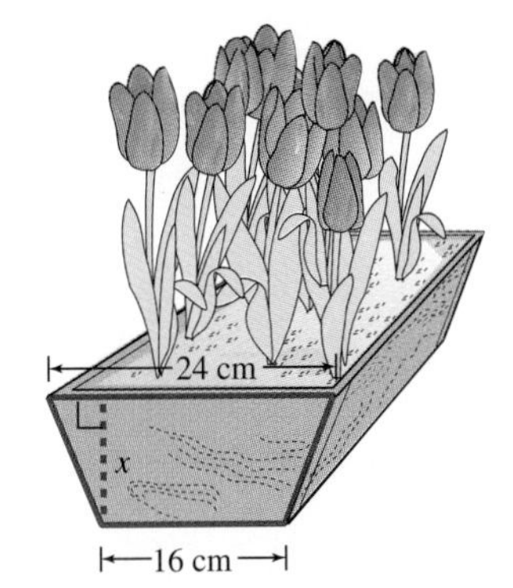

Figure for Exercise 76

77. ***Friend's rule.*** Doctors often prescribe the same drugs for children as they do for adults. The formula $d = 0.08aD$ (Friend's rule) is used to calculate the child's dosage d, where a is the child's age and D is the adult dosage. If a doctor prescribes 1000 milligrams of acetaminophen for an adult, then how many milligrams would the doctor prescribe for an eight-year-old child? Use the bar graph to determine the age at which a child would get the same dosage as an adult.

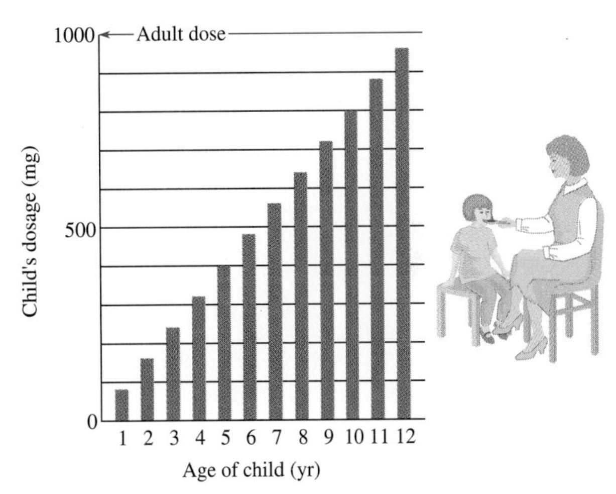

Figure for Exercise 77

78. ***Cowling's rule.*** Cowling's rule is another method for determining the dosage of a drug to prescribe to a child. For this rule, the formula

$$d = \frac{D(a + 1)}{24}$$

gives the child's dosage d, where D is the adult dosage and a is the age of the child in years. If the adult dosage of a drug is 600 milligrams and a doctor uses this formula to determine that a child's dosage is 200 milligrams, then how old is the child?

79. ***Administering Vancomycin.*** A patient is to receive 750 mg of the antibiotic Vancomycin. However, Vancomycin comes in a solution containing 1 gram (available dose) of Vancomycin per 5 milliliters (quantity) of solution. Use the formula

$$\text{Amount} = \frac{\text{desired dose}}{\text{available dose}} \times \text{quantity}$$

to find the amount of this solution that should be administered to the patient.

Getting More Involved

80. ***Writing.*** What is the difference between an equation and a formula?

81. ***Exploration.*** For each table, write a formula using the variables x and y that describes the relationship between the values of x and y.

a)

x	y
3	6
4	8
5	10

b)

x	y
3	6
5	8
7	10

c)

x	y
3	7
4	9
5	11

2.4 Translating Verbal Expressions into Algebraic Expressions

In this section:

- Writing Algebraic Expressions
- Pairs of Numbers
- Consecutive Integers
- Using Formulas
- Writing Equations

You translated some verbal expressions into algebraic expressions in Section 1.6; in this section you will study translating in more detail.

Writing Algebraic Expressions

The following box contains a list of some frequently occurring verbal expressions and their equivalent algebraic expressions.

Translating Words into Algebra

	Verbal Phrase	**Algebraic Expression**
Addition:	The sum of a number and 8	$x + 8$
	Five is added to a number	$x + 5$
	Two more than a number	$x + 2$
	A number increased by 3	$x + 3$
Subtraction:	Four is subtracted from a number	$x - 4$
	Three less than a number	$x - 3$
	The difference between 7 and a number	$7 - x$
	A number decreased by 2	$x - 2$
Multiplication:	The product of 5 and a number	$5x$
	Twice a number	$2x$
	One-half of a number	$\frac{1}{2}x$
	Five percent of a number	$0.05x$
Division:	The ratio of a number to 6	$\frac{x}{6}$
	The quotient of 5 and a number	$\frac{5}{x}$
	Three divided by some number	$\frac{3}{x}$

EXAMPLE 1 Writing algebraic expressions

Translate each verbal expression into an algebraic expression.

a) The sum of a number and 9 **b)** Eighty percent of a number

c) A number divided by 4

Solution

a) If x is the number, then the sum of a number and 9 is $x + 9$.

b) If w is the number, then eighty percent of the number is $0.80w$.

c) If y is the number, then the number divided by 4 is $\frac{y}{4}$. ■

Pairs of Numbers

There is often more than one unknown quantity in a problem, but a relationship between the unknown quantities is given. For example, if one unknown number is 5 more than another unknown number, we can use

$$x \quad \text{and} \quad x + 5,$$

to represent them. Note that x and $x + 5$ can also be used to represent two unknown numbers that differ by 5, for if two numbers differ by 5, one of the numbers is 5 more than the other.

How would you represent two numbers that have a sum of 10? If one of the numbers is 2, the other is certainly $10 - 2$, or 8. Thus if x is one of the numbers, then $10 - x$ is the other. The expressions

$$x \quad \text{and} \quad 10 - x$$

have a sum of 10 for any value of x.

EXAMPLE 2 Algebraic expressions for pairs of numbers

Write algebraic expressions for each pair of numbers.

a) Two numbers that differ by 12 **b)** Two numbers with a sum of -8

Solution

a) The expressions x and $x + 12$ represent two numbers that differ by 12. Of course, x and $x - 12$ also differ by 12.

b) The expressions x and $-8 - x$ have a sum of -8. We can check by addition:

$$-8 - x + x = -8$$ ■

Pairs of numbers occur in geometry in discussing measures of angles. You will need the following facts about degree measures of angles.

Degree Measures of Angles

Two angles are called **complementary** if the sum of their degree measures is 90°.

Two angles are called **supplementary** if the sum of their degree measures is 180°.

The sum of the degree measures of the three angles of any triangle is 180°.

EXAMPLE 3 **Degree measures**

Write algebraic expressions for each pair of angles shown.

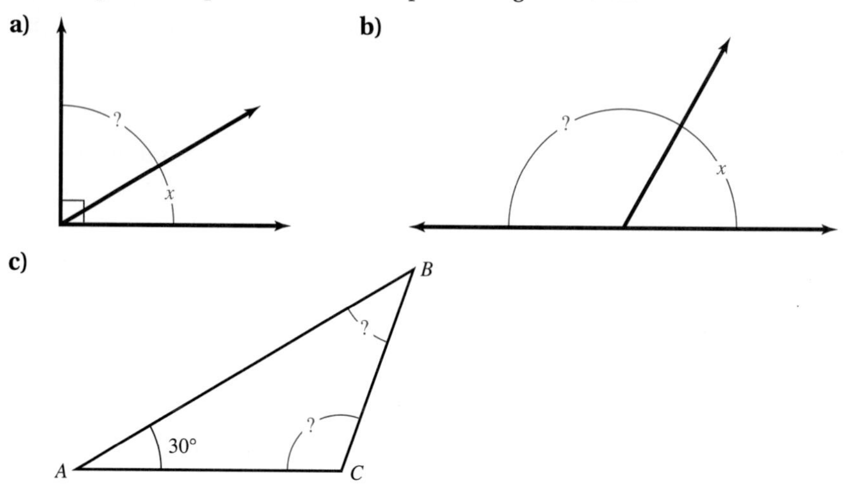

Solution

a) Since the angles shown are complementary, we can use x to represent the degree measure of the smaller angle and $90 - x$ to represent the degree measure of the larger angle.

b) Since the angles shown are supplementary, we can use x to represent the degree measure of the smaller angle and $180 - x$ to represent the degree measure of the larger angle.

c) If we let x represent the degree measure of angle B, then $180 - x - 30$, or $150 - x$, represents the degree measure of angle C. ■

Consecutive Integers

To gain practice in problem solving, we will solve problems about consecutive integers in Section 2.5. Note that each integer is 1 larger than the previous integer, while consecutive even integers as well as consecutive odd integers differ by 2.

EXAMPLE 4 **Expressions for integers**

Write algebraic expressions for the following unknown integers.

a) Three consecutive integers, the smallest of which is w

b) Three consecutive even integers, the largest of which is z

Solution

a) Since each integer is 1 larger than the preceding integer, we can use w, $w + 1$, and $w + 2$ to represent them.

b) Since consecutive even integers differ by 2, these integers can be represented by z, $z - 2$, and $z - 4$. ■

Using Formulas

In writing expressions for unknown quantities, we often use standard formulas such as those given at the back of the book.

EXAMPLE 5 **Writing algebraic expressions using standard formulas**

Find an algebraic expression for

a) the distance if the rate is 30 miles per hour and the time is T hours.

b) the discount if the rate is 40% and the original price is p dollars.

Solution

a) Using the formula $D = RT$, we have $D = 30T$. So $30T$ is an expression that represents the distance in miles.

b) Since the discount is the rate times the original price, an algebraic expression for the discount is $0.40p$ dollars.

Writing Equations

To solve a problem using algebra, we describe or **model** the problem with an equation.

EXAMPLE 6 **Writing equations**

Identify the variable and write an equation that describes each situation.

a) Find two numbers that have a sum of 14 and a product of 45.

b) A coat is on sale for 25% off the list price. If the sale price is $87, then what is the list price?

c) What percent of 8 is 2?

d) The value of x dimes and $x - 3$ quarters is $2.05.

Solution

a) Let $x =$ one of the numbers and $14 - x =$ the other number. Since their product is 45, we have

$$x(14 - x) = 45.$$

b) Let $x =$ the list price and $0.25x =$ the amount of discount. We can write an equation expressing the fact that the selling price is the list price minus the discount:

$$\text{List price} - \text{discount} = \text{selling price}$$
$$x - 0.25x = 87$$

c) If we let x represent the percentage, then the equation is $x \cdot 8 = 2$, or $8x = 2$.

d) The value of x dimes at 10 cents each is $10x$ cents. The value of $x - 3$ quarters at 25 cents each is $25(x - 3)$ cents. We can write an equation expressing the fact that the total value of the coins is 205 cents:

$$\text{Value of dimes} + \text{value of quarters} = \text{total value}$$
$$10x + 25(x - 3) = 205$$

CAUTION The value of the coins in Example 6(d) is either 205 cents or 2.05 dollars. If the total value is expressed in dollars, then all of the values must be expressed in dollars. So we could also write the equation as

$$0.10x + 0.25(x - 3) = 2.05.$$

Warm-ups

True or false? Explain your answer.

1. For any value of x, the numbers x and $x + 6$ differ by 6.
2. For any value of a, a and $10 - a$ have a sum of 10.
3. If Jack ran at x miles per hour for 3 hours, he ran $3x$ miles.
4. If Jill ran at x miles per hour for 10 miles, she ran for $10x$ hours.
5. If the realtor gets 6% of the selling price and the house sells for x dollars, the owner gets $x - 0.06x$ dollars.
6. If the owner got \$50,000 and the realtor got 10% of the selling price, the house sold for \$55,000.
7. Three consecutive odd integers can be represented by x, $x + 1$, and $x + 3$.
8. The value in cents of n nickels and d dimes is $0.05n + 0.10d$.
9. If the sales tax rate is 5% and x represents the amount of goods purchased, then the total bill is $1.05x$.
10. If the length of a rectangle is 4 feet more than the width w, then the perimeter is $w + (w + 4)$ feet.

2.4 EXERCISES

Translate each verbal expression into an algebraic expression. See Example 1.

1. The sum of a number and 3
2. Two more than a number
3. Three less than a number
4. Four subtracted from a number
5. The product of a number and 5
6. Five divided by some number
7. Ten percent of a number
8. Eight percent of a number
9. The ratio of a number and 3
10. The quotient of 12 and a number
11. One-third of a number
12. Three-fourths of a number

Write algebraic expressions for each pair of numbers. See Example 2.

13. Two numbers with a difference of 15
14. Two numbers that differ by 9
15. Two numbers with a sum of 6
16. Two number with a sum of 5
17. Two numbers with a sum of -4
18. Two numbers with a sum of -8
19. Two numbers such that one is 3 larger than the other
20. Two numbers such that one is 8 smaller than the other
21. Two numbers such that one is 5% of the other
22. Two numbers such that one is 40% of the other
23. Two numbers such that one is 30% more than the other
24. Two number such that one is 20% smaller than the other

Write algebraic expressions for the degree measures of each pair of angles. See Example 3.

25.

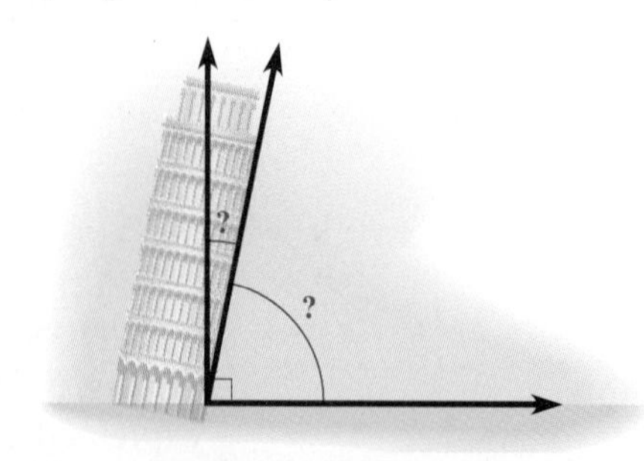

Figure for Exercise 25

26.

Figure for Exercise 26

27.

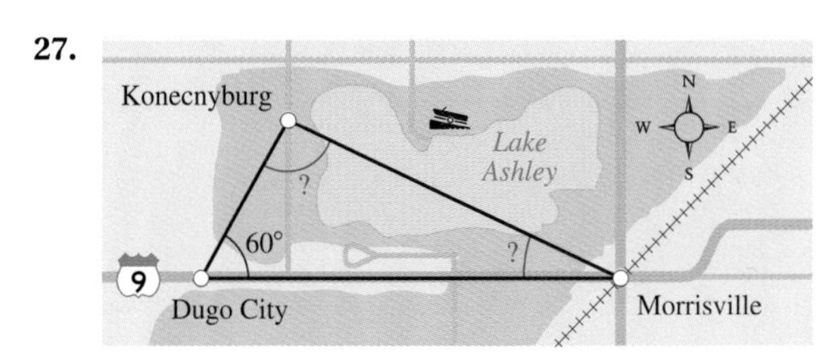

Figure for Exercise 27

28.

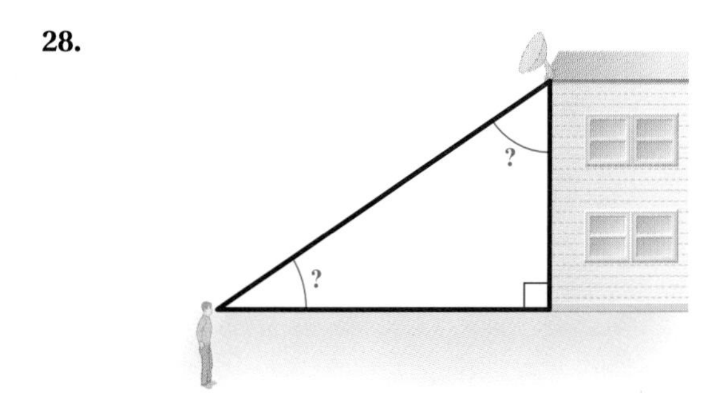

Figure for Exercise 28

Write algebraic expressions for the following unknown integers. See Example 4.

29. Two consecutive even integers

30. Two consecutive odd integers

31. Two consecutive integers

32. Four consecutive odd integers

33. Two consecutive odd integers, the largest of which is m

34. Three consecutive integers, the largest of which is n

35. Three consecutive odd integers, the smallest of which is y

36. Three consecutive even integers, the smallest of which is t

Find an algebraic expression for the quantity in italics using the given information. See Example 5.

37. The *distance,* given that the rate is x miles per hour and the time is 3 hours

38. The *distance,* given that the rate is $x + 10$ miles per hour and the time is 5 hours

39. The *discount,* given that the rate is 25% and the original price is q dollars

40. The *discount,* given that the rate is 10% and the original price is t yen

41. The *time,* given that the distance is x miles and the rate is 20 miles per hour

42. The *time,* given that the distance is 300 kilometers and the rate is $x + 30$ kilometers per hour

43. The *rate,* given that the distance is $x - 100$ meters and the time is 12 seconds

44. The *rate,* given that the distance is 200 feet and the time is $x + 3$ seconds

45. The *area* of a rectangle with length x meters and width 5 meters

46. The *area* of a rectangle with sides b yards and $b - 6$ yards

47. The *perimeter* of a rectangle with length $w + 3$ inches and width w inches

48. The *perimeter* of a rectangle with length r centimeters and width $r - 1$ centimeters

49. The *width* of a rectangle with perimeter 300 feet and length x feet

50. The *length* of a rectangle with area 200 square feet and width w feet

51. The *length* of a rectangle, given that its width is x feet and its length is 1 foot longer than twice the width

52. The *length* of a rectangle, given that its width is w feet and its length is 3 feet shorter than twice the width

53. The *area* of a rectangle, given that the width is x meters and the length is 5 meters longer than the width

54. The *perimeter* of a rectangle, given that the length is x yards and the width is 10 yards shorter

55. The *simple interest,* given that the principal is $x + 1000$, the rate is 18%, and the time is 1 year

56. The *simple interest,* given that the principal is $3x$, the rate is 6%, and the time is 1 year

57. The *price per pound* of peaches, given that x pounds sold for \$16.50

58. The *rate per hour* of a mechanic who gets \$480 for working x hours

59. The *degree measure* of an angle, given that its complementary angle has measure x degrees

60. The *degree measure* of an angle, given that its supplementary angle has measure x degrees

Identify the variable and write an equation that describes each situation. Do not solve the equation. See Example 6.

61. Two numbers differ by 5 and have a product of 8

62. Two numbers differ by 6 and have a product of -9

63. Herman's house sold for x dollars. The real estate agent received 7% of the selling price, and Herman received \$84,532.

64. Gwen sold her car on consignment for x dollars. The saleswoman's commission was 10% of the selling price, and Gwen received \$6570.

65. What percent of 500 is 100?

66. What percent of 40 is 120?

67. The value of x nickels and $x + 2$ dimes is \$3.80.

68. The value of d dimes and $d - 3$ quarters is \$6.75.

69. The sum of a number and 5 is 13.

70. Twelve subtracted from a number is -6.

71. The sum of three consecutive integers is 42.

72. The sum of three consecutive odd integers is 27.

73. The product of two consecutive integers is 182.

74. The product of two consecutive even integers is 168.

75. Twelve percent of Harriet's income is \$3000.

76. If nine percent of the members buy tickets, then we should sell 252 tickets.

77. Thirteen is 5% of what number?

78. Three hundred is 8% percent of what number?

79. The length of a rectangle is 5 feet longer than the width, and the area is 126 square feet.

80. The length of a rectangle is 1 yard shorter than twice the width, and the perimeter is 298 yards.

81. The value of n nickels and $n - 1$ dimes is 95 cents.

82. The value of q quarters, $q + 1$ dimes, and $2q$ nickels is 90 cents.

83. The measure of an angle is 38° smaller than the measure of its supplementary angle.

84. The measure of an angle is 16° larger than the measure of its complementary angle.

Given that the area of each figure is 24 square feet, use the dimensions shown to write an equation expressing this fact. Do not solve the equation.

85.

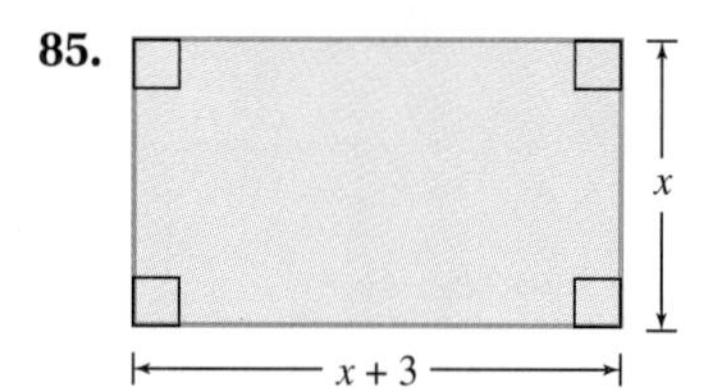

86.

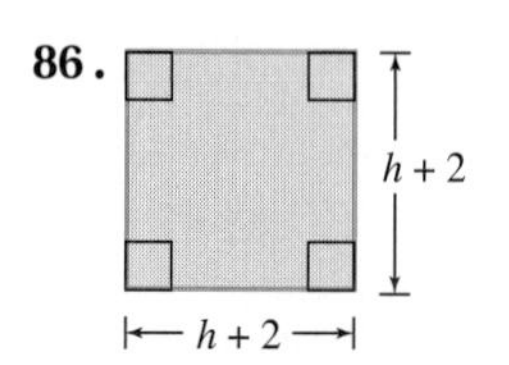

87.

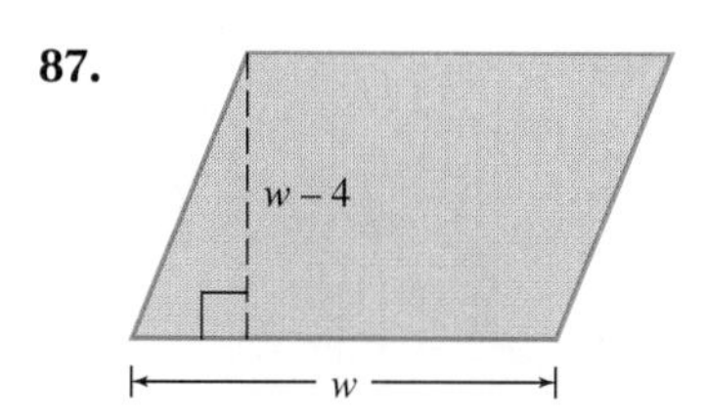

88.

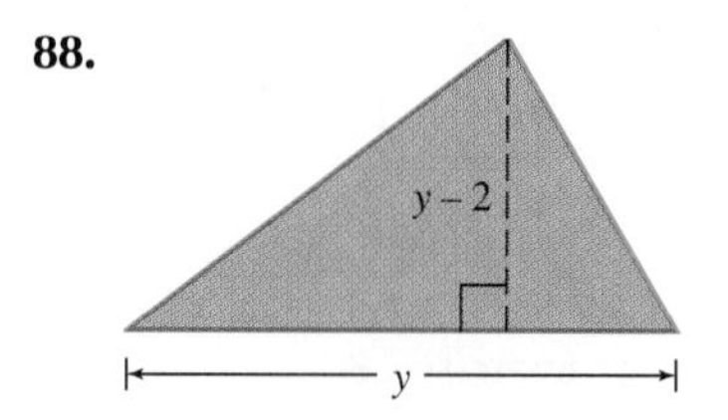

Getting More Involved

89. ***Writing.*** If x represents a rational number, then what expression represents the next larger rational number? Explain your answer.

90. ***Writing.*** Explain why we can use x, $x + 2$, and $x + 4$ to represent three consecutive even integers or three consecutive odd integers.

2.5 Applications

In this section:

- Number Problems
- General Strategy for Solving Verbal Problems
- Geometric Problems
- Uniform Motion Problems

In this section we apply the ideas of Section 2.4 to solving problems. Many of the problems can be solved by using arithmetic only and not algebra. However, remember that we are not just trying to find the answer, we are trying to learn how to apply algebra. So even if the answer is obvious to you, set the problem up and solve it by using algebra as shown in the examples.

Number Problems

Algebra is often applied to problems involving time, rate, distance, interest, or discount. **Number problems** do not involve any physical situation. In number problems we simply find some numbers that satisfy some given conditions. Number problems can provide good practice for solving more complex problems.

EXAMPLE 1 **A consecutive integer problem**

The sum of three consecutive integers is 48. Find the integers.

Solution

We first represent the unknown quantities with variables. Let x, $x + 1$, and $x + 2$ represent the three consecutive integers. We now write an equation that describes the problem and solve it. The equation expresses the fact that the sum of the integers is 48.

$$x + (x + 1) + (x + 2) = 48$$

$$3x + 3 = 48 \quad \text{Combine like terms.}$$

$$3x = 45 \quad \text{Subtract 3 from each side.}$$

$$x = 15 \quad \text{Divide each side by 3.}$$

$$x + 1 = 16 \quad \text{If } x \text{ is 15, then } x + 1 \text{ is 16 and } x + 2 \text{ is 17.}$$

$$x + 2 = 17$$

Because $15 + 16 + 17 = 48$, the three consecutive integers that have a sum of 48 are 15, 16, and 17.

General Strategy for Solving Verbal Problems

You should use the following steps as a guide for solving problems.

Strategy for Solving Problems

1. Read the problem as many times as necessary.
2. If possible, draw a diagram to illustrate the problem.
3. Choose a variable and *write* what it represents.
4. Write algebraic expressions for any other unknowns in terms of that variable.
5. Write an equation that describes the situation.
6. Solve the equation.
7. Check your possible answer in the original problem (not the equation).
8. Answer the original question.

Geometric Problems

Geometric problems involve geometric figures. For these problems you should always draw the figure and label it.

EXAMPLE 2 **A perimeter problem**

The length of a rectangular piece of property is 1 foot less than twice the width. If the perimeter is 748 feet, find the length and width.

Solution

Figure 2.2

Let $x =$ the width. Since the length is one foot less than twice the width, $2x - 1 =$ the length. Draw a diagram as in Fig. 2.2. We know that $2L + 2W = P$ is the formula for perimeter of a rectangle. Substituting $2x - 1$ for L and x for W in this formula yields an equation in x:

$$2L + 2W = P$$

$$2(2x - 1) + 2(x) = 748 \quad \text{Replace } L \text{ by } 2x - 1 \text{ and } W \text{ by } x.$$

$$4x - 2 + 2x = 748 \quad \text{Remove the parentheses.}$$

$$6x - 2 = 748 \quad \text{Combine like terms.}$$

$$6x = 750 \quad \text{Add 2 to each side.}$$

$$x = 125 \quad \text{Divide each side by 6.}$$

$$2x - 1 = 249 \quad \text{If } x = 125, \text{ then } 2x - 1 = 2(125) - 1 = 249.$$

Check these answers by computing $2L + 2W$:

$$2(249) + 2(125) = 748$$

So the width is 125 feet, and the length is 249 feet. ■

The next geometric example involves the degree measures of angles. For this problem, the figure is given.

EXAMPLE 3 **Complementary angles**

In Fig. 2.3 the angle formed by the guy wire and the ground is 3.5 times as large as the angle formed by the guy wire and the antenna. Find the degree measure of each of these angles.

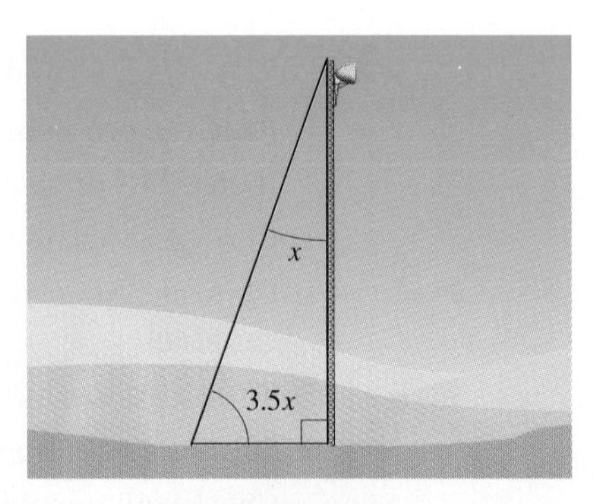

Figure 2.3

Solution

Let x = the degree measure of the smaller angle, and let $3.5x$ = the degree measure of the larger angle. Since the antenna meets the ground at a 90° angle, the sum of the degree measures of the other two angles of the right triangle is 90°. (They are complementary angles.) So we have the following equation:

$$x + 3.5x = 90$$

$$4.5x = 90 \quad \text{Combine like terms.}$$

$$x = 20 \quad \text{Divide each side by 4.5.}$$

$$3.5x = 70 \quad \text{Find the other angle.}$$

Check: 70° is $3.5 \cdot 20°$ and $20° + 70° = 90°$. So the smaller angle is 20°, and the larger angle is 70°. ■

Uniform Motion Problems

Problems involving motion at a constant rate are called **uniform motion problems.** In uniform motion problems we often use an average rate when the actual rate is not constant. For example, you can drive all day and average 50 miles per hour, but you are not driving at a constant 50 miles per hour.

EXAMPLE 4 **Finding the rate**

Bridgette drove her car for 2 hours on an icy road. When the road cleared up, she increased her speed by 35 miles per hour and drove 3 more hours, completing her 255-mile trip. How fast did she travel on the icy road?

Solution

It is helpful to make a table to classify the information given. Remember that $D = RT$.

	Rate	Time	Distance
Icy road	$x \frac{\text{mi}}{\text{hr}}$	2 hr	$2x$ mi
Clear road	$x + 35 \frac{\text{mi}}{\text{hr}}$	3 hr	$3(x + 35)$ mi

The equation expresses the fact that her total distance traveled was 255 miles:

$$\text{Icy road distance} + \text{clear road distance} = \text{total distance}$$

$$2x + 3(x + 35) = 255$$

$$2x + 3x + 105 = 255$$

$$5x + 105 = 255$$

$$5x = 150$$

$$x = 30$$

$$x + 35 = 65$$

If she drove at 30 miles per hour for 2 hours on the icy road, she went 60 miles. If she drove at 65 miles per hour for 3 hours on the clear road, she went 195 miles. Since $60 + 195 = 255$, we can be sure that her speed on the icy road was 30 mph. ■

In the next uniform motion problem we find the time.

EXAMPLE 5 **Finding the time**

Pierce drove from Allentown to Baker, averaging 55 miles per hour. His journey back to Allentown using the same route took 3 hours longer because he averaged only 40 miles per hour. How long did it take him to drive from Allentown to Baker? What is the distance between Allentown and Baker?

Solution

Make a table to classify the information given. Remember that $D = RT$.

	Rate	Time	Distance
Going	$55 \frac{\text{mi}}{\text{hr}}$	x hr	$55x$ mi
Returning	$40 \frac{\text{mi}}{\text{hr}}$	$x + 3$ hr	$40(x + 3)$ mi

We can write an equation expressing the fact that the distance either way is the same:

$$\text{Distance going} = \text{distance returning}$$
$$55x = 40(x + 3)$$
$$55x = 40x + 120$$
$$15x = 120$$
$$x = 8$$

The trip from Allentown to Baker took 8 hours. The distance between Allentown and Baker is $55 \cdot 8$, or 440 miles. ■

Warm-ups

True or false? Explain your answer.

1. The first step in solving a word problem is to write the equation.
2. You should always write down what the variable represents.
3. Diagrams and tables are used as aids in solving problems.
4. To represent two consecutive odd integers, we use x and $x + 1$.
5. If $5x$ is 2 miles more than $3(x + 20)$, then $5x + 2 = 3(x + 20)$.
6. We can represent two numbers with a sum of 6 by x and $6 - x$.
7. Two numbers that differ by 7 can be represented by x and $x + 7$.
8. The degree measures of two complementary angles can be represented by x and $90 - x$.
9. The degree measures of two supplementary angles can be represented by x and $x + 180$.
10. If x is half as large as $x + 50$, then $2x = x + 50$.

2.5 EXERCISES

Show a complete solution to each problem. See Example 1.

1. ***Consecutive integers.*** Find three consecutive integers whose sum is 141.
2. ***Consecutive even integers.*** Find three consecutive even integers whose sum is 114.
3. ***Consecutive odd integers.*** Two consecutive odd integers have a sum of 152. What are the integers?
4. ***Consecutive odd integers.*** Four consecutive odd integers have a sum of 120. What are the integers?
5. ***Consecutive integers.*** Find four consecutive integers whose sum is 194.
6. ***Consecutive even integers.*** Find four consecutive even integers whose sum is 340.

Show a complete solution to each problem. See Examples 2 and 3.

7. ***Olympic swimming.*** If an Olympic swimming pool is twice as long as it is wide and the perimeter is 150 meters, then what are the length and width?

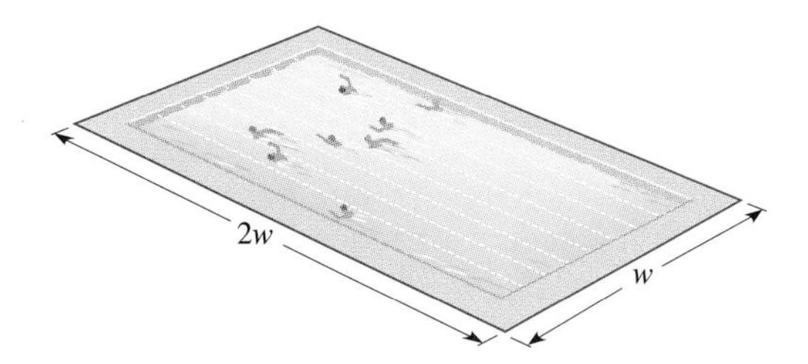

Figure for Exercise 7

8. ***Wimbledon tennis.*** If the perimeter of a tennis court is 228 feet and the length is 6 feet longer than twice the width, then what are the length and width?

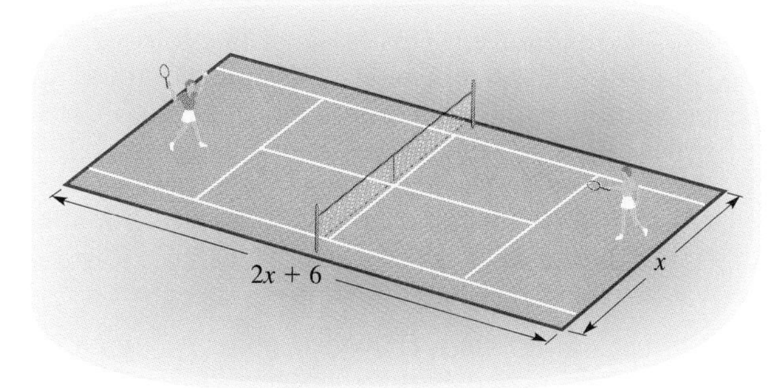

Figure for Exercise 8

9. ***Framed.*** Julia framed an oil painting that her uncle gave her. The painting was 4 inches longer than it was wide, and it took 176 inches of frame molding. What were the dimensions of the picture?
10. ***Industrial triangle.*** Geraldo drove his truck from Indianapolis to Chicago, then to St. Louis, and then back to Indianapolis. He observed that the second side of his triangular route was 81 miles short of being twice as long as the first side and that the third side was 61 miles longer than the first side. If he traveled a total of 720 miles, then how long is each side of this triangular route?

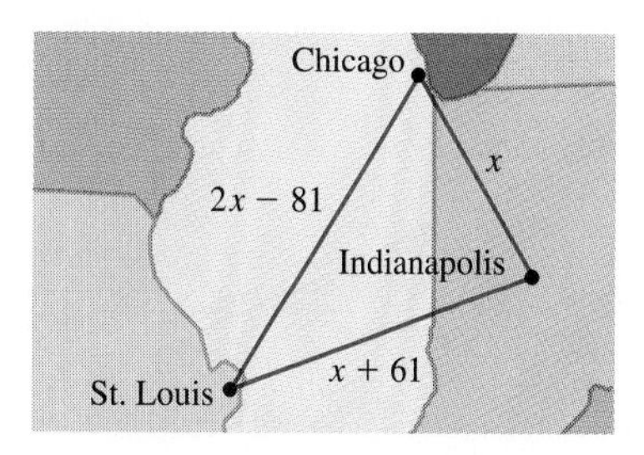

Figure for Exercise 10

11. ***Triangular banner.*** A banner in the shape of an isosceles triangle has a base that is 5 inches shorter than either of the equal sides. If the perimeter of the banner is 34 inches, then what is the length of the equal sides?
12. ***Border paper.*** Dr. Good's waiting room is 8 feet longer than it is wide. When Vincent wallpapered Dr. Good's waiting room, he used 88 feet of border paper. What are the dimensions of Dr. Good's waiting room?

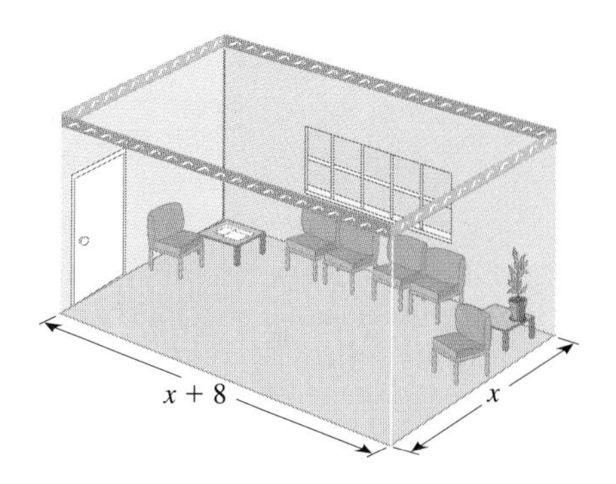

Figure for Exercise 12

13. ***Ramping up.*** A civil engineer is planning a highway overpass as shown in the figure. Find the degree measure of the angle marked w.

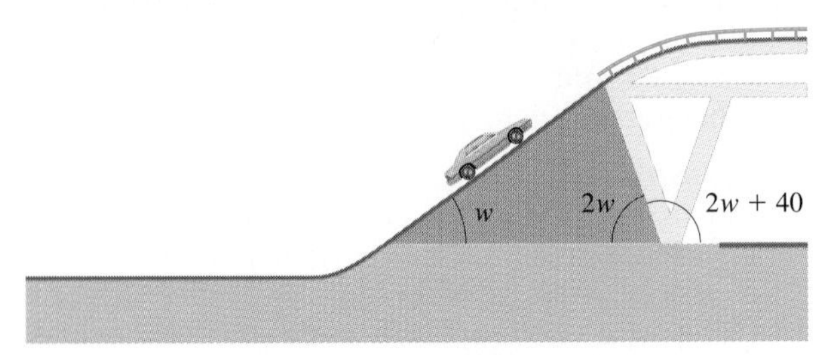

Figure for Exercise 13

14. ***Ramping down.*** For the other side of the overpass, the engineer has drawn the plans shown in the figure. Find the degree measure of the angle marked z.

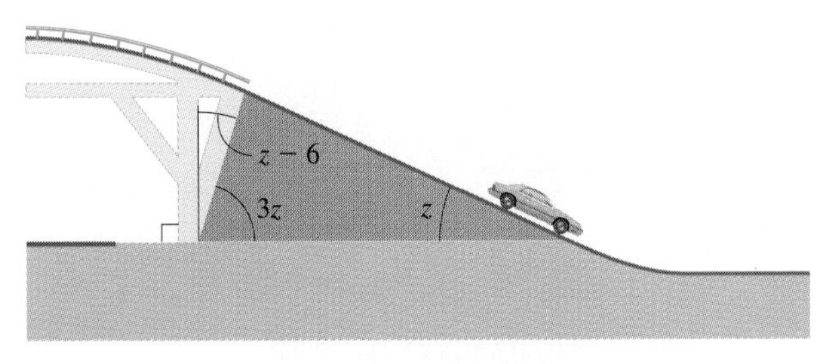

Figure for Exercise 14

Show a complete solution to each problem. See Examples 4 and 5.

15. ***Highway miles.*** Bret drove for 4 hours on the freeway, then decreased his speed by 20 miles per hour and drove for 5 more hours on a country road. If his total trip was 485 miles, then what was his speed on the freeway?

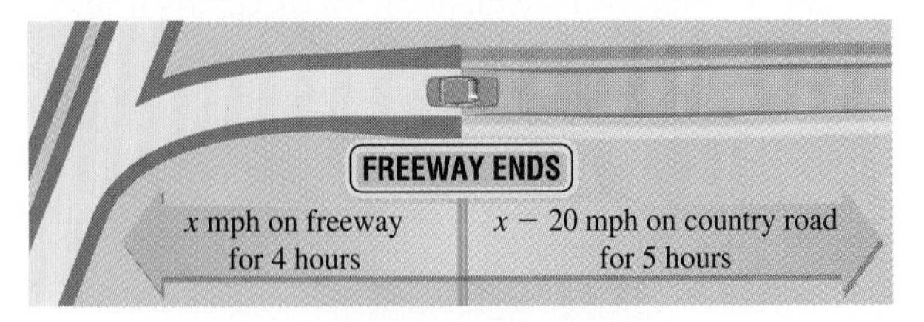

Figure for Exercise 15

16. ***Walking and running.*** On Saturday morning, Lynn walked for 2 hours and then ran for 30 minutes. If she ran twice as fast as she walked and she covered 12 miles altogether, then how fast did she walk?

17. ***Driving all night.*** Kathryn drove her rig 5 hours before dawn and 6 hours after dawn. If her average speed was 5 miles per hour more in the dark and she covered 630 miles altogether, then what was her speed after dawn?

18. ***Commuting to work.*** On Monday, Roger drove to work in 45 minutes. On Tuesday he averaged 12 miles per hour more, and it took him 9 minutes less to get to work. How far does he travel to work?

19. ***Head winds.*** A jet flew at an average speed of 640 mph from Los Angeles to Chicago. Because of head winds the jet averaged only 512 mph on the return trip, and the return trip took 48 minutes longer. How many hours was the flight from Chicago to Los Angeles? How far is it from Chicago to Los Angeles?

20. ***Ride the Peaks.*** Penny's bicycle trip from Colorado Springs to Pikes Peak took 1.5 hours longer than the return trip to Colorado Springs. If she averaged 6 mph on the way to Pikes Peak and 15 mph for the return trip, then how long was the ride from Colorado Springs to Pikes Peak?

Solve each problem.

21. ***Super Bowl score.*** The 1977 Super Bowl was played in the Rose Bowl in Pasadena. In that football game the Oakland Raiders scored 18 more points than the Minnesota Vikings. If the total number of points scored was 46, then what was the final score for the game?

22. ***Burgers, toys, and Mickey Mouse.*** Revenues for McDonald's, Toys Я Us, and Walt Disney totaled $27.9 billion in 1993 (*Money*, April 1994). If Toys Я Us had the same amount of revenue as Walt Disney, and McDonald's revenue was $1.5 billion less than Walt Disney's, then what was the amount of 1993 revenue for each company?

23. ***Idabel to Lawton.*** Before lunch, Sally drove from Idabel to Ardmore, averaging 50 mph. After lunch she continued on to Lawton, averaging 53 mph. If her driving time after lunch was 1 hour less than her driving time before lunch and the total trip was 256 miles, then how many hours did she drive before lunch? How far is to from Ardmore to Lawton?

24. ***Norfolk to Chadron.*** On Monday, Chuck drove from Norfolk to Valentine, averaging 47 mph. On Tuesday, he continued on to Chadron, averaging 69 mph. His driving time on Monday was 2 hours longer than his driving time on Tuesday. If the total distance from Norfolk to Chadron is 326 miles, then how many hours did he drive on Monday? How far is it from Valentine to Chadron?

25. ***Golden oldies.*** Joan Crawford, John Wayne, and James Stewart were born in consecutive years (*Doubleday Almanac*). Joan Crawford was the oldest of the three, and James Stewart was the youngest. In 1950, after all three had their birthdays, the sum of their ages was 129. In what years were they born?

26. ***Leading men.*** Bob Hope was born two years after Clark Gable and two years before Henry Fonda (*Doubleday Almanac*). In 1951, after all three of them had their birthdays, the sum of their ages was 144. In what years were they born?

Getting More Involved

27. ***Writing.*** Write the step-by-step procedure that you used for solving the problems in this section, without looking at the strategy for problem solving given in this section. Explain the reason for each step.

28. ***Discussion.*** Examine the following problem and solution.
Problem: The audience of 300 consisted of twice as many males as females. Find the number of females in the audience.
Solution: Let $M =$ the number of males and $2M =$ the number of females.

$$M + 2M = 300$$
$$3M = 300$$
$$M = 100$$

Check: $100 + 2(100) = 300$. So there were 200 females in the audience.
Is the solution correct? If it is not correct, then where is the error?

2.6 More Applications

In this section:

- Discount Problems
- Commission Problems
- Investment Problems
- Mixture Problems

In this section we continue our study of applications of algebra. The problems in this section involve percents.

Discount Problems

When an item is sold at a discount, the amount of the discount is usually described as being a percentage of the original price.

EXAMPLE 1 **Finding the original price**

Major Motors is discounting all of its deluxe models 12%. When Ralph bought the deluxe model, he got a discount of \$4,500. What was the original price of Ralph's car?

Solution

Let x represent the original price. The discount is found by taking 12% of the original price:

$$12\% \text{ of original price} = \text{discount}$$
$$0.12x = 4500$$
$$x = \frac{4500}{0.12} \quad \text{Divide each side by 0.12.}$$
$$x = 37{,}500$$

To check, we find 12% of \$37,500:

$$0.12(\$37{,}500) = \$4500$$

So the original price of Ralph's car was \$37,500. ■

EXAMPLE 2 **Finding the original price**

When Susan bought her deluxe model at Major Motors, she also got a discount of 12%. She paid \$17,600 for her car. What was the original price of Susan's car?

Solution

Let x represent the original price for Susan's car. The amount of discount is 12% of x, or $0.12x$. We can write an equation expressing the fact that the original price minus the discount is the price Susan paid.

$$\text{Original price} - \text{discount} = \text{sale price}$$
$$x - 0.12x = 17{,}600$$
$$0.88x = 17{,}600 \qquad 1.00x - 0.12x = 0.88x$$
$$x = \frac{17{,}600}{0.88} \qquad \text{Divide each side by } 0.88.$$
$$x = \$20{,}000$$

Check: 12% of \$20,000 is \$2,400, and \$20,000 − \$2,400 = \$17,600. The original price of Susan's car was \$20,000.

Commission Problems

A salesperson's commission for making a sale is often a percentage of the selling price. **Commission problems** are very similar to other problems involving percents.

EXAMPLE 3 **Real estate commission**

Sarah is selling her house through a real estate agent whose commission is 7% of the selling price. What should the selling price be so that Sarah can get the \$83,700 she needs to pay off the mortgage?

Solution

Let x be the selling price. The commission is 7% of x (not 7% of \$83,700). Sarah receives the selling price less the sales commission:

$$\text{Selling price} - \text{commission} = \text{Sarah's share}$$
$$x - 0.07x = 83{,}700$$
$$0.93x = 83{,}700 \qquad 1.00x - 0.07x = 0.93x$$
$$x = \frac{83{,}700}{0.93}$$
$$x = 90{,}000$$

Check: 7% of \$90,000 is \$6,300, and \$90,000 − \$6,300 = \$83,700. So the house should sell for \$90,000

Investment Problems

The interest on an investment is a percentage of the investment, just as the sales commission is a percentage of the sale amount. However, in **investment problems** we must often account for more than one investment at different rates. So it is a good idea to make a table, as in the next example.

EXAMPLE 4 **Diversified investing**

Ruth Ann invested some money in a certificate of deposit with an annual yield of 9%. She invested twice as much in a mutual fund with an annual yield of 10%. Her interest from the two investments at the end of the year was $232. How much was invested at each rate?

Solution

When there are many unknown quantities, it is often helpful to identify them in a table. Since the time is 1 year, the amount of interest is the product of the interest rate and the amount invested.

	Interest rate	Amount invested	Interest for 1 year
CD	9%	x	$0.09x$
Mutual fund	10%	$2x$	$0.10(2x)$

Since the total interest from the investments was $232, we can write the following equation:

$$\begin{aligned} \text{CD interest} + \text{mutual fund interest} &= \text{total interest} \\ 0.09x + 0.10(2x) &= 232 \\ 0.09x + 0.20x &= 232 \\ 0.29x &= 232 \\ x &= \frac{232}{0.29} \\ x &= \$800 \\ 2x &= \$1600 \end{aligned}$$

To check, we find the total interest:

$$\begin{aligned} 0.09(800) + 0.10(1600) &= 72 + 160 \\ &= 232 \end{aligned}$$

So Ruth Ann invested $800 at 9% and $1600 at 10%. ■

Mixture Problems

Mixture problems are concerned with the result of mixing two quantities, each of which contains another substance. Notice how similar the following mixture problem is to the last investment problem.

EXAMPLE 5 **Mixing milk**

How many gallons of milk containing 4% butterfat must be mixed with 80 gallons of 1% milk to obtain 2% milk?

Solution

We again make a table of the unknown quantities:

	Percentage of fat	Amount of milk	Amount of fat
4% milk	4%	x	$0.04x$
1% milk	1%	80	$0.01(80)$
2% milk	2%	$x + 80$	$0.02(x + 80)$

In all mixture problems we write an equation that accounts for one of the quantities being combined. The equation we write here expresses the fact that the total fat from the first two types of milk is the same as the fat in the mixture:

$$\text{Fat in 4\% milk} + \text{fat in 1\% milk} = \text{fat in 2\% milk}$$

$$0.04x + 0.01(80) = 0.02(x + 80)$$

$$0.04x + 0.8 = 0.02x + 1.6 \qquad \text{Simplify.}$$

$$100(0.04x + 0.8) = 100(0.02x + 1.6) \qquad \text{Multiply each side by 100.}$$

$$4x + 80 = 2x + 160 \qquad \text{Distributive property}$$

$$2x + 80 = 160 \qquad \text{Subtract } 2x \text{ from each side.}$$

$$2x = 80 \qquad \text{Subtract 80 from each side.}$$

$$x = 40 \qquad \text{Divide each side by 2.}$$

To check, calculate the total fat:

$$2\% \text{ of 120 gallons} = 0.02(120) = 2.4 \text{ gallons of fat}$$

$$0.04(40) + 0.01(80) = 1.6 + 0.8 = 2.4 \text{ gallons of fat}$$

So we mix 40 gallons of 4% milk with 80 gallons of 1% milk to get 120 gallons of 2% milk. ■

Warm-ups

True or false? Explain your answer.

1. If the original price is w and the discount is 8%, then the selling price is $w - 0.08w$.
2. If x is the selling price and the commission is 8% of the selling price, then the commission is $0.08x$.
3. If you need $40,000 for your house and the agent gets 10% of the selling price, then the agent gets $4,000, and the house sells for $44,000.
4. If you mix 10 liters of a 20% acid solution with x liters of a 30% acid solution, then the total amount of acid is $2 + 0.3x$ liters.
5. A 10% acid solution mixed with a 14% acid solution results in a 24% acid solution.
6. If a TV costs x dollars and sales tax is 5%, then the total bill is $1.05x$ dollars.

2.6 EXERCISES

Show a complete solution to each problem. See Examples 1 and 2.

1. ***Close-out sale.*** At a 25% off sale, Jose saved $80 on a 19-inch Sony TV. What was the original price of the television.

2. ***Big bike.*** A 12% discount on a Giant Perigee saved Melanie $46.68. What was the original price of the bike?

3. ***Circuit city.*** After getting a 20% discount, Robert paid $320 for a Pioneer CD player for his car. What was the original price of the CD player?

4. ***Chrysler Le Baron.*** After getting a 15% discount on the price of a new Chrysler Le Baron convertible, Helen paid $17,000. What was the original price of the convertible?

Show a complete solution to each problem. See Example 3.

5. ***Selling price of a home.*** Kirk wants to get $72,000 for his house. The real estate agent gets a commission equal to 10% of the selling price for selling the house. What should the selling price be?

6. ***Horse trading.*** Gene is selling his palomino at an auction. The auctioneer's commission is 10% of the selling price. If Gene still owes $810 on the horse, then what must the horse sell for so that Gene can pay off his loan?

7. ***Sales tax collection.*** Merilee sells tomatoes at a roadside stand. Her total receipts including the 7% sales tax were $462.24. What amount of sales tax did she collect?

8. ***Toyota Tercel.*** Gwen bought a new Toyota Tercel. The selling price plus the 8% state sales tax was $10,638. What was the selling price?

Show a complete solution to each problem. See Example 4.

9. ***Wise investments.*** Wiley invested some money in the Berger 100 Fund and $3000 more than that amount in the Berger 101 Fund. For the year he has in the fund, the 100 Fund paid 18% simple interest and the 101 Fund paid 15% simple interest. If the income from the two investments totaled $3750 for one year, then how much did he invest in each fund?

10. ***Loan shark.*** Becky lent her brother some money at 8% simple interest, and she lent her sister twice as much at twice the interest rate. If she received a total of 20 cents interest, then how much did she lend to each of them?

11. ***Investing in bonds.*** David split his $25,000 inheritance between Fidelity Short-Term Bond Fund with an annual yield of 5% and T. Rowe Price Tax-Free Short-Intermediate Fund with an annual yield of 4%. If his total income for one year on the two investments was $1140, then how much did he invest in each fund?

12. ***High-risk funds.*** Of the $50,000 that Natasha pocketed on her last real estate deal, $20,000 went to charity. She invested part of the remainder in Dreyfus New Leaders Fund with an annual yield of 16% and the the rest in Templeton Growth Fund with an annual yield of 25%. If she made $6060 on these investments in one year, then how much did she invest in each fund?

Show a complete solution to each problem. See Example 5.

13. ***Mixing milk.*** How many gallons of milk containing 1% butterfat must be mixed with 30 gallons of milk containing 3% butterfat to obtain a mixture containing 2% butterfat?

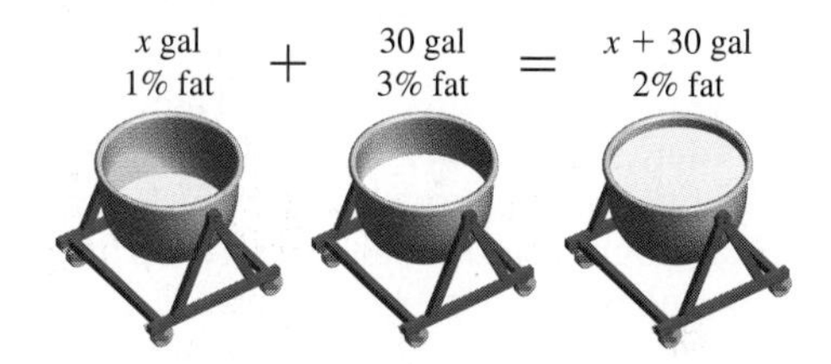

Figure for Exercise 13

14. ***Acid solutions.*** How many gallons of a 5% acid solution should be mixed with 30 gallons of a 10% acid solution to obtain a mixture that is 8% acid?

15. ***Alcohol solutions.*** Gus has on hand a 5% alcohol solution and a 20% alcohol solution. He needs 30 liters of a 10% alcohol solution. How many liters of each solution should he mix together to obtain the 30 liters?

16. ***Adjusting antifreeze.*** Angela needs 20 quarts of 50% antifreeze solution in her radiator. She plans to obtain this by mixing some pure antifreeze with an appropriate amount of a 40% antifreeze solution. How many quarts of each should she use?

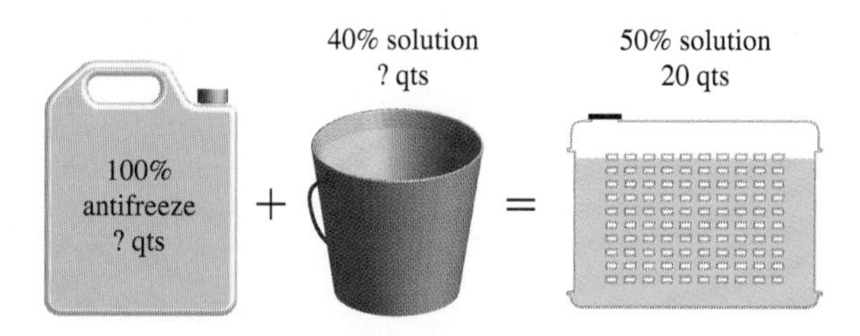

Figure for Exercise 16

Solve each problem.

17. ***Registered voters.*** If 60% of the registered voters of Lancaster County voted in the November election and 33,420 votes were cast, then how many registered voters are there in Lancaster County?

18. ***Tough on crime.*** In a random sample of voters, 594 respondents said that they favored passage of a $33 billion crime bill. If the number in favor of the crime bill was 45% of the number of voters in the sample, then how many voters were in the sample?

19. ***Ford Taurus.*** At an 8% sales tax rate, the sales tax on Peter's new Ford Taurus was $1,200. What was the price of the car?

20. ***Taxpayer blues.*** Last year, Faye paid 24% of her income to taxes. If she paid $9,600 in taxes, then what was her income?

21. ***Making a profit.*** A retail store buys shirts for $8 and sells them for $14. What percent increase is this?

22. ***Monitoring AIDS.*** If 28 new AIDS cases were reported in Landon County this year and 35 new cases were reported last year, then what percent decrease in new cases is this?

23. ***High school integration.*** Wilson High School has 400 students, of whom 20% are African American. The school board plans to merge Wilson High with Jefferson High. This one school will then have a student population that is 44% African American. If Jefferson currently has a student population that is 60% African American, then how many students are at Jefferson?

24. ***Junior high integration.*** The school board plans to merge two junior high schools into one school of 800 students in which 40% of the students will be Caucasian. One of the schools currently has 58% Caucasian students; the other has only 10% Caucasian students. How many students are in each of the two schools?

25. ***Hospital capacity.*** When Memorial Hospital is filled to capacity, it has 18 more people in semiprivate rooms (two patients to a room) than in private rooms. The room rates are $200 per day for a private room and $150 per day for a semiprivate room. If the total receipts for rooms is $17,400 per day when all are full, then how many rooms of each type does the hospital have?

26. ***Public relations.*** Memorial Hospital is planning an advertising campaign. It costs the hospital $3,000 each time a television ad is aired and $2,000 each time a radio ad is aired. The administrator wants to air 60 more television ads than radio ads. If the total cost of airing the ads is $580,000, then how many ads of each type will be aired?

27. ***Mixed nuts.*** Cashews sell for $4.80 per pound, and pistachios sell for $6.40 per pound. How many pounds of pistachios should be mixed with 20 pounds of cashews to get a mixture that sells for $5.40 per pound?

28. ***Premium blend.*** Premium coffee sells for $6.00 per pound, and regular coffee sells for $4.00 per pound. How many pounds of each type of coffee should be blended to obtain 100 pounds of a blend that sells for $4.64 per pound?

29. ***Nickels and dimes.*** Candice paid her library fine with 10 coins consisting of nickels and dimes. If the fine was $0.80, then how many of each type of coin did she use?

30. ***Dimes and quarters.*** Jeremy paid for his breakfast with 36 coins consisting of dimes and quarters. If the bill was $4.50, then how many of each type of coin did he use?

31. ***Cooking oil.*** Crisco Canola Oil is 7% saturated fat. Crisco blends corn oil that is 14% saturated fat with Crisco Canola Oil to get Crisco Canola and Corn Oil, which is 11% saturated fat. How many gallons of corn oil must Crisco mix with 600 gallons of Crisco Canola Oil to get Crisco Canola and Corn Oil?

32. ***Chocolate ripple.*** The Delicious Chocolate Shop makes a dark chocolate that is 35% fat and a white chocolate that is 48% fat. How many kilograms of dark chocolate should be mixed with 50 kilograms of white chocolate to make a ripple blend that is 40% fat?

Getting More Involved

33. ***Discussion.*** How are mixture problems similar to uniform motion problems and investment problems.

34. ***Cooperative learning.*** Working in groups, have each group make up a mixture problem, a uniform motion problem, and an investment problem. Give your problems to members of other groups to solve.

2.7 Inequalities

In this section:

- Basic Ideas
- Graphing Inequalities
- Writing Inequalities

In Chapter 1 we defined inequality in terms of the number line. One number is greater than another number if it lies to the right of the other number on the number line. In this section you will study inequality in greater depth.

Basic Ideas

Recall that we defined an equation to be a statement that two algebraic expressions are equal. An inequality is a statement that two algebraic expressions are not equal. The symbols used to express inequality and their meanings are given in the following box.

Inequality Symbols

Symbol	Meaning
$<$	Is less than
$\le$	Is less than or equal to
$>$	Is greater than
$\ge$	Is greater than or equal to

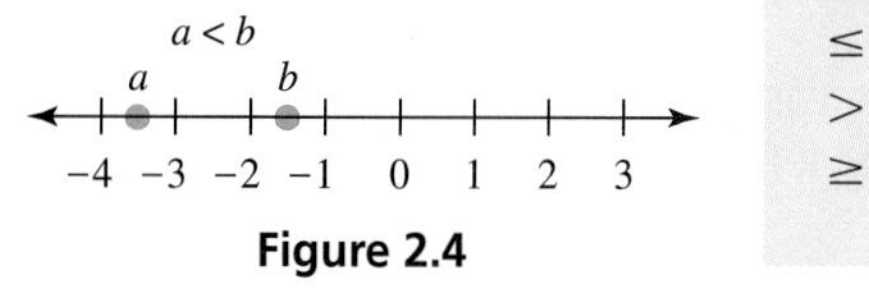
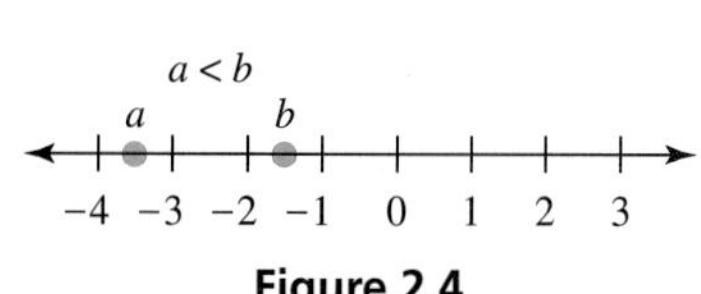

Figure 2.4

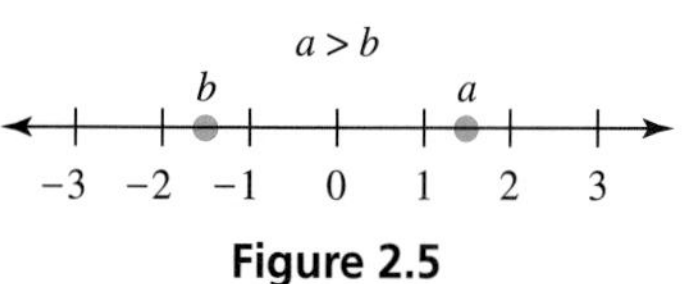

Figure 2.5

The statement $a < b$ means that a is to the left of b on the number line, as shown in Fig. 2.4. The statement $a > b$ means that a is to the right of b on the number line, as shown in Fig. 2.5. Of course, $a < b$ has the same meaning as $b > a$. The statement $a \le b$ is true if a is less than b or if a is equal to b. The statement $a \le b$ has the same meaning as the statement $b \ge a$.

EXAMPLE 1 **Verifying inequalities**

Determine whether each of the following statements is correct.

a) $3 < 4$ **b)** $-1 < -2$ **c)** $-2 \le 0$

d) $0 \ge 0$ **e)** $2(-3) + 8 > 9$ **f)** $(-2)(-5) \le 10$

Solution

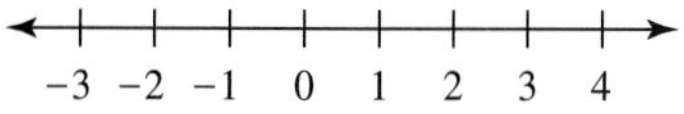

−3 −2 −1 0 1 2 3 4

Figure 2.6

a) Locate 3 and 4 on the number line shown in Fig. 2.6. Because 3 is to the left of 4 on the number line, $3 < 4$ is correct.

b) Locate -1 and -2 on the number line shown in Fig. 2.6. Because -1 is to the right of -2, on the number line, $-1 < -2$ is not correct.

c) Because -2 is to the left of 0 on the number line, $-2 \le 0$ is correct.

d) Because 0 is equal to 0, $0 \ge 0$ is correct.

e) Simplify the left side of the inequality to get $2 > 9$, which is not correct.

f) Simplify the left side of the inequality to get $10 \le 10$, which is correct. ■

Graphing Inequalities

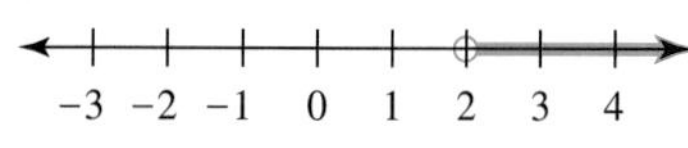

Figure 2.7

A number satisfies the inequality $x > 2$ if it is to the right of 2 on the number line. When we show all of the numbers that satisfy an inequality on a number line, we are **graphing the inequality.** Figure 2.7 shows the graph of $x > 2$. We use an open circle on the number line at 2 to indicate that 2 is not a solution.

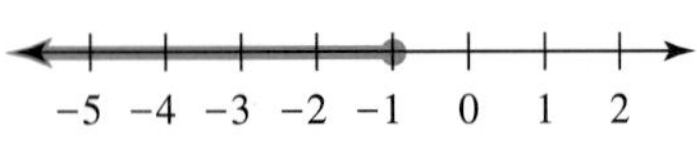

Figure 2.8

A solid circle is used to indicate that a number is a solution. The shading to the right of 2 and the arrow indicate that all real numbers greater than 2 are graphed. Figure 2.8 shows the graph of the inequality $x \le -1$.

A statement involving more than one inequality is called a **compound inequality.** For example, the inequality

$$3 < x < 6$$

indicates that $3 < x$ *and* $x < 6$. The meaning of this compound inequality is clearer if we read the variable first:

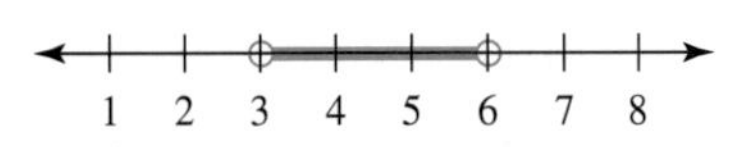

Figure 2.9

"x is greater than 3 *and* x is less than 6."

All real numbers *between* 3 and 6 satisfy $3 < x < 6$. The graph is shown in Fig. 2.9.

EXAMPLE 2 **Graphing inequalities**

Sketch the graph of each inequality on the number line.

a) $x \le 5$ **b)** $-2 < x$ **c)** $-2 \le x < 1$

Solution

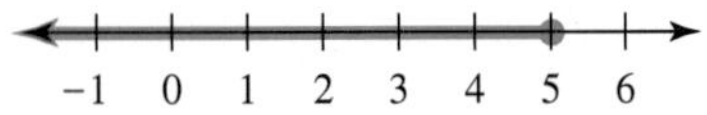

a) The inequality $x \le 5$ indicates that x is less than 5 or equal to 5. To graph this inequality, show all points to the left of 5 and including 5 on the number line, as in Fig. 2.10.

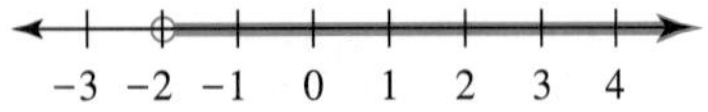

Figure 2.10

b) The inequality $-2 < x$ indicates that x is greater than -2. To graph this inequality, show all points to the right of -2 on the number line, as in Fig. 2.11.

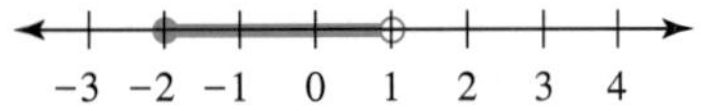

Figure 2.11

c) Numbers that satisfy the inequality $-2 \le x < 1$ are between -2 and 1, including -2 and not including 1. The graph is shown in Fig. 2.12. ■

Figure 2.12

CAUTION When the compound inequality notation $a < x < b$ is used, a must be less than b. An inequality such as $6 < x < 3$ is not correct.

To solve more complicated inequalities, we use steps similar to those used for solving equations. In Section 2.8 we will solve inequalities such as

$$2x - 3 \le -5, \quad x - 5 > 2x + 1, \quad \text{and} \quad 6 < 3x - 5 < 14$$

by reducing them to simpler equivalent inequalities. Here you will learn to determine whether a number is a solution to an inequality of this type.

EXAMPLE 3 **Checking inequalities**

Determine whether the given number satisfies the inequality following it.

a) $0,\ 2x - 3 \le -5$ **b)** $-4,\ x - 5 > 2x + 1$

c) $\frac{13}{3},\ 6 < 3x - 5 < 14$

Solution

a) Replace x by 0 in the inequality and simplify:

$$2x - 3 \le -5$$
$$2 \cdot 0 - 3 \le -5$$
$$-3 \le -5 \quad \text{Incorrect}$$

Since this last inequality is incorrect, 0 is not a solution to the inequality.

b) Replace x by -4 and simplify:

$$x - 5 > 2x + 1$$
$$-4 - 5 > 2(-4) + 1$$
$$-9 > -7 \quad \text{Incorrect}$$

Since this last inequality is incorrect, -4 is not a solution to the inequality.

c) Replace x by $\frac{13}{3}$ and simplify:

$$6 < 3x - 5 < 14$$
$$6 < 3 \cdot \frac{13}{3} - 5 < 14$$
$$6 < 13 - 5 < 14$$
$$6 < 8 < 14 \quad \text{Correct}$$

Since 8 is greater than 6 and less than 14, this inequality is correct. So $\frac{13}{3}$ satisfies the original inequality. ■

Writing Inequalities

Inequalities occur in applications, just as equations do. Certain verbal phrases indicate inequalities. For example, if you must be at least 18 years old to vote, then you can vote if you are 18 or older. The phrase "at least" means "greater than or equal to." If an elevator has a capacity of at most 20 people, then it can hold 20 people or fewer. The phrase "at most" means "less than or equal to."

EXAMPLE 4 **Writing inequalities**

Write an inequality that describes each situation.

a) Lois plans to spend at most \$500 on a washing machine including the 9% sales tax.

b) The length of a certain rectangle must be 4 meters longer than the width, and the perimeter must be at least 120 meters.

c) Fred made a 76 on the midterm exam. To get a B, the average of his midterm and his final exam must be between 80 and 90.

Solution

a) If x is the price of the washing machine, then $0.09x$ is the amount of sales tax. Since the total must be less than or equal to \$500, the inequality is

$$x + 0.09x \le 500.$$

b) If W represents the width of the rectangle, then $W + 4$ represents the length. Since the perimeter $(2W + 2L)$ must be greater than or equal to 120, the inequality is

$$2(W) + 2(W + 4) \ge 120.$$

c) If we let x represent Fred's final exam score, then his average is $\frac{x+76}{2}$.

To indicate that the average is between 80 and 90, we use the compound inequality

$$80 < \frac{x + 76}{2} < 90.$$

■

CAUTION In Example 4(b) you are given that L is 4 meters longer than W. So $L = W + 4$, and you can use $W + 4$ in place of L. If you knew only that L was longer than W, then you would know only that $L > W$. ⊘

Warm-ups

True or false? Explain your answer.

1. $-2 \le -2$ **2.** $-5 < 4 < 6$ **3.** $-3 < 0 < -1$

4. The inequalities $7 < x$ and $x > 7$ have the same graph.

5. The graph of $x < -3$ includes the point at -3.

6. The number 5 satisfies the inequality $x > 2$.

7. The number -3 is a solution to $-2 < x$.

8. The number 4 satisfies the inequality $2x - 1 < 4$.

9. The number 0 is a solution to the inequality $2x - 3 \le 5x - 3$.

10. The inequalities $2x - 1 < x$ and $x < 2x - 1$ have the same solutions.

2.7 EXERCISES

Determine whether each of the following statements is correct. See Example 1.

1. $-3 < 5$ **2.** $-6 < 0$

3. $4 \le 4$ **4.** $-3 \ge -3$

5. $-6 > -5$ **6.** $-2 < -9$

7. $-4 \le -3$ **8.** $-5 \ge -10$

9. $(-3)(4) - 1 < 0 - 3$ **10.** $2(4) - 6 \le -3(5) + 1$

11. $-4(5) - 6 \ge 5(-6)$

12. $4(8) - 30 > 7(5) - 2(17)$

13. $7(4) - 12 \le 3(9) - 2$

14. $-3(4) + 12 \le 2(3) - 6$

Sketch the graph of each inequality on the number line. See Example 2.

15. $x \le 3$ **16.** $x \le -7$ **17.** $x > -2$

18. $x > 4$ **19.** $-1 > x$ **20.** $0 > x$

21. $-2 \le x$ **22.** $-5 \ge x$ **23.** $x \ge \frac{1}{2}$

24. $x \ge -\frac{2}{3}$ **25.** $x \le 5.3$ **26.** $x \le -3.4$

27. $-3 < x < 1$ **28.** $0 < x < 5$ **29.** $3 \le x \le 7$

30. $-3 \le x \le -1$ **31.** $-5 \le x < 0$ **32.** $-2 < x \le 2$

33. $40 < x \le 100$ **34.** $0 \le x < 600$

For each graph, write an inequality that describes the graph.

35.
−3 −2 −1 0 1 2 3 4 5 6 7

36.
−4 −3 −2 −1 0 1 2 3 4 5 6

37.
−6 −5 −4 −3 −2 −1 0 1 2 3 4

38.
−5 −4 −3 −2 −1 0 1 2 3 4 5

39.
−5 −4 −3 −2 −1 0 1 2 3 4 5

40.
−5 −4 −3 −2 −1 0 1 2 3 4 5

41.
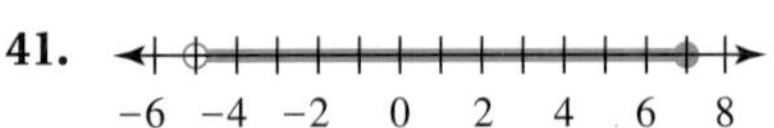

42.
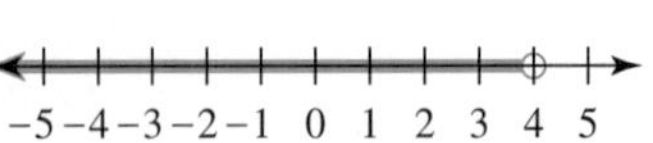

43.
−5 −4 −3 −2 −1 0 1 2 3 4 5

44.
−5 −4 −3 −2 −1 0 1 2 3 4 5

Determine whether the given number satisfies the inequality following it. See Example 3.

45. $-9,\ -x > 3$

46. $5,\ -3 < -x$

47. $-2,\ 5 \le x$

48. $4,\ 4 \ge x$

49. $-6,\ 2x - 3 > -11$

50. $4,\ 3x - 5 < 7$

51. $3,\ -3x + 4 > -7$

52. $-4,\ -5x + 1 > -5$

53. $0,\ 3x - 7 \le 5x - 7$

54. $0,\ 2x + 6 \ge 4x - 9$

55. $2.5,\ -10x + 9 \le 3(x + 3)$

56. $1.5,\ 2x - 3 \le 4(x - 1)$

57. $-7,\ -5 < x < 9$

58. $-9,\ -6 \le x \le 40$

59. $-2,\ -3 \le 2x + 5 \le 9$

60. $-5,\ -3 < -3x - 7 \le 8$

61. $-3.4,\ -4.25x - 13.29 < 0.89$

62. $4.8,\ 3.25x - 14.78 \le 1.3$

For each inequality, determine which of the numbers −5.1, 0, and 5.1 satisfies the inequality.

63. $x > -5$

64. $x \le 0$

65. $5 < x$

66. $-5 > x$

67. $5 < x < 7$

68. $5 < -x < 7$

69. $-6 < -x < 6$

70. $-5 \le x - 0.1 \le 5$

Write an inequality to describe each situation. See Example 4.

71. ***Sales tax.*** At an 8% sales tax rate, Susan paid more than \$1500 sales tax when she purchased her new Camaro.

72. ***Fine dining.*** At Burger Brothers the price of a hamburger is twice the price of an order of French fries, and the price of a Coke is \$0.25 more than the price of the fries. Burger Brothers advertises that you can get a complete meal (burger, fries, and Coke) for under \$2.00.

73. ***Barely passing.*** Travis made 44 and 72 on the first two tests in algebra and has one test remaining. The average on the three tests must be at least 60 for Travis to pass the course.

74. ***Ace the course.*** Florence made 87 on her midterm exam in psychology. The average of her midterm and her final must be at least 90 to get an A in the course.

75. ***Coast to coast.*** On Howard's recent trip from Bangor to San Diego, he drove for 8 hours each day and traveled between 396 and 453 miles each day.

76. ***Mother's Day present.*** Bart and Betty are looking at color televisions that range in price from \$399.99 to \$579.99. Bart can afford more than Betty and has agreed to spend \$100 more than Betty when they purchase this gift for their mother.

77. ***Positioning a ladder.*** Write an inequality in the variable x for the degree measure of the angle at the base of the ladder shown in the figure, given that the angle at the base must be between 60° and 70°.

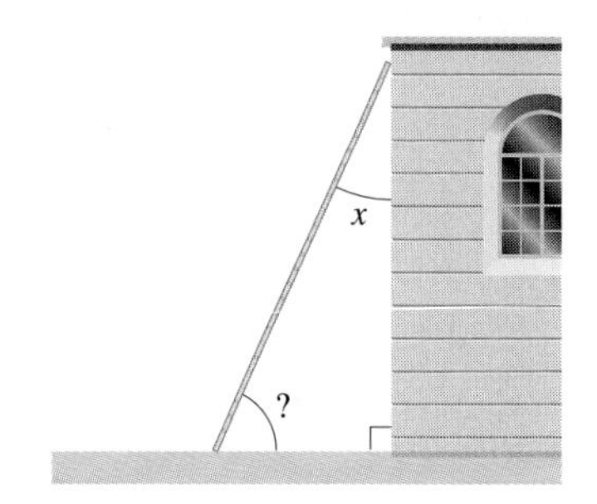

Figure for Exercise 77

78. ***Building a ski ramp.*** Write an inequality in the variable x for the degree measure of the smallest angle of the triangle shown in the figure, given that the degree measure of the smallest angle is at most 30°.

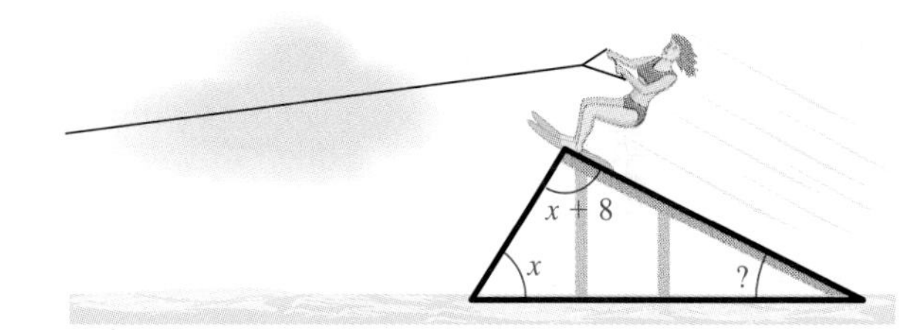

Figure for Exercise 78

Getting More Involved

79. ***Discussion.*** When an instructor asked his students for the smallest number that satisfies $x > 4$, some students immediately gave the answer 5. What is the correct answer? What is the largest number that satisfies $x \le -1$?

80. ***Writing.*** What is the difference between using an open circle or a solid circle on a graph of an inequality? What is indicated by the arrow on the graph of an inequality?

2.8 Solving Inequalities

In this section:

- Rules for Inequalities
- Solving Inequalities
- Applications of Inequalities

In Section 2.1 you learned to solve equations by writing a sequence of equivalent equations that ends in a very simple equation whose solution is obvious. In this section you will learn that the procedure for solving inequalities is the same. However, the rules for performing operations on each side of an inequality are slightly different from the rules for equations.

Rules for Inequalities

Equivalent inequalities are inequalities that have exactly the same solutions. Inequalities such as $x > 3$ and $x + 2 > 5$ are equivalent because any number that is larger than 3 certainly satisfies $x + 2 > 5$ and any number that satisfies $x + 2 > 5$ must certainly be larger than 3.

We can get equivalent inequalities by performing operations on each side of an inequality just as we do for solving equations. If we start with the inequality $6 < 10$ and add 2 to each side, we get the true statement $8 < 12$. Examine the results of performing the same operation on each side of $6 < 10$.

Perform these operations on each side:

	Add 2	Subtract 2	Multiply by 2	Divide by 2
Start with $6 < 10$	$8 < 12$	$4 < 8$	$12 < 20$	$3 < 5$

All of the resulting inequalities are correct. Now if we repeat these operations using -2, we get the following results.

Perform these operations on each side:

	Add -2	Subtract -2	Multiply by -2	Divide by -2
Start with $6 < 10$	$4 < 8$	$8 < 12$	$-12 > -20$	$-3 > -5$

Notice that the direction of the inequality symbol is the same for all of the results except the last two. When we multiplied each side by -2 and when we divided each side by -2, we had to reverse the inequality symbol to get a correct result. These tables illustrate the rules for solving inequalities.

Addition-Subtraction Property of Inequality

If we add the same number to each side of an inequality or subtract the same number from each side of an inequality, we get an equivalent inequality. If $a < b$, then

$$a + c < b + c \quad \text{and} \quad a - c < b - c.$$

Multiplication-Division Property of Inequality

If we multiply or divide each side of an inequality by the same *positive* number, we get an equivalent inequality. If $a < b$ and $c > 0$, then

$$ac < bc \quad \text{and} \quad \frac{a}{c} < \frac{b}{c}.$$

If we multiply or divide each side of an inequality by the same *negative* number and *reverse the inequality symbol,* we get an equivalent inequality. If $a < b$ and $c < 0$, then

$$ac > bc \quad \text{and} \quad \frac{a}{c} > \frac{b}{c}.$$

EXAMPLE 1 **Writing equivalent inequalities**

Write the appropriate inequality symbol in the blank so that the two inequalities are equivalent.

a) $x + 3 > 9$, x _____ 6 **b)** $-2x \leq 6$, x _____ -3

Solution

a) If we subtract 3 from each side of $x + 3 > 9$, we get the equivalent inequality $x > 6$.

b) If we divide each side of $-2x \leq 6$ by -2, we get the equivalent inequality $x \geq -3$. ■

CAUTION We use the properties of inequality just as we use the properties of equality. However, when we multiply or divide each side by a negative number, we must reverse the inequality symbol. ⊘

Solving Inequalities

To solve inequalities, we use the properties of inequality to isolate x on one side.

EXAMPLE 2 **Using the properties of inequality**

Solve and graph the inequality $4x - 5 > 19$.

Solution

$$4x - 5 > 19 \qquad \text{Original inequality}$$
$$4x - 5 + 5 > 19 + 5 \qquad \text{Add 5 to each side.}$$
$$4x > 24 \qquad \text{Simplify.}$$
$$x > 6 \qquad \text{Divide each side by 4.}$$

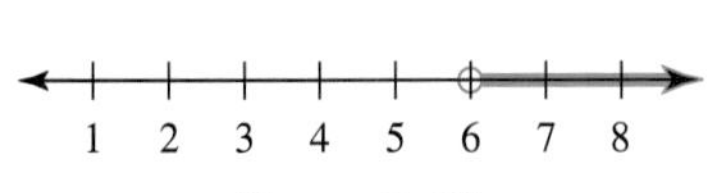

Figure 2.13

Since the last inequality is equivalent to the first, they have the same solutions and the same graph, which is shown in Fig. 2.13. ■

In the next example we divide each side of an inequality by a negative number.

EXAMPLE 3 **Reversing the inequality symbol**

Solve and graph the inequality $5 - 5x \le 1 + 2(5 - x)$.

Solution

$5 - 5x \le 1 + 2(5 - x)$	Original inequality
$5 - 5x \le 11 - 2x$	Simplify the right side.
$5 - 3x \le 11$	Add $2x$ to each side.
$-3x \le 6$	Subtract 5 from each side.
$x \ge -2$	Divide each side by -3, and reverse the inequality.

−6 −5 −4 −3 −2 −1 0 1

Figure 2.14

The inequalities $5 - 5x \le 1 + 2(5 - x)$ and $x \ge -2$ have the same graph, which is shown in Fig. 2.14. ■

We can use the rules for solving inequalities on the compound inequalities that we studied in Section 2.7.

EXAMPLE 4 **Solving a compound inequality**

Solve and graph the inequality $-9 \le \frac{2x}{3} - 7 < 5$.

Solution

$-9 \le \frac{2x}{3} - 7 < 5$	Original inequality
$-9 + 7 \le \frac{2x}{3} - 7 + 7 < 5 + 7$	Add 7 to each part.
$-2 \le \frac{2x}{3} < 12$	Simplify.
$\frac{3}{2}(-2) \le \frac{3}{2} \cdot \frac{2x}{3} < \frac{3}{2} \cdot 12$	Multiply each part by $\frac{3}{2}$.
$-3 \le x < 18$	Simplify.

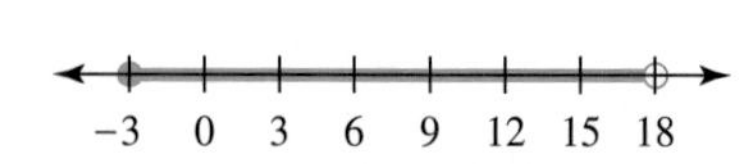

Figure 2.15

Any number that satisfies $-3 \le x < 18$ also satisfies the original inequality. Figure 2.15 shows all of the solutions to the original inequality. ■

CAUTION There are many negative numbers in Example 4, but the inequality was not reversed, since we did not multiply or divide by a negative number. An inequality is reversed only if you multiply or divide by a negative number. ⊘

EXAMPLE 5 **Reversing inequality symbols in a compound inequality**

Solve and graph the inequality $-3 \le 5 - x \le 5$.

Solution

$-3 \le 5 - x \le 5$	Original inequality
$-3 - 5 \le 5 - x - 5 \le 5 - 5$	Subtract 5 from each part.
$-8 \le -x \le 0$	Simplify.
$(-1)(-8) \ge (-1)(-x) \ge (-1)(0)$	Multiply each part by -1, reversing the inequality symbols.
$8 \ge x \ge 0$	

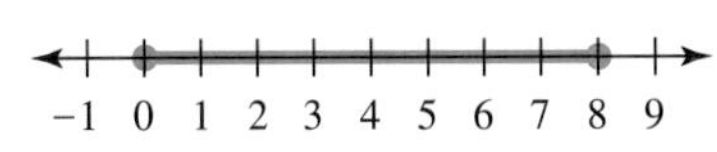

Figure 2.16

It is customary to write $8 \geq x \geq 0$ with the smallest number on the left:

$$0 \leq x \leq 8$$

Figure 2.16 shows all numbers that satisfy $-3 \leq 5 - x \leq 5$. ■

Applications of Inequalities

The following example shows how inequalities can be used in applications.

EXAMPLE 6 **Averaging test scores**

Mei Lin made a 76 on the midterm exam in history. To get a B, the average of her midterm and her final exam must be between 80 and 90. For what range of scores on the final exam will she get a B?

Solution

Let x represent the final exam score. Her average is then $\frac{x + 76}{2}$. The inequality expresses the fact that the average must be between 80 and 90:

$$80 < \frac{x + 76}{2} < 90$$

$$2(80) < 2\left(\frac{x + 76}{2}\right) < 2(90) \quad \text{Multiply each part by 2.}$$

$$160 < x + 76 < 180 \quad \text{Simplify.}$$

$$160 - 76 < x + 76 - 76 < 180 - 76 \quad \text{Subtract 76 from each part.}$$

$$84 < x < 104 \quad \text{Simplify.}$$

The last inequality indicates that Mei Lin's final exam score must be between 84 and 104. ■

Warm-ups

True or false? Explain your answer.

1. The inequality $2x > 18$ is equivalent to $x > 9$.
2. The inequality $x - 5 > 0$ is equivalent to $x < 5$.
3. We can divide each side of an inequality by any real number.
4. The inequality $-2x \leq 6$ is equivalent to $-x \leq 3$.
5. The statement "x is at most 7" is written as $x < 7$.
6. "The sum of x and $0.05x$ is at least 76" is written as $x + 0.05x \geq 76$.
7. The statement "x is not more than 85" is written as $x < 85$.
8. The inequality $-3 > x > -9$ is equivalent to $-9 < x < -3$.
9. If x is the sale price of Glen's truck, the sales tax rate is 8%, and the title fee is \$50, then the total that he pays is $1.08x + 50$ dollars.
10. If the selling price of the house, x, less the sales commission of 6% must be at least \$60,000, then $x - 0.06x \leq 60{,}000$.

2.8 EXERCISES

Write the appropriate inequality symbol in the blank so that the two inequalities are equivalent. See Example 1.

1. $x + 7 > 0$
$x\_\_\_ -7$

2. $x - 6 < 0$
$x\_\_\_ 6$

3. $9 \le 3w$
$w\_\_\_ 3$

4. $10 \ge 5z$
$z\_\_\_ 2$

5. $-4k < -4$
$k\_\_\_ 1$

6. $-9t > 27$
$t\_\_\_ -3$

7. $-\frac{1}{2}y \ge 4$
$y\_\_\_ -8$

8. $-\frac{1}{3}x \le 4$
$x\_\_\_ -12$

Solve and graph each of the following inequalities. See Examples 2 and 3.

9. $x + 3 > 0$

10. $x + 9 \le -8$

11. $-3 < w - 1$

12. $9 > w - 12$

13. $8 > 2b$

14. $35 < 7b$

15. $-4z \le 8$

16. $-5y \ge -20$

17. $3y - 2 < 7$

18. $2y - 5 > -9$

19. $3 - 7z \le 17$

20. $5 - 3z \ge 20$

21. $6 > -r + 3$

22. $6 \le 12 - r$

23. $5 - 4p > -8 - 3p$

24. $7 - 9p > 11 - 8p$

25. $-\frac{5}{6}q \ge -20$

26. $-\frac{2}{3}q \ge -4$

27. $1 - \frac{1}{4}t \ge -2$

28. $2 - \frac{1}{3}t > 0$

29. $2x + 5 < x - 6$

30. $3x - 4 < 2x + 9$

31. $x - 4 < 2(x + 3)$

32. $2x + 3 < 3(x - 5)$

33. $0.52x - 35 < 0.45x + 8$

34. $8455(x - 3.4) > 4320$

Solve and graph each compound inequality. See Examples 4 and 5.

35. $5 < x - 3 < 7$

36. $2 < x - 5 < 6$

37. $3 < 2v + 1 < 10$

38. $-3 < 3v + 4 < 7$

39. $-4 \le 5 - k \le 7$

40. $2 \le 3 - k \le 8$

41. $-2 < 7 - 3y \le 22$

42. $-1 \le 1 - 2y < 3$

43. $5 < \frac{2u}{3} - 3 < 17$

44. $-4 < \frac{3u}{4} - 1 < 11$

45. $-7 < \frac{3m + 1}{2} \le 8$

46. $0 \le \frac{3 - 2m}{2} < 9$

47. $0.02 < 0.54 - 0.0048x < 0.05$

48. $0.44 < \frac{34.55 - 22.3x}{124.5} < 0.76$

Solve and graph each inequality.

49. $\frac{1}{2}x - 1 \le 4 - \frac{1}{3}x$

50. $\frac{y}{4} - \frac{5}{12} \ge \frac{y}{3} + \frac{1}{4}$

51. $\frac{1}{2}\left(x - \frac{1}{4}\right) > \frac{1}{4}\left(6x - \frac{1}{2}\right)$

52. $-\frac{1}{2}\left(z - \frac{2}{5}\right) < \frac{2}{3}\left(\frac{3}{4}z - \frac{6}{5}\right)$

53. $\frac{1}{3} < \frac{1}{4}x - \frac{1}{6} < \frac{7}{12}$

54. $-\frac{3}{5} < \frac{1}{5} - \frac{2}{15}w < -\frac{1}{3}$

Solve each of the following problems by using an inequality. See Example 6.

55. ***Boat storage.*** The length of a rectangular boat storage shed must be 4 meters more than the width, and the perimeter must be at least 120 meters. What is the range of values for the width?

56. ***Fencing a garden.*** Elka is planning a rectangular garden that is to be twice as long as it is wide. If she can afford to buy at most 180 feet of fencing, then what are the possible values for the width?

57. ***Car shopping.*** Harold Ivan is shopping for a new car. In addition to the price of the car, there is a 5% sales tax and a $144 title and license fee. If Harold Ivan decides that he will spend less than $9970 total, then what is the price range for the car?

58. ***Car selling.*** Ronald wants to sell his car through a broker who charges a commission of 10% of the selling price. Ronald still owes $11,025 on the car. Ronald must get enough to at least pay off the loan. What is the range of the selling price?

59. ***Microwave oven.*** Sherie is going to buy a microwave in a city with an 8% sales tax. She has at most $594 to spend. In what price range should she look?

60. ***Dining out.*** At Burger Brothers the price of a hamburger is twice the price of an order of French fries, and the price of a Coke is $0.25 more than the price of the fries. Burger Brothers advertises that you can get a complete meal (burger, fries, and Coke) for under $2.00. What is the price range of an order of fries?

61. ***Averaging test scores.*** Tilak made 44 and 72 on the first two tests in algebra and has one test remaining. For Tilak to pass the course, the average on the three tests must be at least 60. For what range of scores on his last test will Tilak pass the course?

62. ***Averaging income.*** Helen earned \$400 in January, \$450 in February, and \$380 in March. To pay all of her bills, she must average at least \$430 per month. For what income in April would her average for the four months be at least \$430?

63. ***Going for a C.*** Professor Williams gives only a midterm exam and a final exam. The semester average is computed by taking $\frac{1}{3}$ of the midterm exam score plus $\frac{2}{3}$ of the final exam score. To get a C, Stacy must have a semester average between 70 and 79 inclusive. If Stacy scored only 48 on the midterm, then for what range of scores on the final exam will Stacy get a C?

64. ***Different weights.*** Professor Williamson counts his midterm as $\frac{2}{3}$ of the grade and his final as $\frac{1}{3}$ of the grade. Wendy scored only 48 on the midterm. What range of scores on the final exam would put Wendy's average between 70 and 79 inclusive? Compare to the previous exercise.

65. ***Average driving speed.*** On Halley's recent trip from Bangor to San Diego, she drove for 8 hours each day and traveled between 396 and 453 miles each day. In what range was her average speed for each day of the trip?

66. ***Driving time.*** On Halley's trip back to Bangor, she drove at an average speed of 55 mph every day and traveled between 330 and 495 miles per day. In what range was her daily driving time?

67. ***Sailboat navigation.*** As the sloop sailed north along the coast, the captain sighted the lighthouse at points A and B as shown in the figure. If the degree measure of the angle at the lighthouse is less than 30°, then what are the possible values for x?

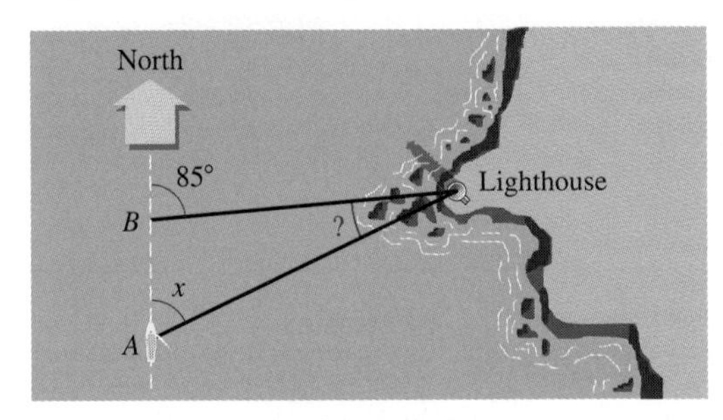

Figure for Exercise 67

68. ***Flight plan.*** A pilot started at point A and flew in the direction shown in the diagram for some time. At point B she made a 110° turn to end up at point C, due east of where she started. If the measure of angle C is less than 85°, then what are the possible values for x?

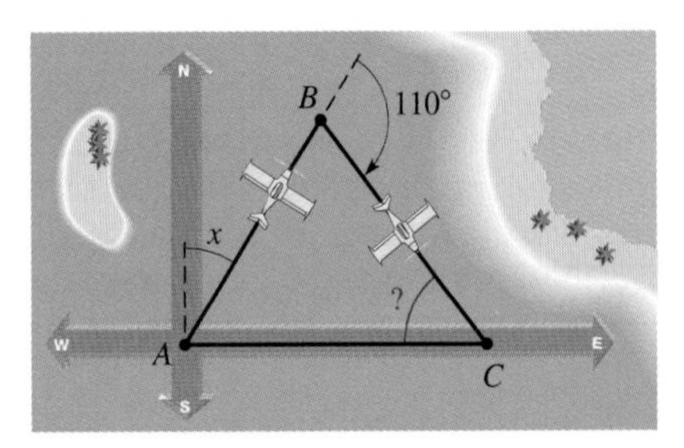

Figure for Exercise 68

Getting More Involved

69. ***Discussion.*** Which of the following inequalities is not equivalent to $1 - x > 0$? Explain.

a) $0 < 1 - x$ **b)** $x - 1 < 0$ **c)** $1 > x$
d) $x < -1$ **e)** $-x > -1$

70. ***Discussion.*** Which of the following inequalities is not equivalent to $-4 \le -2x < 8$? Explain.

a) $4 \ge 2x > -8$ **b)** $\frac{-4}{-2} \le \frac{-2x}{-2} < \frac{8}{-2}$
c) $-8 < 2x \le 4$ **d)** $-4 < x \le 2$
e) $2 \ge x > -4$

71. ***Writing.*** How is solving inequalities similar to solving equations? How is solving inequalities different from solving equations.

72. ***Cooperative learning.*** Work with a group to solve each of the following problems. Explain the difference between them.

a) Two angles are complementary, and the degree measure of one is larger than the degree measure of the other. Find the number of degrees in each angle.

b) Two angles are complementary, and the degree measure of one is 12° larger than the degree measure of the other. Find the number of degrees in each angle.

COLLABORATIVE ACTIVITIES

Finding the Better Deal?

Grouping: 2 to 4 students per group
Topic: Percents

For this activity the students in your group should choose roles. Four standard roles are Moderator (keeps the group on task), Messenger (asks the group's questions to the instructor, tutor, or helper), Quality Manager (checks to see that the work is top quality), and Recorder (records the group's work). See the Instructor's Solution Manual for a description of these roles. After you have chosen roles, read through the activity completely, and answer the questions.

Scenario: You have decided to buy a new car and have asked some friends to help you choose the best deal and the best financing. You have already looked into your finances. You have $1700 from your summer job and $1500 that your parents will give you for a down payment on a car.

You found a car that you really liked that was 10% off the regular $9800 price. Your friends at the student Credit Union tell you it has a 48-month car loan at $7\frac{1}{2}\%$ annual simple interest.

At a second dealership you find a similar car on sale for $9000 if you finance it through the dealership. The dealer said that after the down payment you could pay it off in 5 years with monthly payments of $140. This second deal sounds good! (You have decided you could afford up to $160 a month in payments.) The idea of having an extra $20 a month is appealing. However, you wonder how much you will actually pay for the second car.

Which car should you buy?

Questions: The following questions will help you to work your way through the problem.

For the first car, if you finance it at the Credit Union:

1. How much will it cost after the discount?
2. How much will you need to borrow?
3. What will the total interest be for the 48 months?
4. How much will the monthly payments be?

For the second car:

5. Find the amount to be financed.
6. Find the interest and the interest rate.

For both cars:

7. Find the total cost for each car.

After reviewing the information, decide which car you would buy.

Extension:

1. What other costs would there be? Find these out for your city.
2. Do a comparison of car loans at your local banks.

Wrap-up CHAPTER 2

SUMMARY

Equations		Examples
Linear equation	An equation of the form $ax + b = 0$ with $a \neq 0$	$3x + 7 = 0$
Identity	An equation that is satisfied by every number for which both sides are defined	$x + x = 2x$
Conditional equation	An equation that has at least one solution but is not an identity	$5x - 10 = 0$

Inconsistent equation	An equation that has no solution	$x = x + 1$
Equivalent equations	Equations that have exactly the same solutions	$2x + 1 = 5$ $2x = 4$
Properties of equality	If the same number is added to or subtracted from each side of an equation, the resulting equation is equivalent to the original equation.	$x - 5 = -9$ $x = -4$
	If each side of an equation is multiplied or divided by the same nonzero number, the resulting equation is equivalent to the original equation.	$9x = 27$ $x = 3$
Solving equations	1. Remove parentheses and combine like terms to simplify each side as much as possible. 2. Get like terms from opposite sides onto the same side so that they may be combined. 3. Use the multiplication-division property of equality last to isolate the variable. 4. The equation $-x = a$ is equivalent to $x = -a$. 5. Check that the solution satisfies the original equation.	$2(x - 3) = -7 + 3(x - 1)$ $2x - 6 = -10 + 3x$ $-x - 6 = -10$ $-x = -4$ $x = 4$ *Check:* $2(4 - 3) = -7 + 3(4 - 1)$ $2 = 2$

Applications

Steps in solving applied problems	1. Read the problem. 2. If possible, draw a diagram to illustrate the problem. 3. Choose a variable and write down what it represents. 4. Represent any other unknowns in terms of that variable. 5. Write an equation that describes the situation. 6. Solve the equation. 7. Check your answer by using it to solve the original problem (not the equation). 8. Answer the question posed in the original problem.	

Inequalities

		Examples
Properties of inequality	Addition, subtraction, multiplication, and division may be performed on each side of an inequality, just as we do in solving equations, with one exception. When multiplying or dividing by a negative number, the inequality symbol is reversed.	$-3x + 1 > 7$ $-3x > 6$ $x < -2$

REVIEW EXERCISES

2.1 *Solve each equation.*

1. $2x - 5 = 9$

2. $5x - 8 = 38$

3. $3r - 7 = 0$

4. $3t + 5 = 0$

5. $3 - 4y = 11$

6. $4 - 3k = -8$

7. $2(h - 7) = -14$

8. $2(t - 7) = 0$

9. $3(w - 5) = 6(w + 2) - 3$

10. $2(a - 4) + 4 = 5(9 - a)$

11. $-\frac{2}{3}b = 20$

12. $\frac{3}{4}b = -6$

13. $\frac{1}{3}x = \frac{1}{7}$

14. $\frac{x}{2} = -\frac{2}{5}$

15. $0.24c + 1 = 97$

16. $1.05q = 420$

17. $p - 0.1p = 90$

18. $m + 0.05m = 2.1$

2.2 *Solve each equation. Identify each equation as a conditional equation, an inconsistent equation, or an identity.*

19. $2(x - 7) - 5 = 5 - (3 - 2x)$

20. $2(x - 7) + 5 = -(9 - 2x)$

21. $2(w - w) = 0$

22. $2y - y = 0$

23. $\frac{3r}{3r} = 1$

24. $\frac{3t}{3} = 1$

25. $\frac{1}{2}a - 5 = \frac{1}{3}a - 1$

26. $\frac{1}{2}b - \frac{1}{2} = \frac{1}{4}b$

27. $0.06q + 14 = 0.3q - 5.2$

28. $0.05(z + 20) = 0.1z - 0.5$

29. $0.05(x + 100) + 0.06x = 115$

30. $0.06x + 0.08(x + 1) = 0.41$

Solve each equation.

31. $2x + \frac{1}{2} = 3x + \frac{1}{4}$

32. $5x - \frac{1}{3} = 6x - \frac{1}{2}$

33. $\frac{x}{2} - \frac{3}{4} = \frac{x}{6} + \frac{1}{8}$

34. $\frac{1}{3} - \frac{x}{5} = \frac{1}{2} - \frac{x}{10}$

35. $\frac{5}{6}x = -\frac{2}{3}$

36. $-\frac{2}{3}x = \frac{3}{4}$

37. $-\frac{1}{2}(x - 10) = \frac{3}{4}x$

38. $-\frac{1}{3}(6x - 9) = 23$

39. $3 - 4(x - 1) + 6 = -3(x + 2) - 5$

40. $6 - 5(1 - 2x) + 3 = -3(1 - 2x) - 1$

41. $5 - 0.1(x - 30) = 18 + 0.05(x + 100)$

42. $0.6(x - 50) = 18 - 0.3(40 - 10x)$

2.3 *Solve each equation for x.*

43. $ax + b = 0$

44. $mx + e = t$

45. $ax - 2 = b$

46. $b = 5 - x$

47. $LWx = V$

48. $3xy = 6$

49. $2x - b = 5x$

50. $t - 5x = 4x$

Solve each equation for y. Write the answer in the form $y = mx + b$, where m and b are real numbers.

51. $5x + 2y = 6$

52. $5x - 3y + 9 = 0$

53. $y - 1 = -\frac{1}{2}(x - 6)$

54. $y + 6 = \frac{1}{2}(x + 8)$

55. $\frac{1}{2}x + \frac{1}{4}y = 4$

56. $-\frac{x}{3} + \frac{y}{2} = 1$

Find the value of y in each formula if $x = -3$.

57. $y = 3x - 4$

58. $2x - 3y = -7$

59. $5xy = 6$

60. $3xy - 2x = -12$

61. $y - 3 = -2(x - 4)$

62. $y + 1 = 2(x - 5)$

2.4 *Translate each verbal expression into an algebraic expression.*

63. The sum of a number and 9

64. The product of a number and 7

65. Two numbers that differ by 8

66. Two numbers with a sum of 12

67. Sixty-five percent of a number

68. One half of a number

Identify the variable, and write an equation that describes each situation. Do not solve the equation.

69. One side of a rectangle is 5 feet longer than the other, and the area is 98 square feet.

70. One side of a rectangle is one foot longer than twice the other side, and the perimeter is 56 feet.

71. By driving 10 miles per hour slower than Jim, Barbara travels the same distance in 3 hours as Jim does in 2 hours.

72. Gladys and Ned drove 840 miles altogether, with Gladys averaging 5 miles per hour more in her 6 hours at the wheel than Ned did in his 5 hours at the wheel.

73. The sum of three consecutive even integers is 88.

74. The sum of two consecutive odd integers is 40.

75. The three angles of a triangle have degree measures of t, $2t$, and $t - 10$.

76. Two complementary angles have degree measures p and $3p - 6$.

2.5 and 2.6 *Solve each problem.*

77. ***Odd integers.*** If the sum of three consecutive odd integers is 237, then what are the integers?

78. ***Even integers.*** Find two consecutive even integers that have a sum of 450.

79. ***Driving to the shore.*** Lawanda and Betty both drive the same distance to the shore. By driving 15 miles per hour faster than Betty, Lawanda can get there in 3 hours while Betty takes 4 hours. How fast does each of them drive?

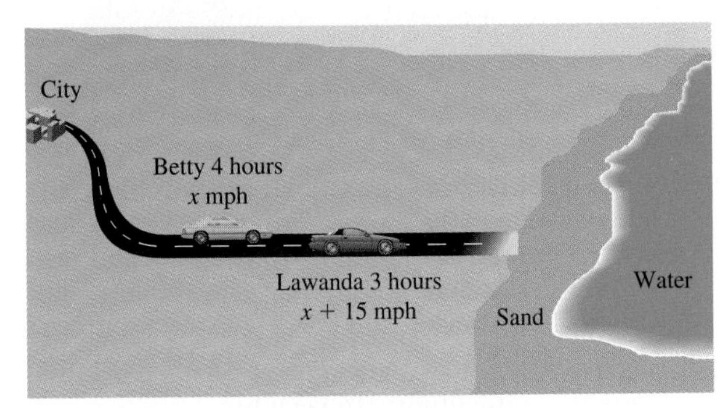

Figure for Exercise 79

80. ***Rectangular lot.*** The length of a rectangular lot is 50 feet more than the width. If the perimeter is 500 feet, then what are the length and width?

81. ***Combined savings.*** Wanda makes \$6000 more per year than her husband does. Wanda saves 10% of her income for retirement, and her husband saves 6%. If together they save \$5400 per year, then how much does each of them make per year?

82. ***Layoffs looming.*** American Products plans to lay off 10% of its employees in its aerospace division and 15% of its employees in its agricultural division. If altogether 12% of the 3,000 employees in these two divisions will be laid off, then how many employees are in each division?

2.7 *Determine whether the given number is a solution to the inequality following it.*

83. $3, -2x + 5 \le x - 6$

84. $-2, 5 - x > 4x + 3$

85. $-1, -2 \le 6 + 4x < 0$

86. $0, 4x + 9 \ge 5(x - 3)$

For each graph, write an inequality that describes the graph.

87. −2 −1 0 1 2 3 4 5 6 7 8

88. −5 −4 −3 −2 −1 0 1 2 3 4 5

89. −2 −1 0 1 2 3 4 5 6 7 8

90. −2 −1 0 1 2 3 4 5 6 7 8

91. −5 −4 −3 −2 −1 0 1 2 3 4 5

92. −6 −5 −4 −3 −2 −1 0 1 2 3 4

93. −8 −7 −6 −5 −4 −3 −2 −1 0 1 2

94. −5 −4 −3 −2 −1 0 1 2 3 4 5

2.8 *Solve and graph each inequality.*

95. $x + 2 > 1$

96. $x - 3 > 7$

97. $3x - 5 < x + 1$

98. $5x - 5 > 9 - 2x$

99. $-\frac{3}{4}x \ge 3$

100. $-\frac{2}{3}x \le 10$

101. $3 - 2x < 11$

102. $5 - 3x > 35$

103. $-3 < 2x - 1 < 9$

104. $2 \le 3x + 2 < 8$

105. $0 \le 1 - 2x < 5$

106. $-5 < 3 - 4x \le 7$

107. $-1 \le \frac{2x - 3}{3} \le 1$

108. $-3 < \frac{4 - x}{2} < 2$

109. $\frac{1}{3} < \frac{1}{3} + \frac{x}{2} < \frac{5}{6}$

110. $-\frac{3}{8} \le -\frac{1}{4}x + \frac{1}{8} < \frac{5}{8}$

Miscellaneous

Use an equation, inequality, or formula to solve each problem.

111. ***Flat yield curve.*** The accompanying graph shows that the *yield curve* for U.S. Treasury bonds was relatively flat from 2 years out to 30 years as of February 16, 1995 (*Fidelity Investments Special Report*, Volume 1, Number 6). In this situation there is not much benefit to be obtained from long-term investing. Use the interest rate in the graph to find the amount of interest earned in the first year on a 2-year bond of $10,000. How much more interest would you earn in the first year on a 30-year bond of $10,000?

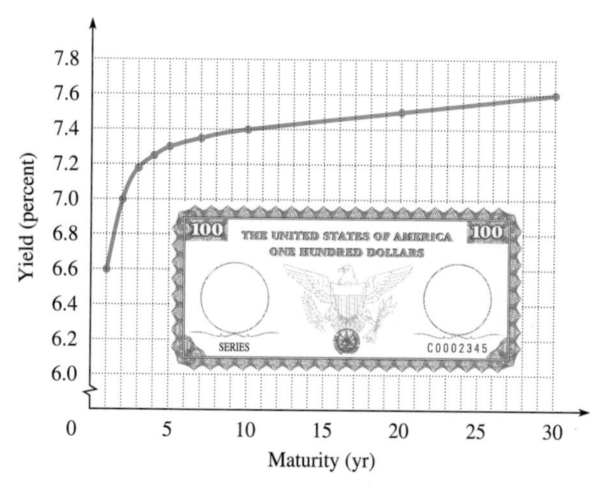

Figure for Exercises 111 and 112

112. ***Reading the curve.*** Use thè accompanying graph to find the maturity of a U.S. Treasury bond that had a yield of 7.4% on February 16, 1995.

113. ***Combined videos.*** The owners of ABC Video discovered that they had no movies in common with XYZ Video and bought XYZ's entire stock. Although XYZ had 200 titles, they had no children's movies, while 60% of ABC's titles were children's movies. If 40% of the movies in the combined stock are children's movies, then how many movies did ABC have before the merger?

114. ***Living comfortably.*** Gary has figured that he needs to take home $30,400 a year to live comfortably. If the government takes 24% of Gary's income, then what must his income be for him to live comfortably?

115. ***Bracing a gate.*** The diagonal brace on a rectangular gate forms an angle with the horizontal side with degree measure x and an angle with the vertical side with degree measure $2x - 3$. Find x.

116. ***Digging up the street.*** A contractor wants to install a pipeline connecting point A with point C on opposite sides of a road as shown in the figure. To save money, the contractor has decided to lay the pipe to point B and then under the road to point C. Find the measure of the angle marked x in the figure.

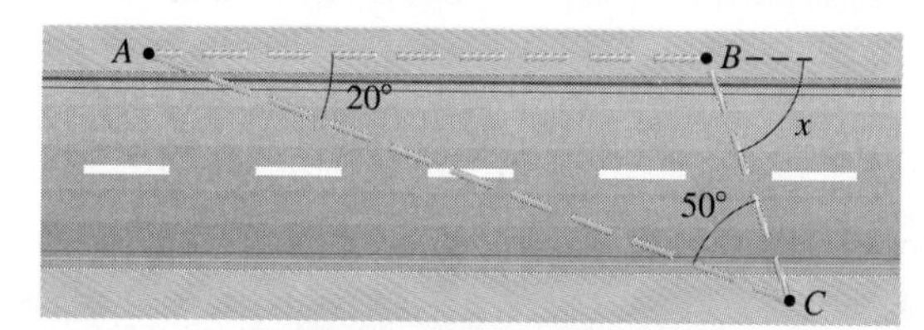

Figure for Exercise 116

117. ***Perimeter of a triangle.*** One side of a triangle is 1 foot longer than the shortest side, and the third side is twice as long as the shortest side. If the perimeter is less than 25 feet, then what is the range of the length of the shortest side?

118. ***Restricted hours.*** Alana makes $5.80 per hour working in the library. To keep her job, she must make at least $116 per week; but to keep her scholarship, she must not earn more than $145 per week. What is the range of the number of hours per week that she may work?

CHAPTER 2 TEST

Solve each equation.

1. $-10x - 6 + 4x = -4x + 8$

2. $5(2x - 3) = x + 3$

3. $-\frac{2}{3}x + 1 = 7$

4. $x + 0.06x = 742$

Solve for the indicated variable.

5. $2x - 3y = 9$ for y

6. $m = aP - w$ for a

Write an inequality that describes the graph.

7.

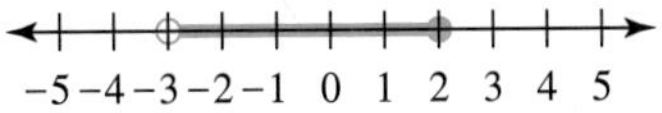

$-5\ -4\ -3\ -2\ -1\ 0\ 1\ 2\ 3\ 4\ 5$

8.

$-2\ -1\ 0\ 1\ 2\ 3\ 4\ 5\ 6\ 7\ 8$

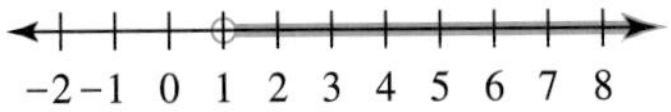

Solve and graph each inequality.

9. $4 - 3(w - 5) < -2w$

10. $1 < \frac{1 - 2x}{3} < 5$

11. $1 < 3x - 2 < 7$

12. $-\frac{2}{3}y < 4$

Solve each equation.

13. $2(x + 6) = 2x - 5$

14. $x + 7x = 8x$

15. $x - 0.03x = 0.97$

16. $6x - 7 = 0$

Write a complete solution to each problem.

17. The perimeter of a rectangle is 72 meters. If the width is 8 meters less than the length, then what is the width of the rectangle?

18. If the area of a triangle is 54 square inches and the base is 12 inches, then what is the height?

19. How many liters of a 20% alcohol solution should Maria mix with 50 liters of a 60% alcohol solution to obtain a 30% solution?

20. Brandon gets a 40% discount on loose diamonds where he works. The cost of the setting is \$250. If he plans to spend at most \$1450, then what is the price range (list price) of the diamonds that he can afford?

21. If the degree measure of the smallest angle of a triangle is one-half of the degree measure of the second largest angle and one-third of the degree measure of the largest angle, then what is the degree measure of each angle?

Tying It All Together CHAPTERS 1 – 2

Simplify each expression.

1. $3x + 5x$

2. $3x \cdot 5x$

3. $\dfrac{4x + 2}{2}$

4. $5 - 4(3 - x)$

5. $3x + 8 - 5(x - 1)$

6. $(-6)^2 - 4(-3)2$

7. $3^2 \cdot 2^3$

8. $4(-7) - (-6)(3)$

9. $-2x \cdot x \cdot x$

10. $(-1)(-1)(-1)(-1)(-1)$

Solve each equation.

11. $3x + 5x = 8$

12. $3x + 5x = 8x$

13. $3x + 5x = 7x$

14. $3x + 5 = 8$

15. $3x + 1 = 7$

16. $5 - 4(3 - x) = 1$

17. $3x + 8 = 5(x - 1)$

18. $x - 0.05x = 190$

Solve the problem.

19. ***Linear Depreciation.*** In computing income taxes, a company is allowed to depreciate a $20,000 computer system over five years. Using *linear depreciation,* the value V of the computer system at any year t from 0 through 5 is given by

$$V = C - \frac{(C - S)}{5}t,$$

where C is the initial cost of the system and S is the scrap value of the system. What is the value of the computer system after 2 years if its scrap value is estimated to be $4,000? If the value of the system after three years is claimed to be $14,000, then what is the scrap value of the company's system?

CHAPTER 3

Polynomials and Exponents

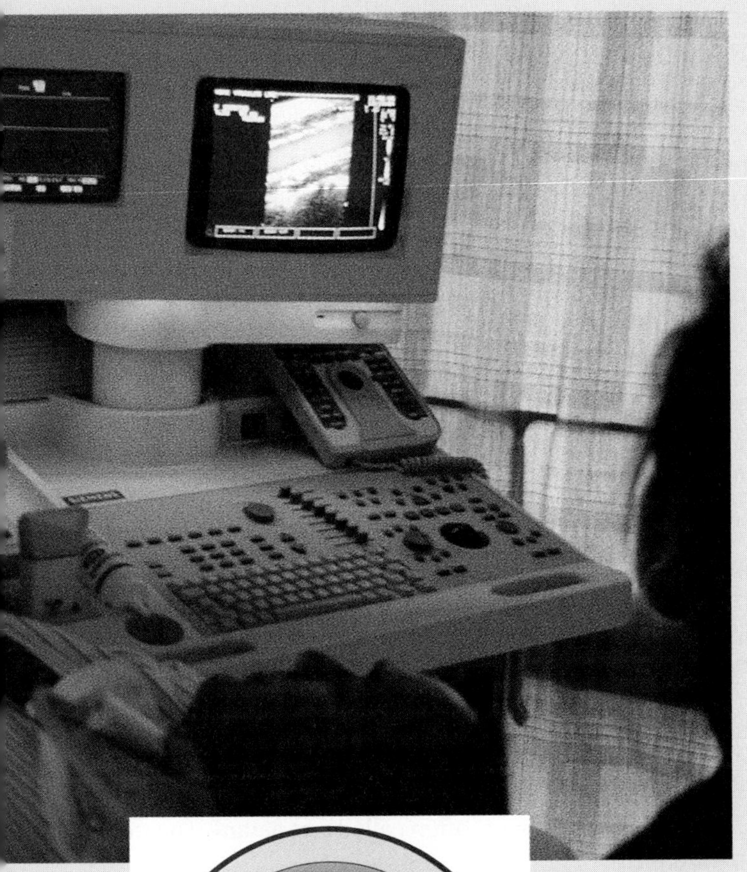

The nineteenth century physician and physicist, Jean Louis Marie Poiseuille (1799–1869) is given credit for discovering a formula associated with the circulation of blood through arteries. Poiseuille's law, as it is known, can be used to determine the velocity of blood in an artery at a given distance from the center of the artery. The formula states that the flow of blood in an artery is faster toward the center of the blood vessel and is slower toward the outside. Blood flow can also be affected by a person's blood pressure, the length of the blood vessel, and the viscosity of the blood itself.

In later years, Poiseuille's continued interest in blood circulation led him to experiments to show that blood pressure rises and falls when a person exhales and inhales. In modern medicine, physicians can use Poiseuille's law to determine how much the radius of a blocked blood vessel must be widened to create a healthy flow of blood.

In this chapter you will study polynomials, the fundamental expressions of algebra. Polynomials are to algebra what integers are to arithmetic. We use polynomials to represent quantities in general, such as perimeter, area, revenue, and the volume of blood flowing through an artery. In Exercise 79 of Section 3.4, you will see Poiseuille's law represented by a polynomial.

3.1 Addition and Subtraction of Polynomials

In this section:

- Polynomials
- Value of a Polynomial
- Addition of Polynomials
- Subtraction of Polynomials
- Applications

We first used polynomials in Chapter 1 but did not identify them as polynomials. Polynomials also occurred in the equations and inequalities of Chapter 2. In this section we will define polynomials and begin a thorough study of polynomials.

Polynomials

In Chapter 1 we defined a **term** as an expression containing a number or the product of a number and one or more variables raised to powers. Some examples of terms are

$$4x^3, \quad -x^2y^3, \quad 6ab, \quad \text{and} \quad -2.$$

The **degree of the term** $4x^3$ is 3. If a term has more than one variable, then the degree of the term is the sum of the exponents of the variables. The degree of $-x^2y^3$ is 5, and the degree of $6ab$ is 2. The term -2 has **degree 0** because -2 can be written as $-2x^0$. (The definition $x^0 = 1$ is given in Section 3.5.)

A **polynomial** can be defined as a single term or a finite sum of terms. The powers of the variables in the terms of a polynomial must be whole numbers. For example,

$$4x^3 + (-15x^2) + 7x + (-2)$$

is a polynomial. Because it is simpler to write addition of a negative as subtraction, this polynomial is written as

$$4x^3 - 15x^2 + 7x - 2.$$

The **degree of a polynomial** is the highest degree of any of its terms. So the degree of $4x^3 - 15x^2 + 7x - 2$ is 3, and the degree of $4w - w^2$ is 2. The degree of a polynomial consisting of a single nonzero number such as 8 is 0. The number 0 is considered a polynomial without degree because $0 = 0 \cdot x^n$ for any value of n.

The number preceding the variable in each term is called the **coefficient** of that variable or the coefficient of that term. In the polynomial $4x^3 - 15x^2 + 7x - 2$, 4 is the coefficient of x^3, -15 is the coefficient of x^2, and 7 is the coefficient of x. In algebra a number is often called a **constant.** The term -2 is called the **constant term.**

EXAMPLE 1 **Identifying coefficients**

Determine the coefficients of x^3 and x^2 in each polynomial:

a) $x^3 + 5x^2 - 6$ **b)** $4x^6 - x^3 + x$

Solution

a) Write the polynomial as $1 \cdot x^3 + 5x^2 - 6$ to see that the coefficient of x^3 is 1 and the coefficient of x^2 is 5.

b) The x^2-term is missing in $4x^6 - x^3 + x$. Because $4x^6 - x^3 + x$ can be written as

$$4x^6 - 1 \cdot x^3 + 0 \cdot x^2 + x,$$

the coefficient of x^3 is -1 and the coefficient of x^2 is 0. ■

For simplicity we generally write polynomials with the exponents decreasing from left to right and the constant term last. So we write

$$x^3 - 4x^2 + 5x + 1 \qquad \text{rather than} \qquad -4x^2 + 1 + 5x + x^3.$$

When a polynomial is written with decreasing exponents, the coefficient of the first term is called the **leading coefficient.**

Certain polynomials are given special names. A **monomial** is a polynomial that has one term, a **binomial** is a polynomial that has two terms, and a **trinomial** is a polynomial that has three terms. For example, $3x^5$ is a monomial, $2x - 1$ is a binomial, and $4x^6 - 3x + 2$ is a trinomial.

EXAMPLE 2 **Types of polynomials**

Identify each polynomial as a monomial, binomial, or trinomial and state its degree.

a) $5x^2 - 7x^3 + 2$ **b)** $x^{43} - x^2$

c) $5x$ **d)** -12

Solution

a) The polynomial $5x^2 - 7x^3 + 2$ is a third-degree trinomial.

b) The polynomial $x^{43} - x^2$ is a binomial with degree 43.

c) Because $5x = 5x^1$, this polynomial is a monomial with degree 1.

d) The polynomial -12 is a monomial with degree 0. ■

Value of a Polynomial

A polynomial is an algebraic expression. Like other algebraic expressions, a polynomial has no specific value unless the variables are replaced by numbers.

EXAMPLE 3 **Evaluating polynomials**

Find the value of the polynomial $-3x^4 - x^3 + 20x + 3$ when $x = -2$ and when $x = 1$.

Solution

Replace x by -2:

$$\begin{aligned} -3x^4 - x^3 + 20x + 3 &= -3(-2)^4 - (-2)^3 + 20(-2) + 3 \\ &= -3(16) - (-8) - 40 + 3 \\ &= -48 + 8 - 40 + 3 \\ &= -77 \end{aligned}$$

Now replace x by 1:

$$\begin{aligned} -3x^4 - x^3 + 20x + 3 &= -3(1)^4 - (1)^3 + 20(1) + 3 \\ &= -3 - 1 + 20 + 3 \\ &= 19 \end{aligned}$$

■

Calculator Close-up

A calculator can be used to evaluate a polynomial when the calculations get more difficult. For example, to evaluate $-2x^2 - 3x + 5$ for $x = 1.3$ using a scientific calculator, try the following:

2 [+/−] [×] 1.3 [x^2] [−] 3 [×] 1.3 [+] 5 [=]

The display should read

−2.28

On a graphing calculator you can use parentheses to make the display appear as follows:

⁻2(1.3)²−3(1.3)+5 [ENTER]

⁻2.28

Addition of Polynomials

You learned how to combine like terms in Chapter 1. Also, you combined like terms when solving equations in Chapter 2. Addition of polynomials is done simply by adding the like terms.

Addition of Polynomials

To add two polynomials, add the like terms.

Polynomials can be added horizontally or vertically, as shown in the next example.

EXAMPLE 4 **Adding polynomials**

Perform the indicated operation.

a) $(x^2 - 6x + 5) + (-3x^2 + 5x - 9)$

b) $(-5a^3 + 3a - 7) + (4a^2 - 3a + 7)$

Solution

a) We can use the commutative and associative properties to get the like terms next to each other and then combine them:

$$\begin{aligned}(x^2 - 6x + 5) + (-3x^2 + 5x - 9) &= x^2 - 3x^2 - 6x + 5x + 5 - 9 \\ &= -2x^2 - x - 4\end{aligned}$$

b) When adding vertically, we line up the like terms:

$$\begin{array}{rrrr} -5a^3 & & +3a & -7 \\ & 4a^2 & -3a & +7 \\ \hline -5a^3 & +4a^2 & & \end{array}$$

Add. ■

Subtraction of Polynomials

When we subtract polynomials, we subtract the like terms. Because $a - b = a + (-b)$, we can subtract by adding the opposite of the second polynomial to the first polynomial. Remember that a negative sign in front of parentheses changes the sign of each term in the parentheses. For example,

$$-(x^2 - 2x + 8) = -x^2 + 2x - 8.$$

Polynomials can be subtracted horizontally or vertically, as shown in the next example .

EXAMPLE 5 **Subtracting polynomials**

Perform the indicated operation.

a) $(x^2 - 5x - 3) - (4x^2 + 8x - 9)$ **b)** $(4y^3 - 3y + 2) - (5y^2 - 7y - 6)$

Solution

a) $(x^2 - 5x - 3) - (4x^2 + 8x - 9) = x^2 - 5x - 3 - 4x^2 - 8x + 9$ Change signs.

$= -3x^2 - 13x + 6$ Add.

b) To subtract $5y^2 - 7y - 6$ from $4y^3 - 3y + 2$ vertically, we line up the like terms as we do for addition:

$$\begin{array}{r} 4y^3 \qquad\quad - 3y + 2 \\ - \qquad (5y^2 - 7y - 6) \\ \hline \end{array}$$

Now change the signs of $5y^2 - 7y - 6$ and add the like terms:

$$\begin{array}{r} 4y^3 \qquad\quad - 3y + 2 \\ -5y^2 + 7y + 6 \\ \hline 4y^3 - 5y^2 + 4y + 8 \end{array}$$

CAUTION When adding or subtracting polynomials vertically, be sure to line up the like terms.

In the next example we combine addition and subtraction of polynomials.

EXAMPLE 6 **Adding and subtracting**

Perform the indicated operations:

$$(2x^2 - 3x) + (x^3 + 6) - (x^4 - 6x^2 - 9)$$

Solution

Remove the parentheses and combine the like terms:

$$\begin{aligned}(2x^2 - 3x) + (x^3 + 6) - (x^4 - 6x^2 - 9) &= 2x^2 - 3x + x^3 + 6 - x^4 + 6x^2 + 9 \\ &= -x^4 + x^3 + 8x^2 - 3x + 15\end{aligned}$$

Applications

Polynomials are often used to represent unknown quantities. In certain situations it is necessary to add or subtract such polynomials.

EXAMPLE 7 **Profit from prints**

Trey pays \$60 per day for a permit to sell famous art prints in the Student Union Mall. Each print costs him \$4, so the polynomial $4x + 60$ represents his daily cost in dollars for x prints sold. He sells the prints for \$10 each. So the polynomial $10x$ represents his daily revenue for x prints sold. Find a polynomial that represents his daily profit from selling x prints. Evaluate the profit polynomial for $x = 30$.

Solution

Because profit is revenue minus cost, we can subtract the corresponding polynomials to get a polynomial that represents the daily profit:

$$\begin{aligned} 10x - (4x + 60) &= 10x - 4x - 60 \\ &= 6x - 60 \end{aligned}$$

His daily profit from selling x prints is $6x - 60$ dollars. Evaluate this profit polynomial for $x = 30$:

$$6x - 60 = 6(30) - 60 = 120$$

So for $x = 30$, Trey's profit is \$120. ■

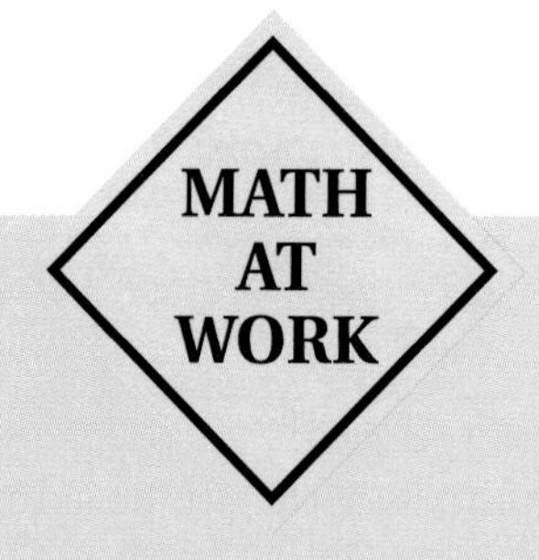

Nutritionist

The message we hear today about healthy eating is "more fiber and less fat." But healthy eating can be challenging. Jill Brown, Registered Dietitian and owner of Healthy Habits Nutritional Counseling and Consulting, provides nutritional counseling for people who are basically healthy but interested in improving their eating habits.

On a client's first visit, an extensive nutrition, medical, and family history is taken. Behavior patterns are discussed, and nutritional goals are set. Ms. Brown strives for a realistic rather than an idealistic approach. Whenever possible, a client should eat foods that are high in vitamin content rather than take vitamin pills. Moreover, certain frame types and family history make a slight and slender look difficult to achieve. Here, Ms. Brown might use the Harris-Benedict formula to determine the daily basal energy expenditure based on age, gender, and size. In addition to diet, exercise is discussed and encouraged. Finally, Ms. Brown provides a support system and serves as a "nutritional coach." By following her advice, many of her clients lower their blood pressure, reduce their cholesterol, have more energy, and look and feel better.

In Exercises 95 and 96 of this section you will use the Harris-Benedict formula to calculate the basal energy expenditure for a female and a male.

Warm-ups

True or false? Explain your answer.

1. In the polynomial $2x^2 - 4x + 7$ the coefficient of x is 4.
2. The degree of the polynomial $x^2 + 5x - 9x^3 + 6$ is 2.
3. In the polynomial $x^2 - x$ the coefficient of x is -1.
4. The degree of the polynomial $x^2 - x$ is 2.
5. A binomial always has a degree of 2.
6. The polynomial $3x^2 - 5x + 9$ is a trinomial.
7. Every trinomial has degree 2.
8. $x^2 - 7x^2 = -6x^2$ for any value of x.
9. $(3x^2 - 8x + 6) + (x^2 + 4x - 9) = 4x^2 - 4x - 3$ for any value of x.
10. $(x^2 - 4x) - (x^2 - 3x) = -7x$ for any value of x.

3.1 EXERCISES

Determine the coefficients of x^3 and x^2 in each polynomial. See Example 1.

1. $-3x^3 + 7x^2$ **2.** $10x^3 - x^2$

3. $x^4 + 6x^2 - 9$ **4.** $x^5 - x^3 + 3$

5. $\frac{x^3}{3} + \frac{7x^2}{2} - 4$ **6.** $\frac{x^3}{2} - \frac{x^2}{4} + 2x + 1$

Identify each polynomial as a monomial, binomial, or trinomial and state its degree. See Example 2.

7. -1 **8.** 5 **9.** m^3

10. $3a^8$ **11.** $4x + 7$ **12.** $a + 6$

13. $x^{10} - 3x^2 + 2$ **14.** $y^6 - 6y^3 + 9$

15. $x^6 + 1$ **16.** $b^2 - 4$

17. $a^3 - a^2 + 5$ **18.** $-x^2 + 4x - 9$

Evaluate each polynomial for $x = -1$ and $x = 3$. See Example 3.

19. $2x^2 - 3x + 1$ **20.** $3x^2 - x + 2$

21. $-3x^3 - x^2 + 3x - 4$ **22.** $-2x^4 - 3x^2 + 5x - 9$

Use a calculator to evaluate each polynomial for $x = 1.45$ and $x = -2.36$.

23. $x^2 - 6x + 3$ **24.** $x^3 - 3x + 2$

25. $1.2x^3 - 4.3x - 2.4$ **26.** $-3.5x^4 - 4.6x^3 + 5.5$

Perform the indicated operation. See Example 4.

27. $(x - 3) + (3x - 5)$ **28.** $(x - 2) + (x + 3)$

29. $(q - 3) + (q + 3)$ **30.** $(q + 4) + (q + 6)$

31. $(3x + 2) + (x^2 - 4)$

32. $(5x^2 - 2) + (-3x^2 - 1)$

33. $(4x - 1) + (x^3 + 5x - 6)$

34. $(3x - 7) + (x^2 - 4x + 6)$

35. $(a^2 - 3a + 1) + (2a^2 - 4a - 5)$

36. $(w^2 - 2w + 1) + (2w - 5 + w^2)$

37. $(w^2 - 9w - 3) + (w - 4w^2 + 8)$

38. $(a^3 - a^2 - 5a) + (6 - a - 3a^2)$

39. $(5.76x^2 - 3.14x - 7.09) + (3.9x^2 + 1.21x + 5.6)$

40. $(8.5x^2 + 3.27x - 9.33) + (x^2 - 4.39x - 2.32)$

Perform the indicated operation. See Example 5.

41. $(x - 2) - (5x - 8)$ **42.** $(x - 7) - (3x - 1)$

43. $(m - 2) - (m + 3)$ **44.** $(m + 5) - (m + 9)$

45. $(2z^2 - 3z) - (3z^2 - 5z)$

46. $(z^2 - 4z) - (5z^2 - 3z)$

47. $(w^5 - w^3) - (-w^4 + w^2)$

48. $(w^6 - w^3) - (-w^2 + w)$

49. $(t^2 - 3t + 4) - (t^2 - 5t - 9)$

50. $(t^2 - 6t + 7) - (5t^2 - 3t - 2)$

51. $(9 - 3y - y^2) - (2 + 5y - y^2)$

52. $(4 - 5y + y^3) - (2 - 3y + y^2)$

53. $(3.55x - 879) - (26.4x - 455.8)$

54. $(345.56x - 347.4) - (56.6x + 433)$

Add or subtract the polynomials as indicated. See Examples 4 and 5.

55. Add:
$$\begin{array}{r} 3a - 4 \\ \underline{a + 6} \end{array}$$

56. Add:
$$\begin{array}{r} 2w - 8 \\ \underline{w + 3} \end{array}$$

57. Subtract:
$$\begin{array}{r} 3x + 11 \\ \underline{5x + 7} \end{array}$$

58. Subtract:
$$\begin{array}{r} 4x + 3 \\ \underline{2x + 9} \end{array}$$

59. Add:
$$\begin{array}{r} a - b \\ \underline{a + b} \end{array}$$

60. Add:
$$\begin{array}{r} s - 6 \\ \underline{s - 1} \end{array}$$

61. Subtract:
$$\begin{array}{r} -3m + 1 \\ \underline{2m - 6} \end{array}$$

62. Subtract:
$$\begin{array}{r} -5n + 2 \\ \underline{3n - 4} \end{array}$$

63. Add:
$$\begin{array}{r} 2x^2 - x - 3 \\ \underline{2x^2 + x + 4} \end{array}$$

64. Add:
$$\begin{array}{r} -x^2 + 4x - 6 \\ \underline{3x^2 - x - 5} \end{array}$$

65. Subtract:
$$\begin{array}{r} 3a^3 - 5a^2 + 7 \\ \underline{2a^3 + 4a^2 - 2a} \end{array}$$

66. Subtract:
$$\begin{array}{r} -2b^3 + 7b^2 - 9 \\ \underline{b^3 - 4b - 2} \end{array}$$

67. Subtract:
$$\begin{array}{r} x^2 - 3x + 6 \\ \underline{x^2 - 3} \end{array}$$

68. Subtract:
$$\begin{array}{r} x^4 - 3x^2 + 2 \\ \underline{3x^4 - 2x^2 } \end{array}$$

69. Add:
$$\begin{array}{r} y^3 + 4y^2 - 6y - 5 \\ \underline{y^3 + 3y^2 + 2y - 9} \end{array}$$

70. Add:
$$\begin{array}{r} q^2 - 4q + 9 \\ \underline{-3q^2 - 7q + 5} \end{array}$$

Perform the operations indicated.

71. Find the sum of $2m - 9$ and $3m + 4$.

72. Find the sum of $-3n - 2$ and $6m - 3$.

73. Find the difference when $7y - 3$ is subtracted from $9y - 2$.

74. Find the difference when $-2y - 1$ is subtracted from $3y - 4$.

75. Subtract $x^2 - 3x - 1$ from the sum of $2x^2 - x + 3$ and $x^2 + 5x - 9$.

76. Subtract $-2y^2 + 3y - 8$ from the sum of $-3y^2 - 2y + 6$ and $7y^2 + 8y - 3$.

Perform the indicated operations. Write down only the answer. See Example 6.

77. $(4m - 2) + (2m + 4) - (9m - 1)$

78. $(-5m - 6) + (8m - 3) - (-5m + 3)$

79. $(6y - 2) - (8y + 3) - (9y - 2)$

80. $(-5y - 1) - (8y - 4) - (y + 3)$

81. $(-x^2 - 5x + 4) + (6x^2 - 8x + 9) - (3x^2 - 7x + 1)$

82. $(-8x^2 + 5x - 12) + (-3x^2 - 9x + 18) - (-3x^2 + 9x - 4)$

83. $(-6z^4 - 3z^3 + 7z^2) - (5z^3 + 3z^2 - 2) + (z^4 - z^2 + 5)$

84. $(-v^3 - v^2 - 1) - (v^4 - v^2 - v - 1) + (v^3 - 3v^2 + 6)$

Solve each problem. See Example 7.

85. ***Profitable pumps.*** Walter Waterman, of Walter's Water Pumps, has found that when he produces x water pumps per month, his revenue is $x^2 + 400x + 300$ dollars. His cost for producing x water pumps per month is $x^2 + 300x - 200$ dollars. Write a polynomial that represents his monthly profit for x water pumps. Evaluate this profit polynomial for $x = 50$.

86. ***Manufacturing costs.*** Ace manufacturing has determined that the cost of labor for producing x constant velocity joints is $300x^2 + 400x - 550$ dollars, while the cost of materials is $100x^2 - 50x + 800$ dollars. Write a polynomial that represents the total cost of materials and labor for producing x constant velocity joints. Evaluate this total cost polynomial for $x = 10$.

87. ***Perimeter of a triangle.*** The shortest side of a triangle is x meters, and the other two sides are $3x - 1$ and $2x + 4$ meters. Write a polynomial that represents the perimeter and then evaluate the perimeter polynomial if x is 4 meters.

88. ***Perimeter of a rectangle.*** The width of a rectangular playground is $2x - 5$ feet, and the length is $3x + 9$ feet. Write a polynomial that represents the perimeter and then evaluate this perimeter polynomial if x is 4 feet.

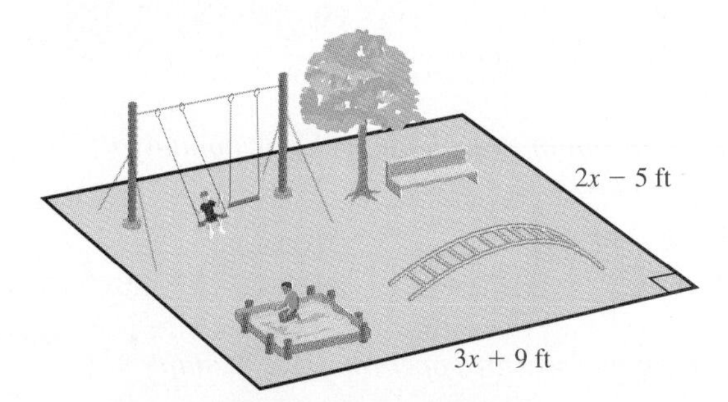

Figure for Exercise 88

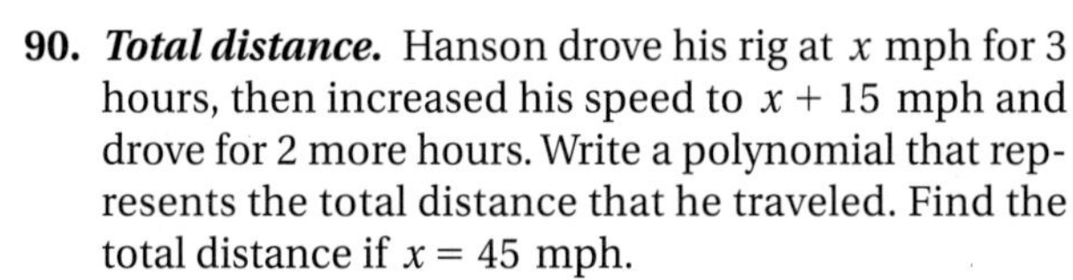

89. ***Before and after.*** Jessica traveled $2x + 50$ miles in the morning and $3x - 10$ miles in the afternoon. Write a polynomial that represents the total distance that she traveled. Find the total distance if $x = 20$.

90. ***Total distance.*** Hanson drove his rig at x mph for 3 hours, then increased his speed to $x + 15$ mph and drove for 2 more hours. Write a polynomial that represents the total distance that he traveled. Find the total distance if $x = 45$ mph.

91. ***Sky divers.*** Bob and Betty simultaneously jump from two airplanes at different altitudes. Bob's altitude t seconds after leaving the plane is $-16t^2 + 6600$ feet. Betty's altitude t seconds after leaving the plane is $-16t^2 + 7400$ feet. Write a polynomial that represents the difference between their altitudes t seconds after leaving the planes. What is the difference between their altitudes 3 seconds after leaving the planes?

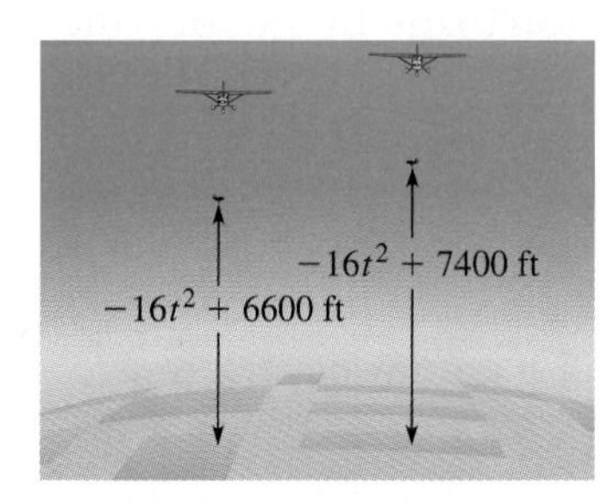

Figure for Exercise 91

92. ***Height difference.*** A red ball and a green ball are simultaneously tossed into the air. The red ball is given an initial velocity of 48 feet per second, and its height t seconds after it is tossed is $-16t^2 + 48t$ feet. The green ball is given an initial velocity of 30 feet per second, and its height t seconds after it is tossed is $-16t^2 + 30t$ feet. Find a polynomial that represents the difference in the heights of the two balls. How much higher is the red ball 2 seconds after the balls are tossed?

93. ***Total interest.*** Donald received $0.08(x + 554)$ dollars interest on one investment and $0.09(x + 335)$ interest on another investment. Write a polynomial that represents the total interest he received. What is the total interest if $x = 1000$?

94. ***Total acid.*** Deborah figured that the amount of acid in one bottle of solution is $0.12x$ milliliters and the amount of acid in another bottle of solution is $0.22(75 - x)$ milliliters. Find a polynomial that represents the total amount of acid? What is the total amount of acid if $x = 50$?

95. ***Harris-Benedict for females.*** The Harris-Benedict polynomial

$$655.1 + 9.56w + 1.85h - 4.68a$$

represents the number of calories needed to maintain a female at rest for 24 hours, where w is her weight in kilograms, h is her height in centimeters, and a is her age. Find the number of calories needed by a 30-year-old 54-kilogram female who is 157 centimeters tall.

96. ***Harris-Benedict for males.*** The Harris-Benedict polynomial

$$66.5 + 13.75w + 5.0h - 6.78a$$

represents the number of calories needed to maintain a male at rest for 24 hours, where w is his weight in kilograms, h is his height in centimeters, and a is his age. Find the number of calories needed by a 40-year-old 90-kilogram male who is 185 centimeters tall.

Getting More Involved

97. ***Discussion.*** Is the sum of two natural numbers always a natural number? Is the sum of two integers always an integer? Is the sum of two polynomials always a polynomial? Explain.

98. ***Discussion.*** Is the difference of two natural numbers always a natural number? Is the difference of two rational numbers always a rational number? Is the difference of two polynomials always a polynomial? Explain.

99. ***Writing.*** Explain why the polynomial $2^4 - 7x^3 + 5x^2 - x$ has degree 3 and not degree 4.

100. ***Discussion.*** Which of the following polynomials does not have degree 2? Explain.

a) πr^2 **b)** $\pi^2 - 4$
c) $y^2 - 4$ **d)** $x^2 - x^4$
e) $a^2 - 3a + 9$

3.2 Multiplication of Polynomials

In this section:
- Multiplying Monomials with the Product Rule
- Multiplying Polynomials
- Applications

You learned to multiply some polynomials in Chapter 1. In this section you will learn how to multiply any two polynomials.

Multiplying Monomials with the Product Rule

To multiply two monomials, such as x^3 and x^5, recall that

$$x^3 = x \cdot x \cdot x \quad \text{and} \quad x^5 = x \cdot x \cdot x \cdot x \cdot x.$$

So

$$x^3 \cdot x^5 = \underbrace{\overbrace{(x \cdot x \cdot x)}^{\text{3 factors}}\overbrace{(x \cdot x \cdot x \cdot x \cdot x)}^{\text{5 factors}}}_{\text{8 factors}} = x^8.$$

The exponent of the product of x^3 and x^5 is the sum of the exponents 3 and 5. This example illustrates the **product rule** for multiplying exponential expressions.

> **Product Rule**
>
> If a is any real number and m and n are any positive integers, then
>
> $$a^m \cdot a^n = a^{m+n}.$$

EXAMPLE 1 **Multiplying monomials**

Find the indicated products.

a) $x^2 \cdot x^4 \cdot x$ **b)** $(-2ab)(-3ab)$ **c)** $-4x^2y^2 \cdot 3xy^5$ **d)** $(3a)^2$

Solution

a) $x^2 \cdot x^4 \cdot x = x^2 \cdot x^4 \cdot x^1$

$= x^7$ Product rule

b) $(-2ab)(-3ab) = (-2)(-3) \cdot a \cdot a \cdot b \cdot b$

$= 6a^2b^2$ Product rule

c) $(-4x^2y^2)(3xy^5) = (-4)(3)x^2 \cdot x \cdot y^2 \cdot y^5$

$= -12x^3y^7$ Product rule

d) $(3a)^2 = 3a \cdot 3a$

$= 9a^2$

CAUTION Be sure to distinguish between adding and multiplying monomials. You can add like terms to get $3x^4 + 2x^4 = 5x^4$, but you cannot combine the terms in $3w^5 + 6w^2$. However, you can multiply any two monomials: $3x^4 \cdot 2x^4 = 6x^8$ and $3w^5 \cdot 6w^2 = 18w^7$.

Multiplying Polynomials

To multiply a monomial and a polynomial, we use the distributive property.

EXAMPLE 2 **Multiplying monomials and polynomials**

Find each product.

a) $3x^2(x^3 - 4x)$ **b)** $(y^2 - 3y + 4)(-2y)$ **c)** $-a(b - c)$

Solution

a) $3x^2(x^3 - 4x) = 3x^2 \cdot x^3 - 3x^2 \cdot 4x$ Distributive property

$= 3x^5 - 12x^3$

b) $(y^2 - 3y + 4)(-2y) = y^2(-2y) - 3y(-2y) + 4(-2y)$ Distributive property

$= -2y^3 - (-6y^2) + (-8y)$

$= -2y^3 + 6y^2 - 8y$

c) $-a(b - c) = (-a)b - (-a)c$ Distributive property

$= -ab + ac$

$= ac - ab$

Note in part (c) that either of the last two binomials is the correct answer. The last one is just a little simpler to read. ■

Just as we use the distributive property to find the product of a monomial and a polynomial, we can use the distributive property to find the product of two binomials and the product of a binomial and a trinomial.

EXAMPLE 3 **Using the distributive property**

Use the distributive property to find each product.

a) $(x + 2)(x + 5)$ **b)** $(x + 3)(x^2 + 2x - 7)$

Solution

a) First multiply $x + 2$ by each term of $x + 5$:

$(x + 2)(x + 5) = (x + 2)x + (x + 2)5$ Distributive property

$= x^2 + 2x + 5x + 10$ Distributive property

$= x^2 + 7x + 10$ Combine like terms.

b) First multiply $x + 3$ by each term of the trinomial:

$(x + 3)(x^2 + 2x - 7) = (x + 3)x^2 + (x + 3)2x + (x + 3)(-7)$ Distributive property

$= x^3 + 3x^2 + 2x^2 + 6x - 7x - 21$ Distributive property

$= x^3 + 5x^2 - x - 21$ Combine like terms. ■

Products of polynomials can also be found by arranging the multiplication vertically like multiplication of whole numbers.

EXAMPLE 4 **Multiplying vertically**

Find each product.

a) $(x - 2)(3x + 7)$ **b)** $(x + y)(a + 3)$

Solution

a)

$$\begin{array}{r} 3x + 7 \\ x - 2 \\ \hline -6x - 14 \\ 3x^2 + 7x \qquad \\ \hline 3x^2 + x - 14 \end{array}$$

$\leftarrow -2$ times $3x + 7$

$\leftarrow x$ times $3x + 7$

Add.

b)

$$\begin{array}{r} x + y \\ a + 3 \\ \hline 3x + 3y \\ ax + ay \qquad \qquad \\ \hline ax + ay + 3x + 3y \end{array}$$

These examples illustrate the following rule.

Multiplication of Polynomials

To multiply polynomials, multiply each term of the first polynomial by every term of the second polynomial, then combine like terms.

Note the result of multiplying the difference $a - b$ by -1:

$$-1(a - b) = -a + b = b - a$$

Because multiplying by -1 is the same as taking the opposite, we can write

$$-(a - b) = b - a.$$

So $a - b$ and $b - a$ are opposites or additive inverses of each other. If a and b are replaced by numbers, the values of $a - b$ and $b - a$ are additive inverses. For example, $3 - 7 = -4$ and $7 - 3 = 4$.

CAUTION The opposite of $a + b$ is $-a - b$, *not* $a - b$.

EXAMPLE 5 **Opposite of a polynomial**

Find the opposite of each polynomial.

a) $x - 2$ **b)** $9 - y^2$ **c)** $a + 4$ **d)** $-x^2 + 6x - 3$

Solution

a) $-(x - 2) = 2 - x$

b) $-(9 - y^2) = y^2 - 9$

c) $-(a + 4) = -a - 4$

d) $-(-x^2 + 6x - 3) = x^2 - 6x + 3$

Applications

EXAMPLE 6 **Multiplying polynomials**

A commuter parking lot is 20 yards wide and 30 yards long. If the university increases the length and width by the same amount to handle an increasing number of commuters, then what polynomial represents the area of the new lot? What is the new area if the increase is 15 yards?

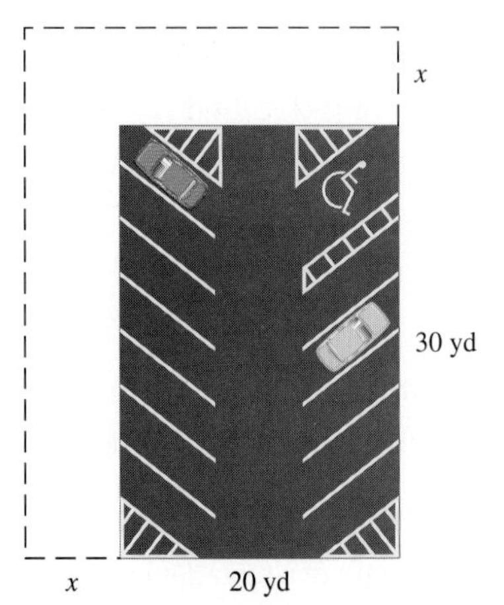

Figure 3.1

Solution

If x is the amount of increase, then the new lot will be $x + 20$ yards wide and $x + 30$ yards long as shown in Fig. 3.1. Multiply the length and width to get the area:

$$\begin{aligned}(x + 20)(x + 30) &= (x + 20)x + (x + 20)30 \\ &= x^2 + 20x + 30x + 600 \\ &= x^2 + 50x + 600\end{aligned}$$

The polynomial $x^2 + 50x + 600$ represents the area of the new lot. If $x = 15$, then

$$x^2 + 50x + 600 = (15)^2 + 50(15) + 600 = 1575.$$

If the increase is 15 yards, then the area of the lot will be 1575 square yards.

Warm-ups

True or false? Explain your answer.

1. $3x^3 \cdot 5x^4 = 15x^{12}$ for any value of x.
2. $3x^2 \cdot 2x^7 = 5x^9$ for any value of x.
3. $(3y^3)^2 = 9y^6$ for any value of y.
4. $-3x(5x - 7x^2) = -15x^3 + 21x^2$ for any value of x.
5. $2x(x^2 - 3x + 4) = 2x^3 - 6x^2 + 8x$ for any number x.
6. $-2(3 - x) = 2x - 6$ for any number x.
7. $(a + b)(c + d) = ac + ad + bc + bd$ for any values of a, b, c, and d.
8. $-(x - 7) = 7 - x$ for any value of x.
9. $83 - 37 = -(37 - 83)$
10. The opposite of $x + 3$ is $x - 3$ for any number x.

3.2 EXERCISES

Find each product. See Example 1.

1. $3x^2 \cdot 9x^3$
2. $5x^7 \cdot 3x^5$
3. $2a^3 \cdot 7a^8$
4. $3y^{12} \cdot 5y^{15}$
5. $-6x^2 \cdot 5x^2$
6. $-2x^2 \cdot 8x^5$
7. $(-9x^{10})(-3x^7)$
8. $(-2x^2)(-8x^9)$
9. $-6st \cdot 9st$
10. $-12sq \cdot 3s$
11. $3wt \cdot 8w^7t^6$
12. $h^8k^3 \cdot 5h$
13. $(5y)^2$
14. $(6x)^2$
15. $(2x^3)^2$
16. $(3y^5)^2$

Find each product. See Example 2.

17. $4y^2(y^5 - 2y)$
18. $6t^3(t^5 + 3t^2)$
19. $-3y(6y - 4)$
20. $-9y(y^2 - 1)$
21. $(y^2 - 5y + 6)(-3y)$
22. $(x^3 - 5x^2 - 1)7x^2$
23. $-x(y^2 - x^2)$
24. $-ab(a^2 - b^2)$
25. $(3ab^3 - a^2b^2 - 2a^3b)5a^3$
26. $(3c^2d - d^3 + 1)8cd^2$
27. $-\frac{1}{2}t^2v(4t^3v^2 - 6tv - 4v)$
28. $-\frac{1}{3}m^2n^3(-6mn^2 + 3mn - 12)$

Use the distributive property to find each product. See Example 3.

29. $(x + 1)(x + 2)$

30. $(x + 6)(x + 3)$

31. $(x - 3)(x + 5)$

32. $(y - 2)(y + 4)$

33. $(t - 4)(t - 9)$

34. $(w - 3)(w - 5)$

35. $(x + 1)(x^2 + 2x + 2)$

36. $(x - 1)(x^2 + x + 1)$

37. $(3y + 2)(2y^2 - y + 3)$

38. $(4y + 3)(y^2 + 3y + 1)$

39. $(y^2z - 2y^4)(y^2z + 3z^2 - y^4)$

40. $(m^3 - 4mn^2)(6m^4n^2 - 3m^6 + m^2n^4)$

Find each product vertically. See Example 4.

41. $\begin{array}{r} 2a - 3 \\ \underline{a + 5} \end{array}$

42. $\begin{array}{r} 2w - 6 \\ \underline{w + 5} \end{array}$

43. $\begin{array}{r} 7x + 30 \\ \underline{2x + 5} \end{array}$

44. $\begin{array}{r} 5x + 7 \\ \underline{3x + 6} \end{array}$

45. $\begin{array}{r} 5x + 2 \\ \underline{4x - 3} \end{array}$

46. $\begin{array}{r} 4x + 3 \\ \underline{2x - 6} \end{array}$

47. $\begin{array}{r} m - 3n \\ \underline{2a + b} \end{array}$

48. $\begin{array}{r} 3x + 7 \\ \underline{a - 2b} \end{array}$

49. $\begin{array}{r} x^2 + 3x - 2 \\ \underline{x + 6} \end{array}$

50. $\begin{array}{r} -x^2 + 3x - 5 \\ \underline{x - 7} \end{array}$

51. $\begin{array}{r} 2a^3 - 3a^2 + 4 \\ \underline{-2a - 3} \end{array}$

52. $\begin{array}{r} -3x^2 + 5x - 2 \\ \underline{-5x - 6} \end{array}$

53. $\begin{array}{r} x - y \\ \underline{x + y} \end{array}$

54. $\begin{array}{r} a^2 + b^2 \\ \underline{a^2 - b^2} \end{array}$

55. $\begin{array}{r} x^2 - xy + y^2 \\ \underline{x + y} \end{array}$

56. $\begin{array}{r} 4w^2 + 2wv + v^2 \\ \underline{2w - v} \end{array}$

Find the opposite of each polynomial. See Example 5.

57. $3t - u$

58. $-3t - u$

59. $3x + y$

60. $x - 3y$

61. $-3a^2 - a + 6$

62. $3b^2 - b - 6$

63. $3v^2 + v - 6$

64. $-3t^2 + t - 6$

Perform the indicated operations.

65. $-3x(2x - 9)$

66. $-1(2 - 3x)$

67. $2 - 3x(2x - 9)$

68. $6 - 3(4x - 8)$

69. $(2 - 3x) + (2x - 9)$

70. $(2 - 3x) - (2x - 9)$

71. $(6x^6)^2$

72. $(-3a^3b)^2$

73. $3ab^3(-2a^2b^7)$

74. $-4xst \cdot 8xs$

75. $(5x + 6)(5x + 6)$

76. $(5x - 6)(5x - 6)$

77. $(5x - 6)(5x + 6)$

78. $(2x - 9)(2x + 9)$

79. $2x^2(3x^5 - 4x^2)$

80. $4a^3(3ab^3 - 2ab^3)$

81. $(m - 1)(m^2 + m + 1)$

82. $(a + b)(a^2 - ab + b^2)$

83. $(3x - 2)(x^2 - x - 9)$

84. $(5 - 6y)(3y^2 - y - 7)$

Solve each problem. See Example 6.

85. ***Office space.*** The length of a professor's office is x feet, and the width is $x + 4$ feet. Write a polynomial that represents the area. Find the area if $x = 10$ ft.

86. ***Swimming space.*** The length of a rectangular swimming pool is $2x - 1$ meters, and the width is $x + 2$ meters. Write a polynomial that represents the area. Find the area if x is 5 meters.

87. ***Area of a truss.*** A roof truss is in the shape of a triangle with a height of x feet and a base of $2x + 1$ feet. Write a polynomial that represents the area of the triangle. What is the area if x is 5 feet?

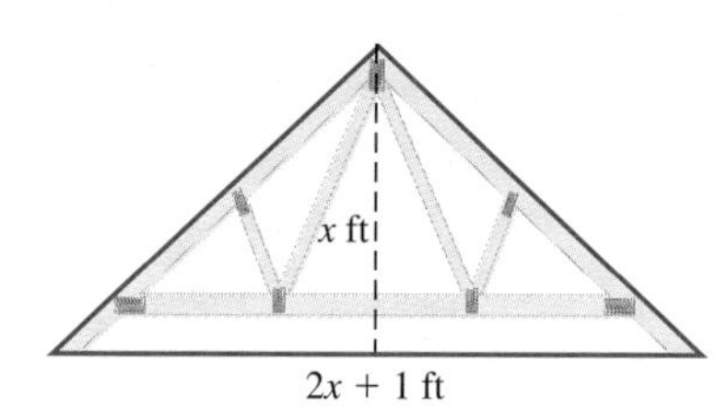

Figure for Exercise 87

88. ***Volume of a box.*** The length, width, and height of a box are x, $2x$, and $3x - 5$ inches, respectively. Write a polynomial that represents its volume.

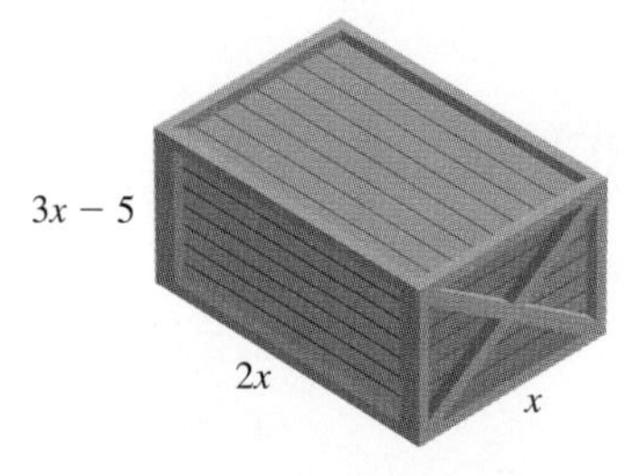

Figure for Exercise 88

89. ***Number pairs.*** If two numbers differ by 5, then what polynomial represents their product?

90. ***Number pairs.*** If two numbers have a sum of 9, then what polynomial represents their product?

91. ***Area of a rectangle.*** The length of a rectangle is $2.3x + 1.2$ meters, and its width is $3.5x + 5.1$ meters. What polynomial represents its area?

92. ***Patchwork.*** A quilt patch cut in the shape of a triangle has a base of $5x$ inches and a height of $1.732x$ inches. What polynomial represents its area?

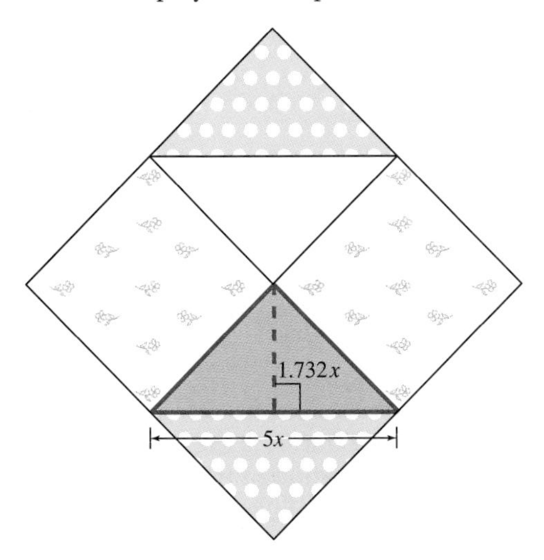

Figure for Exercise 92

93. ***Total revenue.*** If a promoter charges p dollars per ticket for a concert in Tulsa, then she expects to sell $40{,}000 - 1000p$ tickets to the concert. How many tickets will she sell if the tickets are \$10 each? Find the total revenue when the tickets are \$10 each. What polynomial represents the total revenue expected for the concert when the tickets are p dollars each?

94. ***Manufacturing shirts.*** If a manufacturer charges p dollars each for rugby shirts, then he expects to sell $2000 - 100p$ shirts per week. What polynomial represents the total revenue expected for a week? How many shirts will be sold if the manufacturer charges \$20 each for the shirts? Find the total revenue when the shirts are sold for \$20 each. Use the bar graph to determine the price that will give the maximum total revenue.

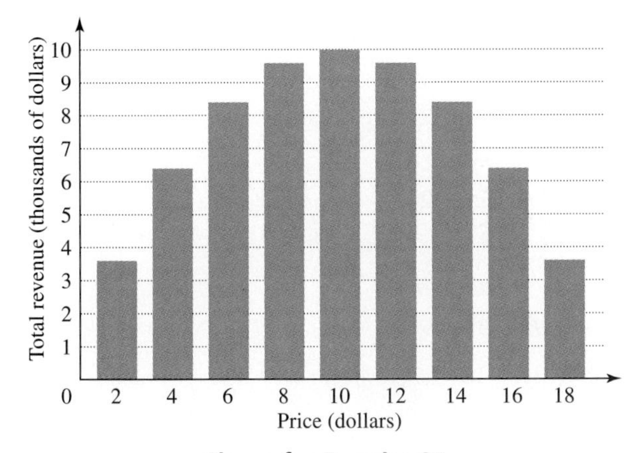

Figure for Exercise 94

95. ***Periodic deposits.*** At the beginning of each year for 5 years, an investor invests \$10 in a mutual fund with an average annual return of r. If we let $x = 1 + r$, then at the end of the first year (just before the next investment) the value is $10x$ dollars. Because \$10 is then added to the $10x$ dollars, the amount at the end of the second year is $(10x + 10)x$ dollars. Find a polynomial that represents the value of the investment at the end of the fifth year. Evaluate this polynomial if $r = 10\%$.

96. ***Increasing deposits.*** At the beginning of each year for 5 years, an investor invests in a mutual fund with an average annual return of r. The first year, she invests \$10; the second year, she invests \$20; the third year, she invests \$30; the fourth year, she invests \$40; the fifth year, she invests \$50. Let $x = 1 + r$ as in Exercise 95 and write a polynomial in x that represents the value of the investment at the end of the fifth year. Evaluate this polynomial for $r = 8\%$.

Getting More Involved

97. ***Discussion.*** Name all properties of the real numbers that are used in finding the following products:

a) $-2ab^3c^2 \cdot 5a^2bc$ **b)** $(x^2 + 3)(x^2 - 8x - 6)$

98. ***Discussion.*** Find the product of 27 and 436 without using a calculator. Then use the distributive property to find the product $(20 + 7)(400 + 30 + 6)$ as you would find the product of a binomial and a trinomial. Explain how the two methods are related.

3.3 Multiplication of Binomials

In this section:

- The FOIL Method
- Multiplying Binomials Quickly

In Section 3.2 you learned to multiply polynomials. In this section you will learn a rule that makes multiplication of binomials simpler.

The FOIL Method

We can use the distributive property to find the product of two binomials. For example,

$$
\begin{aligned}
(x+2)(x+3) &= x(x+3) + 2(x+3) && \text{Distributive property}\\
&= x^2 + 3x + 2x + 6 && \text{Distributive property}\\
&= x^2 + 5x + 6 && \text{Combine like terms.}
\end{aligned}
$$

There are four terms in $x^2 + 3x + 2x + 6$. The term x^2 is the product of the *first* term of each binomial, x and x. The term $3x$ is the product of the two *outer* terms, 3 and x. The term $2x$ is the product of the two *inner* terms, 2 and x. The term 6 is the product of the last term of each binomial, 2 and 3. We can connect the terms multiplied by lines as follows:

$$(x+2)(x+3)$$

F = First terms
O = Outer terms
I = Inner terms
L = Last terms

If you remember the word FOIL, you can get the product of the two binomials much faster than writing out all of the steps above. This method is called the **FOIL method.** The name should make it easier to remember.

EXAMPLE 1 **Using the FOIL method**

Find each product.

a) $(x+2)(x-4)$ **b)** $(2x+5)(3x-4)$
c) $(a-b)(2a-b)$ **d)** $(x+3)(y+5)$

Solution

a)
$$
\begin{aligned}
(x+2)(x-4) &= \overset{F}{x^2} \overset{O}{-\,4x} \overset{I}{+\,2x} \overset{L}{-\,8}\\
&= x^2 - 2x - 8 && \text{Combine the like terms.}
\end{aligned}
$$

b)
$$
\begin{aligned}
(2x+5)(3x-4) &= 6x^2 - 8x + 15x - 20\\
&= 6x^2 + 7x - 20 && \text{Combine the like terms.}
\end{aligned}
$$

c)
$$
\begin{aligned}
(a-b)(2a-b) &= 2a^2 - ab - 2ab + b^2\\
&= 2a^2 - 3ab + b^2
\end{aligned}
$$

d) $(x+3)(y+5) = xy + 5x + 3y + 15$ There are no like terms to combine. ■

FOIL can be used to multiply any two binomials. The binomials in the next example have higher powers than those of Example 1.

EXAMPLE 2 **Using the FOIL method**

Find each product.

a) $(x^3 - 3)(x^3 + 6)$ **b)** $(2a^2 + 1)(a^2 + 5)$

Solution

a) $(x^3 - 3)(x^3 + 6) = x^6 + 6x^3 - 3x^3 - 18$

$= x^6 + 3x^3 - 18$

b) $(2a^2 + 1)(a^2 + 5) = 2a^4 + 10a^2 + a^2 + 5$

$= 2a^4 + 11a^2 + 5$

Multiplying Binomials Quickly

The outer and inner products in the FOIL method are often like terms, and we can combine them without writing them down. Once you become proficient at using FOIL, you can find the product of two binomials without writing anything except the answer.

EXAMPLE 3 **Using FOIL to find a product quickly**

Find each product. Write down only the answer.

a) $(x + 3)(x + 4)$ **b)** $(2x - 1)(x + 5)$ **c)** $(a - 6)(a + 6)$

Solution

a) $(x + 3)(x + 4) = x^2 + 7x + 12$ Combine like terms: $3x + 4x = 7x$.

b) $(2x - 1)(x + 5) = 2x^2 + 9x - 5$ Combine like terms: $10x - x = 9x$.

c) $(a - 6)(a + 6) = a^2 - 36$ Combine like terms: $6a - 6a = 0$.

Warm-ups

True or false? Answer true only if the equation is true for all values of the variable or variables. Explain your answer.

1. $(x + 3)(x + 2) = x^2 + 6$

2. $(x + 2)(y + 1) = xy + x + 2y + 2$

3. $(3a - 5)(2a + 1) = 6a^2 + 3a - 10a - 5$

4. $(y + 3)(y - 2) = y^2 + y - 6$

5. $(x^2 + 2)(x^2 + 3) = x^4 + 5x^2 + 6$

6. $(3a^2 - 2)(3a^2 + 2) = 9a^2 - 4$

7. $(t + 3)(t + 5) = t^2 + 8t + 15$

8. $(y - 9)(y - 2) = y^2 - 11y - 18$

9. $(x + 4)(x - 7) = x^2 + 4x - 28$

10. It is not necessary to learn FOIL as long as you can get the answer.

3.3 EXERCISES

Use FOIL to find each product. See Example 1.

1. $(x + 2)(x + 4)$

2. $(x + 3)(x + 5)$

3. $(a - 3)(a + 2)$

4. $(b - 1)(b + 2)$

5. $(2x - 1)(x - 2)$

6. $(2y - 5)(y - 2)$

7. $(2a - 3)(a + 1)$

8. $(3x - 5)(x + 4)$

9. $(w - 50)(w - 10)$

10. $(w - 30)(w - 20)$

11. $(y - a)(y + 5)$

12. $(a + t)(3 - y)$

13. $(5 - w)(w + m)$

14. $(a - h)(b + t)$

15. $(2m - 3t)(5m + 3t)$

16. $(2x - 5y)(x + y)$

17. $(5a + 2b)(9a + 7b)$

18. $(11x + 3y)(x + 4y)$

Use FOIL to find each product. See Example 2.

19. $(x^2 - 5)(x^2 + 2)$

20. $(y^2 + 1)(y^2 - 2)$

21. $(h^3 + 5)(h^3 + 5)$

22. $(y^6 + 1)(y^6 - 4)$

23. $(3b^3 + 2)(b^3 + 4)$

24. $(5n^4 - 1)(n^4 + 3)$

25. $(y^2 - 3)(y - 2)$

26. $(x - 1)(x^2 - 1)$

27. $(3m^3 - n^2)(2m^3 + 3n^2)$

28. $(6y^4 - 2z^2)(6y^4 - 3z^2)$

29. $(3u^2v - 2)(4u^2v + 6)$

30. $(5y^3w^2 + z)(2y^3w^2 + 3z)$

Find each product. Try to write only the answer. See Example 3.

31. $(b + 4)(b + 5)$

32. $(y + 8)(y + 4)$

33. $(x - 3)(x + 9)$

34. $(m + 7)(m - 8)$

35. $(a + 5)(a + 5)$

36. $(t - 4)(t - 4)$

37. $(2x - 1)(2x - 1)$

38. $(3y + 4)(3y + 4)$

39. $(z - 10)(z + 10)$

40. $(3h - 5)(3h + 5)$

41. $(a + b)(a + b)$

42. $(x - y)(x - y)$

43. $(a - 1)(a - 2)$

44. $(b - 8)(b - 1)$

45. $(2x - 1)(x + 3)$

46. $(3y + 5)(y - 3)$

47. $(5t - 2)(t - 1)$

48. $(2t - 3)(2t - 1)$

49. $(h - 7)(h - 9)$

50. $(h - 7w)(h - 7w)$

51. $(h + 7w)(h + 7w)$

52. $(h - 7q)(h + 7q)$

53. $(2h^2 - 1)(2h^2 - 1)$

54. $(3h^2 + 1)(3h^2 + 1)$

Perform the indicated operations.

55. $\left(2a + \frac{1}{2}\right)\left(4a - \frac{1}{2}\right)$

56. $\left(3b + \frac{2}{3}\right)\left(6b - \frac{1}{3}\right)$

57. $\left(\frac{1}{2}x - \frac{1}{3}\right)\left(\frac{1}{4}x + \frac{1}{2}\right)$

58. $\left(\frac{2}{3}t - \frac{1}{4}\right)\left(\frac{1}{2}t - \frac{1}{2}\right)$

59. $-2x^4(3x - 1)(2x + 5)$

60. $4xy^3(2x - y)(3x + y)$

61. $(x - 1)(x + 1)(x + 3)$

62. $(a - 3)(a + 4)(a - 5)$

63. $(3x - 2)(3x + 2)(x + 5)$

64. $(x - 6)(9x + 4)(9x - 4)$

65. $(x - 1)(x + 2) - (x + 3)(x - 4)$

66. $(k - 4)(k + 9) - (k - 3)(k + 7)$

Solve each problem.

67. ***Area of a rug.*** Find a trinomial that represents the area of a rectangular rug whose sides are $x + 3$ feet and $2x - 1$ feet.

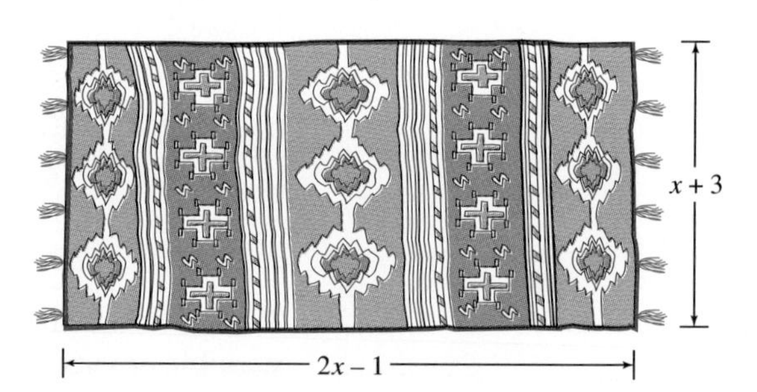

Figure for Exercise 67

68. ***Area of a parallelogram.*** Find a trinomial that represents the area of a parallelogram whose base is $3x + 2$ meters and whose height is $2x + 3$ meters.

69. ***Area of a sail.*** The sail of a tall ship is triangular in shape with a base of $4.57x + 3$ meters and a height of $2.3x - 1.33$ meters. Find a polynomial that represents the area of the triangle.

70. ***Area of a square.*** A square has a side of length $1.732x + 1.414$ meters. Find a polynomial that represents its area.

Getting More Involved

71. ***Exploration.*** Find the area of each of the four regions shown in the figure. What is the total area of the four regions? What does this exercise illustrate?

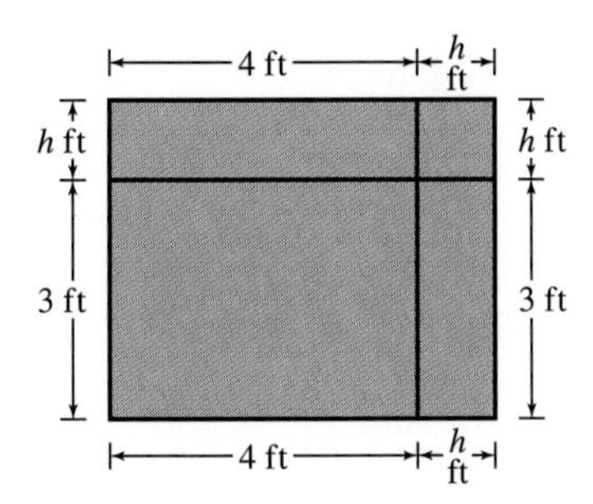

Figure for Exercise 71

72. ***Exploration.*** Find the area of each of the four regions shown in the figure. What is the total area of the four regions? What does this exercise illustrate?

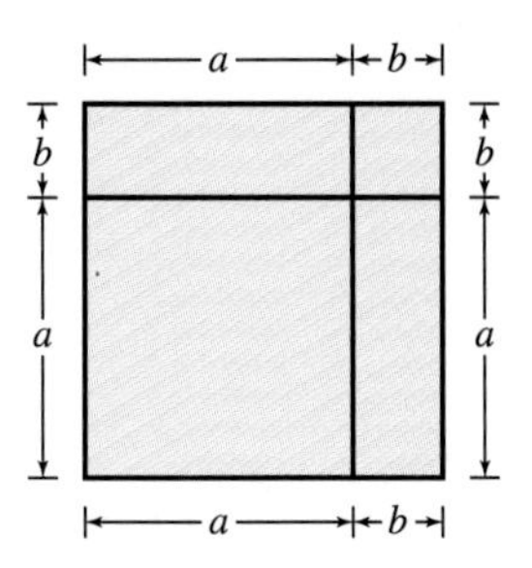

Figure for Exercise 72

3.4 Special Products

In this section:

- The Square of a Binomial
- Product of a Sum and a Difference
- Higher Powers of Binomials

In Section 3.3 you learned the FOIL method to make multiplying binomials simpler. In this section you will learn rules for squaring binomials and for finding the product of a sum and a difference. These products are called **special products.**

The Square of a Binomial

To compute $(a + b)^2$, the square of a binomial, we can write it as $(a + b)(a + b)$ and use FOIL:

$$\begin{aligned}(a + b)^2 &= (a + b)(a + b)\\ &= a^2 + ab + ab + b^2\\ &= a^2 + 2ab + b^2\end{aligned}$$

So to square $a + b$, *we square the first term* (a^2), *add twice the product of the two terms* $(2ab)$, *then add the square of the last term* (b^2). The square of a binomial occurs so frequently that it is helpful to learn this new rule to find it. The rule for squaring a sum is given symbolically as follows.

The Square of a Sum

$$(a + b)^2 = a^2 + 2ab + b^2$$

EXAMPLE 1 **Using the rule for squaring a sum**

Find the square of each sum.

a) $(x + 3)^2$ **b)** $(2a + 5)^2$

Solution

a) $(x + 3)^2 = x^2 + 2(x)(3) + 3^2 = x^2 + 6x + 9$

Square of first (x^2); Twice the product ($2(x)(3)$); Square of last (3^2)

b)
$$(2a + 5)^2 = (2a)^2 + 2(2a)(5) + 5^2$$
$$= 4a^2 + 20a + 25$$

CAUTION Do not forget the middle term when squaring a sum. The equation $(x + 3)^2 = x^2 + 6x + 9$ is an identity, but $(x + 3)^2 = x^2 + 9$ is not an identity. For example, if $x = 1$ in $(x + 3)^2 = x^2 + 9$, then we get $4^2 = 1^2 + 9$, which is false.

When we use FOIL to find $(a - b)^2$, we see that

$$(a - b)^2 = (a - b)(a - b)$$
$$= a^2 - ab - ab + b^2$$
$$= a^2 - 2ab + b^2.$$

So to square $a - b$, *we square the first term* (a^2), *subtract twice the product of the two terms* $(-2ab)$, *and add the square of the last term* (b^2). The rule for squaring a difference is given symbolically as follows.

The Square of a Difference

$$(a - b)^2 = a^2 - 2ab + b^2$$

EXAMPLE 2 **Using the rule for squaring a difference**

Find the square of each difference.

a) $(x - 4)^2$ **b)** $(4b - 5y)^2$

Solution

a)
$$(x - 4)^2 = x^2 - 2(x)(4) + 4^2$$
$$= x^2 - 8x + 16$$

b)
$$(4b - 5y)^2 = (4b)^2 - 2(4b)(5y) + (5y)^2$$
$$= 16b^2 - 40by + 25y^2$$

Product of a Sum and a Difference

If we multiply the sum $a + b$ and the difference $a - b$ by using FOIL, we get

$$\begin{aligned}(a + b)(a - b) &= a^2 - ab + ab - b^2 \\ &= a^2 - b^2.\end{aligned}$$

The inner and outer products have a sum of 0. So *the product of a sum and a difference of the same two terms is equal to the difference of two squares.*

The Product of a Sum and a Difference

$$(a + b)(a - b) = a^2 - b^2$$

EXAMPLE 3 **Product of a sum and a difference**

Find each product.

a) $(x + 2)(x - 2)$ **b)** $(b + 7)(b - 7)$ **c)** $(3x - 5)(3x + 5)$

Solution

a) $(x + 2)(x - 2) = x^2 - 4$

b) $(b + 7)(b - 7) = b^2 - 49$

c) $(3x - 5)(3x + 5) = 9x^2 - 25$ ■

Higher Powers of Binomials

To find a power of a binomial that is higher than 2, we can use the rule for squaring a binomial along with the method of multiplying binomials using the distributive property.

EXAMPLE 4 **Higher powers of a binomial**

Find each product.

a) $(x + 4)^3$ **b)** $(y - 2)^4$

Solution

a)
$$\begin{aligned}(x + 4)^3 &= (x + 4)^2(x + 4) \\ &= (x^2 + 8x + 16)(x + 4) \\ &= (x^2 + 8x + 16)x + (x^2 + 8x + 16)4 \\ &= x^3 + 8x^2 + 16x + 4x^2 + 32x + 64 \\ &= x^3 + 12x^2 + 48x + 64\end{aligned}$$

b)
$$\begin{aligned}(y - 2)^4 &= (y - 2)^2(y - 2)^2 \\ &= (y^2 - 4y + 4)(y^2 - 4y + 4) \\ &= (y^2 - 4y + 4)(y^2) + (y^2 - 4y + 4)(-4y) + (y^2 - 4y + 4)(4) \\ &= y^4 - 8y^3 + 24y^2 - 32y + 16\end{aligned}$$
■

Warm-ups

True or false? Explain your answer.

1. $(2 + 3)^2 = 2^2 + 3^2$

2. $(x + 3)^2 = x^2 + 6x + 9$ for any value of x.

3. $(3 + 5)^2 = 9 + 25 + 30$

4. $(2x + 7)^2 = 4x^2 + 28x + 49$ for any value of x.

5. $(y + 8)^2 = y^2 + 64$ for any value of y.

6. The product of a sum and a difference of the same two terms is equal to the difference of two squares.

7. $(40 - 1)(40 + 1) = 1599$

8. $49 \cdot 51 = 2499$

9. $(x - 3)^2 = x^2 - 3x + 9$ for any value of x.

10. The square of a sum is equal to a sum of two squares.

3.4 EXERCISES

Square each binomial. See Example 1.

1. $(x + 1)^2$ **2.** $(y + 2)^2$ **3.** $(y + 4)^2$

4. $(z + 3)^2$ **5.** $(3x + 8)^2$ **6.** $(2m + 7)^2$

7. $(s + t)^2$ **8.** $(x + z)^2$ **9.** $(2x + y)^2$

10. $(3t + v)^2$ **11.** $(2t + 3h)^2$ **12.** $(3z + 5k)^2$

Square each binomial. See Example 2.

13. $(a - 3)^2$ **14.** $(w - 4)^2$ **15.** $(t - 1)^2$

16. $(t - 6)^2$ **17.** $(3t - 2)^2$ **18.** $(5a - 6)^2$

19. $(s - t)^2$ **20.** $(r - w)^2$ **21.** $(3a - b)^2$

22. $(4w - 7)^2$ **23.** $(3z - 5y)^2$ **24.** $(2z - 3w)^2$

Find each product. See Example 3.

25. $(a - 5)(a + 5)$ **26.** $(x - 6)(x + 6)$

27. $(y - 1)(y + 1)$ **28.** $(p + 2)(p - 2)$

29. $(3x - 8)(3x + 8)$ **30.** $(6x + 1)(6x - 1)$

31. $(r + s)(r - s)$ **32.** $(b - y)(b + y)$

33. $(8y - 3a)(8y + 3a)$ **34.** $(4u - 9v)(4u + 9v)$

35. $(5x^2 - 2)(5x^2 + 2)$ **36.** $(3y^2 + 1)(3y^2 - 1)$

Find each product. See Example 4.

37. $(x + 1)^3$ **38.** $(y - 1)^3$ **39.** $(2a - 3)^3$

40. $(3w - 1)^3$ **41.** $(a - 3)^4$ **42.** $(2b + 1)^4$

43. $(a + b)^4$ **44.** $(2a - 3b)^4$

Find each product.

45. $(a - 20)(a + 20)$ **46.** $(1 - x)(1 + x)$

47. $(x + 8)(x + 7)$ **48.** $(x - 9)(x + 5)$

49. $(4x - 1)(4x + 1)$ **50.** $(9y - 1)(9y + 1)$

51. $(9y - 1)^2$ **52.** $(4x - 1)^2$

53. $(2t - 5)(3t + 4)$ **54.** $(2t + 5)(3t - 4)$

55. $(2t - 5)^2$ **56.** $(2t + 5)^2$

57. $(2t + 5)(2t - 5)$ **58.** $(3t - 4)(3t + 4)$

59. $(x^2 - 1)(x^2 + 1)$ **60.** $(y^3 - 1)(y^3 + 1)$

61. $(2y^3 - 9)^2$ **62.** $(3z^4 - 8)^2$

63. $(2x^3 + 3y^2)^2$ **64.** $(4y^5 + 2w^3)^2$

65. $\left(\frac{1}{2}x + \frac{1}{3}\right)^2$ **66.** $\left(\frac{2}{3}y - \frac{1}{2}\right)^2$

67. $(0.2x - 0.1)^2$ **68.** $(0.1y + 0.5)^2$

69. $(a + b)^3$ **70.** $(2a - 3b)^3$

71. $(1.5x + 3.8)^2$ **72.** $(3.45a - 2.3)^2$

73. $(3.5t - 2.5)(3.5t + 2.5)$

74. $(4.5h + 5.7)(4.5h - 5.7)$

In Exercises 75–84, solve each problem.

75. ***Shrinking garden.*** Rose's garden is a square with sides of length x feet. Next spring she plans to make it rectangular by lengthening one side 5 feet and shortening the other side by 5 feet. What polynomial represents the new area? By how much will the area of the new garden differ from that of the old garden?

76. ***Square lot.*** Sam lives on a lot that he thought was a square, 157 feet by 157 feet. When he had it surveyed, he discovered that one side was actually 2 feet longer than he thought and the other was actually 2 feet shorter than he thought. How much less area does he have than he thought he had?

77. ***Area of a circle.*** Find a polynomial that represents the area of a circle whose radius is $b + 1$ meters. Use the value 3.14 for π.

78. ***Comparing dart boards.*** A toy store sells two sizes of circular dartboards. The larger of the two has a radius that is 3 inches greater than that of the other. The radius of the smaller dartboard is t inches. Find a polynomial that represents the difference in area between the two dartboards?

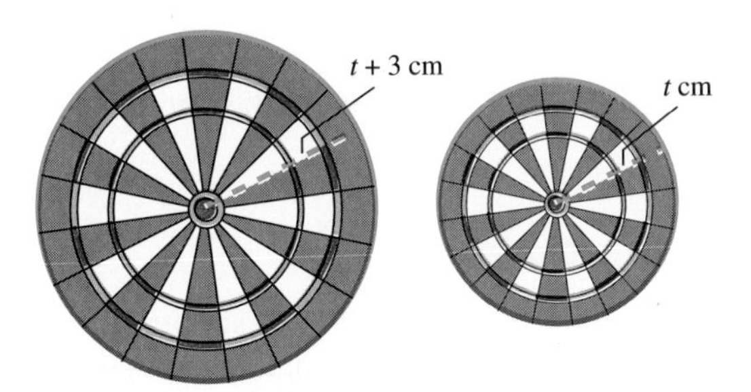

Figure for Exercise 78

79. ***Poiseuille's law.*** According to the nineteenth-century physician Poiseuille, the velocity (in centimeters per second) of blood r centimeters from the center of an artery of radius R centimeters is given by

$$v = k(R - r)(R + r),$$

where k is a constant. Rewrite the formula using a special product rule.

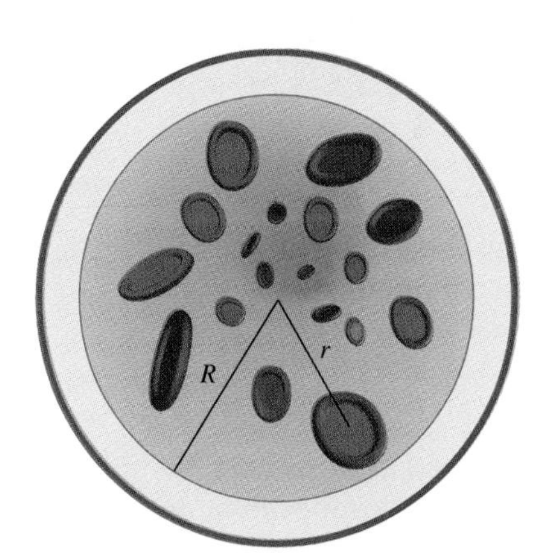

Figure for Exercise 79

80. ***Going in circles.*** A promoter is planning a circular race track with an inside radius of r feet and a width of w feet. The cost in dollars for paving the track is given by the formula

$$C = 1.2\pi[(r + w)^2 - r^2].$$

Use a special product rule to simplify this formula. What is the cost of paving the track if the inside radius is 1000 feet and the width of the track is 40 feet?

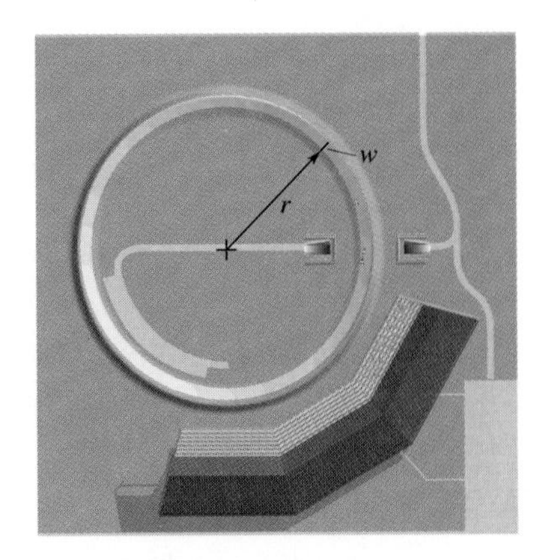

Figure for Exercise 80

81. ***Compounded annually.*** P dollars is invested at annual interest rate r for 2 years. If the interest is compounded annually, then the polynomial $P(1 + r)^2$ represents the value of the investment after 2 years. Rewrite this expression without parentheses. Evaluate the polynomial if $P = \$200$ and $r = 10\%$.

82. ***Compounded semiannually.*** P dollars is invested at annual interest rate r for 1 year. If the interest is compounded semiannually, then the polynomial $P\left(1 + \frac{r}{2}\right)^2$ represents the value of the investment after 1 year. Rewrite this expression without parentheses. Evaluate the polynomial if $P = \$200$ and $r = 10\%$.

83. ***Investing in treasury bills.*** An investment advisor uses the polynomial $P(1 + r)^{10}$ to predict the value in 10 years of a client's investment of P dollars with average annual return r. The accompanying graph shows historic average annual returns for various investments *(Stocks, Bonds, Bills, and Inflation: 1995 Yearbook)*. What is the predicted value in 10 years of an investment of \$10,000 in U.S. treasury bills?

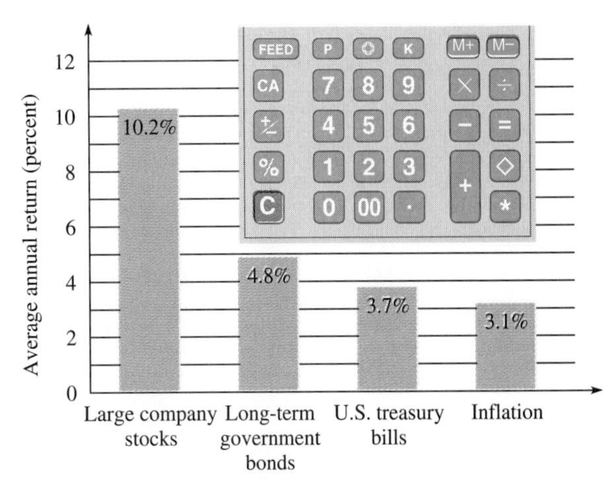

Figure for Exercises 83 and 84

84. ***Comparing investments.*** How much more would the investment in Exercise 83 be worth in 10 years if the client invests in large company stocks rather than U.S. treasury bills?

Getting More Involved

85. ***Writing.*** What is the difference between the equations $(x + 5)^2 = x^2 + 10x + 25$ and $(x + 5)^2 = x^2 + 25$?

86. ***Writing.*** Is it possible to square a sum or a difference without using the rules presented in this section? Why should you learn the rules given in this section?

3.5 Division of Polynomials

In this section:

- Dividing Monomials Using the Quotient Rule
- Dividing a Polynomial by a Monomial
- Dividing a Polynomial by a Binomial

You multiplied polynomials in Section 3.2. In this section you will learn to divide polynomials.

Dividing Monomials Using the Quotient Rule

In Chapter 1 we used the definition of division to divide signed numbers. Because the definition of division applies to any division, we restate it here.

Division of Real Numbers

If a, b, and c are any numbers with $b \neq 0$, then

$$a \div b = c \qquad \text{provided that} \qquad c \cdot b = a.$$

If $a \div b = c$, we call a the **dividend,** b the **divisor,** and c (or $a \div b$) the **quotient.**

You can find the quotient of two monomials by writing the quotient as a fraction and then reducing the fraction. For example,

$$x^5 \div x^2 = \frac{x^5}{x^2} = \frac{x \cdot x \cdot x \cdot \cancel{x} \cdot \cancel{x}}{\cancel{x} \cdot \cancel{x}} = x^3.$$

You can be sure that x^3 is correct by checking that $x^3 \cdot x^2 = x^5$. You can also divide x^2 by x^5, but the result is not a monomial:

$$x^2 \div x^5 = \frac{x^2}{x^5} = \frac{1 \cdot \cancel{x} \cdot \cancel{x}}{x \cdot x \cdot x \cdot \cancel{x} \cdot \cancel{x}} = \frac{1}{x^3}$$

Note that the exponent 3 can be obtained in either case by subtracting 5 and 2. These examples illustrate the quotient rule for exponents.

Quotient Rule

Suppose $a \neq 0$, and m and n are positive integers.

If $m \geq n$, then $\dfrac{a^m}{a^n} = a^{m-n}$.

If $n > m$, then $\dfrac{a^m}{a^n} = \dfrac{1}{a^{n-m}}$.

Note that if you use the quotient rule to subtract the exponents in $x^4 \div x^4$, you get the expression x^{4-4}, or x^0, which has not been defined yet. Because we must have $x^4 \div x^4 = 1$ if $x \neq 0$, we define the zero power of a nonzero real number to be 1. We do not define the expression 0^0.

Zero Exponent

For any nonzero real number a,

$$a^0 = 1.$$

EXAMPLE 1 **Using the definition of zero exponent**

Simplify each expression. Assume that all variables are nonzero real numbers.

a) 5^0 **b)** $(3xy)^0$ **c)** $a^0 + b^0$

Solution

a) $5^0 = 1$ **b)** $(3xy)^0 = 1$ **c)** $a^0 + b^0 = 1 + 1 = 2$ ■

With the definition of zero exponent the quotient rule is valid for all positive integers as stated.

EXAMPLE 2 **Using the quotient rule in dividing monomials**

Find each quotient.

a) $\dfrac{y^9}{y^5}$ **b)** $\dfrac{12b^2}{3b^7}$ **c)** $-6x^3 \div (2x^9)$ **d)** $\dfrac{x^8y^2}{x^2y^2}$

Solution

a) $\dfrac{y^9}{y^5} = y^{9-5} = y^4$

Use the definition of division to check that $y^4 \cdot y^5 = y^9$.

b) $\dfrac{12b^2}{3b^7} = \dfrac{12}{3} \cdot \dfrac{b^2}{b^7} = 4 \cdot \dfrac{1}{b^{7-2}} = \dfrac{4}{b^5}$

Use the definition of division to check that

$$\frac{4}{b^5} \cdot 3b^7 = \frac{12b^7}{b^5} = 12b^2.$$

c) $-6x^3 \div (2x^9) = \dfrac{-6x^3}{2x^9} = \dfrac{-3}{x^6}$

Use the definition of division to check that

$$\frac{-3}{x^6} \cdot 2x^9 = \frac{-6x^9}{x^6} = -6x^3.$$

d) $\dfrac{x^8y^2}{x^2y^2} = \dfrac{x^8}{x^2} \cdot \dfrac{y^2}{y^2} = x^6 \cdot y^0 = x^6$

Use the definition of division to check that $x^6 \cdot x^2y^2 = x^8y^2$. ■

We showed more steps in Example 2 than are necessary. For division problems like these you should try to write down only the quotient.

Dividing a Polynomial by a Monomial

We divided some simple polynomials by monomials in Chapter 1. For example,

$$\frac{6x + 8}{2} = \frac{1}{2}(6x + 8) = 3x + 4.$$

We use the distributive property to take one-half of $6x$ and one-half of 8 to get $3x + 4$. So both $6x$ and 8 are divided by 2. To divide any polynomial by a monomial, we divide each term of the polynomial by the monomial.

EXAMPLE 3 **Dividing a polynomial by a monomial**

Find the quotient for $(-8x^6 + 12x^4 - 4x^2) \div (4x^2)$.

Solution

$$\begin{aligned}\frac{-8x^6 + 12x^4 - 4x^2}{4x^2} &= \frac{-8x^6}{4x^2} + \frac{12x^4}{4x^2} - \frac{4x^2}{4x^2} \\ &= -2x^4 + 3x^2 - 1\end{aligned}$$

The quotient is $-2x^4 + 3x^2 - 1$. We can check by multiplying.

$$4x^2(-2x^4 + 3x^2 - 1) = -8x^6 + 12x^4 - 4x^2.$$

■

Because division by zero is undefined, we will always assume that the divisor is nonzero in any quotient involving variables. For example, the division in Example 3 is valid only if $4x^2 \neq 0$, or $x \neq 0$.

Dividing a Polynomial by a Binomial

Division of whole numbers is often done with a procedure called **long division.** For example, 253 is divided by 7 as follows:

$$\begin{array}{rrl} & 36 & \leftarrow \text{Quotient} \\ \text{Divisor} \rightarrow & 7\overline{)253} & \leftarrow \text{Dividend} \\ & \underline{21} & \\ & 43 & \\ & \underline{42} & \\ & 1 & \leftarrow \text{Remainder} \end{array}$$

Note that $36 \cdot 7 + 1 = 253$. It is always true that

$$(\text{quotient})(\text{divisor}) + (\text{remainder}) = \text{dividend}.$$

To divide a polynomial by a binomial, we perform the division like long division of whole numbers. For example, to divide $x^2 - 3x - 10$ by $x + 2$, we get the first term of the quotient by dividing the first term of $x + 2$ into the first term of $x^2 - 3x - 10$. So divide x^2 by x to get x:

$$\begin{array}{lrl} & x & x^2 \div x = x \\ \text{Multiply:} & x + 2\overline{)x^2 - 3x - 10} & \\ & \underline{x^2 + 2x} & x \cdot (x + 2) = x^2 + 2x \\ \text{Subtract:} & -5x & -3x - 2x = -5x \end{array}$$

Now bring down -10 and continue the process. We get the second term of the quotient (below) by dividing the first term of $x + 2$ into the first term of $-5x - 10$. So divide $-5x$ by x to get -5:

$$
\begin{array}{lrl}
 & x - 5 & -5x \div x = -5 \\
 & x + 2\overline{)x^2 - 3x - 10} & \\
 & \underline{x^2 + 2x} & \\
 & -5x - 10 & \\
\text{Multiply:} & \underline{-5x - 10} & -5(x + 2) = -5x - 10 \\
\text{Subtract:} & 0 & -10 - (-10) = 0
\end{array}
$$

So the quotient is $x - 5$, and the remainder is 0.

In the next example we must rearrange the dividend before dividing.

EXAMPLE 4 **Dividing a polynomial by a binomial**

Divide $2x^3 - 4 - 7x^2$ by $2x - 3$, and identify the quotient and the remainder.

Solution

Rearrange the dividend as $2x^3 - 7x^2 - 4$. Because the x-term in the dividend is missing, we write $0 \cdot x$ for it:

$$
\begin{array}{rl}
x^2 - 2x - 3 & \\
2x - 3\overline{)2x^3 - 7x^2 + 0 \cdot x - 4} & \\
\underline{2x^3 - 3x^2} & \\
-4x^2 + 0 \cdot x & -7x^2 - (-3x^2) = -4x^2 \\
\underline{-4x^2 + 6x} & \\
-6x - 4 & 0 \cdot x - 6x = -6x \\
\underline{-6x + 9} & \\
-13 & -4 - (9) = -13
\end{array}
$$

The quotient is $x^2 - 2x - 3$, and the remainder is -13. Note that the degree of the remainder is 0 and the degree of the divisor is 1. To check, we must verify that

$$(2x - 3)(x^2 - 2x - 3) - 13 = 2x^3 - 7x^2 - 4.$$

CAUTION To avoid errors, always write the terms of the divisor and the dividend in descending order of the exponents and insert a zero for any term that is missing.

If we divide both sides of the equation

$$\text{dividend} = (\text{quotient})(\text{divisor}) + (\text{remainder})$$

by the divisor, we get the equation

$$\frac{\text{dividend}}{\text{divisor}} = \text{quotient} + \frac{\text{remainder}}{\text{divisor}}.$$

This fact is used in expressing improper fractions as mixed numbers. For example, if 19 is divided by 5, the quotient is 3 and the remainder is 4. So

$$\frac{19}{5} = 3 + \frac{4}{5} = 3\frac{4}{5}.$$

We can also use this form to rewrite algebraic fractions.

EXAMPLE 5 **Rewriting algebraic fractions**

Express $\frac{-3x}{x-2}$ in the form

$$\text{quotient} + \frac{\text{remainder}}{\text{divisor}}.$$

Solution

Use long division to get the quotient and remainder:

$$\begin{array}{r} -3 \\ x-2\overline{)-3x+0} \\ \underline{-3x+6} \\ -6 \end{array}$$

Because the quotient is -3 and the remainder is -6, we can write

$$\frac{-3x}{x-2} = -3 + \frac{-6}{x-2}.$$

To check, we must verify that $-3(x-2) - 6 = -3x$. ■

CAUTION When dividing polynomials by long division, we do not stop until the remainder is 0 or the degree of the remainder is smaller than the degree of the divisor. For example, we stop dividing in Example 5 because the degree of the remainder -6 is 0 and the degree of the divisor $x - 2$ is 1. ⊘

Warm-ups

True or false? Explain your answer.

1. $y^{10} \div y^2 = y^5$ for any nonzero value of y.
2. $\frac{7x+2}{7} = x + 2$ for any value of x.
3. $\frac{7x^2}{7} = x^2$ for any value of x.
4. If $3x^2 + 6$ is divided by 3, the quotient is $x^2 + 6$.
5. If $4y^2 - 6y$ is divided by $2y$, the quotient is $2y - 3$.
6. The quotient times the remainder plus the dividend equals the divisor.
7. $(x+2)(x+1) + 3 = x^2 + 3x + 5$ for any value of x.
8. If $x^2 + 3x + 5$ is divided by $x + 2$, then the quotient is $x + 1$.
9. If $x^2 + 3x + 5$ is divided by $x + 2$, the remainder is 3.
10. If the remainder is zero, then (divisor)(quotient) = dividend.

3.5 EXERCISES

Simplify each expression. See Example 1.

1. 9^0 **2.** m^0 **3.** $(-2x^3)^0$

4. $(5a^3b)^0$ **5.** $2 \cdot 5^0 - 3^0$ **6.** $-4^0 - 8^0$

7. $(2x - y)^0$ **8.** $(a^2 + b^2)^0$

Find each quotient. Try to write only the answer. See Example 2.

9. $\frac{x^8}{x^2}$ **10.** $\frac{y^9}{y^3}$ **11.** $\frac{6a^7}{2a^{12}}$ **12.** $\frac{30b^2}{3b^6}$

13. $-12x^5 \div (3x^9)$ **14.** $-6y^5 \div (-3y^{10})$

15. $-6y^2 \div (6y)$

16. $-3a^2b \div (3ab)$

17. $\dfrac{-6x^3y^2}{2x^2y^2}$

18. $\dfrac{-4h^2k^4}{-2hk^3}$

19. $\dfrac{-9x^2y^2}{3x^5y^2}$

20. $\dfrac{-12z^4y^2}{-2z^{10}y^2}$

Find the quotients. See Example 3.

21. $\dfrac{3x-6}{3}$

22. $\dfrac{5y-10}{-5}$

23. $\dfrac{x^5+3x^4-x^3}{x^2}$

24. $\dfrac{6y^6-9y^4+12y^2}{3y^2}$

25. $\dfrac{-8x^2y^2+4x^2y-2xy^2}{-2xy}$

26. $\dfrac{-9ab^2-6a^3b^3}{-3ab^2}$

27. $(x^2y^3-3x^3y^2) \div (x^2y)$

28. $(4h^5k-6h^2k^2) \div (-2h^2k)$

Find the quotient and remainder for each division. Check by using the fact that dividend = (divisor)(quotient) + remainder. See Example 4.

29. $(x^2+5x+13) \div (x+3)$

30. $(x^2+3x+6) \div (x+3)$

31. $(2x) \div (x+5)$

32. $(5x) \div (x-1)$

33. $(a^3+4a-3) \div (a-2)$

34. $(w^3+2w^2-3) \div (w-2)$

35. $(x^2-3x) \div (x+1)$

36. $(3x^2) \div (x+1)$

37. $(h^3-27) \div (h-3)$

38. $(w^3+1) \div (w+1)$

39. $(6x^2-13x+7) \div (3x-2)$

40. $(4b^2+25b-3) \div (4b+1)$

41. $(x^3-x^2+x-2) \div (x-1)$

42. $(a^3-3a^2+4a-4) \div (a-2)$

Write each expression in the form

$$\textit{quotient} + \frac{\textit{remainder}}{\textit{divisor}}.$$

See Example 5.

43. $\dfrac{3x}{x-5}$

44. $\dfrac{2x}{x-1}$

45. $\dfrac{-x}{x+3}$

46. $\dfrac{-3x}{x+1}$

47. $\dfrac{x-1}{x}$

48. $\dfrac{a-5}{a}$

49. $\dfrac{3x+1}{x}$

50. $\dfrac{2y+1}{y}$

51. $\dfrac{x^2}{x+1}$

52. $\dfrac{x^2}{x-1}$

53. $\dfrac{x^2+4}{x+2}$

54. $\dfrac{x^2+1}{x-1}$

55. $\dfrac{x^3}{x-2}$

56. $\dfrac{x^3-1}{x+1}$

57. $\dfrac{x^3+3}{x}$

58. $\dfrac{2x^2+4}{2x}$

Find each quotient.

59. $-6a^3b \div (2a^2b)$

60. $-14x^7 \div (-7x^2)$

61. $-8w^4t^7 \div (-2w^9t^3)$

62. $-9y^7z^4 \div (3y^3z^{11})$

63. $(3a-12) \div (-3)$

64. $(-6z+3z^2) \div (-3z)$

65. $(3x^2-9x) \div (3x)$

66. $(5x^3+15x^2-25x) \div (5x)$

67. $(12x^4-4x^3+6x^2) \div (-2x^2)$

68. $(-9x^3+3x^2-15x) \div (-3x)$

69. $(t^2-5t-36) \div (t-9)$

70. $(b^2+2b-35) \div (b-5)$

71. $(6w^2-7w-5) \div (3w-5)$

72. $(4z^2+23z-6) \div (4z-1)$

73. $(8x^3+27) \div (2x+3)$

74. $(8y^3-1) \div (2y-1)$

75. $(t^3-3t^2+5t-6) \div (t-2)$

76. $(2u^3-13u^2-8u+7) \div (u-7)$

77. $(-6v^2-4+9v+v^3) \div (v-4)$

78. $(14y+8y^2+y^3+12) \div (6+y)$

Solve each problem.

79. ***Area of a rectangle.*** The area of a rectangle is x^2+x-30 square meters. If the length is $x+6$ meters, find a binomial that represents the width.

80. ***Perimeter of a rectangle.*** The perimeter of a rectangle is $6x+6$ yards. If the width is x yards, find a binomial that represents the length.

Getting More Involved

81. ***Exploration.*** Divide x^3-1 by $x-1$, x^4-1 by $x-1$, and x^5-1 by $x-1$. What is the quotient when x^9-1 is divided by $x-1$?

82. ***Exploration.*** Divide a^3-b^3 by $a-b$ and a^4-b^4 by $a-$ b. What is the quotient when a^8-b^8 is divided by $a-b$?

83. ***Discussion.*** Are the expressions $\frac{10x}{5x}$, $10x \div 5x$, and $(10x) \div (5x)$ equivalent? Before you answer, review the order of operations in Section 1.5 and evaluate each expression for $x=3$.

3.6 Positive Integral Exponents

In this section:

- The Product and Quotient Rules
- Raising an Exponential Expression to a Power
- Power of a Product
- Power of a Quotient
- Summary of Rules

The product rule for positive integral exponents was presented in Section 3.2, and the quotient rule was presented in Section 3.5. In this section we review those rules and then further investigate the properties of exponents.

The Product and Quotient Rules

The product rule allows us to write the product of exponential expressions with the same base as a single exponential expression. The exponent of the new expression is the sum of the exponents of the factors.

Product Rule

If m and n are positive integers, then $a^m \cdot a^n = a^{m+n}$.

CAUTION The product rule applies only if the bases of the expressions are identical. For example, $3^2 \cdot 3^4 = 3^6$, but the product rule cannot be applied to $5^2 \cdot 3^4$. Note also that the bases are not multiplied: $3^2 \cdot 3^4 \neq 9^6$. ⊘

The quotient rule from Section 3.5 allows us to write the quotient of two exponential expressions with the same base as a single exponential expression. The exponent in the new expression is the difference of the exponents in the quotient.

Quotient Rule

Suppose $a \neq 0$, and m and n are positive integers.

$$\text{If } m \geq n, \text{ then } \frac{a^m}{a^n} = a^{m-n}.$$

$$\text{If } n > m, \text{ then } \frac{a^m}{a^n} = \frac{1}{a^{n-m}}.$$

In Section 3.5 we defined the zero power of any nonzero number to be 1.

Zero Exponent

If a is any nonzero real number, then $a^0 = 1$.

EXAMPLE 1 **Using the product and quotient rules**

Use the rules of exponents to simplify each expression. Assume that all variables represent nonzero real numbers.

a) $2^3 \cdot 3^2$ **b)** $(3x)^0(5x^2)(4x)$ **c)** $\dfrac{(3a^2b)b^9}{(6a^5)a^3b^2}$

Solution

a) $2^3 \cdot 3^2 = 8 \cdot 9 = 72$

Note that since the bases are different (2 and 3), we cannot use the product rule to add the exponents.

b) $(3x)^0(5x^2)(4x) = 1 \cdot 5x^2 \cdot 4x$ Definition of zero exponent

$= 20x^3$ Product rule

c) $\dfrac{(3a^2b)b^9}{(6a^5)a^3b^2} = \dfrac{3a^2b^{10}}{6a^8b^2}$ Product rule

$= \dfrac{b^8}{2a^6}$ Quotient rule ■

Raising an Exponential Expression to a Power

When we raise an exponential expression to a power, we can use the product rule to find the result, as shown in the following example:

$(w^4)^3 = w^4 \cdot w^4 \cdot w^4$ Three factors of w^4 because of the exponent 3

$= w^{12}$ Product rule

By the product rule we add the three 4's to get 12, but 12 is also the product of 4 and 3. This example illustrates the **power rule** for exponents.

Power Rule

If m and n are nonnegative integers and $a \neq 0$, then

$$(a^m)^n = a^{mn}.$$

In the next example we use the new rule along with the other rules.

EXAMPLE 2 **Using the power rule**

Use the rules of exponents to simplify each expression. Assume that all variables represent nonzero real numbers.

a) $3x^2(x^3)^5$ **b)** $\dfrac{(2^3)^4 \cdot 2^7}{2^5 \cdot 2^9}$ **c)** $\dfrac{3(x^5)^4}{15x^{22}}$

Solution

a) $3x^2(x^3)^5 = 3x^2x^{15}$ Power rule

$= 3x^{17}$ Product rule

b) $\dfrac{(2^3)^4 \cdot 2^7}{2^5 \cdot 2^9} = \dfrac{2^{12} \cdot 2^7}{2^{14}}$ Power rule and product rule

$= \dfrac{2^{19}}{2^{14}}$ Product rule

$= 2^5$ Quotient rule

$= 32$ Evaluate 2^5.

c) $\dfrac{3(x^5)^4}{15x^{22}} = \dfrac{3x^{20}}{15x^{22}} = \dfrac{1}{5x^2}$ ■

Power of a Product

Consider an example of raising a monomial to a power. We will use known rules to rewrite the expression.

$$\begin{aligned}(2x)^3 &= 2x \cdot 2x \cdot 2x && \text{Definition of exponent 3}\\ &= 2 \cdot 2 \cdot 2 \cdot x \cdot x \cdot x && \text{Commutative and associative properties}\\ &= 2^3x^3 && \text{Definition of exponents}\end{aligned}$$

Note that the power 3 is distributed over each factor of the product. This example illustrates the **power of a product rule.**

Power of a Product Rule

If a and b are real numbers and n is a positive integer, then

$$(ab)^n = a^n b^n.$$

EXAMPLE 3 **Using the power of a product rule**

Simplify. Assume that the variables are nonzero.

a) $(xy^3)^5$ **b)** $(-3m)^3$ **c)** $(2x^3y^2z^7)^3$

Solution

a) $(xy^3)^5 = x^5(y^3)^5$ Power of a product rule

$= x^5y^{15}$ Power rule

b) $(-3m)^3 = (-3)^3m^3$ Power of a product rule

$= -27m^3$ $(-3)(-3)(-3) = -27$

c) $(2x^3y^2z^7)^3 = 2^3(x^3)^3(y^2)^3(z^7)^3 = 8x^9y^6z^{21}$ ■

Power of a Quotient

Raising a quotient to a power is similar to raising a product to a power:

$$\begin{aligned}\left(\frac{x}{5}\right)^3 &= \frac{x}{5} \cdot \frac{x}{5} \cdot \frac{x}{5} && \text{Definition of exponent 3}\\ &= \frac{x \cdot x \cdot x}{5 \cdot 5 \cdot 5} && \text{Definition of multiplication of fractions}\\ &= \frac{x^3}{5^3} && \text{Definition of exponents}\end{aligned}$$

The power is distributed over each term of the quotient. This example illustrates the **power of a quotient rule.**

Power of a Quotient Rule

If a and b are real numbers, $b \neq 0$, and n is a positive integer, then

$$\left(\frac{a}{b}\right)^n = \frac{a^n}{b^n}.$$

EXAMPLE 4 **Using the power of a quotient rule**

Simplify. Assume that the variables are nonzero.

a) $\left(\frac{2}{5x^3}\right)^2$ **b)** $\left(\frac{3x^4}{2y^3}\right)^3$ **c)** $\left(\frac{-12a^5b}{4a^2b^7}\right)^3$

Solution

a) $\left(\frac{2}{5x^3}\right)^2 = \frac{2^2}{(5x^3)^2}$ Power of a quotient rule

$= \frac{4}{25x^6}$ $(5x^3)^2 = 5^2(x^3)^2 = 25x^6$

b) $\left(\frac{3x^4}{2y^3}\right)^3 = \frac{3^3x^{12}}{2^3y^9}$ Power of a quotient and power of a product rules

$= \frac{27x^{12}}{8y^9}$ Simplify.

c) $\left(\frac{-12a^5b}{4a^2b^7}\right)^3 = \left(\frac{-3a^3}{b^6}\right)^3$ Use the quotient rule first.

$= \frac{-27a^9}{b^{18}}$ Power of a quotient rule ■

Summary of Rules

The rules for exponents are summarized in the following box.

Rules for Nonnegative Integral Exponents

The following rules hold for nonzero real numbers a and b and nonnegative integers m and n.

1. $a^0 = 1$ Definition of zero exponent
2. $a^m \cdot a^n = a^{m+n}$ Product rule
3. $\frac{a^m}{a^n} = a^{m-n}$ for $m \geq n$,

   $\frac{a^m}{a^n} = \frac{1}{a^{n-m}}$ for $n > m$ Quotient rule
4. $(a^m)^n = a^{mn}$ Power rule
5. $(ab)^n = a^n \cdot b^n$ Power of a product rule
6. $\left(\frac{a}{b}\right)^n = \frac{a^n}{b^n}$ Power of a quotient rule

Warm-ups

True or false? Assume that all variables represent nonzero real numbers. A statement involving variables is to be marked true only if it is an identity. Explain your answer.

1. $-3^0 = 1$ **2.** $2^5 \cdot 2^8 = 4^{13}$ **3.** $2^3 \cdot 3^2 = 6^5$ **4.** $(2x)^4 = 2x^4$

5. $(q^3)^5 = q^8$ **6.** $(-3x^2)^3 = 27x^6$ **7.** $(ab^3)^4 = a^4b^{12}$

8. $\dfrac{a^{12}}{a^4} = a^3$ **9.** $\dfrac{6w^4}{3w^9} = 2w^5$ **10.** $\left(\dfrac{2y^3}{9}\right)^2 = \dfrac{4y^6}{81}$

3.6 EXERCISES

For all exercises in this section, assume that the variables represent nonzero real numbers.

Simplify the exponential expressions. See Example 1.

1. $2^3 \cdot 5^2$ **2.** $2^2 \cdot 10^3$

3. $2^5 \cdot 2^{10}$ **4.** $x^6 \cdot x^7$

5. $(-3u^8v^7)(-2u^2v)$ **6.** $(3r^4t^3)(-6r^2t^{10})$

7. $a^3b^4 \cdot ab^6(ab)^0$ **8.** $x^2y \cdot x^3y^6(x + y)^0$

9. $\dfrac{-2a^3}{4a^7}$ **10.** $\dfrac{-3t^9}{6t^{18}}$

11. $\dfrac{2a^5 \cdot 3a^7}{15a^6}$ **12.** $\dfrac{3y^8 \cdot 5y^9}{20y^{14}}$

Simplify. See Example 2.

13. $(x^2)^3$ **14.** $(y^2)^4$ **15.** $2x^2 \cdot (x^2)^5$

16. $(y^2)^6 \cdot 3y^5$ **17.** $\dfrac{(t^2)^5}{(t^3)^4}$ **18.** $\dfrac{(r^4)^2}{(r^5)^3}$

19. $\dfrac{3x(x^5)^2}{6x^3(x^2)^4}$ **20.** $\dfrac{5y^3(y^5)^2}{10y^5(y^2)^6}$

Simplify. See Example 3.

21. $(xy^2)^3$ **22.** $(wy^2)^6$ **23.** $(-2t^5)^3$

24. $(-3r^3)^3$ **25.** $(-2x^2y^5)^3$ **26.** $(-3y^2z^3)^3$

27. $\dfrac{(a^4b^2c^5)^3}{a^3b^4c}$ **28.** $\dfrac{(2ab^2c^3)^5}{(2a^3bc)^4}$

Simplify. See Example 4.

29. $\left(\dfrac{x^4}{4}\right)^3$ **30.** $\left(\dfrac{y^2}{2}\right)^3$ **31.** $\left(\dfrac{-2a^2}{b^3}\right)^4$

32. $\left(\dfrac{-9r^3}{t^5}\right)^2$ **33.** $\left(\dfrac{2x^2y}{-4y^2}\right)^3$ **34.** $\left(\dfrac{3y^8}{2zy^2}\right)^4$

35. $\left(\dfrac{-6x^2y^4z^9}{3x^6y^4z^3}\right)^2$ **36.** $\left(\dfrac{-10rs^9t^4}{2rs^2t^7}\right)^3$

Simplify each expression.

37. $2^3 + 3^3$ **38.** $5^2 - 3^2$ **39.** $(2 + 3)^3$

40. $(5 - 3)^2$ **41.** $\left(\dfrac{2}{3}\right)^3$ **42.** $\left(\dfrac{3}{4}\right)^2$

43. $5^2 \cdot 2^3$ **44.** $10^3 \cdot 10^2$ **45.** $3x^4 \cdot 5x^7$

46. $-2y^3(3y)$ **47.** $(-5x^4)^3$ **48.** $(4z^3)^3$

49. $-3y^5z^{12} \cdot 9yz^7$ **50.** $2a^4b^5 \cdot 2a^9b^2$

51. $\dfrac{-9u^4v^9}{-3u^5v^8}$ **52.** $\dfrac{-20a^5b^{13}}{5a^4b^{13}}$

53. $(-xt^2)(-2x^2t)^4$ **54.** $(-ab)^3(-3ba^2)^4$

55. $\left(\dfrac{2x^2}{x^4}\right)^3$ **56.** $\left(\dfrac{3y^8}{y^5}\right)^2$

57. $\left(\dfrac{-8a^3b^4}{4c^5}\right)^5$ **58.** $\left(\dfrac{-10a^5c}{5a^5b^4}\right)^5$

59. $\left(\dfrac{-8x^4y^7}{-16x^5y^6}\right)^5$ **60.** $\left(\dfrac{-5x^2yz^3}{-5x^2yz}\right)^5$

Solve each problem.

61. ***Long-term investing.*** Sheila invested P dollars at annual rate r for 10 years. At the end of 10 years her investment was worth $P(1 + r)^{10}$ dollars. She then reinvested this money for another 5 years at annual rate r. At the end of the second time period her investment was worth $P(1 + r)^{10}(1 + r)^5$ dollars. Which law of exponents can be used to simplify the last expression? Simplify it.

62. ***CD rollover.*** Ronnie invested P dollars in a 2-year CD with an annual rate of return of r. After the CD rolled over two times, its value was $P((1 + r)^2)^3$. Which law of exponents can be used to simplify the expression? Simplify it.

Getting More Involved

63. ***Writing.*** When we square a product, we square each factor in the product. For example, $(3b)^2 = 9b^2$. Explain why we cannot square a sum by simply squaring each term of the sum.

64. ***Writing.*** Explain why we define 2^0 to be 1. Explain why $-2^0 \neq 1$.

COLLABORATIVE ACTIVITIES

Area as a Model of FOIL

Grouping: Pairs
Topic: Multiplying polynomials

Sometimes we can use drawings to represent mathematical operations. The area of a rectangle can represent the process we use when multiplying binomials. The rectangle below represents the multiplication of the binomials $(x + 3)$ and $(x + 5)$:

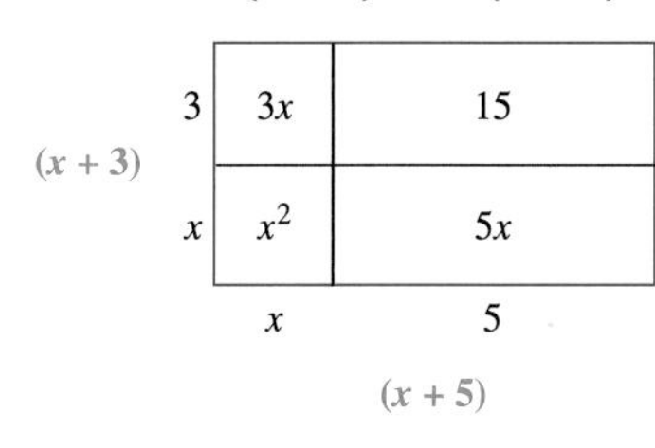

The areas of the inner rectangles are x^2, $3x$, $5x$, and 15.

The area of the red rectangle equals the sum of the areas of the four inner rectangles.

Area of red rectangle:

$$\begin{aligned}(x + 3)(x + 5) &= x^2 + 3x + 5x + 15\\ &= x^2 + 8x + 15\end{aligned}$$

1. a. With your partner, find the areas of the inner rectangles to find the product $(x + 2)(x + 7)$ below:

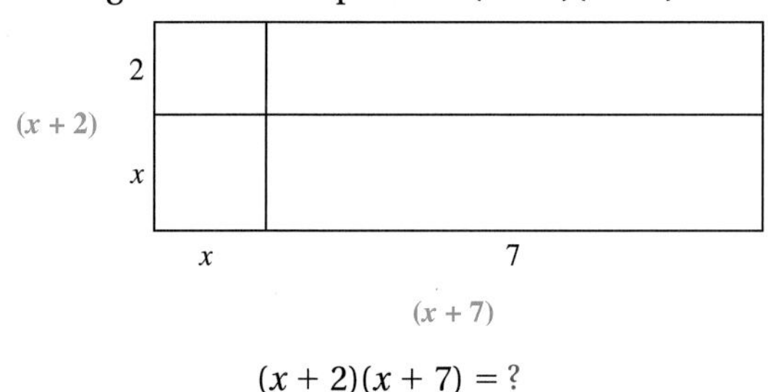

$$(x + 2)(x + 7) = ?$$

b. Find the same product $(x + 2)(x + 7)$ using FOIL.

For problem 2 student A uses FOIL to find the given product while student B finds the area with the diagram.

2.

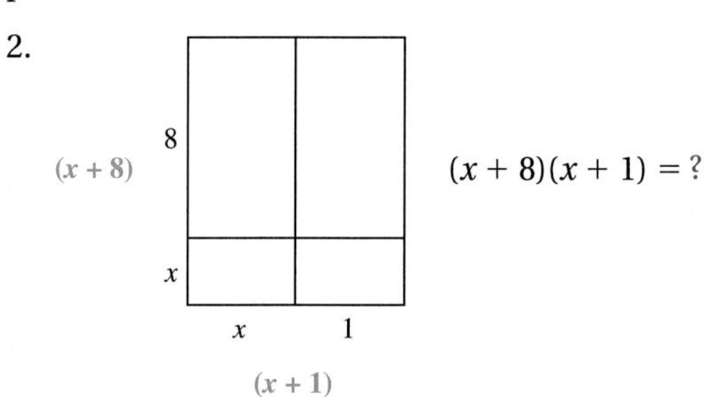

For problem 3 student B uses FOIL and A uses the diagram.

3.

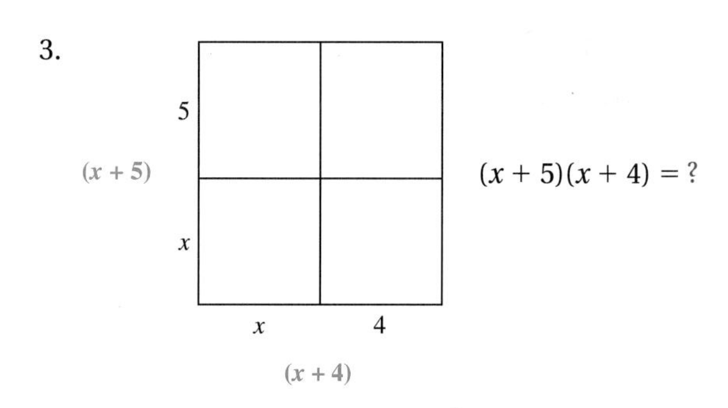

4. Student A draws a diagram to find the product $(x + 3)(x + 7)$. Student B finds $(x + 3)(x + 7)$ using FOIL.
5. Student B draws a diagram to find the product $(x + 2)(x + 1)$. Student A finds $(x + 2)(x + 1)$ using FOIL.

Thinking in reverse: Work together to complete the product that is represented by the given diagram.

6.

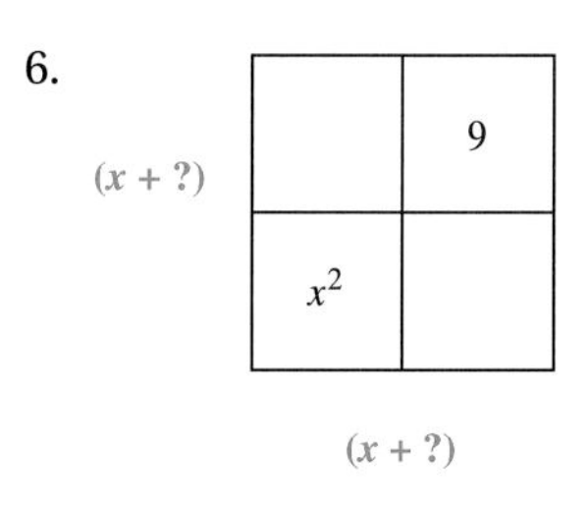

7.

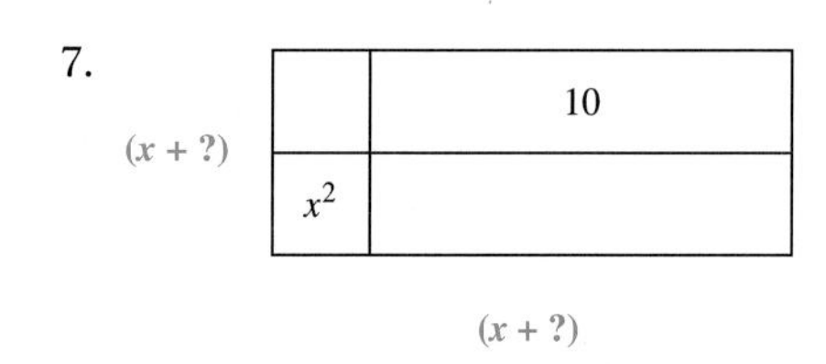

Extension: Make up a FOIL problem, then have your partner draw a diagram of it.

Wrap-up CHAPTER 3

SUMMARY

Polynomials		Examples
Term	A number or the product of a number and one or more variables raised to powers	$5x^3$, $-4x$, 7
Polynomial	A single term or a finite sum of terms	$2x^5 - 9x^2 + 11$
Degree of a polynomial	The highest degree of any of the terms	Degree of $2x - 9$ is 1. Degree of $5x^3 - x^2$ is 3.

Adding, Subtracting, and Multiplying Polynomials		Examples
Add or subtract polynomials	Add or subtract the like terms.	$(x + 1) + (x - 4) = 2x - 3$ $(x^2 - 3x) - (4x^2 - x)$ $= -3x^2 - 2x$
Multiply monomials	Use the product rule for exponents	$-2x^5 \cdot 6x^8 = -12x^{13}$
Multiply polynomials	Multiply each term of one polynomial by every term of the other polynomial, then combine like terms.	$\begin{array}{r} x^2 + 2x + 5 \\ x - 1 \\ \hline -x^2 - 2x - 5 \\ x^3 + 2x^2 + 5x \\ \hline x^3 + x^2 + 3x - 5 \end{array}$

Binomials		Examples
FOIL	A method for multiplying two binomials quickly	$(x - 2)(x + 3) = x^2 + x - 6$
Square of a sum	$(a + b)^2 = a^2 + 2ab + b^2$	$(x + 3)^2 = x^2 + 6x + 9$
Square of a difference	$(a - b)^2 = a^2 - 2ab + b^2$	$(m - 5)^2 = m^2 - 10m + 25$
Product of a sum and a difference	$(a - b)(a + b) = a^2 - b^2$	$(x + 2)(x - 2) = x^2 - 4$

Dividing Polynomials		Examples
Dividing monomials	Use the quotient rule for exponents	$8x^5 \div (2x^2) = 4x^3$
Divide a polynomial by a monomial	Divide each term of the polynomial by the monomial.	$\dfrac{3x^5 + 9x}{3x} = x^4 + 3$
Divide a polynomial by a binomial	If the divisor is a binomial, use long division. (divisor)(quotient) + (remainder) = dividend	$\begin{array}{r} x - 7 \\ x + 2 \overline{) x^2 - 5x - 4} \\ x^2 + 2x \\ \hline -7x - 4 \\ -7x - 14 \\ \hline 10 \end{array}$ Divisor: $x + 2$; Quotient: $x - 7$; Dividend: $x^2 - 5x - 4$; Remainder: 10

Rules of Exponents		**Examples**
The following rules hold for any positive integers m and n, and nonzero real numbers a and b.		
Zero exponent	$a^0 = 1$	$2^0 = 1,\ (-34)^0 = 1$
Product rule	$a^m \cdot a^n = a^{m+n}$	$a^2 \cdot a^3 = a^5$ $3x^6 \cdot 4x^9 = 12x^{15}$
Quotient rule	If $m \geq n$, then $\dfrac{a^m}{a^n} = a^{m-n}$.	$x^8 \div x^2 = x^6,\ \dfrac{y^3}{y^3} = y^0 = 1$
	If $n > m$, then $\dfrac{a^m}{a^n} = \dfrac{1}{a^{n-m}}$.	$\dfrac{c^7}{c^9} = \dfrac{1}{c^2}$
Power rule	$(a^m)^n = a^{mn}$	$(2^2)^3 = 2^6,\ (w^5)^3 = w^{15}$
Power of a product rule	$(ab)^n = a^n b^n$	$(2t)^3 = 8t^3$
Power of a quotient rule	$\left(\dfrac{a}{b}\right)^n = \dfrac{a^n}{b^n}$	$\left(\dfrac{x}{3}\right)^3 = \dfrac{x^3}{27}$

REVIEW EXERCISES

3.1 *Perform the indicated operations.*

1. $(2w - 6) + (3w + 4)$

2. $(1 - 3y) + (4y - 6)$

3. $(x^2 - 2x - 5) - (x^2 + 4x - 9)$

4. $(3 - 5x - x^2) - (x^2 - 7x + 8)$

5. $(5 - 3w + w^2) + (w^2 - 4w - 9)$

6. $(-2t^2 + 3t - 4) + (t^2 - 7t + 2)$

7. $(4 - 3m - m^2) - (m^2 - 6m + 5)$

8. $(n^3 - n^2 + 9) - (n^4 - n^3 + 5)$

3.2 *Perform the indicated operations.*

9. $5x^2 \cdot (-10x^9)$

10. $3h^3t^2 \cdot 2h^2t^5$

11. $(-11a^7)^2$

12. $(12b^3)^2$

13. $x - 5(x - 3)$

14. $x - 4(x - 9)$

15. $5x + 3(x^2 - 5x + 4)$

16. $5 + 4x^2(x - 5)$

17. $3m^2(5m^3 - m + 2)$

18. $-4a^4(a^2 + 2a + 4)$

19. $(x - 5)(x^2 - 2x + 10)$

20. $(x + 2)(x^2 - 2x + 4)$

21. $(x^2 - 2x + 4)(3x - 2)$

22. $(5x + 3)(x^2 - 5x + 4)$

3.3 *Perform the indicated operations.*

23. $(q - 6)(q + 8)$

24. $(w + 5)(w + 12)$

25. $(2t - 3)(t - 9)$

26. $(5r + 1)(5r + 2)$

27. $(4y - 3)(5y + 2)$

28. $(11y + 1)(y + 2)$

29. $(3x^2 + 5)(2x^2 + 1)$

30. $(x^3 - 7)(2x^3 + 7)$

3.4 *Perform the indicated operations. Try to write only the answers.*

31. $(z - 7)(z + 7)$

32. $(a - 4)(a + 4)$

33. $(y + 7)^2$

34. $(a + 5)^2$

35. $(w - 3)^2$

36. $(a - 6)^2$

37. $(x^2 - 3)(x^2 + 3)$

38. $(2b^2 - 1)(2b^2 + 1)$

39. $(3a + 1)^2$

40. $(1 - 3c)^2$

41. $(4 - y)^2$

42. $(9 - t)^2$

3.5 *In Exercises 43–54, find each quotient.*

43. $-10x^5 \div (2x^3)$

44. $-6x^4y^2 \div (-2x^2y^2)$

45. $\dfrac{6a^5b^7c^6}{-3a^3b^9c^6}$

46. $\dfrac{-9h^5t^9r^2}{3h^7t^6r^2}$

47. $\dfrac{3x - 9}{-3}$

48. $\dfrac{7 - y}{-1}$

49. $\dfrac{9x^3 - 6x^2 + 3x}{-3x}$

50. $\dfrac{-8x^3y^5 + 4x^2y^4 - 2xy^3}{2xy^2}$

51. $(a - 1) \div (1 - a)$

52. $(t - 3) \div (3 - t)$

53. $(m^4 - 16) \div (m - 2)$

54. $(x^4 - 1) \div (x - 1)$

Find the quotient and remainder.

55. $(3m^3 - 9m^2 + 18m) \div (3m)$

56. $(8x^3 - 4x^2 - 18x) \div (2x)$

57. $(b^2 - 3b + 5) \div (b + 2)$

58. $(r^2 - 5r + 9) \div (r - 3)$

59. $(4x^2 - 9) \div (2x + 1)$

60. $(9y^3 + 2y) \div (3y + 2)$

61. $(x^3 + x^2 - 11x + 10) \div (x - 1)$

62. $(y^3 - 9y^2 + 3y - 6) \div (y + 1)$

Write each expression in the form

$$\textit{quotient} + \frac{\textit{remainder}}{\textit{divisor}}.$$

63. $\dfrac{2x}{x - 3}$

64. $\dfrac{3x}{x - 4}$

65. $\dfrac{2x}{1 - x}$

66. $\dfrac{3x}{5 - x}$

67. $\dfrac{x^2 - 3}{x + 1}$

68. $\dfrac{x^2 + 3x + 1}{x - 3}$

69. $\dfrac{x^2}{x + 1}$

70. $\dfrac{-2x^2}{x - 3}$

3.6 *Simplify each expression.*

71. $2y^{10} \cdot 3y^{20}$

72. $(-3a^5)(5a^3)$

73. $\dfrac{-10b^5c^3}{2b^5c^9}$

74. $\dfrac{-30k^3y^9}{15k^3y^2}$

75. $(b^5)^6$

76. $(y^5)^8$

77. $(-2x^3y^2)^3$

78. $(-3a^4b^6)^4$

79. $\left(\dfrac{2a}{b}\right)^3$

80. $\left(\dfrac{3y}{2}\right)^3$

81. $\left(\dfrac{-6x^2y^5}{-3z^6}\right)^3$

82. $\left(\dfrac{-3a^4b^8}{6a^3b^{12}}\right)^4$

Miscellaneous

Perform the indicated operations.

83. $(x + 3)(x + 7)$

84. $(k + 5)(k + 4)$

85. $(t - 3y)(t - 4y)$

86. $(t + 7z)(t + 6z)$

87. $(2x^3)^0 + (2y)^0$

88. $(4y^2 - 9)^0$

89. $(-3ht^6)^3$

90. $(-9y^3c^4)^2$

91. $(2w + 3)(w - 6)$

92. $(3x + 5)(2x - 6)$

93. $(3u - 5v)(3u + 5v)$

94. $(9x^2 - 2)(9x^2 + 2)$

95. $(3h + 5)^2$

96. $(4v - 3)^2$

97. $(x + 3)^3$

98. $(k - 10)^3$

99. $(-7s^2t)(-2s^3t^5)$

100. $-5w^3r^2 \cdot 2w^4r^8$

101. $\left(\dfrac{k^4m^2}{2k^2m^2}\right)^4$

102. $\left(\dfrac{-6h^3y^5}{2h^7y^2}\right)^4$

103. $(5x^2 - 8x - 8) - (4x^2 + x - 3)$

104. $(4x^2 - 6x - 8) - (9x^2 - 5x + 7)$

105. $(2x^2 - 2x - 3) + (3x^2 + x - 9)$

106. $(x^2 - 3x - 1) + (x^2 - 2x + 1)$

107. $(x + 4)(x^2 - 5x + 1)$

108. $(2x^2 - 7x + 4)(x + 3)$

109. $(x^2 + 4x - 12) \div (x - 2)$

110. $(a^2 - 3a - 10) \div (a - 5)$

Solve each problem.

111. ***Roundball court.*** The length of a basketball court is 44 feet larger than its width. Find polynomials that represent its perimeter and area. The actual width of a basketball court is 50 feet. Evaluate these polynomials to find the actual perimeter and area of the court.

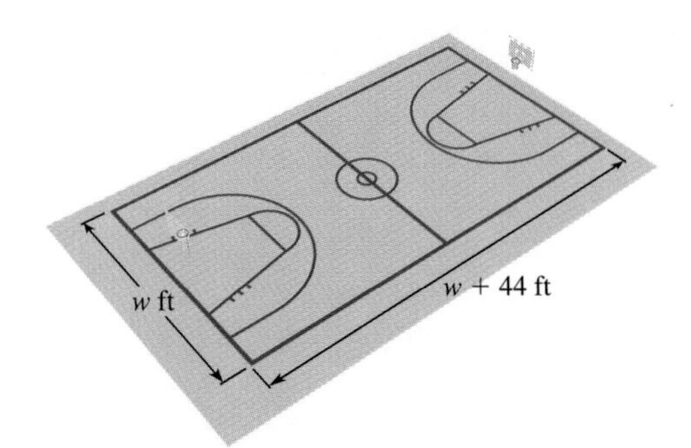

Figure for Exercise 111

112. ***Badminton court.*** The width of a badminton court is 24 feet less than its length. Find polynomials that represent its perimeter and area. The actual length of a badminton court is 44 feet. Evaluate these polynomials to find the perimeter and area of the court.

113. ***Smoke alert.*** A retailer of smoke alarms knows that at a price of p dollars each, she can sell $600 - 15p$ smoke alarms per week. Find a polynomial that represents the weekly revenue for the smoke alarms. Find the revenue for a week in which the price is $12 per smoke alarm. Use the bar graph to find the price per smoke alarm that gives the maximum weekly revenue.

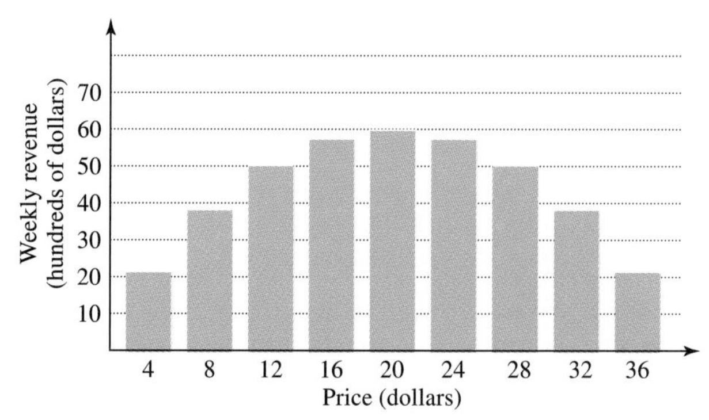

Figure for Exercise 113

114. ***Boom box sales.*** A retailer of boom boxes knows that at a price of q dollars each, he can sell $900 - 3q$ boom boxes per month. Find a polynomial that represents the monthly revenue for the boom boxes? How many boom boxes will he sell if the price is \$300 each?

CHAPTER 3 TEST

Perform the indicated operations.

1. $(7x^3 - x^2 - 6) + (5x^2 + 2x - 5)$

2. $(x^2 - 3x - 5) - (2x^2 + 6x - 7)$

3. $-5x^3 \cdot 7x^5$

4. $3x^3y \cdot (2xy^4)^2$

5. $-4a^6b^5 \div (-2a^5b)$

6. $(x - 2) \div (2 - x)$

7. $\dfrac{6y^3 - 9y^2}{-3y}$

8. $\dfrac{-6a^7b^6c^2}{-2a^3b^8c^2}$

9. $\left(\dfrac{-2a}{b}\right)^5$

10. $\left(\dfrac{12u^3v^9}{-3u^8v^6}\right)^2$

11. $(x^3 - 2x^2 - 4x + 3) \div (x - 3)$

12. $3x^2(5x^3 - 7x^2 + 4x - 1)$

Find the products.

13. $(x + 5)(x - 2)$

14. $(3a - 7)(2a + 5)$

15. $(a - 7)^2$

16. $(4x + 3y)^2$

17. $(b - 3)(b + 3)$

18. $(3t^2 - 7)(3t^2 + 7)$

19. $(4x^2 - 3)(x^2 + 2)$

20. $(x - 2)(x + 3)(x - 4)$

Write each expression in the form

$$quotient + \frac{remainder}{divisor}.$$

21. $\dfrac{2x}{x - 3}$

22. $\dfrac{x^2 - 3x + 5}{x + 2}$

Solve each problem.

23. Find the quotient and remainder when $x^2 - 5x + 9$ is divided by $x - 3$.

24. Subtract $3x^2 - 4x - 9$ from $x^2 - 3x + 6$.

25. The width of a pool table is x feet, and the length is 4 feet longer than the width. Find polynomials that represent the area and the perimeter of the pool table. Evaluate these polynomials for a width of 4 feet.

26. If a manufacturer charges q dollars each for footballs, then he can sell $3000 - 150q$ footballs per week. Find a polynomial that represents the revenue for one week. Find the weekly revenue if the price is \$8 for each football.

Tying It All Together CHAPTERS 1 – 3

Simplify each expression.

1. $-16 \div (-2)$

2. $(-2)^3 - 1$

3. $(-5)^2 - 3(-5) + 1$

4. $2^{10} \cdot 2^{15}$

5. $2^{15} \div 2^{10}$

6. $2^{10} - 2^5$

7. $3^2 \cdot 4^2$

8. $(172 - 85) \div (85 - 172)$

9. $(5 + 3)^2$

10. $5^2 + 3^2$

11. $(30 - 1)(30 + 1)$

12. $(30 + 1)^2$

Perform the indicated operations.

13. $(x + 3)(x + 5)$

14. $(x^2 + 8x + 15) \div (x + 5)$

15. $x + 3(x + 5)$

16. $(x^2 + 8x + 15)(x + 5)$

17. $-5t^3v \cdot 3t^2v^6$

18. $(-10t^3v^2) \div (-2t^2v)$

19. $(-6y^3 + 8y^2) \div (-2y^2)$

20. $(y^2 - 3y - 9) - (-3y^2 + 2y - 6)$

Solve each equation.

21. $2x + 1 = 0$

22. $x - 7 = 0$

23. $2x - 3 = 0$

24. $3x - 7 = 5$

25. $8 - 3x = x + 20$

26. $4 - 3(x + 2) = 0$

Solve the problem.

27. ***Average cost.*** Pineapple Recording plans to spend \$100,000 to record a new CD by the Woozies and \$2.25 per CD to manufacture the disks. The polynomial $2.25n + 100{,}000$ represents the total cost in dollars for recording and manufacturing n disks. Find an expression that represents the average cost per disk by dividing the total cost by n. Find the average cost per disk for $n = 1000$, 100,000, and 1,000,000. What happens to the large initial investment of \$100,000 if the company sells one million CDs?

CHAPTER 4

Factoring

The sport of skydiving was born in the 1930s soon after the military began using parachutes as a means of deploying troops. Today, skydiving is a popular sport around the world.

With as little as 8 hours of ground instruction, first-time jumpers can be ready to make a solo jump. Without the assistance of oxygen, sky divers can jump from as high as 14,000 feet and reach speeds of more than 100 miles per hour as they fall toward the earth. Jumpers usually open their parachutes between 2000 and 3000 feet and then gradually glide down to their landing area. If the jump and the parachute are handled correctly, the landing can be as gentle as jumping off two steps.

Making a jump and floating to earth are only part of the sport of skydiving. For example, in an activity called "relative work skydiving," a team of as many as 920 free-falling sky divers join together to make geometrically shaped formations. In a related exercise called "canopy relative work," the team members form geometric patterns after their parachutes or canopies have opened. This kind of skydiving takes skill and practice, and teams are not always successful in their attempts.

The amount of time a sky diver has for a free fall depends on the height of the jump and how much the sky diver uses the air to slow the fall. In Exercises 57 and 58 of Section 4.6 we find the amount of time that it takes a sky diver to fall from a given height.

4.1 Factoring Out Common Factors

In this section:

- Prime Factorization of Integers
- Greatest Common Factor
- Finding the Greatest Common Factor for Monomials
- Factoring Out the Greatest Common Factor
- Factoring Out the Opposite of the GCF

In Chapter 3 you learned how to multiply a monomial and a polynomial. In this section you will learn how to reverse that multiplication by finding the greatest common factor for the terms of a polynomial and then factoring the polynomial.

Prime Factorization of Integers

To **factor** an expression means to write the expression as a product. If we start with 12 and write $12 = 4 \cdot 3$, we have factored 12. Both 4 and 3 are **factors,** or **divisors,** of 12. The number 3 is a prime number, but 4 is not a prime number.

> **Prime Number**
>
> A positive integer larger than 1 that has no integral factors other than itself and 1 is called a **prime number.**

The numbers 2, 3, 5, 7, 11, 13, 17, 19, and 23 are the first nine prime numbers.

There are other factorizations of 12:

$$12 = 2 \cdot 6 \qquad 12 = 1 \cdot 12 \qquad 12 = 2 \cdot 2 \cdot 3 = 2^2 \cdot 3$$

The one that is most useful to us is $12 = 2^2 \cdot 3$. This factorization is called the **prime factorization** because it expresses 12 as a product of prime numbers.

EXAMPLE 1 **Prime factorization**

Find the prime factorization for 36.

Solution

We start by writing 36 as a product of two integers:

$$\begin{aligned} 36 &= 2 \cdot 18 && \text{Write 36 as } 2 \cdot 18. \\ &= 2 \cdot 2 \cdot 9 && \text{Replace 18 by } 2 \cdot 9. \\ &= 2 \cdot 2 \cdot 3 \cdot 3 && \text{Replace 9 by } 3 \cdot 3. \\ &= 2^2 \cdot 3^2 && \text{Use exponential notation.} \end{aligned}$$

The prime factorization of 36 is $2^2 \cdot 3^2$. ■

For larger numbers it is helpful to use the method shown in the next example.

EXAMPLE 2 **Factoring a large number**

Find the prime factorization for 420.

Solution

Start by dividing 420 by the smallest prime number that will divide into it evenly (without remainder). The smallest prime divisor of 420 is 2.

$$2\overline{)420}^{\;210}$$

Now find the smallest prime that will divide evenly into the quotient, 210. The smallest prime divisor of 210 is 2. Continue this procedure, as follows, until the quotient is a prime number:

$$\begin{array}{rl} 7 & 35 \div 5 = 7 \\ 5\overline{)35} & 105 \div 3 = 35 \\ 3\overline{)105} & 210 \div 2 = 105 \\ 2\overline{)210} & \\ \text{Start here} \rightarrow \quad 2\overline{)420} & \end{array}$$

The prime factorization of 420 is $2 \cdot 2 \cdot 3 \cdot 5 \cdot 7$, or $2^2 \cdot 3 \cdot 5 \cdot 7$. Note that it is really not necessary to divide by the smallest prime divisor at each step. We obtain the same factorization if we divide by any prime divisor at each step. ■

Greatest Common Factor

The largest integer that is a factor of two or more integers is called the **greatest common factor (GCF)** of the integers. For example, 1, 2, 3, and 6 are common factors of 18 and 24. Because 6 is the largest, 6 is the GCF of 18 and 24. We can use prime factorizations to find the GCF. For example, to find the GCF of 8 and 12, we first factor 8 and 12:

$$8 = 2 \cdot 2 \cdot 2 = 2^3 \qquad 12 = 2 \cdot 2 \cdot 3 = 2^2 \cdot 3$$

We see that the factor 2 appears twice in both 8 and 12. So 2^2, or 4, is the GCF of 8 and 12. Notice that 2 is a factor in both 2^3 and $2^2 \cdot 3$ and that 2^2 is the smallest power of 2 in these factorizations. In general, *the GCF is the product of the common prime factors in which the exponent on each common factor is the smallest power that appears on that factor in the prime factorizations.*

EXAMPLE 3 **Greatest common factor**

Find the GCF for each group of numbers.

a) 150, 225 **b)** 216, 360, 504 **c)** 55, 168

Solution

a) First find the prime factorization for each number:

$$150 = 2 \cdot 3 \cdot 5^2 \qquad 225 = 3^2 \cdot 5^2$$

Because 2 is not a factor of 225, it is not a common factor of 150 and 225. Only 3 and 5 appear in both factorizations. Looking at both $2 \cdot 3 \cdot 5^2$ and $3^2 \cdot 5^2$, we see that the smallest power of 5 is 2 and the smallest power of 3 is 1. So the GCF of 150 and 225 is $3 \cdot 5^2$, or 75.

b) First find the prime factorization for each number:

$$216 = 2^3 \cdot 3^3 \qquad 360 = 2^3 \cdot 3^2 \cdot 5 \qquad 504 = 2^3 \cdot 3^2 \cdot 7$$

The only common prime factors are 2 and 3. The smallest power of 2 in the factorizations is 3, and the smallest power of 3 is 2. So the GCF is $2^3 \cdot 3^2$, or 72.

c) First find the prime factorization for each number:

$$55 = 5 \cdot 11 \qquad 168 = 2^3 \cdot 3 \cdot 7$$

Because there are no common factors other than 1, the GCF is 1. ■

Finding the Greatest Common Factor for Monomials

To find the GCF for a group of monomials, we use the same procedure as that used for integers.

Strategy for Finding the GCF for Monomials

1. Find the GCF for the coefficients of the monomials.
2. Form the product of the GCF of the coefficients and each variable that is common to all of the monomials, where the exponent on each variable is the smallest power of that variable in any of the monomials.

EXAMPLE 4 **Greatest common factor of monomials**

Find the greatest common factor for each group of monomials.

a) $15x^2, 9x^3$ **b)** $12x^2y^2, 30x^2yz, 42x^3y$

Solution

a) The GCF for 15 and 9 is 3, and the smallest power of x is 2. So the GCF for the monomials is $3x^2$. If we write these monomials as

$$15x^2 = 5 \cdot 3 \cdot x \cdot x \quad \text{and} \quad 9x^3 = 3 \cdot 3 \cdot x \cdot x \cdot x,$$

we can see that $3x^2$ is the GCF.

b) The GCF for 12, 30, and 42 is 6. For the common variables x and y, 2 is the smallest power of x and 1 is the smallest power of y. So the GCF for the monomials is $6x^2y$. ■

Factoring Out the Greatest Common Factor

In Chapter 3 we used the distributive property to multiply monomials and polynomials. For example,

$$6(5x - 3) = 30x - 18.$$

If we start with $30x - 18$ and write

$$30x - 18 = 6(5x - 3),$$

we have factored $30x - 18$. Because 6 is the GCF of 30 and 18, we have **factored out** the GCF.

EXAMPLE 5 **Factoring out the greatest common factor**

Factor the following polynomials by factoring out the GCF.

a) $25a^2 + 40a$ **b)** $6x^4 - 12x^3 + 3x^2$

c) $x^2y^5 + x^6y^3$ **d)** $(a + b)w + (a + b)6$

Solution

a) The GCF of the coefficients 25 and 40 is 5. Because the smallest power of the common factor a is 1, we can factor $5a$ out of each term:

$$\begin{aligned}25a^2 + 40a &= 5a \cdot 5a + 5a \cdot 8\\ &= 5a(5a + 8)\end{aligned}$$

b) The GCF of 6, 12, and 3 is 3. We can factor x^2 out of each term, since the smallest power of x in the three terms is 2. So factor $3x^2$ out of each term as follows:

$$\begin{aligned}6x^4 - 12x^3 + 3x^2 &= 3x^2 \cdot 2x^2 - 3x^2 \cdot 4x + 3x^2 \cdot 1\\ &= 3x^2(2x^2 - 4x + 1)\end{aligned}$$

Check by multiplying: $3x^2(2x^2 - 4x + 1) = 6x^4 - 12x^3 + 3x^2$.

c) The GCF of the numerical coefficients is 1. Both x and y are common to each term. Using the lowest powers of x and y, we get

$$\begin{aligned}x^2y^5 + x^6y^3 &= x^2y^3 \cdot y^2 + x^2y^3 \cdot x^4\\ &= x^2y^3(y^2 + x^4).\end{aligned}$$

Check by multiplying.

d) Even though this expression looks different from the rest, we can factor it in the same way. The binomial $a + b$ is a common factor, and we can factor it out just as we factor out a monomial:

$$(a + b)w + (a + b)6 = (a + b)(w + 6)$$

CAUTION If the GCF is one of the terms of the polynomial, then you must remember to leave a 1 in place of that term when the GCF is factored out. For example,

$$ab + b = a \cdot b + 1 \cdot b = b(a + 1).$$

You should always check your answer by multiplying the factors.

Factoring Out the Opposite of the GCF

Because the greatest common factor for $-4x + 2xy$ is $2x$, we write

$$-4x + 2xy = 2x(-2 + y).$$

We could factor out $-2x$, the opposite of the greatest common factor:

$$-4x + 2xy = -2x(2 - y).$$

It will be necessary to factor out the opposite of the greatest common factor when you learn factoring by grouping in Section 4.2. Remember that you can check all factoring by multiplying the factors to see whether you get the original polynomial.

EXAMPLE 6 Factoring out the opposite of the GCF

Factor each polynomial twice. First factor out the greatest common factor, and then factor out the opposite of the GCF.

a) $3x - 3y$ **b)** $a - b$ **c)** $-x^3 + 2x^2 - 8x$

Solution

a) $3x - 3y = 3(x - y)$ Factor out 3.

$= -3(-x + y)$ Factor out -3.

Note that the signs of the terms in parentheses change when -3 is factored out. Check the answers by multiplying.

b) $a - b = 1(a - b)$ Factor out 1, the GCF of a and b.

$= -1(-a + b)$ Factor out -1.

We can also write $a - b = -1(b - a)$.

c) $-x^3 + 2x^2 - 8x = x(-x^2 + 2x - 8)$ Factor out x.

$= -x(x^2 - 2x + 8)$ Factor out $-x$.

CAUTION Be sure to change *all* of the signs in the polynomial when you factor out the opposite of the greatest common factor.

Warm-ups

True or false? Explain your answer.

1. There are only nine prime numbers.
2. The prime factorization of 32 is $2^3 \cdot 3$.
3. The integer 51 is a prime number.
4. The GCF of the integers 12 and 16 is 4.
5. The GCF of the integers 10 and 21 is 1.
6. The GCF of the polynomial $x^5y^3 - x^4y^7$ is x^4y^3.
7. For the polynomial $2x^2y - 6xy^2$ we can factor out either $2xy$ or $-2xy$.
8. The greatest common factor of the polynomial $8a^3b - 12a^2b$ is $4ab$.
9. $x - 7 = 7 - x$ for any real number x.
10. $-3x^2 + 6x = -3x(x - 2)$ for any real number x.

4.1 EXERCISES

Find the prime factorization of each integer. See Examples 1 and 2.

1. 18 **2.** 20 **3.** 52

4. 76 **5.** 98 **6.** 100

7. 460 **8.** 345 **9.** 924

10. 585

Find the greatest common factor (GCF) for each group of integers. See Example 3.

11. 8, 20 **12.** 18, 42 **13.** 36, 60

14. 42, 70 **15.** 40, 48, 88 **16.** 15, 35, 45

17. 76, 84, 100 **18.** 66, 72, 120 **19.** 39, 68, 77

20. 81, 200, 539

Find the greatest common factor (GCF) for each group of monomials. See Example 4.

21. $6x$, $8x^3$

22. $12x^2$, $4x^3$

23. $12x^3$, $4x^2$, $6x$

24. $3y^5$, $9y^4$, $15y^3$

25. $3x^2y$, $2xy^2$

26. $7a^2x^3$, $5a^3x$

27. $24a^2bc$, $60ab^2$

28. $30x^2yz^3$, $75x^3yz^6$

29. $12u^3v^2$, $25s^2t^4$

30. $45m^2n^5$, $56a^4b^8$

31. $18a^3b$, $30a^2b^2$, $54ab^3$

32. $16x^2z$, $40xz^2$, $72z^3$

Complete the factoring of each monomial.

33. $27x = 9(\ \ )$

34. $51y = 3y(\ \ )$

35. $24t^2 = 8t(\ \ )$

36. $18u^2 = 3u(\ \ )$

37. $36y^5 = 4y^2(\ \ )$

38. $42z^4 = 3z^2(\ \ )$

39. $u^4v^3 = uv(\ \ )$

40. $x^5y^3 = x^2y(\ \ )$

41. $-14m^4n^3 = 2m^4(\ \ )$

42. $-8y^3z^4 = 4z^3(\ \ )$

43. $-33x^4y^3z^2 = -3x^3yz(\ \ )$

44. $-96a^3b^4c^5 = -12ab^3c^3(\ \ )$

Factor out the GCF in each expression. See Example 5.

45. $x^3 - 6x$

46. $10y^4 - 30y^2$

47. $5ax + 5ay$

48. $6wz + 15wa$

49. $h^5 - h^3$

50. $y^6 + y^5$

51. $-2k^7m^4 + 4k^3m^6$

52. $-6h^5t^2 + 3h^3t^6$

53. $2x^3 - 6x^2 + 8x$

54. $6x^3 + 18x^2 - 24x$

55. $12x^4t + 30x^3t - 24x^2t^2$

56. $15x^2y^2 - 9xy^2 + 6x^2y$

57. $(x - 3)a + (x - 3)b$

58. $(y + 4)3 + (y + 4)z$

59. $a(y + 1)^2 + b(y + 1)^2$

60. $w(w + 2)^2 + 8(w + 2)^2$

61. $36a^3b^5 - 27a^2b^4 + 18a^2b^9$

62. $56x^3y^5 - 40x^2y^6 + 8x^2y^3$

First factor out the GCF, and then factor out the opposite of the GCF. See Example 6.

63. $8x - 8y$

64. $2a - 6b$

65. $-4x + 8x^2$

66. $-5x^2 + 10x$

67. $x - 5$

68. $a - 6$

69. $4 - 7a$

70. $7 - 5b$

71. $-24a^3 + 16a^2$

72. $-30b^4 + 75b^3$

73. $-12x^2 - 18x$

74. $-20b^2 - 8b$

75. $-2x^3 - 6x^2 + 14x$

76. $-8x^4 + 6x^3 - 2x^2$

77. $4a^3b - 6a^2b^2 - 4ab^3$

78. $12u^5v^6 + 18u^2v^3 - 15u^4v^5$

Solve each problem by factoring.

79. ***Uniform motion.*** Helen traveled a distance of $20x + 40$ miles at 20 miles per hour. Find a binomial that represents the time that she traveled.

80. ***Area of a painting.*** A rectangular painting with a width of x centimeters has an area of $x^2 + 50x$ square centimeters. Find a binomial that represents the length.

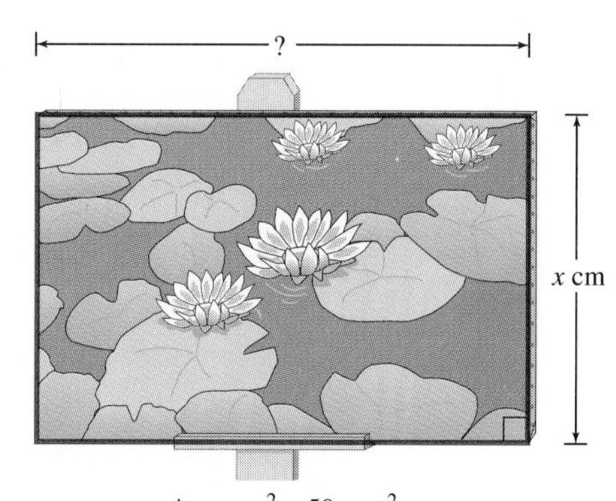

Figure for Exercise 80

81. ***Tomato soup.*** The amount of metal (in square inches) that it takes to make the can for tomato soup shown in the figure is given by the formula $S = 2\pi r^2 + 2\pi rh$. Rewrite this formula by factoring out the greatest common factor on the right-hand side.

Figure for Exercise 81

82. ***Amount of an investment.*** The amount of an investment of P dollars for t years at simple interest rate r is given by $A = P + Prt$. Rewrite this formula by factoring out the greatest common factor on the right-hand side.

Getting More Involved

83. ***Discussion.*** Is the greatest common factor of $-6x^2 + 3x$ positive or negative? Explain.

84. ***Writing.*** Explain in your own words why you use the smallest power of each common prime factor when finding the GCF of two or more integers.

4.2 Factoring the Special Products

In this section:

- Factoring a Difference of Two Squares
- Factoring a Perfect Square Trinomial
- Factoring Completely
- Factoring by Grouping

In Section 3.4 you learned how to find the special products: the square of a sum, the square of a difference, and the product of a sum and a difference. In this section you will learn how to reverse those operations.

Factoring a Difference of Two Squares

In Section 3.4 you learned that the product of a sum and a difference is a difference of two squares:

$$(a + b)(a - b) = a^2 - b^2$$

So a difference of two squares can be factored as a product of a sum and a difference, using the following rule.

Factoring a Difference of Two Squares

For any real numbers a and b,

$$a^2 - b^2 = (a + b)(a - b).$$

Note that the square of an integer is a perfect square. For example, 64 is a perfect square because $64 = 8^2$. The square of a monomial in which the coefficient is an integer is also called a **perfect square** or simply a **square.** For example, $9m^2$ is a perfect square because $9m^2 = (3m)^2$.

EXAMPLE 1 **Factoring a difference of two squares**

Factor each polynomial.

a) $y^2 - 81$ **b)** $9m^2 - 16$ **c)** $4x^2 - 9y^2$

Solution

a) Because $81 = 9^2$, the binomial $y^2 - 81$ is a difference of two squares:

$$y^2 - 81 = y^2 - 9^2 \quad \text{Rewrite as a difference of two squares.}$$
$$= (y + 9)(y - 9) \quad \text{Factor.}$$

Check by multiplying.

b) Because $9m^2 = (3m)^2$ and $16 = 4^2$, the binomial $9m^2 - 16$ is a difference of two squares:

$$9m^2 - 16 = (3m)^2 - 4^2 \quad \text{Rewrite as a difference of two squares.}$$
$$= (3m + 4)(3m - 4) \quad \text{Factor.}$$

Check by multiplying.

c) Because $4x^2 = (2x)^2$ and $9y^2 = (3y)^2$, the binomial $4x^2 - 9y^2$ is a difference of two squares:

$$4x^2 - 9y^2 = (2x + 3y)(2x - 3y)$$

■

Factoring a Perfect Square Trinomial

In Section 3.4 you learned how to square a binomial using the rule

$$(a + b)^2 = a^2 + 2ab + b^2.$$

You can reverse this rule to factor a trinomial such as $x^2 + 6x + 9$. Notice that

$$x^2 + 6x + 9 = \underset{\substack{\uparrow \\ a^2}}{x^2} + \underbrace{2 \cdot x \cdot 3}_{2ab} + \underset{\substack{\uparrow \\ b^2}}{3^2}.$$

So if $a = x$ and $b = 3$, then $x^2 + 6x + 9$ fits the form $a^2 + 2ab + b^2$, and

$$x^2 + 6x + 9 = (x + 3)^2.$$

A trinomial that is of the form $a^2 + 2ab + b^2$ is called a **perfect square trinomial.** A perfect square trinomial is the square of a binomial. Perfect square trinomials can be identified by using the following strategy.

Strategy for Identifying a Perfect Square Trinomial

A trinomial is a perfect square trinomial if:

1. the first and last terms are of the form a^2 and b^2 (perfect squares)
2. the middle term is $2ab$ or $-2ab$.

EXAMPLE 2 **Identifying the special products**

Determine whether each binomial is a difference of two squares and whether each trinomial is a perfect square trinomial.

a) $x^2 - 14x + 49$ **b)** $4x^2 - 81$

c) $4a^2 + 24a + 25$ **d)** $9y^2 - 24y - 16$

Solution

a) The first term is x^2, and the last term is 7^2. The middle term, $-14x$, is $-2 \cdot x \cdot 7$. So this trinomial is a perfect square trinomial.

b) Both terms of $4x^2 - 81$ are perfect squares, $(2x)^2$ and 9^2. So $4x^2 - 81$ is a difference of two squares.

c) The first term of $4a^2 + 24a + 25$ is $(2a)^2$ and the last term is 5^2. However, $2 \cdot 2a \cdot 5$ is $20a$. Because the middle term is $24a$, this trinomial is not a perfect square trinomial.

d) The first and last terms in a perfect square trinomial are both positive. Because the last term in $9y^2 - 24y - 16$ is negative, the trinomial is not a perfect square trinomial. ■

Note that the middle term in a perfect square trinomial may have a positive or a negative coefficient, while the first and last terms must be positive. Any perfect square trinomial can be factored as the square of a binomial by using the following rule.

Factoring Perfect Square Trinomials

For any real numbers a and b,

$$a^2 + 2ab + b^2 = (a + b)^2$$
$$a^2 - 2ab + b^2 = (a - b)^2.$$

EXAMPLE 3 **Factoring perfect square trinomials**

Factor.

a) $x^2 - 4x + 4$ b) $a^2 + 16a + 64$ c) $4x^2 - 12x + 9$

Solution

a) The first term is x^2, and the last term is 2^2. Because the middle term is $-2 \cdot 2 \cdot x$, or $-4x$, this polynomial is a perfect square trinomial:

$$x^2 - 4x + 4 = (x - 2)^2$$

Check by finding $(x - 2)^2$.

b) $a^2 + 16a + 64 = (a + 8)^2$
Check by finding $(a + 8)^2$.

c) The first term is $(2x)^2$, and the last term is 3^2. Because $-2 \cdot 2x \cdot 3 = -12x$, the polynomial is a perfect square trinomial. So

$$4x^2 - 12x + 9 = (2x - 3)^2.$$

Check by finding $(2x - 3)^2$. ■

Factoring Completely

To factor a polynomial means to write it as a product of simpler polynomials. A polynomial that cannot be factored is called a **prime polynomial,** and any monomial is a prime polynomial. For example, $6x + 1$ is a prime polynomial because it cannot be factored, and $12x^3$ is a prime polynomial because it is a monomial. A polynomial is **factored completely** when it is written as a product of prime polynomials.

Some polynomials have a factor common to all terms. To factor such polynomials completely, it is simpler to factor out the greatest common factor (GCF) and then factor the remaining polynomial. The following example illustrates factoring completely.

EXAMPLE 4 **Factoring completely**

Factor each polynomial completely.

a) $2x^3 - 50x$ b) $8x^2y - 32xy + 32y$

Solution

a) The greatest common factor of $2x^3$ and $50x$ is $2x$:

$$2x^3 - 50x = 2x(x^2 - 25) \quad \text{Check this step by multiplying.}$$
$$= 2x(x + 5)(x - 5) \quad \text{Difference of two squares}$$

b) $8x^2y - 32xy + 32y = 8y(x^2 - 4x + 4)$ Check this step by multiplying.

$= 8y(x - 2)^2$ Perfect square trinomial

Even though 8 is not a prime number, $8y$ is a prime polynomial because it is a monomial. So the polynomial is factored completely. ■

Remember that factoring reverses multiplication and *every step of factoring can be checked by multiplication.*

Factoring by Grouping

The product of two binomials may be a polynomial with four terms. For example,

$$(x + a)(x + 3) = (x + a)x + (x + a)3$$
$$= x^2 + ax + 3x + 3a.$$

We can factor a polynomial of this type by simply reversing the steps we used to find the product. To reverse these steps, we factor out common factors from the first two terms and from the last two terms. This procedure is called **factoring by grouping.**

EXAMPLE 5 **Factoring by grouping**

Use grouping to factor each polynomial completely.

a) $xy + 2y + 3x + 6$ **b)** $2x^3 - 3x^2 - 2x + 3$ **c)** $ax + 3y - 3x - ay$

Solution

a) Notice that the first two terms have a common factor of y and the last two terms have a common factor of 3:

$xy + 2y + 3x + 6 = (xy + 2y) + (3x + 6)$ Use the associative property to group the terms.

$= y(x + 2) + 3(x + 2)$ Factor out the common factors in each group.

$= (y + 3)(x + 2)$ Factor out $x + 2$.

b) We can factor x^2 out of the first two terms and 1 out of the last two terms:

$2x^3 - 3x^2 - 2x + 3 = (2x^3 - 3x^2) + (-2x + 3)$ Group the terms.

$= x^2(2x - 3) + 1(-2x + 3)$

However, we cannot proceed any further because $2x - 3$ and $-2x + 3$ are not the same. To get $2x - 3$ as a common factor, we must factor out -1 from the last two terms:

$2x^3 - 3x^2 - 2x + 3 = x^2(2x - 3) - 1(2x - 3)$ Factor out the common factors.

$= (x^2 - 1)(2x - 3)$ Factor out $2x - 3$.

$= (x - 1)(x + 1)(2x - 3)$ Difference of two squares

c) In $ax + 3y - 3x - ay$ there are no common factors in the first two or the last two terms. However, if we rewrite the polynomial as $ax - 3x - ay + 3y$, then we can factor by grouping:

$ax + 3y - 3x - ay = ax - 3x - ay + 3y$ Rearrange the terms.

$= x(a - 3) - y(a - 3)$ Factor out x and $-y$.

$= (x - y)(a - 3)$ Factor out $a - 3$. ■

Warm-ups

True or false? Explain your answer.

1. The polynomial $x^2 + 16$ is a difference of two squares.
2. The polynomial $x^2 - 8x + 16$ is a perfect square trinomial.
3. The polynomial $9x^2 + 21x + 49$ is a perfect square trinomial.
4. $4x^2 + 4 = (2x + 2)^2$ for any real number x.
5. A difference of two squares is equal to a product of a sum and a difference.
6. The monomial $16y^2$ is a prime polynomial.
7. The polynomial $x^2 + 9$ can be factored as $(x + 3)(x + 3)$.
8. The polynomial $4x^2 - 4$ is factored completely as $4(x^2 - 1)$.
9. $y^2 - 2y + 1 = (y - 1)^2$ for any real number y.
10. $2x^2 - 18 = 2(x - 3)(x + 3)$ for any real number x.

4.2 EXERCISES

Factor each polynomial. See Example 1.

1. $a^2 - 4$ **2.** $h^2 - 9$

3. $x^2 - 49$ **4.** $y^2 - 36$

5. $y^2 - 9x^2$ **6.** $16x^2 - y^2$

7. $25a^2 - 49b^2$ **8.** $9a^2 - 64b^2$

9. $121m^2 - 1$ **10.** $144n^2 - 1$

11. $9w^2 - 25c^2$ **12.** $144w^2 - 121a^2$

Determine whether each binomial is a difference of two squares and whether each trinomial is a perfect square trinomial. See Example 2.

13. $x^2 - 20x + 100$ **14.** $x^2 - 10x - 25$

15. $y^2 - 40$ **16.** $a^2 - 49$

17. $4y^2 + 12y + 9$ **18.** $9a^2 - 30a - 25$

19. $x^2 - 8x + 64$ **20.** $x^2 + 4x + 4$

21. $9y^2 - 25c^2$ **22.** $9x^2 + 4$

23. $9a^2 + 6ab + b^2$ **24.** $4x^2 - 4xy + y^2$

Factor each perfect square trinomial. See Example 3.

25. $x^2 + 12x + 36$ **26.** $y^2 + 14y + 49$

27. $a^2 - 4a + 4$ **28.** $b^2 - 6b + 9$

29. $4w^2 + 4w + 1$ **30.** $9m^2 + 6m + 1$

31. $16x^2 - 8x + 1$ **32.** $25y^2 - 10y + 1$

33. $4t^2 + 20t + 25$ **34.** $9y^2 - 12y + 4$

35. $9w^2 + 42w + 49$ **36.** $144x^2 + 24x + 1$

37. $n^2 + 2nt + t^2$ **38.** $x^2 - 2xy + y^2$

Factor each polynomial completely. See Example 4.

39. $5x^2 - 125$ **40.** $3y^2 - 27$

41. $-2x^2 + 18$ **42.** $-5y^2 + 20$

43. $a^3 - ab^2$ **44.** $x^2y - y$

45. $3x^2 + 6x + 3$ **46.** $12a^2 + 36a + 27$

47. $-5y^2 + 50y - 125$ **48.** $-2a^2 - 16a - 32$

49. $x^3 - 2x^2y + xy^2$ **50.** $x^3y + 2x^2y^2 + xy^3$

51. $-3x^2 + 3y^2$ **52.** $-8a^2 + 8b^2$

53. $2ax^2 - 98a$ **54.** $32x^2y - 2y^3$

55. $3ab^2 - 18ab + 27a$ **56.** $-2a^2b + 8ab - 8b$

57. $-4m^3 + 24m^2n - 36mn^2$

58. $10a^3 - 20a^2b + 10ab^2$

Use grouping to factor each polynomial completely. See Example 5.

59. $bx + by + cx + cy$ **60.** $3x + 3z + ax + az$

61. $x^3 + x^2 - 4x - 4$ **62.** $x^3 + x^2 - x - 1$

63. $3a - 3b - xa + xb$ **64.** $ax - bx - 4a + 4b$

65. $a^3 + 3a^2 + a + 3$ **66.** $y^3 - 5y^2 + 8y - 40$

67. $xa + ay + 3y + 3x$ **68.** $x^3 + ax + 3a + 3x^2$

69. $abc - 3 + c - 3ab$ **70.** $xa + tb + ba + tx$

71. $x^2a - b + bx^2 - a$

72. $a^2m - b^2n + a^2n - b^2m$

73. $y^2 + y + by + b$

74. $ac + mc + aw^2 + mw^2$

Factor each polynomial completely.

75. $6a^3y + 24a^2y^2 + 24ay^3$

76. $8b^5c - 8b^4c^2 + 2b^3c^3$

77. $24a^3y - 6ay^3$

78. $27b^3c - 12bc^3$

79. $2a^3y^2 - 6a^2y$

80. $9x^3y - 18x^2y^2$

81. $ab + 2bw - 4aw - 8w^2$

82. $3am - 6n - an + 18m$

Solve each problem.

83. ***Skydiving.*** The height (in feet) above the earth for a sky diver t seconds after jumping from an airplane at 6400 ft is approximated by the formula $h = -16t^2 + 6400$, provided that $t < 5$. Rewrite the formula with the right-hand side factored completely. Use your revised formula to find h when $t = 2$.

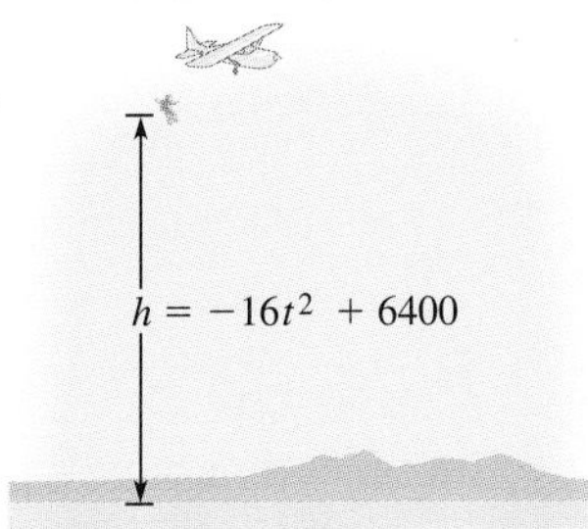

Figure for Exercise 83

84. ***Demand for pools.*** Tropical Pools sells an aboveground model for p dollars each. The monthly revenue from the sale of this model depends on the price and is given by $R = -0.08p^2 + 300p$. Revenue is the product of the price p and the demand (quantity sold). Factor out the price p on the right-hand side of the formula to find the monthly demand. What is the demand for this pool when the price is $3000?

85. ***Volume of a tank.*** The volume of a fish tank with a square base and height y is $y^3 - 6y^2 + 9y$ cubic inches. Find the length of a side of the square base.

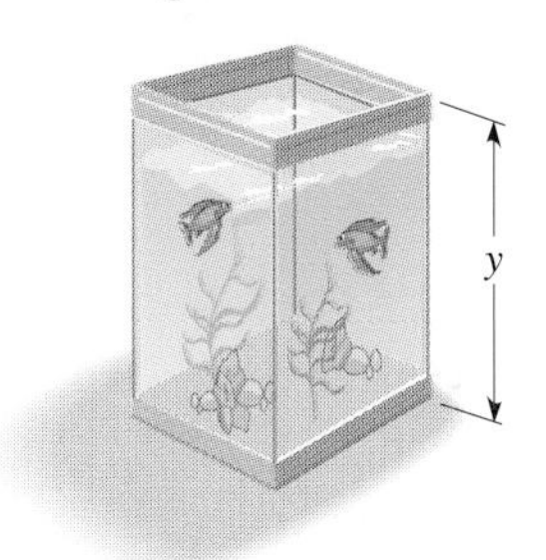

Figure for Exercise 85

Getting More Involved

86. ***Discussion.*** For what real number k, does $3x^2 - k$ factor as $3(x - 2)(x + 2)$?

87. ***Writing.*** Explain in your own words how to factor a four-term polynomial by grouping.

88. ***Writing.*** Explain how you know that $x^2 + 1$ is a prime polynomial.

4.3 Factoring Trinomials

In this section:

- Factoring $ax^2 + bx + c$ with $a = 1$
- Factoring Completely

In this section we will factor the type of trinomials that result from multiplying two different binomials. We will do this only for trinomials in which the coefficient of x^2, the leading coefficient, is 1. Factoring trinomials with leading coefficient not equal to 1 will be done in Section 4.4.

Factoring $ax^2 + bx + c$ with $a = 1$

Let's look closely at an example of finding the product of two binomials using the distributive property:

$$\begin{aligned}(x + 2)(x + 3) &= (x + 2)x + (x + 2)3 && \text{Distributive property}\\ &= x^2 + 2x + 3x + 6 && \text{Distributive property}\\ &= x^2 + 5x + 6 && \text{Combine like terms.}\end{aligned}$$

To factor $x^2 + 5x + 6$, we need to reverse these steps. First observe that the coefficient 5 is the sum of two numbers that have a product of 6. The only numbers that have a product of 6 and a sum of 5 are 2 and 3. So write $5x$ as $2x + 3x$:

$$x^2 + 5x + 6 = x^2 + 2x + 3x + 6$$

Now we can factor the polynomial $x^2 + 2x + 3x + 6$ by grouping:

$$\begin{aligned} x^2 + 5x + 6 &= (x^2 + 2x) + (3x + 6) && \text{Group terms together.} \\ &= (x + 2)x + (x + 2)3 && \text{Factor out the common factors.} \\ &= (x + 2)(x + 3) && \text{Factor out } x + 2. \end{aligned}$$

EXAMPLE 1 **Factoring a trinomial**

Factor.

a) $x^2 + 8x + 12$

b) $a^2 - 9a + 20$

Solution

a) To factor $x^2 + 8x + 12$, we must find two integers that have a product of 12 and a sum of 8. The pairs of integers with a product of 12 are 1 and 12, 2 and 6, and 3 and 4. Only 2 and 6 have a sum of 8. So write $8x$ as $2x + 6x$ and factor by grouping:

$$\begin{aligned} x^2 + 8x + 12 &= x^2 + 2x + 6x + 12 \\ &= (x + 2)x + (x + 2)6 && \text{Factor out the common factors.} \\ &= (x + 2)(x + 6) && \text{Factor out } x + 2. \end{aligned}$$

Check by using FOIL: $(x + 2)(x + 6) = x^2 + 8x + 12$.

b) To factor $a^2 - 9a + 20$, we need two integers that have a product of 20 and a sum of -9. The integers are -4 and -5. Now replace $-9a$ by $-4a - 5a$ and factor by grouping:

$$\begin{aligned} a^2 - 9a + 20 &= a^2 - 4a - 5a + 20 && \text{Replace } -9a \text{ by } -4a - 5a. \\ &= a(a - 4) - 5(a - 4) && \text{Factor by grouping.} \\ &= (a - 5)(a - 4) && \text{Factor out } a - 4. \end{aligned}$$

■

After sufficient practice factoring trinomials, you may be able to skip most of the steps shown in these examples. For example, to factor $x^2 + x - 6$, simply find a pair of integers with a product of -6 and a sum of 1. The integers are 3 and -2, so we can write

$$x^2 + x - 6 = (x + 3)(x - 2)$$

and check by using FOIL.

EXAMPLE 2 **Factoring trinomials**

Factor.

a) $x^2 + 5x + 4$

b) $w^2 - 5w - 24$

Solution

a) To get a product of 4 and a sum of 5, use 1 and 4:

$$x^2 + 5x + 4 = (x + 1)(x + 4)$$

Check by using FOIL on $(x + 1)(x + 4)$.

b) To get a product of -24 and a sum of -5, use -8 and 3:

$$w^2 - 5w - 24 = (w - 8)(w + 3)$$

Check by using FOIL.

■

If a trinomial is not in the form $ax^2 + bx + c$, rewrite it in this form before factoring. In the next example we factor a polynomial that must be rewritten and one that has two variables.

EXAMPLE 3 **Factoring trinomials**

Factor.

a) $2x - 8 + x^2$ **b)** $a^2 - 7ab + 10b^2$

Solution

a) Before factoring, write the trinomial as $x^2 + 2x - 8$. Now, to get a product of -8 and a sum of 2, use -2 and 4:

$$\begin{aligned} 2x - 8 + x^2 &= x^2 + 2x - 8 && \text{Write in } ax^2 + bx + c \text{ form.} \\ &= (x + 4)(x - 2) && \text{Factor and check by multiplying.} \end{aligned}$$

b) To get a product of 10 and a sum of -7, use -5 and -2. So to get a product of $10b^2$, we use $-5b$ and $-2b$:

$$a^2 - 7ab + 10b^2 = (a - 5b)(a - 2b)$$

To factor $x^2 + bx + c$, we try pairs of integers that have a product of c until we find a pair that has a sum of b. If there is no such pair of integers, then the polynomial cannot be factored and it is a prime polynomial. Before you can conclude that a polynomial is prime, you must try *all* possibilities.

EXAMPLE 4 **Prime polynomials**

Factor.

a) $x^2 + 7x - 6$ **b)** $x^2 + 9$

Solution

a) Because the last term is -6, we want a positive integer and a negative integer that have a product of -6 and a sum of 7. Check all possible pairs of integers:

Product	Sum
$-6 = (-1)(6)$	$-1 + 6 = 5$
$-6 = (1)(-6)$	$1 + (-6) = -5$
$-6 = (2)(-3)$	$2 + (-3) = -1$
$-6 = (-2)(3)$	$-2 + 3 = 1$

None of these possible factors of -6 have a sum of 7, so we can be certain that $x^2 + 7x - 6$ cannot be factored. It is a prime polynomial.

b) Because the x-term is missing in $x^2 + 9$, its coefficient, b, is 0. So we seek two positive integers or two negative integers that have a product of 9 and a sum of 0. Check all possibilities:

Product	Sum
$9 = (3)(3)$	$3 + 3 = 6$
$9 = (-3)(-3)$	$-3 + (-3) = -6$
$9 = (9)(1)$	$9 + 1 = 10$
$9 = (-9)(-1)$	$-9 + (-1) = -10$

None of these pairs of integers have a sum of 0, so we can conclude that $x^2 + 9$ is a prime polynomial. Note that $x^2 + 9$ does not factor as $(x + 3)^2$ because $(x + 3)^2 = x^2 + 6x + 9$.

The prime polynomial $x^2 + 9$ in Example 4(b) is a sum of two squares. It can be shown that any sum of two squares (in which there are no common factors) is a prime polynomial.

Sum of Two Squares

If a sum of two squares, $a^2 + b^2$, has no common factor other than 1, then it is a prime polynomial.

Factoring Completely

In Section 4.2 you learned that any monomial is a prime polynomial. You also learned that binomials such as $3x - 5$ (with no common factor) are prime polynomials. In Example 4 of this section we saw a trinomial that is a prime polynomial. There are infinitely many prime trinomials. When factoring a polynomial completely, we could have a factor that is a prime trinomial.

EXAMPLE 5 **Factoring completely**

Factor each polynomial completely.

a) $x^3 - 6x^2 - 16x$ **b)** $4x^3 + 4x^2 + 4x$

Solution

a)
$$\begin{aligned} x^3 - 6x^2 - 16x &= x(x^2 - 6x - 16) && \text{Factor out the GCF.} \\ &= x(x - 8)(x + 2) && \text{Factor } x^2 - 6x - 16. \end{aligned}$$

b) First factor out $4x$, the greatest common factor:

$$4x^3 + 4x^2 + 4x = 4x(x^2 + x + 1)$$

To factor $x^2 + x + 1$, we would need two integers with a product of 1 and a sum of 1. Because there are no such integers, $x^2 + x + 1$ is prime, and the factorization is complete. ■

Warm-ups

True or false? Answer true if the correct factorization is given and false if the factorization is incorrect. Explain your answer.

1. $x^2 - 6x + 9 = (x - 3)^2$
2. $x^2 + 6x + 9 = (x + 3)^2$
3. $x^2 + 10x + 9 = (x - 9)(x - 1)$
4. $x^2 - 8x - 9 = (x - 8)(x - 9)$
5. $x^2 + 8x - 9 = (x + 9)(x - 1)$
6. $x^2 + 8x + 9 = (x + 3)^2$
7. $x^2 - 10xy + 9y^2 = (x - y)(x - 9y)$
8. $x^2 + x + 1 = (x + 1)(x + 1)$
9. $x^2 + xy + 20y^2 = (x + 5y)(x - 4y)$
10. $x^2 + 1 = (x + 1)(x + 1)$

4.3 EXERCISES

Factor each trinomial. Write out all of the steps as shown in Example 1.

1. $x^2 + 4x + 3$

2. $y^2 + 6y + 5$

3. $x^2 + 9x + 18$

4. $w^2 + 6w + 8$

5. $a^2 - 7a + 12$

6. $m^2 - 9m + 14$

7. $b^2 - 5b - 6$

8. $a^2 + 5a - 6$

Factor each polynomial. See Examples 2–4. If the polynomial is prime, say so.

9. $y^2 + 7y + 10$

10. $x^2 + 8x + 15$

11. $a^2 - 6a + 8$

12. $b^2 - 8b + 15$

13. $m^2 - 10m + 16$

14. $m^2 - 17m + 16$

15. $w^2 + 9w - 10$

16. $m^2 + 6m - 16$

17. $w^2 - 2w - 8$

18. $m^2 - 6m - 16$

19. $a^2 - 2a - 12$

20. $x^2 + 3x + 3$

21. $m^2 + 15m - 16$

22. $y^2 + 3y - 10$

23. $a^2 - 4a + 12$

24. $y^2 - 6y - 8$

25. $z^2 - 25$

26. $p^2 - 1$

27. $h^2 + 49$

28. $q^2 + 4$

29. $m^2 + 12m + 20$

30. $m^2 + 21m + 20$

31. $t^2 - 3t + 10$

32. $x^2 - 5x - 3$

33. $m^2 - 17m - 18$

34. $h^2 + 5h - 36$

35. $m^2 - 23m + 24$

36. $m^2 + 23m + 24$

37. $5t - 24 + t^2$

38. $t^2 - 24 - 10t$

39. $t^2 - 2t - 24$

40. $t^2 + 14t + 24$

41. $t^2 - 10t - 200$

42. $t^2 + 30t + 200$

43. $x^2 - 5x - 150$

44. $x^2 - 25x + 150$

45. $13y + 30 + y^2$

46. $18z + 45 + z^2$

47. $x^2 - 4xy - 12y^2$

48. $y^2 + yt - 12t^2$

49. $x^2 - 13xy + 12y^2$

50. $h^2 - 9hs + 9s^2$

51. $x^2 - 5xs - 24s^2$

52. $x^2 + 4xz - 32z^2$

Factor each polynomial completely. Use the methods discussed in Sections 4.1 through 4.3. See Example 5.

53. $w^2 - 8w$

54. $x^4 - x^3$

55. $2w^2 - 162$

56. $6w^4 - 54w^2$

57. $x^2w^2 + 9x^2$

58. $a^4b + a^2b^3$

59. $w^2 - 18w + 81$

60. $w^2 + 30w + 81$

61. $6w^2 - 12w - 18$

62. $9w - w^3$

63. $32x^2 - 2x^4$

64. $20w^2 + 100w + 40$

65. $3w^2 + 27w + 54$

66. $w^3 - 3w^2 - 18w$

67. $18w^2 + w^3 + 36w$

68. $18a^2 + 3a^3 + 36a$

69. $8vw^2 + 32vw + 32v$

70. $3h^2t + 6ht + 3t$

71. $6x^3y + 30x^2y^2 + 36xy^3$

72. $3x^3y^2 - 3x^2y^2 + 3xy^2$

Use factoring to solve each problem.

73. ***Area of a deck.*** A rectangular deck has an area of $x^2 + 6x + 8$ square feet and a width of $x + 2$ feet. Find the length of the deck.

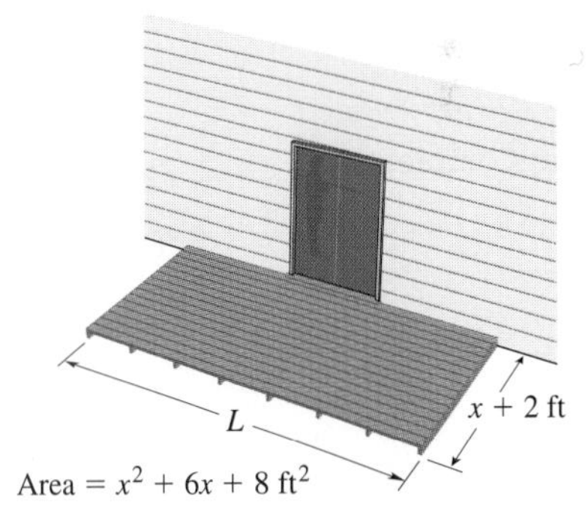

Figure for Exercise 73

74. ***Area of a sail.*** A triangular sail has an area of $x^2 + 5x + 6$ square meters and a height of $x + 3$ meters. Find the length of the sail's base.

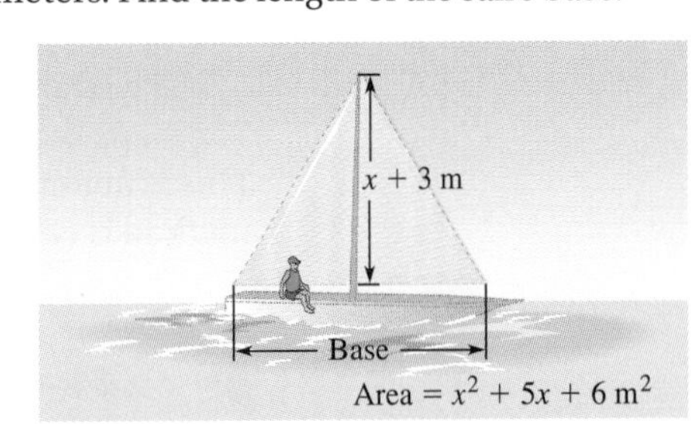

Figure for Exercise 74

75. ***Volume of a cube.*** Hector designed a cubic box with volume x^3 cubic feet. After increasing the dimensions of the bottom, the box has a volume of $x^3 + 8x^2 + 15x$ cubic feet. If each of the dimensions of the bottom was increased by a whole number of feet, then how much was each increase?

76. ***Volume of a container.*** A cubic shipping container had a volume of a^3 cubic meters. The height was decreased by a whole number of meters and the width was increased by a whole number of meters so that the volume of the container is now $a^3 + 2a^2 - 3a$ cubic meters. By how many meters were the height and width changed?

Getting More Involved

77. ***Discussion.*** Which of the following products is not equivalent to the others. Explain your answer.
    a) $(2x - 4)(x + 3)$ **b)** $(x - 2)(2x + 6)$
    c) $2(x - 2)(x + 3)$ **d)** $(2x - 4)(2x + 6)$

78. ***Discussion.*** When asked to factor completely a certain polynomial, four students gave the following answers. Only one student gave the correct answer. Which one must it be? Explain your answer.
    a) $3(x^2 - 2x - 15)$ **b)** $(3x - 5)(5x - 15)$
    c) $3(x - 5)(x - 3)$ **d)** $(3x - 15)(x - 3)$

4.4 Factoring More Trinomials

In this section:
- Factoring $ax^2 + bx + c$ with $a \neq 1$
- Trial and Error
- Factoring Completely

In Section 4.3 we factored trinomials with a leading coefficient of 1. In this section we will use a slightly different technique to factor trinomials with leading coefficients not equal to 1.

Factoring $ax^2 + bx + c$ with $a \neq 1$

If the leading coefficient of a trinomial is not 1, we can again use grouping to factor the trinomial. However, the procedure is slightly different.

Consider the trinomial $2x^2 + 7x + 6$. First find the product of the leading coefficient and the constant term. In this case it is $2 \cdot 6$, or 12. Now find two numbers with a product of 12 and a sum of 7. The pairs of numbers with a product of 12 are 1 and 12, 2 and 6, and 3 and 4. Only 3 and 4 have a product of 12 and a sum of 7. Now replace $7x$ by $3x + 4x$ and factor by grouping:

$$2x^2 + 7x + 6 = 2x^2 + 3x + 4x + 6 \quad \text{Replace } 7x \text{ by } 3x + 4x.$$
$$= (2x + 3)x + (2x + 3)2 \quad \text{Factor out the common factors.}$$
$$= (2x + 3)(x + 2) \quad \text{Factor out } 2x + 3.$$

The strategy for factoring a trinomial is summarized in the following box. The steps listed here actually work whether or not the leading coefficient is 1. This is actually the method that you learned in Section 4.3 with $a = 1$. This method is called the ***ac* method.**

Strategy for Factoring $ax^2 + bx + c$ by the ac Method

To factor the trinomial $ax^2 + bx + c$:

1. Find two numbers that have a product equal to ac and a sum equal to b.
2. Replace bx by two terms using the two new numbers as coefficients.
3. Factor the resulting four-term polynomial by grouping.

EXAMPLE 1 **The *ac* method**

Factor each trinomial.

a) $2x^2 + x - 6$ **b)** $10x^2 + 13x - 3$

Solution

a) Because $2 \cdot (-6) = -12$, we need two integers with a product of -12 and a sum of 1. We can list the possible pairs of integers with a product of -12:

1 and -12	2 and -6	3 and -4
-1 and 12	-2 and 6	-3 and 4

Only -3 and 4 have a sum of 1. Replace x by $-3x + 4x$ and factor by grouping:

$$\begin{aligned} 2x^2 + x - 6 &= 2x^2 - 3x + 4x - 6 && \text{Replace } x \text{ by } -3x + 4x. \\ &= (2x - 3)x + (2x - 3)2 && \text{Factor out the common factors.} \\ &= (2x - 3)(x + 2) && \text{Factor out } 2x - 3. \end{aligned}$$

Check by FOIL.

b) Because $10 \cdot (-3) = -30$, we need two integers with a product of -30 and a sum of 13. The product is negative, so the integers must have opposite signs. We can list all pairs of factors of -30 as follows:

1 and -30	2 and -15	3 and -10	5 and -6
-1 and 30	-2 and 15	-3 and 10	-5 and 6

The only pair that has a sum of 13 is -2 and 15:

$$\begin{aligned} 10x^2 + 13x - 3 &= 10x^2 - 2x + 15x - 3 && \text{Replace } 13x \text{ by } -2x + 15x. \\ &= (5x - 1)2x + (5x - 1)3 && \text{Factor out the common factors.} \\ &= (5x - 1)(2x + 3) && \text{Factor out } 5x - 1. \end{aligned}$$

Check by FOIL. ■

Trial and Error

After you have gained some experience at factoring by the *ac* method, you can often find the factors without going through the steps of grouping. For example, consider the polynomial

$$3x^2 + 7x - 6.$$

The factors of $3x^2$ can only be $3x$ and x. The factors of 6 could be 2 and 3 or 1 and 6. We can list all of the possibilities that give the correct first and last terms, without regard to the signs:

$$(3x \quad 3)(x \quad 2) \qquad (3x \quad 2)(x \quad 3) \qquad (3x \quad 6)(x \quad 1) \qquad (3x \quad 1)(x \quad 6)$$

Because the factors of -6 have unlike signs, one binomial factor is a sum and the other binomial is a difference. Now try each product to see which has a middle term of $7x$:

$$\begin{aligned} (3x + 3)(x - 2) &= 3x^2 - 3x - 6 && \text{Incorrect.} \\ (3x - 3)(x + 2) &= 3x^2 + 3x - 6 && \text{Incorrect.} \\ (3x + 2)(x - 3) &= 3x^2 - 7x - 6 && \text{Incorrect.} \\ (3x - 2)(x + 3) &= 3x^2 + 7x - 6 && \text{Correct.} \end{aligned}$$

Even though there may be many possibilities in some factoring problems, it is often possible to find the correct factors without writing down every possibility. We can use a bit of guesswork in factoring trinomials. *Try* whichever possibility you think might work. *Check* it by multiplying. If it is not right, then *try again.* That is why this method is called **trial and error.**

EXAMPLE 2 **Trial and error**

Factor each trinomial using trial and error.

a) $2x^2 + 5x - 3$ **b)** $3x^2 - 11x + 6$

Solution

a) Because $2x^2$ factors only as $2x \cdot x$ and 3 factors only as $1 \cdot 3$, there are only two possible ways to get the correct first and last terms, without regard to the signs:

$$(2x \quad 1)(x \quad 3) \qquad \text{and} \qquad (2x \quad 3)(x \quad 1)$$

Because the last term of the trinomial is negative, one of the missing signs must be $+$, and the other must be $-$. The trinomial is factored correctly as

$$2x^2 + 5x - 3 = (2x - 1)(x + 3).$$

Check by using FOIL.

b) There are four possible ways to factor $3x^2 - 11x + 6$:

$$(3x \quad 1)(x \quad 6) \qquad (3x \quad 2)(x \quad 3)$$
$$(3x \quad 6)(x \quad 1) \qquad (3x \quad 3)(x \quad 2)$$

Because the last term in $3x^2 - 11x + 6$ is positive and the middle term is negative, both signs in the factors must be negative. The trinomial is factored correctly as

$$3x^2 - 11x + 6 = (3x - 2)(x - 3).$$

Check by using FOIL. ■

In the next example we factor a trinomial that has two variables.

EXAMPLE 3 **Factoring a trinomial with two variables**

Factor $6x^2 - 7xy + 2y^2$.

Solution

We list the possible ways to factor the trinomial:

$$(3x \quad 2y)(2x \quad y) \qquad (3x \quad y)(2x \quad 2y) \qquad (6x \quad 2y)(x \quad y) \qquad (6x \quad y)(x \quad 2y)$$

Because the last term of the trinomial is positive and the middle term is negative, both factors must contain subtraction symbols. To get the middle term of $-7xy$, we use the first possibility listed:

$$6x^2 - 7xy + 2y^2 = (3x - 2y)(2x - y)$$ ■

Factoring Completely

You can use the latest factoring technique along with the techniques that you learned earlier to factor polynomials completely. Remember always to first factor out the greatest common factor (if it is not 1).

EXAMPLE 4 **Factoring completely**

Factor each polynomial completely.

a) $4x^3 + 14x^2 + 6x$ **b)** $12x^2y + 6xy + 6y$

Solution

a) $4x^3 + 14x^2 + 6x = 2x(2x^2 + 7x + 3)$ Factor out the GCF, $2x$.

$= 2x(2x + 1)(x + 3)$ Factor $2x^2 + 7x + 3$.

Check by multiplying.

b) $12x^2y + 6xy + 6y = 6y(2x^2 + x + 1)$ Factor out the GCF, $6y$.

To factor $2x^2 + x + 1$ by the ac method, we need two numbers with a product of 2 and a sum of 1. Because there are no such numbers, $2x^2 + x + 1$ is prime and the factorization is complete. ■

Usually, our first step in factoring is to factor out the greatest common factor (if it is not 1). If the first term of a polynomial has a negative coefficient, then it is better to factor out the opposite of the GCF.

EXAMPLE 5 **Factoring out the opposite of the GCF**

Factor each polynomial completely.

a) $-18x^3 + 51x^2 - 15x$ **b)** $-3a^2 + 2a + 21$

Solution

a) The GCF is $3x$. Because the first term has a negative coefficient, we factor out $-3x$:

$-18x^3 + 51x^2 - 15x = -3x(6x^2 - 17x + 5)$ Factor out $-3x$.

$= -3x(3x - 1)(2x - 5)$ Factor $6x^2 - 17x + 5$.

b) The GCF for $-3a^2 + 2a + 21$ is 1. Because the first term has a negative coefficient, factor out -1:

$-3a^2 + 2a + 21 = -1(3a^2 - 2a - 21)$ Factor out -1.

$= -1(3a + 7)(a - 3)$ Factor $3a^2 - 2a - 21$. ■

Warm-ups

True or false? Answer true if the correct factorization is given and false if the factorization is incorrect. Explain your answer.

1. $2x^2 + 3x + 1 = (2x + 1)(x + 1)$

2. $2x^2 + 5x + 3 = (2x + 1)(x + 3)$

3. $3x^2 + 10x + 3 = (3x + 1)(x + 3)$

4. $15x^2 + 31x + 14 = (3x + 7)(5x + 2)$

5. $2x^2 - 7x - 9 = (2x - 9)(x + 1)$

6. $2x^2 + 3x - 9 = (2x + 3)(x - 3)$

7. $2x^2 - 16x - 9 = (2x - 9)(2x + 1)$

8. $8x^2 - 22x - 5 = (4x - 1)(2x + 5)$

9. $9x^2 + x - 1 = (5x - 1)(4x + 1)$

10. $12x^2 - 13x + 3 = (3x - 1)(4x - 3)$

4.4 EXERCISES

Find the following. See Example 1.

1. Two integers that have a product of 20 and a sum of 12
2. Two integers that have a product of 36 and a sum of −20
3. Two integers that have a product of −12 and a sum of −4
4. Two integers that have a product of −8 and a sum of 7

Each of the following trinomials is in the form $ax^2 + bx + c$. For each trinomial, find two integers that have a product of ac and a sum of b. Do not factor the trinomials. See Example 1.

5. $6x^2 + 7x + 2$
6. $5x^2 + 17x + 6$
7. $6y^2 - 11y + 3$
8. $6z^2 - 19z + 10$
9. $12w^2 + w - 1$
10. $15t^2 - 17t - 4$

Factor each trinomial using the ac method. See Example 1.

11. $2x^2 + 3x + 1$
12. $2x^2 + 11x + 5$
13. $2x^2 + 9x + 4$
14. $2h^2 + 7h + 3$
15. $3t^2 + 7t + 2$
16. $3t^2 + 8t + 5$
17. $2x^2 + 5x - 3$
18. $3x^2 - x - 2$
19. $6x^2 + 7x - 3$
20. $21x^2 + 2x - 3$
21. $2x^2 - 7x + 6$
22. $3a^2 - 14a + 15$
23. $5b^2 - 13b + 6$
24. $7y^2 + 16y - 15$
25. $4y^2 - 11y - 3$
26. $35x^2 - 2x - 1$
27. $3x^2 + 2x + 1$
28. $6x^2 - 4x - 5$
29. $8x^2 - 2x - 1$
30. $8x^2 - 10x - 3$
31. $9t^2 - 9t + 2$
32. $9t^2 + 5t - 4$
33. $15x^2 + 13x + 2$
34. $15x^2 - 7x - 2$
35. $15x^2 - 13x + 2$
36. $15x^2 + x - 2$

Complete the factoring.

37. $3x^2 + 7x + 2 = (x + 2)(\quad)$
38. $2x^2 - x - 15 = (x - 3)(\quad)$
39. $5x^2 + 11x + 2 = (5x + 1)(\quad)$
40. $4x^2 - 19x - 5 = (4x + 1)(\quad)$
41. $6a^2 - 17a + 5 = (3a - 1)(\quad)$
42. $4b^2 - 16b + 15 = (2b - 5)(\quad)$

Factor each trinomial using trial and error. See Examples 2 and 3.

43. $5a^2 + 11a + 2$
44. $3y^2 + 10y + 7$
45. $4w^2 + 8w + 3$
46. $6z^2 + 13z + 5$
47. $15x^2 - x - 2$
48. $15x^2 + 13x - 2$
49. $8x^2 - 6x + 1$
50. $8x^2 - 22x + 5$
51. $15x^2 - 31x + 2$
52. $15x^2 + 31x + 2$
53. $2x^2 + 18x - 90$
54. $3x^2 + 11x + 10$
55. $3x^2 + x - 10$
56. $3x^2 - 17x + 10$
57. $10x^2 - 3xy - y^2$
58. $8x^2 - 2xy - y^2$
59. $42a^2 - 13ab + b^2$
60. $10a^2 - 27ab + 5b^2$

Factor each polynomial completely. See Examples 4 and 5.

61. $81w^3 - w$
62. $81w^3 - w^2$
63. $4w^2 + 2w - 30$
64. $2x^2 - 28x + 98$
65. $27 + 12x^2 + 36x$
66. $24y + 12y^2 + 12$
67. $6w^2 - 11w - 35$
68. $18x^2 - 6x + 6$
69. $3x^2z - 3zx - 18z$
70. $a^2b + 2ab - 15b$
71. $10x^2y^2 + xy^2 - 9y^2$
72. $2x^2y^2 + xy^2 + 3y^2$
73. $a^2 + 2ab - 15b^2$
74. $a^2b^2 - 2a^2b - 15a^2$
75. $-6t^3 - t^2 + 2t$
76. $-36t^2 - 6t + 12$
77. $12t^4 - 2t^3 - 4t^2$
78. $12t^3 + 14t^2 + 4t$
79. $4x^2y - 8xy^2 + 3y^3$
80. $9x^2 + 24xy - 9y^2$
81. $-4w^2 + 7w - 3$
82. $-30w^2 + w + 1$
83. $-12a^3 + 22a^2b - 6ab^2$
84. $-36a^2b + 21ab^2 - 3b^3$

Solve each problem.

85. ***Height of a ball.*** If a ball is thrown upward at 40 feet per second from a rooftop 24 feet above the ground, then its height above the ground t seconds after it is thrown is given by $h = -16t^2 + 40t + 24$. Rewrite this formula with the polynomial on the right-hand side factored completely. Use the factored version of the formula to find h when $t = 3$.

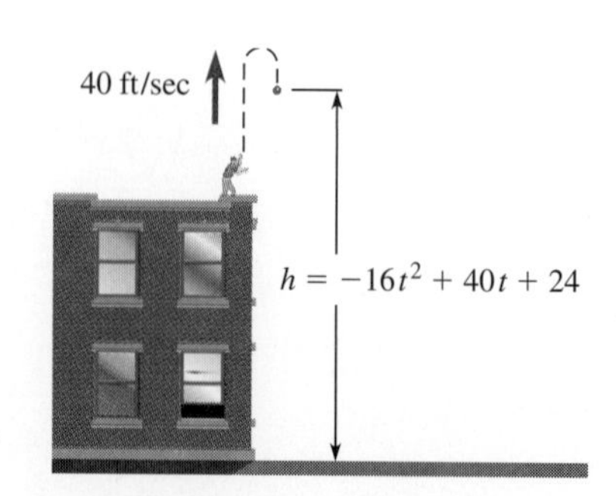

Figure for Exercise 85

86. ***Worker efficiency.*** In a study of efficiency at Wong Laboratories it was found that the number of components assembled per hour by the average worker t hours after starting work is given by

$$N = -3t^3 + 23t^2 + 8t.$$

Rewrite the formula by factoring the right-hand side completely. Use the factored version of the formula to find N when $t = 3$.

Getting More Involved

87. ***Exploration.*** Find all positive and negative integers b for which each polynomial can be factored.

a) $x^2 + bx + 3$ **b)** $3x^2 + bx + 5$
c) $2x^2 + bx - 15$

88. ***Exploration.*** Find two integers c (positive or negative) for which each polynomial can be factored. Many answers are possible.

a) $x^2 + x + c$ **b)** $x^2 - 2x + c$
c) $2x^2 - 3x + c$

89. ***Cooperative learning.*** Working in groups, cut two large squares, three rectangles, and one small square out of paper that are exactly the same size as shown in the accompanying figure. Then try to place the six figures next to one another so that they form a large rectangle. Do not overlap the pieces or leave any gaps. Explain how factoring $2x^2 + 3x + 1$ can help you solve this puzzle.

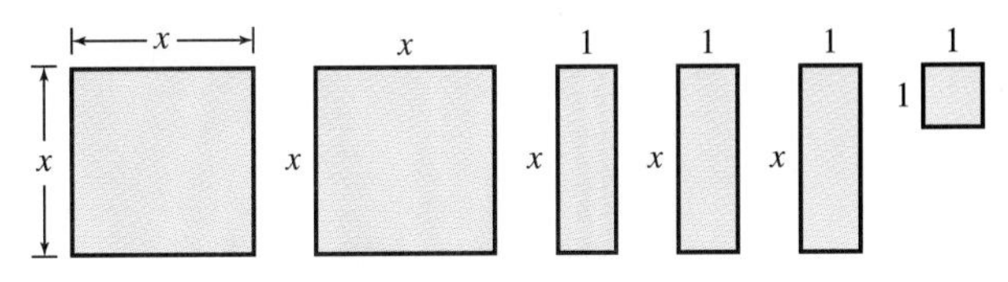

Figure for Exercise 89

90. ***Cooperative learning.*** Working in groups, cut four squares and eight rectangles out of paper as in the previous exercise to illustrate the trinomial $4x^2 + 7x + 3$. Select one group to demonstrate how to arrange the 12 pieces to form a large rectangle. Have another group explain how factoring the trinomial can help you solve this puzzle.

4.5 The Factoring Strategy

In this section:

- Using Division in Factoring
- Factoring a Difference or Sum of Two Cubes
- The Factoring Strategy

In previous sections we established the general idea of factoring and some special cases. In this section we will see how division relates to factoring and see two more special cases. We will then summarize all of the factoring that we have done with a factoring strategy.

Using Division in Factoring

To find the prime factorization for a large integer such as 1001, you could divide possible factors (prime numbers) into 1001 until you find one that leaves no remainder. If you are told that 13 is a factor (or make a lucky guess), then you could divide 1001 by 13 to get the quotient 77. With this information you can factor 1001:

$$1001 = 77 \cdot 13$$

Now you can factor 77 to get the prime factorization of 1001:

$$1001 = 7 \cdot 11 \cdot 13$$

We can use this same idea with polynomials that are of higher degree than the ones we have been factoring. If we can guess a factor or if we are given a factor, we can use division to find the other factor and then proceed to factor the polynomial completely. Of course, it is harder to guess a factor of a polynomial than it is to guess a factor of an integer. In the next example we will factor a third-degree polynomial completely, given one factor.

EXAMPLE 1 **Using division in factoring**

Factor the polynomial $x^3 + 2x^2 - 5x - 6$ completely, given that the binomial $x + 1$ is a factor of the polynomial.

Solution

Divide the polynomial by the binomial:

$$\begin{array}{r|l} & x^2 + x - 6 \\ \hline x + 1 & x^3 + 2x^2 - 5x - 6 \\ & \underline{x^3 + x^2} \\ & \quad x^2 - 5x \\ & \quad \underline{x^2 + x} \\ & \qquad -6x - 6 \qquad -5x - x = -6x \\ & \qquad \underline{-6x - 6} \\ & \qquad\qquad 0 \qquad -6 - (-6) = 0 \end{array}$$

Because the remainder is 0, the dividend is the divisor times the quotient:

$$x^3 + 2x^2 - 5x - 6 = (x + 1)(x^2 + x - 6)$$

Now we factor the remaining trinomial to get the complete factorization:

$$x^3 + 2x^2 - 5x - 6 = (x + 1)(x + 3)(x - 2)$$

Factoring a Difference or Sum of Two Cubes

We can use division to discover that $a - b$ is a factor of $a^3 - b^3$ (a difference of two cubes) and $a + b$ is a factor of $a^3 + b^3$ (a sum of two cubes):

$$\begin{array}{r|l} & a^2 + ab + b^2 \\ \hline a - b & a^3 + 0a^2b + 0ab^2 - b^3 \\ & \underline{a^3 - a^2b} \\ & \quad a^2b + 0ab^2 \\ & \quad \underline{a^2b - ab^2} \\ & \qquad ab^2 - b^3 \\ & \qquad \underline{ab^2 - b^3} \\ & \qquad\qquad 0 \end{array} \qquad \begin{array}{r|l} & a^2 - ab + b^2 \\ \hline a + b & a^3 + 0a^2b + 0ab^2 + b^3 \\ & \underline{a^3 + a^2b} \\ & \quad -a^2b + 0ab^2 \\ & \quad \underline{-a^2b - ab^2} \\ & \qquad ab^2 + b^3 \\ & \qquad \underline{ab^2 + b^3} \\ & \qquad\qquad 0 \end{array}$$

So $a - b$ is a factor of $a^3 - b^3$, and $a + b$ is a factor of $a^3 + b^3$. These results give us two more factoring rules.

Factoring a Difference or Sum of Two Cubes

$$a^3 - b^3 = (a - b)(a^2 + ab + b^2)$$
$$a^3 + b^3 = (a + b)(a^2 - ab + b^2)$$

Note that $a^2 + ab + b^2$ and $a^2 - ab + b^2$ are both prime polynomials.

EXAMPLE 2 **Factoring a difference or sum of two cubes**

Factor each polynomial.

a) $w^3 - 8$ **b)** $x^3 + 1$ **c)** $8y^3 - 27$

Solution

a) Because $8 = 2^3$, $w^3 - 8$ is a difference of two cubes. To factor $w^3 - 8$, let $a = w$ and $b = 2$ in the formula $a^3 - b^3 = (a - b)(a^2 + ab + b^2)$:

$$w^3 - 8 = (w - 2)(w^2 + 2w + 4)$$

b) Because $1 = 1^3$, the binomial $x^3 + 1$ is a sum of two cubes. Let $a = x$ and $b = 1$ in the formula $a^3 + b^3 = (a + b)(a^2 - ab + b^2)$:

$$x^3 + 1 = (x + 1)(x^2 - x + 1)$$

c) $8y^3 - 27 = (2y)^3 - 3^3$ This is a difference of two cubes.

$= (2y - 3)(4y^2 + 6y + 9)$ Let $a = 2y$ and $b = 3$ in the formula.

CAUTION The polynomial $(a - b)^3$ is not equivalent to $a^3 - b^3$ because if $a = 2$ and $b = 1$, then

$$(a - b)^3 = (2 - 1)^3 = 1^3 = 1$$

and

$$a^3 - b^3 = 2^3 - 1^3 = 8 - 1 = 7.$$

Likewise, $(a + b)^3$ is not equivalent to $a^3 + b^3$.

The Factoring Strategy

The following is a summary of the ideas that we use to factor a polynomial completely.

Strategy for Factoring Polynomials Completely

1. If there are any common factors, factor them out first.
2. When factoring a binomial, check to see whether it is a difference of two squares, a difference of two cubes, or a sum of two cubes. *A sum of two squares does not factor.*
3. When factoring a trinomial, check to see whether it is a perfect square trinomial.
4. When factoring a trinomial that is not a perfect square, use the *ac* method or the trial-and-error method.
5. If the polynomial has four terms, try factoring by grouping.
6. Check to see whether any of the factors can be factored again.

We will use the factoring strategy in the next example.

EXAMPLE 3 **Factoring polynomials**

Factor each polynomial completely.

a) $2a^2b - 24ab + 72b$

b) $3x^3 + 6x^2 - 75x - 150$

Solution

a) $2a^2b - 24ab + 72b = 2b(a^2 - 12a + 36)$ First factor out the GCF, $2b$.

$= 2b(a - 6)^2$ Factor the perfect square trinomial.

b) $3x^3 + 6x^2 - 75x - 150 = 3[x^3 + 2x^2 - 25x - 50]$ Factor out the GCF, 3.

$= 3[x^2(x + 2) - 25(x + 2)]$ Factor out common factors.

$= 3(x^2 - 25)(x + 2)$ Factor by grouping.

$= 3(x + 5)(x - 5)(x + 2)$ Factor the difference of two squares.

Warm-ups

True or false? Explain your answer.

1. $x^2 - 4 = (x - 2)^2$ for any real number x.
2. The trinomial $4x^2 + 6x + 9$ is a perfect square trinomial.
3. The polynomial $4y^2 + 25$ is a prime polynomial.
4. $3y + ay + 3x + ax = (x + y)(3 + a)$ for any values of the variables.
5. The polynomial $3x^2 + 51$ cannot be factored.
6. If the GCF is not 1, then you should factor it out first.
7. $x^2 + 9 = (x + 3)^2$ for any real number x.
8. The polynomial $x^2 - 3x - 5$ is a prime polynomial.
9. The polynomial $y^2 - 5y - my + 5m$ can be factored by grouping.
10. The polynomial $x^2 + ax - 3x + 3a$ can be factored by grouping.

4.5 EXERCISES

Factor each polynomial completely, given that the binomial following it is a factor of the polynomial. See Example 1.

1. $x^3 + 3x^2 - 10x - 24,\ x + 4$
2. $x^3 - 7x + 6,\ x - 1$
3. $x^3 + 4x^2 + x - 6,\ x - 1$
4. $x^3 - 5x^2 - 2x + 24,\ x + 2$
5. $x^3 - 8,\ x - 2$
6. $x^3 + 27,\ x + 3$
7. $x^3 + 4x^2 - 3x + 10,\ x + 5$
8. $2x^3 - 5x^2 - x - 6,\ x - 3$
9. $x^3 + 2x^2 + 2x + 1,\ x + 1$
10. $x^3 + 2x^2 - 5x - 6,\ x + 3$

Factor each difference or sum of cubes. See Example 2.

11. $m^3 - 1$
12. $z^3 - 27$
13. $x^3 + 8$
14. $y^3 + 27$
15. $8w^3 + 1$
16. $125m^3 + 1$
17. $8t^3 - 27$
18. $125n^3 - 8$
19. $x^3 - y^3$
20. $m^3 + n^3$
21. $8t^3 + y^3$
22. $u^3 - 125v^3$

Factor each polynomial completely. If a polynomial is prime, say so. See Example 3.

23. $2x^2 - 18$
24. $3x^3 - 12x$
25. $4x^2 + 8x - 60$
26. $3x^2 + 18x + 27$
27. $x^3 + 4x^2 + 4x$
28. $a^3 - 5a^2 + 6a$
29. $5max^2 + 20ma$
30. $3bmw^2 - 12bm$
31. $9x^2 + 6x + 1$
32. $9x^2 + 6x + 3$

33. $6x^2y + xy - 2y$
34. $5x^2y^2 - xy^2 - 6y^2$
35. $y^2 + 10y - 25$
36. $8b^2 + 24b + 18$
37. $16m^2 - 4m - 2$
38. $32a^2 + 4a - 6$
39. $9a^2 + 24a + 16$
40. $3x^2 - 18x - 48$
41. $24x^2 - 26x + 6$
42. $4x^2 - 6x - 12$
43. $3a^2 - 27a$
44. $a^2 - 25a$
45. $8 - 2x^2$
46. $x^3 + 6x^2 + 9x$
47. $6x^3 - 5x^2 + 12x$
48. $x^3 + 2x^2 - x - 2$
49. $a^3b - 4ab$
50. $2m^2 - 1800$
51. $x^3 + 2x^2 - 4x - 8$
52. $m^2a + 2ma^2 + a^3$
53. $2w^4 - 16w$
54. $m^4n + mn^4$
55. $3a^2w - 18aw + 27w$
56. $8a^3 + 4a$
57. $5x^2 - 500$
58. $25x^2 - 16y^2$
59. $2m + 2n - wm - wn$
60. $aw - 5b - bw + 5a$
61. $3x^4 + 3x$
62. $3a^5 - 81a^2$
63. $4w^2 + 4w - 4$
64. $4w^2 + 8w - 5$
65. $a^4 + 7a^3 - 30a^2$
66. $2y^5 + 3y^4 - 20y^3$
67. $4aw^3 - 12aw^2 + 9aw$
68. $9bn^3 + 15bn^2 - 14bn$
69. $t^2 + 6t + 9$
70. $t^3 + 12t^2 + 36t$

Solve each problem.

71. ***Increasing cube.*** Each of the three dimensions of a cube with a volume of x^3 cubic centimeters is increased by a whole number of centimeters. If the new volume is $x^3 + 10x^2 + 31x + 30$ cubic centimeters and the new height is $x + 2$ centimeters, then what are the new length and width?

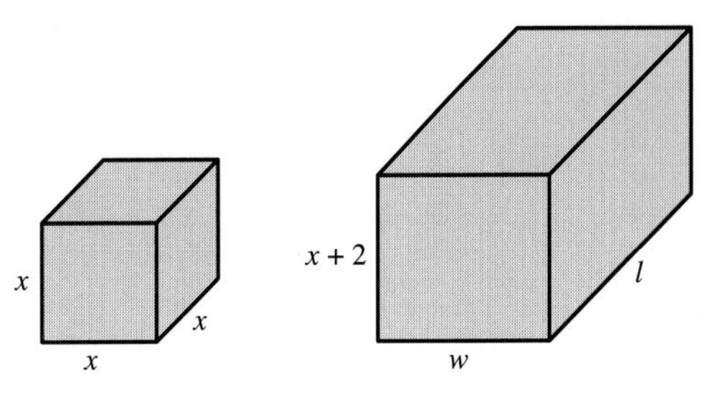

Figure for Exercise 71

72. ***Decreasing cube.*** Each of the three dimensions of a cube with a volume of y^3 cubic centimeters is decreased by a whole number of centimeters. If the new volume is $y^3 - 13y^2 + 54y - 72$ cubic centimeters and the new width is $y - 6$ centimeters, then what are the new length and height?

Getting More Involved

73. ***Discussion.*** Are there any values for a and b for which $(a + b)^3 = a^3 + b^3$? Find a pair of values for a and b for which $(a + b)^3 \neq a^3 + b^3$. Is $(a + b)^3$ equivalent to $a^3 + b^3$? Explain your answers.

74. ***Writing.*** Explain why $a^2 + ab + b^2$ and $a^2 - ab + b^2$ are prime polynomials.

4.6 Using Factoring to Solve Equations

In this section:

- The Zero Factor Property
- Applications

The techniques of factoring can be used to solve equations involving polynomials. These equations cannot be solved by the other methods that you have learned. After you learn to solve equations by factoring, you will use this technique to solve some new types of problems.

The Zero Factor Property

In this chapter you learned to factor polynomials such as $x^2 + x - 6$. The equation $x^2 + x - 6 = 0$ is called a *quadratic equation.*

Quadratic Equation

If a, b, and c are real numbers with $a \neq 0$, then

$$ax^2 + bx + c = 0$$

is called a **quadratic equation.**

To solve $x^2 + x - 6 = 0$, we factor the trinomial on the left side to get

$$(x + 3)(x - 2) = 0.$$

This equation states that the product of $x + 3$ and $x - 2$ is 0. The product of two numbers is 0 only when one or the other of the numbers is 0. Therefore the equation is equivalent to the *compound equation*

$$x + 3 = 0 \quad \text{or} \quad x - 2 = 0.$$

A sentence made up of two or more equations connected with the word "or" is called a **compound equation.** The compound equation above is equivalent to

$$x = -3 \quad \text{or} \quad x = 2.$$

We now check that -3 and 2 satisfy the original equation.

For $x = -3$:

$$\begin{aligned} x^2 + x - 6 &= (-3)^2 + (-3) - 6 \\ &= 9 - 3 - 6 \\ &= 0 \end{aligned}$$

For $x = 2$:

$$\begin{aligned} x^2 + x - 6 &= (2)^2 + (2) - 6 \\ &= 4 + 2 - 6 \\ &= 0 \end{aligned}$$

We solved the quadratic equation $x^2 + x - 6 = 0$ by obtaining two linear equations. The main idea used here is called the **zero factor property.**

The Zero Factor Property

The equation $a \cdot b = 0$ is equivalent to the compound equation

$$a = 0 \quad \text{or} \quad b = 0.$$

EXAMPLE 1 **Using the zero factor property**

Solve the equation $3x^2 = -3x$.

Solution

First rewrite the equation with 0 on the right-hand side:

$$\begin{aligned} 3x^2 &= -3x & \\ 3x^2 + 3x &= 0 & \text{Add } 3x \text{ to each side.} \\ 3x(x + 1) &= 0 & \text{Factor the left-hand side.} \\ 3x = 0 \quad \text{or} \quad x + 1 &= 0 & \text{Zero factor property} \\ x = 0 \quad \text{or} \quad x &= -1 & \text{Solve each equation.} \end{aligned}$$

Check 0 and -1 in the original equation $3x^2 = -3x$.

For $x = 0$:

$$\begin{aligned} 3(0)^2 &= -3(0) \\ 0 &= 0 \end{aligned}$$

For $x = -1$:

$$\begin{aligned} 3(-1)^2 &= -3(-1) \\ 3 &= 3 \end{aligned}$$

There are two solutions to the original equation, 0 and -1. ■

CAUTION If in Example 1 you divide each side of $3x^2 = -3x$ by $3x$, you would get $x = -1$ but not the solution $x = 0$. For this reason we usually do not divide each side of an equation by a variable.

The basic strategy for solving an equation by factoring follows.

Strategy for Solving an Equation by Factoring

1. Rewrite the equation with 0 on the right-hand side.
2. Factor the left-hand side completely.
3. Use the zero factor property to get simple linear equations.
4. Solve the linear equations.
5. Check the answers in the original equation.
6. State the solution(s) to the original equation.

EXAMPLE 2 **Using the zero factor property**

Solve $(2x + 1)(x - 1) = 14$.

Solution

To write the equation with 0 on the right-hand side, multiply the binomials on the left and then subtract 14 from each side:

$$(2x + 1)(x - 1) = 14 \quad \text{Original equation}$$
$$2x^2 - x - 1 = 14 \quad \text{Multiply the binomials.}$$
$$2x^2 - x - 15 = 0 \quad \text{Subtract 14 from each side.}$$
$$(2x + 5)(x - 3) = 0 \quad \text{Factor.}$$
$$2x + 5 = 0 \quad \text{or} \quad x - 3 = 0 \quad \text{Zero factor property}$$
$$2x = -5 \quad \text{or} \quad x = 3$$
$$x = -\frac{5}{2} \quad \text{or} \quad x = 3$$

Check $-\frac{5}{2}$ and 3 in the original equation. The solutions are $-\frac{5}{2}$ and 3.

CAUTION In Example 2 we started with a product of two factors equal to 14. Because there are many pairs of factors that have a product of 14, we *cannot make any conclusion about the factors.* If the product of two factors is 0, then we can conclude that one or the other factor is 0.

If a perfect square trinomial occurs in a quadratic equation, then there are two identical factors of the trinomial. In this case it is not necessary to set both factors equal to zero. The solution can be found from one factor.

EXAMPLE 3 **An equation with a repeated factor**

Solve $5x^2 - 30x + 45 = 0$.

Solution

Notice that the trinomial on the left-hand side has a common factor:

$$
\begin{aligned}
5x^2 - 30x + 45 &= 0 \\
5(x^2 - 6x + 9) &= 0 && \text{Factor out the GCF.} \\
5(x - 3)^2 &= 0 && \text{Factor the perfect square trinomial.} \\
(x - 3)^2 &= 0 && \text{Divide each side by 5.} \\
x - 3 &= 0 && \text{Zero factor property} \\
x &= 3
\end{aligned}
$$

You should check that 3 satisfies the original equation. Even though $x - 3$ occurs twice as a factor, it is not necessary to write $x - 3 = 0$ or $x - 3 = 0$. The only solution to the equation is 3. ■

CAUTION To solve $7(3x - 4) = 0$, we do not write $7 = 0$ or $3x - 4 = 0$. Do not include 7 as a solution to the equation. If one of the factors is a nonzero number, divide each side of the equation by that number to eliminate that factor. ⊘

If the left-hand side of the equation has more than two factors, we can write an equivalent equation by setting each factor equal to zero.

EXAMPLE 4 **An equation with three solutions**

Solve $2x^3 - x^2 - 8x + 4 = 0$.

Solution

We can factor the four-term polynomial by grouping:

$$
\begin{aligned}
2x^3 - x^2 - 8x + 4 &= 0 \\
x^2(2x - 1) - 4(2x - 1) &= 0 && \text{Factor out the common factors.} \\
(x^2 - 4)(2x - 1) &= 0 && \text{Factor out } 2x - 1. \\
(x - 2)(x + 2)(2x - 1) &= 0 && \text{Difference of two squares}
\end{aligned}
$$

$$x - 2 = 0 \quad \text{or} \quad x + 2 = 0 \quad \text{or} \quad 2x - 1 = 0 \qquad \text{Zero factor property}$$

$$x = 2 \quad \text{or} \quad x = -2 \quad \text{or} \quad x = \frac{1}{2} \qquad \text{Solve each equation.}$$

You should check that all three numbers satisfy the original equation. The solutions to this equation are -2, $\frac{1}{2}$, and 2. ■

Applications

There are many problems that can be solved by equations like those we have just discussed.

EXAMPLE 5 **Area of a garden**

Merida's garden has a rectangular shape with a length that is 1 foot longer than twice the width. If the area of the garden is 55 square feet, then what are the dimensions of the garden?

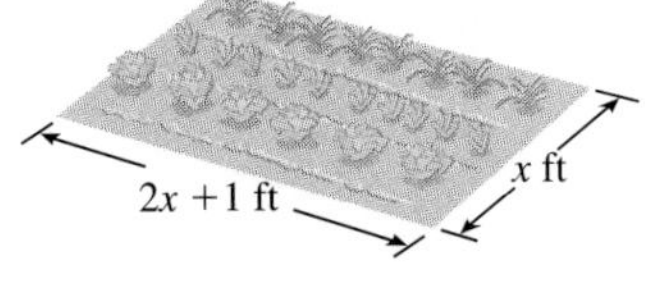

Figure 4.1

Solution

If x represents the width of the garden, then $2x + 1$ represents the length. See Fig. 4.1. Because the area of a rectangle is the length times the width, we can write the equation

$$x(2x + 1) = 55.$$

We must have zero on the right-hand side of the equation to use the zero factor property. So we rewrite the equation and then factor:

$$2x^2 + x - 55 = 0$$

$$(2x + 11)(x - 5) = 0 \quad \text{Factor.}$$

$$2x + 11 = 0 \quad \text{or} \quad x - 5 = 0 \quad \text{Zero factor property}$$

$$x = -\frac{11}{2} \quad \text{or} \quad x = 5$$

The width is certainly not $-\frac{11}{2}$. So we use $x = 5$ to get the length:

$$2x + 1 = 2(5) + 1 = 11$$

We check by multiplying 11 feet and 5 feet to get the area of 55 square feet. So the width is 5 ft, and the length is 11 ft. ■

The next application involves a theorem from geometry called the **Pythagorean theorem.** This theorem says that in any right triangle the sum of the squares of the lengths of the legs is equal to the square of the length of the hypotenuse.

The Pythagorean Theorem

The triangle shown in Fig. 4.2 is a right triangle if and only if

$$a^2 + b^2 = c^2.$$

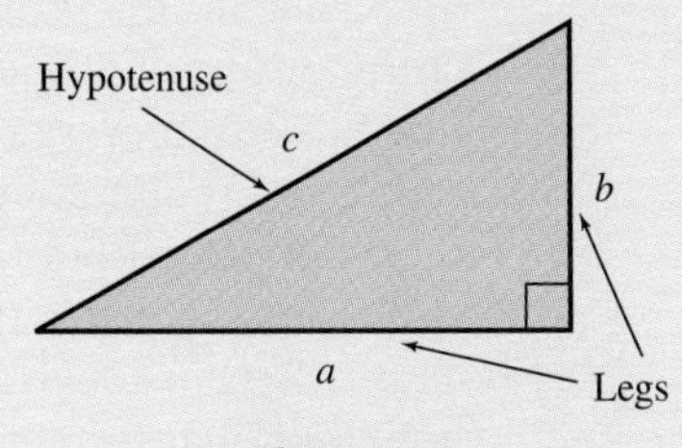

Figure 4.2

EXAMPLE 6 Using the Pythagorean theorem

The length of a rectangle is 1 meter longer than the width, and the diagonal measures 5 meters. What are the length and width?

Solution

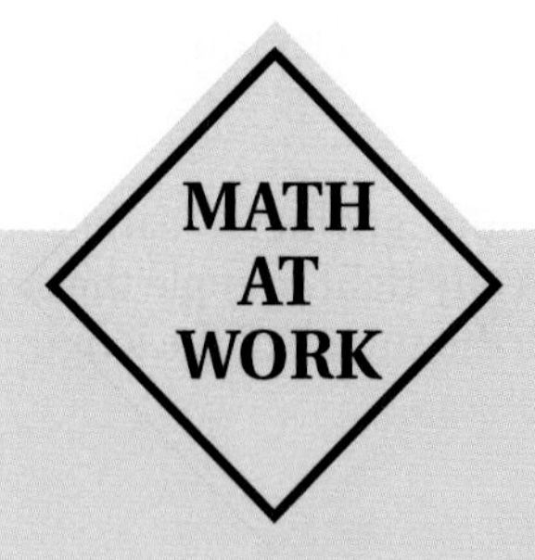

Figure 4.3

If x represents the width of the rectangle, then $x + 1$ represents the length. Because the two sides are the legs of a right triangle, we can use the Pythagorean theorem to get a relationship between the length, width, and diagonal. See Fig. 4.3.

$$x^2 + (x + 1)^2 = 5^2 \quad \text{Pythagorean theorem}$$
$$x^2 + x^2 + 2x + 1 = 25 \quad \text{Simplify.}$$
$$2x^2 + 2x - 24 = 0$$
$$x^2 + x - 12 = 0 \quad \text{Divide each side by 2.}$$
$$(x - 3)(x + 4) = 0$$
$$x - 3 = 0 \quad \text{or} \quad x + 4 = 0 \quad \text{Zero factor property}$$
$$x = 3 \quad \text{or} \quad x = -4 \quad \text{The length cannot be negative.}$$
$$x + 1 = 4$$

To check this answer, we compute $3^2 + 4^2 = 5^2$, or $9 + 16 = 25$. So the rectangle is 3 meters by 4 meters. ■

CAUTION The hypotenuse is the longest side of a right triangle. So if the lengths of the sides of a right triangle are 5 meters, 12 meters, and 13 meters, then the length of the hypotenuse is 13 meters, and $5^2 + 12^2 = 13^2$. ⊘

MATH AT WORK

Investment Advisor

Can you successfully invest money and at the same time be socially responsible? Geeta Bhide, president and founder of Walden Capital Management, answers with an emphatic "yes." Ms. Bhide helps clients integrate their social values with a portfolio of stocks, bonds, cash, and cash equivalents.

With the client's consent, Ms. Bhide might invest in bonds backed by the Department of Housing and Urban Development for housing in specific inner-city areas. Other choices might be environmentally conscious companies or companies that have a proven record of equal employment practices. In many cases, categorizing a particular company as socially conscious is a judgment call. For example, many oil companies provide a good return on investment, but they might not have an unblemished record on oil spills. In this instance, picking the best of the worst might be the correct choice. Because such trade-offs are necessary, clients are encouraged to define both their investment goals and the social ideals to which they subscribe.

As any investment advisor would, Ms. Bhide tries to minimize risk and maximize reward, or return on investment. In Exercises 71 and 72 of this section you will see how solving a quadratic equation by factoring can be used to find the average annual return on an investment.

Warm-ups

True or false? Explain your answer.

1. The equation $x(x + 2) = 3$ is equivalent to $x = 3$ or $x + 2 = 3$.
2. Equations solved by factoring always have two different solutions.
3. The equation $a \cdot d = 0$ is equivalent to $a = 0$ or $d = 0$.
4. If x is the width in feet of a rectangular room and the length is 5 feet longer than the width, then the area is $x^2 + 5x$ square feet.
5. Both 1 and -4 are solutions to the equation $(x - 1)(x + 4) = 0$.
6. If a, b, and c are the sides of any triangle, then $a^2 + b^2 = c^2$.
7. If the perimeter of a rectangular room is 50 feet, then the sum of the length and width is 25 feet.
8. Equations solved by factoring may have more than two solutions.
9. Both 0 and 2 are solutions to the equation $x(x - 2) = 0$.
10. The solutions to $3(x - 2)(x + 5) = 0$ are 3, 2, and -5.

4.6 EXERCISES

Solve each equation. See Examples 1 and 2.

1. $(x + 5)(x + 4) = 0$
2. $(a + 6)(a + 5) = 0$
3. $(2x + 5)(3x - 4) = 0$
4. $(3k - 8)(4k + 3) = 0$
5. $w^2 - 9w + 14 = 0$
6. $t^2 + 6t - 27 = 0$
7. $m^2 = -7m$
8. $h^2 = -5h$
9. $a^2 + a = 20$
10. $p^2 + p = 42$
11. $2x^2 + 5x = 3$
12. $3x^2 - 10x = -7$
13. $(x + 2)(x + 6) = 12$
14. $(x + 2)(x - 6) = 20$
15. $(a + 3)(2a - 1) = 15$
16. $(b - 3)(3b + 4) = 10$
17. $2(4 - 5h) = 3h^2$
18. $2w(4w + 1) = 1$

Solve each equation. See Examples 3 and 4.

19. $2x^2 + 50 = 20x$
20. $3x^2 + 48 = 24x$
21. $4m^2 - 12m + 9 = 0$
22. $25y^2 + 20y + 4 = 0$
23. $x^3 - 9x = 0$
24. $25x - x^3 = 0$
25. $w^3 + 4w^2 - 4w = 16$
26. $a^3 + 2a^2 - a = 2$
27. $n^3 - 3n^2 + 3 = n$
28. $w^3 + w^2 - 25w = 25$
29. $y^3 - 9y^2 + 20y = 0$
30. $m^3 + 2m^2 - 3m = 0$

Solve each equation.

31. $x^2 - 16 = 0$
32. $x^2 - 36 = 0$
33. $x^2 = 9$
34. $x^2 = 25$
35. $a^3 = a$
36. $x^3 = 4x$
37. $3x^2 + 15x + 18 = 0$
38. $-2x^2 - 2x + 24 = 0$
39. $z^2 + \frac{11}{2}z = -6$
40. $m^2 + \frac{8}{3}m = 1$
41. $(t - 3)(t + 5) = 9$
42. $3x(2x + 1) = 18$
43. $(x - 2)^2 + x^2 = 10$
44. $(x - 3)^2 + (x + 2)^2 = 17$
45. $\frac{1}{16}x^2 + \frac{1}{8}x = \frac{1}{2}$
46. $\frac{1}{18}h^2 - \frac{1}{2}h + 1 = 0$
47. $a^3 + 3a^2 - 25a = 75$
48. $m^4 + m^3 = 100m^2 + 100m$

Solve each problem. See Examples 5 and 6.

49. ***Dimensions of a rectangle.*** The perimeter of a rectangle is 34 feet, and the diagonal is 13 feet long. What are the length and width of the rectangle?

50. ***Address book.*** The perimeter of the cover of an address book is 14 inches, and the diagonal measures 5 inches. What are the length and width of the cover?

Figure for Exercise 50

51. ***Violla's bathroom.*** The length of Violla's bathroom is 2 feet longer than twice the width. If the diagonal measures 13 feet, then what are the length and width?

52. ***Rectangular stage.*** One side of a rectangular stage is 2 meters longer than the other. If the diagonal is 10 meters, then what are the lengths of the sides?

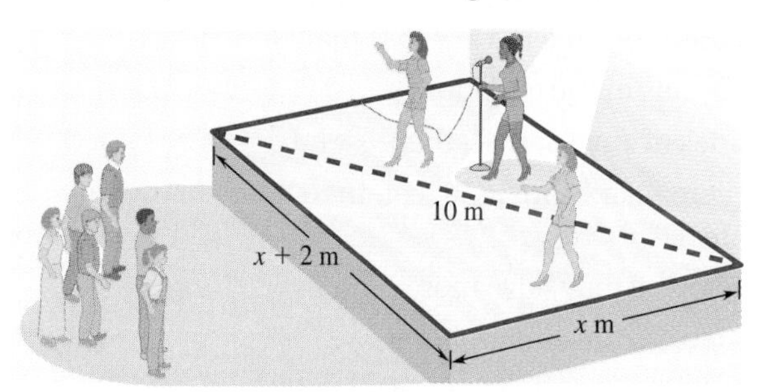

Figure for Exercise 52

53. ***Consecutive integers.*** The sum of the squares of two consecutive integers is 13. Find the integers.

54. ***Consecutive integers.*** The sum of the squares of two consecutive even integers is 52. Find the integers.

55. ***Two numbers.*** The sum of two numbers is 11, and their product is 30. Find the numbers.

56. ***Missing ages.*** Molly's age is twice Anita's. If the sum of the squares of their ages is 80, then what are their ages?

57. ***Skydiving.*** If there were no air resistance, then the height (in feet) above the earth for a sky diver t seconds after jumping from an airplane at 10,000 ft would be given by $h = -16t^2 + 10{,}000$. Find the time it would take to fall to the earth with no air resistance, that is, find t for $h = 0$. A sky diver actually gets about twice as much free fall time due to air resistance.

58. ***Skydiving.*** If a sky diver jumps from an airplane at a height of 8256 feet, then for the first five seconds, her height above the earth is approximated by the formula $h = -16t^2 + 8256$. How many seconds does it take her to reach 8000 feet.

59. ***Throwing a sandbag.*** If a balloonist throws a sandbag downward at 24 feet per second from an altitude of 720 feet, then its height (in feet) above the ground after t seconds is given by $S = -16t^2 - 24t + 720$. How long does it take for the sandbag to reach the earth? (On the ground, $S = 0$.)

60. ***Throwing a sandbag.*** If the balloonist of the previous exercise throws his sandbag downward from an altitude of 128 feet with an initial velocity of 32 feet per second, then its altitude after t seconds is given by the formula $S = -16t^2 - 32t + 128$. How long does it take for the sandbag to reach the earth?

61. ***Glass prism.*** One end of a glass prism is in the shape of a triangle with a height that is 1 inch longer than twice the base. If the area of the triangle is 39 square inches, then how long are the base and height?

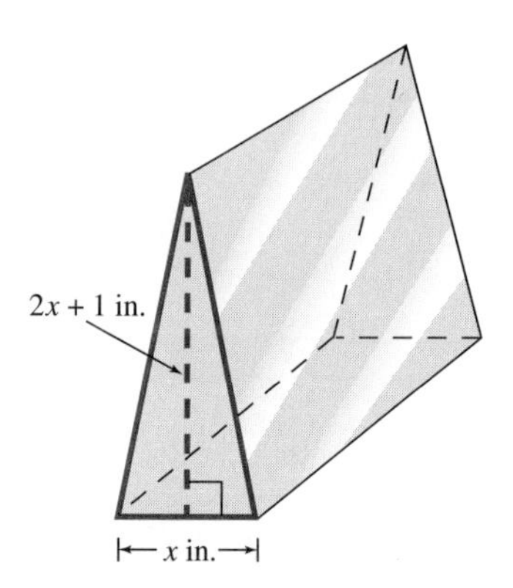

Figure for Exercise 61

62. ***Areas of two circles.*** The radius of a circle is 1 meter longer than the radius of another circle. If their areas differ by 5π square meters, then what is the radius of each?

63. ***Changing area.*** Last year Otto's garden was square. This year he plans to make it smaller by shortening one side 5 feet and the other 8 feet. If the area of the smaller garden will be 180 square feet, then what was the size of Otto's garden last year?

64. ***Dimensions of a box.*** Rosita's Christmas present from Carlos is in a box that has a width that is 3 inches shorter than the height. The length of the base is 5 inches longer than the height. If the area of the base is 84 square inches, then what is the height of the package?

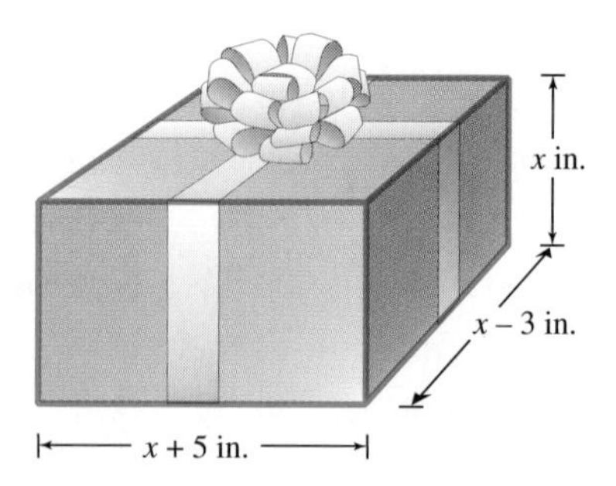

Figure for Exercise 64

65. ***Flying a kite.*** Imelda and Gordon have designed a new kite. While Imelda is flying the kite, Gordon is standing directly below it. The kite is designed so that its altitude is always 20 feet larger than the distance between Imelda and Gordon. What is the altitude of the kite when it is 100 feet from Imelda?

66. ***Avoiding a collision.*** A car is traveling on a road that is perpendicular to a railroad track. When the car is 30 meters from the crossing, the car's new collision detector warns the driver that there is a train 50 meters from the car and heading toward the same crossing. How far is the train from the crossing?

67. ***Carpeting two rooms.*** Virginia is buying carpet for two square rooms. One room is 3 yards wider than the other. If she needs 45 square yards of carpet, then what are the dimensions of each room?

68. ***Winter wheat.*** While finding the amount of seed needed to plant his three square wheat fields, Hank observed that the side of one field was 1 kilometer longer than the side of the smallest field and that the side of the largest field was 3 kilometers longer than the side of the smallest field. If the total area of the three fields is 38 square kilometers, then what is the area of each field?

69. ***Sailing to Miami.*** At point A the captain of a ship determined that the distance to Miami was 13 miles. If she sailed north to point B and then west to Miami, the distance would be 17 miles. If the distance from point A to point B is greater than the distance from point B to Miami, then how far is it from point A to point B?

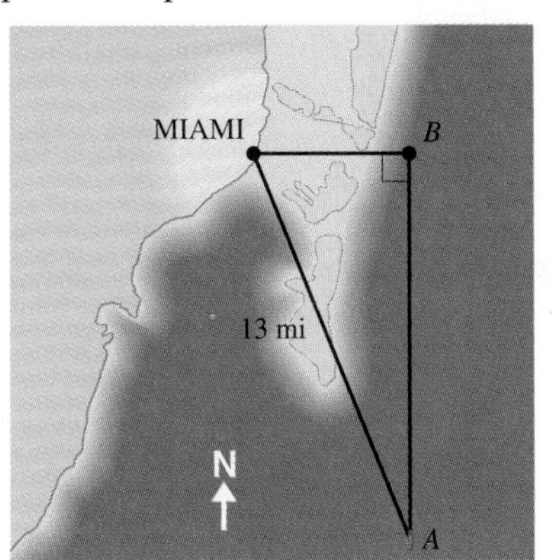

Figure for Exercise 69

70. ***Buried treasure.*** Ahmed has half of a treasure map, which indicates that the treasure is buried in the desert $2x + 6$ paces from Castle Rock. Vanessa has the other half of the map. Her half indicates that to find the treasure, one must get to Castle Rock, walk x paces to the north, and then walk $2x + 4$ paces to the east. If they share their information, then they can find x and save a lot of digging. What is x?

71. ***Emerging markets.*** Catarina's investment of \$16,000 in an emerging market fund grew to \$25,000 in two years. Find the average annual rate of return by solving the equation $16{,}000(1 + r)^2 = 25{,}000$.

72. ***Venture capital.*** Henry invested \$12,000 in a new restaurant. When the restaurant was sold two years later, he received \$27,000. Find his average annual return by solving the equation $12{,}000(1 + r)^2 = 27{,}000$.

COLLABORATIVE ACTIVITIES

The Puzzle Box

Grouping: 4 students
Topic: Multiplying and factoring polynomials

After graduating from college, you get a job working for a small business that makes jigsaw puzzles. You are on the team to design the cover of a puzzle box. Your design will use 525 square centimeters. The production manager has found a great deal on the price of cardboard. The cardboard is precut to 29 cm by 33 cm. The production manager tells your team that the depth of the box can vary to fit the needs of your design. He asks your team not only to come up with a design, but also to determine what the depth of the box should be so that production can fold it correctly to allow for your design.

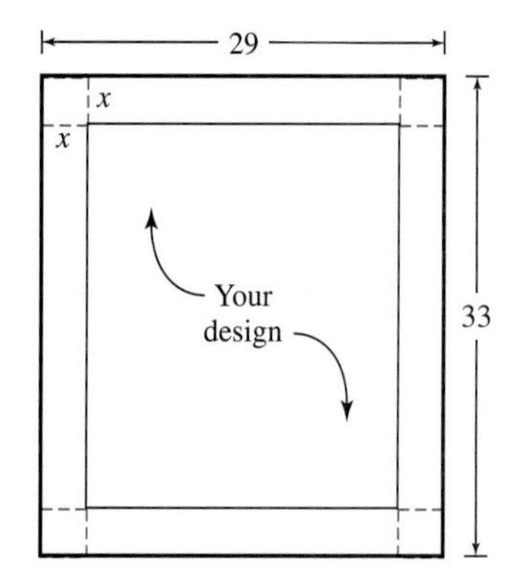

Getting started: Choose roles (consult your instructor). Decide on a design and a name for your puzzle.

Part I: Finding the depth and dimensions

1. Write an equation using the given information and the diagram.
2. Solve the equation to find the depth of the box.
3. Use the depth of the box to find the dimensions for your design.

Part II: Presenting your final design. For shipping purposes, management has decided to make the boxes only 3 cm deep. This will change the dimensions of the box top. There isn't time for you to come up with a new design. Your team must find a way to incorporate your present design with its original dimensions onto the new box top. You will need to make an exact drawing in centimeters of your solution, using the new dimensions for the box top.

1. Using the same precut cardboard with the new depth, what will be the dimensions of the top of the box?
2. Make an *exact* drawing of your puzzle box top using the new dimensions.
3. Include the puzzle name in your drawing.

Report: Write all your work and solutions as a report. List the names of the members of your group. Include answers and work for all questions.

Wrap-up CHAPTER 4

SUMMARY

Factoring		Examples
Prime number	A positive integer larger than 1 that has no integral factors other than 1 and itself	2, 3, 5, 7, 11
Prime polynomial	Any monomial is prime, and a polynomial of more than one term that cannot be factored is prime.	$6x^3$, $x^2 + 3$, and $x^2 - x + 5$ are prime.

Strategy for finding the GCF for monomials	1. Find the GCF for the coefficients of the monomials. 2. Form the product of the GCF of the coefficients and each variable that is common to all of the monomials, where the exponent on each variable equals the smallest power of that variable in any of the monomials.	$12x^3yz,\ 8x^2y^3$ $\text{GCF} = 4x^2y$
Factoring out the GCF	Use the distributive property to factor out the GCF from all terms of a polynomial.	$2x^3 - 4x = 2x(x^2 - 2)$

Special Cases		**Examples**
Difference of two squares	$a^2 - b^2 = (a + b)(a - b)$	$m^2 - 9 = (m - 3)(m + 3)$
Perfect square trinomial	$a^2 + 2ab + b^2 = (a + b)^2$ $a^2 - 2ab + b^2 = (a - b)^2$	$x^2 + 6x + 9 = (x + 3)^2$ $4h^2 - 12h + 9 = (2h - 3)^2$
Difference or sum of two cubes	$a^3 - b^3 = (a - b)(a^2 + ab + b^2)$ $a^3 + b^3 = (a + b)(a^2 - ab + b^2)$	$t^3 - 8 = (t - 2)(t^2 + 2t + 4)$ $p^3 + 1 = (p + 1)(p^2 - p + 1)$

Factoring Polynomials		**Examples**
Factoring by grouping	Factor out common factors from groups of terms.	$6x + 6w + ax + aw$ $= 6(x + w) + a(x + w)$ $= (6 + a)(x + w)$
Strategy for factoring $ax^2 + bx + c$ by the *ac* method	1. Find two numbers that have a product equal to *ac* and a sum equal to *b*. 2. Replace *bx* by two terms using the two new numbers as coefficients. 3. Factor the resulting four-term polynomial by grouping.	$6x^2 + 17x + 12$ $= 6x^2 + 9x + 8x + 12$ $= (2x + 3)3x + (2x + 3)4$ $= (2x + 3)(3x + 4)$
Factoring by trial and error	Try possible factors of the trinomial and check by using FOIL. If incorrect, try again.	$2x^2 + 5x - 12$ $= (2x - 3)(x + 4)$
Strategy for factoring polynomials completely	1. First factor out any common factors. 2. When factoring a binomial, check to see whether it is a difference of two squares, a difference of two cubes, or a sum of two cubes. Remember that a sum of two squares (with no common factor) is prime. 3. When factoring a trinomial, check to see whether it is a perfect square trinomial. 4. When factoring a trinomial that is not a perfect square, use the *ac* method or trial-and-error. 5. If the polynomial has four terms, try factoring by grouping. 6. Check to see whether any factors can be factored again.	

Solving Equations		Examples
Zero factor property	The equation $a \cdot b = 0$ is equivalent to $a = 0$ or $b = 0$.	$x(x-1) = 0$ $x = 0$ or $x - 1 = 0$
Strategy for solving an equation by factoring	1. Rewrite the equation with 0 on the right-hand side. 2. Factor the left-hand side completely. 3. Set each factor equal to zero to get linear equations. 4. Solve the linear equations. 5. Check the answers in the original equation. 6. State the solution(s) to the original equation.	$x^2 + 3x = 18$ $x^2 + 3x - 18 = 0$ $(x+6)(x-3) = 0$ $x + 6 = 0$ or $x - 3 = 0$ $x = -6$ or $x = 3$

REVIEW EXERCISES

4.1 *Find the prime factorization for each integer.*

1. 144 **2.** 121 **3.** 58

4. 76 **5.** 150 **6.** 200

Find the greatest common factor for each group.

7. 36, 90 **8.** 30, 42, 78 **9.** $8x$, $12x^2$

10. $6a^2b$, $9ab^2$, $15a^2b^2$

Complete the factorization of each binomial.

11. $3x + 6 = 3(\quad)$ **12.** $7x^2 + x = x(\quad)$

13. $2a - 20 = -2(\quad)$ **14.** $a^2 - a = -a(\quad)$

Factor each polynomial by factoring out the GCF.

15. $2a - a^2$ **16.** $9 - 3b$

17. $6x^2y^2 - 9x^5y$ **18.** $a^3b^5 + a^3b^2$

19. $3x^2y - 12xy - 9y^2$ **20.** $2a^2 - 4ab^2 - ab$

4.2 *Factor each polynomial completely.*

21. $y^2 - 400$ **22.** $4m^2 - 9$

23. $w^2 - 8w + 16$ **24.** $t^2 + 20t + 100$

25. $4y^2 + 20y + 25$ **26.** $2a^2 - 4a - 2$

27. $r^2 - 4r + 4$ **28.** $3m^2 - 75$

29. $8t^3 - 24t^2 + 18t$ **30.** $t^2 - 9w^2$

31. $x^2 + 12xy + 36y^2$ **32.** $9y^2 - 12xy + 4x^2$

33. $x^2 + 5x - xy - 5y$ **34.** $x^2 + xy + ax + ay$

4.3 *Factor each polynomial.*

35. $b^2 + 5b - 24$ **36.** $a^2 - 2a - 35$

37. $r^2 - 4r - 60$ **38.** $x^2 + 13x + 40$

39. $y^2 - 6y - 55$ **40.** $a^2 + 6a - 40$

41. $u^2 + 26u + 120$ **42.** $v^2 - 22v - 75$

Factor completely.

43. $3t^3 + 12t^2$ **44.** $-4m^4 - 36m^2$

45. $5w^3 + 25w^2 + 25w$ **46.** $-3t^3 + 3t^2 - 6t$

47. $2a^3b + 3a^2b^2 + ab^3$ **48.** $6x^2y^2 - xy^3 - y^4$

49. $9x^3 - xy^2$ **50.** $h^4 - 100h^2$

4.4 *Factor each polynomial completely.*

51. $14t^2 + t - 3$ **52.** $15x^2 - 22x - 5$

53. $6x^2 - 19x - 7$ **54.** $2x^2 - x - 10$

55. $6p^2 + 5p - 4$ **56.** $3p^2 + 2p - 5$

57. $-30p^3 + 8p^2 + 8p$ **58.** $-6q^2 - 40q - 50$

59. $6x^2 - 29xy - 5y^2$ **60.** $10a^2 + ab - 2b^2$

61. $32x^2 + 16xy + 2y^2$ **62.** $8a^2 + 40ab + 50b^2$

4.5 *Factor completely.*

63. $5x^3 + 40x$ **64.** $w^2 + 6w + 9$

65. $9x^2 + 3x - 2$ **66.** $ax^3 + ax$

67. $x^3 + 2x^2 - x - 2$ **68.** $16x^2 - 2x - 3$

69. $x^2y - 16xy^2$ **70.** $-3x^2 + 27$

71. $a^2 + 2a + 1$ **72.** $-2w^2 - 12w - 18$

73. $x^3 - x^2 + x - 1$ **74.** $9x^2y^2 - 9y^2$

75. $a^2 + ab + 2a + 2b$ **76.** $4m^2 + 20m + 25$

77. $-2x^2 + 16x - 24$ **78.** $6x^2 + 21x - 45$

79. $m^3 - 1000$ **80.** $8p^3 + 1$

Factor each polynomial completely, given that the binomial following it is a factor of the polynomial.

81. $x^3 + x + 10,\ x + 2$

82. $x^3 - 5x - 12,\ x - 3$

83. $x^3 + 6x^2 - 7x - 60,\ x + 4$

84. $x^3 - 4x^2 - 3x - 10,\ x - 5$

4.6 *Solve each equation.*

85. $x^3 = 5x^2$

86. $2m^2 + 10m = -12$

87. $(a - 2)(a - 3) = 6$

88. $(w - 2)(w + 3) = 50$

89. $2m^2 - 9m - 5 = 0$

90. $12x^2 + 5x - 3 = 0$

91. $m^3 + 4m^2 - 9m = 36$

92. $w^3 + 5w^2 - w = 5$

93. $(x + 3)^2 + x^2 = 5$

94. $(h - 2)^2 + (h + 1)^2 = 9$

95. $p^2 + \frac{1}{4}p - \frac{1}{8} = 0$

96. $t^2 + 1 = \frac{13}{6}t$

Solve each problem.

97. ***Positive numbers.*** Two positive numbers differ by 6, and their squares differ by 96. Find the numbers.

98. ***Consecutive integers.*** Find three consecutive integers such that the sum of their squares is 77.

99. ***Dimensions of a notebook.*** The perimeter of a notebook is 28 inches, and the diagonal measures 10 inches. What are the length and width of the notebook?

100. ***Two numbers.*** The sum of two numbers is 8.5, and their product is 18. Find the numbers.

101. ***Poiseuille's law.*** According to the nineteenth century physician Poiseuille, the velocity (in centimeters per second) of blood r centimeters from the center of an artery of radius R centimeters is given by $v = kR^2 - kr^2$, where k is a constant. Rewrite the formula by factoring the right-hand side completely.

102. ***Racquetball.*** The volume of rubber (in cubic centimeters) in a hollow rubber ball used in racquetball is given by

$$V = \frac{4}{3}\pi R^3 - \frac{4}{3}\pi r^3,$$

where the inside radius is r centimeters and the outside radius is R centimeters. Rewrite the formula by factoring the right-hand side completely.

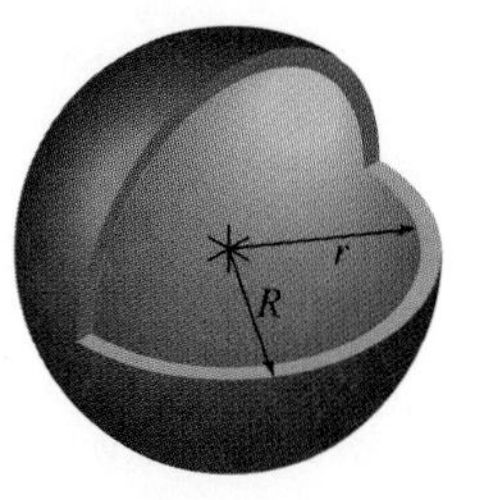

Figure for Exercise 102

103. ***Leaning ladder.*** A 10-foot ladder is placed against a building so that the distance from the bottom of the ladder to the building is 2 feet less than the distance from the top of the ladder to the ground. What is the distance from the bottom of the ladder to the building?

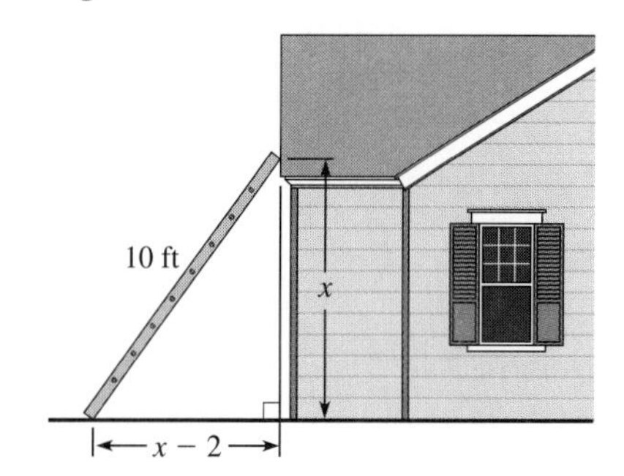

Figure for Exercise 103

104. ***Towering antenna.*** A guy wire of length 50 feet is attached to the ground and to the top of an antenna. The height of the antenna is 10 feet larger than the distance from the base of the antenna to the point where the guy wire is attached to the ground. What is the height of the antenna?

CHAPTER 4 TEST

Give the prime factorization for each integer.

1. 66

2. 336

Find the greatest common factor (GCF) for each group.

3. 48, 80

4. 42, 66, 78

5. $6y^2$, $15y^3$

6. $12a^2b$, $18ab^2$, $24a^3b^3$

Factor each polynomial completely.

7. $5x^2 - 10x$

8. $6x^2y^2 + 12xy^2 + 12y^2$

9. $3a^3b - 3ab^3$

10. $a^2 + 2a - 24$

11. $4b^2 - 28b + 49$

12. $3m^3 + 27m$

13. $ax - ay + bx - by$

14. $ax - 2a - 5x + 10$

15. $6b^2 - 7b - 5$

16. $m^2 + 4mn + 4n^2$

17. $2a^2 - 13a + 15$

18. $z^3 + 9z^2 + 18z$

Factor the polynomial completely, given that $x - 1$ is a factor.

19. $x^3 - 6x^2 + 11x - 6$

Solve each equation.

20. $2x^2 + 5x - 12 = 0$

21. $3x^3 = 12x$

22. $(2x - 1)(3x + 5) = 5$

Write a complete solution to each problem.

23. If the length of a rectangle is 3 feet longer than the width and the diagonal is 15 feet, then what are the length and width?

24. The sum of two numbers is 4, and their product is -32. Find the numbers.

Tying It All Together CHAPTERS 1–4

Simplify each expression.

1. $\dfrac{91 - 17}{17 - 91}$

2. $\dfrac{4 - 18}{-6 - 1}$

3. $5 - 2(7 - 3)$

4. $3^2 - 4(6)(-2)$

5. $2^5 - 2^4$

6. $0.07(37) + 0.07(63)$

Perform the indicated operations.

7. $x \cdot 2x$

8. $x + 2x$

9. $\dfrac{6 + 2x}{2}$

10. $\dfrac{6 \cdot 2x}{2}$

11. $2 \cdot 3y \cdot 4z$

12. $2(3y + 4z)$

13. $2 - (3 - 4z)$

14. $t^8 \div t^2$

15. $t^8 \cdot t^2$

16. $\dfrac{8t^8}{2t^2}$

Solve and graph each inequality.

17. $2x - 5 > 3x + 4$

18. $4 - 5x \leq -11$

19. $-\dfrac{2}{3}x + 3 < -5$

20. $0.05(x - 120) - 24 < 0$

Solve each equation.

21. $2x - 3 = 0$

22. $2x + 1 = 0$

23. $(x - 3)(x + 5) = 0$

24. $(2x - 3)(2x + 1) = 0$

25. $3x(x - 3) = 0$

26. $x^2 = x$

27. $3x - 3x = 0$

28. $3x - 3x = 1$

29. $0.01x - x + 14.9 = 0.5x$

30. $0.05x + 0.04(x - 40) = 2$

31. $2x^2 = 18$

32. $2x^2 + 7x - 15 = 0$

Solve the problem.

33. ***Another ace.*** Professional tennis players can serve a tennis ball at speeds over 120 mph into a rectangular region that has a perimeter of 69 feet and an area of 283.5 square feet. Find the length and width of the service region.

CHAPTER 5

Rational Expressions

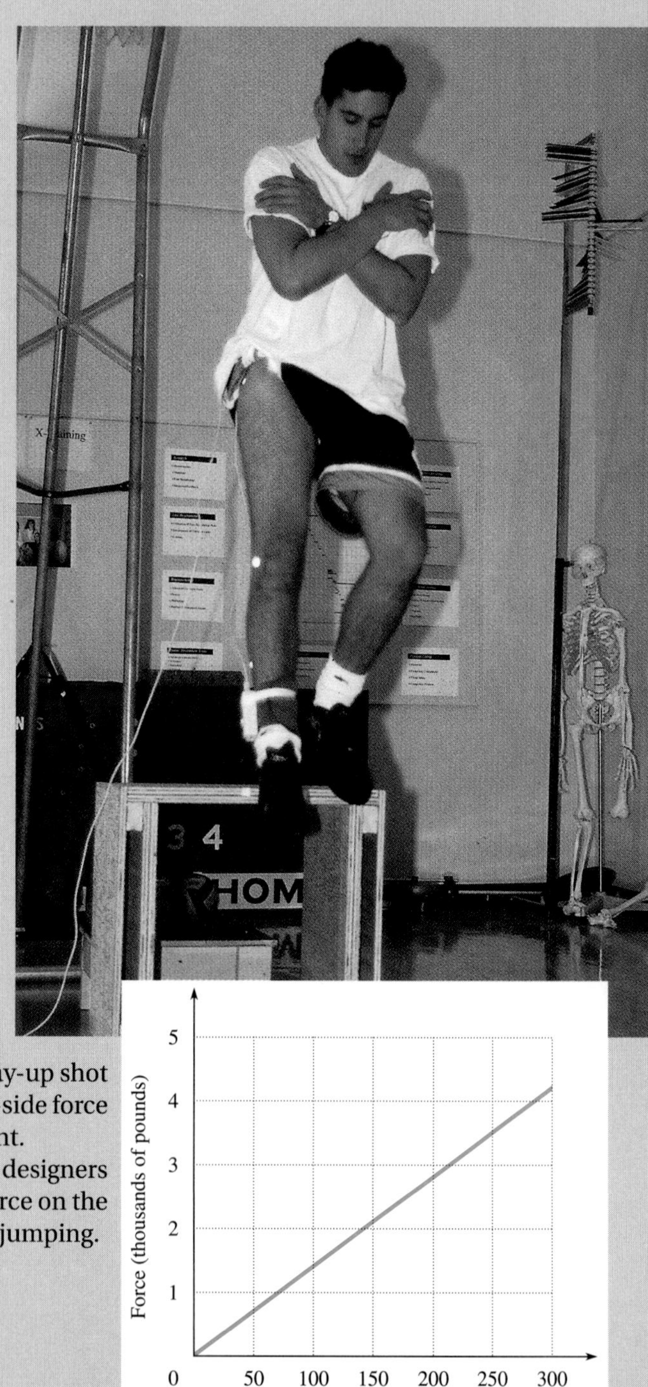

Advanced technical developments have made sports equipment faster, lighter, and more responsive to the human body. Behind the more flexible skis, lighter bats, and comfortable athletic shoes lies the science of biomechanics, which is the study of human movement and the factors that influence it.

Designing and testing an athletic shoe go hand in hand. While a shoe is being designed, it is tested in a multitude of ways, including long-term wear, rear foot stability, and strength of materials. Testing basketball shoes usually includes an evaluation of the force applied to the ground by the foot during running, jumping, and landing. Many biomechanics laboratories have a special platform that can measure the force exerted when a player cuts from side to side as well as the force against the bottom of the shoe. Force exerted in landing from a lay-up shot can be as high as 14 times the weight of the body. Side-to-side force is usually about 1 to 2 body weights in a cutting movement.

In Exercises 47 and 48 of Section 5.7 you will see how designers of athletic shoes use proportions to find the amount of force on the foot and soles of shoes for activities such as running and jumping.

5.1 Properties of Rational Expressions

In this section:

Rational expressions in algebra are similar to the rational numbers in arithmetic. In this section you will learn the basic ideas of rational expressions.

Rational Expressions

A rational number is the ratio of two integers with the denominator not equal to 0. For example,

$$\frac{3}{4}, \quad \frac{-9}{-6}, \quad 7, \quad \text{and} \quad 0$$

are rational numbers. A **rational expression** is the ratio of two polynomials with the denominator not equal to 0. Because an integer is a monomial, a rational number is a rational expression. The following expressions are rational expressions:

$$\frac{x^2 - 1}{x + 8}, \quad \frac{3a^2 + 5a - 3}{a - 9}, \quad \frac{3}{7}, \quad w$$

We say that w is a rational expression because w can be written as $\frac{w}{1}$.

Because the denominator cannot be zero, any number can be used in place of the variable *except* numbers that cause the denominator to be zero.

EXAMPLE 1 **Rational expressions**

Which numbers cannot be used in place of x in each rational expression?

a) $\dfrac{x^2 - 1}{x + 8}$ **b)** $\dfrac{x + 2}{2x + 1}$ **c)** $\dfrac{x + 5}{x^2 - 4}$

Solution

a) The denominator is 0 if $x + 8 = 0$, or $x = -8$. So -8 cannot be used in place of x.

b) The denominator is zero if $2x + 1 = 0$, or $x = -\frac{1}{2}$. So we cannot use $-\frac{1}{2}$ in place of x.

c) The denominator is zero if $x^2 - 4 = 0$. Solve this equation:

$$x^2 - 4 = 0$$

$$(x - 2)(x + 2) = 0 \quad \text{Factor.}$$

$$x - 2 = 0 \quad \text{or} \quad x + 2 = 0 \quad \text{Zero factor property}$$

$$x = 2 \quad \text{or} \quad x = -2$$

So 2 and -2 cannot be used in place of x. ■

When dealing with rational expressions in this book, we will generally assume that the variables represent numbers for which the denominator is not zero.

Reducing to Lowest Terms

Rational expressions are a generalization of rational numbers. The operations that we perform on rational numbers can be performed on rational expressions in exactly the same manner.

Each rational number can be written in infinitely many equivalent forms. For example,

$$\frac{3}{5} = \frac{6}{10} = \frac{9}{15} = \frac{12}{20} = \frac{15}{25} = \cdots.$$

Each equivalent form of $\frac{3}{5}$ is obtained from $\frac{3}{5}$ by multiplying both numerator and denominator by the same nonzero number. This is equivalent to multiplying the fraction by 1, which does not change its value. For example,

$$\frac{3}{5} = \frac{3}{5} \cdot 1 = \frac{3}{5} \cdot \frac{2}{2} = \frac{6}{10} \quad \text{and} \quad \frac{3}{5} = \frac{3 \cdot 3}{5 \cdot 3} = \frac{9}{15}.$$

If we start with $\frac{6}{10}$ and convert it into $\frac{3}{5}$, we say that we are **reducing $\frac{6}{10}$ to lowest terms.** We reduce by dividing the numerator and denominator by the common factor 2:

$$\frac{6}{10} = \frac{\cancel{2} \cdot 3}{\cancel{2} \cdot 5} = \frac{3}{5}$$

A rational number is expressed in lowest terms when the numerator and the denominator have no common factors other than 1.

CAUTION We can reduce fractions only by dividing the numerator and the denominator by a common factor. Although it is true that

$$\frac{6}{10} = \frac{2+4}{2+8},$$

we cannot eliminate the 2's, because they are not factors. Removing them from the sums in the numerator and denominator would not result in $\frac{3}{5}$. ⊘

Reducing Fractions

If $a \neq 0$ and $c \neq 0$, then

$$\frac{ab}{ac} = \frac{b}{c}.$$

To reduce rational expressions to lowest terms, we use exactly the same procedure as with fractions:

Reducing Rational Expressions

1. Factor the numerator and denominator completely.
2. Divide the numerator and denominator by the greatest common factor.

Dividing the numerator and denominator by the GCF is often referred to as **dividing out the GCF.**

EXAMPLE 2 **Reducing**

Reduce to lowest terms.

a) $\frac{30}{42}$ **b)** $\frac{x^2 - 9}{6x + 18}$

Solution

a) $\frac{30}{42} = \frac{\cancel{2} \cdot \cancel{3} \cdot 5}{\cancel{2} \cdot \cancel{3} \cdot 7}$ Factor.

$= \frac{5}{7}$ Divide out the GCF: $2 \cdot 3$ or 6.

b) $\frac{x^2 - 9}{6x + 18} = \frac{(x - 3)\cancel{(x + 3)}}{6\cancel{(x + 3)}}$ Factor.

$= \frac{x - 3}{6}$ Divide out the GCF: $x + 3$.

If two rational expressions are equivalent, then they have the same numerical value for any replacement of the variables. Of course, the replacement must not give us an undefined expression (0 in the denominator). So in Example 2(b) the equation

$$\frac{x^2 - 9}{6x + 18} = \frac{x - 3}{6}$$

is satisfied by all real numbers except $x = -3$. The equation is an identity.

Reducing with the Quotient Rule

To reduce rational expressions involving exponential expressions, we use the quotient rule for exponents from Chapter 3. We restate it here for reference.

Quotient Rule

Suppose $a \neq 0$, and m and n are positive integers.

If $m \geq n$, then $\frac{a^m}{a^n} = a^{m-n}$.

If $m < n$, then $\frac{a^m}{a^n} = \frac{1}{a^{n-m}}$.

EXAMPLE 3 **Using the quotient rule in reducing**

Reduce to lowest terms.

a) $\frac{3a^{15}}{6a^7}$ b) $\frac{6x^4y^2}{4xy^5}$

Solution

a)
$$\begin{aligned}\frac{3a^{15}}{6a^7} &= \frac{\cancel{3}a^{15}}{\cancel{3} \cdot 2a^7} && \text{Factor.}\\ &= \frac{a^{15-7}}{2} && \text{Quotient rule}\\ &= \frac{a^8}{2}\end{aligned}$$

b)
$$\begin{aligned}\frac{6x^4y^2}{4xy^5} &= \frac{\cancel{2} \cdot 3x^4y^2}{\cancel{2} \cdot 2xy^5} && \text{Factor.}\\ &= \frac{3x^{4-1}}{2y^{5-2}} && \text{Quotient rule}\\ &= \frac{3x^3}{2y^3}\end{aligned}$$

■

The essential part of reducing is getting a complete factorization for the numerator and denominator. To get a complete factorization, you must use the techniques for factoring from Chapter 4. If there are large integers in the numerator and denominator, you can use the technique shown in Section 4.1 to get a prime factorization of each integer.

EXAMPLE 4 **Reducing expressions involving large integers**

Reduce $\frac{420}{616}$ to lowest terms.

Solution

Use the method of Section 4.1 to get a prime factorization of 420 and 616:

$$\begin{array}{rr} 7 & 11 \\ 5\overline{)35} & 7\overline{)77} \\ 3\overline{)105} & 2\overline{)154} \\ 2\overline{)210} & 2\overline{)308} \\ \text{Start here} \rightarrow 2\overline{)420} & 2\overline{)616} \end{array}$$

The complete factorization for 420 is $2^2 \cdot 3 \cdot 5 \cdot 7$, and the complete factorization for 616 is $2^3 \cdot 7 \cdot 11$. To reduce the fraction, we divide out the common factors:

$$\begin{aligned}\frac{420}{616} &= \frac{2^2 \cdot 3 \cdot 5 \cdot 7}{2^3 \cdot 7 \cdot 11}\\ &= \frac{3 \cdot 5}{2 \cdot 11}\\ &= \frac{15}{22}\end{aligned}$$

■

Dividing $a - b$ by $b - a$

In Section 3.2 you learned that $a - b = -(b - a) = -1(b - a)$. So if $a - b$ is divided by $b - a$, the quotient is -1:

$$\frac{a-b}{b-a} = \frac{-1(b-a)}{b-a}$$
$$= -1$$

We will use this fact in the next example.

EXAMPLE 5 **Expressions with $a - b$ and $b - a$**

Reduce to lowest terms.

a) $\dfrac{5x - 5y}{4y - 4x}$ **b)** $\dfrac{m^2 - n^2}{n - m}$

Solution

a) $\dfrac{5x - 5y}{4y - 4x} = \dfrac{5(x - y)}{4(y - x)}$ Factor.

$= \dfrac{5}{4} \cdot (-1)$ $\quad \dfrac{x - y}{y - x} = -1$

$= -\dfrac{5}{4}$

b) $\dfrac{m^2 - n^2}{n - m} = \dfrac{\overset{-1}{\cancel{(m - n)}}(m + n)}{\cancel{n - m}}$ Factor.

$= -1(m + n)$ $\quad \dfrac{m - n}{n - m} = -1$

$= -m - n$

CAUTION We can reduce $\dfrac{a-b}{b-a}$ to -1, but we cannot reduce $\dfrac{a-b}{a+b}$. There is no factor that is common to the numerator and denominator of $\dfrac{a-b}{a+b}$.

Factoring Out the Opposite of a Common Factor

If we can factor out a common factor, we can also factor out the opposite of that common factor. For example, from $-3x - 6y$ we can factor out the common factor 3 or the common factor -3:

$$-3x - 6y = 3(-x - 2y) \quad \text{or} \quad -3x - 6y = -3(x + 2y)$$

To reduce an expression, it is sometimes necessary to factor out the opposite of a common factor.

EXAMPLE 6 **Factoring out the opposite of a common factor**

Reduce $\frac{-3w - 3w^2}{w^2 - 1}$ to lowest terms.

Solution

We can factor $3w$ or $-3w$ from the numerator. If we factor out $-3w$, we get a common factor in the numerator and denominator:

$$\frac{-3w - 3w^2}{w^2 - 1} = \frac{-3w(1 + w)}{(w - 1)(w + 1)} \quad \text{Factor.}$$

$$= \frac{-3w}{w - 1} \quad \text{Since } 1 + w = w + 1 \text{, we divide out } w + 1.$$

$$= \frac{3w}{1 - w} \quad \text{Multiply numerator and denominator by } -1.$$

The last step in this reduction is not absolutely necessary, but we usually perform it to make the answer look a little simpler.

The main points to remember for reducing rational expressions are summarized in the following reducing strategy.

Strategy for Reducing Rational Expressions

1. Reducing is done by dividing out all common factors.
2. Factor the numerator and denominator completely to see the common factors.
3. Use the quotient rule to reduce a ratio of two monomials.
4. You may have to factor out a common factor with a negative sign to get identical factors in the numerator and denominator.
5. The quotient of $a - b$ and $b - a$ is -1.

Evaluating Rational Expressions

In Chapter 3 we evaluated a polynomial for a specified value of the variable. For example, if $x = 2$, then the value of the polynomial $x^2 + 3$ is $2^2 + 3$, or 7. We can evaluate a rational expression in the same manner.

EXAMPLE 7 **Evaluating a rational expression**

Find the value of $\frac{4x - 1}{x + 2}$ for $x = -3$.

Solution

To find the value of the rational expression for $x = -3$, replace x by -3 in the rational expression:

$$\frac{4(-3) - 1}{-3 + 2} = \frac{-13}{-1} = 13$$

So the value of the rational expression is 13.

Calculator Close-up

When you evaluate a rational expression with a calculator, you must be sure to divide the value of the numerator by the value of the denominator. The simplest way to accomplish this is to enclose the numerator and denominator in parentheses. For example, to evaluate the expression

$$\frac{4(-3) - 1}{-3 + 2}$$

from Example 7 with a scientific calculator, try the following:

(4 × 3 +/− − 1) ÷ (3 +/− + 2) =

The display should read

13

With a graphing calculator the display should look like the following:

(4∗⁻3−1)/(⁻3+2) ENTER

13

Warm-ups

True or false? Explain your answer.

1. A complete factorization of 3003 is $2 \cdot 3 \cdot 7 \cdot 11 \cdot 13$.
2. A complete factorization of 120 is $2^3 \cdot 3 \cdot 5$.
3. Any number can be used in place of x in the expression $\frac{x - 2}{5}$.
4. We cannot replace x by -1 or 3 in the expression $\frac{x + 1}{x - 3}$.
5. The rational expression $\frac{x + 2}{2}$ reduces to x.
6. $\frac{2x}{2} = x$ for any real number x.
7. $\frac{x^{13}}{x^{20}} = \frac{1}{x^7}$ for any nonzero value of x.
8. $\frac{a^2 + b^2}{a + b}$ reduced to lowest terms is $a + b$.
9. If $a \neq b$, then $\frac{a - b}{b - a} = 1$.
10. The expression $\frac{-3x - 6}{x + 2}$ reduces to -3.

5.1 EXERCISES

Which numbers cannot be used in place of the variable in each rational expression? See Example 1.

1. $\frac{x}{x + 1}$
2. $\frac{3x}{x - 7}$
3. $\frac{7a}{3a - 5}$
4. $\frac{84}{3 - 2a}$
5. $\frac{2x + 3}{x^2 - 16}$
6. $\frac{2y + 1}{y^2 - y - 6}$
7. $\frac{p - 1}{2}$
8. $\frac{m + 31}{5}$

Reduce each rational expression to lowest terms. Assume that the variables represent only numbers for which the denominators are nonzero. See Example 2.

9. $\frac{6}{27}$
10. $\frac{14}{21}$
11. $\frac{42}{90}$
12. $\frac{42}{54}$
13. $\frac{36a}{90}$
14. $\frac{56y}{40}$
15. $\frac{78}{30w}$
16. $\frac{68}{44y}$
17. $\frac{6x+2}{6}$
18. $\frac{2w+2}{2w}$
19. $\frac{2x+4y}{6y+3x}$
20. $\frac{3m+9w}{3m-6w}$
21. $\frac{w^2-49}{w+7}$
22. $\frac{a^2-b^2}{a-b}$
23. $\frac{a^2-1}{a^2+2a+1}$
24. $\frac{x^2-y^2}{x^2+2xy+y^2}$
25. $\frac{2x^2+4x+2}{4x^2-4}$
26. $\frac{2x^2+10x+12}{3x^2-27}$
27. $\frac{3x^2+18x+27}{21x+63}$
28. $\frac{x^3-3x^2-4x}{x^2-4x}$

Reduce each expression to lowest terms. Assume that all denominators are nonzero. See Example 3.

29. $\frac{x^{10}}{x^7}$
30. $\frac{y^8}{y^5}$
31. $\frac{z^3}{z^8}$
32. $\frac{w^9}{w^{12}}$
33. $\frac{4x^7}{-2x^5}$
34. $\frac{-6y^3}{3y^9}$
35. $\frac{-12m^9n^{18}}{8m^6n^{16}}$
36. $\frac{-9u^9v^{19}}{6u^9v^{14}}$
37. $\frac{6b^{10}c^4}{-8b^{10}c^7}$
38. $\frac{9x^{20}y}{-6x^{25}y^3}$
39. $\frac{30a^3bc}{18a^7b^{17}}$
40. $\frac{15m^{10}n^3}{24m^{12}np}$

Reduce each expression to lowest terms. See Example 4.

41. $\frac{210}{264}$
42. $\frac{616}{660}$
43. $\frac{231}{168}$
44. $\frac{936}{624}$
45. $\frac{630x^5}{300x^9}$
46. $\frac{96y^2}{108y^5}$
47. $\frac{924a^{23}}{448a^{19}}$
48. $\frac{270b^{75}}{165b^{12}}$

Reduce each expression to lowest terms. See Example 5.

49. $\frac{3a-2b}{2b-3a}$
50. $\frac{5m-6n}{6n-5m}$
51. $\frac{h^2-t^2}{t-h}$
52. $\frac{r^2-s^2}{s-r}$
53. $\frac{2g-6h}{9h^2-g^2}$
54. $\frac{5a-10b}{4b^2-a^2}$
55. $\frac{x^2-x-6}{9-x^2}$
56. $\frac{1-a^2}{a^2+a-2}$

Reduce each expression to lowest terms. See Example 6.

57. $\frac{-x-6}{x+6}$
58. $\frac{-5x-20}{3x+12}$
59. $\frac{-2y-6y^2}{3+9y}$
60. $\frac{y^2-16}{-8-2y}$
61. $\frac{-3x-6}{3x-6}$
62. $\frac{8-4x}{-8x-16}$
63. $\frac{-12a-6}{2a^2+7a+3}$
64. $\frac{-2b^2-6b-4}{b^2-1}$

Reduce each expression to lowest terms.

65. $\frac{2x^{12}}{4x^8}$
66. $\frac{4x^2}{2x^9}$
67. $\frac{2x+4}{4x}$
68. $\frac{2x+4x^2}{4x}$
69. $\frac{a-4}{4-a}$
70. $\frac{2b-4}{2b+4}$
71. $\frac{2c-4}{4-c^2}$
72. $\frac{-2t-4}{4-t^2}$
73. $\frac{x^2+4x+4}{x^2-4}$
74. $\frac{3x-6}{x^2-4x+4}$
75. $\frac{-2x-4}{x^2+5x+6}$
76. $\frac{-2x-8}{x^2+2x-8}$
77. $\frac{2q^8+q^7}{2q^6+q^5}$
78. $\frac{8s^{12}}{12s^6-16s^5}$
79. $\frac{u^2-6u-16}{u^2-16u+64}$
80. $\frac{v^2+3v-18}{v^2+12v+36}$
81. $\frac{a^3-8}{2a-4}$
82. $\frac{4w^2-12w+36}{2w^3+54}$
83. $\frac{y^3-2y^2-4y+8}{y^2-4y+4}$
84. $\frac{mx+3x+my+3y}{m^2-3m-18}$

Evaluate each rational expression for the given values of x. See Example 7.

85. $\frac{x-5}{x+3}$ for $x = 2$, $x = -4$, $x = -3.02$, $x = -2.96$
86. $\frac{x^2-2x-3}{x-2}$ for $x = 3$, $x = 5$, $x = 2.05$, $x = 1.999$

Find a rational expression that answers each question.

87. If Sergio drove 300 miles at $x + 10$ miles per hour, then how many hours did he drive?
88. If Carrie walked 40 miles in x hours, then how fast did she walk?
89. If $x + 4$ pounds of peaches cost \$4.50, then what is the cost per pound?
90. If nine pounds of pears cost x dollars, then what is the price per pound?

91. If Ayesha can clean the entire swimming pool in x hours, then how much of the pool does she clean per hour?

92. If Ramon can mow the entire lawn in $x - 3$ hours, then how much of the lawn does he mow per hour?

Solve each problem.

93. ***Annual reports.*** The Crest Meat Company found that the cost per report for printing x annual reports at Peppy Printing is given by the formula

$$C = \frac{150 + 0.60x}{x},$$

where C is in dollars.

a) Use the accompanying graph to estimate the cost per report for printing 1000 reports.

b) Use the formula to find the cost per report for printing 1000, 5000, and 10,000 reports.

c) What happens to the cost per report as the number of reports gets very large?

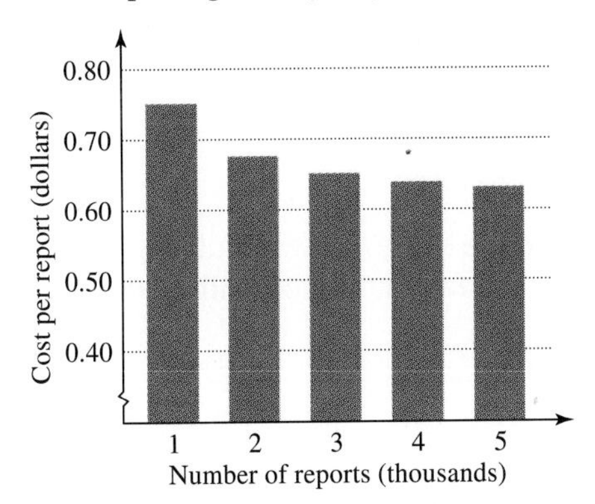

Figure for Exercise 93

94. ***Toxic pollutants.*** The annual cost in dollars for removing $p\%$ of the toxic chemicals from a town's water supply is given by the formula

$$C = \frac{500{,}000}{100 - p}.$$

a) Use the accompanying graph to estimate the cost for removing 90% and 95% of the toxic chemicals.

b) Use the formula to determine the cost for removing 99.5% of the toxic chemicals.

c) What happens to the cost as the percentage of pollutants removed approaches 100%?

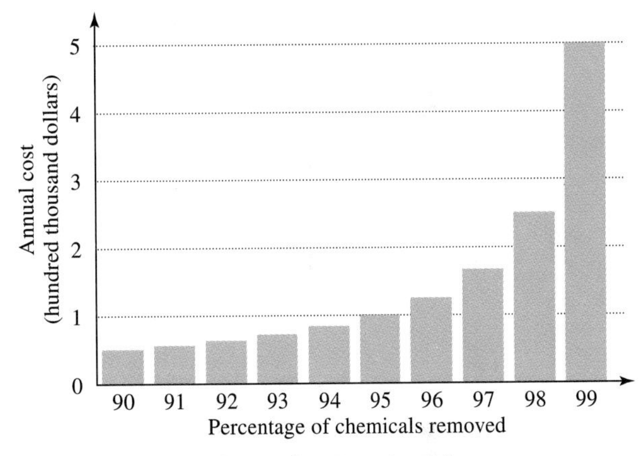

Figure for Exercise 94

5.2 Multiplication and Division

In this section:

- Multiplication of Rational Numbers
- Multiplication of Rational Expressions
- Division of Rational Numbers
- Division of Rational Expressions

In Section 5.1 you learned to reduce rational expressions in the same way that we reduce rational numbers. In this section we will multiply and divide rational expressions using the same procedures that we use for rational numbers.

Multiplication of Rational Numbers

Two rational numbers are multiplied by multiplying their numerators and multiplying their denominators.

Multiplication of Rational Numbers

If $b \neq 0$ and $d \neq 0$, then

$$\frac{a}{b} \cdot \frac{c}{d} = \frac{ac}{bd}.$$

EXAMPLE 1 **Multiplying rational numbers**

Find the product $\frac{6}{7} \cdot \frac{14}{15}$.

Solution

The product is found by multiplying the numerators and multiplying the denominators:

$$\begin{aligned} \frac{6}{7} \cdot \frac{14}{15} &= \frac{84}{105} \\ &= \frac{21 \cdot 4}{21 \cdot 5} && \text{Factor the numerator and denominator.} \\ &= \frac{4}{5} && \text{Divide out the GCF 21.} \end{aligned}$$

The reducing that we did after multiplying is easier to do before multiplying. First factor all terms, reduce, and then multiply:

$$\begin{aligned} \frac{6}{7} \cdot \frac{14}{15} &= \frac{2 \cdot \cancel{3}}{\cancel{7}} \cdot \frac{2 \cdot \cancel{7}}{\cancel{3} \cdot 5} \\ &= \frac{4}{5} \end{aligned}$$

■

Multiplication of Rational Expressions

We multiply rational expressions in the same way we multiply rational numbers. As with rational numbers, we can factor, reduce, and then multiply.

EXAMPLE 2 **Multiplying rational expressions**

Find the indicated products.

a) $\frac{9x}{5y} \cdot \frac{10y}{3xy}$ **b)** $\frac{-8xy^4}{3z^3} \cdot \frac{15z}{2x^5y^3}$

Solution

a)
$$\begin{aligned} \frac{9x}{5y} \cdot \frac{10y}{3xy} &= \frac{3 \cdot \cancel{3x}}{\cancel{5y}} \cdot \frac{2 \cdot \cancel{5y}}{\cancel{3xy}} && \text{Factor.} \\ &= \frac{6}{y} \end{aligned}$$

b)
$$\begin{aligned} \frac{-8xy^4}{3z^3} \cdot \frac{15z}{2x^5y^3} &= \frac{-2 \cdot 2 \cdot \cancel{2}xy^4}{\cancel{3}z^3} \cdot \frac{\cancel{3} \cdot 5z}{\cancel{2}x^5y^3} && \text{Factor.} \\ &= \frac{-20xy^4z}{z^3x^5y^3} && \text{Reduce.} \\ &= \frac{-20y}{z^2x^4} && \text{Quotient rule} \end{aligned}$$

■

EXAMPLE 3 **Multiplying rational expressions**

Find the indicated products.

a) $\frac{2x - 2y}{4} \cdot \frac{2x}{x^2 - y^2}$

b) $\frac{x^2 + 7x + 12}{2x + 6} \cdot \frac{x}{x^2 - 16}$

c) $\frac{a + b}{6a} \cdot \frac{8a^2}{a^2 + 2ab + b^2}$

Solution

a) $\frac{2x - 2y}{4} \cdot \frac{2x}{x^2 - y^2} = \frac{\cancel{2}\cancel{(x - y)}}{\cancel{2} \cdot 2} \cdot \frac{\cancel{2} \cdot x}{\cancel{(x - y)}(x + y)}$ Factor.

$= \frac{x}{x + y}$ Reduce.

b) $\frac{x^2 + 7x + 12}{2x + 6} \cdot \frac{x}{x^2 - 16} = \frac{\cancel{(x + 3)}\cancel{(x + 4)}}{2\cancel{(x + 3)}} \cdot \frac{x}{(x - 4)\cancel{(x + 4)}}$ Factor.

$= \frac{x}{2(x - 4)}$ Reduce.

$= \frac{x}{2x - 8}$

c) $\frac{a + b}{6a} \cdot \frac{8a^2}{a^2 + 2ab + b^2} = \frac{a + b}{2 \cdot 3a} \cdot \frac{2 \cdot 4a^2}{(a + b)^2}$ Factor.

$= \frac{4a}{3(a + b)}$ Reduce.

$= \frac{4a}{3a + 3b}$ ■

Division of Rational Numbers

Division of rational numbers can be accomplished by multiplying by the reciprocal of the divisor.

Division of Rational Numbers

If $b \neq 0$, $c \neq 0$, and $d \neq 0$, then

$$\frac{a}{b} \div \frac{c}{d} = \frac{a}{b} \cdot \frac{d}{c}.$$

EXAMPLE 4 **Dividing rational numbers**

Find each quotient.

a) $5 \div \frac{1}{2}$

b) $\frac{6}{7} \div \frac{3}{14}$

Solution

a) $5 \div \frac{1}{2} = 5 \cdot 2 = 10$

b) $\frac{6}{7} \div \frac{3}{14} = \frac{6}{7} \cdot \frac{14}{3} = \frac{2 \cdot \cancel{3}}{\cancel{7}} \cdot \frac{2 \cdot \cancel{7}}{\cancel{3}} = 4$ ■

Division of Rational Expressions

We divide rational expressions in the same way we divide rational numbers: Invert the divisor and multiply.

EXAMPLE 5 **Dividing rational expressions**

Find each quotient.

a) $\frac{5}{3x} \div \frac{5}{6x}$

b) $\frac{x^7}{2} \div (2x^2)$

c) $\frac{4 - x^2}{x^2 + x} \div \frac{x - 2}{x^2 - 1}$

Solution

a)
$$\frac{5}{3x} \div \frac{5}{6x} = \frac{5}{3x} \cdot \frac{6x}{5} \quad \text{Invert the divisor and multiply.}$$
$$= \frac{\cancel{5}}{\cancel{3x}} \cdot \frac{2 \cdot \cancel{3x}}{\cancel{5}} \quad \text{Factor.}$$
$$= 2 \quad \text{Divide out the common factors.}$$

b)
$$\frac{x^7}{2} \div (2x^2) = \frac{x^7}{2} \cdot \frac{1}{2x^2} \quad \text{Invert and multiply.}$$
$$= \frac{x^5}{4} \quad \text{Quotient rule}$$

c)
$$\frac{4 - x^2}{x^2 + x} \div \frac{x - 2}{x^2 - 1} = \frac{4 - x^2}{x^2 + x} \cdot \frac{x^2 - 1}{x - 2} \quad \text{Invert and multiply.}$$
$$= \frac{\overset{-1}{\cancel{(2 - x)}}(2 + x)}{x\cancel{(x + 1)}} \cdot \frac{\cancel{(x + 1)}(x - 1)}{\cancel{x - 2}} \quad \text{Factor.}$$
$$= \frac{-1(2 + x)(x - 1)}{x} \quad \frac{2 - x}{x - 2} = -1$$
$$= \frac{-1(x^2 + x - 2)}{x} \quad \text{Simplify.}$$
$$= \frac{-x^2 - x + 2}{x}$$ ■

We sometimes write division of rational expressions using the fraction bar. For example, we can write

$$\frac{a+b}{3} \div \frac{1}{6} \quad \text{as} \quad \frac{\frac{a+b}{3}}{\frac{1}{6}}.$$

No matter how division is expressed, we invert the divisor and multiply.

EXAMPLE 6 **Division expressed with a fraction bar**

Find each quotient.

a) $\dfrac{\frac{a+b}{3}}{\frac{1}{6}}$ **b)** $\dfrac{\frac{x^2-1}{2}}{\frac{x-1}{3}}$ **c)** $\dfrac{\frac{a^2+5}{3}}{2}$

Solution

a)
$$\frac{\frac{a+b}{3}}{\frac{1}{6}} = \frac{a+b}{3} \div \frac{1}{6} \qquad \text{Rewrite as division.}$$
$$= \frac{a+b}{3} \cdot \frac{6}{1} \qquad \text{Invert and multiply.}$$
$$= \frac{a+b}{\cancel{3}} \cdot \frac{2 \cdot \cancel{3}}{1} \qquad \text{Factor.}$$
$$= (a+b)2 \qquad \text{Reduce.}$$
$$= 2a + 2b$$

b)
$$\frac{\frac{x^2-1}{2}}{\frac{x-1}{3}} = \frac{x^2-1}{2} \div \frac{x-1}{3} \qquad \text{Rewrite as division.}$$
$$= \frac{x^2-1}{2} \cdot \frac{3}{x-1} \qquad \text{Invert and multiply.}$$
$$= \frac{\cancel{(x-1)}(x+1)}{2} \cdot \frac{3}{\cancel{x-1}} \qquad \text{Factor.}$$
$$= \frac{3x+3}{2} \qquad \text{Reduce.}$$

c)
$$\frac{\frac{a^2+5}{3}}{2} = \frac{a^2+5}{3} \div 2 \qquad \text{Rewrite as division.}$$
$$= \frac{a^2+5}{3} \cdot \frac{1}{2} = \frac{a^2+5}{6}$$

■

Warm-ups

True or false? Explain your answer.

1. $\frac{2}{3} \cdot \frac{5}{3} = \frac{10}{9}$
2. The product of $\frac{x-7}{3}$ and $\frac{6}{7-x}$ is -2.
3. Dividing by 2 is equivalent to multiplying by $\frac{1}{2}$.
4. $3 \div x = \frac{1}{3} \cdot x$ for any nonzero number x.
5. Factoring polynomials is essential in multiplying rational expressions.
6. One-half of one-fourth is one-sixth.
7. One-half divided by three is three-halves.
8. The quotient of $(839 - 487)$ and $(487 - 839)$ is -1.
9. $\frac{a}{3} \div 3 = \frac{a}{9}$ for any value of a.
10. $\frac{a}{b} \cdot \frac{b}{a} = 1$ for any nonzero values of a and b.

5.2 EXERCISES

Perform the indicated operation. See Example 1.

1. $\frac{8}{15} \cdot \frac{35}{24}$
2. $\frac{3}{4} \cdot \frac{8}{21}$
3. $\frac{12}{17} \cdot \frac{51}{10}$
4. $\frac{25}{48} \cdot \frac{56}{35}$
5. $24 \cdot \frac{7}{20}$
6. $\frac{3}{10} \cdot 35$

Perform the indicated operation. See Example 2.

7. $\frac{5a}{12b} \cdot \frac{3ab}{55a}$
8. $\frac{3m}{7p} \cdot \frac{35p}{6mp}$
9. $\frac{-2x^6}{7a^5} \cdot \frac{21a^2}{6x}$
10. $\frac{5z^3w}{-9y^3} \cdot \frac{-6y^5}{20z^9}$
11. $\frac{15t^3y^5}{20w^7} \cdot 24t^5w^3y^2$
12. $22x^2y^3z \cdot \frac{6x^5}{33y^3z^4}$

Perform the indicated operation. See Example 3.

13. $\frac{3a+3b}{15} \cdot \frac{10a}{a^2-b^2}$
14. $\frac{b^3+b}{5} \cdot \frac{10}{b^2+b}$
15. $(x^2 - 6x + 9) \cdot \frac{3}{x-3}$
16. $\frac{12}{4x+10} \cdot (4x^2 + 20x + 25)$
17. $\frac{16a+8}{5a^2+5} \cdot \frac{2a^2+a-1}{4a^2-1}$
18. $\frac{6x-18}{2x^2-5x-3} \cdot \frac{4x^2+4x+1}{6x+3}$

Perform the indicated operation. See Example 4.

19. $12 \div \frac{2}{5}$
20. $32 \div \frac{1}{4}$
21. $\frac{5}{7} \div \frac{15}{14}$
22. $\frac{3}{4} \div \frac{15}{2}$
23. $\frac{40}{3} \div 12$
24. $\frac{22}{9} \div 9$

Perform the indicated operation. See Example 5.

25. $\frac{5x^2}{3} \div \frac{10x}{21}$
26. $\frac{4u^2}{3v} \div \frac{14u}{15v^6}$
27. $\frac{8m^3}{n^4} \div (12mn^2)$
28. $\frac{2p^4}{3q^3} \div (4pq^5)$
29. $\frac{y-6}{2} \div \frac{6-y}{6}$
30. $\frac{4-a}{5} \div \frac{a^2-16}{3}$
31. $\frac{x^2+4x+4}{8} \div \frac{(x+2)^3}{16}$
32. $\frac{a^2+2a+1}{3} \div \frac{a^2-1}{a}$
33. $\frac{t^2+3t-10}{t^2-25} \div (4t-8)$

34. $\dfrac{w^2 - 7w + 12}{w^2 - 4w} \div (w^2 - 9)$

35. $(2x^2 - 3x - 5) \div \dfrac{2x - 5}{x - 1}$

36. $(6y^2 - y - 2) \div \dfrac{2y + 1}{3y - 2}$

Perform the indicated operation. See Example 6.

37. $\dfrac{\frac{x - 2y}{5}}{\frac{1}{10}}$

38. $\dfrac{\frac{3m + 6n}{8}}{\frac{3}{4}}$

39. $\dfrac{\frac{x^2 - 4}{12}}{\frac{x - 2}{6}}$

40. $\dfrac{\frac{6a^2 + 6}{5}}{\frac{6a + 6}{5}}$

41. $\dfrac{\frac{x^2 + 9}{3}}{5}$

42. $\dfrac{\frac{1}{a - 3}}{4}$

43. $\dfrac{x^2 - y^2}{\frac{x - y}{9}}$

44. $\dfrac{x^2 + 6x + 8}{\frac{x + 2}{x + 1}}$

Perform the indicated operation.

45. $\dfrac{x - 1}{3} \cdot \dfrac{9}{1 - x}$

46. $\dfrac{2x - 2y}{3} \cdot \dfrac{1}{y - x}$

47. $\dfrac{3a + 3b}{a} \cdot \dfrac{1}{3}$

48. $\dfrac{a - b}{2b - 2a} \cdot \dfrac{2}{5}$

49. $\dfrac{\frac{b}{a}}{\frac{1}{2}}$

50. $\dfrac{\frac{2g}{3h}}{\frac{1}{h}}$

51. $\dfrac{6y}{3} \div (2x)$

52. $\dfrac{8x}{9} \div (18x)$

53. $\dfrac{a^3b^4}{-2ab^2} \cdot \dfrac{a^5b^7}{ab}$

54. $\dfrac{-2a^2}{3a^2} \cdot \dfrac{20a}{15a^3}$

55. $\dfrac{2mn^4}{6mn^2} \div \dfrac{3m^5n^7}{m^2n^4}$

56. $\dfrac{rt^2}{rt^2} \div \dfrac{rt^2}{r^3t^2}$

57. $\dfrac{3x^2 + 16x + 5}{x} \cdot \dfrac{x^2}{9x^2 - 1}$

58. $\dfrac{x^2 + 6x + 5}{x} \cdot \dfrac{x^4}{3x + 3}$

59. $\dfrac{a^2 - 2a + 4}{a^2 - 4} \cdot \dfrac{(a + 2)^3}{2a + 4}$

60. $\dfrac{w^2 - 1}{(w - 1)^2} \cdot \dfrac{w - 1}{w^2 + 2w + 1}$

61. $\dfrac{2x^2 + 19x - 10}{x^2 - 100} \div \dfrac{4x^2 - 1}{2x^2 - 19x - 10}$

62. $\dfrac{x^3 - 1}{x^2 + 1} \div \dfrac{9x^2 + 9x + 9}{x^2 - x}$

63. $\dfrac{9 + 6m + m^2}{9 - 6m + m^2} \cdot \dfrac{m^2 - 9}{m^2 + mk + 3m + 3k}$

64. $\dfrac{3x + 3w + bx + bw}{x^2 - w^2} \cdot \dfrac{6 - 2b}{9 - b^2}$

Solve each problem.

65. ***Area of a rectangle.*** If the length of a rectangle is x meters and its width is $\frac{5}{x}$ meters, then what is the area of the rectangle?

66. ***Area of a triangle.*** If the base of a triangle is $8x + 16$ yards and its height is $\frac{1}{x + 2}$ yards, then what is the area of the triangle?

Getting More Involved

67. ***Discussion.*** Evaluate each expression.

a) One-half of $\frac{1}{4}$ **b)** One-third of 4

c) One-half of $\frac{4x}{3}$ **d)** One-half of $\frac{3x}{2}$

68. ***Exploration.*** Let $R = \dfrac{6x^2 + 23x + 20}{24x^2 + 29x - 4}$ and $H = \dfrac{2x + 5}{8x - 1}$.

a) Find R when $x = 2$ and $x = 3$. Find H when $x = 2$ and $x = 3$.

b) How are these values of R and H related and why?

5.3 Building Up the Denominator

In this section:

- Building Up the Denominator
- Finding the Least Common Denominator
- Converting to the LCD

Every rational expression can be written in infinitely many equivalent forms. Because we can add or subtract only fractions with identical denominators, we must be able to change the denominator of a fraction. You have already learned how to change the denominator of a fraction by reducing. In this section you will learn the opposite of reducing, which is called **building up the denominator.**

Building Up the Denominator

To convert the fraction $\frac{2}{3}$ into an equivalent fraction with a denominator of 21, we factor 21 as $21 = 3 \cdot 7$. Because $\frac{2}{3}$ already has a 3 in the denominator, multiply the numerator and denominator of $\frac{2}{3}$ by the missing factor 7 to get a denominator of 21:

$$\frac{2}{3} = \frac{2}{3} \cdot \frac{7}{7} = \frac{14}{21}$$

For rational expressions the process is the same. To convert the rational expression

$$\frac{5}{x + 3}$$

into an equivalent rational expression with a denominator of $x^2 - x - 12$, first factor $x^2 - x - 12$:

$$x^2 - x - 12 = (x + 3)(x - 4)$$

From the factorization we can see that the denominator $x + 3$ needs only a factor of $x - 4$ to have the required denominator. So multiply the numerator and denominator by the missing factor $x - 4$:

$$\frac{5}{x + 3} = \frac{5(x - 4)}{(x + 3)(x - 4)} = \frac{5x - 20}{x^2 - x - 12}$$

EXAMPLE 1 **Building up the denominator**

Build each rational expression into an equivalent rational expression with the indicated denominator.

a) $3 = \frac{?}{12}$ **b)** $\frac{3}{w} = \frac{?}{wx}$ **c)** $\frac{2}{3y^3} = \frac{?}{12y^8}$

Solution

a) Because $3 = \frac{3}{1}$, we get a denominator of 12 by multiplying the numerator and denominator by 12:

$$3 = \frac{3}{1} = \frac{3 \cdot 12}{1 \cdot 12} = \frac{36}{12}$$

b) Multiply the numerator and denominator by x:

$$\frac{3}{w} = \frac{3 \cdot x}{w \cdot x} = \frac{3x}{wx}$$

c) To build the denominator $3y^3$ up to $12y^8$, multiply by $4y^5$:

$$\frac{2}{3y^3} = \frac{2 \cdot 4y^5}{3y^3 \cdot 4y^5} = \frac{8y^5}{12y^8}$$

■

In the next example we must factor the original denominator before building up the denominator.

EXAMPLE 2 **Building up the denominator**

Build each rational expression into an equivalent rational expression with the indicated denominator.

a) $\frac{7}{3x - 3y} = \frac{?}{6y - 6x}$ **b)** $\frac{x - 2}{x + 2} = \frac{?}{x^2 + 8x + 12}$

Solution

a) Because $3x - 3y = 3(x - y)$, we factor -6 out of $6y - 6x$. This will give a factor of $x - y$ in each denominator:

$$3x - 3y = 3(x - y)$$
$$6y - 6x = -6(x - y) = -2 \cdot 3(x - y)$$

To get the required denominator, we multiply the numerator and denominator by -2 only:

$$\frac{7}{3x - 3y} = \frac{7(-2)}{(3x - 3y)(-2)}$$
$$= \frac{-14}{6y - 6x}$$

b) Because $x^2 + 8x + 12 = (x + 2)(x + 6)$, we multiply the numerator and denominator by $x + 6$, the missing factor:

$$\frac{x - 2}{x + 2} = \frac{(x - 2)(x + 6)}{(x + 2)(x + 6)}$$
$$= \frac{x^2 + 4x - 12}{x^2 + 8x + 12}$$

■

CAUTION When building up a denominator, *both* the numerator and the denominator must be multiplied by the appropriate expression. ⊘

Finding the Least Common Denominator

We can use the idea of building up the denominator to convert two fractions with different denominators into fractions with identical denominators. For example,

$$\frac{5}{6} \text{ and } \frac{1}{4}$$

can both be converted into fractions with a denominator of 12, since $12 = 2 \cdot 6$ and $12 = 3 \cdot 4$:

$$\frac{5}{6} = \frac{5 \cdot 2}{6 \cdot 2} = \frac{10}{12} \qquad \frac{1}{4} = \frac{1 \cdot 3}{4 \cdot 3} = \frac{3}{12}$$

The smallest number that is a multiple of all of the denominators is called the **least common denominator (LCD).** The LCD for the denominators 6 and 4 is 12.

To find the LCD in a systematic way, we look at a complete factorization of each denominator. Consider the denominators 24 and 30:

$$24 = 2 \cdot 2 \cdot 2 \cdot 3 = 2^3 \cdot 3$$
$$30 = 2 \cdot 3 \cdot 5$$

Any multiple of 24 must have three 2's in its factorization, and any multiple of 30 must have one 2 as a factor. So a number with three 2's in its factorization will have enough to be a multiple of both 24 and 30. The LCD must also have one 3 and one 5 in its factorization. *We use each factor the maximum number of times it appears in either factorization.* So the LCD is $2^3 \cdot 3 \cdot 5$:

$$2^3 \cdot 3 \cdot 5 = \overbrace{2 \cdot 2 \cdot 2 \cdot 3}^{24} \cdot 5 = 120, \quad 2 \cdot 2 \cdot \underbrace{2 \cdot 3 \cdot 5}_{30}$$

If we omitted any one of the factors in $2 \cdot 2 \cdot 2 \cdot 3 \cdot 5$, we would not have a multiple of both 24 and 30. That is what makes 120 the *least* common denominator. To find the LCD for two polynomials, we use the same strategy.

Strategy for Finding the LCD for Polynomials

1. Factor each denominator completely. Use exponent notation for repeated factors.
2. Write the product of all of the different factors that appear in the denominators.
3. On each factor, use the highest power that appears on that factor in any of the denominators.

EXAMPLE 3 **Finding the LCD**

If the given expressions were used as denominators of rational expressions, then what would be the LCD for each group of denominators?

a) 20, 50 **b)** x^3yz^2, x^5y^2z, xyz^5 **c)** $a^2 + 5a + 6,\ a^2 + 4a + 4$

Solution

a) First factor each number completely:

$$20 = 2^2 \cdot 5 \qquad 50 = 2 \cdot 5^2$$

The highest power of 2 is 2, and the highest power of 5 is 2. So the LCD of 20 and 50 is $2^2 \cdot 5^2$, or 100.

b) The expressions x^3yz^2, x^5y^2z, and xyz^5 are already factored. For the LCD, use the highest power of each variable. So the LCD is $x^5y^2z^5$.

c) First factor each polynomial.

$$a^2 + 5a + 6 = (a + 2)(a + 3) \qquad a^2 + 4a + 4 = (a + 2)^2$$

The highest power of $(a + 3)$ is 1, and the highest power of $(a + 2)$ is 2. So the LCD is $(a + 3)(a + 2)^2$. ■

Converting to the LCD

When adding or subtracting rational expressions, we must convert the expressions into expressions with identical denominators. To keep the computations as simple as possible, we use the least common denominator.

EXAMPLE 4 **Converting to the LCD**

Find the LCD for the rational expressions, and convert each expression into an equivalent rational expression with the LCD as the denominator.

a) $\dfrac{4}{9xy}, \dfrac{2}{15xz}$ **b)** $\dfrac{5}{6x^2}, \dfrac{1}{8x^3y}, \dfrac{3}{4y^2}$

Solution

a) Factor each denominator completely:

$$9xy = 3^2xy \qquad 15xz = 3 \cdot 5xz$$

The LCD is $3^2 \cdot 5xyz$. Now convert each expression into an expression with this denominator. We must multiply the numerator and denominator of the first rational expression by $5z$ and the second by $3y$:

$$\left.\begin{aligned} \frac{4}{9xy} &= \frac{4 \cdot 5z}{9xy \cdot 5z} = \frac{20z}{45xyz} \\ \frac{2}{15xz} &= \frac{2 \cdot 3y}{15xz \cdot 3y} = \frac{6y}{45xyz} \end{aligned}\right\} \text{Same denominator}$$

b) Factor each denominator completely:

$$6x^2 = 2 \cdot 3x^2 \qquad 8x^3y = 2^3x^3y \qquad 4y^2 = 2^2y^2$$

The LCD is $2^3 \cdot 3 \cdot x^3y^2$ or $24x^3y^2$. Now convert each expression into an expression with this denominator:

$$\frac{5}{6x^2} = \frac{5 \cdot 4xy^2}{6x^2 \cdot 4xy^2} = \frac{20xy^2}{24x^3y^2}$$

$$\frac{1}{8x^3y} = \frac{1 \cdot 3y}{8x^3y \cdot 3y} = \frac{3y}{24x^3y^2}$$

$$\frac{3}{4y^2} = \frac{3 \cdot 6x^3}{4y^2 \cdot 6x^3} = \frac{18x^3}{24x^3y^2}$$

■

EXAMPLE 5 **Converting to the LCD**

Find the LCD for the rational expressions

$$\frac{5x}{x^2 - 4} \quad \text{and} \quad \frac{3}{x^2 + x - 6}$$

and convert each into an equivalent rational expression with that denominator.

Solution

First factor the denominators:

$$x^2 - 4 = (x - 2)(x + 2)$$
$$x^2 + x - 6 = (x - 2)(x + 3)$$

The LCD is $(x - 2)(x + 2)(x + 3)$. Now we multiply the numerator and denominator of the first rational expression by $(x + 3)$ and those of the second rational expression by $(x + 2)$. Because each denominator already has one factor of $(x - 2)$, there is no reason to multiply by $(x - 2)$. We multiply each denominator by the factors in the LCD that are missing from that denominator:

$$\frac{5x}{x^2 - 4} = \frac{5x(x + 3)}{(x - 2)(x + 2)(x + 3)} = \frac{5x^2 + 15x}{(x - 2)(x + 2)(x + 3)}$$
$$\frac{3}{x^2 + x - 6} = \frac{3(x + 2)}{(x - 2)(x + 3)(x + 2)} = \frac{3x + 6}{(x - 2)(x + 2)(x + 3)}$$

Same denominator

Note that in Example 5 we multiplied the expressions in the numerators but left the denominators in factored form. The numerators are simplified because it is the numerators that must be added when we add rational expressions in the next section. Because we can add rational expressions with identical denominators, there is no need to multiply the denominators.

Warm-ups

True or false? Explain your answer.

1. To convert $\frac{2}{3}$ into an equivalent fraction with a denominator of 18, we would multiply only the denominator of $\frac{2}{3}$ by 6.
2. Factoring has nothing to do with finding the least common denominator.
3. $\frac{3}{2ab^2} = \frac{15a^2b^2}{10a^3b^4}$ for any nonzero values of a and b.
4. The LCD for the denominators $2^5 \cdot 3$ and $2^4 \cdot 3^2$ is $2^5 \cdot 3^2$.
5. The LCD for the fractions $\frac{1}{6}$ and $\frac{1}{10}$ is 60.
6. The LCD for the denominators $6a^2b$ and $4ab^3$ is $2ab$.
7. The LCD for the denominators $a^2 + 1$ and $a + 1$ is $a^2 + 1$.
8. $\frac{x}{2} = \frac{x + 7}{2 + 7}$ for any real number x.
9. The LCD for the rational expressions $\frac{1}{x - 2}$ and $\frac{3}{x + 2}$ is $x^2 - 4$.
10. $x = \frac{3x}{3}$ for any real number x.

5.3 EXERCISES

Build each rational expression into an equivalent rational expression with the indicated denominator. See Example 1.

1. $\frac{1}{3} = \frac{?}{27}$

2. $\frac{2}{5} = \frac{?}{35}$

3. $7 = \frac{?}{2x}$

4. $6 = \frac{?}{4y}$

5. $\frac{5}{b} = \frac{?}{3bt}$

6. $\frac{7}{2ay} = \frac{?}{2ayz}$

7. $\frac{-9z}{2aw} = \frac{?}{8awz}$

8. $\frac{-7yt}{3x} = \frac{?}{18xyt}$

9. $\frac{2}{3a} = \frac{?}{15a^3}$

10. $\frac{7b}{12c^5} = \frac{?}{36c^8}$

11. $\frac{4}{5xy^2} = \frac{?}{10x^2y^5}$

12. $\frac{5y^2}{8x^3z} = \frac{?}{24x^5z^3}$

Build each rational expression into an equivalent rational expression with the indicated denominator. See Example 2.

13. $\frac{5}{2x+2} = \frac{?}{-8x-8}$

14. $\frac{3}{m-n} = \frac{?}{2n-2m}$

15. $\frac{8a}{5b^2-5b} = \frac{?}{20b^2-20b^3}$

16. $\frac{5x}{-6x-9} = \frac{?}{18x^2+27x}$

17. $\frac{3}{x+2} = \frac{?}{x^2-4}$

18. $\frac{a}{a+3} = \frac{?}{a^2-9}$

19. $\frac{3x}{x+1} = \frac{?}{x^2+2x+1}$

20. $\frac{-7x}{2x-3} = \frac{?}{4x^2-12x+9}$

21. $\frac{y-6}{y-4} = \frac{?}{y^2+y-20}$

22. $\frac{z-6}{z+3} = \frac{?}{z^2-2z-15}$

If the given expressions were used as denominators of rational expressions, then what would be the LCD for each group of denominators? See Example 3.

23. 12, 16

24. 28, 42

25. 12, 18, 20

26. 24, 40, 48

27. $6a^2$, $15a$

28. $18x^2$, $20xy$

29. $2a^4b$, $3ab^6$, $4a^3b^2$

30. $4m^3nw$, $6mn^5w^8$, $9m^6nw$

31. $x^2 - 16$, $x^2 + 8x + 16$

32. $x^2 - 9$, $x^2 + 6x + 9$

33. x, $x + 2$, $x - 2$

34. y, $y - 5$, $y + 2$

35. $x^2 - 4x$, $x^2 - 16$, $2x$

36. y, $y^2 - 3y$, $3y$

Find the LCD for the given rational expressions, and convert each rational expression into an equivalent rational expression with the LCD as the denominator. See Example 4.

37. $\frac{1}{6}, \frac{3}{8}$

38. $\frac{5}{12}, \frac{3}{20}$

39. $\frac{3}{84a}, \frac{5}{63b}$

40. $\frac{4b}{75a}, \frac{6}{105ab}$

41. $\frac{1}{3x^2}, \frac{3}{2x^5}$

42. $\frac{3}{8a^3b^9}, \frac{5}{6a^2c}$

43. $\frac{x}{9y^5z}, \frac{y}{12x^3}, \frac{1}{6x^2y}$

44. $\frac{5}{12a^6b}, \frac{3b}{14a^3}, \frac{1}{2ab^3}$

In Exercises 45–56, find the LCD for the given rational expressions, and convert each rational expression into an equivalent rational expression with the LCD as the denominator. See Example 5.

45. $\frac{2x}{x-3}, \frac{5x}{x+2}$

46. $\frac{2a}{a-5}, \frac{3a}{a+2}$

47. $\frac{4}{a-6}, \frac{5}{6-a}$

48. $\frac{4}{x-y}, \frac{5x}{2y-2x}$

49. $\frac{x}{x^2-9}, \frac{5x}{x^2-6x+9}$

50. $\frac{5x}{x^2-1}, \frac{-4}{x^2-2x+1}$

51. $\frac{w+2}{w^2-2w-15}, \frac{-2w}{w^2-4w-5}$

52. $\frac{z-1}{z^2+6z+8}, \frac{z+1}{z^2+5z+6}$

53. $\dfrac{-5}{6x-12}, \dfrac{x}{x^2-4}, \dfrac{3}{2x+4}$

54. $\dfrac{3}{4b^2-9}, \dfrac{2b}{2b+3}, \dfrac{-5}{2b^2-3b}$

55. $\dfrac{2}{2q^2-5q-3}, \dfrac{3}{2q^2+9q+4}, \dfrac{4}{q^2+q-12}$

56. $\dfrac{-3}{2p^2+7p-15}, \dfrac{p}{2p^2-11p+12}, \dfrac{2}{p^2+p-20}$

Getting More Involved

57. ***Discussion.*** Why do we learn how to convert two rational expressions into equivalent rational expressions with the same denominator?

58. ***Discussion.*** Which expression is the LCD for

$$\frac{3x-1}{2^2 \cdot 3 \cdot x^2(x+2)} \text{ and } \frac{2x+7}{2 \cdot 3^2 \cdot x(x+2)^2}?$$

a) $2 \cdot 3 \cdot x(x+2)$ **b)** $36x(x+2)$
c) $36x^2(x+2)^2$ **d)** $2^3 \cdot 3^3x^3(x+2)^2$

5.4 Addition and Subtraction

In this section:

- Addition and Subtraction of Rational Numbers
- Addition and Subtraction of Rational Expressions

In Section 5.3 you learned how to find the LCD and build up the denominators of rational expressions. In this section we will use that knowledge to add and subtract rational expressions with different denominators.

Addition and Subtraction of Rational Numbers

We can add or subtract rational numbers (or fractions) only with identical denominators according to the following definition.

Addition and Subtraction of Rational Numbers

If $b \neq 0$, then

$$\frac{a}{b} + \frac{c}{b} = \frac{a+c}{b} \quad \text{and} \quad \frac{a}{b} - \frac{c}{b} = \frac{a-c}{b}.$$

EXAMPLE 1 **Adding or subtracting fractions with the same denominator**

Perform the indicated operations. Reduce answers to lowest terms.

a) $\dfrac{1}{12} + \dfrac{7}{12}$

b) $\dfrac{1}{4} - \dfrac{3}{4}$

Solution

a) $\dfrac{1}{12} + \dfrac{7}{12} = \dfrac{8}{12} = \dfrac{\cancel{4} \cdot 2}{\cancel{4} \cdot 3} = \dfrac{2}{3}$

b) $\dfrac{1}{4} - \dfrac{3}{4} = \dfrac{-2}{4} = -\dfrac{1}{2}$

■

If the rational numbers have different denominators, we must convert them to equivalent rational numbers that have identical denominators and then add or subtract. Of course, it is most efficient to use the least common denominator (LCD), as in the following example.

EXAMPLE 2 **Adding or subtracting fractions with different denominators**

Find each sum or difference.

a) $\frac{3}{20} + \frac{7}{12}$

b) $\frac{1}{6} - \frac{4}{15}$

Solution

a) Because $20 = 2^2 \cdot 5$ and $12 = 2^2 \cdot 3$, the LCD is $2^2 \cdot 3 \cdot 5$, or 60. Convert each fraction to an equivalent fraction with a denominator of 60:

$$\frac{3}{20} + \frac{7}{12} = \frac{3 \cdot 3}{20 \cdot 3} + \frac{7 \cdot 5}{12 \cdot 5} \quad \text{Build up the denominators.}$$

$$= \frac{9}{60} + \frac{35}{60} \quad \text{Simplify numerators and denominators.}$$

$$= \frac{44}{60} \quad \text{Add the fractions.}$$

$$= \frac{4 \cdot 11}{4 \cdot 15} \quad \text{Factor.}$$

$$= \frac{11}{15} \quad \text{Reduce.}$$

b) Because $6 = 2 \cdot 3$ and $15 = 3 \cdot 5$, the LCD is $2 \cdot 3 \cdot 5$ or 30:

$$\frac{1}{6} - \frac{4}{15} = \frac{1}{2 \cdot 3} - \frac{4}{3 \cdot 5} \quad \text{Factor the denominators.}$$

$$= \frac{1 \cdot 5}{2 \cdot 3 \cdot 5} - \frac{4 \cdot 2}{3 \cdot 5 \cdot 2} \quad \text{Build up the denominators.}$$

$$= \frac{5}{30} - \frac{8}{30} \quad \text{Simplify the numerators and denominators.}$$

$$= \frac{-3}{30} \quad \text{Subtract.}$$

$$= \frac{-1 \cdot 3}{10 \cdot 3} \quad \text{Factor.}$$

$$= -\frac{1}{10} \quad \text{Reduce.}$$

■

Addition and Subtraction of Rational Expressions

Rational expressions are added or subtracted just like rational numbers. We can add or subtract only rational expressions that have identical denominators.

EXAMPLE 3 **Rational expressions with the same denominator**

Perform the indicated operations and reduce answers to lowest terms.

a) $\frac{2}{3y} + \frac{4}{3y}$

b) $\frac{2x}{x+2} + \frac{4}{x+2}$

c) $\frac{x^2 + 2x}{(x-1)(x+3)} - \frac{2x+1}{(x-1)(x+3)}$

Solution

a) $\frac{2}{3y} + \frac{4}{3y} = \frac{6}{3y}$ Add the fractions.

$= \frac{2}{y}$ Reduce.

b) $\frac{2x}{x+2} + \frac{4}{x+2} = \frac{2x+4}{x+2}$ Add the fractions.

$= \frac{2(x+2)}{x+2}$ Factor the numerator.

$= 2$ Reduce.

c) $\frac{x^2 + 2x}{(x-1)(x+3)} - \frac{2x+1}{(x-1)(x+3)} = \frac{x^2 + 2x - (2x+1)}{(x-1)(x+3)}$ Subtract the fractions.

$= \frac{x^2 + 2x - 2x - 1}{(x-1)(x+3)}$ Remove parentheses.

$= \frac{x^2 - 1}{(x-1)(x+3)}$ Combine like terms.

$= \frac{(x-1)(x+1)}{(x-1)(x+3)}$ Factor.

$= \frac{x+1}{x+3}$ Reduce. ■

CAUTION When subtracting a numerator containing more than one term, be sure to enclose it in parentheses, as in Example 3(c). Because that numerator is a binomial, the sign of each of its terms must be changed for the subtraction.

In the next example the rational expressions have different denominators.

EXAMPLE 4 **Rational expressions with different denominators**

Perform the indicated operations.

a) $\dfrac{5}{2x} + \dfrac{2}{3}$

b) $\dfrac{4}{x^3y} + \dfrac{2}{xy^3}$

c) $\dfrac{a+1}{6} - \dfrac{a-2}{8}$

Solution

a) The LCD for $2x$ and 3 is $6x$:

$$\frac{5}{2x} + \frac{2}{3} = \frac{5 \cdot 3}{2x \cdot 3} + \frac{2 \cdot 2x}{3 \cdot 2x} \quad \text{Build up both denominators to } 6x.$$

$$= \frac{15}{6x} + \frac{4x}{6x} \quad \text{Simplify numerators and denominators.}$$

$$= \frac{15 + 4x}{6x} \quad \text{Add the rational expressions.}$$

b) The LCD is x^3y^3.

$$\frac{4}{x^3y} + \frac{2}{xy^3} = \frac{4 \cdot y^2}{x^3y \cdot y^2} + \frac{2 \cdot x^2}{xy^3 \cdot x^2} \quad \text{Build up both denominators to the LCD.}$$

$$= \frac{4y^2}{x^3y^3} + \frac{2x^2}{x^3y^3} \quad \text{Simplify numerators and denominators.}$$

$$= \frac{4y^2 + 2x^2}{x^3y^3} \quad \text{Add the rational expressions.}$$

c) Because $6 = 2 \cdot 3$ and $8 = 2^3$, the LCD is $2^3 \cdot 3$, or 24:

$$\frac{a+1}{6} - \frac{a-2}{8} = \frac{(a+1)4}{6 \cdot 4} - \frac{(a-2)3}{8 \cdot 3} \quad \text{Build up both denominators to the LCD 24.}$$

$$= \frac{4a+4}{24} - \frac{3a-6}{24} \quad \text{Simplify numerators and denominators.}$$

$$= \frac{4a + 4 - (3a - 6)}{24} \quad \text{Subtract the rational expressions.}$$

$$= \frac{4a + 4 - 3a + 6}{24} \quad \text{Remove the parentheses.}$$

$$= \frac{a + 10}{24} \quad \text{Combine like terms.}$$

■

EXAMPLE 5 **Rational expressions with different denominators**

Perform the indicated operations:

a) $\dfrac{1}{x^2 - 9} + \dfrac{2}{x^2 + 3x}$

b) $\dfrac{4}{5 - a} - \dfrac{2}{a - 5}$

Solution

a) $$\frac{1}{x^2 - 9} + \frac{2}{x^2 + 3x} = \underbrace{\frac{1}{(x-3)(x+3)}}_{\text{Needs } x} + \underbrace{\frac{2}{x(x+3)}}_{\text{Needs } x-3}$$ The LCD is $x(x - 3)(x + 3)$.

$$= \frac{1 \cdot x}{(x-3)(x+3)x} + \frac{2(x-3)}{x(x+3)(x-3)}$$

$$= \frac{x}{x(x-3)(x+3)} + \frac{2x-6}{x(x-3)(x+3)}$$

$$= \frac{3x-6}{x(x-3)(x+3)}$$ We usually leave the denominator in factored form.

b) Because $-1(5 - a) = a - 5$, we can get identical denominators by multiplying only the first expression by -1 in the numerator and denominator:

$$\frac{4}{5-a} - \frac{2}{a-5} = \frac{4(-1)}{(5-a)(-1)} - \frac{2}{a-5}$$

$$= \frac{-4}{a-5} - \frac{2}{a-5}$$

$$= \frac{-6}{a-5}$$ $-4 - 2 = -6$

$$= -\frac{6}{a-5}$$ ■

In the next example we combine three rational expressions by addition and subtraction.

EXAMPLE 6 **Rational expressions with different denominators**

Perform the indicated operations.

$$\frac{x+1}{x^2+2x} + \frac{2x+1}{6x+12} - \frac{1}{6}$$

Solution

The LCD for $x(x + 2)$, $6(x + 2)$, and 6 is $6x(x + 2)$.

$$\frac{x+1}{x^2+2x} + \frac{2x+1}{6x+12} - \frac{1}{6} = \frac{x+1}{x(x+2)} + \frac{2x+1}{6(x+2)} - \frac{1}{6}$$ Factor denominators.

$$= \frac{6(x+1)}{6x(x+2)} + \frac{x(2x+1)}{6x(x+2)} - \frac{1x(x+2)}{6x(x+2)}$$ Build up to the LCD.

$$= \frac{6x+6}{6x(x+2)} + \frac{2x^2+x}{6x(x+2)} - \frac{x^2+2x}{6x(x+2)}$$ Simply numerators.

$$= \frac{6x+6+2x^2+x-x^2-2x}{6x(x+2)}$$ Combine the numerators.

$$= \frac{x^2+5x+6}{6x(x+2)}$$ Combine like terms.

$$= \frac{(x+3)(x+2)}{6x(x+2)}$$ Factor.

$$= \frac{x+3}{6x}$$ Reduce. ■

Warm-ups

True or false? Explain your answer.

1. $\frac{1}{2} + \frac{1}{3} = \frac{2}{5}$ **2.** $\frac{7}{12} - \frac{1}{12} = \frac{1}{2}$ **3.** $\frac{3}{5} + \frac{4}{3} = \frac{29}{15}$

4. $\frac{4}{5} - \frac{5}{7} = \frac{3}{35}$ **5.** $\frac{5}{20} + \frac{3}{4} = 1$

6. $\frac{2}{x} + 1 = \frac{3}{x}$ for any nonzero value of x.

7. $1 + \frac{1}{a} = \frac{a+1}{a}$ for any nonzero value of a.

8. $a - \frac{1}{4} = \frac{3}{4}a$ for any value of a.

9. $\frac{a}{2} + \frac{b}{3} = \frac{3a+2b}{6}$ for any values of a and b.

10. The LCD for the rational expressions $\frac{1}{x}$ and $\frac{3x}{x-1}$ is $x^2 - 1$.

5.4 EXERCISES

Perform the indicated operation. Reduce each answer to lowest terms. See Example 1.

1. $\frac{1}{10} + \frac{1}{10}$ **2.** $\frac{1}{8} + \frac{3}{8}$ **5.** $\frac{1}{6} - \frac{5}{6}$ **6.** $-\frac{3}{8} - \frac{7}{8}$

3. $\frac{7}{8} - \frac{1}{8}$ **4.** $\frac{4}{9} - \frac{1}{9}$ **7.** $-\frac{7}{8} + \frac{1}{8}$ **8.** $-\frac{9}{20} + \left(-\frac{3}{20}\right)$

Perform the indicated operation. Reduce each answer to lowest terms. See Example 2.

9. $\frac{1}{3}+\frac{2}{9}$

10. $\frac{1}{4}+\frac{5}{6}$

11. $\frac{7}{16}+\frac{5}{18}$

12. $\frac{7}{6}+\frac{4}{15}$

13. $\frac{1}{8}-\frac{9}{10}$

14. $\frac{2}{15}-\frac{5}{12}$

15. $-\frac{1}{6}-\left(-\frac{3}{8}\right)$

16. $-\frac{1}{5}-\left(-\frac{1}{7}\right)$

Perform the indicated operation. Reduce each answer to lowest terms. See Example 3.

17. $\frac{3}{2w}+\frac{7}{2w}$

18. $\frac{5x}{3y}+\frac{7x}{3y}$

19. $\frac{3a}{a+5}+\frac{15}{a+5}$

20. $\frac{a+7}{a-4}+\frac{9-5a}{a-4}$

21. $\frac{q-1}{q-4}-\frac{3q-9}{q-4}$

22. $\frac{3-a}{3}-\frac{a-5}{3}$

23. $\frac{4h-3}{h(h+1)}-\frac{h-6}{h(h+1)}$

24. $\frac{2t-9}{t(t-3)}-\frac{t-9}{t(t-3)}$

25. $\frac{x^2-x-5}{(x+1)(x+2)}+\frac{1-2x}{(x+1)(x+2)}$

26. $\frac{2x-5}{(x-2)(x+6)}+\frac{x^2-2x+1}{(x-2)(x+6)}$

Perform the indicated operation. Reduce each answer to lowest terms. See Example 4.

27. $\frac{3}{2a}+\frac{1}{5a}$

28. $\frac{5}{6y}-\frac{3}{8y}$

29. $\frac{w-3}{9}-\frac{w-4}{12}$

30. $\frac{y+4}{10}-\frac{y-2}{14}$

31. $\frac{b^2}{4a}-c$

32. $y+\frac{3}{7b}$

33. $\frac{2}{wz^2}+\frac{3}{w^2z}$

34. $\frac{1}{a^5b}-\frac{5}{ab^3}$

Perform the indicated operation. Reduce each answer to lowest terms. See Examples 5 and 6.

35. $\frac{2}{x+1}-\frac{3}{x}$

36. $\frac{1}{a-1}-\frac{2}{a}$

37. $\frac{2}{a-b}+\frac{1}{a+b}$

38. $\frac{3}{x+1}+\frac{2}{x-1}$

39. $\frac{3}{x^2+x}-\frac{4}{5x+5}$

40. $\frac{3}{a^2+3a}-\frac{2}{5a+15}$

41. $\frac{2a}{a^2-9}+\frac{a}{a-3}$

42. $\frac{x}{x^2-1}+\frac{3}{x-1}$

43. $\frac{4}{a-b}+\frac{4}{b-a}$

44. $\frac{2}{x-3}+\frac{3}{3-x}$

45. $\frac{3}{2a-2}-\frac{2}{1-a}$

46. $\frac{5}{2x-4}-\frac{3}{2-x}$

47. $\frac{1}{x^2-4}-\frac{3}{x^2-3x-10}$

48. $\frac{2x}{x^2-9}+\frac{3x}{x^2+4x+3}$

49. $\frac{3}{x^2+x-2}+\frac{4}{x^2+2x-3}$

50. $\frac{x-1}{x^2-x-12}+\frac{x+4}{x^2+5x+6}$

51. $\frac{2}{x}-\frac{1}{x-1}+\frac{1}{x+2}$

52. $\frac{1}{a}-\frac{2}{a+1}+\frac{3}{a-1}$

53. $\frac{5}{3a-9}-\frac{3}{2a}+\frac{4}{a^2-3a}$

54. $\frac{3}{4c+2}-\frac{c-4}{2c^2+c}-\frac{5}{6c}$

Solve each problem.

55. ***Perimeter of a rectangle.*** Suppose that the length of a rectangle is $\frac{3}{x}$ feet and its width is $\frac{5}{2x}$ feet. Find a rational expression for the perimeter of the rectangle.

56. ***Perimeter of a triangle.*** The lengths of the sides of a triangle are $\frac{1}{x}$, $\frac{1}{2x}$, and $\frac{2}{3x}$ meters. Find a rational expression for the perimeter of the triangle.

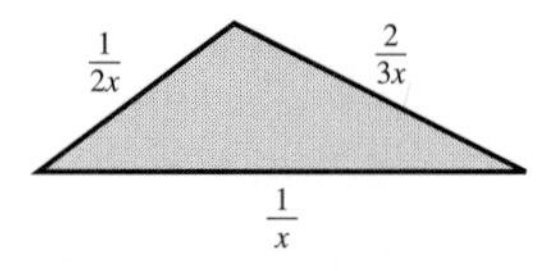

Figure for Exercise 56

57. ***Traveling time.*** Janet drove 120 miles at x mph before 6:00 A.M. After 6:00 A.M. she increased her speed by 5 mph and drove 195 additional miles. Use the fact that $T=\frac{D}{R}$ to complete the table on the next page.

	Rate	Time	Distance
Before	$x \dfrac{\text{mi}}{\text{hr}}$		120 mi
After	$x + 5 \dfrac{\text{mi}}{\text{hr}}$		195 mi

Write a rational expression for her total traveling time. Evaluate the expression for $x = 60$.

58. ***Traveling time.*** Hanson drove 200 kilometers at x km/hr and then decreased his speed by 20 km/hr and drove 240 additional kilometers. Make a table like the one in Exercise 57. Write a rational expression for his total traveling time. Evaluate the expression for $x = 100$.

59. ***House painting.*** Kent can paint a certain house by himself in x days. His helper Keith can paint the same house by himself in $x + 3$ days. Suppose that they work together on the job for 2 days. Use the fact that the work completed is the product of the rate and the time to complete the table.

	Rate	Time	Work
Kent	$\dfrac{1}{x} \dfrac{\text{job}}{\text{day}}$	2 days	
Keith	$\dfrac{1}{x+3} \dfrac{\text{job}}{\text{day}}$	2 days	

Write a rational expression for the fraction of the house that they complete by working together for 2 days. Evaluate the expression for $x = 6$.

60. ***Barn painting.*** Melanie can paint a certain barn by herself in x days. Her helper Melissa can paint the same barn by herself in $2x$ days. Write a rational expression for the fraction of the barn that they complete in one day by working together. Evaluate the expression for $x = 5$.

Getting More Involved

61. ***Writing.*** Write a step-by-step procedure for adding rational expressions.

62. ***Writing.*** Explain why fractions must have the same denominator to be added. Use real-life examples.

5.5 Complex Fractions

In this section:

- Complex Fractions
- Using the LCD to Simplify Complex Fractions
- Applications

In this section we will use the idea of least common denominator to simplify complex fractions. Also we will see how complex fractions can arise in applications.

Complex Fractions

A **complex fraction** is a fraction having rational expressions in the numerator, denominator, or both. Consider the following complex fraction:

$$\frac{\dfrac{1}{2} + \dfrac{2}{3}}{\dfrac{1}{4} - \dfrac{5}{8}} \quad \begin{matrix} \leftarrow \text{Numerator of complex fraction} \\ \leftarrow \text{Denominator of complex fraction} \end{matrix}$$

To simplify it, we can combine the fractions in the numerator as follows:

$$\frac{1}{2} + \frac{2}{3} = \frac{1 \cdot 3}{2 \cdot 3} + \frac{2 \cdot 2}{3 \cdot 2} = \frac{3}{6} + \frac{4}{6} = \frac{7}{6}$$

We can combine the fractions in the denominator as follows:

$$\frac{1}{4} - \frac{5}{8} = \frac{1 \cdot 2}{4 \cdot 2} - \frac{5}{8} = \frac{2}{8} - \frac{5}{8} = -\frac{3}{8}$$

Now divide the numerator by the denominator:

$$\frac{\frac{1}{2} + \frac{2}{3}}{\frac{1}{4} - \frac{5}{8}} = \frac{\frac{7}{6}}{-\frac{3}{8}} = \frac{7}{6} \div \left(-\frac{3}{8}\right)$$

$$= \frac{7}{6} \cdot \left(-\frac{8}{3}\right)$$

$$= -\frac{56}{18}$$

$$= -\frac{28}{9}$$

Using the LCD to Simplify Complex Fractions

A complex fraction can be simplified by writing the numerator and denominator as single fractions and then dividing, as we just did. However, there is a better method. The next example shows how to simplify a complex fraction by using the LCD of all of the single fractions in the complex fraction.

EXAMPLE 1 **Using the LCD to simplify a complex fraction**

Use the LCD to simplify

$$\frac{\frac{1}{2} + \frac{2}{3}}{\frac{1}{4} - \frac{5}{8}}.$$

Solution

The LCD of 2, 3, 4, and 8 is 24. Now multiply the numerator and denominator of the complex fraction by the LCD:

$$\frac{\frac{1}{2} + \frac{2}{3}}{\frac{1}{4} - \frac{5}{8}} = \frac{\left(\frac{1}{2} + \frac{2}{3}\right)24}{\left(\frac{1}{4} - \frac{5}{8}\right)24}$$ Multiply the numerator and denominator by the LCD.

$$= \frac{\frac{1}{2} \cdot 24 + \frac{2}{3} \cdot 24}{\frac{1}{4} \cdot 24 - \frac{5}{8} \cdot 24}$$ Distributive property

$$= \frac{12 + 16}{6 - 15}$$ Simplify.

$$= \frac{28}{-9}$$

$$= -\frac{28}{9}$$

■

CAUTION We simplify a complex fraction by multiplying the numerator and denominator of the *complex fraction* by the LCD. Do not multiply the numerator and denominator of each fraction in the complex fraction by the LCD.

In the next example we simplify a complex fraction involving variables.

EXAMPLE 2 **A complex fraction with variables**

Simplify

$$\frac{2 - \frac{1}{x}}{\frac{1}{x^2} - \frac{1}{2}}.$$

Solution

The LCD of the denominators x, x^2, and 2 is $2x^2$:

$$\frac{2 - \frac{1}{x}}{\frac{1}{x^2} - \frac{1}{2}} = \frac{\left(2 - \frac{1}{x}\right)(2x^2)}{\left(\frac{1}{x^2} - \frac{1}{2}\right)(2x^2)} \quad \text{Multiply the numerator and denominator by } 2x^2.$$

$$= \frac{2 \cdot 2x^2 - \frac{1}{x} \cdot 2x^2}{\frac{1}{x^2} \cdot 2x^2 - \frac{1}{2} \cdot 2x^2} \quad \text{Distributive property}$$

$$= \frac{4x^2 - 2x}{2 - x^2} \quad \text{Simplify.}$$

The numerator of this answer can be factored, but the rational expression cannot be reduced.

The general strategy for simplifying a complex fraction is stated as follows.

Strategy for Simplifying a Complex Fraction

1. Find the LCD for all the denominators in the complex fraction.
2. Multiply both the numerator and the denominator of the complex fraction by the LCD. Use the distributive property if necessary.
3. Combine like terms if possible.
4. Reduce to lowest terms when possible.

EXAMPLE 3 **Simplifying a complex fraction**

Simplify

$$\frac{\frac{1}{x-2} - \frac{2}{x+2}}{\frac{3}{2-x} + \frac{4}{x+2}}.$$

Solution

Because $x - 2$ and $2 - x$ are opposites, we can use $(x-2)(x+2)$ as the LCD. Multiply the numerator and denominator by $(x-2)(x+2)$:

$$\frac{\frac{1}{x-2} - \frac{2}{x+2}}{\frac{3}{2-x} + \frac{4}{x+2}} = \frac{\frac{1}{x-2}(x-2)(x+2) - \frac{2}{x+2}(x-2)(x+2)}{\frac{3}{2-x}(x-2)(x+2) + \frac{4}{x+2}(x-2)(x+2)}$$

$$= \frac{x+2-2(x-2)}{3(-1)(x+2)+4(x-2)} \qquad \frac{x-2}{2-x} = -1$$

$$= \frac{x+2-2x+4}{-3x-6+4x-8} \qquad \text{Distributive property}$$

$$= \frac{-x+6}{x-14} \qquad \text{Combine like terms.}$$

■

Applications

As their name suggests, complex fractions arise in some fairly complex situations.

EXAMPLE 4 **Fast-food workers**

A survey of college students found that $\frac{1}{2}$ of the female students had jobs and $\frac{2}{3}$ of the male students had jobs. It was also found that $\frac{1}{4}$ of the female students worked in fast-food restaurants and $\frac{1}{6}$ of the male students worked in fast-food restaurants. If equal numbers of male and female students were surveyed, then what fraction of the working students worked in fast-food restaurants?

Solution

Let x represent the number of males surveyed. The number of females surveyed is also x. The total number of students working in fast-food restaurants is

$$\frac{1}{4}x + \frac{1}{6}x.$$

The total number of working students in the survey is

$$\frac{1}{2}x + \frac{2}{3}x.$$

So the fraction of working students who work in fast-food restaurants is

$$\frac{\frac{1}{4}x + \frac{1}{6}x}{\frac{1}{2}x + \frac{2}{3}x}.$$

The LCD of the denominators 2, 3, 4, and 6 is 12. Multiply the numerator and denominator by 12 to eliminate the fractions as follows:

$$\frac{\frac{1}{4}x + \frac{1}{6}x}{\frac{1}{2}x + \frac{2}{3}x} = \frac{\left(\frac{1}{4}x + \frac{1}{6}x\right)12}{\left(\frac{1}{2}x + \frac{2}{3}x\right)12} \quad \text{Multiply numerator and denominator by 12.}$$

$$= \frac{3x + 2x}{6x + 8x} \quad \text{Distributive property}$$

$$= \frac{5x}{14x} \quad \text{Combine like terms.}$$

$$= \frac{5}{14} \quad \text{Reduce.}$$

So $\frac{5}{14}$ (or about 36%) of the working students work in fast-food restaurants. ■

Warm-ups

True or false? Explain your answer.

1. The LCD for the denominators 4, x, 6, and x^2 is $12x^3$.
2. The LCD for the denominators $a - b$, $2b - 2a$, and 6 is $6a - 6b$.
3. The fraction $\frac{4117}{7983}$ is a complex fraction.
4. The LCD for the denominators $a - 3$ and $3 - a$ is $a^2 - 9$.
5. The largest common denominator for the fractions $\frac{1}{2}$, $\frac{1}{3}$, and $\frac{1}{4}$ is 24.

Questions 6–10 refer to the following complex fractions:

a) $\dfrac{\frac{1}{2} + \frac{x}{3}}{\frac{1}{4} + \frac{1}{5}}$ **b)** $\dfrac{1 + \frac{2}{b}}{\frac{2}{a} + 5}$ **c)** $\dfrac{x - \frac{1}{2}}{x + \frac{3}{2}}$ **d)** $\dfrac{\frac{1}{2} + \frac{1}{3}}{1 + \frac{1}{2}}$

6. To simplify (a), we multiply the numerator and denominator by $60x$.
7. To simplify (b), we multiply the numerator and denominator by $\frac{ab}{ab}$.
8. The complex fraction (c) is equivalent to $\frac{2x - 1}{2x + 3}$.
9. If $x \neq -\frac{3}{2}$, then (c) represents a real number.
10. The complex fraction (d) can be written as $\frac{5}{6} \div \frac{3}{2}$.

5.5 EXERCISES

Simplify each complex fraction. See Example 1.

1. $\dfrac{\frac{1}{2}+\frac{1}{3}}{\frac{1}{4}-\frac{1}{2}}$

2. $\dfrac{\frac{1}{3}-\frac{1}{4}}{\frac{1}{3}+\frac{1}{6}}$

3. $\dfrac{\frac{2}{5}+\frac{5}{6}-\frac{1}{2}}{\frac{1}{2}-\frac{1}{3}+\frac{1}{15}}$

4. $\dfrac{\frac{2}{5}-\frac{2}{9}-\frac{1}{3}}{\frac{1}{3}+\frac{1}{5}+\frac{2}{15}}$

5. $\dfrac{3+\frac{1}{2}}{5-\frac{3}{4}}$

6. $\dfrac{1+\frac{1}{12}}{1-\frac{1}{12}}$

7. $\dfrac{1-\frac{1}{6}+\frac{2}{3}}{1+\frac{1}{15}-\frac{3}{10}}$

8. $\dfrac{3-\frac{2}{9}-\frac{1}{6}}{\frac{5}{18}-\frac{1}{3}-2}$

Simplify each complex fraction. See Example 2.

9. $\dfrac{\frac{1}{a}+\frac{3}{b}}{\frac{1}{b}-\frac{3}{a}}$

10. $\dfrac{\frac{1}{x}-\frac{3}{2}}{\frac{3}{4}+\frac{1}{x}}$

11. $\dfrac{5-\frac{3}{a}}{3+\frac{1}{a}}$

12. $\dfrac{4+\frac{3}{y}}{1-\frac{2}{y}}$

13. $\dfrac{\frac{1}{2}-\frac{2}{x}}{3-\frac{1}{x^2}}$

14. $\dfrac{\frac{2}{a}+\frac{5}{3}}{\frac{3}{a}-\frac{3}{a^2}}$

15. $\dfrac{\frac{3}{2b}+\frac{1}{b}}{\frac{3}{4}-\frac{1}{b^2}}$

16. $\dfrac{\frac{3}{2w}+\frac{4}{3w}}{\frac{1}{4w}-\frac{5}{9w}}$

Simplify each complex fraction. See Example 3.

17. $\dfrac{1-\frac{3}{y+1}}{3+\frac{1}{y+1}}$

18. $\dfrac{2-\frac{1}{a-3}}{3-\frac{1}{a-3}}$

19. $\dfrac{x+\frac{4}{x-2}}{x-\frac{x+1}{x-2}}$

20. $\dfrac{x-\frac{x-6}{x-1}}{x-\frac{x+15}{x-1}}$

21. $\dfrac{\frac{1}{3-x}-5}{\frac{1}{x-3}-2}$

22. $\dfrac{\frac{2}{x-5}-x}{\frac{3x}{5-x}-1}$

23. $\dfrac{1-\frac{5}{a-1}}{3-\frac{2}{1-a}}$

24. $\dfrac{\frac{1}{3}-\frac{2}{9-x}}{\frac{1}{6}-\frac{1}{x-9}}$

25. $\dfrac{\frac{1}{m-3}-\frac{4}{m}}{\frac{3}{m-3}+\frac{1}{m}}$

26. $\dfrac{\frac{1}{y+3}-\frac{4}{y}}{\frac{1}{y}-\frac{2}{y+3}}$

27. $\dfrac{\frac{2}{w-1}-\frac{3}{w+1}}{\frac{4}{w+1}+\frac{5}{w-1}}$

28. $\dfrac{\frac{1}{x+2}-\frac{3}{x+3}}{\frac{2}{x+3}+\frac{3}{x+2}}$

29. $\dfrac{\frac{1}{a-b}-\frac{1}{a+b}}{\frac{1}{b-a}+\frac{1}{b+a}}$

30. $\dfrac{\frac{1}{2+x}-\frac{1}{2-x}}{\frac{1}{x+2}-\frac{1}{x-2}}$

Simplify each complex fraction.

31. $\dfrac{\frac{2x-9}{6}}{\frac{2x-3}{9}}$

32. $\dfrac{\frac{a-5}{12}}{\frac{a+2}{15}}$

33. $\dfrac{\frac{2x-4y}{xy^2}}{\frac{3x-6y}{x^3y}}$

34. $\dfrac{\frac{ab+b^2}{4ab^5}}{\frac{a+b}{6a^2b^4}}$

35. $\dfrac{\frac{a^2+2a-24}{a+1}}{\frac{a^2-a-12}{(a+1)^2}}$

36. $\dfrac{\frac{y^2-3y-18}{y^2-4}}{\frac{y^2+5y+6}{y-2}}$

37. $\dfrac{\dfrac{x}{x+1}}{\dfrac{1}{x^2-1} - \dfrac{1}{x-1}}$

38. $\dfrac{\dfrac{a}{a^2-b^2}}{\dfrac{1}{a+b} + \dfrac{1}{a-b}}$

Solve each problem. See Example 4.

39. ***Sophomore math.*** A survey of college sophomores showed that $\frac{5}{6}$ of the males were taking a mathematics class and $\frac{3}{4}$ of the females were taking a mathematics class. One-third of the males were enrolled in calculus, and $\frac{1}{5}$ of the females were enrolled in calculus. If just as many males as females were surveyed, then what fraction of the surveyed students taking mathematics were enrolled in calculus? Rework this problem assuming that the number of females in the survey was twice the number of males.

40. ***Commuting students.*** At a well-known university, $\frac{1}{4}$ of the undergraduate students commute, and $\frac{1}{3}$ of the graduate students commute. One-tenth of the undergraduate students drive more than 40 miles daily, and $\frac{1}{6}$ of the graduate students drive more than 40 miles daily. If there are twice as many undergraduate students as there are graduate students, then what fraction of the commuters drive more than 40 miles daily?

Getting More Involved

41. ***Exploration.*** Simplify

$$\frac{1}{1+\frac{1}{2}}, \quad \frac{1}{1+\dfrac{1}{1+\frac{1}{2}}}, \quad \text{and} \quad \frac{1}{1+\dfrac{1}{1+\dfrac{1}{1+\frac{1}{2}}}}.$$

a) Are these fractions getting larger or smaller as the fractions become more complex?

b) Continuing the pattern, find the next two complex fractions and simplify them.

c) Now what can you say about the values of all five complex fractions?

42. ***Discussion.*** A complex fraction can be simplified by writing the numerator and denominator as single fractions and then dividing them or by multiplying the numerator and denominator by the LCD. Simplify the complex fraction

$$\frac{\dfrac{4}{xy^2} - \dfrac{6}{xy}}{\dfrac{2}{x^2} + \dfrac{4}{x^2y}}$$

by using each of these methods. Compare the number of steps used in each method, and determine which method requires fewer steps.

5.6 Solving Equations

In this section:

- Equations with Rational Expressions
- Extraneous Solutions

Many problems in algebra can be solved by using equations involving rational expressions. In this section you will learn how to solve equations that involve rational expressions, and in Section 5.7 and 5.8 you will solve problems using these equations.

Equations with Rational Expressions

We solved some equations involving fractions in Section 2.2. In that section the equations had only integers in the denominators. Our first step in solving those equations was to multiply by the LCD to eliminate all of the denominators.

EXAMPLE 1 **Integers in the denominators**

Solve $\frac{1}{2} - \frac{x-2}{3} = \frac{1}{6}$.

Solution

The LCD for 2, 3, and 6 is 6. Multiply each side of the equation by 6:

$$\frac{1}{2} - \frac{x-2}{3} = \frac{1}{6} \quad \text{Original equation}$$

$$6\left(\frac{1}{2} - \frac{x-2}{3}\right) = 6 \cdot \frac{1}{6} \quad \text{Multiply each side by 6.}$$

$$6 \cdot \frac{1}{2} - \overset{2}{\cancel{6}} \cdot \frac{x-2}{\cancel{3}} = \cancel{6} \cdot \frac{1}{\cancel{6}} \quad \text{Distributive property}$$

$$3 - 2(x-2) = 1 \quad \text{Simplify.}$$

$$3 - 2x + 4 = 1 \quad \text{Distributive property}$$

$$-2x = -6 \quad \text{Subtract 7 from each side.}$$

$$x = 3 \quad \text{Divide each side by } -2.$$

Check $x = 3$ in the original equation:

$$\frac{1}{2} - \frac{3-2}{3} = \frac{1}{2} - \frac{1}{3} = \frac{3}{6} - \frac{2}{6} = \frac{1}{6}$$

The solution to the equation is 3. ■

CAUTION When a numerator contains a binomial, as in Example 1, the numerator must be enclosed in parentheses when the denominator is eliminated. ⊘

To solve an equation involving rational expressions, we usually multiply each side of the equation by the LCD for all the denominators involved, just as we do for an equation with fractions.

EXAMPLE 2 **Variables in the denominators**

Solve $\frac{1}{x} + \frac{1}{6} = \frac{1}{4}$.

Solution

We multiply each side of the equation by $12x$, the LCD for 4, 6, and x:

$$\frac{1}{x} + \frac{1}{6} = \frac{1}{4} \quad \text{Original equation}$$

$$12x\left(\frac{1}{x} + \frac{1}{6}\right) = 12x\left(\frac{1}{4}\right) \quad \text{Multiply each side by } 12x.$$

$$12\cancel{x} \cdot \frac{1}{\cancel{x}} + \overset{2}{\cancel{12}}x \cdot \frac{1}{\cancel{6}} = \overset{3}{\cancel{12}}x \cdot \frac{1}{\cancel{4}} \quad \text{Distributive property}$$

$$12 + 2x = 3x \quad \text{Simplify.}$$

$$12 = x \quad \text{Subtract } 2x \text{ from each side.}$$

Check that 12 satisfies the original equation:

$$\frac{1}{12} + \frac{1}{6} = \frac{1}{12} + \frac{2}{12} = \frac{3}{12} = \frac{1}{4}$$ ■

EXAMPLE 3 **An equation with two solutions**

Solve the equation $\frac{100}{x} + \frac{100}{x+5} = 9$.

Solution

The LCD for the denominators x and $x + 5$ is $x(x + 5)$:

$$\frac{100}{x} + \frac{100}{x+5} = 9 \quad \text{Original equation}$$

$$x(x+5)\frac{100}{x} + x(x+5)\frac{100}{x+5} = x(x+5)9 \quad \text{Multiply each side by } x(x+5).$$

$$(x+5)100 + x(100) = (x^2 + 5x)9 \quad \text{All denominators are eliminated.}$$

$$100x + 500 + 100x = 9x^2 + 45x \quad \text{Simplify.}$$

$$500 + 200x = 9x^2 + 45x$$

$$0 = 9x^2 - 155x - 500 \quad \text{Get 0 on one side.}$$

$$0 = (9x + 25)(x - 20) \quad \text{Factor.}$$

$$9x + 25 = 0 \quad \text{or} \quad x - 20 = 0 \quad \text{Zero factor property}$$

$$x = -\frac{25}{9} \quad \text{or} \quad x = 20$$

A check will show that both $-\frac{25}{9}$ and 20 satisfy the original equation. ■

Extraneous Solutions

In a rational expression we can replace the variable only by real numbers that do not cause the denominator to be 0. When solving equations involving rational expressions, we must check every solution to see whether it causes 0 to appear in a denominator. If a number causes the denominator to be 0, then it cannot be a solution to the equation. A number that appears to be a solution but causes 0 in a denominator is called an **extraneous solution.**

EXAMPLE 4 **An equation with an extraneous solution**

Solve the equation $\frac{1}{x-2} = \frac{x}{2x-4} + 1$.

Solution

Because the denominator $2x - 4$ factors as $2(x - 2)$, the LCD is $2(x - 2)$.

$$2(x-2)\frac{1}{x-2} = 2(x-2)\frac{x}{2(x-2)} + 2(x-2)\cdot 1 \quad \text{Multiply each side of the original equation by } 2(x-2).$$

$$2 = x + 2x - 4 \quad \text{Simplify.}$$

$$2 = 3x - 4$$

$$6 = 3x$$

$$2 = x$$

Check 2 in the original equation:

$$\frac{1}{2-2} = \frac{2}{2\cdot 2 - 4} + 1$$

The denominator $2 - 2$ is 0. So 2 does not satisfy the equation, and it is an extraneous solution. The equation has no solutions. ■

EXAMPLE 5 **Another extraneous solution**

Solve the equation $\dfrac{1}{x} + \dfrac{1}{x-3} = \dfrac{x-2}{x-3}$.

Solution

The LCD for the denominators x and $x - 3$ is $x(x - 3)$:

$$\frac{1}{x} + \frac{1}{x-3} = \frac{x-2}{x-3} \quad \text{Original equation}$$

$$x(x-3)\cdot\frac{1}{x} + x(x-3)\cdot\frac{1}{x-3} = x(x-3)\cdot\frac{x-2}{x-3} \quad \text{Multiply each side by } x(x-3).$$

$$x - 3 + x = x(x - 2)$$

$$2x - 3 = x^2 - 2x$$

$$0 = x^2 - 4x + 3$$

$$0 = (x - 3)(x - 1)$$

$$x - 3 = 0 \quad \text{or} \quad x - 1 = 0$$

$$x = 3 \quad \text{or} \quad x = 1$$

If $x = 3$, then the denominator $x - 3$ has a value of 0. If $x = 1$, the original equation is satisfied. The only solution to the equation is 1. ■

CAUTION Be sure to always check your answers in the original equation to determine whether they are extraneous solutions.

Warm-ups

True or false? Explain your answers.

1. The LCD is not used in solving equations with rational expressions.
2. To solve the equation $x^2 = 8x$, we divide each side by x.
3. An extraneous solution is an irrational number.

Use the following equations for Questions 4–10.

a) $\dfrac{3}{x} + \dfrac{5}{x-2} = \dfrac{2}{3}$ **b)** $\dfrac{1}{x} + \dfrac{1}{2} = \dfrac{3}{4}$ **c)** $\dfrac{1}{x-1} + 2 = \dfrac{1}{x+1}$

4. To solve Eq. (a), we must add the expressions on the left-hand side.
5. Both 0 and 2 satisfy Eq. (a).
6. To solve Eq. (a), we multiply each side by $3x^2 - 6x$.
7. The only solution to Eq. (b) is 4.
8. Equation (b) is equivalent to $4 + 2x = 3x$.
9. To solve Eq. (c), we multiply each side by $x^2 - 1$.
10. The numbers 1 and -1 do not satisfy Eq. (c).

5.6 EXERCISES

Solve each equation. See Example 1.

1. $\frac{x}{3} - 5 = \frac{x}{2} - 7$

2. $\frac{x}{3} - \frac{x}{2} = \frac{x}{5} - 11$

3. $\frac{y}{5} - \frac{2}{3} = \frac{y}{6} + \frac{1}{3}$

4. $\frac{z}{6} + \frac{5}{4} = \frac{z}{2} - \frac{3}{4}$

5. $\frac{3}{4} - \frac{t-4}{3} = \frac{t}{12}$

6. $\frac{4}{5} - \frac{v-1}{10} = \frac{v-5}{30}$

7. $\frac{1}{5} - \frac{w+10}{15} = \frac{1}{10} - \frac{w+1}{6}$

8. $\frac{q}{5} - \frac{q-1}{2} = \frac{13}{20} - \frac{q+1}{4}$

Solve each equation. See Example 2.

9. $\frac{1}{x} + \frac{1}{2} = \frac{3}{4}$

10. $\frac{3}{x} + \frac{1}{4} = \frac{5}{8}$

11. $\frac{2}{3x} + \frac{1}{2x} = \frac{7}{24}$

12. $\frac{1}{6x} - \frac{1}{8x} = \frac{1}{72}$

13. $\frac{1}{2} + \frac{a-2}{a} = \frac{a+2}{2a}$

14. $\frac{1}{b} + \frac{1}{5} = \frac{b-1}{5b} + \frac{3}{10}$

15. $\frac{1}{3} - \frac{k+3}{6k} = \frac{1}{3k} - \frac{k-1}{2k}$

16. $\frac{3}{p} - \frac{p+3}{3p} = \frac{2p-1}{2p} - \frac{5}{6}$

Solve each equation. See Example 3.

17. $\frac{x}{2} = \frac{5}{x+3}$

18. $\frac{x}{3} = \frac{4}{x+1}$

19. $\frac{2}{x+1} = \frac{1}{x} + \frac{1}{6}$

20. $\frac{1}{w+1} - \frac{1}{2w} = \frac{3}{40}$

21. $\frac{a-1}{a^2-4} + \frac{1}{a-2} = \frac{a+4}{a+2}$

22. $\frac{b+17}{b^2-1} - \frac{1}{b+1} = \frac{b-2}{b-1}$

Solve each equation. Watch for extraneous solutions. See Examples 4 and 5.

23. $\frac{1}{x-1} + \frac{2}{x} = \frac{x}{x-1}$

24. $\frac{4}{x} + \frac{3}{x-3} = \frac{x}{x-3} - \frac{1}{3}$

25. $\frac{5}{x+2} + \frac{2}{x-3} = \frac{x-1}{x-3}$

26. $\frac{6}{y-2} + \frac{7}{y-8} = \frac{y-1}{y-8}$

27. $1 + \frac{3y}{y-2} = \frac{6}{y-2}$

28. $\frac{5}{y-3} = \frac{y+7}{2y-6} + 1$

29. $\frac{z}{z+1} - \frac{1}{z+2} = \frac{2z+5}{z^2+3z+2}$

30. $\frac{z}{z-2} - \frac{1}{z+5} = \frac{7}{z^2+3z-10}$

In Exercises 31–52, solve each equation.

31. $\frac{a}{4} = \frac{5}{2}$

32. $\frac{y}{3} = \frac{6}{5}$

33. $\frac{w}{6} = \frac{3w}{11}$

34. $\frac{2m}{3} = \frac{3m}{2}$

35. $\frac{5}{x} = \frac{x}{5}$

36. $\frac{-3}{x} = \frac{x}{-3}$

37. $\frac{x-3}{5} = \frac{x-3}{x}$

38. $\frac{a+4}{2} = \frac{a+4}{a}$

39. $\frac{1}{x+2} = \frac{x}{x+2}$

40. $\frac{-3}{w+2} = \frac{w}{w+2}$

41. $\frac{1}{2x-4} + \frac{1}{x-2} = \frac{3}{2}$

42. $\frac{7}{3x-9} - \frac{1}{x-3} = \frac{4}{3}$

43. $\frac{3}{a^2-a-6} = \frac{2}{a^2-4}$

44. $\frac{8}{a^2+a-6} = \frac{6}{a^2-9}$

45. $\frac{4}{c-2} - \frac{1}{2-c} = \frac{25}{c+6}$

46. $\frac{3}{x+1} - \frac{1}{1-x} = \frac{10}{x^2-1}$

47. $\frac{1}{x^2-9} + \frac{3}{x+3} = \frac{4}{x-3}$

48. $\frac{3}{x-2} - \frac{5}{x+3} = \frac{1}{x^2+x-6}$

49. $\frac{3}{2x+4} - \frac{1}{x+2} = \frac{1}{3x+1}$

50. $\dfrac{5}{2m+6} - \dfrac{1}{m+1} = \dfrac{1}{m+3}$

51. $\dfrac{2t-1}{3t+3} + \dfrac{3t-1}{6t+6} = \dfrac{t}{t+1}$

52. $\dfrac{4w-1}{3w+6} - \dfrac{w-1}{3} = \dfrac{w-1}{w+2}$

Solve each problem.

53. ***Lens equation.*** The focal length f for a camera lens is related to the object distance o and the image distance i by the formula

$$\frac{1}{f} = \frac{1}{o} + \frac{1}{i}.$$

See the accompanying figure. The image is in focus at distance i from the lens. For an object that is 600mm from a 50-mm lens, use $f = 50$-mm and $o = 600$ mm to find i.

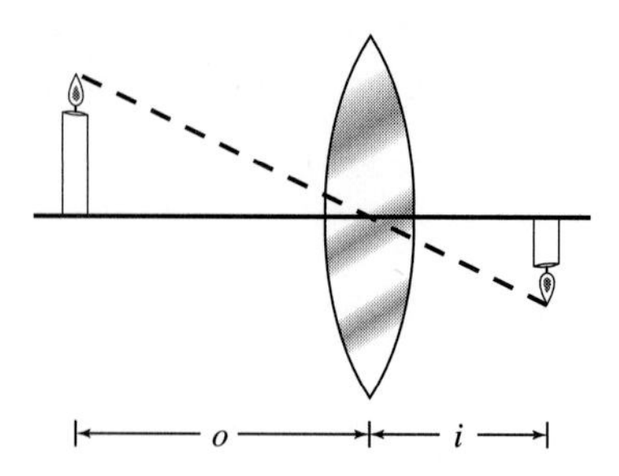

Figure for Exercise 53

54. ***Telephoto lens.*** Use the formula from Exercise 53 to find the image distance i for an object that is 2,000,000 mm from a 250-mm telephoto lens.

5.7 Ratios and Proportions

In this section:
- Ratios
- Proportions

In this section we will use the ideas of rational expressions in ratio and proportion problems. We will solve proportions in the same way we solved equations in Section 5.6.

Ratios

In Chapter 1 we defined a rational number as the *ratio of two integers.* We will now give a more general definition of ratio. If a and b are any real numbers (not just integers), with $b \neq 0$, then the expression $\frac{a}{b}$ is called the **ratio of a and b** or the **ratio of a to b.** The ratio of a to b is also written as $a:b$. A ratio is a comparison of two numbers. Some examples of ratios are

$$\frac{3}{4}, \quad \frac{4.2}{2.1}, \quad \frac{\frac{1}{4}}{\frac{1}{2}}, \quad \frac{3.6}{5}, \quad \text{and} \quad \frac{100}{1}.$$

Ratios are treated just like fractions. We can reduce ratios, and we can build them up. We generally express ratios as ratios of integers. When possible, we will convert a ratio into an equivalent ratio of integers in lowest terms.

EXAMPLE 1 **Finding equivalent ratios**

Find an equivalent ratio of integers in lowest terms for each ratio.

a) $\dfrac{4.2}{2.1}$ **b)** $\dfrac{\frac{1}{4}}{\frac{1}{2}}$ **c)** $\dfrac{3.6}{5}$

Solution

a) Because both the numerator and the denominator have one decimal place, we will multiply the numerator and denominator by 10 to eliminate the decimals:

$$\frac{4.2}{2.1} = \frac{4.2(10)}{2.1(10)} = \frac{42}{21} = \frac{21 \cdot 2}{21 \cdot 1} = \frac{2}{1} \quad \text{Do not omit the 1 in a ratio.}$$

So the ratio of 4.2 to 2.1 is equivalent to the ratio 2 to 1.

b) This ratio is a complex fraction. We can simplify this expression using the LCD method as shown in Section 5.5. Multiply the numerator and denominator of this ratio by 4:

$$\frac{\frac{1}{4}}{\frac{1}{2}} = \frac{\frac{1}{4} \cdot 4}{\frac{1}{2} \cdot 4} = \frac{1}{2}$$

c) We can get a ratio of integers if we multiply the numerator and denominator by 10.

$$\frac{3.6}{5} = \frac{3.6(10)}{5(10)} = \frac{36}{50}$$

$$= \frac{18}{25} \quad \text{Reduce to lowest terms.}$$

In the next example a ratio is used to compare quantities.

EXAMPLE 2 **Nitrogen to potash**

In a 50-pound bag of lawn fertilizer there are 8 pounds of nitrogen and 12 pounds of potash. What is the ratio of nitrogen to potash?

Solution

The nitrogen and potash occur in this fertilizer in the ratio of 8 pounds to 12 pounds:

$$\frac{8}{12} = \frac{2 \cdot \not{4}}{3 \cdot \not{4}} = \frac{2}{3}$$

So the ratio of nitrogen to potash is 2 to 3.

EXAMPLE 3 **Males to females**

In a class of 50 students, there were exactly 20 male students. What was the ratio of males to females in this class?

Solution

Because there were 20 males in the class of 50, there were 30 females. The ratio of males to females was 20 to 30, or 2 to 3.

Ratios give us a means of comparing the size of two quantities. For this reason *the numbers compared in a ratio should be expressed in the same units.* For example, if one dog is 24 inches high and another is 1 foot high, then the ratio of their heights is 2 to 1, not 24 to 1.

EXAMPLE 4 **Quantities with different units**

What is the ratio of length to width for a poster with a length of 30 inches and a width of 2 feet?

Solution

Because the width is 2 feet, or 24 inches, the ratio of length to width is 30 to 24. Reduce as follows:

$$\frac{30}{24} = \frac{5 \cdot 6}{4 \cdot 6} = \frac{5}{4}$$

So the ratio of length to width is 5 to 4. ■

Proportions

A **proportion** is any statement expressing the equality of two ratios. The statement

$$\frac{a}{b} = \frac{c}{d} \qquad \text{or} \qquad a{:}b = c{:}d$$

is a proportion. In any proportion the numbers in the positions of a and d above are called the **extremes.** The numbers in the positions of b and c above are called the **means.** In the proportion

$$\frac{30}{24} = \frac{5}{4},$$

the means are 24 and 5, and the extremes are 30 and 4.

If we multiply each side of the proportion

$$\frac{a}{b} = \frac{c}{d}$$

by the LCD, bd, we get

$$\frac{a}{b} \cdot bd = \frac{c}{d} \cdot bd$$

or

$$a \cdot d = b \cdot c.$$

We can express this result by saying that *the product of the extremes is equal to the product of the means.* We call this fact the **extremes-means property** or **cross-multiplying.**

Extremes-Means Property (Cross-Multiplying)

Suppose a, b, c, and d are real numbers with $b \neq 0$ and $d \neq 0$. If

$$\frac{a}{b} = \frac{c}{d}, \text{ then } ad = bc.$$

We use the extremes-means property to solve proportions.

EXAMPLE 5 **Using the extremes-means property**

Solve the proportion $\frac{3}{x} = \frac{5}{x+5}$ for x.

Solution

Instead of multiplying each side by the LCD, we use the extremes-means property:

$$\frac{3}{x} = \frac{5}{x+5} \quad \text{Original proportion}$$

$$3(x+5) = 5x \quad \text{Extremes-means property}$$

$$3x + 15 = 5x \quad \text{Distributive property}$$

$$15 = 2x$$

$$\frac{15}{2} = x$$

Check:

$$\frac{3}{\frac{15}{2}} = 3 \cdot \frac{2}{15} = \frac{2}{5}$$

$$\frac{5}{\frac{15}{2} + 5} = \frac{5}{\frac{25}{2}} = 5 \cdot \frac{2}{25} = \frac{2}{5}$$

So $\frac{15}{2}$ is the solution to the equation or the solution to the proportion. ■

EXAMPLE 6 **Solving a proportion**

The ratio of men to women at Brighton City College is 2 to 3. If there are 894 men, then how many women are there?

Solution

Because the ratio of men to women is 2 to 3, we have

$$\frac{\text{Number of men}}{\text{Number of women}} = \frac{2}{3}.$$

If x represents the number of women, then we have the following proportion:

$$\frac{894}{x} = \frac{2}{3}$$

$$2x = 2682 \quad \text{Extremes-means property}$$

$$x = 1341$$

The number of women is 1341. ■

Note that any proportion can be solved by multiplying each side by the LCD as we did when we solved other equations involving rational expressions. The extremes-means property gives us a shortcut for solving proportions.

EXAMPLE 7 **Solving a proportion**

In a conservative portfolio the ratio of the amount invested in bonds to the amount invested in stocks should be 3 to 1. A conservative investor invested \$2850 more in bonds than she did in stocks. How much did she invest in each category?

Solution

Because the ratio of the amount invested in bonds to the amount invested in stocks is 3 to 1, we have

$$\frac{\text{Amount invested in bonds}}{\text{Amount invested in stocks}} = \frac{3}{1}.$$

If x represents the amount invested in stocks and $x + 2850$ represents the amount invested in bonds, then we can write and solve the following proportion:

$$\frac{x + 2850}{x} = \frac{3}{1}$$

$$3x = x + 2850 \quad \text{Extremes-means property}$$

$$2x = 2850$$

$$x = 1425$$

$$x + 2850 = 4275$$

So she invested \$4275 in bonds and \$1425 in stocks. Note that these amounts are in the ratio of 3 to 1. ■

The next example shows how conversions from one unit of measurement to another can be done by using proportions.

EXAMPLE 8 **Converting measurements**

There are 3 feet in 1 yard. How many feet are there in 12 yards?

Solution

Let x represent the number of feet in 12 yards. There are two proportions that we can write to solve the problem:

$$\frac{3 \text{ feet}}{x \text{ feet}} = \frac{1 \text{ yard}}{12 \text{ yards}} \qquad \frac{3 \text{ feet}}{1 \text{ yard}} = \frac{x \text{ feet}}{12 \text{ yards}}$$

The ratios in the second proportion violate the rule of comparing only measurements that are expressed in the same units. Note that each side of the second proportion is actually the ratio 1 to 1, since 3 feet = 1 yard and x feet = 12 yards. For doing conversions we can use ratios like this to compare measurements in different units. Applying the extremes-means property to either proportion gives

$$3 \cdot 12 = x \cdot 1,$$

or

$$x = 36.$$

So there are 36 feet in 12 yards. ■

MATH AT WORK

Sales Analyst

Did you ever wonder how your local store calculates how much of your favorite cosmetic to stock on the shelf? Mike Pittman, National Account Manager for a major cosmetic company, is responsible for providing more than 2000 stores across the United States with personal care products such as skin lotions, fragrances, and cosmetics.

Data on what has been sold is transmitted from the point of sale across a number of satellite dishes and computers to Mr. Pittman. The data usually includes size, color, and other pertinent facts. The information is then combined with demographics for certain geographic areas and movement data, as well as advertising and promotional information to answer questions such as: What color is selling best? Is it time to stock sunscreen? Is this a trend-setting area of the country? On the basis of his analysis of these and many other questions, Mr. Pittman recommends changes in packaging, promotional programs, and the quantities of products to be shipped.

Mr. Pittman's job requires a unique blend of sales, marketing, and quantitative skills. Of course, knowledge of computers and an understanding of people help. So the next time you see a whole aisle of personal care and cosmetic products, think of all the information that has been analyzed to put it there.

In Exercise 53 of this section you will see how Mr. Pittman uses a proportion to determine the quantity of mascara needed in a warehouse.

Warm-ups

True or false? Explain your answer.

1. The ratio of 40 men to 30 women can be expressed as the ratio 4 to 3.
2. The ratio of 3 feet to 2 yards can be expressed as the ratio 3 to 2.
3. If the ratio of men to women in the Chamber of Commerce is 3 to 2 and there are 20 men, then there must be 30 women.
4. The ratio of 1.5 to 2 is equivalent to the ratio of 3 to 4.
5. A statement that two ratios are equal is called a proportion.
6. The product of the extremes is equal to the product of the means.
7. If $\frac{2}{x} = \frac{3}{5}$, then $5x = 6$.
8. The ratio of the height of a 12-inch cactus to the height of a 3-foot cactus is 4 to 1.
9. If 30 out of 100 lawyers preferred aspirin and the rest did not, then the ratio of lawyers that preferred aspirin to those who did not is 30 to 100.
10. If $\frac{x + 5}{x} = \frac{2}{3}$, then $3x + 15 = 2x$.

5.7 EXERCISES

For each ratio, find an equivalent ratio of integers in lowest terms. See Example 1.

1. $\frac{2.5}{3.5}$ **2.** $\frac{4.8}{1.2}$ **3.** $\frac{0.32}{0.6}$

4. $\frac{0.05}{0.8}$ **5.** $\frac{35}{10}$ **6.** $\frac{88}{33}$

7. $\frac{4.5}{7}$ **8.** $\frac{3}{2.5}$ **9.** $\frac{\frac{1}{2}}{\frac{1}{5}}$

10. $\frac{\frac{2}{3}}{\frac{3}{4}}$ **11.** $\frac{5}{\frac{1}{3}}$ **12.** $\frac{4}{\frac{1}{4}}$

Find a ratio for each of the following, and write it as a ratio of integers in lowest terms. See Examples 2–4.

13. ***Men and women.*** Find the ratio of men to women in a bowling league containing 12 men and 8 women.

14. ***Coffee drinkers.*** Among 100 coffee drinkers, 36 said that they preferred their coffee black and the rest did not prefer their coffee black. Find the ratio of those who prefer black coffee to those who prefer nonblack coffee.

15. ***Smokers.*** A life insurance company found that among its last 200 claims, there were six dozen smokers.What is the ratio of smokers to nonsmokers in this group of claimants?

16. ***Hits and misses.*** A woman threw 60 darts and hit the target a dozen times. What is her ratio of hits to misses?

17. ***Violence and kindness.*** While watching television for one week, a consumer group counted 1240 acts of violence and 40 acts of kindness. What is the violence to kindness ratio for television, according to this group?

18. ***Length to width.*** What is the ratio of length to width for the given rectangle?

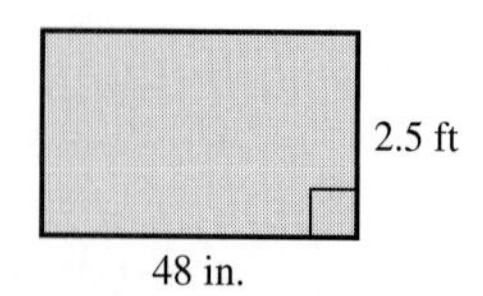

Figure for Exercise 18

19. ***Rise to run.*** What is the ratio of rise to run for the stairway shown in the accompanying figure?

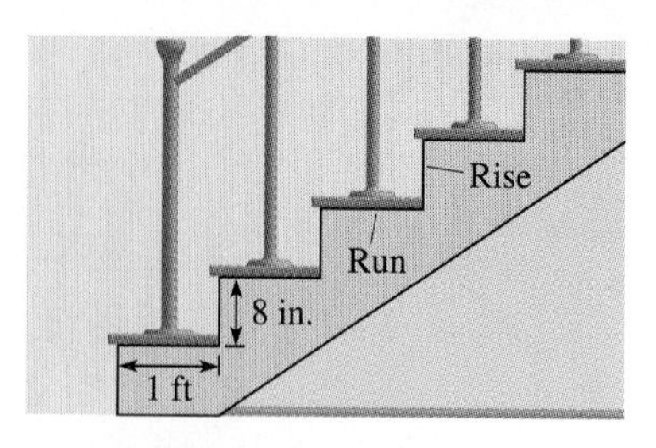

Figure for Exercise 19

20. ***Rise and run.*** If the rise is $\frac{3}{2}$ and the run is 5, then what is the ratio of the rise to the run?

Solve each proportion. See Example 5.

21. $\frac{4}{x} = \frac{2}{3}$ **22.** $\frac{9}{x} = \frac{3}{2}$ **23.** $\frac{a}{2} = \frac{-1}{5}$

24. $\frac{b}{3} = \frac{-3}{4}$ **25.** $-\frac{5}{9} = \frac{3}{x}$ **26.** $-\frac{3}{4} = \frac{5}{x}$

27. $\frac{10}{x} = \frac{34}{x + 12}$ **28.** $\frac{x}{3} = \frac{x + 1}{2}$

29. $\frac{a}{a + 1} = \frac{a + 3}{a}$ **30.** $\frac{c + 3}{c - 1} = \frac{c + 2}{c - 3}$

31. $\frac{m - 1}{m - 2} = \frac{m - 3}{m + 4}$ **32.** $\frac{h}{h - 3} = \frac{h}{h - 9}$

Use a proportion to solve each problem. See Examples 6–8.

33. ***New shows and reruns.*** The ratio of new shows to reruns on cable TV is 2 to 27. If Frank counted only eight new shows one evening, then how many reruns were there?

34. ***Fast food.*** If four out of five doctors prefer fast food, then at a convention of 445 doctors, how many prefer fast food?

35. ***Voting.*** If 220 out of 500 voters surveyed said that they would vote for the incumbent, then how many votes could the incumbent expect out of the 400,000 voters in the state?

36. ***New product.*** A taste test with 200 randomly selected people found that only three of them said that they would buy a box of new Sweet Wheats cereal. How many boxes could the manufacturer expect to sell in a country of 280 million people?

37. ***Basketball blowout.*** As the final buzzer signaled the end of the basketball game, the Lions were 34 points ahead of the Tigers. If the Lions scored 5 points for every 3 scored by the Tigers, then what was the final score?

38. ***The golden ratio.*** The ancient Greeks thought that the most pleasing shape for a rectangle was one for which the ratio of the length to the width was 8 to 5, the golden ratio. If the length of a rectangular painting is 2 ft longer than its width, then for what dimensions would the length and width have the golden ratio?

39. ***Automobile sales.*** The ratio of sports cars to luxury cars sold in Wentworth one month was 3 to 2. If 20 more sports cars were sold than luxury cars, then how many of each were sold that month?

40. ***Foxes and rabbits.*** The ratio of foxes to rabbits in the Deerfield Forest Preserve is 2 to 9. If there are 35 fewer foxes than rabbits, then how many of each are there?

41. ***Inches and feet.*** If there are 12 inches in 1 foot, then how many inches are there in 7 feet?

42. ***Feet and yards.*** If there are 3 feet in 1 yard, then how many yards are there in 28 feet?

43. ***Minutes and hours.*** If there are 60 minutes in 1 hour, then how many minutes are there in 0.25 hour?

44. ***Meters and kilometers.*** If there are 1000 meters in 1 kilometer, then how many meters are there in 2.33 kilometers.

45. ***Miles and hours.*** If Alonzo travels 230 miles in 3 hours, then how many miles does he travel in 7 hours?

46. ***Hiking time.*** If Evangelica can hike 19 miles in 2 days on the Appalachian Trail, then how many days will it take her to hike 63 miles?

47. ***Force on basketball shoes.*** The designers of Converse shoes know that the force exerted on shoe soles in a jump shot is proportional to the weight of the person jumping. If a 70-pound boy exerts a force of 980 pounds on his shoe soles when he returns to the court after a jump, then what force does a 6 ft 8 in. professional ball player weighing 280 pounds exert on the soles of his shoes when he returns to the court after a jump? Use the accompanying graph to estimate the force for a 150-pound player.

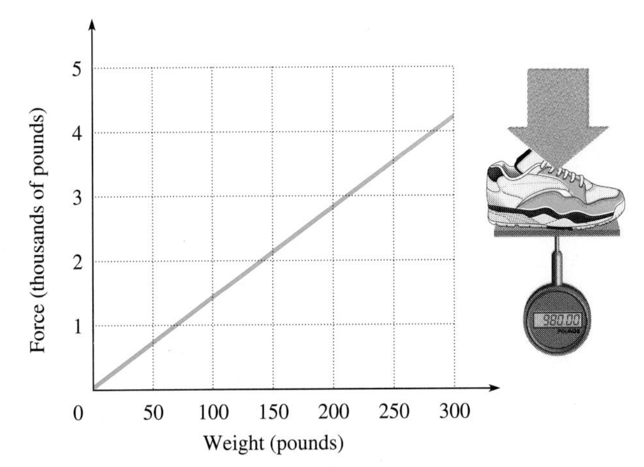

Figure for Exercise 47

48. ***Force on running shoes.*** The designers of Converse shoes know that the ratio of the force on the shoe soles to the weight of a runner is 3 to 1. What force does a 130-pound jogger exert on the soles of her shoes.

49. ***Capture-recapture.*** To estimate the number of trout in Trout Lake, rangers used the capture-recapture method. They caught, tagged, and released 200 trout. One week later, they caught a sample of 150 trout and found that 5 of them were tagged. Assuming that the ratio of tagged trout to the total number of trout in the lake is the same as the ratio of tagged trout in the sample to the number of trout in the sample, find the number of trout in the lake.

50. ***Bear population.*** To estimate the size of the bear population on the Keweenaw Peninsula, conservationists captured, tagged, and released 50 bears. One year later, a random sample of 100 bears included only 2 tagged bears. What is the conservationist's estimate of the size of the bear population?

51. ***Public college graduation rate.*** The accompanying graph shows five-year graduation rates for private and public colleges for students entering in 1987 (*Boston Globe*, Feb. 23, 1994). What is the ratio of graduates to nongraduates for students starting college in 1987 at Northern State College, a public college with an average SAT of 745? If the number of graduates was 200 less than the number of nongraduates in the group that started at Northern State in 1987, then how many students were in that group?

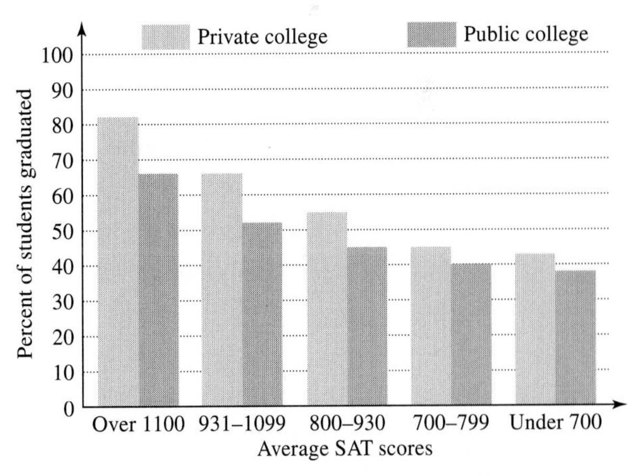

Figure for Exercises 51 and 52

52. ***Private college graduation rate.*** Use the accompanying graph to find the five-year graduation rate for a private college with an average SAT of 960 for students entering in 1987. If 330 of the students entering this private college in 1987 graduated in five years, then how many of those entering students did not graduate in five years?

53. ***Mascara needs.*** In determining warehouse needs for a particular mascara for a chain of 2000 stores, Mike Pittman first determines a need B based on sales figures for the past 52 weeks. He then determines the actual need A from the equation $\frac{A}{B} = k$, where

$$k = 1 + V + C + X - D.$$

He uses $V = 0.22$ if there is a national TV ad and $V = 0$ if not, $C = 0.26$ if there is a national coupon and $C = 0$ if not, $X = 0.36$ if there is a chain-specific ad and $X = 0$ if not, and $D = 0.29$ if there is a special display in the chain and $D = 0$ if not. (D is subtracted because less product is needed in the warehouse when more is on display in the store.) If $B = 4200$ units and there is a special display and a national coupon but no national TV ad and no chain-specific ad, then what is the value of A?

Getting More Involved

54. ***Discussion.*** Which of the following equations is not a proportion? Explain.

a) $\frac{1}{2} = \frac{1}{2}$ **b)** $\frac{x}{x+2} = \frac{4}{5}$

c) $\frac{x}{4} = \frac{9}{x}$ **d)** $\frac{8}{x+2} - 1 = \frac{5}{x+2}$

55. ***Discussion.*** Find all of the errors in the following solution to an equation.

$$\frac{7}{x} = \frac{8}{x+3} + 1$$
$$7(x + 3) = 8x + 1$$
$$7x + 3 = 8x$$
$$-x = -3$$
$$x = 3$$

5.8 Applications

In this section:

- Formulas
- Uniform Motion Problems
- Work Problems
- Miscellaneous Problems

In this section we will study additional applications of rational expressions.

Formulas

Many formulas involve rational expressions. When solving a formula of this type for a certain variable, we usually multiply each side by the LCD to eliminate the denominators.

EXAMPLE 1 **An equation of a line**

The formula

$$\frac{y-4}{x+2} = \frac{3}{2}$$

is the equation for a certain straight line. We will study equations of this type further in Chapter 6. Solve this equation for y.

Solution

To isolate y on the left-hand side of the equation, we multiply each side by $x + 2$:

$$\frac{y-4}{x+2} = \frac{3}{2} \quad \text{Original equation}$$

$$(x+2) \cdot \frac{y-4}{x+2} = (x+2) \cdot \frac{3}{2} \quad \text{Multiply by } x + 2.$$

$$y - 4 = \frac{3}{2}x + 3 \quad \text{Simplify.}$$

$$y = \frac{3}{2}x + 7 \quad \text{Add 4 to each side.}$$

Because the original equation is a proportion, we could have used the extremes-means property to solve it for y. ■

EXAMPLE 2 **Distance, rate, and time**

Solve the formula $\frac{D}{T} = R$ for T.

Solution

Because the only denominator is T, we multiply each side by T:

$\frac{D}{T} = R$	Original formula
$T \cdot \frac{D}{T} = T \cdot R$	Multiply each side by T.
$D = TR$	
$\frac{D}{R} = \frac{TR}{R}$	Divide each side by R.
$\frac{D}{R} = T$	Simplify.

The formula solved for T is $T = \frac{D}{R}$. ■

In the next example, different subscripts are used on a variable to indicate that they are different variables. Think of R_1 as the first resistance, R_2 as the second resistance, and R as a combined resistance.

EXAMPLE 3 **Total resistance**

The formula

$$\frac{1}{R} = \frac{1}{R_1} + \frac{1}{R_2}$$

(from physics) expresses the relationship between different amounts of resistance in a parallel circuit. Solve it for R_2.

Solution

The LCD for R, R_1, and R_2 is RR_1R_2:

$\frac{1}{R} = \frac{1}{R_1} + \frac{1}{R_2}$	Original formula
$RR_1R_2 \cdot \frac{1}{R} = RR_1R_2 \cdot \frac{1}{R_1} + RR_1R_2 \cdot \frac{1}{R_2}$	Multiply each side by the LCD, RR_1R_2.
$R_1R_2 = RR_2 + RR_1$	All denominators are eliminated.
$R_1R_2 - RR_2 = RR_1$	Get all terms involving R_2 onto the left side.
$R_2(R_1 - R) = RR_1$	Factor out R_2.
$R_2 = \frac{RR_1}{R_1 - R}$	Divide each side by $R_1 - R$.

■

EXAMPLE 4 **Finding the value of a variable**

In the formula of Example 1, find x if $y = -3$.

Solution

Substitute $y = -3$ into the formula, then solve for x:

$$\frac{y-4}{x+2} = \frac{3}{2} \quad \text{Original formula}$$

$$\frac{-3-4}{x+2} = \frac{3}{2} \quad \text{Replace } y \text{ by } -3.$$

$$\frac{-7}{x+2} = \frac{3}{2} \quad \text{Simplify.}$$

$$3x + 6 = -14 \quad \text{Extremes-means property}$$

$$3x = -20$$

$$x = -\frac{20}{3}$$

Uniform Motion Problems

In uniform motion problems we use the formula $D = RT$. In some problems in which the time is unknown, we can use the formula $T = \frac{D}{R}$ to get an equation involving rational expressions.

EXAMPLE 5 **Driving to Florida**

Susan drove 1500 miles to Daytona Beach for spring break. On the way back she averaged 10 miles per hour less, and the drive back took her 5 hours longer. Find Susan's average speed on the way to Daytona Beach.

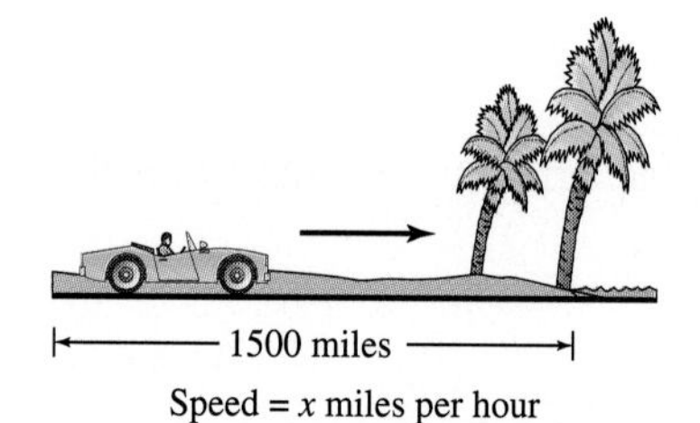

Speed = $x - 10$ miles per hour

Figure 5.1

Solution

If x represents her average speed going there, then $x - 10$ is her average speed for the return trip. See Fig. 5.1. We use the formula $T = \frac{D}{R}$ to make the following table.

	D	**R**	**T**	
Going	1500	x	$\frac{1500}{x}$	← Shorter time
Returning	1500	$x - 10$	$\frac{1500}{x-10}$	← Longer time

Because the difference between the two times is 5 hours, we have

$$\text{longer time} - \text{shorter time} = 5.$$

Using the time expressions from the table, we get the following equation:

$$\frac{1500}{x-10} - \frac{1500}{x} = 5$$

$$x(x-10)\frac{1500}{x-10} - x(x-10)\frac{1500}{x} = x(x-10)5 \quad \text{Multiply by } x(x-10).$$

$$1500x - 1500(x-10) = 5x^2 - 50x$$

$$15{,}000 = 5x^2 - 50x \quad \text{Simplify.}$$

$$3000 = x^2 - 10x \quad \text{Divide each side by 5.}$$

$$0 = x^2 - 10x - 3000$$

$$(x+50)(x-60) = 0 \quad \text{Factor.}$$

$$x + 50 = 0 \quad \text{or} \quad x - 60 = 0$$

$$x = -50 \quad \text{or} \quad x = 60$$

The answer $x = -50$ is a solution to the equation, but it cannot indicate the average speed of the car. Her average speed going to Daytona Beach was 60 mph. ■

Work Problems

If you can complete a job in 3 hours, then you are working at the rate of $\frac{1}{3}$ of the job per hour. If you work for 2 hours at the rate of $\frac{1}{3}$ of the job per hour, then you will complete $\frac{2}{3}$ of the job. The product of the rate and time is the amount of work completed. For problems involving work, we will always assume that the work is done at a constant rate. So if a job takes x hours to complete, then the rate is $\frac{1}{x}$ of the job per hour.

EXAMPLE 6 **Shoveling snow**

After a heavy snowfall, Brian can shovel all of the driveway in 30 minutes. If his younger brother Allen helps, the job takes only 20 minutes. How long would it take Allen to do the job by himself?

Solution

Let x represent the number of minutes it would take Allen to do the job by himself. Brian's rate for shoveling is $\frac{1}{30}$ of the driveway per minute, and Allen's rate for shoveling is $\frac{1}{x}$ of the driveway per minute. We organize all of the information in a table like the table in Example 5.

	Rate	Time	Work
Brian	$\frac{1}{30}\frac{\text{job}}{\text{min}}$	20 min	$\frac{2}{3}$ job
Allen	$\frac{1}{x}\frac{\text{job}}{\text{min}}$	20 min	$\frac{20}{x}$ job

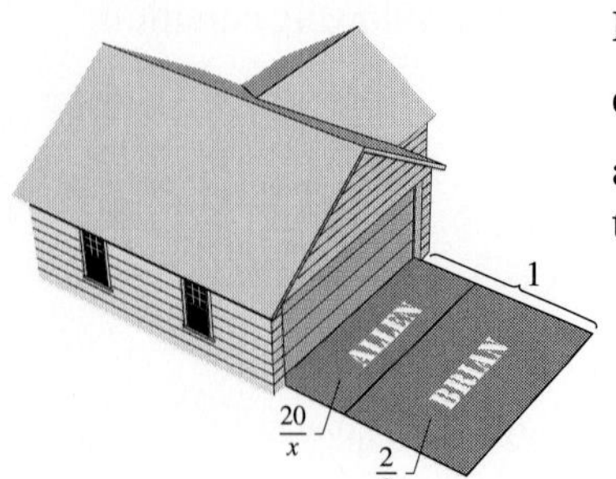

Figure 5.2

If Brian works for 20 min at the rate $\frac{1}{30}$ of the job per minute, then he does $\frac{20}{30}$ or $\frac{2}{3}$ of the job, as shown in Fig. 5.2. The amount of work that each boy does is a fraction of the whole job. So the expressions for work in the last column of the table have a sum of 1:

$$\frac{2}{3} + \frac{20}{x} = 1$$

$$3x \cdot \frac{2}{3} + 3x \cdot \frac{20}{x} = 3x \cdot 1 \quad \text{Multiply each side by } 3x.$$

$$2x + 60 = 3x$$

$$60 = x$$

If it takes Allen 60 min to do the job by himself, then he works at the rate of $\frac{1}{60}$ of the job per minute. In 20 minutes he does $\frac{1}{3}$ of the job while Brian does $\frac{2}{3}$. So it would take Allen 60 minutes to shovel the driveway by himself. ■

Miscellaneous Problems

If you bought 5 pounds of meat for \$20, then you paid 4 dollars per pound. If you bought x pounds of meat for \$20, then you paid $\frac{20}{x}$ dollars per pound.

EXAMPLE 7 **Oranges and grapefruit**

Tamara bought 50 pounds of fruit consisting of Florida oranges and Texas grapefruit. She paid twice as much per pound for the grapefruit as she did for the oranges. If Tamara bought \$12 worth of oranges and \$16 worth of grapefruit, then how many pounds of each did she buy?

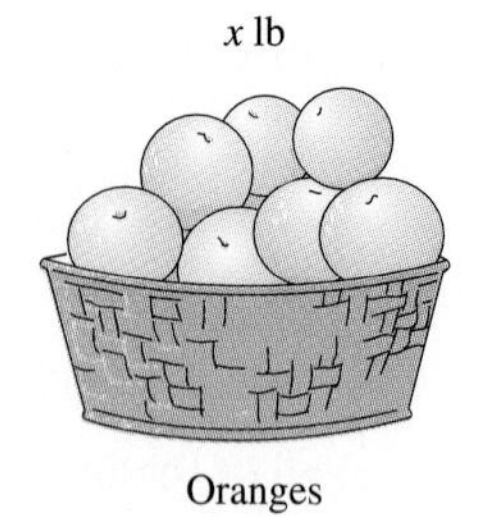

$50 - x$ lb

Grapefruit

Figure 5.3

Solution

If x represents the number of pounds of oranges, then $50 - x$ is the number of pounds of grapefruit. See Fig. 5.3. So she paid $\frac{12}{x}$ dollars per pound for the oranges and $\frac{16}{50 - x}$ dollars per pound for the grapefruit. Because the price per pound of the grapefruit is twice that of the oranges, we have

2(price per pound of oranges) = price per pound of grapefruit.

Rewrite this equation using the rational expressions for the prices per pound:

$$2\left(\frac{12}{x}\right) = \frac{16}{50 - x}$$

$$\frac{24}{x} = \frac{16}{50 - x}$$

$$16x = 1200 - 24x \quad \text{Extremes-means property}$$

$$40x = 1200$$

$$x = 30$$

$$50 - x = 20$$

If Tamara purchased 20 pounds of grapefruit for \$16, then she paid \$0.80 per pound. If she purchased 30 pounds of oranges for \$12, then she paid \$0.40 per pound. Because \$0.80 is twice \$0.40, we can be sure that she purchased 20 pounds of grapefruit and 30 pounds of oranges. ■

Warm-ups

True or false? Explain your answer.

1. The formula $t = \frac{1-t}{m}$, solved for m, is $m = \frac{1-t}{t}$.
2. To solve $\frac{1}{m} + \frac{1}{n} = \frac{1}{2}$ for m, we multiply each side by $2mn$.
3. If Fiona drives 300 miles in x hours, then her average speed is $\frac{x}{300}$ mph.
4. If Miguel drives 20 hard bargains in x hours, then he is driving $\frac{20}{x}$ hard bargains per hour.
5. If Fred can paint a house in y days, then he paints $\frac{1}{y}$ of the house per day.
6. If $\frac{1}{x}$ is 1 less than $\frac{2}{x+3}$, then $\frac{1}{x} - 1 = \frac{2}{x+3}$.
7. If a and b are nonzero and $a = \frac{m}{b}$, then $b = am$.
8. If $D = RT$, then $T = \frac{D}{R}$.
9. Solving $P + Prt = I$ for P gives $P = I - Prt$.
10. To solve $3R + yR = m$ for R, we must first factor the left-hand side.

5.8 EXERCISES

Solve each equation for y. See Example 1.

1. $\frac{y-1}{x-3} = 2$
2. $\frac{y-2}{x-4} = -2$
3. $\frac{y-1}{x+6} = -\frac{1}{2}$
4. $\frac{y+5}{x-2} = -\frac{1}{2}$
5. $\frac{y+a}{x-b} = m$
6. $\frac{y-h}{x+k} = a$
7. $\frac{y-1}{x+4} = -\frac{1}{3}$
8. $\frac{y-1}{x+3} = -\frac{3}{4}$

Solve each formula for the indicated variable. See Examples 2 and 3.

9. $A = \frac{B}{C}$ for C
10. $P = \frac{A}{C+D}$ for A
11. $\frac{1}{a} + m = \frac{1}{p}$ for p
12. $\frac{2}{f} + t = \frac{3}{m}$ for m
13. $F = k\frac{m_1 m_2}{r^2}$ for m_1
14. $F = \frac{mv^2}{r}$ for r
15. $\frac{1}{a} + \frac{1}{b} = \frac{1}{f}$ for a
16. $\frac{1}{R} = \frac{1}{R_1} + \frac{1}{R_2}$ for R
17. $S = \frac{a}{1-r}$ for r
18. $I = \frac{E}{R+r}$ for R
19. $\frac{P_1V_1}{T_1} = \frac{P_2V_2}{T_2}$ for P_2
20. $\frac{P_1V_1}{T_1} = \frac{P_2V_2}{T_2}$ for T_1
21. $V = \frac{4}{3}\pi r^2 h$ for h
22. $h = \frac{S - 2\pi r^2}{2\pi r}$ for S

Find the value of the indicated variable. See Example 4.

23. In the formula of Exercise 9, if $A = 12$ and $B = 5$, find C.

24. In the formula of Exercise 10, if $A = 500$, $P = 100$, and $C = 2$, find D.

25. In the formula of Exercise 11, if $p = 6$ and $m = 4$, find a.

26. In the formula of Exercise 12, if $m = 4$ and $t = 3$, find f.

27. In the formula of Exercise 13, if $F = 32$, $r = 4$, $m_1 = 2$, and $m_2 = 6$, find k.

28. In the formula of Exercise 14, if $F = 10$, $v = 8$, and $r = 6$, find m.

29. In the formula of Exercise 15, if $f = 3$ and $a = 2$, find b.

30. In the formula of Exercise 16, if $R = 3$ and $R_1 = 5$, find R_2.

31. In the formula of Exercise 17, if $S = \frac{3}{2}$ and $r = \frac{1}{5}$, find a.

32. In the formula of Exercise 18, if $I = 15$, $E = 3$, and $R = 2$, find r.

Show a complete solution to each problem. See Example 5.

33. ***Fast walking.*** Marcie can walk 8 miles in the same time as Frank walks 6 miles. If Marcie walks 1 mile per hour faster than Frank, then how fast does each person walk?

34. ***Upstream, downstream.*** Junior's boat will go 15 miles per hour in still water. If he can go 12 miles downstream in the same amount of time as it takes to go 9 miles upstream, then what is the speed of the current?

35. ***Delivery routes.*** Pat travels 70 miles on her milk route, and Bob travels 75 miles on his route. Pat travels 5 miles per hour slower than Bob, and her route takes her one-half hour longer than Bob's. How fast is each one traveling?

36. ***Ride the peaks.*** Smith bicycled 45 miles going east from Durango, and Jones bicycled 70 miles. Jones averaged 5 miles per hour more than Smith, and his trip took one-half hour longer than Smith's. How fast was each one traveling?

37. ***Walking and running.*** Raffaele ran 8 miles and then walked 6 miles. If he ran 5 miles per hour faster than he walked and the total time was 2 hours, then how fast did he walk?

38. ***Triathlon.*** Luisa participated in a triathlon in which she swam 3 miles, ran 5 miles, and then bicycled 10 miles. Luisa ran twice as fast as she swam, and she cycled three times as fast as she swam. If her total time for the triathlon was 1 hour and 46 minutes, then how fast did she swim?

Show a complete solution to each problem. See Example 6.

39. ***Fence painting.*** Kiyoshi can paint a certain fence in 3 hours by himself. If Red helps, the job takes only 2 hours. How long would it take Red to paint the fence by himself?

40. ***Envelope stuffing.*** Every week, Linda must stuff 1000 envelopes. She can do the job by herself in 6 hours. If Laura helps, they get the job done in $5\frac{1}{2}$ hours. How long would it take Laura to do the job by herself?

41. ***Garden destroying.*** Mr. McGregor has discovered that a large dog can destroy his entire garden in 2 hours and that a small boy can do the same job in 1 hour. How long would it take the large dog and the small boy working together to destroy Mr. McGregor's garden?

42. ***Draining the vat.*** With only the small valve open, all of the liquid can be drained from a large vat in 4 hours. With only the large valve open, all of the liquid can be drained from the same vat in 2 hours. How long would it take to drain the vat with both valves open?

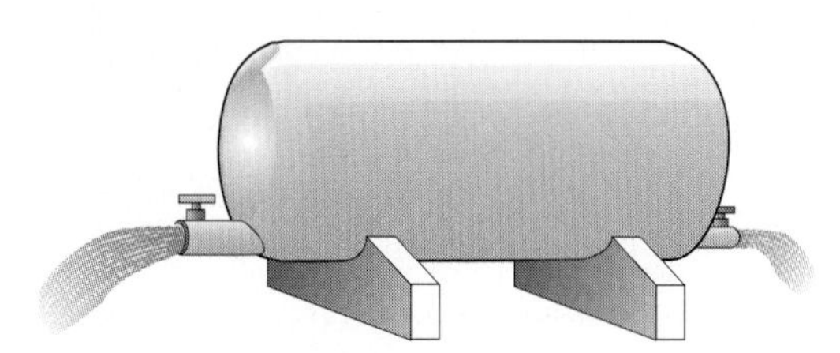

Figure for Exercise 42

43. ***Cleaning sidewalks.*** Edgar can blow the leaves off the sidewalks around the capitol building in 2 hours using a gasoline-powered blower. Ellen can do the same job in 8 hours using a broom. How long would it take them working together?

44. ***Computer time.*** It takes a computer 8 days to print all of the personalized letters for a national sweepstakes. A new computer is purchased that can do the same job in 5 days. How long would it take to do the job with both computers working on it?

Show a complete solution to each problem. See Example 7.

45. ***Apples and bananas.*** Bertha bought 18 pounds of fruit consisting of apples and bananas. She paid \$9 for the apples and \$2.40 for the bananas. If the price per pound of the apples was 3 times that of the bananas, then how many pounds of each type of fruit did she buy?

46. ***Running backs.*** In the playoff game the ball was carried by either Anderson or Brown on 21 plays. Anderson gained 36 yards, and Brown gained 54 yards. If Brown averaged twice as many yards per carry as Anderson, then on how many plays did Anderson carry the ball?

47. ***Fuel efficiency.*** Last week, Joe's Electric Service used 110 gallons of gasoline in its two trucks. The large truck was driven 800 miles, and the small truck was driven 600 miles. If the small truck gets twice as many miles per gallon as the large truck, then how many gallons of gasoline did the large truck use?

48. ***Repair work.*** Sally received a bill for a total of 8 hours labor on the repair of her bulldozer. She paid $50 to the master mechanic and $90 to his apprentice. If the master mechanic gets $10 more per hour than his apprentice, then how many hours did each work on the bulldozer?

COLLABORATIVE ACTIVITIES

How Do I Get There from Here?

Grouping: 3 students
Topic: Distance formula

You want to decide whether to ride your bicycle, drive your car, or take the bus to school this year. The best thing to do is to analyze each of your options. Read all the information given below. Then have each person in your group pick one mode of transportation. Working individually, answer the questions using the given information, the map, and the distance formula $d = rt$. Present a case to your group for your type of transportation.

Information available (see the map for distances):

Bicycle: You would need to buy a new bike lock for $15 and two new tubes at $2.50 apiece. Determine how fast you would have to bike to beat the car.

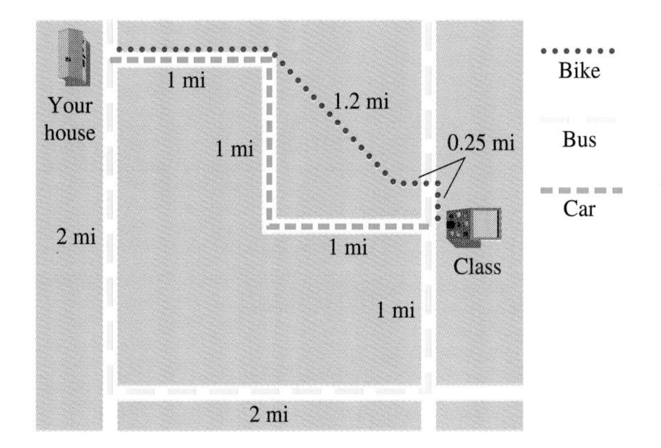

Car: You would need to pay for a parking permit, which costs $40. Traffic has increased, so it takes 12 minutes to get to school. What is your average speed?

Bus: The bus stops at the end of your block and has a new student rate of $1 a week. It leaves at 8:30 A.M. and will get to the college at 8:55 A.M. On Mondays and Wednesdays you have a 9:00 A.M. class, four blocks from the bus stop. Find the average speed of the bus and figure the cost for the 16-week semester.

When presenting your case: Include the time needed to get there, speed, cost, and convenience. Have at least three reasons why this would be the best way to travel. Consider unique features of your area such as traffic, weather, and terrain.

After each of you has presented your case, decide as a group which type of transportation you would choose.

Wrap-up CHAPTER 5

SUMMARY

Rational Expressions

Rational Expressions		Examples
Rational expression	The ratio of two polynomials with the denominator not equal to 0	$\frac{x-1}{x-3}$ $(x \neq 3)$
Rule for reducing rational expressions	If $a \neq 0$ and $c \neq 0$, then $\frac{ab}{ac} = \frac{b}{c}$. (Divide out the common factors.)	$\frac{8x+2}{4x} = \frac{2(4x+1)}{2(2x)} = \frac{4x+1}{2x}$
Quotient rule for exponents	Suppose $a \neq 0$ and m and n are positive integers. If $m \geq n$, then $\frac{a^m}{a^n} = a^{m-n}$. If $m < n$, then $\frac{a^m}{a^n} = \frac{1}{a^{n-m}}$.	$\frac{x^7}{x^5} = x^2$ $\frac{x^2}{x^5} = \frac{1}{x^3}$

Multiplication and Division of Rational Expressions

Multiplication and Division of Rational Expressions		Examples
Multiplication	If $b \neq 0$ and $d \neq 0$, then $\frac{a}{b} \cdot \frac{c}{d} = \frac{ac}{bd}$.	$\frac{3}{x^3} \cdot \frac{6}{x^5} = \frac{18}{x^8}$
Division	If $b \neq 0$, $c \neq 0$, and $d \neq 0$, then $\frac{a}{b} \div \frac{c}{d} = \frac{a}{b} \cdot \frac{d}{c}$. (Invert the divisor and multiply.)	$\frac{a}{x^3} \div \frac{5}{x^9} = \frac{a}{x^3} \cdot \frac{x^9}{5} = \frac{ax^6}{5}$

Addition and Subtraction of Rational Expressions

Addition and Subtraction of Rational Expressions		Examples
Least common denominator	The LCD of a group of denominators is the smallest number that is a multiple of all of them.	8, 12 LCD = 24
Finding the least common denominator	1. Factor each denominator completely. Use exponent notation for repeated factors. 2. Write the product of all of the different factors that appear in the denominators. 3. On each factor, use the highest power that appears on that factor in any of the denominators.	$6a^2b$, $4ab^3$ $4ab^3 = 2^2ab^3$ $6a^2b = 2 \cdot 3a^2b$ $\text{LCD} = 2^2 \cdot 3a^2b^3 = 12a^2b^3$

Addition and subtraction of rational expressions	If $b \neq 0$, then $\frac{a}{b} + \frac{c}{b} = \frac{a+c}{b}$ and $\frac{a}{b} - \frac{c}{b} = \frac{a-c}{b}$. If the denominators are not identical, change each fraction to an equivalent fraction so that all denominators are identical.	$\frac{2x}{x-3} + \frac{7x}{x-3} = \frac{9x}{x-3}$ $\frac{2}{x} + \frac{1}{3x} = \frac{6}{3x} + \frac{1}{3x} = \frac{7}{3x}$
Complex fraction	A rational expression that has fractions in the numerator and/or the denominator	$\dfrac{\frac{1}{2} + \frac{1}{3}}{\frac{1}{3} - \frac{3}{4}}$
Simplifying complex fractions	Multiply the numerator and denominator by the LCD.	$\dfrac{\left(\frac{1}{2} + \frac{1}{3}\right)12}{\left(\frac{1}{3} - \frac{3}{4}\right)12} = \frac{6+4}{4-9} = -2$

Equations with Rational Expressions

Examples

Solving equations	Multiply each side by the LCD.	$\frac{1}{x} - \frac{1}{3} = \frac{1}{2x} - \frac{1}{6}$ $6x\left(\frac{1}{x} - \frac{1}{3}\right) = 6x\left(\frac{1}{2x} - \frac{1}{6}\right)$ $6 - 2x = 3 - x$
Proportion	An equation expressing the equality of two ratios	$\frac{2}{x-3} = \frac{5}{6}$
Extremes-means property (Cross-multiply)	If $b \neq 0$ and $d \neq 0$, then $\frac{a}{b} = \frac{c}{d}$ is equivalent to $ad = bc$. Cross-multiplying is a quick way to eliminate the fractions in a proportion.	$12 = 5x - 15$

REVIEW EXERCISES

5.1 *Reduce each rational expression to lowest terms.*

1. $\frac{24}{28}$
2. $\frac{42}{18}$
3. $\frac{2a^3c^3}{8a^5c}$
4. $\frac{39x^6}{15x}$
5. $\frac{6w-9}{9w-12}$
6. $\frac{3t-6}{8-4t}$
7. $\frac{x^2-1}{3-3x}$
8. $\frac{3x^2-9x+6}{10-5x}$

5.2 *Perform the indicated operation.*

9. $\frac{1}{6k} \cdot 3k^2$
10. $\frac{1}{15abc} \cdot 5a^3b^5c^2$
11. $\frac{2xy}{3} \div y^2$
12. $4ab \div \frac{1}{2a^4}$
13. $\frac{a^2-9}{a-2} \cdot \frac{a^2-4}{a+3}$
14. $\frac{x^2-1}{3x} \cdot \frac{6x}{2x-2}$
15. $\frac{w-2}{3w} \div \frac{4w-8}{6w}$
16. $\frac{2y+2x}{x-xy} \div \frac{x^2+2xy+y^2}{y^2-y}$

5.3 *Find the least common denominator for each group of denominators.*

17. 36, 54

18. 10, 15, 35

19. $6ab^3$, $8a^7b^2$

20. $20u^4v$, $18uv^5$, $12u^2v^3$

21. $4x$, $6x - 6$

22. $8a$, $6a$, $2a^2 + 2a$

23. $x^2 - 4$, $x^2 - x - 2$

24. $x^2 - 9$, $x^2 + 6x + 9$

Convert each rational expression into an equivalent rational expression with the indicated denominator.

25. $\frac{5}{12} = \frac{?}{36}$

26. $\frac{2a}{15} = \frac{?}{45}$

27. $\frac{2}{3xy} = \frac{?}{15x^2y}$

28. $\frac{3z}{7x^2y} = \frac{?}{42x^3y^8}$

29. $\frac{5}{y-6} = \frac{?}{12-2y}$

30. $\frac{-3}{2-t} = \frac{?}{2t-4}$

31. $\frac{x}{x-1} = \frac{?}{x^2-1}$

32. $\frac{t}{t-3} = \frac{?}{t^2+2t-15}$

5.4 *Perform the indicated operation.*

33. $\frac{5}{36} + \frac{9}{28}$

34. $\frac{7}{30} - \frac{11}{42}$

35. $3 - \frac{4}{x}$

36. $1 + \frac{3a}{2b}$

37. $\frac{2}{ab^2} - \frac{1}{a^2b}$

38. $\frac{3}{4x^3} + \frac{5}{6x^2}$

39. $\frac{9a}{2a-3} + \frac{5}{3a-2}$

40. $\frac{3}{x-2} - \frac{5}{x+3}$

41. $\frac{1}{a-8} - \frac{2}{8-a}$

42. $\frac{5}{x-14} + \frac{4}{14-x}$

43. $\frac{3}{2x-4} + \frac{1}{x^2-4}$

44. $\frac{x}{x^2-2x-3} - \frac{3x}{x^2-9}$

5.5 *Simplify each complex fraction.*

45. $\dfrac{\frac{1}{2} - \frac{3}{4}}{\frac{2}{3} + \frac{1}{2}}$

46. $\dfrac{\frac{2}{3} + \frac{5}{8}}{\frac{1}{2} - \frac{3}{8}}$

47. $\dfrac{\frac{1}{a} + \frac{2}{3b}}{\frac{1}{2b} - \frac{3}{a}}$

48. $\dfrac{\frac{3}{xy} - \frac{1}{3y}}{\frac{1}{6x} - \frac{3}{5y}}$

49. $\dfrac{\frac{1}{x-2} - \frac{3}{x+3}}{\frac{2}{x+3} + \frac{1}{x-2}}$

50. $\dfrac{\frac{4}{a+1} + \frac{5}{a^2-1}}{\frac{1}{a^2-1} - \frac{3}{a-1}}$

51. $\dfrac{\frac{x-1}{x-3}}{\frac{1}{x^2-x-6} - \frac{4}{x+2}}$

52. $\dfrac{\frac{6}{a^2+5a+6} - \frac{8}{a+2}}{\frac{2}{a+3} - \frac{4}{a+2}}$

5.6 *Solve each equation.*

53. $\frac{-2}{5} = \frac{3}{x}$

54. $\frac{3}{x} + \frac{5}{3x} = 1$

55. $\frac{14}{a^2-1} + \frac{1}{a-1} = \frac{3}{a+1}$

56. $2 + \frac{3}{y-5} = \frac{2y}{y-5}$

57. $z - \frac{3z}{2-z} = \frac{6}{z-2}$

58. $\frac{1}{x} + \frac{1}{3} = \frac{1}{2}$

5.7 *Solve each proportion.*

59. $\frac{3}{x} = \frac{2}{7}$

60. $\frac{4}{x} = \frac{x}{4}$

61. $\frac{2}{w-3} = \frac{5}{w}$

62. $\frac{3}{t-3} = \frac{5}{t+4}$

Solve each problem by using a proportion.

63. ***Taxis in Times Square.*** The ratio of taxis to private automobiles in Times Square at 6:00 P.M. on New Year's Eve was estimated to be 15 to 2. If there were 60 taxis, then how many private automobiles were there?

64. ***Student-teacher ratio.*** The student-teacher ratio for Washington High was reported to be 27.5 to 1. If there are 42 teachers, then how many students are there?

65. ***Water and rice.*** At Wong's Chinese Restaurant the secret recipe for white rice calls for a 2 to 1 ratio of water to rice. In one batch the chef used 28 more cups of water than rice. How many cups of each did he use?

66. ***Oil and gas.*** An outboard motor calls for a fuel mixture that has a gasoline-to-oil ratio of 50 to 1. How many pints of oil should be added to 6 gallons of gasoline?

5.8 *Solve each formula for the indicated variable.*

67. $\frac{y-b}{m} = x$ for y

68. $\frac{A}{h} = \frac{a+b}{2}$ for a

69. $F = \frac{mv+1}{m}$ for m

70. $m = \frac{r}{1+rt}$ for r

71. $\frac{y+1}{x-3} = 4$ for y

72. $\frac{y-3}{x+2} = \frac{-1}{3}$ for y

Solve each problem.

73. ***Making a puzzle.*** Tracy, Stacy, and Fred assembled a very large puzzle together in 40 hours. If Stacy worked twice as fast as Fred and Tracy worked just as fast as Stacy, then how long would it have taken Fred to assemble the puzzle alone?

74. ***Going skiing.*** Leon drove 270 miles to the lodge in the same time as Pat drove 330 miles to the lodge. If Pat drove 10 miles per hour faster than Leon, then how fast did each of them drive?

75. ***Merging automobiles.*** When Bert and Ernie merged their automobile dealerships, Bert had 10 more cars than Ernie. While 36% of Ernie's stock consisted of new cars, only 25% of Bert's stock consisted of new cars. If they had 33 new cars on the lot after the merger, then how many cars did each one have before the merger?

76. ***Magazine sales.*** A company specializing in magazine sales over the telephone found that in 2,500 phone calls, 360 resulted in sales and were made by male callers, and 480 resulted in sales and were made by female callers. If the company gets twice as many sales per call with a women's voice than with a man's voice, then how many of the 2500 calls were made by females?

77. ***Chlorofluorocarbons used worldwide.*** The accompanying graph shows the percentages of chlorofluorocarbons used annually in various products worldwide (*Time*, Feb. 17, 1992). If the amount used in cleaning agents is 30,000 metric tons more than the amount used in vehicle air conditioning, then what amount is used in vehicle air conditioning?

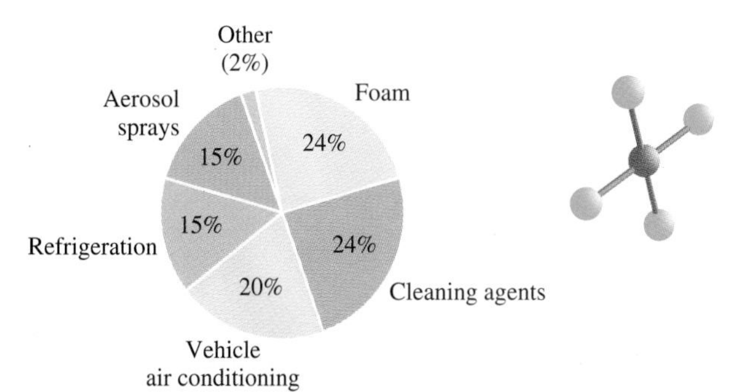

Figure for Exercises 77 and 78

78. ***Ozone depletion.*** Chlorofluorocarbons are blamed for causing a hole in the earth's ozone layer that is as big as Alaska. Use the accompanying graph and the information given in Exercise 77 to find the total amount of chlorofluorocarbons used annually worldwide and the amount used in refrigeration.

Miscellaneous

In place of each question mark, put an expression that makes each equation an identity.

79. $\frac{5}{x} = \frac{?}{2x}$

80. $\frac{?}{a} = \frac{6}{3a}$

81. $\frac{2}{a-5} = \frac{?}{5-a}$

82. $\frac{-1}{a-7} = \frac{1}{?}$

83. $3 = \frac{?}{x}$

84. $2a = \frac{?}{b}$

85. $m \div \frac{1}{2} = ?$

86. $5x \div \frac{1}{x} = ?$

87. $2a \div ? = 12a$

88. $10x \div ? = 20x^2$

89. $\frac{a-1}{a^2-1} = \frac{1}{?}$

90. $\frac{?}{x^2-9} = \frac{1}{x-3}$

91. $\frac{1}{a} - \frac{1}{5} = ?$

92. $\frac{3}{7} - \frac{2}{b} = ?$

93. $\frac{a}{2} - 1 = \frac{?}{2}$

94. $\frac{1}{a} - 1 = \frac{?}{a}$

95. $(a-b) \div (-1) = ?$

96. $(a-7) \div (7-a) = ?$

97. $\frac{\frac{1}{5a}}{2} = ?$

98. $\frac{3a}{\frac{1}{2}} = ?$

For each expression in Exercises 99–118, either perform the indicated operation or solve the equation, whichever is appropriate.

99. $\frac{1}{x} + \frac{1}{2x}$

100. $\frac{1}{y} + \frac{1}{3y} = 2$

101. $\frac{2}{3xy} + \frac{1}{6x}$

102. $\frac{3}{x-1} - \frac{3}{x}$

103. $\frac{5}{a-5} - \frac{3}{5-a}$

104. $\frac{2}{x-2} - \frac{3}{x} = \frac{-1}{x}$

105. $\frac{2}{x-1} - \frac{2}{x} = 1$

106. $\frac{2}{x-2} \cdot \frac{6x-12}{14}$

107. $\frac{-3}{x+2} \cdot \frac{5x+10}{9}$

108. $\frac{3}{10} = \frac{5}{x}$

109. $\frac{1}{-3} = \frac{-2}{x}$

110. $\frac{x^2 - 4}{x} \div \frac{4x - 8}{x}$

111. $\frac{ax + am + 3x + 3m}{a^2 - 9} \div \frac{2x + 2m}{a - 3}$

112. $\frac{-2}{x} = \frac{3}{x + 2}$

113. $\frac{2}{x^2 - 25} + \frac{1}{x^2 - 4x - 5}$

114. $\frac{4}{a^2 - 1} + \frac{1}{2a + 2}$

115. $\frac{-3}{a^2 - 9} - \frac{2}{a^2 + 5a + 6}$

116. $\frac{-5}{a^2 - 4} - \frac{2}{a^2 - 3a + 2}$

117. $\frac{1}{a^2 - 1} + \frac{2}{1 - a} = \frac{3}{a + 1}$

118. $3 + \frac{1}{x - 2} = \frac{2x - 3}{x - 2}$

CHAPTER 5 TEST

What numbers cannot be used for x in each rational expression?

1. $\frac{2x - 1}{x^2 - 1}$

2. $\frac{5}{2 - 3x}$

3. $\frac{1}{x}$

Perform the indicated operation. Write each answer in lowest terms.

4. $\frac{2}{15} - \frac{4}{9}$

5. $\frac{1}{y} + 3$

6. $\frac{3}{a - 2} - \frac{1}{2 - a}$

7. $\frac{2}{x^2 - 4} - \frac{3}{x^2 + x - 2}$

8. $\frac{m^2 - 1}{(m - 1)^2} \cdot \frac{2m - 2}{3m + 3}$

9. $\frac{a - b}{3} \div \frac{b^2 - a^2}{6}$

10. $\frac{5a^2b}{12a} \cdot \frac{2a^3b}{15ab^6}$

Simplify each complex fraction.

11. $\frac{\frac{2}{3} + \frac{4}{5}}{\frac{2}{5} - \frac{3}{2}}$

12. $\frac{\frac{2}{x} + \frac{1}{x - 2}}{\frac{1}{x - 2} - \frac{3}{x}}$

Solve each equation.

13. $\frac{3}{x} = \frac{7}{5}$

14. $\frac{x}{x - 1} - \frac{3}{x} = \frac{1}{2}$

15. $\frac{1}{x} + \frac{1}{6} = \frac{1}{4}$

Solve each formula for the indicated variable.

16. $\frac{y - 3}{x + 2} = \frac{-1}{5}$ for y

17. $M = \frac{1}{3}b(c + d)$ for c

Solve each problem.

18. When all of the grocery carts escape from the supermarket, it takes Reginald 12 minutes to round them up and bring them back. Because Norman doesn't make as much per hour as Reginald, it takes Norman 18 minutes to do the same job. How long would it take them working together to complete the roundup?

19. Brenda and her husband Randy bicycled cross-country together. One morning, Brenda rode 30 miles. By traveling only 5 miles per hour faster and putting in one more hour, Randy covered twice the distance Brenda covered. What was the speed of each cyclist?

20. For a certain time period the ratio of the dollar value of exports to the dollar value of imports for this country was 2 to 3. If the value of exports during that time period was 48 billion dollars, then what was the value of imports?

Tying It All Together CHAPTERS 1–5

Solve each equation.

1. $3x - 2 = 5$
2. $\frac{3}{5}x = -2$
3. $2(x - 2) = 4x$
4. $2(x - 2) = 2x$
5. $2(x + 3) = 6x + 6$
6. $2(3x + 4) + x^2 = 0$
7. $4x - 4x^3 = 0$
8. $\frac{3}{x} = \frac{-2}{5}$
9. $\frac{3}{x} = \frac{x}{12}$
10. $\frac{x}{2} = \frac{4}{x - 2}$
11. $\frac{w}{18} - \frac{w - 1}{9} = \frac{4 - w}{6}$
12. $\frac{x}{x + 1} + \frac{1}{2x + 2} = \frac{7}{8}$

Solve each equation for y.

13. $2x + 3y = c$
14. $\frac{y - 3}{x - 5} = \frac{1}{2}$
15. $2y = ay + c$
16. $\frac{A}{y} = \frac{C}{B}$
17. $\frac{A}{y} + \frac{1}{3} = \frac{B}{y}$
18. $\frac{A}{y} - \frac{1}{2} = \frac{1}{3}$
19. $3y - 5ay = 8$
20. $y^2 - By = 0$
21. $A = \frac{1}{2}h(b + y)$
22. $2(b + y) = b$

Calculate the value of $b^2 - 4ac$ for each choice of a, b, and c.

23. $a = 1,\ b = 2,\ c = -15$
24. $a = 1,\ b = 8,\ c = 12$
25. $a = 2,\ b = 5,\ c = -3$
26. $a = 6,\ b = 7,\ c = -3$

Perform each indicated operation.

27. $(3x - 5) - (5x - 3)$
28. $(2a - 5)(a - 3)$
29. $x^7 \div x^3$
30. $\frac{x - 3}{5} + \frac{x + 4}{5}$
31. $\frac{1}{2} \cdot \frac{1}{x}$
32. $\frac{1}{2} + \frac{1}{x}$
33. $\frac{1}{2} \div \frac{1}{x}$
34. $\frac{1}{2} - \frac{1}{x}$
35. $\frac{x - 3}{5} - \frac{x + 4}{5}$
36. $\frac{3a}{2} \div 2$
37. $(x - 8)(x + 8)$
38. $3x(x^2 - 7)$
39. $2a^5 \cdot 5a^9$
40. $x^2 \cdot x^8$
41. $(k - 6)^2$
42. $(j + 5)^2$
43. $(g - 3) \div (3 - g)$
44. $(6x^3 - 8x^2) \div (2x)$

Solve.

45. ***Present value.*** An investor is interested in the amount or present value that she would have to invest today to receive periodic payments in the future. The present value of \$1 in one year and \$1 in two years with interest rate r compounded annually is given by the formula

$$P = \frac{1}{1 + r} + \frac{1}{(1 + r)^2}.$$

a) Rewrite the formula so that the right-hand side is a single rational expression.

b) Find P if $r = 7\%$.

CHAPTER 6

Linear Equations and Their Graphs

If you pick up any package of food and read the label, you will find a long list that usually ends with some mysterious looking names. Many of these strange elements are food additives. A food additive is a substance or a mixture of substances other than basic foodstuffs that is present in food as a result of production, processing, storage, or packaging. They can be natural or synthetic and are categorized in many ways: preservatives, coloring agents, processing aids, and nutritional supplements, to name a few.

Food additives have been around since prehistoric humans discovered that salt would help to preserve meat. Today, food additives can include simple ingredients such as red color from Concord grape skins, calcium, or an enzyme. Throughout the centuries there have been lively discussions on what is healthy to eat. At the present time the food industry is working to develop foods that have less cholesterol, fats, and other unhealthy ingredients.

Although they frequently have different viewpoints, the food industry and the Food and Drug Administration (FDA) are working to provide consumers with information on a healthier diet. Recent developments such as the synthetically engineered tomato stirred great controversy, even though the FDA declared the tomato safe to eat.

In Exercise 29 of Section 6.5 you will see how a food chemist uses a linear equation in testing the concentration of an enzyme in a fruit juice.

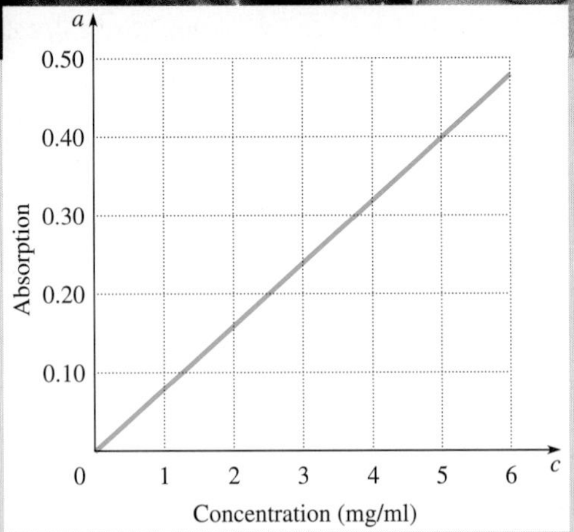

6.1 Graphing in the Coordinate Plane

In this section:

In Chapter 1 you learned to graph numbers on a number line. We also used number lines to illustrate the solution to inequalities in Chapter 2. In this section you will learn to graph pairs of numbers in a coordinate system made up of a pair of number lines. We will use this coordinate system to illustrate the solution to equations and inequalities in two variables.

Ordered Pairs

The equation $y = 2x - 1$ is an equation in two variables. This equation is satisfied if we choose a value for x and a value for y that make it true. If we choose $x = 2$ and $y = 3$, then $y = 2x - 1$ becomes

$$\overset{\overset{y}{\downarrow}}{3} = 2(\overset{\overset{x}{\downarrow}}{2}) - 1.$$

$$3 = 3$$

Because the last statement is true, we say that the pair of numbers 2 and 3 **satisfies the equation** or is a **solution to the equation.** We use the **ordered pair** (2, 3) to represent $x = 2$ and $y = 3$. The format is always to write the value for x first and the value for y second. The first number of the ordered pair is called the ***x*-coordinate,** and the second number is called the ***y*-coordinate.** Note that the ordered pair (3, 2) does not satisfy the equation $y = 2x - 1$, because for $x = 3$ and $y = 2$ we have

$$2 \neq 2(3) - 1.$$

The ordered pair (2, 3) is a solution to $y = 2x - 1$. We can find as many solutions as we please by simply choosing any value for x or y and then using the equation to find the other coordinate of the ordered pair.

EXAMPLE 1 **Finding solutions to an equation**

Each of the ordered pairs below is missing one coordinate. Complete each ordered pair so that it satisfies the equation $y = -3x + 4$.

a) (2,) **b)** (, −5) **c)** (0,)

Solution

a) The x-coordinate of (2,) is 2. Let $x = 2$ in the equation $y = -3x + 4$:

$$\begin{aligned} y &= -3 \cdot 2 + 4 \\ &= -6 + 4 \\ &= -2 \end{aligned}$$

The ordered pair (2, −2) satisfies the equation.

b) The y-coordinate of (, −5) is −5. Let $y = -5$ in the equation $y = -3x + 4$:

$$\begin{aligned} -5 &= -3x + 4 \\ -9 &= -3x \\ 3 &= x \end{aligned}$$

The ordered pair (3, −5) satisfies the equation.

c) Replace x by 0 in the equation $y = -3x + 4$:

$$y = -3 \cdot 0 + 4$$
$$= 4$$

So $(0, 4)$ satisfies the equation. ■

The Rectangular Coordinate System

We use the **rectangular (or Cartesian) coordinate system** to get a visual image of ordered pairs of real numbers. The rectangular coordinate system consists of two number lines drawn at a right angle to one another, intersecting at zero on each number line, as shown in Fig. 6.1. The plane containing these number lines is called the **coordinate plane.** On the horizontal number line the positive numbers are to the right of zero, and on the vertical number line the positive numbers are above zero.

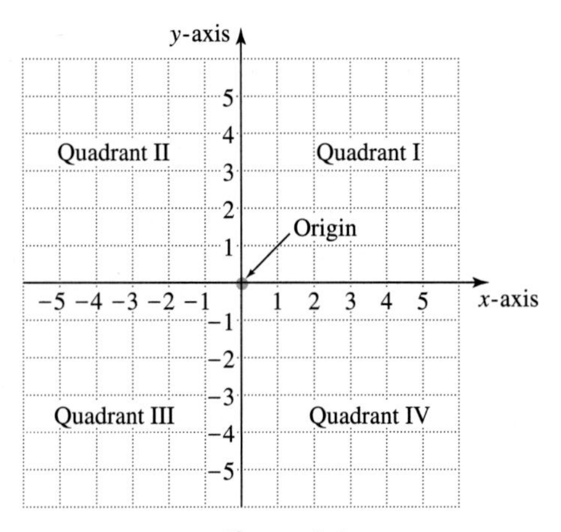

Figure 6.1

The horizontal number line is called the ***x*-axis,** and the vertical number line is called the ***y*-axis.** The point at which they intersect is called the **origin.** The two number lines divide the plane into four regions called **quadrants.** They are numbered as shown in Fig. 6.1. The quadrants do not include any points on the axes.

Plotting Points

Just as every real number corresponds to a point on the number line, *every pair of real numbers corresponds to a point in the rectangular coordinate system.* For example, the point corresponding to the pair $(2, 3)$ is found by starting at the origin and moving two units to the right and then three units up. The point corresponding to the pair $(-3, -2)$ is found by starting at the origin and moving three units to the left and then two units down. Both of these points are shown in Fig. 6.2.

When we locate a point in the rectangular coordinate system, we are **plotting** or **graphing** the point. Because ordered pairs of numbers correspond to points in the coordinate plane, we frequently refer to an ordered pair as a point.

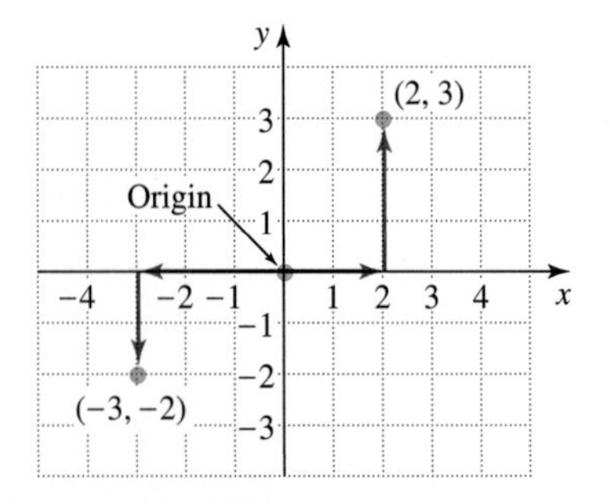

Figure 6.2

EXAMPLE 2 **Plotting points**

Plot the points $(2, 5)$, $(-1, 4)$, $(-3, -4)$, and $(3, -2)$.

Solution

To locate $(2, 5)$, start at the origin, move two units to the right, and then move up five units. To locate $(-1, 4)$, start at the origin, move one unit to the left, and then move up four units. All four points are shown in Fig. 6.3. ■

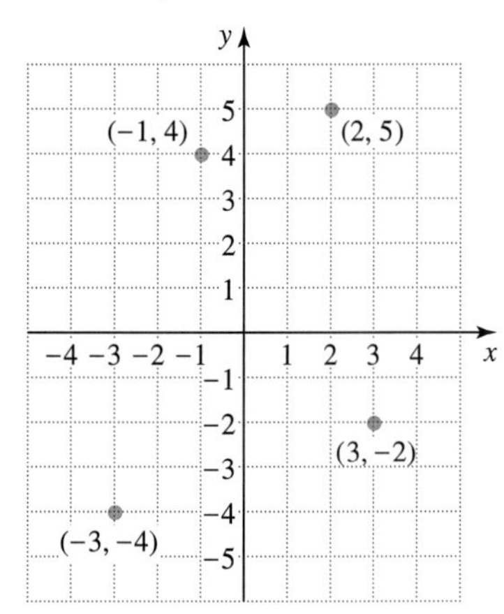

Figure 6.3

Graphing a Linear Equation

The **graph** of an equation is an illustration in the coordinate plane that shows all of the ordered pairs that satisfy the equation. When we draw the graph, we are **graphing the equation.**

Consider again the equation $y = 2x - 1$. The following table shows some ordered pairs that satisfy this equation.

x	-3	-2	-1	0	1	2	3
$y = 2x - 1$	-7	-5	-3	-1	1	3	5

The ordered pairs in this table are graphed in Fig. 6.4. Notice that the points lie in a straight line. If we choose any real number for x and find the point (x, y) that satisfies $y = 2x - 1$, we get another point along this line. Likewise, any point along this line satisfies the equation. So the graph of $y = 2x - 1$ is the straight line shown in Fig. 6.5. Because it is not possible to actually show all of the line, the arrows on the ends of the line indicate that it goes on indefinitely in both directions. The equation $y = 2x - 1$ is an example of a linear equation in two variables.

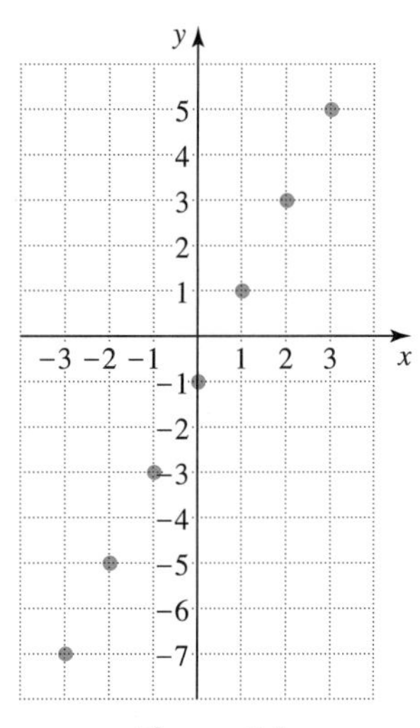

Figure 6.4

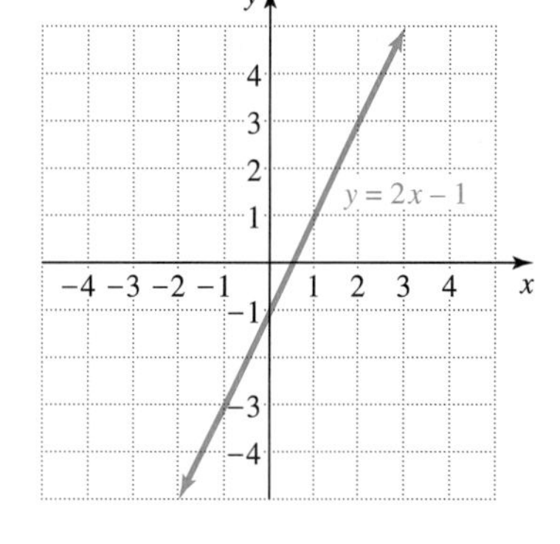

Figure 6.5

Linear Equation in Two Variables

A **linear equation in two variables** is an equation of the form

$$Ax + By = C,$$

where A, B, and C are real numbers, with A and B not both equal to zero.

The graph of any linear equation in two variables is a straight line.

EXAMPLE 3 **Graphing an equation**

Graph the equation $3x + y = 2$. Plot at least five points.

Solution

First solve the equation for y:

$$y = -3x + 2$$

Next, arbitrarily select values for x and then calculate the corresponding value for y. The following table of values shows five ordered pairs that satisfy the equation:

x	-2	-1	0	1	2
$y = -3x + 2$	8	5	2	-1	-4

The graph of the line through these points is shown in Fig. 6.6. ■

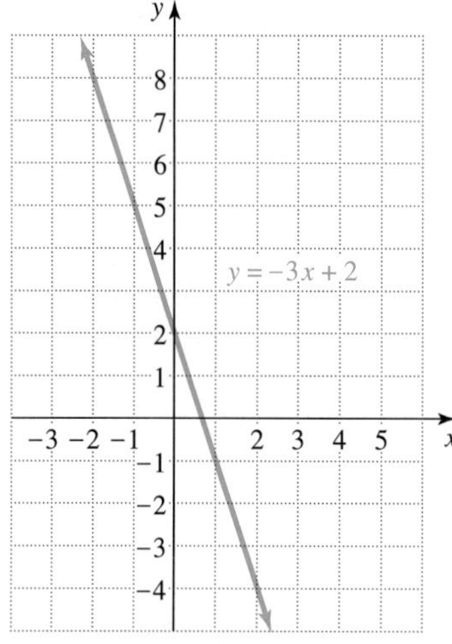

Figure 6.6

Figure 6.7

EXAMPLE 4 **A vertical line**

Graph the equation $x + 0 \cdot y = 3$. Plot at least five points.

Solution

If we choose a value of 3 for x, then we can choose any number for y, since y is multiplied by 0. The equation $x + 0 \cdot y = 3$ is usually written simply as $x = 3$. The following table shows five ordered pairs that satisfy the equation:

$x = 3$	3	3	3	3	3
y	-2	-1	0	1	2

Figure 6.7 shows the line through these points. ■

CAUTION If an equation such as $x = 3$ is discussed in the context of equations in two variables, then we assume that it is a simplified form of $x + 0 \cdot y = 3$, and there are infinitely many ordered pairs that satisfy the equation. If the equation $x = 3$ is discussed in the context of equations in a single variable, then $x = 3$ has only one solution, 3. ⊘

All of the equations we have considered so far have involved single-digit numbers. If an equation involves large numbers, then we must change the scale on the x-axis, the y-axis, or both to accommodate the numbers involved. The change of scale is arbitrary, and the graph will look different for different scales.

EXAMPLE 5 **Adjusting the scale**

Graph the equation $y = 20x + 500$. Plot at least five points.

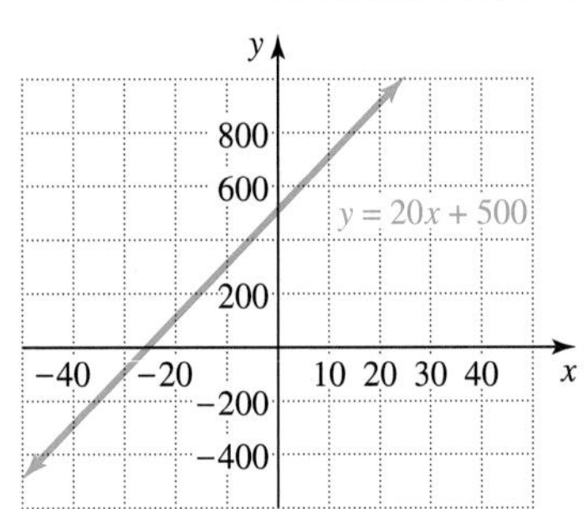

Figure 6.8

Solution

The following table shows five ordered pairs that satisfy the equation.

x	-20	-10	0	10	20
$y = 20x + 500$	100	300	500	700	900

To fit these points onto a graph, we change the scale on the x-axis to let each division represent 10 units and change the scale on the y-axis to let each division represent 200 units. The graph is shown in Fig. 6.8. ■

Graphing a Line Using Intercepts

We know that the graph of a linear equation is a straight line. Because it takes only two points to determine a line, we can graph a linear equation using only two points. The two points that are the easiest to locate are usually the points where the line crosses the axes. The point where the graph crosses the x-axis is the ***x*-intercept,** and the point where the graph crosses the y-axis is the ***y*-intercept.**

EXAMPLE 6 **Graphing a line using intercepts**

Graph the equation $2x - 3y = 6$ by using the x- and y-intercepts.

Solution

To find the x-intercept, let $y = 0$ in the equation $2x - 3y = 6$:

$$2x - 3 \cdot 0 = 6$$
$$2x = 6$$
$$x = 3$$

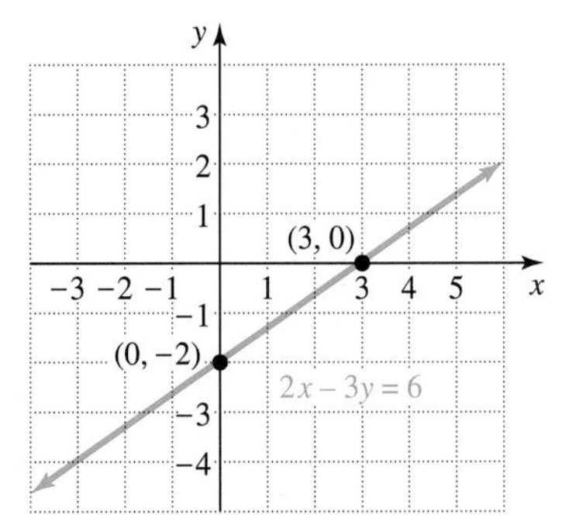

Figure 6.9

The x-intercept is $(3, 0)$. To find the y-intercept, let $x = 0$ in $2x - 3y = 6$:

$$2 \cdot 0 - 3y = 6$$
$$-3y = 6$$
$$y = -2$$

The y-intercept is $(0, -2)$. Locate the intercepts and draw a line through them as shown in Fig. 6.9. To check, find one additional point that satisfies the equation, say $(6, 2)$, and see whether the line goes through that point. ■

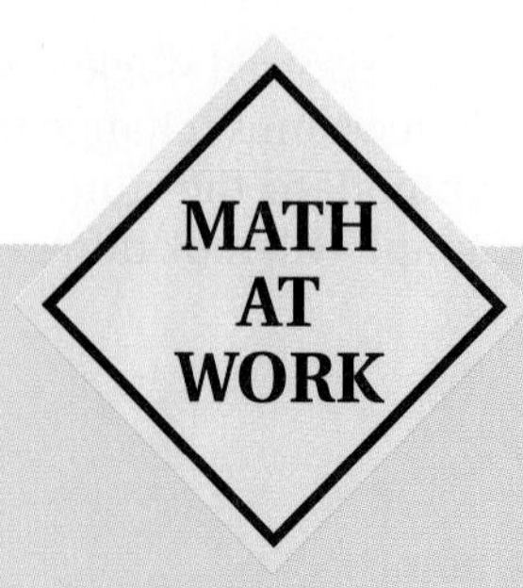

Christopher J. Edington, Manager of the Biomechanics Laboratory at Converse, Inc., is a specialist in studying human movements from the hip down. In the past he has worked with diabetics, helping to educate them about the role of shoes and stress points in the shoes and their relationship to preventing foot injuries. More recently, he has helped to design and run tests for Converse's new athletic and leisure shoes. The latest development is a new basketball shoe that combines the lightweight characteristic of a running shoe with the support and durability of a standard basketball sneaker.

Information on how the foot strikes the ground, the length of contact time, and movements of the foot, knee, and hip can be recorded by using high-speed video equipment. This information is then used to evaluate the performance and design requirements of a lightweight, flexible, and well-fitting shoe. To meet the requirements of a good basketball shoe, Mr. Edington helped design and test the "React" shock-absorbing technology that is in Converse's latest sneakers.

In Exercise 80 of this section you will see the motion of a runner's heel as Mr. Edington does.

Warm-ups

True or false? Explain your answer.

1. The point $(2, 4)$ satisfies the equation $2y - 3x = -8$.
2. If $(1, 5)$ satisfies an equation, then $(5, 1)$ also satisfies the equation.
3. The origin is in quadrant I.
4. The point $(4, 0)$ is on the y-axis.
5. The graph of $x + 0 \cdot y = 9$ is the same as the graph of $x = 9$.
6. The graph of $x = -5$ is a vertical line.
7. The graph of $0 \cdot x + y = 6$ is a horizontal line.
8. The y-intercept for the line $x + 2y = 5$ is $(5, 0)$.
9. The point $(5, -3)$ is in quadrant II.
10. The point $(-349, 0)$ is on the x-axis.

6.1 EXERCISES

Complete each ordered pair so that it satisfies the given equation. See Example 1.

1. $y = 3x + 9$: (0,), (, 24), (2,)

2. $y = 2x + 5$: (8,), (−1,), (, −1)

3. $y = -3x - 7$: (0,), (−4,), (, −1)

4. $y = -5x - 3$: (, 2), (−3,), (0,)

5. $y = -12x + 5$: (0,), (10,), (, 17)

6. $y = 18x + 200$: (1,), (−10,), (, 200)

7. $2x - 3y = 6$: (3,), (, −2), (12,)

8. $3x + 5y = 0$: (−5,), (, −3), (10,)

9. $0 \cdot y + x = 5$: (, −3), (, 5), (, 0)

10. $0 \cdot x + y = -6$: (3,), (−1,), (4,)

Plot the points on a rectangular coordinate system. See Example 2.

11. $(1, 5)$ **12.** $(4, 3)$ **13.** $(-2, 1)$

14. $(-3, 5)$ **15.** $\left(3, -\frac{1}{2}\right)$ **16.** $\left(2, -\frac{1}{3}\right)$

17. $(-2, -4)$ **18.** $(-3, -5)$ **19.** $(0, 3)$

20. $(0, -2)$ **21.** $(-3, 0)$ **22.** $(5, 0)$

23. $(\pi, 1)$ **24.** $(-2, \pi)$ **25.** $(1.4, 4)$

26. $(-3, 1.4)$

Graph each equation. Plot at least five points for each. See Examples 3 and 4.

27. $y = x + 1$ **28.** $y = x - 1$

29. $y = 2x + 1$ **30.** $y = 3x - 1$

31. $y = 3x - 2$ **32.** $y = 2x + 3$

33. $y = x$ **34.** $y = -x$

35. $y = 1 - x$ **36.** $y = 2 - x$

37. $y = -2x + 3$ **38.** $y = -3x + 2$

39. $y = -3$ **40.** $y = 2$

41. $x = 2$ **42.** $x = -4$

43. $2x + y = 5$ **44.** $3x + y = 5$

45. $x + 2y = 4$ **46.** $x - 2y = 6$

47. $x - 3y = 6$ **48.** $x + 4y = 5$

49. $y = 0.36x + 0.4$ **50.** $y = 0.27x - 0.42$

For each point, name the quadrant in which it lies or the axis on which it lies.

51. $(-3, 45)$ **52.** $(-33, 47)$ **53.** $(-3, 0)$

54. $(0, -9)$ **55.** $(-2.36, -5)$ **56.** $(89.6, 0)$

57. $(3.4, 8.8)$ **58.** $(4.1, 44)$ **59.** $\left(-\frac{1}{2}, 50\right)$

60. $\left(-6, -\frac{1}{2}\right)$ **61.** $(0, -99)$ **62.** $(\pi, 0)$

Graph each equation. Plot at least five points for each equation. See Example 5.

63. $y = x + 1200$ **64.** $y = 2x - 3000$

65. $y = 50x - 2000$ **66.** $y = -300x + 4500$

67. $y = -400x + 2000$ **68.** $y = 500x + 3$

Graph each equation using the x- and y-intercepts. See Example 6.

69. $3x + 2y = 6$ **70.** $2x + y = 6$

71. $x - 4y = 4$ **72.** $-2x + y = 4$

73. $y = \frac{3}{4}x - 9$ **74.** $y = -\frac{1}{2}x + 5$

75. $\frac{1}{2}x + \frac{1}{4}y = 1$ **76.** $\frac{1}{3}x - \frac{1}{2}y = 3$

Solve each problem.

77. ***Advertising blitz.*** Furniture City had \$24,000 to spend on advertising a year-end clearance sale. A 30-second radio ad costs \$300, and a 30-second local television ad costs \$400. To model this situation, the advertising manager wrote the equation $300x + 400y = 24{,}000$. What do x and y represent? Graph the equation. How many solutions are there to the equation, given that the number of ads of each type must be a whole number?

78. ***Material allocation.*** A tent maker had 4500 square yards of nylon tent material available. It takes 45 square yards of nylon to make an 8×10 tent and 50 square yards to make a 9×12 tent. To model this situation, the manager wrote the equation $45x + 50y = 4500$. What do x and y represent? Graph the equation. How many solutions are there to the equation, given that the number of tents of each type must be a whole number?

79. ***Percentage of full benefit.*** The age at which you retire affects your Social Security benefits. The accompanying graph gives the percentage of full benefit for each age from 62 through 70, based on current legislation and retirement after the year 2005 (Source: Social Security Administration). What percentage of full benefit does a person receive if that person retires at age 63? At what age will a retiree receive the full benefit? For what ages do you receive more than the full benefit?

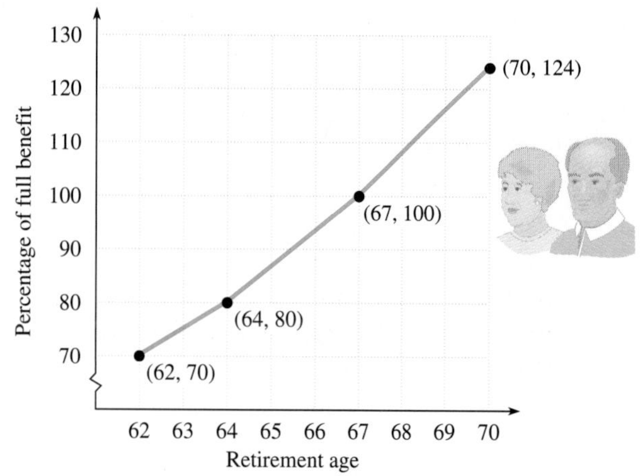

Figure for Exercise 79

80. ***Heel motion.*** When designing running shoes, Chris Edington studies the motion of a runner's foot. The following data gives the coordinates of the heel (in centimeters) at intervals of 0.05 millisecond during one cycle of level treadmill running at 3.8 meters per second *(Sagittal Plane Kinematics, Milliron and Cavanagh)*:

$$(31.7, 5.7), (48.0, 5.7), (68.3, 5.8), (88.9, 6.9), (107.2, 13.3), (119.4, 24.7), (127.2, 37.8), (125.7, 52.0), (116.1, 60.2), (102.2, 59.5), (88.7, 50.2), (73.9, 35.8), (52.6, 20.6), (29.6, 10.7), (22.4, 5.9).$$

Graph these ordered pairs to see the heel motion.

Getting More Involved

81. ***Exploration.*** An electronics repair shop uses the following table to determine the labor charge C (in dollars) for a repair that takes n hours. However, the table has one error in it. Assuming that the correct ordered pairs satisfy a linear equation, find the ordered pair that is in error and correct it. Explain how you knew which ordered pair was in error.

n	1	2	3	4	5
C	70	110	150	180	230

82. ***Writing.*** Find examples of tables of ordered pairs used in real life, and explain what they are used for.

6.2 Slope

In this section:

- Slope Concepts
- Slope Using Coordinates
- Graphing a Line Given a Point and Its Slope
- Parallel Lines
- Perpendicular Lines
- Geometric Applications

In Section 6.1 you learned that the graph of a linear equation is a straight line. In this section we will continue our study of lines in the coordinate plane.

Slope Concepts

If a highway rises 6 feet in a horizontal run of 100 feet, then the grade is $\frac{6}{100}$, or 6%. See Fig. 6.10. The grade of a road is a measurement of the steepness of the road. We measure the steepness of a line in a coordinate system in a similar way.

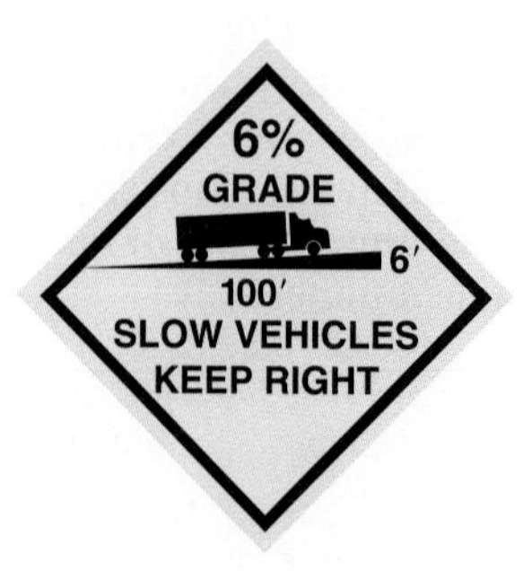

Figure 6.10

The steepness of a line is called the *slope* of the line. To find the slope of the line in Fig. 6.11, imagine that you are to move from (1, 1) to (4, 3) but you cannot travel along the line. You can move horizontally and vertically only. To go from (1, 1) to (4, 3), move three units to the right and then two units upward. There is a change of +2 in the y-coordinate and a change of +3 in the x-coordinate. See Fig. 6.11. The slope of the line is the ratio of these two changes:

$$\text{Slope} = \frac{\text{change in } y\text{-coordinate}}{\text{change in } x\text{-coordinate}} = \frac{+2}{+3} = \frac{2}{3}$$

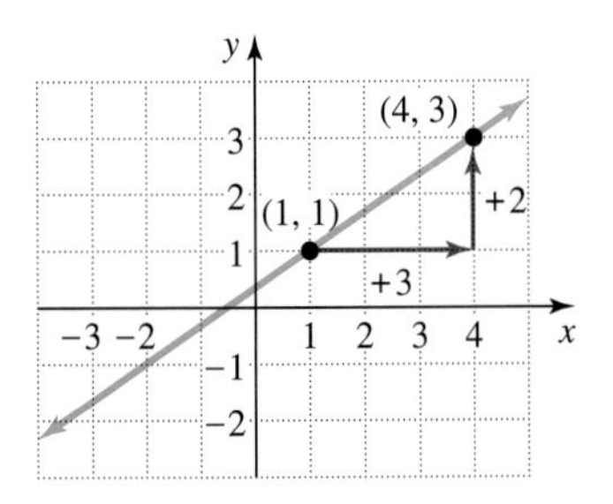

Figure 6.11

So the slope of the line is $\frac{2}{3}$. In general, we have the following definition.

Slope

$$\text{Slope} = \frac{\text{change in } y\text{-coordinate}}{\text{change in } x\text{-coordinate}}$$

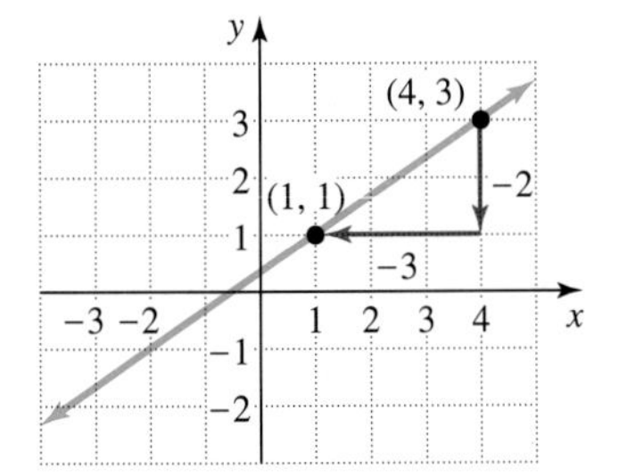

Figure 6.12

If we move from the point $(4, 3)$ to the point $(1, 1)$, there is a change of -2 in the y-coordinate and a change of -3 in the x-coordinate. See Fig. 6.12. In this case we get

$$\text{Slope} = \frac{-2}{-3} = \frac{2}{3}.$$

Note that going from $(4, 3)$ to $(1, 1)$ gives the same slope as going from $(1, 1)$ to $(4, 3)$.

We call the change in y-coordinate the **rise** and the change in x-coordinate the **run.** Moving up is a positive rise, and moving down is a negative rise. Moving to the right is a positive run, and moving to the left is a negative run. We usually use the letter m to stand for slope. So we have

$$m = \frac{\text{change in } y}{\text{change in } x} = \frac{\text{rise}}{\text{run}}.$$

EXAMPLE 1 **Finding the slope of a line**

Find the slopes of the given lines by going from point A to point B.

a)
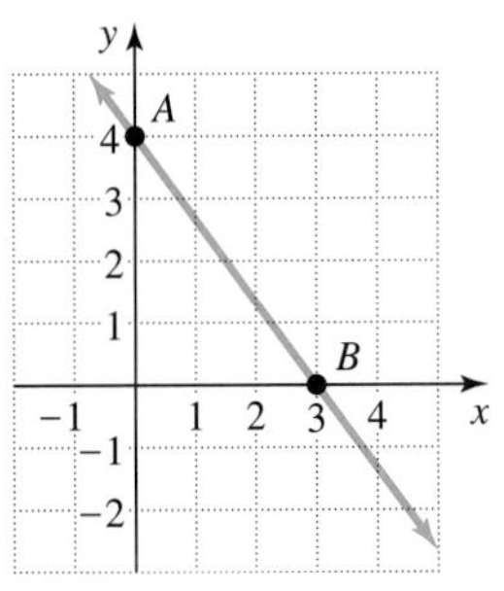

b)
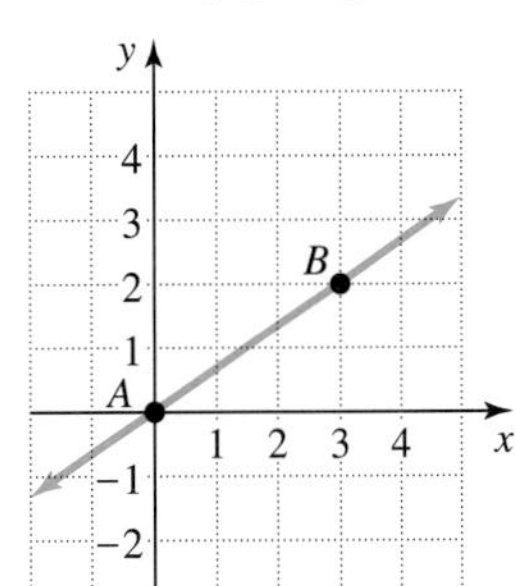

c)
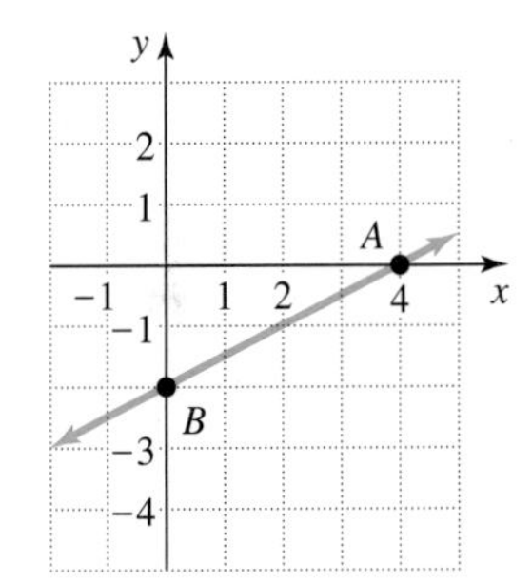

Solution

a) The coordinates of point A are $(0, 4)$, and the coordinates of point B are $(3, 0)$. Going from A to B, the change in y is -4, and the change in x is $+3$. So

$$m = \frac{-4}{3} = -\frac{4}{3}.$$

b) Going from A to B, the rise is 2, and the run is 3. So

$$m = \frac{2}{3}.$$

c) Going from A to B, the rise is -2, and the run is -4. So

$$m = \frac{-2}{-4} = \frac{1}{2}.$$

■

CAUTION The change in y is always in the numerator, and the change in x is always in the denominator.

The ratio of rise to run is the ratio of the lengths of the two legs of any right triangle whose hypotenuse is on the line. As long as one leg is vertical and the other is horizontal, all such triangles for a certain line have the same shape. These triangles are similar triangles. The ratio of the length of the vertical side to the length of the horizontal side for any two such triangles is the same number. So we get the same value for the slope no matter which two points of the line are used to calculate it or in which order the points are used.

EXAMPLE 2 **Finding slope**

Find the slope of the line shown here using points A and B, points A and C, and points B and C.

Solution

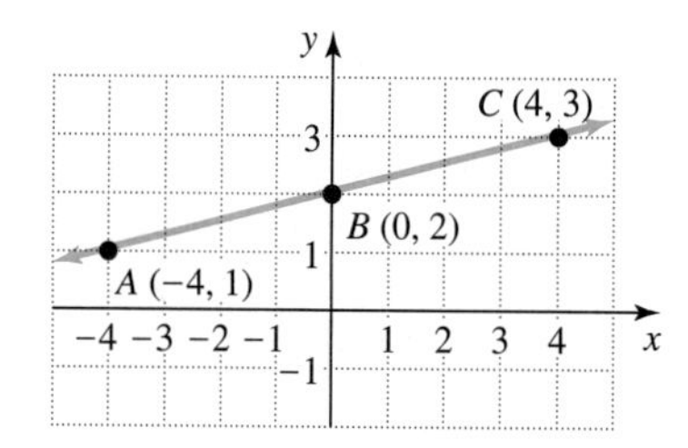

Using A and B, we get

$$m = \frac{\text{rise}}{\text{run}} = \frac{1}{4}.$$

Using A and C, we get

$$m = \frac{\text{rise}}{\text{run}} = \frac{2}{8} = \frac{1}{4}.$$

Using B and C, we get

$$m = \frac{\text{rise}}{\text{run}} = \frac{1}{4}.$$

Slope Using Coordinates

One way to obtain the rise and run is from a graph. The rise and run can also be found by using the coordinates of two points on the line as shown in Fig. 6.13.

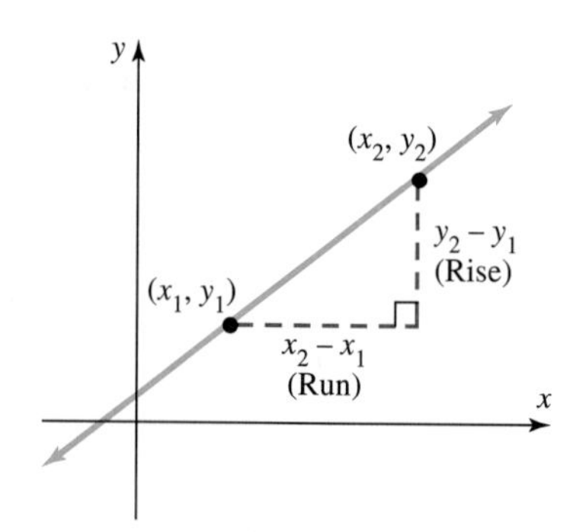

Figure 6.13

Coordinate Formula for Slope

The slope of the line containing the points (x_1, y_1) and (x_2, y_2) is given by

$$m = \frac{y_2 - y_1}{x_2 - x_1},$$

provided that $x_2 - x_1 \neq 0$.

EXAMPLE 3 **Using coordinates to find slope**

Find the slope of each of the following lines.

a) The line through $(0, 5)$ and $(6, 3)$

b) The line through $(-3, 4)$ and $(-5, -2)$

c) The line through $(-4, 2)$ and the origin

Solution

a) Let $(x_1, y_1) = (0, 5)$ and $(x_2, y_2) = (6, 3)$. Which point is called (x_1, y_1) is arbitrary.

$$m = \frac{y_2 - y_1}{x_2 - x_1} = \frac{3 - 5}{6 - 0}$$

$$= \frac{-2}{6} = -\frac{1}{3}$$

b) Let $(x_1, y_1) = (-3, 4)$ and $(x_2, y_2) = (-5, -2)$:

$$m = \frac{y_2 - y_1}{x_2 - x_1} = \frac{-2 - 4}{-5 - (-3)}$$

$$= \frac{-6}{-2} = 3$$

c) Let $(x_1, y_1) = (0, 0)$ and $(x_2, y_2) = (-4, 2)$:

$$m = \frac{2 - 0}{-4 - 0} = \frac{2}{-4} = -\frac{1}{2}$$

■

CAUTION It does not matter which point is called (x_1, y_1) and which is called (x_2, y_2), but if you divide $y_2 - y_1$ by $x_1 - x_2$, the slope will have the wrong sign. ⊘

Note that slope is not defined if $x_2 - x_1 = 0$. So slope is not defined if the x-coordinates of the two points are equal. The x-coordinates for two points are equal only for points on a vertical line. So *slope is undefined for vertical lines.* See Fig. 6.14.

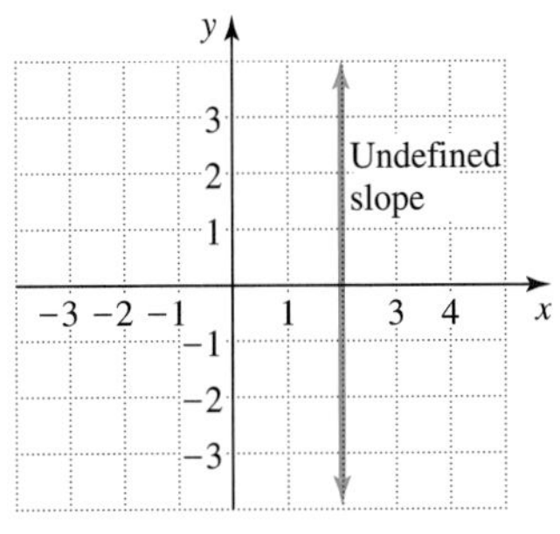

Vertical line

Figure 6.14

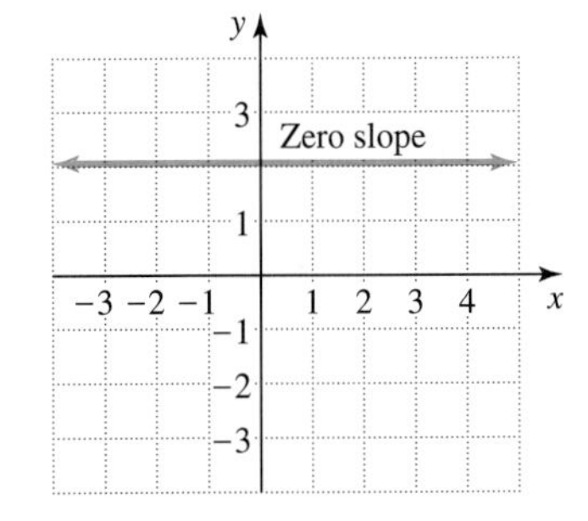

Horizontal line

Figure 6.15

If $y_2 - y_1 = 0$, then the points have equal y-coordinates and lie on a horizontal line. *The slope for any horizontal line is zero.* See Fig. 6.15.

Note that for a line with *positive slope*, the y-values increase as the x-values increase. For a line with *negative slope*, the y-values decrease as the x-values increase. See Fig. 6.16.

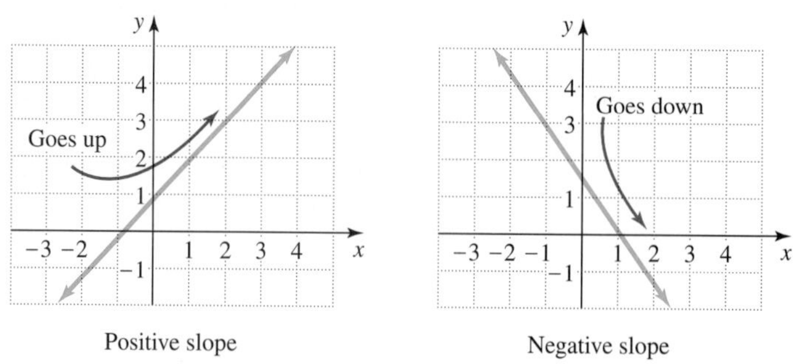

Figure 6.16

Graphing a Line Given a Point and Its Slope

We can find the slope of a line by examining its graph. We can also draw the graph of a line if we know its slope and a point on the line.

EXAMPLE 4 **Graphing a line given a point and its slope**

Graph each line.

a) The line through $(2, 1)$ with slope $\frac{3}{4}$

b) The line through $(-2, 4)$ with slope -3

Solution

a) First locate the point $(2, 1)$. Because the slope is $\frac{3}{4}$, we can find another point on the line by going up three units and to the right four units to get the point $(6, 4)$, as shown in Fig. 6.17. Now draw the line through $(2, 1)$ and $(6, 4)$.

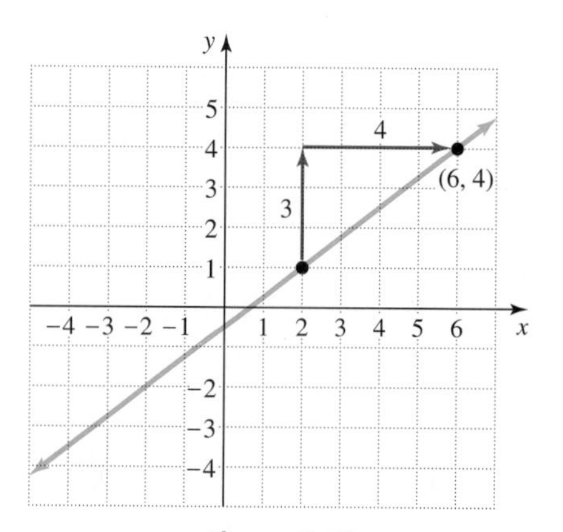

Figure 6.17

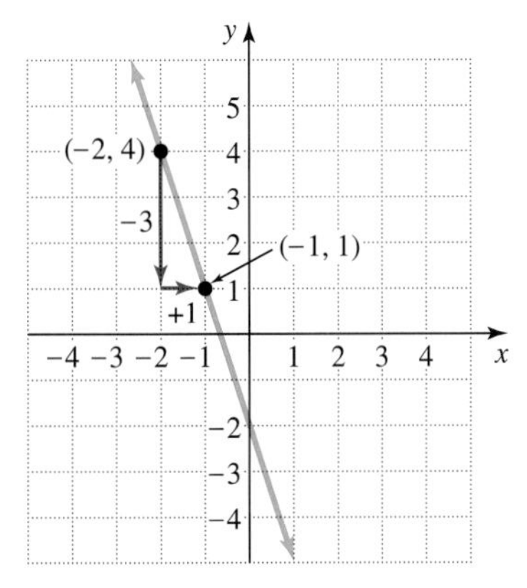

Figure 6.18

b) First locate the point $(-2, 4)$. Because the slope is -3, or $\frac{-3}{1}$, we can locate another point on the line by starting at $(-2, 4)$ and moving down three units and then one unit to the right to get the point $(-1, 1)$. Now draw a line through $(-2, 4)$ and $(-1, 1)$, as shown in Fig. 6.18. ■

Parallel Lines

Every nonvertical line has a unique slope, but there are infinitely many lines with a given slope. All lines that have a given slope are parallel.

> **Parallel Lines**
>
> Nonvertical lines are parallel if and only if they have equal slopes. Any two vertical lines are parallel to each other.

EXAMPLE 5 **Graphing parallel lines**

Draw a line through the point $(-2, 1)$ with slope $\frac{1}{2}$ and a line through $(3, 0)$ with slope $\frac{1}{2}$.

Solution

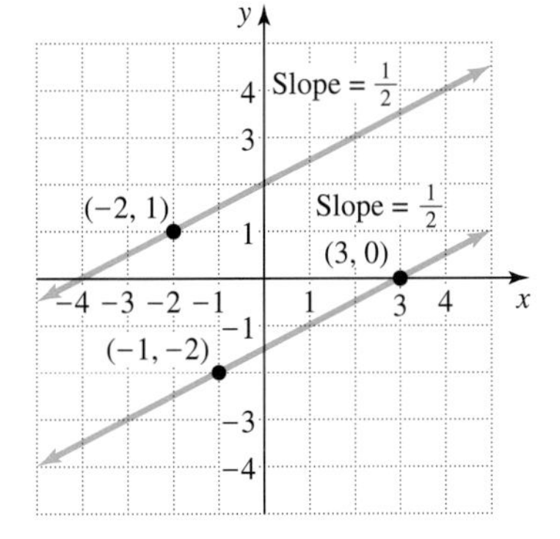

Figure 6.19

Because slope is the ratio of rise to run, a slope of $\frac{1}{2}$ means that we can locate a second point of the line by starting at $(-2, 1)$ and going up one unit and to the right two units. For the line through $(3, 0)$ we start at $(3, 0)$ and go up one unit and to the right two units. See Fig. 6.19. ■

Perpendicular Lines

Slope can also be used to determine whether lines are perpendicular. If the slope of one line is the opposite of the reciprocal of the slope of another line, then the lines are perpendicular. For example, lines with slopes $\frac{3}{4}$ and $-\frac{4}{3}$ are perpendicular.

> **Perpendicular Lines**
>
> Two lines with slopes m_1 and m_2 are perpendicular if and only if
>
> $$m_1 = -\frac{1}{m_2}.$$
>
> Any vertical line is perpendicular to any horizontal line.

EXAMPLE 6 **Graphing perpendicular lines**

Draw two lines through the point $(-1, 2)$, one with slope $-\frac{1}{3}$ and the other with slope 3.

Solution

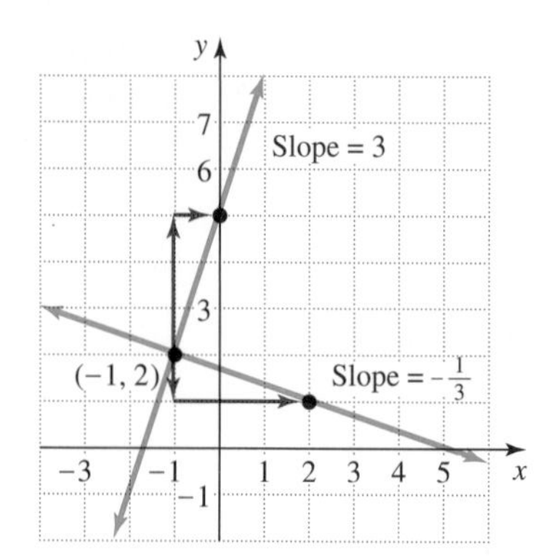

Figure 6.20

Because slope is the ratio of rise to run, a slope of $-\frac{1}{3}$ means that we can locate a second point on the line by starting at $(-1, 2)$ and going down one unit and to the right three units. For the line with slope 3, we start at $(-1, 2)$ and go up three units and to the right one unit. See Fig. 6.20. ■

Geometric Applications

If the opposite sides of a four-sided figure are parallel, then the figure is a **parallelogram.** If the adjacent sides of a parallelogram are perpendicular, then the figure is a **rectangle.**

EXAMPLE 7 **A rectangle**

Use slope to determine whether the four points $(-4, 1)$, $(-3, -3)$, $(5, -1)$, and $(4, 3)$ are the vertices of a rectangle.

Solution

First we sketch the figure determined by these points. See Fig. 6.21. It appears to be a rectangle, but to prove that it is, we find the slope of each side:

$$m_{AB} = \frac{1 - (-3)}{-4 - (-3)} = \frac{4}{-1} = -4$$

$$m_{BC} = \frac{-3 - (-1)}{-3 - 5} = \frac{-2}{-8} = \frac{1}{4}$$

$$m_{CD} = \frac{-1 - 3}{5 - 4} = \frac{-4}{1} = -4$$

$$m_{AD} = \frac{1 - 3}{-4 - 4} = \frac{-2}{-8} = \frac{1}{4}$$

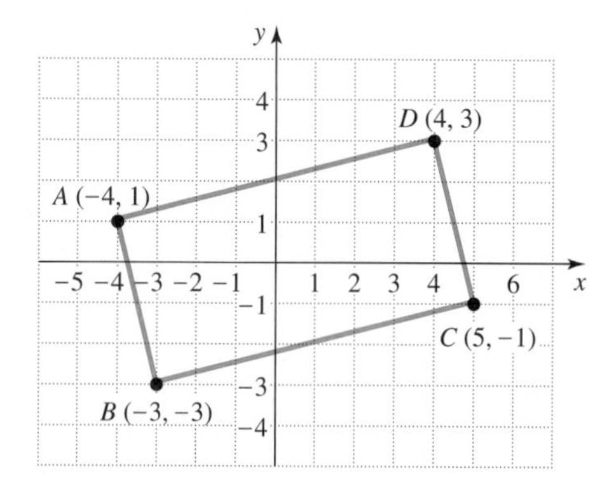

Figure 6.21

Because the opposite sides have the same slope, the opposite sides are parallel, and the figure is a parallelogram. Because -4 is the opposite of the reciprocal of $\frac{1}{4}$, the adjacent sides are perpendicular and the figure is a rectangle. ■

Warm-ups

True or false? Explain your answer.

1. Slope is a measurement of the steepness of a line.
2. Slope is rise divided by run.
3. Every line in the coordinate plane has a slope.
4. The line through the point $(1, 1)$ and the origin has slope 1.
5. Slope can never be negative.
6. A line with slope 2 is perpendicular to any line with slope -2.
7. The slope of the line through $(0, 3)$ and $(4, 0)$ is $\frac{3}{4}$.
8. Two different lines cannot have the same slope.
9. The line through $(1, 3)$ and $(-5, 3)$ has zero slope.
10. A quadrilateral whose opposite sides are parallel is a parallelogram.

6.2 EXERCISES

In Exercises 1–12, find the slope of each line. See Examples 1 and 2.

1.

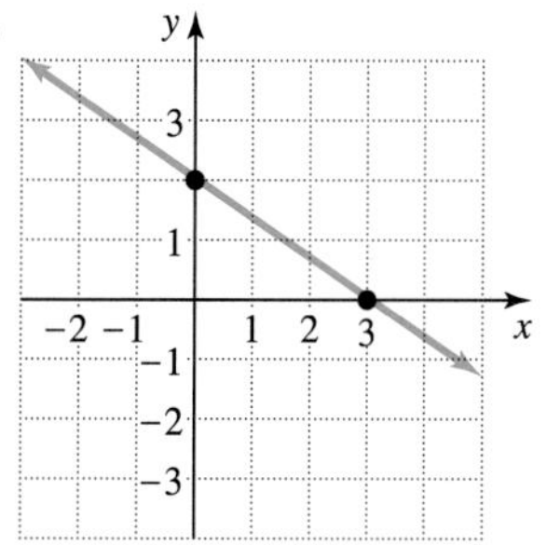

2.

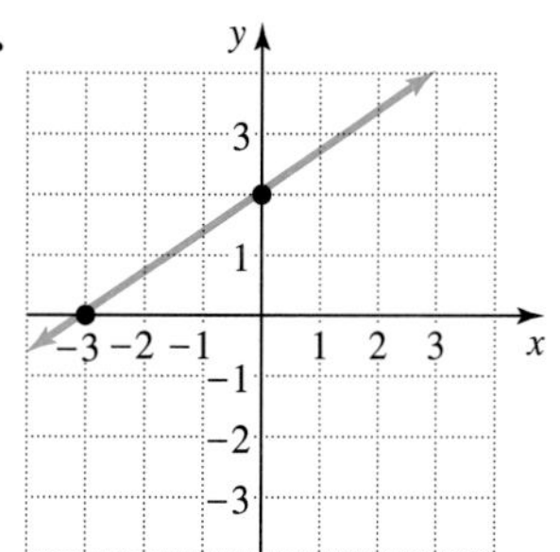

3.

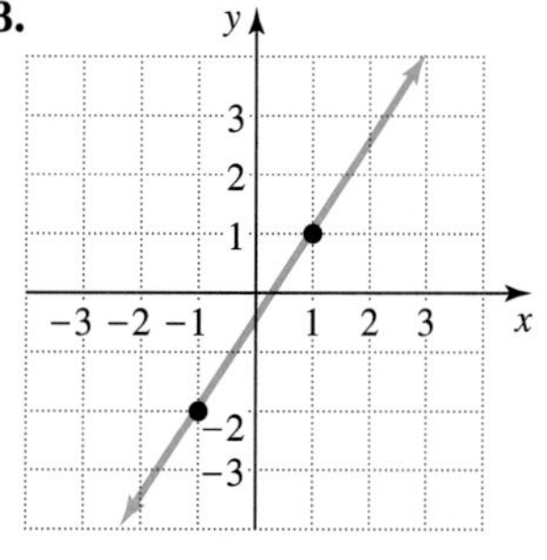

4.

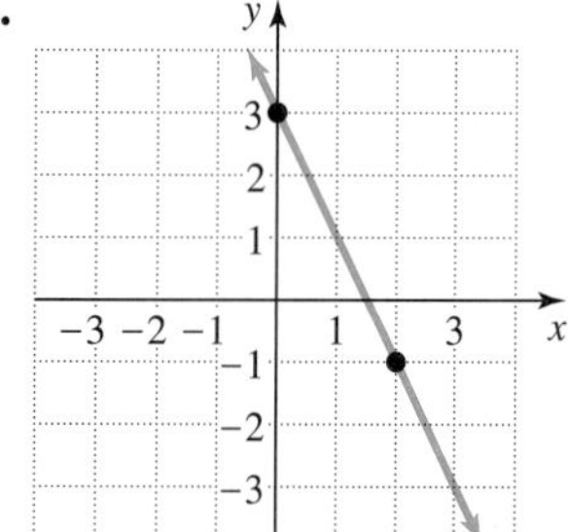

5.

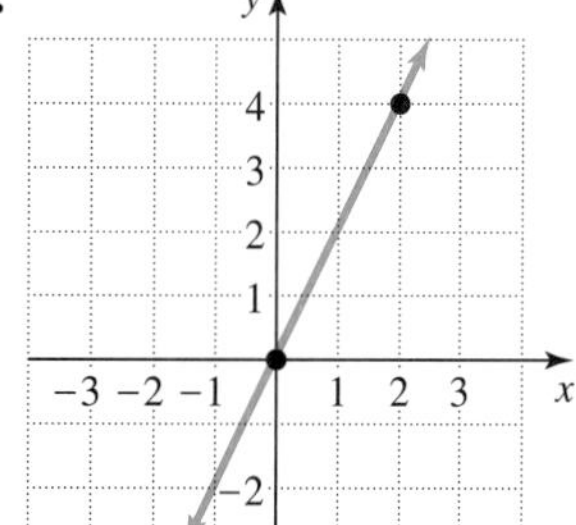

6.

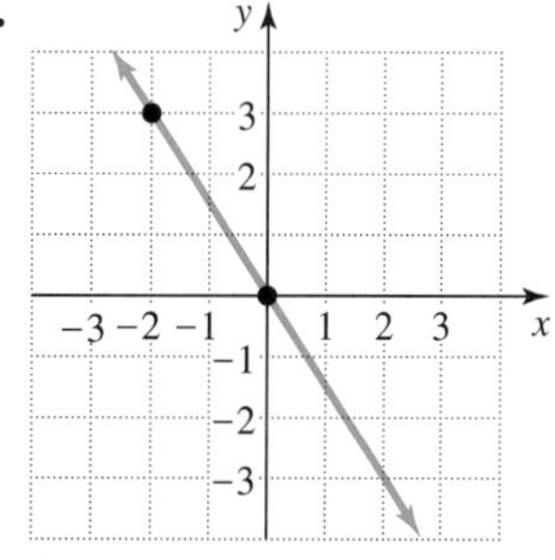

7.

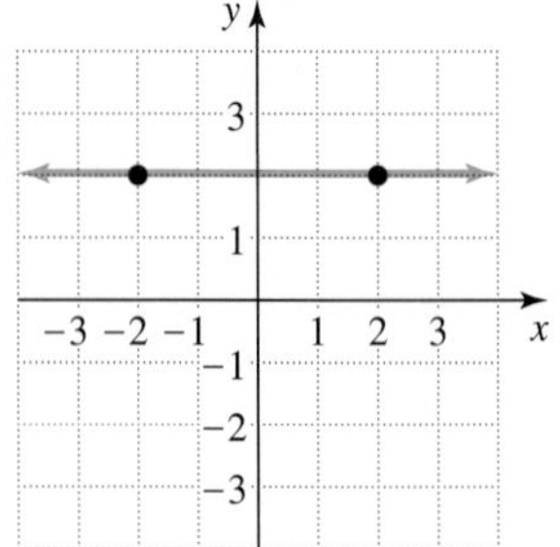

8.

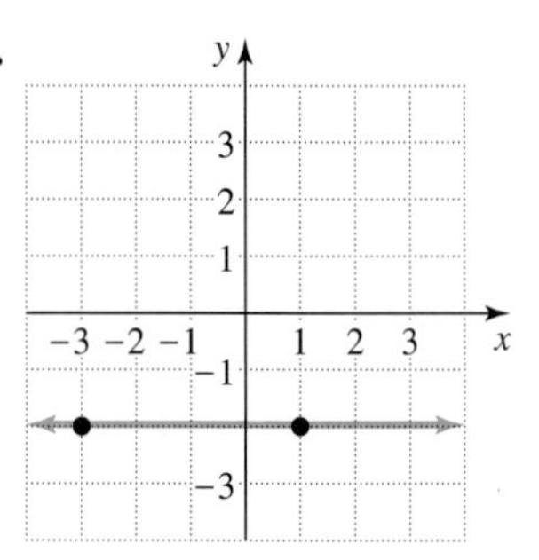

9.

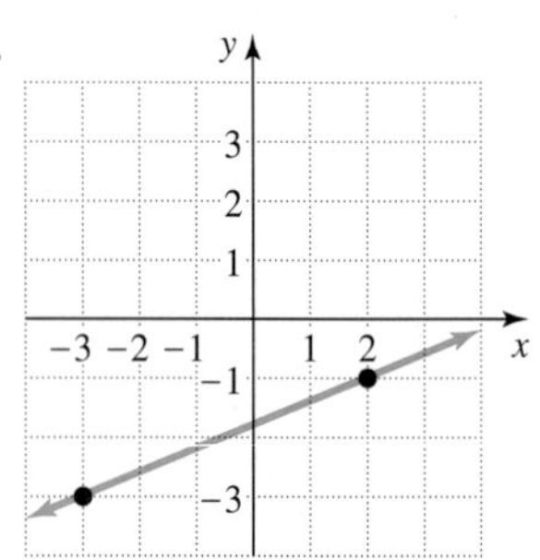

10.

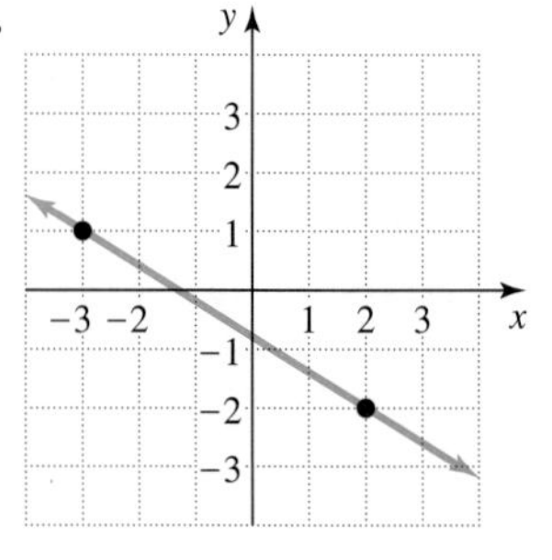

11.

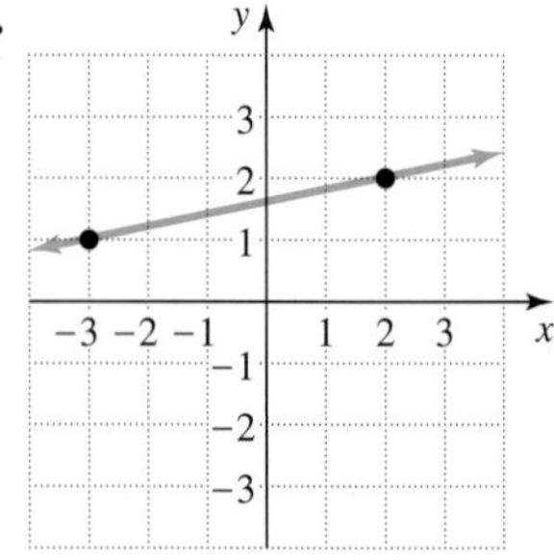

12.

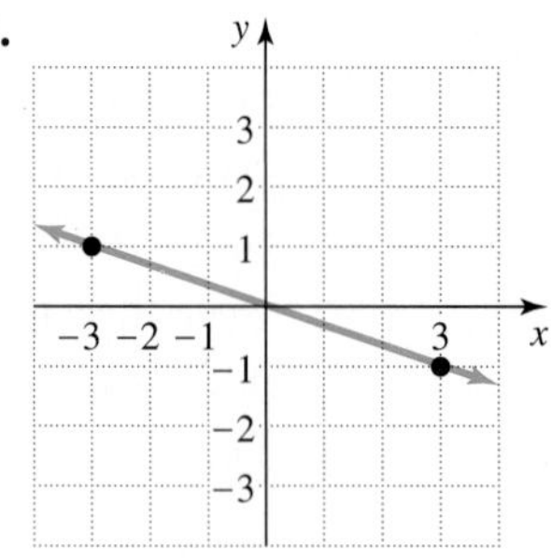

Find the slope of the line that goes through each pair of points. See Example 3.

13. $(1, 2), (3, 6)$

14. $(2, 5), (6, 10)$

15. $(2, 4), (5, -1)$

16. $(3, 1), (6, -2)$

17. $(-2, 4), (5, 9)$

18. $(-1, 3), (3, 5)$

19. $(-2, -3), (-5, 1)$

20. $(-6, -3), (-1, 1)$

21. $(-3, 4), (3, -2)$

22. $(-1, 3), (5, -2)$

23. $\left(\frac{1}{2}, 2\right), \left(-1, \frac{1}{2}\right)$

24. $\left(\frac{1}{3}, 2\right), \left(-\frac{1}{3}, 1\right)$

25. $(2, 3), (2, -9)$

26. $(-3, 6), (8, 6)$

27. $(-2, -5), (9, -5)$

28. $(4, -9), (4, 6)$

29. $(0.3, 0.9), (-0.1, -0.3)$

30. $(-0.1, 0.2), (0.5, 0.8)$

Graph the line with the given point and slope. See Example 4.

31. The line through $(1, 1)$ with slope $\frac{2}{3}$

32. The line through $(2, 3)$ with slope $\frac{1}{2}$

33. The line through $(-2, 3)$ with slope -2

34. The line through $(-2, 5)$ with slope -1

35. The line through $(0, 0)$ with slope $-\frac{2}{5}$

36. The line through $(-1, 4)$ with slope $-\frac{2}{3}$

Solve each problem. See Examples 5 and 6.

37. Draw line l_1 through $(1, -2)$ with slope $\frac{1}{2}$ and line l_2 through $(-1, 1)$ with slope $\frac{1}{2}$.

38. Draw line l_1 through $(0, 3)$ with slope 1 and line l_2 through $(0, 0)$ with slope 1.

39. Draw l_1 through $(1, 2)$ with slope $\frac{1}{2}$, and draw l_2 through $(1, 2)$ with slope -2.

40. Draw l_1 through $(-2, 1)$ with slope $\frac{2}{3}$, and draw l_2 through $(-2, 1)$ with slope $-\frac{3}{2}$.

41. Draw any line l_1 with slope $\frac{3}{4}$. What is the slope of any line perpendicular to l_1? Draw any line l_2 perpendicular to l_1.

42. Draw any line l_1 with slope -1. What is the slope of any line perpendicular to l_1? Draw any line l_2 perpendicular to l_1.

43. Draw l_1 through $(-2, -3)$ and $(4, 0)$. What is the slope of any line parallel to l_1? Draw l_2 through $(1, 2)$ so that it is parallel to l_1.

44. Draw l_1 through $(-4, 0)$ and $(0, 6)$. What is the slope of any line parallel to l_1? Draw l_2 through the origin and parallel to l_1.

45. Draw l_1 through $(-2, 4)$ and $(3, -1)$. What is the slope of any line perpendicular to l_1? Draw l_2 through $(1, 3)$ so that it is perpendicular to l_1.

46. Draw l_1 through $(0, -3)$ and $(3, 0)$. What is the slope of any line perpendicular to l_1? Draw l_2 through the origin so that it is perpendicular to l_1.

Use slope to solve each problem. See Example 7.

47. Show that the points $(-3, 2)$, $(2, 3)$, $(3, 0)$, and $(-2, -1)$ are the vertices of a parallelogram.

48. Show that the points $(-3, 2)$, $(0, 1)$, $(1, 0)$, and $(-2, 1)$ are the vertices of a parallelogram.

49. Show that the points $(0, 1)$, $(-2, 0)$, $(1, -1)$, and $(-1, -2)$ are the vertices of a rectangle.

50. Show that the points $(-4, 1)$, $(2, 3)$, $(3, 0)$, and $(-3, -2)$ are the vertices of a rectangle.

51. Show that the points $(-4, -3)$, $(-3, -4)$, and $(5, 4)$ are the vertices of a right triangle.

52. Show that the points $(-4, -3)$, $(-3, -4)$, and $(1, 2)$ are the vertices of a right triangle.

53. Determine whether the points $(-3, -8)$, $(4, 9)$, and $(1, 2)$ are all on the same straight line.

54. Determine whether the points $(2, -7)$, $(0, -1)$, and $(-3, 8)$ are all on the same straight line.

Solve each problem.

55. ***Super cost.*** The average cost of an ad during the Super Bowl in 1990 was \$700,000; in 1995 it was \$1 million (*USA Today*, November 3, 1994). Find the slope of the line through the points (90, 700,000) and (95, 1,000,000), and interpret your result. Use the bar graph to estimate the average cost of an ad in 1993. What do you think the average cost will be in the year 2000?

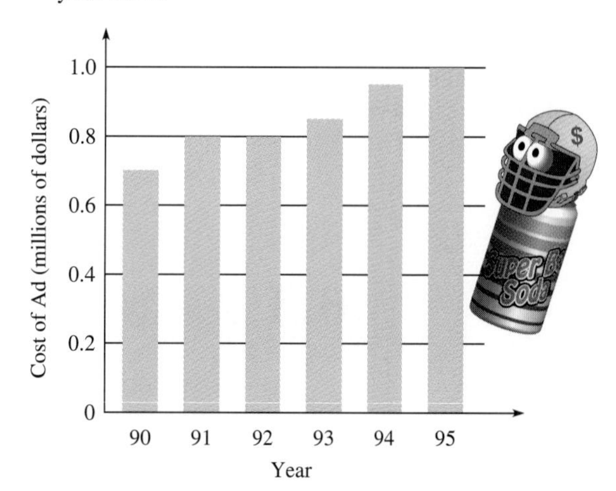

Figure for Exercise 55

56. ***Retirement pay.*** The annual Social Security benefit of a retiree depends on the age at the time of retirement. The accompanying graph gives the annual benefit for persons retiring at ages 62 through 70 in the year 2005 or later (Source: Social Security Administration). What is the annual benefit for a person who retires at age 64? At what retirement age does a person receive an annual benefit of \$11,600? Find the slope of each line segment on the graph, and interpret your results. Why do people who postpone retirement until 70 years of age get the highest benefit?

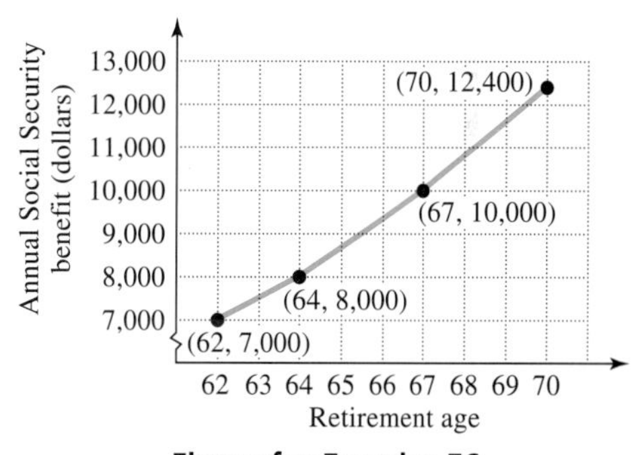

Figure for Exercise 56

Getting More Involved

57. ***Writing.*** Explain how builders measure steepness of a roof in terms of pitch. What are the most common pitches used in roof construction? How is pitch related to slope? What is the pitch of the roof that you live under?

58. ***Cooperative learning.*** According to the 1990 Americans with Disabilities Act, all new or substantially renovated public buildings must be accessible to physically challenged people. Working in a group, locate three wheelchair ramps in your local area and measure their slopes. What do you think is the maximum slope for a wheelchair ramp?

59. ***Exploration.*** The following is a table of ordered pairs that satisfy a linear equation. However, the table has one error in it. Find the ordered pair that is in error and correct it. Explain how you found the error. Can you think of another way to find the error?

x	-2	3	1	-1	0	2
y	8	-7	-1	5	3	-4

6.3 Equations of Lines

In this section:

- Slope-Intercept Form
- Standard Form
- Using Slope-Intercept Form for Graphing
- Writing the Equation for a Line

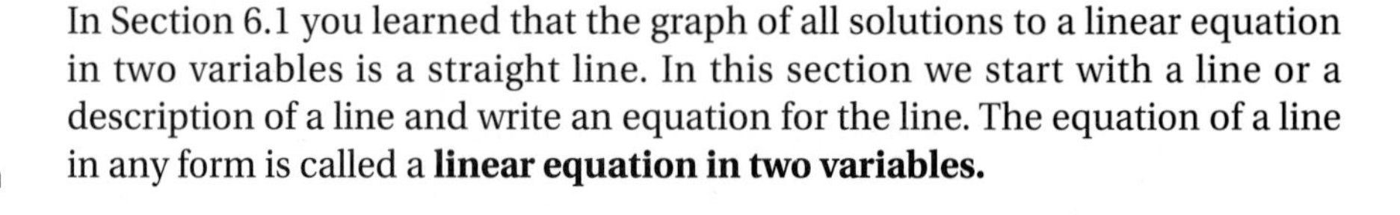

In Section 6.1 you learned that the graph of all solutions to a linear equation in two variables is a straight line. In this section we start with a line or a description of a line and write an equation for the line. The equation of a line in any form is called a **linear equation in two variables.**

Slope-Intercept Form

Consider the line through $(0, 1)$ with slope $\frac{2}{3}$ shown in Fig. 6.22. If we use the points (x, y) and $(0, 1)$ in the slope formula, we get an equation that is satisfied by every point on the line:

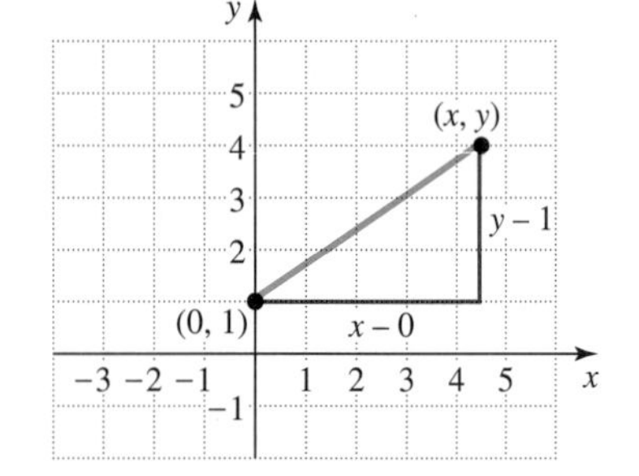

Figure 6.22

$$\frac{y_2 - y_1}{x_2 - x_1} = m \quad \text{Slope formula}$$

$$\frac{y - 1}{x - 0} = \frac{2}{3} \quad \text{Let } (x_1, y_1) = (0, 1) \text{ and } (x_2, y_2) = (x, y).$$

$$\frac{y - 1}{x} = \frac{2}{3}$$

Now solve the equation for y:

$$x \cdot \frac{y - 1}{x} = \frac{2}{3} \cdot x \quad \text{Multiply each side by } x.$$

$$y - 1 = \frac{2}{3}x$$

$$y = \frac{2}{3}x + 1 \quad \text{Add 1 to each side.}$$

Because $(0, 1)$ is on the y-axis, it is called the ***y*-intercept** of the line. Note how the slope $\frac{2}{3}$ and the y-coordinate of the y-intercept $(0, 1)$ appear in $y = \frac{2}{3}x + 1$. For this reason it is called the **slope-intercept form** of the equation of the line.

Slope-Intercept Form

The equation of the line with y-intercept $(0, b)$ and slope m is

$$y = mx + b.$$

EXAMPLE 1 **Using slope-intercept form**

Write the equation of each line in slope-intercept form.

a)

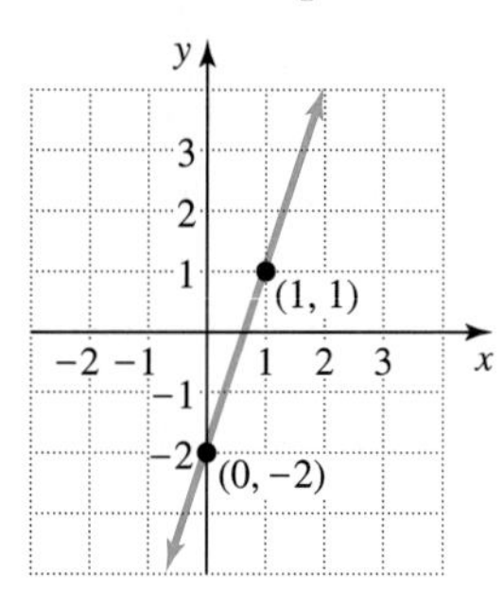

b)

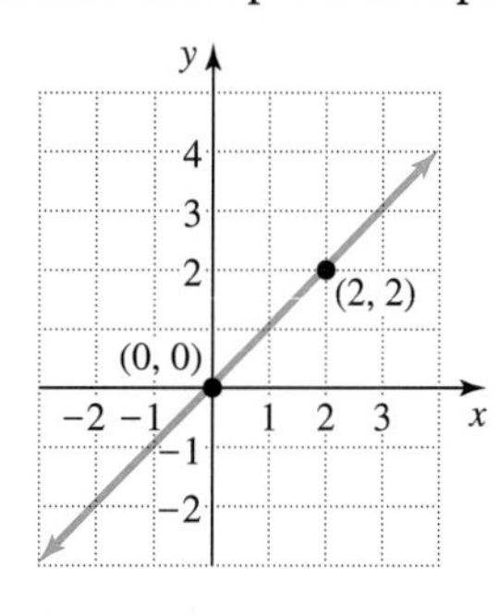

c)

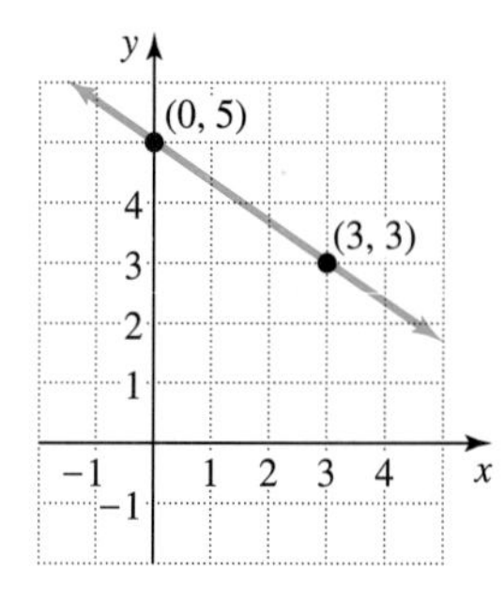

Solution

a) The y-intercept is $(0, -2)$, and the slope is 3. Use the form $y = mx + b$ with $b = -2$ and $m = 3$. The equation in slope-intercept form is

$$y = 3x - 2.$$

b) The y-intercept is $(0, 0)$, and the slope is 1. So the equation is

$$y = x.$$

c) The y-intercept is $(0, 5)$, and the slope is $-\frac{2}{3}$. So the equation is

$$y = -\frac{2}{3}x + 5.$$

■

The equation of a line may take many different forms. The easiest way to find the slope and y-intercept for a line is to rewrite the equation in slope-intercept form.

EXAMPLE 2 **Finding slope and y-intercept**

Determine the slope and y-intercept of the line $3x - 2y = 6$.

Solution

Solve for y to get slope-intercept form:

$$3x - 2y = 6$$
$$-2y = -3x + 6$$
$$y = \frac{3}{2}x - 3$$

The slope is $\frac{3}{2}$, and the y-intercept is $(0, -3)$. ■

Standard Form

The graph of the equation $x = 3$ is a vertical line. Because slope is not defined for vertical lines, this line does not have an equation in slope-intercept form. Only nonvertical lines have equations in slope-intercept form. However, there is a form that includes all lines. It is called **standard form.**

> **Standard Form**
>
> Every line has an equation in the form
>
> $$Ax + By = C$$
>
> where A, B, and C are real numbers with A and B not both zero.

To write the equation $x = 3$ in this form, let $A = 1$, $B = 0$, and $C = 3$. We get

$$1 \cdot x + 0 \cdot y = 3,$$

which is equivalent to

$$x = 3.$$

In Example 2 we converted an equation in standard form to slope-intercept form. Any linear equation in standard form with $B \neq 0$ can be written in slope-intercept form by solving for y. In the next example we convert an equation in slope-intercept form to standard form.

EXAMPLE 3 **Converting to standard form**

Write the equation of the line $y = \frac{2}{5}x + 3$ in standard form using only integers.

Solution

To get standard form, first subtract $\frac{2}{5}x$ from each side:

$$y = \frac{2}{5}x + 3$$

$$-\frac{2}{5}x + y = 3$$

$$-5\left(-\frac{2}{5}x + y\right) = -5 \cdot 3 \quad \text{Multiply each side by } -5 \text{ to eliminate the fraction and get positive } 2x.$$

$$2x - 5y = -15$$

The answer $2x - 5y = -15$ in Example 3 is not the only answer using only integers. Equations such as $-2x + 5y = 15$ and $4x - 10y = -30$ are equivalent equations in standard form. We prefer to write $2x - 5y = -15$ because the greatest common factor of 2, 5, and 15 is 1 and the coefficient of x is positive.

Using Slope-Intercept Form for Graphing

One way to graph a linear equation is to find several points that satisfy the equation and then draw a straight line through them. We can also graph a linear equation by using the y-intercept and the slope.

Strategy for Graphing a Line Using Slope and y-Intercept

1. Write the equation in slope-intercept form if necessary.
2. Plot the y-intercept.
3. Starting from the y-intercept, use the rise and run to locate a second point.
4. Draw a line through the two points.

EXAMPLE 4 **Graphing a line using y-intercept and slope**

Graph the line $2x - 3y = 3$.

Solution

First write the equation in slope-intercept form:

$$2x - 3y = 3$$

$$-3y = -2x + 3 \quad \text{Subtract } 2x \text{ from each side.}$$

$$y = \frac{2}{3}x - 1 \quad \text{Divide each side by } -3.$$

The slope is $\frac{2}{3}$, and the y-intercept is $(0, -1)$. A slope of $\frac{2}{3}$ means a rise of 2 and a run of 3. Start at $(0, -1)$ and go up two units and to the right three units to locate a second point on the line. Now draw a line through the two points. See Fig. 6.23 for the graph of $2x - 3y = 3$. ■

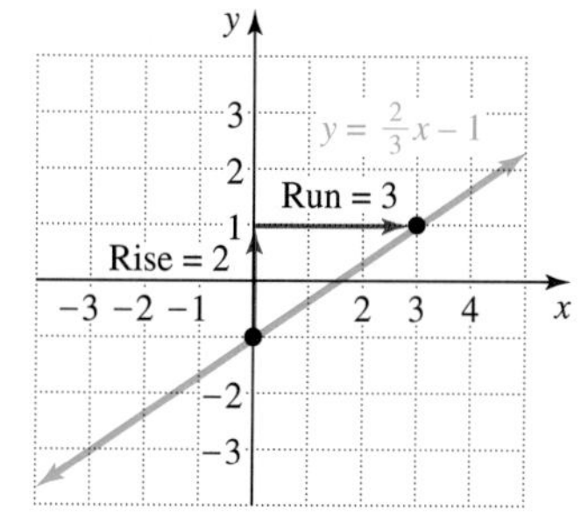

Figure 6.23

CAUTION When using the slope to find a second point on the line, be sure to start at the y-intercept, not at the origin.

EXAMPLE 5 **Graphing a line using y-intercept and slope**

Graph the line $y = -3x + 4$.

Solution

The slope is -3, and the y-intercept is $(0, 4)$. Because $-3 = \frac{-3}{1}$, we use a rise of -3 and a run of 1. To locate a second point on the line, start at $(0, 4)$ and go down three units and to the right one unit. Draw a line through the two points. See Fig. 6.24. ■

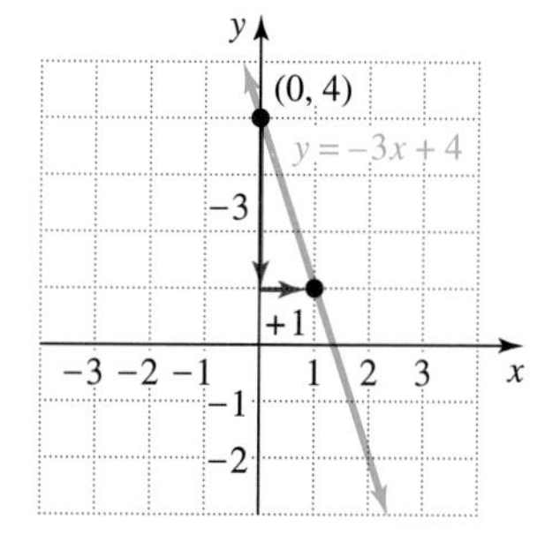

Figure 6.24

Writing the Equation for a Line

In Example 1 we wrote the equation of a line by finding its slope and y-intercept from a graph. In the next example we write the equation of a line from a description of the line.

EXAMPLE 6 **Writing an equation**

Write the equation in slope-intercept form for the line through $(0, 4)$ that is perpendicular to the line $2x - 4y = 1$.

Solution

First find the slope of $2x - 4y = 1$:

$$2x - 4y = 1$$
$$-4y = -2x + 1$$
$$y = \frac{1}{2}x - \frac{1}{4} \quad \text{The slope of this line is } \tfrac{1}{2}.$$

The slope of the line that we are interested in is the opposite of the reciprocal of $\frac{1}{2}$.

So the line has slope -2 and y-intercept $(0, 4)$. Its equation is $y = -2x + 4$. ■

Warm-ups

True or false? Explain your answer.

1. There is only one line with y-intercept $(0, 3)$ and slope $-\frac{4}{3}$.
2. The equation of the line through $(1, 2)$ with slope 3 is $y = 3x + 2$.
3. The vertical line $x = -2$ has no y-intercept.
4. The equation $x = 5$ has a graph that is a vertical line.
5. The line $y = x - 3$ is perpendicular to the line $y = 5 - x$.
6. The line $y = 2x - 3$ is parallel to the line $y = 4x - 3$.
7. The line $2y = 3x - 8$ has a slope of 3.
8. Every straight line in the coordinate plane has an equation in standard form.
9. The line $x = 2$ is perpendicular to the line $y = 5$.
10. The line $y = x$ has no y-intercept.

6.3 EXERCISES

Write an equation for each line. Use slope-intercept form if possible. See Example 1.

1.

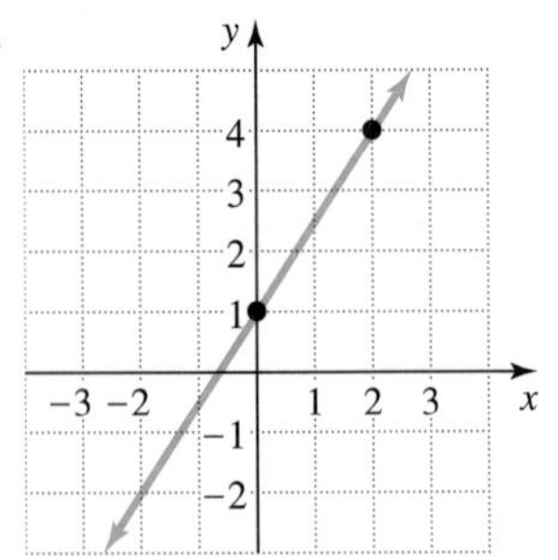

2.

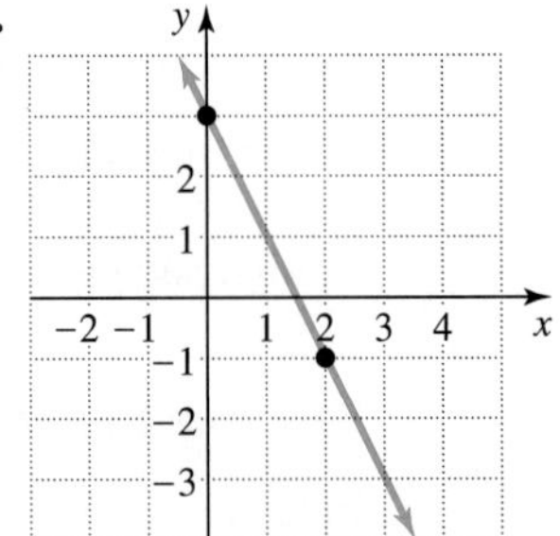

3.

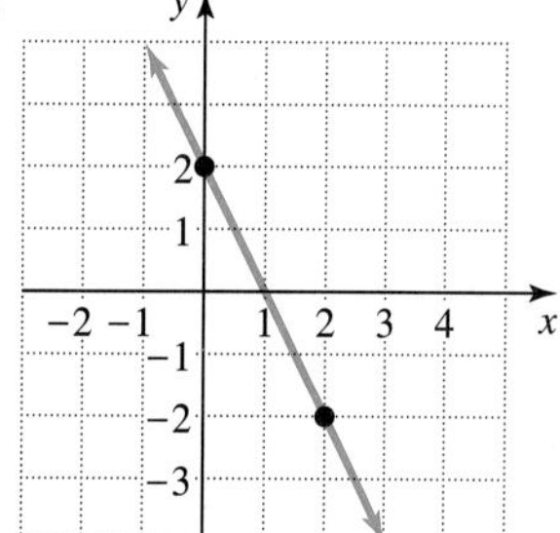

4.

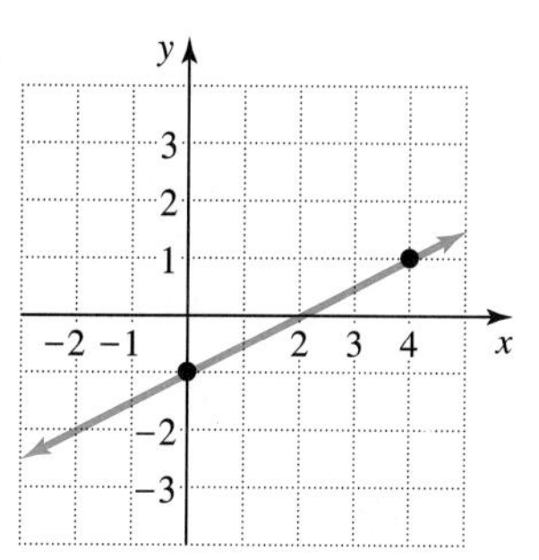

5.

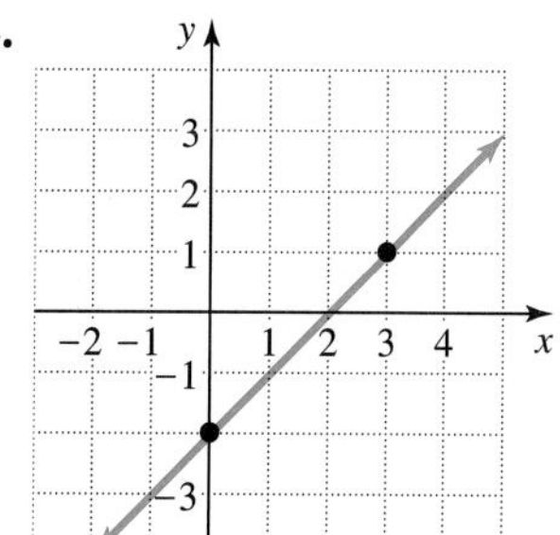

6.

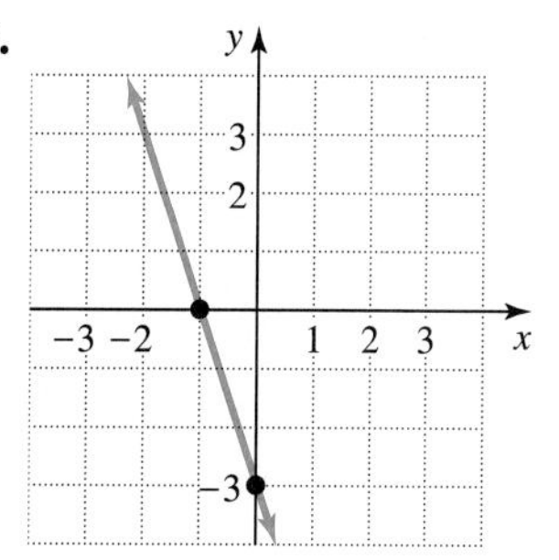

7.

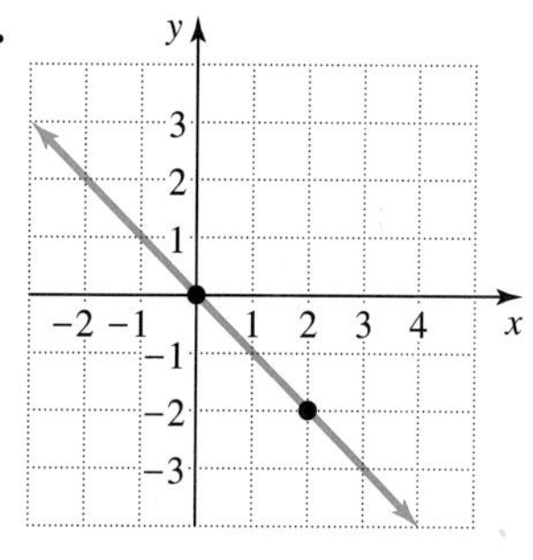

8.

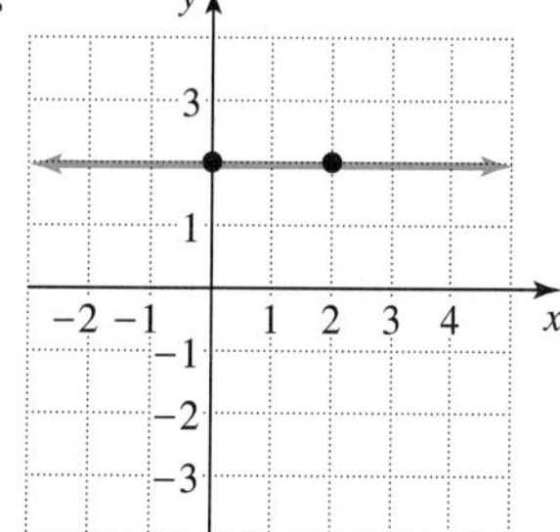

9.

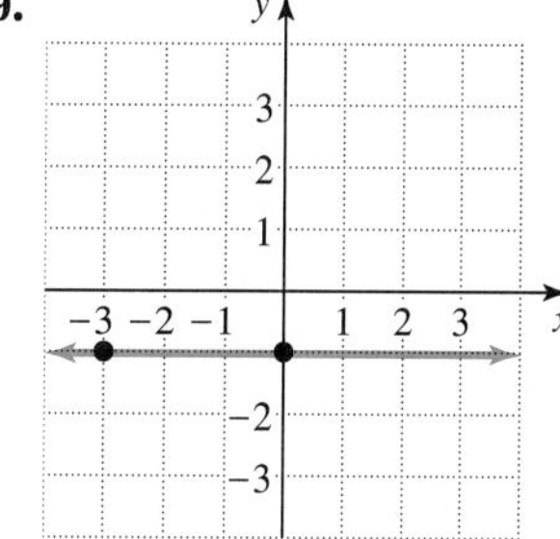

10.

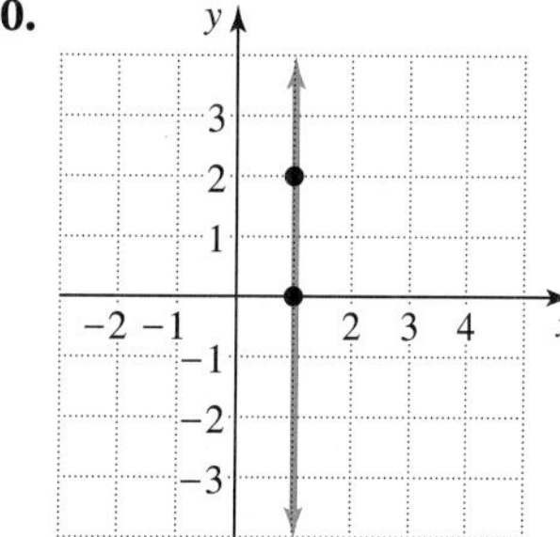

11.

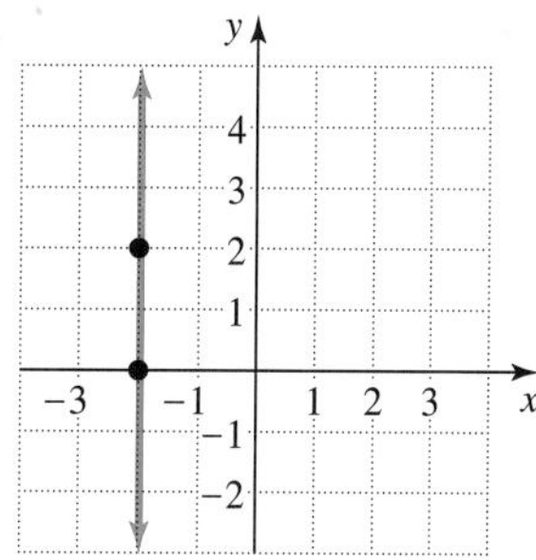

12.

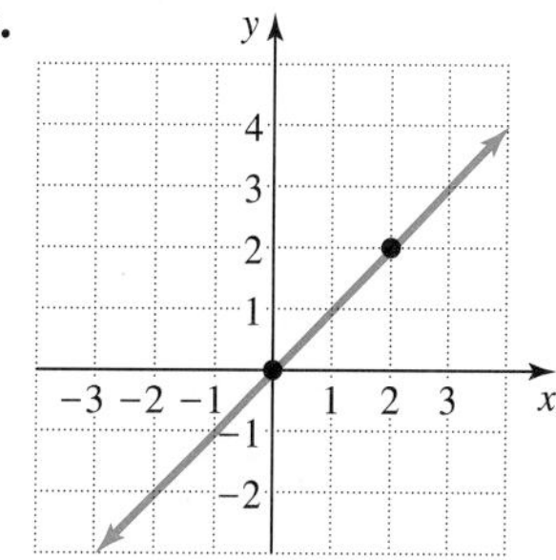

Find the slope and y-intercept for each line that has a slope and y-intercept. See Example 2.

13. $y = 3x - 9$

14. $y = -5x + 4$

15. $y = 4$

16. $y = -5$

17. $y = -3x$

18. $y = 2x$

19. $x + y = 5$

20. $x - y = 4$

21. $x - 2y = 4$

22. $x + 2y = 3$

23. $2x - 5y = 10$

24. $2x + 3y = 9$

25. $2x - y + 3 = 0$

26. $3x - 4y - 8 = 0$

27. $x = -3$

28. $\frac{2}{3}x = 4$

Write each equation in standard form using only integers. See Example 3.

29. $y = -x + 2$

30. $y = 3x - 5$

31. $y = \frac{1}{2}x + 3$

32. $y = \frac{2}{3}x - 4$

33. $y = \frac{3}{2}x - \frac{1}{3}$

34. $y = \frac{4}{5}x + \frac{2}{3}$

35. $y = -\frac{3}{5}x + \frac{7}{10}$

36. $y = -\frac{2}{3}x - \frac{5}{6}$

37. $\frac{3}{5}x + 6 = 0$

38. $\frac{1}{2}x - 9 = 0$

39. $\frac{3}{4}y = \frac{5}{2}$

40. $\frac{2}{3}y = \frac{1}{9}$

41. $\frac{x}{2} = \frac{3y}{5}$

42. $\frac{x}{8} = -\frac{4y}{5}$

43. $y = 0.02x + 0.5$

44. $0.2x = 0.03y - 0.1$

Draw the graph of each line using its y-intercept and its slope. See Examples 4 and 5.

45. $y = 2x - 1$

46. $y = 3x - 2$

47. $y = -3x + 5$

48. $y = -4x + 1$

49. $y = \frac{3}{4}x - 2$

50. $y = \frac{3}{2}x - 4$

51. $2y + x = 0$

52. $2x + y = 0$

53. $3x - 2y = 10$

54. $4x + 3y = 9$

55. $y - 2 = 0$

56. $y + 5 = 0$

Write an equation in slope-intercept form, if possible, for each line. See Example 6. In each case, make a sketch.

57. The line through $(0, 6)$ that is perpendicular to the line $y = 3x - 5$

58. The line through $(0, -1)$ that is perpendicular to the line $y = x$

59. The line with y-intercept $(0, 3)$ that is parallel to the line $2x + y = 5$

60. The line through the origin that is parallel to the line $2x - 5y = 8$

61. The line through $(2, 3)$ that runs parallel to the x-axis

62. The line through $(-3, 5)$ that runs parallel to the y-axis

63. The line through $(0, 4)$ and $(5, 0)$

64. The line through $(0, -3)$ and $(4, 0)$

Solve each problem.

65. ***Marginal cost.*** A manufacturer plans to spend \$150,000 on research and development for a new lawn mower and then \$200 to manufacture each mower. The formula $C = 200n + 150{,}000$ gives the cost in dollars of n mowers. What is the cost of 5000 mowers? What is the cost of 5001 mowers? By how much did the one extra lawn mower increase the cost? (The increase in cost is called the *marginal cost* of the 5001st lawn mower.)

66. ***Marginal revenue.*** A defense attorney charges her client \$4000 plus \$120 per hour. The formula $R = 120n + 4000$ gives her revenue in dollars for n hours of work. What is her revenue for 100 hours of work? What is her revenue for 101 hours of work? By how much did the one extra hour of work increase the revenue? (The increase in revenue is called the *marginal revenue* for the 101st hour.)

Getting More Involved

67. ***Discussion.*** Which of the following lines cannot be graphed by using the y-intercept and the slope? Explain your answer.

a) $y = -3x - 100$ **b)** $3x - 50y = 0$

c) $4x - 4 = 0$ **d)** $2x - 3y - 6 = 0$

68. ***Discussion.*** Which of the following lines is not parallel to the others? Explain your answer.

a) $3x - 2y = 6$ **b)** $-3x + 2y = -6$

c) $y = \frac{3}{2}x - 3$ **d)** $y = -\frac{3}{2}x + 3$

69. ***Discussion.*** Which of the following lines is perpendicular to $y = -0.03x + 5$?

a) $y = -\frac{100}{3}x + 5$ **b)** $y = 33.3x + 9$

c) $y = \frac{3}{100}x - 12$ **d)** $y = \frac{100}{3}x - 2.6$

6.4 More on Equations of Lines

In this section:

- Point-Slope Form
- Parallel Lines
- Perpendicular Lines

In Section 6.3 we wrote the equation of a line given its slope and y-intercept. In this section you will learn to write the equation of a line given the slope and *any* other point on the line.

Point-Slope Form

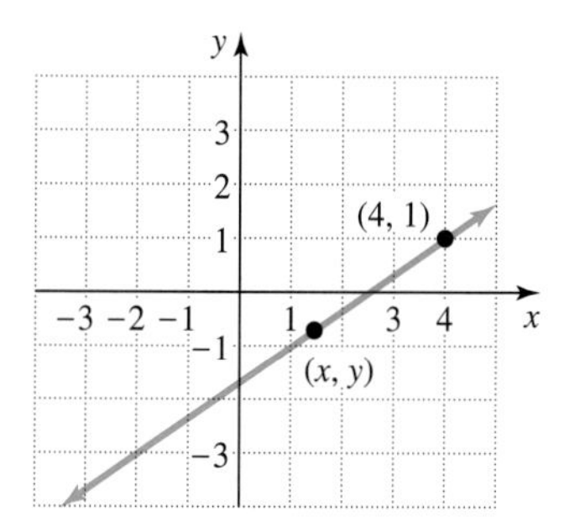

Figure 6.25

Consider a line through the point $(4, 1)$ with slope $\frac{2}{3}$ as shown in Fig. 6.25. Because the slope can be found by using any two points on the line, we use $(4, 1)$ and an arbitrary point (x, y) in the formula for slope:

$$\frac{y_2 - y_1}{x_2 - x_1} = m \qquad \text{Slope formula}$$

$$\frac{y - 1}{x - 4} = \frac{2}{3} \qquad \text{Let } m = \tfrac{2}{3},\ (x_1, y_1) = (4, 1), \text{ and } (x_2, y_2) = (x, y).$$

$$y - 1 = \frac{2}{3}(x - 4) \qquad \text{Multiply each side by } x - 4.$$

Note how the coordinates of the point $(4, 1)$ and the slope $\frac{2}{3}$ appear in the above equation. We can use the same procedure to get the equation of any line given one point on the line and the slope. The resulting equation is called the **point-slope form** of the equation of the line.

Point-Slope Form

The equation of the line through the point (x_1, y_1) with slope m is

$$y - y_1 = m(x - x_1).$$

EXAMPLE 1 **Writing an equation given a point and a slope**

Find the equation of the line through $(-2, 3)$ with slope $\frac{1}{2}$, and write it in slope-intercept form.

Solution

Because we know a point and the slope, we can use the point-slope form:

$$y - y_1 = m(x - x_1) \qquad \text{Point-slope form}$$

$$y - 3 = \frac{1}{2}[x - (-2)] \qquad \text{Substitute } m = \tfrac{1}{2} \text{ and } (x_1, y_1) = (-2, 3).$$

$$y - 3 = \frac{1}{2}(x + 2) \qquad \text{Simplify.}$$

$$y - 3 = \frac{1}{2}x + 1 \qquad \text{Distributive property}$$

$$y = \frac{1}{2}x + 4 \qquad \text{Slope-intercept form}$$

■

CAUTION The point-slope form can be used to find the equation of a line for *any* given point and slope. However, if the given point is the y-intercept, then it is simpler to use the slope-intercept form.

EXAMPLE 2 Writing an equation given two points

Find the equation of the line that contains the points $(-3, -2)$ and $(4, -1)$, and write it in standard form.

Solution

First find the slope using the two given points:

$$m = \frac{-2 - (-1)}{-3 - 4} = \frac{-1}{-7} = \frac{1}{7}$$

Now use one of the points, say $(-3, -2)$, and slope $\frac{1}{7}$ in the point-slope form:

$$y - y_1 = m(x - x_1) \quad \text{Point-slope form}$$
$$y - (-2) = \frac{1}{7}[x - (-3)] \quad \text{Substitute.}$$
$$y + 2 = \frac{1}{7}(x + 3) \quad \text{Simplify.}$$
$$7(y + 2) = 7 \cdot \frac{1}{7}(x + 3) \quad \text{Multiply each side by 7.}$$
$$7y + 14 = x + 3$$
$$7y = x - 11 \quad \text{Subtract 14 from each side.}$$
$$-x + 7y = -11 \quad \text{Subtract } x \text{ from each side.}$$
$$x - 7y = 11 \quad \text{Multiply each side by } -1.$$

The equation in standard form is $x - 7y = 11$. Using the other given point, $(4, -1)$, would give the same final equation in standard form. Try it.

Parallel Lines

In Section 6.2 you learned that parallel lines have the same slope. For example, the lines $y = 6x - 4$ and $y = 6x + 7$ are parallel because each has slope 6. In the next example we write the equation of a line that is parallel to a given line and contains a given point.

EXAMPLE 3 Writing an equation given a point and a parallel line

Write the equation of the line that is parallel to the line $3x + y = 9$ and contains the point $(2, -1)$. Give the answer in slope-intercept form.

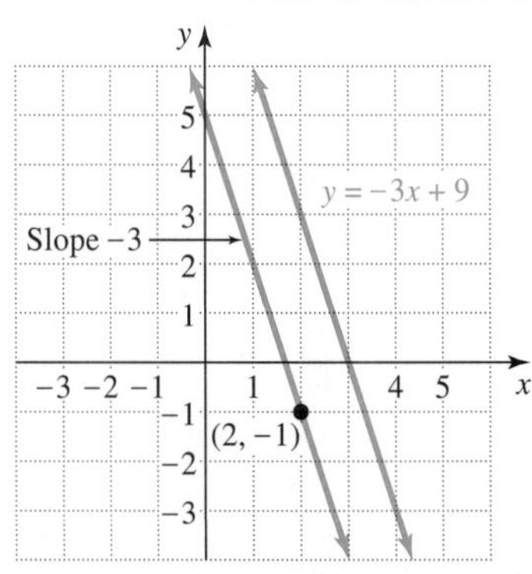

Figure 6.26

Solution

We want the equation of the line through $(2, -1)$ that is parallel to $3x + y = 9$, as shown in Fig. 6.26. First write $3x + y = 9$ in slope-intercept form to determine its slope:

$$3x + y = 9$$
$$y = -3x + 9$$

The slope of $3x + y = 9$ and any line parallel to it is -3. So we want the equation of the line that has slope -3 and contains $(2, -1)$. Use the point-slope form:

$$y - y_1 = m(x - x_1) \quad \text{Point-slope form}$$
$$y - (-1) = -3(x - 2) \quad \text{Substitute.}$$
$$y + 1 = -3x + 6 \quad \text{Simplify.}$$
$$y = -3x + 5 \quad \text{Slope-intercept form}$$

The line $y = -3x + 5$ has slope -3 and contains the point $(2, -1)$. Check that $(2, -1)$ satisfies $y = -3x + 5$. ■

Perpendicular Lines

In Section 6.2 you learned that lines with slopes m and $-\frac{1}{m}$ (for $m \neq 0$) are perpendicular to each other. For example, the lines

$$y = -2x + 7 \qquad \text{and} \qquad y = \frac{1}{2}x - 8$$

are perpendicular to each other. In the next example we will write the equation of a line that is perpendicular to a given line and contains a given point.

EXAMPLE 4 **Writing an equation given a point and a perpendicular line**

Write the equation of the line that is perpendicular to $3x + 2y = 8$ and contains the point $(1, -3)$. Write the answer in slope-intercept form.

Solution

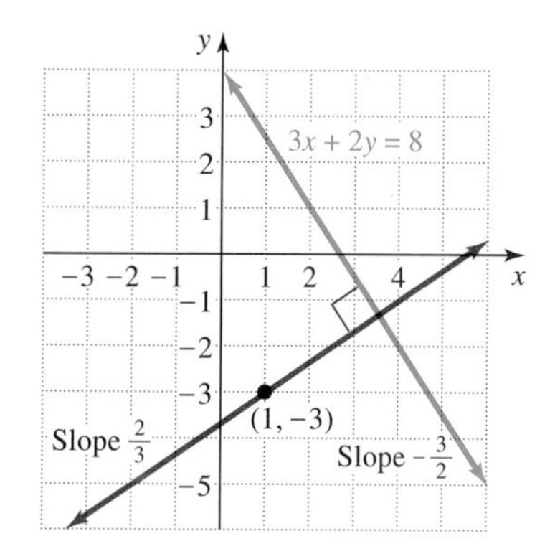

Figure 6.27

We want the equation of the line containing $(1, -3)$ that is perpendicular to $3x + 2y = 8$ as shown in Fig. 6.27. First write $3x + 2y = 8$ in slope-intercept form to determine the slope:

$$3x + 2y = 8$$
$$2y = -3x + 8$$
$$y = -\frac{3}{2}x + 4 \quad \text{Slope-intercept form}$$

The slope of the given line is $-\frac{3}{2}$. The slope of any line perpendicular to it is $\frac{2}{3}$. Now we use the point-slope form with the point $(1, -3)$ and the slope $\frac{2}{3}$:

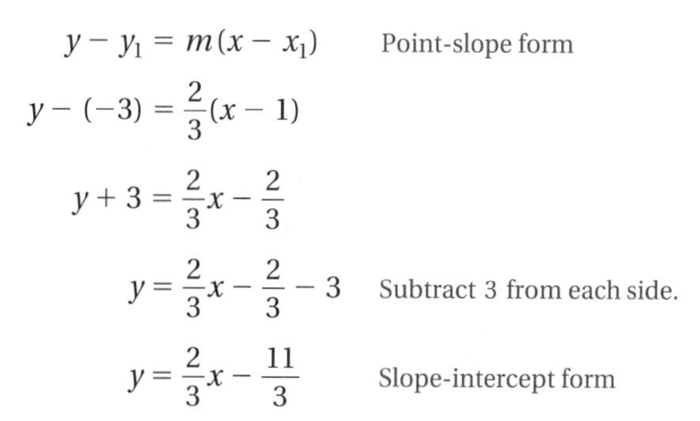

$$y - y_1 = m(x - x_1) \quad \text{Point-slope form}$$
$$y - (-3) = \frac{2}{3}(x - 1)$$
$$y + 3 = \frac{2}{3}x - \frac{2}{3}$$
$$y = \frac{2}{3}x - \frac{2}{3} - 3 \quad \text{Subtract 3 from each side.}$$
$$y = \frac{2}{3}x - \frac{11}{3} \quad \text{Slope-intercept form}$$

So $y = \frac{2}{3}x - \frac{11}{3}$ is the equation of the line that contains $(1, -3)$ and is perpendicular to $3x + 2y = 8$. Check that $(1, -3)$ satisfies $y = \frac{2}{3}x - \frac{11}{3}$. ■

Calculator Close-up

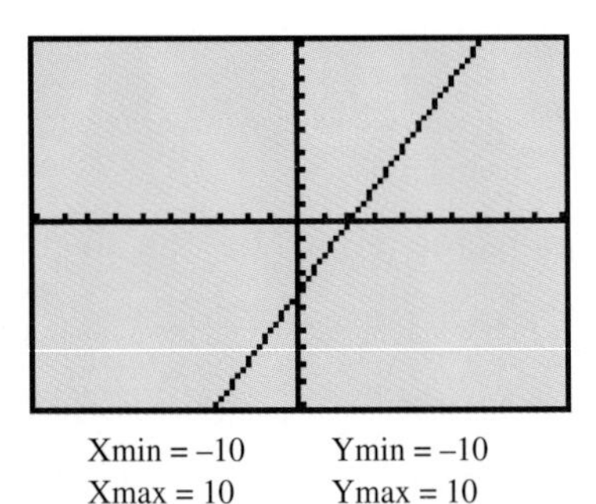

Xmin = –10 Ymin = –10
Xmax = 10 Ymax = 10

Figure 6.28

A graphing calculator (or a computer graphing utility) can find many ordered pairs that satisfy an equation and then plot them on its screen. Because a graphing calculator screen is made up of a finite number of pixels, a graphing calculator can plot only a finite number of ordered pairs, even if there are infinitely many ordered pairs that satisfy the equation. However, the graph shown on the graphing calculator can usually give you a good idea what the graph of the equation looks like.

To plot some ordered pairs that satisfy $y = 2x - 4$, enter the equation into the calculator using the Y= key and then press the GRAPH key. (Because there are many differences in how different brands of graphing calculators graph equations, you should consult your manual at this point.) Your calculator might show ordered pairs as in Fig. 6.28. This output is consistent with the fact that $y = 2x - 4$ is a linear equation and its graph is a straight line.

The viewing window used in Fig. 6.28 has x-values ranging from a minimum of -10 to a maximum of 10 and y-values ranging from a minimum of -10 to a maximum of 10, the **standard viewing window.** If the points satisfying $y = 2x - 4$ do not appear on your calculator, your viewing window might be set so that it does not contain any of the points. You can change the viewing window by pressing the RANGE key. Consult your manual. Exercises for graphing calculators are included in this section and in others for the remainder of the text.

Warm-ups

True or false? Explain your answer.

1. The formula $y = m(x - x_1)$ is the point-slope form for a line.
2. It is impossible to find the equation of a line through $(2, 5)$ and $(-3, 1)$.
3. The point-slope form will not work for the line through $(3, 4)$ and $(3, 6)$.
4. The equation of the line through the origin with slope 1 is $y = x$.
5. The slope of the line $5x + y = 4$ is 5.
6. The slope of any line perpendicular to the line $y = 4x - 3$ is $-\frac{1}{4}$.
7. The slope of any line parallel to the line $x + y = 1$ is -1.
8. The line $2x - y = -1$ goes through the point $(-2, -3)$.
9. The lines $2x + y = 4$ and $y = -2x + 7$ are parallel.
10. The equation of the line through $(0, 0)$ perpendicular to $y = x$ is $y = -x$.

6.4 EXERCISES

Write each equation in slope-intercept form. See Example 1.

1. $y - 1 = 5(x + 2)$
2. $y + 3 = -3(x - 6)$
3. $3x - 4y = 80$
4. $2x + 3y = 90$
5. $y - \frac{1}{2} = \frac{2}{3}\left(x - \frac{1}{4}\right)$
6. $y + \frac{2}{3} = -\frac{1}{2}\left(x - \frac{2}{5}\right)$

Find the equation of each line. Write each answer in slope-intercept form. See Example 1.

7. The line through $(2, 3)$ with slope $\frac{1}{3}$
8. The line through $(1, 4)$ with slope $\frac{1}{4}$

9. The line through $(-2, 5)$ with slope $-\frac{1}{2}$

10. The line through $(-3, 1)$ with slope $-\frac{1}{3}$

11. The line with slope -6 that goes through $(-1, -7)$

12. The line with slope -8 that goes through $(-1, -5)$

Write each equation in standard form using only integers. See Example 2.

13. $y - 3 = 2(x - 5)$

14. $y + 2 = -3(x - 1)$

15. $y = \frac{1}{2}x - 3$

16. $y = \frac{1}{3}x + 5$

17. $y - 2 = \frac{2}{3}(x - 4)$

18. $y + 1 = \frac{3}{2}(x + 4)$

Find the equation of each line. Write each answer in standard form using only integers. See Example 2.

19. The line through the points $(1, 2)$ and $(5, 8)$

20. The line through the points $(3, 5)$ and $(8, 15)$

21. The line through the points $(-2, -1)$ and $(3, -4)$

22. The line through the points $(-1, -3)$ and $(2, -1)$

23. The line through the points $(-2, 0)$ and $(0, 2)$

24. The line through the points $(0, 3)$ and $(5, 0)$

Find the equation of the solid line in each figure. Write the equation in slope-intercept form.

25.

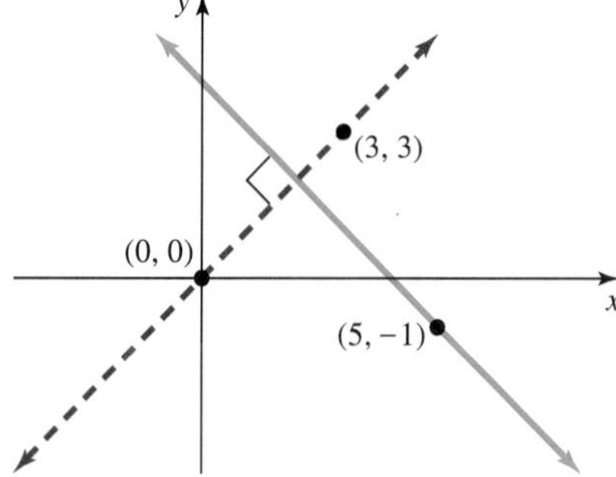

26.

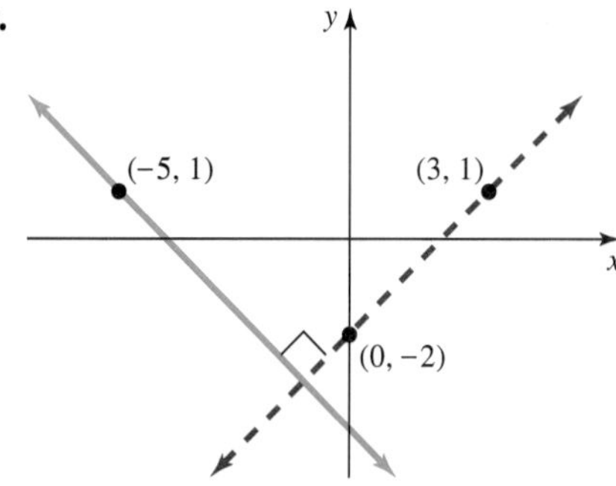

27.

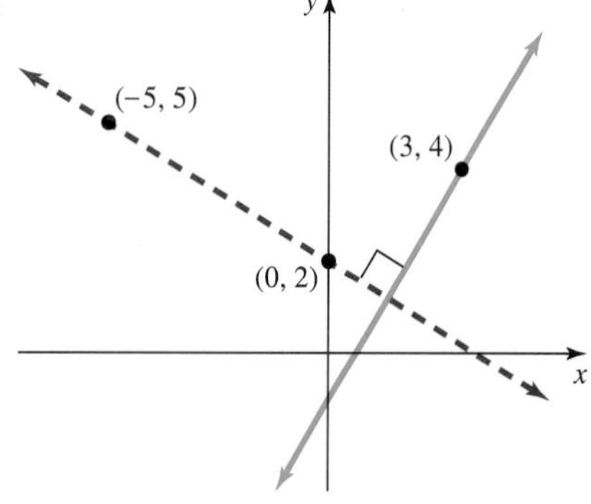

28.

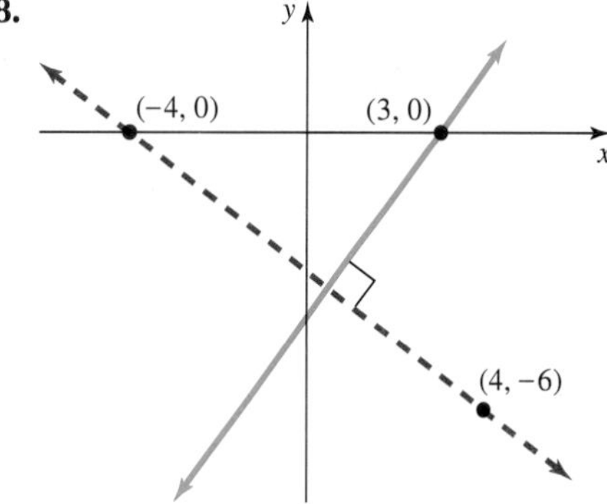

Find the equation of each line. Write each answer in slope-intercept form. See Examples 3 and 4.

29. The line contains the point $(3, 4)$ and is perpendicular to $y = 3x - 1$.

30. The line contains the point $(-2, 3)$ and is perpendicular to $y = 2x + 7$.

31. The line is parallel to $y = x - 9$ and goes through the point $(7, 10)$.

32. The line is parallel to $y = -x + 5$ and goes through the point $(-3, 6)$.

33. The line is perpendicular to $3x - 2y = 10$ and passes through the point $(1, 1)$.

34. The line is perpendicular to $x - 5y = 4$ and passes through the point $(-1, 1)$.

35. The line is parallel to $2x + y = 8$ and contains the point $(-1, -3)$.

36. The line is parallel to $-3x + 2y = 9$ and contains the point $(-2, 1)$.

37. The line goes through $(-1, 2)$ and is perpendicular to $3x + y = 5$.

38. The line goes through $(1, 2)$ and is perpendicular to $y = \frac{1}{2}x - 3$.

39. The line goes through $(2, 3)$ and is parallel to $-2x + y = 6$.

40. The line goes through $(1, 4)$ and is parallel to $x - 2y = 6$.

Solve each problem.

41. ***Automated tellers.*** Banks are pleased that automated tellers are growing in popularity because they cost banks less per transaction than human tellers (*New York Times*, October 23, 1994). The accompanying graph shows the steady growth of automated tellers. Write the equation for the line through (88, 4.5) and (92, 7). Use the equation to predict the number of transactions at automated teller machines in the year 2000.

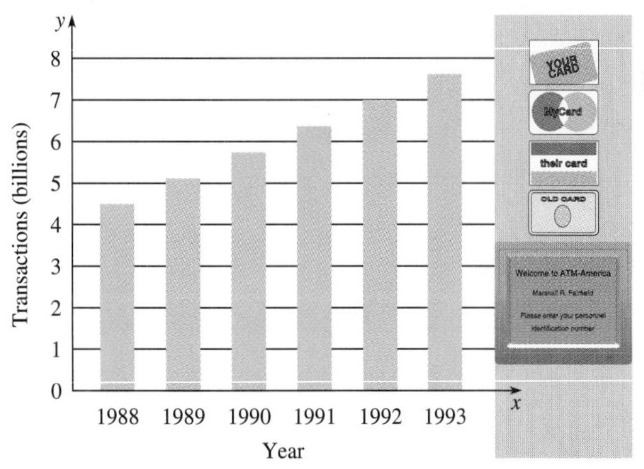

Figure for Exercise 41

42. ***Direct deposit.*** In 1988, only 8% of employees in private industry had direct deposit of their paychecks; in 1993 the percentage was 35% (*New York Times*, October 23, 1994). Write the equation of the line through (88, 0.08) and (93, 0.35). Use the equation to predict the year in which 100% of all employees in private industry will have direct deposit of their paychecks.

Getting More Involved

43. ***Writing.*** Write a summary of all of the different ways that we have used to describe a line. Explain how to get the equation of the line from each description.

44. ***Exploration.*** What is the slope of the line $2x + 3y = 9$? What is the slope of $4x - 5y = 6$? Write a formula for the slope of $Ax + By = C$, where $B \neq 0$.

Graphing Calculator Exercises

45. Graph each equation on a graphing calculator. Choose a viewing window that includes both the x- and y-intercepts. Use the calculator output to help you draw the graph on paper.

a) $y = 20x - 300$ **b)** $y = -30x + 500$
c) $2x - 3y = 6000$

46. Graph $y = 2x + 1$ and $y = 1.99x - 1$ on a graphing calculator. Are these lines parallel? Explain your answer.

47. Graph $y = 0.5x + 0.8$ and $y = 0.5x + 0.7$ on a graphing calculator. Find a viewing window in which the two lines are separate.

48. Graph $y = 3x + 1$ and $y = -\frac{1}{3}x + 2$ on a graphing calculator. Do the lines look perpendicular? Explain.

6.5 Applications of Linear Equations

In this section:

- Applied Examples
- Graphing
- Finding a Formula

The linear equation $y = mx + b$ is a formula that determines a value of y for each given value of x. In this section you will study linear equations used as formulas.

Applied Examples

The daily rental charge for renting a 1988 Buick at Wrenta-Wreck is \$30 plus 25 cents per mile. The rental charge depends on the number of miles you drive. If x represents the number of miles driven in one day, then $0.25x + 30$ will give the rental charge in dollars for that day. If we let y represent the rental charge, then we can write the equation

$$y = 0.25x + 30.$$

Because the value of y depends on the value of x, we call x the **independent variable** and y the **dependent variable.**

For variables in applications we generally use letters that help us to remember what the variables represent. In the rental example above, we could let m represent the number of miles and R the rental charge. Then R is determined from m by the formula

$$R = 0.25m + 30.$$

There are many examples in which the value of one variable is determined from the value of another variable by means of a linear equation. When this is the case, we say that the dependent variable is a *linear function* of the independent variable. For example, the formula

$$F = \frac{9}{5}C + 32$$

is a linear equation that expresses Fahrenheit temperature in terms of Celsius temperature. In other words, the Fahrenheit temperature F is a linear function of the Celsius temperature C.

EXAMPLE 1 **Distance as a function of time**

A car is averaging 50 miles per hour. Write an equation that expresses the distance it travels as a linear function of the time it travels.

Solution

If the speed is 50 miles per hour, then from the formula $D = RT$, we can write

$$D = 50T.$$

This linear equation expresses D as a linear function of T. ■

Graphing

The graph of a linear equation is a straight line. A linear equation used as a formula is graphed the same way that any linear equation is graphed. If a formula is in slope-intercept form, then we can graph it using the slope and intercept as in Section 6.3. If we use letters other than x and y for the variables, then we label the axes with these letters. The horizontal axis is always the axis of the independent variable. The vertical axis is the axis of the dependent variable.

EXAMPLE 2 **Graphing a formula**

Graph the linear equation $R = 0.25m + 30$ for $0 \le m \le 500$. R represents the rental charge in dollars, and m represents the number of miles. Find the rental charge for driving 200 miles.

Solution

We label the x-axis with the letter m and the y-axis with the letter R. See Fig. 6.29. We adjust the scale on the m-axis to graph the values from 0 to 500. The slope of this line is 0.25, or $\frac{25}{100}$. To sketch the graph, start at the R-intercept (0, 30). Move 100 units to the right and up 25 units to locate a second point on the line. Draw the line as in Fig. 6.29. To find the rental charge for 200 miles, let $m = 200$ in $R = 0.25m + 30$:

$$R = 0.25(200) + 30 = 80$$

So the rental charge for 200 miles is \$80. ■

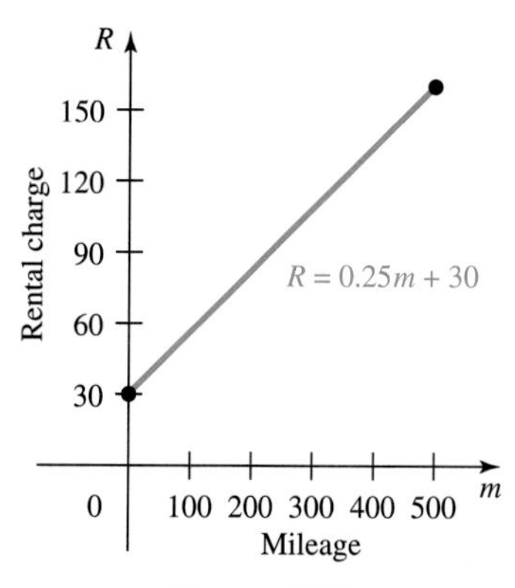

Figure 6.29

Finding a Formula

If one variable is a linear function of another, then there is a linear equation expressing one variable in terms of the other. In Section 6.4 we used the point-slope form to find the equation of a line given two points on the line. We can use that same procedure to find a linear function for two variables.

EXAMPLE 3 **Writing a linear function given two points**

A contractor found that his labor cost for installing 100 feet of pipe was \$30. He also found that his labor cost for installing 500 feet of pipe was \$120. If the cost C in dollars is a linear function of the length L in feet, then what is the formula for this function? What would his labor cost be for installing 240 feet of pipe?

Solution

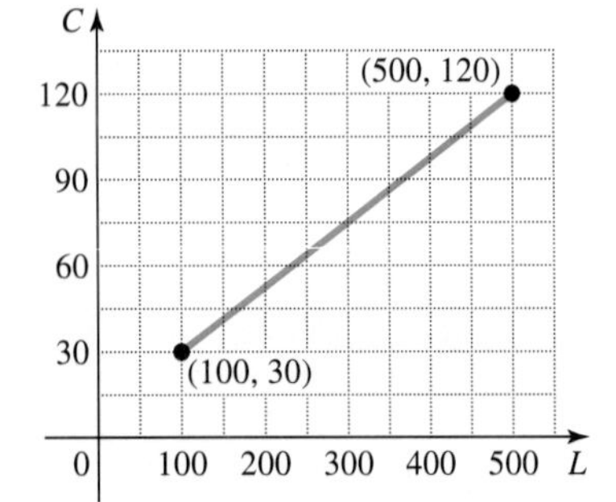

Figure 6.30

Because C is determined from L, we let C take the place of the dependent variable y and let L take the place of the independent variable x. So the ordered pairs are in the form (L, C). We can use the slope formula to find the slope of the line through the two points $(100, 30)$ and $(500, 120)$ shown in Fig. 6.30:

$$m = \frac{120 - 30}{500 - 100} = \frac{90}{400} = \frac{9}{40}$$

Now we use the point-slope form with the point $(100, 30)$ and slope $\frac{9}{40}$:

$$y - y_1 = m(x - x_1)$$

$$C - 30 = \frac{9}{40}(L - 100)$$

$$C - 30 = \frac{9}{40}L - \frac{45}{2}$$

$$C = \frac{9}{40}L - \frac{45}{2} + 30$$

$$C = \frac{9}{40}L + \frac{15}{2} \quad \text{C is a linear function of } L.$$

Now that we have a formula for C in terms of L, we can find C for any value of L. If $L = 240$ feet, then

$$C = \frac{9}{40} \cdot 240 + \frac{15}{2}$$

$$C = 54 + 7.5$$

$$C = 61.5$$

The labor cost to install 240 feet of pipe would be \$61.50. ■

Warm-ups

True or false? Explain your answer.

1. If $z = 3r - 9$, then z is a linear function of r.
2. The circumference of a circle is a linear function of its radius.
3. The area of a circle is a linear function of its radius.
4. The distance driven in 8 hours is a linear function of your average speed.
5. Celsius temperature is a linear function of Fahrenheit temperature.
6. The slope of the line through (1980, 3000) and (1990, 2000) is 100.
7. If your lawyer charges $90 per hour, then your bill is a linear function of the time spent on the case.
8. The area of a square is a linear function of the length of a side.
9. The perimeter of a square is a linear function of the length of a side.
10. The perimeter of a rectangle with a length of 5 meters is a linear function of its width.

6.5 EXERCISES

Write an equation that expresses one variable as a linear function of the other. See Example 1.

1. Express length in feet as a linear function of length in yards.
2. Express length in yards as a linear function of length in feet.
3. For a car averaging 65 miles per hour, express the distance it travels as a linear function of the time spent traveling.
4. For a car traveling 6 hours, express the distance it travels as a linear function of its average speed.
5. Express the circumference of a circle as a linear function of its diameter.
6. For a rectangle with a fixed width of 12 feet, express the perimeter as a linear function of its length.
7. Rodney makes $7.80 per hour. Express his weekly pay as a linear function of the number of hours he works.
8. A triangle has a base of 5 feet. Express its area as a linear function of its height.

Graph each formula for the given values of the independent variable. See Example 2.

9. $P = 40n + 300,\ 0 \le n \le 200$
10. $C = -50r + 500,\ 0 \le r \le 10$
11. $R = 30t + 1000,\ 100 \le t \le 900$
12. $W = 3m - 4000,\ 1000 \le m \le 5000$
13. $C = 2\pi r,\ 1 \le r \le 10$
14. $P = 4s,\ 100 \le s \le 500$
15. $h = -7.5d + 350,\ 0 \le d \le 40$
16. $a = -50g + 2500,\ 0 \le g \le 50$

Solve each problem. See Example 2.

17. ***Profit per share.*** In the 1980s, People's Gas had a profit per share, P, that was determined by the equation $P = 0.35x + 4.60$, where x ranges from 0 to 9 corresponding to the years 1980 to 1989. What was the profit per share in 1987? Sketch the graph of this formula for x ranging from 0 to 9.
18. ***Loan value of a car.*** For the first 6 years the loan value of a $30,000 Corvette is determined by the formula $V = -4{,}000a + 30{,}000$, where a is the age in years of the Corvette. What is the loan value of this automobile when it is 5 years old? Sketch the graph of this formula for a between 0 and 6 inclusive.

In Exercises 19–30, solve each problem. See Example 3.

19. ***Plumbing problems.*** When Millie called Pete's Plumbing, Pete worked 2 hours and charged her $70. When her neighbor Rosalee called Pete, he worked 4 hours and charged her $110. Pete's charge is a linear function of the number of hours he works. Find a formula for this function. How much will Pete charge for working 7 hours at Fred's house?

20. ***Interior angles.*** The sum of the measures of the interior angles of a triangle is 180°. The sum of the measures of the interior angles of a square is 360°. The sum S of the measures of the interior angles of any n-sided polygon is a linear function of the number of sides n. Express S as a linear function of n. What is the sum of the measures of the interior angles of an octagon?

Figure for Exercise 20

21. ***If the shoe fits.*** If a child's foot is 7.75 inches long, then the child wears a size 13 shoe. If the child has a foot that is 5.75 inches long, then the child wears a size 7 shoe. The shoe size S is a linear function of the length of the foot L. Write a linear equation expressing S as a function of L. What size shoe fits a child with a 6.25-inch foot?

22. ***Celsius to Fahrenheit.*** Fahrenheit temperature F is a linear function of Celsius temperature C. When $C = 0$, $F = 32$. When $C = 100$, $F = 212$. Use the point-slope form to write F as a linear function of C. What is the Fahrenheit temperature when $C = 45$?

23. ***Velocity of a projectile.*** The velocity v of a projectile is a linear function of the time t that it is in the air. A ball is thrown downward from the top of a tall building. Its velocity is 42 feet per second after 1 second and 74 feet per second after 2 seconds. Write v as a linear function of t. What is the velocity when $t = 3.5$ seconds?

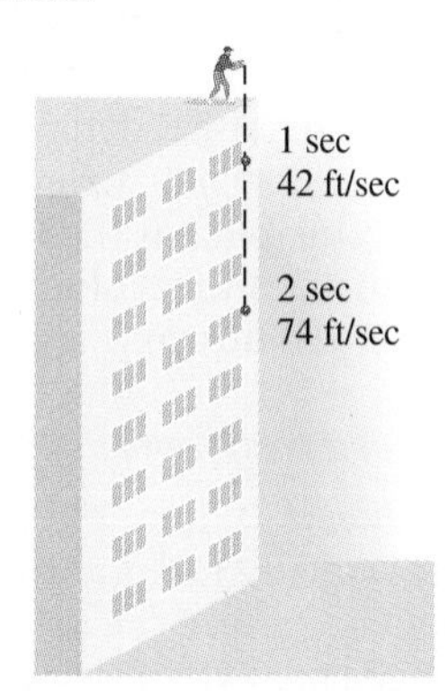

Figure for Exercise 23

24. ***Natural gas.*** The cost C of natural gas is a linear function of the number n of cubic feet of gas used. The cost of 1000 cubic feet of gas is \$39, and the cost of 3000 cubic feet of gas is \$99. Express C as a linear function of n. What is the cost of 2400 cubic feet of gas?

25. ***Expansion joint.*** The width of an expansion joint on the Carl T. Hull bridge is a linear function of the temperature of the roadway. When the temperature is 90°F, the width is 0.75 inch. When the temperature is 30°F, the width is 1.25 inches. Express w as a linear function of t. What is the width of the joint when the temperature is 80°F?

26. ***Perimeter of a rectangle.*** The perimeter P of a rectangle with a fixed width is a linear function of its length. The perimeter is 28 inches when the length is 6.5 inches, and the perimeter is 36 inches when the length is 10.5 inches. Write P as a linear function of L. What is the perimeter when $L = 40$ feet? What is the fixed width of the rectangle?

27. ***Stretching a spring.*** The amount A that a spring stretches beyond its natural length is a linear function of the weight w placed on the spring. A weight of 3 pounds stretches a certain spring 1.8 inches and a weight of 5 pounds stretches the same spring 3 inches. Express A as a linear function of w. How much will the spring stretch with a weight of 6 pounds?

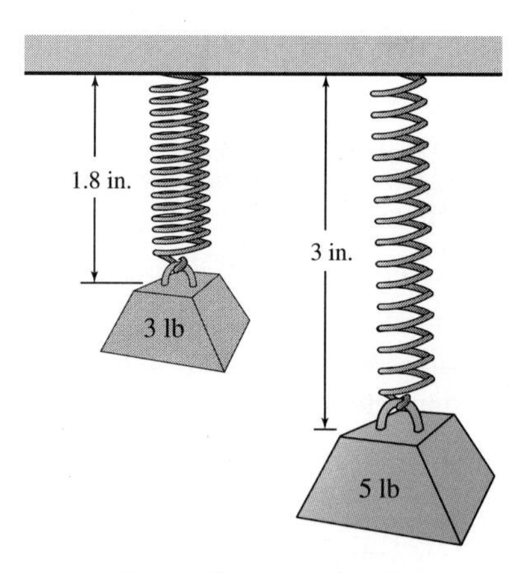

Figure for Exercise 27

28. ***Velocity of a bullet.*** If a gun is fired straight upward, then the velocity v of the bullet is a linear function of the time t that has elapsed since the gun was fired. Suppose that the bullet leaves the gun at 100 feet per second (time $t = 0$) and that after 2 seconds its velocity is 36 feet per second. Express v as a linear function of t. What is the velocity after 3 seconds?

29. ***Enzyme concentration.*** A food chemist tests enzymes for their ability to break down pectin in fruit juices. Excess pectin makes the juice cloudy. The clarity of the juice is measured by the fraction of light that it absorbs. The absorption of the liquid is a linear function of the concentration of the enzyme. Suppose that a concentration of 2 mg/ml (milligrams per milliliter) produces an absorption of 0.16 and a concentration of 5 mg/ml produces an absorption of 0.40. Express the absorption a as a linear function of the concentration c. What should the absorption be if the concentration is 3 mg/ml? Use the accompanying graph to estimate the concentration when the absorption is 0.50.

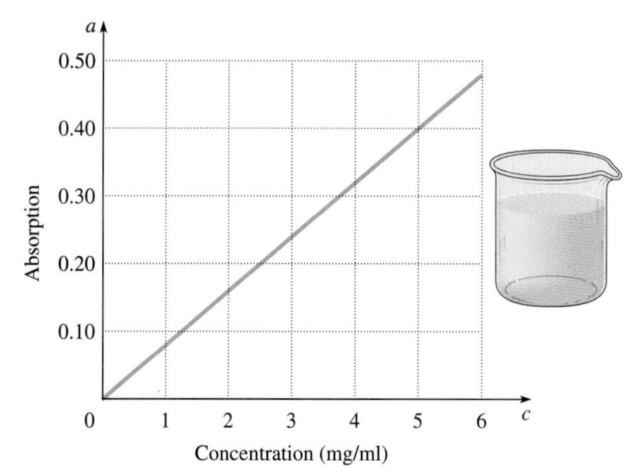

Figure for Exercise 29

30. ***Basal energy requirement.*** The basal energy requirement B is the number of calories that a person needs to maintain the life processes. B depends on the height, weight, and age of the person. For a 28-year-old female with a height of 160 cm, B is a linear function of the person's weight w (in kilograms). For a weight of 45 kg, B is 1300 calories. For a weight of 50 kg, B is 1365 calories. Express B as a linear function of w. What is B for a 28-year-old 160-cm female who weighs 53.2 kg?

Graphing Calculator Exercises

31. ***Energy decreasing with age.*** The basal energy requirement (in calories) for a 55-kg 160-cm male at age A is given by $B = 1481 - 4.7A$; the basal energy requirement for a female with the same weight and height is given by $B = 1623 - 6.9A$. Graph these functions on your calculator, and use the graphs to answer the following questions.

a) Which person has a higher basal energy requirement at age 25?

b) Which person has a higher basal energy requirement at age 72?

c) Use the graph to estimate the age at which the basal energy requirements are equal.

d) At what age does the female require no calories?

32. ***Equality of energy.*** The basal energy requirement B for a 70-kg 160-cm male at age A is given by $B = 1620 - 4.7A$. The basal energy requirement B for a 65-kg 165-cm female at age A is given by $B = 1786 - 6.8A$. Graph these linear functions on your calculator, and use the graphs to estimate the age at which these two people have the same basal energy requirement.

6.6 Functions

In this section:

- Functions Defined by Formulas
- Functions Defined by Tables
- Functions Defined by Ordered Pairs
- Graphs of Functions
- Domain and Range
- The f-Notation

In Section 6.5 you learned that if y is determined from the value of x by an equation of the form $y = mx + b$, then y is a linear function of x. In this section we will discuss other types of functions, but the idea is the same.

Functions Defined by Formulas

If you get a speeding ticket, then your speed determines the cost of the ticket. You may not know exactly how the judge determines the cost, but the judge is using some rule to determine a cost from knowing your speed. The cost of the ticket is a function of your speed.

Function (as a Rule)

A function is a rule by which any allowable value of one variable (the **independent variable**) determines a *unique* value of a second variable (the **dependent variable**).

One way to express a function is to use a formula. For example, the formula

$$A = \pi r^2$$

gives the area of a circle as a function of its radius. The formula gives us a rule for finding a *unique* area for any given radius. A is the dependent variable, and r is the independent variable. The formula

$$S = -16t^2 + v_0 t + s_0$$

expresses altitude S of a projectile as a function of time t, where v_0 is the initial velocity and s_0 is the initial altitude. S is the dependent variable, and t is the independent variable.

In many areas of study, formulas are used to describe relationships between variables. In the next example we write a formula that describes or **models** a real situation.

EXAMPLE 1 **Writing a formula for a function**

A carpet layer charges \$25 plus \$4 per square yard for installing carpet. Write the total charge C as a function of the number n of square yards of carpet installed.

Solution

At \$4 per square yard, n square yards installed cost $4n$ dollars. If we include the \$25 charge, then the total cost is $4n + 25$ dollars. Thus the equation

$$C = 4n + 25$$

expresses C as a function of n. ■

In the next example, we modify a well-known geometric formula.

EXAMPLE 2 **A function in geometry**

Express the area of a circle as a function of its diameter.

Solution

The area of a circle is given by $A = \pi r^2$. Because the radius of a circle is one-half of the diameter, we have $r = \frac{d}{2}$. Now replace r by $\frac{d}{2}$ in the formula $A = \pi r^2$:

$$\begin{aligned} A &= \pi\left(\frac{d}{2}\right)^2 \\ &= \frac{\pi d^2}{4} \end{aligned}$$

So $A = \frac{\pi d^2}{4}$ expresses the area of a circle as a function of its diameter. ■

Functions Defined by Tables

Another way to define a function is with a table. For example, Table 6.1 can be used to determine the cost at United Freight Service for shipping a package that weighs under 100 pounds. For any *allowable* weight, the table gives us a rule for finding the unique shipping cost. The weight is the independent variable, and the cost is the dependent variable.

Weight in Pounds	Cost
0 to 10	\$4.60
11 to 30	\$12.75
31 to 79	\$32.90
80 to 99	\$55.82

Table 6.1

Now consider Table 6.2. It does not look much different from Table 6.1, but there is an important difference. The cost for shipping a 12-pound package according to Table 6.2 is either \$4.60 or \$12.75. Either the table has an error or perhaps \$4.60 and \$12.75 are costs for shipping to different destinations. In any case the weight does not determine a unique cost. So Table 6.2 does not define the cost as a function of the weight.

Weight in Pounds	Cost
0 to 15	\$4.60
10 to 30	\$12.75
31 to 79	\$32.90
80 to 99	\$55.82

Table 6.2

EXAMPLE 3 **Functions defined by tables**

Which of the following tables defines y as a function of x?

a)

x	y
1	3
2	6
3	9
4	12
5	15

b)

x	y
1	1
−1	1
2	2
−2	2
3	3
−3	3

c)

x	y
1988	27,000
1989	27,000
1990	28,500
1991	29,000
1992	30,000
1993	30,750

d)

x	y
23	48
35	27
19	28
23	37
41	56
22	34

Solution

In Tables (a), (b), and (c), every value of x corresponds to only one value of y. Tables (a), (b), and (c) each express y as a function of x. Notice that different values of x may correspond to the same value of y. In Table (d) we have the value of 23 for x corresponding to two different values of y, 48 and 37. So Table (d) does not define y as a function of x. ■

Functions Defined by Ordered Pairs

If the value of the independent variable is written as the first coordinate of an ordered pair and the value of the dependent variable is written as the second coordinate of an ordered pair, then a function can be defined by a set of ordered pairs. In a function, each value of the independent variable corresponds to a unique value of the dependent variable. So no two ordered pairs can have the same first coordinate and different second coordinates.

Function (as a Set of Ordered Pairs)

A function is a set of ordered pairs of real numbers such that no two ordered pairs have the same first coordinates and different second coordinates.

EXAMPLE 4 **Functions defined by a set of ordered pairs**

Determine whether each set of ordered pairs is a function.

a) $\{(1, 2), (1, 5), (-4, 6)\}$ **b)** $\{(-1, 3), (0, 3), (6, 3), (-3, 2)\}$

Solution

a) This set of ordered pairs is not a function because $(1, 2)$ and $(1, 5)$ have the same first coordinates but different second coordinates.

b) This set of ordered pairs is a function. Note that the same second coordinate with different first coordinates is permitted in a function. ■

If there are infinitely many ordered pairs in a function, then we can use set-builder notation from Chapter 1 along with an equation to define the function. For example,

$$\{(x, y) \mid y = x^2\}$$

is the set of ordered pairs in which the y-coordinate is the square of the x-coordinate. Ordered pairs such as $(0, 0)$, $(2, 4)$, and $(-2, 4)$ belong to this set. This set is a function because every value of x determines only one value of y.

EXAMPLE 5 **Functions defined by set-builder notation**

Determine whether each set of ordered pairs is a function.

a) $\{(x, y) \mid y = 3x^2 - 2x + 1\}$ **b)** $\{(x, y) \mid y^2 = x\}$ **c)** $\{(x, y) \mid x + y = 6\}$

Solution

a) This set is a function because each value we select for x determines only one value for y.

b) If $x = 9$, then we have $y^2 = 9$. Because both 3 and -3 satisfy $y^2 = 9$, both $(9, 3)$ and $(9, -3)$ belong to this set. So the set is not a function.

c) If we solve $x + y = 6$ for y, we get $y = -x + 6$. Because each value of x determines only one value for y, this set is a function. In fact, this set is a linear function. ■

We often omit the set notation when discussing functions. For example, the equation

$$y = 3x^2 - 2x + 1$$

expresses y as a function of x because the set of ordered pairs determined by the equation is a function. However, the equation

$$y^2 = x$$

does not express y as a function of x because ordered pairs such as $(9, 3)$ and $(9, -3)$ satisfy the equation.

EXAMPLE 6 **Functions defined by equations**

Determine whether each equation defines y as a function of x.

a) $y = |x|$ **b)** $y = x^3$ **c)** $x = |y|$

Solution

a) Because every number has a unique absolute value, $y = |x|$ is a function.

b) Because every number has a unique cube, $y = x^3$ is a function.

c) The equation $x = |y|$ does not define a function because both $(4, -4)$ and $(4, 4)$ satisfy this equation. These ordered pairs have the same first coordinate but different second coordinates. ■

Graphs of Functions

Every function determines a set of ordered pairs, and any set of ordered pairs has a graph in the rectangular coordinate system. For example, the set of ordered pairs determined by the linear function $y = 2x - 1$ is shown in Fig. 6.31.

Every graph illustrates a set of ordered pairs, but not every graph is a graph of a function. For example, the circle in Fig. 6.32 is not a graph of a function because the ordered pairs $(0, 4)$ and $(0, -4)$ are both on the graph, and these two ordered pairs have the same first coordinate and different second coordinates. Whether a graph has such ordered pairs can be determined by a simple visual test called the **vertical-line test.**

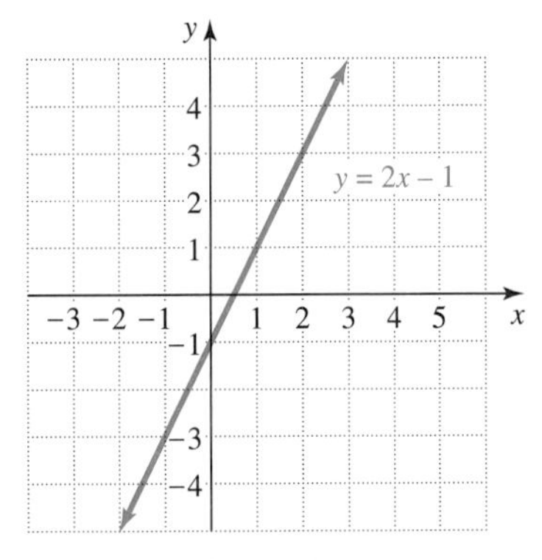

Figure 6.31

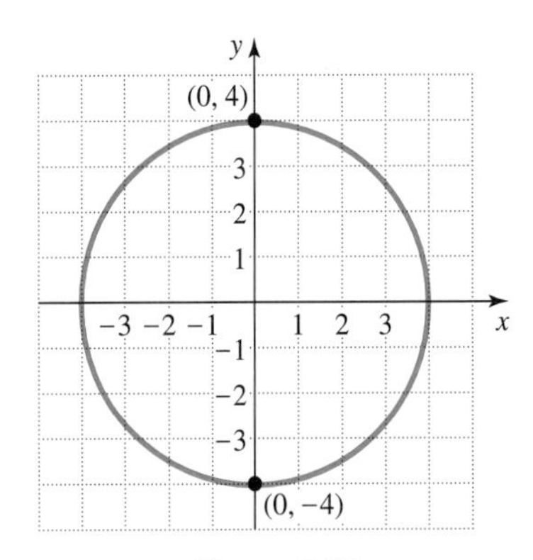

Figure 6.32

Vertical-Line Test

If it is possible to draw a vertical line that crosses a graph two or more times, then the graph is not the graph of a function.

If there is a vertical line that crosses a graph twice (or more), then we have two points (or more) with the same x-coordinate and different y-coordinates, and so the graph is not the graph of a function. If you mentally consider every possible vertical line and none of them cross the graph more than once, then you can conclude that the graph is the graph of a function.

EXAMPLE 7 **Using the vertical-line test**

Which of the following graphs are graphs of functions?

a)

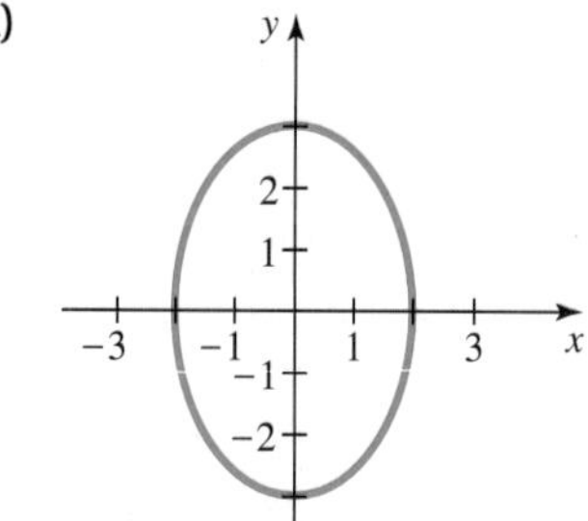

b)

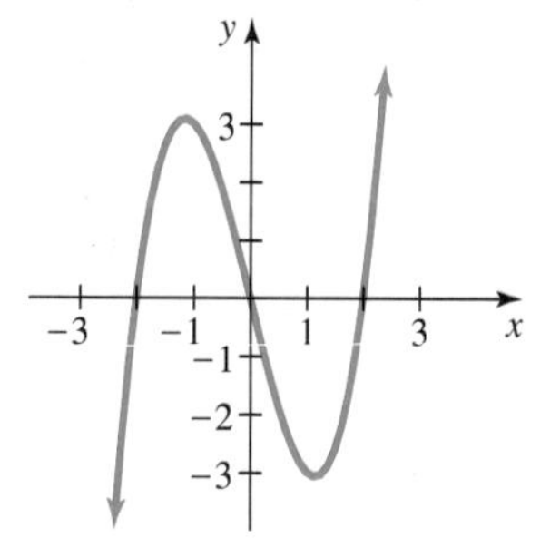

c)

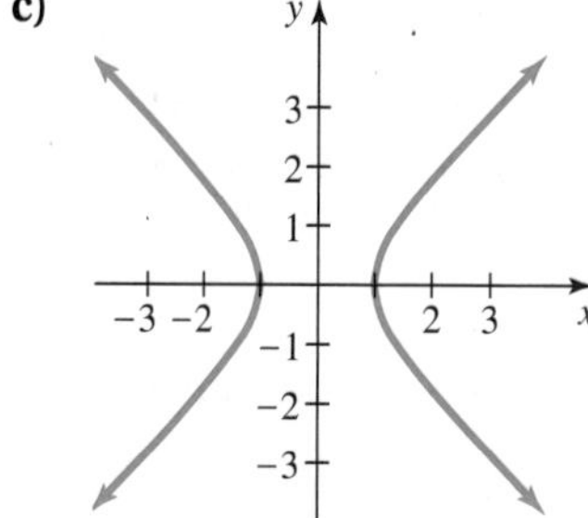

Solution

Neither (a) nor (c) is the graph of a function, since we can draw vertical lines that cross these graphs twice. Graph (b) is the graph of a function, since no vertical line crosses it twice. ■

Domain and Range

The set of all possible numbers that can be used for the independent variable is called the **domain** of the function. For example, the domain of the function

$$y = \frac{1}{x}$$

is the set of all nonzero real numbers because $\frac{1}{x}$ is undefined for $x = 0$. For some functions the domain is clearly stated when the function is given. The set of all values of the dependent variable is called the **range** of the function.

EXAMPLE 8 **Domain and range**

State the domain and range of each function.

a) $\{(3, -1), (2, 5), (1, 5)\}$ **b)** $y = |x|$ **c)** $A = \pi r^2$ for $r > 0$

Solution

a) The domain is the set of numbers used as first coordinates, $\{1, 2, 3\}$. The range is the set of second coordinates, $\{-1, 5\}$.

b) Because $|x|$ is a real number for any real number x, the domain is the set of all real numbers. The range is the set of numbers that result from taking the absolute value of every real number. Thus the range is the set of nonnegative real numbers, $\{y \mid y \geq 0\}$.

c) The condition $r > 0$ specifies the domain of the function. The domain is $\{r \mid r > 0\}$, the positive real numbers. Because $A = \pi r^2$, the value of A is also greater than zero. So the range is also the set of positive real numbers.

The *f*-Notation

When the variable y is a function of x, we may use the notation $f(x)$ to represent y. The symbol $f(x)$ is read as "f of x." So if x is the independent variable, we may use y or $f(x)$ to represent the dependent variable. For example, the function

$$y = 2x + 3$$

can also be written as

$$f(x) = 2x + 3.$$

We use y and $f(x)$ interchangeably. We think of f as the name of the function, and we may use letters other than f.

The expression $f(x)$ represents the second coordinate when the first coordinate is x; it does not mean f times x. For example, if we replace x by 4 in $f(x) = 2x + 3$, we get

$$f(4) = 2 \cdot 4 + 3 = 11.$$

So if the first coordinate is 4, then the second coordinate is $f(4)$, or 11. The ordered pair $(4, 11)$ belongs to the function f. This statement means that the function f pairs 4 with 11. We can use the diagram in Fig. 6.33 to picture this situation.

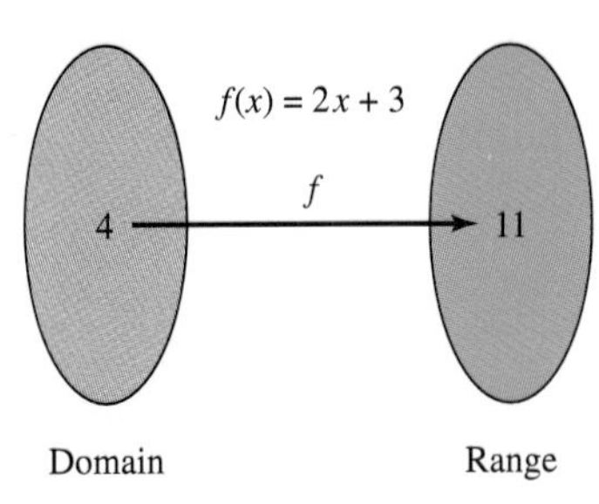

Figure 6.33

EXAMPLE 9 **Using f-notation**

Suppose $f(x) = x^2 - 1$ and $g(x) = -3x + 2$. Find the following:

a) $f(-2)$

b) $f(-1)$

c) $g(0)$

d) $g(6)$

Solution

a) Replace x by -2 in the formula $f(x) = x^2 - 1$:

$$\begin{aligned} f(-2) &= (-2)^2 - 1 \\ &= 4 - 1 \\ &= 3 \end{aligned}$$

So $f(-2) = 3$.

b) Replace x by -1 in the formula $f(x) = x^2 - 1$:

$$\begin{aligned} f(-1) &= (-1)^2 - 1 \\ &= 1 - 1 \\ &= 0 \end{aligned}$$

So $f(-1) = 0$.

c) Replace x by 0 in the formula $g(x) = -3x + 2$:

$$g(0) = -3 \cdot 0 + 2 = 2$$

So $g(0) = 2$.

d) Replace x by 6 in $g(x) = -3x + 2$ to get $g(6) = -16$. ■

Warm-ups

True or false? Explain your answer.

1. Any set of ordered pairs is a function.
2. The area of a square is a function of the length of a side.
3. The set $\{(-1, 3), (-3, 1), (-1, -3)\}$ is a function.
4. The set $\{(1, 5), (3, 5), (7, 5)\}$ is a function.
5. The domain of $f(x) = x^3$ is the set of all real numbers.
6. The domain of $y = |x|$ is the set of nonnegative real numbers.
7. The range of $y = |x|$ is the set of all real numbers.
8. The set $\{(x, y) \mid x = 2y\}$ is a function.
9. The set $\{(x, y) \mid x = y^2\}$ is a function.
10. If $f(x) = x^2 - 5$, then $f(-2) = -1$.

6.6 EXERCISES

Write a formula that describes the function for each of the following. See Examples 1 and 2.

1. A small pizza costs \$5.00 plus \$0.50 for each topping. Express the total cost C as a function of the number of toppings t.

2. A developer prices condominiums in Florida at \$20,000 plus \$40 per square foot of living area. Express the cost C as a function of the number of square feet of living area s.

3. The sales tax rate on groceries in Mayberry is 9%. Express the total cost T (including tax) as a function of the total price of the groceries S.

4. With a GM MasterCard, 5% of the amount charged is credited toward a rebate on the purchase of a new car. Express the rebate R as a function of the amount charged A.

5. Express the circumference of a circle as a function of its radius.

6. Express the circumference of a circle as a function of its diameter.

7. Express the perimeter P of a square as a function of the length s of a side.

8. Express the perimeter P of a rectangle with width 10 ft as a function of its length L.

9. Express the area A of a triangle with a base of 10 m as a function of its height h.

10. Express the area A of a trapezoid with bases 12 cm and 10 cm as a function of its height h.

Determine whether each table defines the second variable as a function of the first variable. See Example 3.

11.

x	y
1	1
4	2
9	3
16	4
25	5
36	6
49	8

12.

x	y
2	4
3	9
4	16
5	25
8	36
9	49
10	100

13.

t	V
2	2
−2	2
3	3
−3	3
4	4
−4	4
5	5

14.

s	W
5	17
6	17
−1	17
−2	17
−3	17
7	17
8	17

15.

a	P
2	2
2	−2
3	3
3	−3
4	4
4	−4
5	5

16.

n	r
17	5
17	6
17	−1
17	−2
17	−3
17	−4
17	−5

17.

b	q
1970	0.14
1972	0.18
1974	0.18
1976	0.22
1978	0.25
1980	0.28

18.

c	h
345	0.3
350	0.4
355	0.5
360	0.6
365	0.7
370	0.8
375	0.9

Determine whether each set of ordered pairs is a function. See Example 4.

19. $\{(1, 2), (2, 3), (3, 4)\}$

20. $\{(1, -3), (1, 3), (2, 12)\}$

21. $\{(-1, 4), (2, 4), (3, 4)\}$

22. $\{(1, 7), (7, 1)\}$

23. $\{(0, -1), (0, 1)\}$

24. $\{(1, 7), (-2, 7), (3, 7), (4, 7)\}$

25. $\{(50, 50)\}$

26. $\{(0, 0)\}$

Determine whether each set is a function. See Example 5.

27. $\{(x, y) \mid y = x - 3\}$

28. $\{(x, y) \mid y = x^2 - 2x - 1\}$

29. $\{(x, y) \mid x = |y|\}$

30. $\{(x, y) \mid x = y^2 + 1\}$

31. $\{(x, y) \mid x = y + 1\}$

32. $\left\{(x, y) \middle| y = \dfrac{1}{x}\right\}$

33. $\{(x, y) \mid x = y^2 - 1\}$

34. $\{(x, y) \mid x = 3y\}$

Determine whether each equation defines a function. See Example 6.

35. $x = 4y$ **36.** $x = -3y$ **37.** $y = \dfrac{2}{x}$

38. $y = \dfrac{x}{2}$ **39.** $y = x^3 - 1$ **40.** $y = |x - 1|$

41. $x^2 + y^2 = 25$ **42.** $x^2 - y^2 = 9$

Which of the following graphs are graphs of functions? See Example 7.

43.

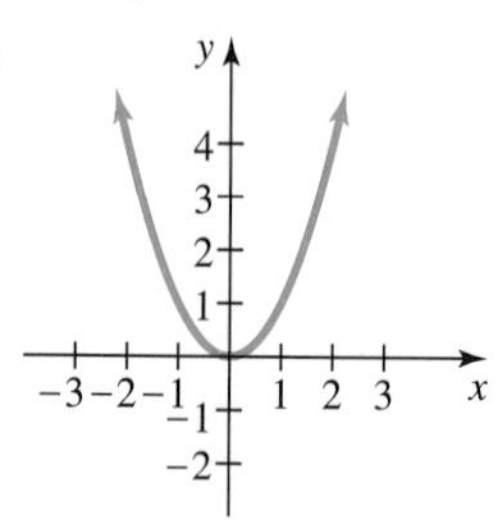

44.

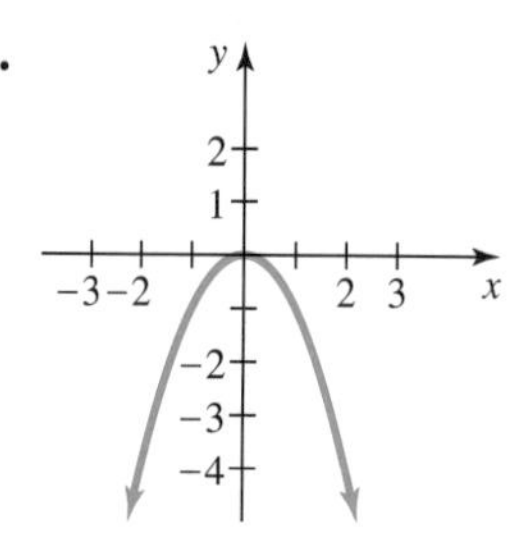

45.

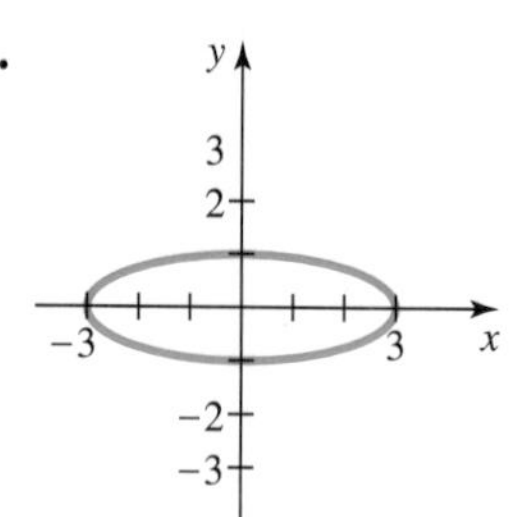

46.

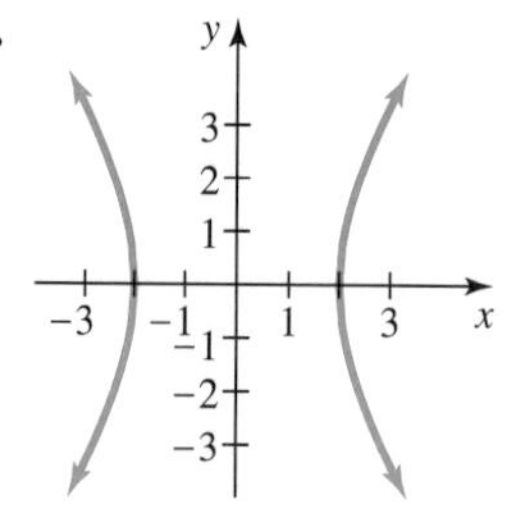

47.

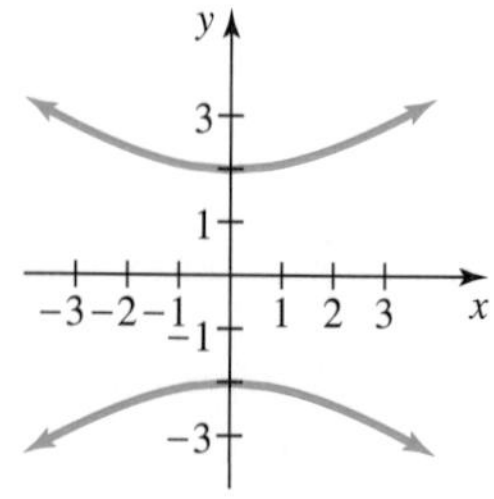

48.

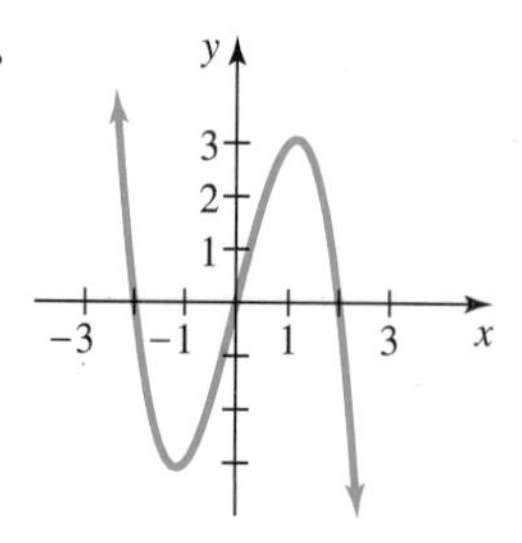

Determine the domain and range of each function. See Example 8.

49. $\{(3, 3), (2, 5), (1, 7)\}$ **50.** $\{(0, 1), (2, 1), (4, 1)\}$

51. $y = |x + 3|$ **52.** $y = |x - 1|$

53. $y = x$ **54.** $y = 2x + 1$

55. $y = x^2$ **56.** $y = x^3$

57. $A = s^2$ for $s > 0$ **58.** $S = -16t^2$ for $t \geq 0$

Let $f(x) = 2x - 1$, $g(x) = x^2 - 3$, and $h(x) = |x - 1|$. Find the following. See Example 9.

59. $f(0)$ **60.** $f(-1)$ **61.** $f\left(\dfrac{1}{2}\right)$

62. $f\left(\dfrac{3}{4}\right)$ **63.** $g(4)$ **64.** $g(-4)$

65. $g(0.5)$ **66.** $g(-1.5)$ **67.** $h(3)$

68. $h(-1)$ **69.** $h(0)$ **70.** $h(1)$

Let $f(x) = x^3 - x^2$ and $g(x) = x^2 - 4.2x + 2.76$. Find the following. Round each answer to three decimal places.

71. $f(5.68)$ **72.** $g(-2.7)$

73. $g(3.5)$ **74.** $f(67.2)$

Solve each problem.

75. ***Threshold weight.*** The threshold weight for an individual is the weight beyond which the risk of death increases significantly. For middle-aged males the function $W(h) = 0.000534h^3$ expresses the threshold weight in pounds as a function of the height h in inches. Find $W(70)$. Find the threshold weight for a 6'2" middle-aged male.

76. ***Pole vaulting.*** The height a pole vaulter attains is a function of the vaulter's velocity on the runway. The function $h(v) = \frac{1}{64}v^2$ gives the height in feet as a function of the velocity v in feet per second. Find $h(35)$ to the nearest tenth of an inch.

77. ***Credit card fees.*** A certain credit card company gets 4% of each charge, and the retailer receives the rest. At the end of a billing period the retailer receives a statement showing only the retailer's portion of each transaction. Express the original amount charged C as a function of the retailer's portion r.

78. ***More credit card fees.*** Suppose that the amount charged on the credit card in the previous exercise includes 8% sales tax. The credit card company does not get any of the sales tax. In this case the retailer's portion of each transaction includes sales tax on the original cost of the goods. Express the original amount charged C as a function of the retailers portion.

Getting More Involved

79. ***Writing.*** What are the three ways to define a function? What are the advantages of each method?

80. ***Discussion.*** Is the score on the last test in this class a function of the number of hours spent studying the night before the test? Is the test score a function of the number of classes attended? A variable such as a test score might be a function of many variables. Discuss the variables that might have had an effect on your score for the last test.

6.7 Variation

In this section:

- Direct Variation
- Finding the Constant
- Inverse Variation
- Joint Variation

If $y = 5x$, the value of y depends on the value of x. As x varies, so does y. Certain functions are customarily expressed in terms of variation. In this section you will learn to write formulas for those functions from verbal descriptions of the functions.

Direct Variation

Suppose you average 60 miles per hour on the freeway. The distance D that you travel depends on the amount of time T that you travel. Using the formula $D = R \cdot T$, we can write

$$D = 60T.$$

Consider the possible values for T and D given in Table 6.3.

T (hours)	1	2	3	4	5	6
D (miles)	60	120	180	240	300	360

Table 6.3

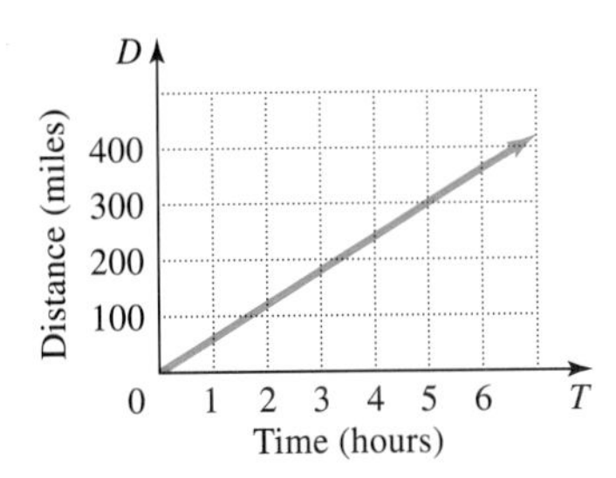

Figure 6.34

The graph of $D = 60T$ is shown in Fig. 6.34. Note that as T gets larger, so does D. In this situation we say that D *varies directly with* T, or D is *directly proportional* to T. The constant rate of 60 miles per hour is called the **variation constant** or **proportionality constant.** Notice that D is simply a linear function of T. We are just introducing some new terms to express an old idea.

> **Direct Variation**
>
> The statement "**y varies directly as x**" or "**y is directly proportional to x**" means that
>
> $$y = kx$$
>
> for some constant k. The constant k is a fixed nonzero real number.

Finding the Constant

If we know one ordered pair in a direct variation, then we can find the constant of variation.

EXAMPLE 1 **Finding a constant of variation**

Natasha is traveling by car, and the distance D that she travels varies directly as the rate R at which she drives. At 45 miles per hour, Natasha travels 135 miles. Find the constant of variation, and write D as a function of R.

Solution

Because D varies directly as R, there is a constant k such that

$$D = kR.$$

Because $D = 135$ when $R = 45$, we can write

$$135 = k \cdot 45$$

or

$$3 = k.$$

So $D = 3R$. ■

In the next example we find the constant of variation and use it to solve a variation problem.

EXAMPLE 2 **A direct variation problem**

Your electric bill at Middle States Electric Co-op varies directly with the amount of electricity that you use. If the bill for 2800 kilowatts of electricity is \$196, then what is the bill for 4000 kilowatts of electricity?

Solution

Because the amount A of the electric bill varies directly as the amount E of electricity used, we have

$$A = kE$$

for some constant k. Because 2800 kilowatts cost \$196, we have

$$196 = k2800$$

or

$$0.07 = k.$$

So $A = 0.07E$. Now if $E = 4000$ we get

$$A = 0.07(4000) = 280.$$

The bill for 4000 kilowatts would be \$280. ■

Inverse Variation

If you plan to make a 400-mile trip by car, the time it will take depends on your rate of speed. Using the formula $D = RT$, we can write

$$T = \frac{400}{R}.$$

Consider the possible values for R and T given in the following table:

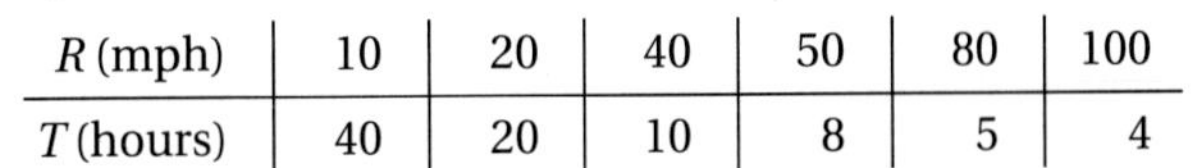

R (mph)	10	20	40	50	80	100
T (hours)	40	20	10	8	5	4

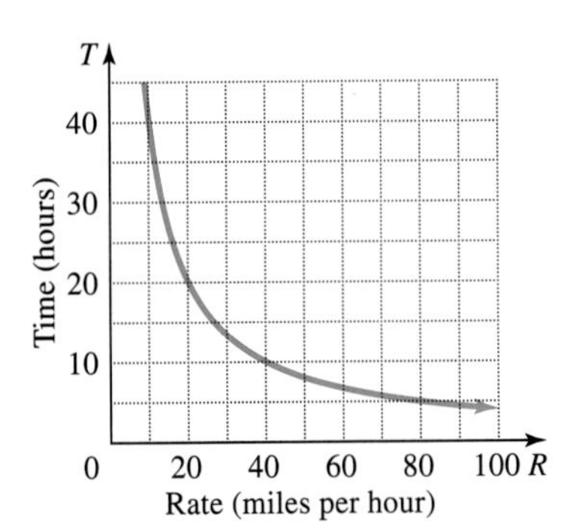

Figure 6.35

The graph of $T = \frac{400}{R}$ is shown in Fig. 6.35. As your rate increases, the time for the trip decreases. In this situation we say that the time is *inversely proportional* to the speed. In general, we make the following definition.

Inverse Variation

The statement **"y varies inversely as x"**, or **"y is inversely proportional to x"** means that

$$y = \frac{k}{x}$$

for some nonzero constant k.

CAUTION The constant of variation is usually positive because most physical examples involve positive quantities. However, the definitions of direct and inverse variation do not rule out a negative constant.

EXAMPLE 3 **An inverse variation problem**

The volume of a gas in a cylinder is inversely proportional to the pressure on the gas. If the volume is 12 cubic centimeters when the pressure on the gas is 200 kilograms per square centimeter, then what is the volume when the pressure is 150 kilograms per square centimeter? See Fig. 6.36.

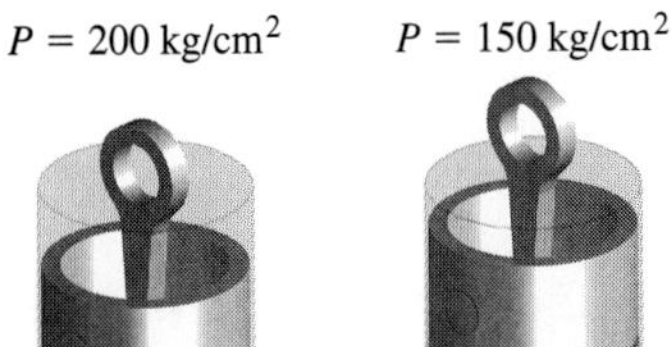

Figure 6.36

Solution

Because the volume V is inversely proportional to the pressure P, we have

$$V = \frac{k}{P}$$

for some constant k. Because $V = 12$ when $P = 200$, we can find k by substituting these values into the above formula:

$$12 = \frac{k}{200}$$

$$200 \cdot 12 = 200 \cdot \frac{k}{200} \quad \text{Multiply each side by 200.}$$

$$2400 = k$$

Now to find V when $P = 150$, we can use the formula $V = \frac{2400}{P}$:

$$V = \frac{2400}{150}$$

$$= 16$$

So the volume is 16 cubic centimeters when the pressure is 150 kilograms per square centimeter.

Joint Variation

If the price of carpet is \$30 per square yard, then the cost C of carpeting a rectangular room depends on the width W (in yards) and the length L (in yards). As the width or length of the room increases, so does the cost. We can write the cost as a function of the two variables L and W:

$$C = 30LW$$

We say that C *varies jointly* as L and W.

Joint Variation

The statement **"y varies jointly as x and z"** or **"y is jointly proportional to x and z"** means that

$$y = kxz$$

for some nonzero constant k.

EXAMPLE 4 **A joint variation problem**

The cost of shipping a piece of machinery by truck varies jointly with the weight of the machinery and the distance that it is shipped. It costs \$3000 to ship a 2500-lb milling machine a distance of 600 miles. Find the cost for shipping a 1500-lb lathe a distance of 800 miles.

Solution

Because the cost C varies jointly with the weight w and the distance d, we have

$$C = kwd$$

where k is the constant of proportionality. To find k, we use $C = 3000$, $w = 2500$, and $d = 600$:

$$3000 = k \cdot 2500 \cdot 600$$

$$\frac{3000}{2500 \cdot 600} = k \quad \text{Divide each side by } 2500 \cdot 600.$$

$$0.002 = k$$

Now use $w = 1500$ and $d = 800$ in the formula $C = 0.002wd$:

$$C = 0.002 \cdot 1500 \cdot 800$$
$$= 2400$$

So the cost of shipping the lathe is \$2400. ■

CAUTION The variation terms (directly, inversely, or jointly) are never used to indicate addition or subtraction. We use multiplication in the formula unless we see the word "inversely." We use division only for inverse variation.

Warm-ups

True or false? Explain your answer.

1. If y varies directly as z, then $y = kz$ for some constant k.
2. If a varies inversely as b, then $a = \frac{b}{k}$ for some constant k.
3. If y varies directly as x and $y = 8$ when $x = 2$, then the variation constant is 4.
4. If y varies inversely as x and $y = 8$ when $x = 2$, then the variation constant is $\frac{1}{4}$.
5. If C varies jointly as h and t, then $C = ht$.
6. The amount of sales tax on a new car varies directly with the purchase price of the car.
7. If z varies inversely as w and $z = 10$ when $w = 2$, then $z = \frac{20}{w}$.
8. The time that it takes to travel a fixed distance varies inversely with the rate.
9. If m varies directly as w, then $m = w + k$ for some constant k.
10. If y varies jointly as x and z, then $y = k(x + z)$ for some constant k.

6.7 EXERCISES

Write a formula that expresses the relationship described by each statement. Use k for the constant in each case. See Examples 1–4.

1. T varies directly as h.
2. m varies directly as p.
3. y varies inversely as r.
4. u varies inversely as n.
5. R is jointly proportional to t and s.
6. W varies jointly as u and v.
7. i is directly proportional to b.
8. p is directly proportional to x.
9. A is jointly proportional to y and m.
10. t is inversely proportional to e.

Find the variation constant, and write a formula that expresses the indicated variation. See Example 1.

11. y varies directly as x, and $y = 5$ when $x = 3$.
12. m varies directly as w, and $m = \frac{1}{2}$ when $w = \frac{1}{4}$.
13. A varies inversely as B, and $A = 3$ when $B = 2$.
14. c varies inversely as d, and $c = 5$ when $d = 2$.
15. m varies inversely as p, and $m = 22$ when $p = 9$.
16. s varies inversely as v, and $s = 3$ when $v = 4$.
17. A varies jointly as t and u, and $A = 24$ when $t = 6$ and $u = 2$.
18. N varies jointly as p and q, and $N = 720$ when $p = 3$ and $q = 2$.
19. T varies directly as u, and $T = 9$ when $u = 2$.
20. R varies directly as p, and $R = 30$ when $p = 6$.

Solve each variation problem. See Examples 2–4.

21. Y varies directly as x, and $Y = 100$ when $x = 20$. Find Y when $x = 5$.
22. n varies directly as q, and $n = 39$ when $q = 3$. Find n when $q = 8$.
23. a varies inversely as b, and $a = 3$, when $b = 4$. Find a when $b = 12$.
24. y varies inversely as w, and $y = 9$ when $w = 2$. Find y when $w = 6$.
25. P varies jointly as s and t, and $P = 56$ when $s = 2$ and $t = 4$. Find P when $s = 5$ and $t = 3$.
26. B varies jointly as u and v, and $B = 12$ when $u = 4$ and $v = 6$. Find B when $u = 5$ and $v = 8$.

Solve each problem.

27. ***Aluminum flatboat.*** The weight of an aluminum flatboat varies directly with the length of the boat. If a 12-foot boat weighs 86 pounds, then what is the weight of a 14-foot boat?

28. ***Christmas tree.*** The price of a Christmas tree varies directly with the height. If a 5-foot tree costs \$20, then what is the price of a 6-foot tree?

29. ***Sharing the work.*** The time it takes to erect the big circus tent varies inversely as the number of elephants working on the job. If it takes four elephants 75 minutes, then how long would it take six elephants?

30. ***Gas laws.*** The volume of a gas is inversely proportional to the pressure on the gas. If the volume is 6 cubic centimeters when the pressure on the gas is 8 kilograms per square centimeter, then what is the volume when the pressure is 12 kilograms per square centimeter?

31. ***Steel tubing.*** The cost of steel tubing is jointly proportional to its length and diameter. If a 10-foot tube with a 1-inch diameter costs \$5.80, then what is the cost of a 15-foot tube with a 2-inch diameter?

32. ***Sales tax.*** The amount of sales tax varies jointly with the number of Cokes purchased and the price per Coke. If the sales tax on eight Cokes at 65 cents each is 26 cents, then what is the sales tax on six cokes at 90 cents each?

33. ***Approach speed.*** The approach speed of an airplane is directly proportional to its landing speed. If the approach speed for a Piper Cheyenne is 90 mph with a landing speed of 75 mph, then what is the landing speed for an airplane with an approach speed of 96 mph?

34. ***Ideal waist size.*** According to Dr. Aaron R. Folsom of the University of Minnesota School of Public Health, your maximum ideal waist size is directly proportional to your hip size. For a woman with 40-inch hips, the maximum ideal waist size is 32 inches. What is the maximum ideal waist size for a woman with 35-inch hips?

Getting More Involved

35. ***Discussion.*** If y varies directly as x, then the graph of the equation is a straight line. What is its slope? What is the y-intercept? If $y = 3x + 2$, then does y vary directly as x? Which straight lines correspond to direct variations?

36. ***Writing.*** Write a summary of the three types of variation. Include an example of each type that is not found in this text.

COLLABORATIVE ACTIVITIES

Inches or Centimeters?

Grouping: 3 to 4 students
Topic: Plotting points, graphing lines

In this activity you will generate data by measuring in both inches and centimeters the height of each member of your group. Then you will plot the points on a graph and use any two of your points to find the conversion formula for converting inches to centimeters.

Part I: Measure the height of each person in your group and fill out a table like the one below:

Name	Height in inches	Height in centimeters

Part II: The numbers for inches and centimeters from the table will give you three or four ordered pairs to graph. Plot these points on a graph. Let inches be the horizontal x-axis and centimeters be the vertical y-axis. Let each mark on the axes represent 10 units. When graphing, you will need to estimate the place to plot fractional values.

Part III: Use any two of your points to find an equation of the line you have graphed. What is the slope of your line? Where does it cross the horizontal axis?

Extension: Look up the conversion formula for converting inches to centimeters. Is it the same as the one you found by measuring? If it is different, what could account for the difference?

Wrap-up CHAPTER 6

SUMMARY

Slope of a Line

		Examples
Slope	The slope of the line through (x_1, y_1) and (x_2, y_2) is given by $$m = \frac{y_2 - y_1}{x_2 - x_1}, \text{ provided that } x_2 - x_1 \neq 0.$$ Slope is the ratio of the rise to the run for any two points on the line: $$m = \frac{\text{change in } y}{\text{change in } x} = \frac{\text{rise}}{\text{run}}$$	$(0, 1), (3, 5)$ $m = \frac{5 - 1}{3 - 0} = \frac{4}{3}$
Types of slope	Positive slope; Negative slope; Zero slope; Undefined slope	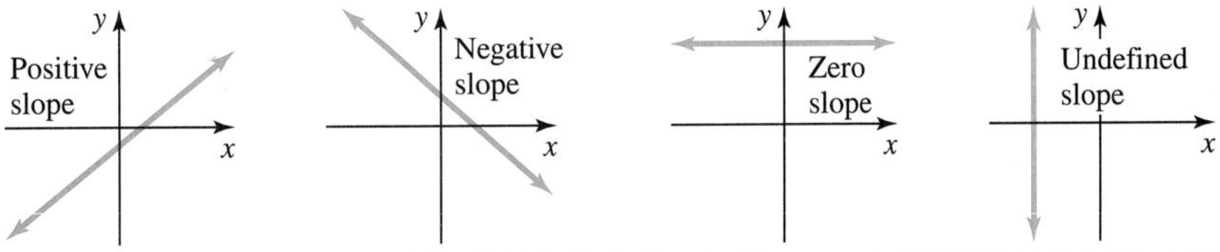
Parallel lines	Nonvertical parallel lines have equal slopes. Two vertical lines are parallel.	The lines $y = 3x - 9$ and $y = 3x + 7$ are parallel lines.
Perpendicular lines	Lines with slopes m and $-\frac{1}{m}$ are perpendicular. Any vertical line is perpendicular to any horizontal line.	The lines $y = -5x + 7$ and $y = \frac{1}{5}x$ are perpendicular.

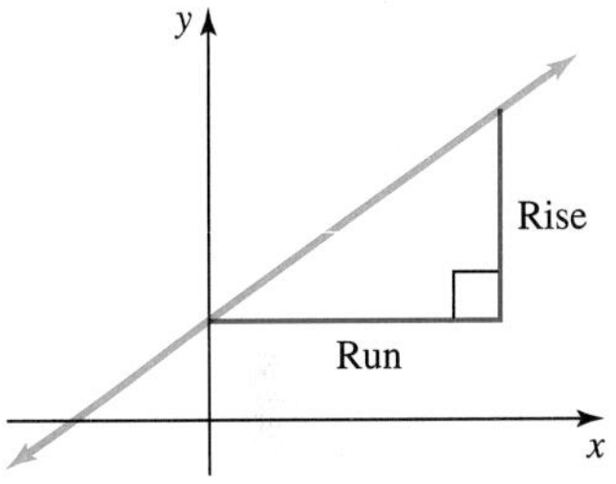

Equations of Lines

		Examples
Slope-intercept form	The equation of the line with y-intercept $(0, b)$ and slope m is $y = mx + b$.	$y = 3x - 1$ has slope 3 and y-intercept $(0, -1)$.
Point-slope form	The equation of the line with slope m that contains the point (x_1, y_1) is $y - y_1 = m(x - x_1)$.	The line through $(2, -1)$ with slope -5 is $y + 1 = -5(x - 2)$.
Standard form	Every line has an equation of the form $Ax + By = C$, where A, B, and C are real numbers with A and B not both equal to zero.	$4x - 9y = 15$ $x = 5$ (vertical line) $y = -7$ (horizontal line)
Graphing a line using y-intercept and slope	1. Write the equation in slope-intercept form. 2. Plot the y-intercept. 3. Use the rise and run to locate a second point. 4. Draw a line through the two points.	

Functions		**Examples**
Definition of a function	A function is a rule by which any allowable value of one variable (the independent variable) determines a unique value of a second variable (the dependent variable).	$A = \pi r^2$
Equivalent definition of a function	A function is a set of ordered pairs such that no two ordered pairs have the same first coordinates and different second coordinates. To say that y is a function of x means that y is determined uniquely by x.	$\{(1, 0), (3, 8)\}$ $\{(x, y) \mid y = x^2\}$
Domain	The set of values of the independent variable, x	$y = x^2$ Domain: all real numbers
Range	The set of values of the dependent variable, y	$y = x^2$ Range: nonnegative real numbers
Linear functions	If $y = mx + b$, we say that y is a linear function of x.	$F = \frac{9}{5}C + 32$
f-notation	If x is the independent variable, then we use the notation $f(x)$ to represent the dependent variable.	$y = 2x + 3$ $f(x) = 2x + 3$
Variation	If $y = kx$, then y varies directly as x.	$D = 50T$
	If $y = \frac{k}{x}$, then y varies inversely as x.	$R = \frac{400}{T}$
	If $y = kxz$, then y varies jointly as x and z.	$V = 6LW$

REVIEW EXERCISES

6.1 *For each point, name the quadrant in which it lies or the axis on which it lies.*

1. $(-2, 5)$ **2.** $(-3, -5)$ **3.** $(3, 0)$

4. $(9, 10)$ **5.** $(0, -6)$ **6.** $(0, \pi)$

7. $(1.414, -3)$ **8.** $(-4, 1.732)$

Complete the given ordered pairs so that each ordered pair satisfies the given equation.

9. $y = 3x - 5$: $(0,\ )$, $(-3,\ )$, $(4,\ )$

10. $y = -2x + 1$: $(9,\ )$, $(3,\ )$, $(-1,\ )$

11. $2x - 3y = 8$: $(0,\ )$, $(3,\ )$, $(-6,\ )$

12. $x + 2y = 1$: $(0,\ )$, $(-2,\ )$, $(2,\ )$

Sketch the graph of each equation by finding three ordered pairs that satisfy each equation.

13. $y = -3x + 4$ **14.** $y = 2x - 6$

15. $x + y = 7$ **16.** $x - y = 4$

6.2 *Determine the slope of the line that goes through each pair of points.*

17. $(0, 0)$ and $(1, 1)$ **18.** $(-1, 1)$ and $(2, -2)$

19. $(-2, -3)$ and $(0, 0)$ **20.** $(-1, -2)$ and $(4, -1)$

21. $(-4, -2)$ and $(3, 1)$ **22.** $(0, 4)$ and $(5, 0)$

Use slope to solve each geometric problem.

23. Determine whether the points $(-4, 1)$, $(-3, -3)$, and $(5, -1)$ are the vertices of a right triangle.

24. Determine whether the points $(-3, 4)$, $(-1, 2)$, $(0, 3)$, and $(2, 1)$ are the vertices of a parallelogram.

25. Show that the diagonals of the quadrilateral with vertices $(2, 1)$, $(6, 2)$, $(7, 6)$, and $(3, 5)$ are perpendicular.

26. Show that the opposite sides of the quadrilateral of Exercise 25 are parallel.

6.3 *Find the slope and y-intercept for each line.*

27. $y = 3x - 18$

28. $y = -x + 5$

29. $2x - y = 3$

30. $x - 2y = 1$

31. $4x - 2y - 8 = 0$

32. $3x + 5y + 10 = 0$

Sketch the graph of each equation.

33. $y = \frac{2}{3}x - 5$

34. $y = \frac{3}{2}x + 1$

35. $2x + y = -6$

36. $-3x - y = 2$

37. $y = -4$

38. $x = 9$

Determine the equation of each line. Write the answer in standard form using only integers as the coefficients.

39. The line through $(0, 4)$ with slope $\frac{1}{3}$

40. The line through $(-2, 0)$ with slope $-\frac{3}{4}$

41. The line through the origin that is perpendicular to the line $y = 2x - 1$

42. The line through $(0, 9)$ that is parallel to the line $3x + 5y = 15$

43. The line through $(3, 5)$ that is parallel to the x-axis

44. The line through $(-2, 4)$ that is perpendicular to the x-axis

6.4 *Write each equation in slope-intercept form.*

45. $y - 3 = \frac{2}{3}(x + 6)$

46. $y + 2 = -6(x - 1)$

47. $3x - 7y - 14 = 0$

48. $1 - x - y = 0$

49. $y - 5 = -\frac{3}{4}(x + 1)$

50. $y + 8 = -\frac{2}{5}(x - 2)$

Determine the equation of each line. Write the answer in slope-intercept form.

51. The line through $(-4, 7)$ with slope -2

52. The line through $(9, 0)$ with slope $\frac{1}{2}$

53. The line through the two points $(-2, 1)$ and $(3, 7)$

54. The line through the two points $(4, 0)$ and $(-3, -5)$

55. The line through $(3, -5)$ that is parallel to the line $y = 3x - 1$

56. The line through $(4, 0)$ that is perpendicular to the line $x + y = 3$

6.5 *Graph each linear equation for the indicated values of the independent variable.*

57. $P = -3t + 400,\ 10 \le t \le 90$

58. $R = 40w - 300,\ 20 \le w \le 80$

59. $v = 50n + 30,\ 0.1 \le n \le 0.9$

60. $w = -40q + 8000,\ 0 \le q \le 1000$

Solve each problem.

61. ***Rental charge.*** The charge C for renting an air hammer from Taylor and Son Equipment Rental is a linear function of the number n of days in the rental period. The charge is \$113 for two days and \$209 for five days. Write C as a linear function of n. What would the charge be for four days?

62. ***Time on a treadmill.*** After 2 minutes on a treadmill, Jenny has a heart rate of 82. After 3 minutes she has a heart rate of 86. Assuming that Jenny's heart rate h is a linear function of the time t on the treadmill, write h as a linear function of t. What heart rate could be expected for Jenny after 10 minutes on the treadmill?

63. ***Probability of rain.*** When the probability p of rain is 90%, the probability q that it does not rain is 10%. When the probability p of rain is 80%, the probability q that it does not rain is 20%. Assuming that the probability q that it does not rain is a linear function of the probability p of rain, write q as a function of p.

64. ***Social Security benefits.*** On the basis of current legislation, if you earned an average salary of \$25,000 over your working life and you retire after the year 2005 at age 62, 63, or 64, then your annual Social Security benefit will be \$7000, \$7500, or \$8000, respectively (Source: Social Security Administration). Write a formula that gives annual benefit as a linear function of age for these three ages.

6.6 *Determine whether each set of ordered pairs is a function.*

65. $\{(4, 3), (5, 3)\}$

66. $\{(0, 0), (0, 1), (0, 2)\}$

67. $\{(3, 4), (3, 5)\}$

68. $\{(1, 2), (2, 3), (3, 4)\}$

69. $\{(x, y) \mid y = 45x\}$

70. $\{(x, y) \mid x = y^3\}$

Determine whether each equation defines y as a function of x.

71. $y = x^2 + 10$

72. $y = 2x - 7$

73. $x^2 + y^2 = 1$

74. $x^2 = y^2$

Determine the domain and range of each function.

75. $f(x) = 2x - 3$

76. $g(x) = -|x|$

77. $\{(1, 2), (2, 0), (3, 0)\}$

78. $\{(2, 3), (4, 3), (6, 3)\}$

79. $\{(x, y) \mid y = x^2\}$

80. $\{(x, y) \mid y = x^4\}$

6.7 *Solve each variation problem.*

81. Suppose y varies directly as w. If $y = 48$ when $w = 4$, then what is y when $w = 11$?

82. Suppose m varies directly as t. If $m = 13$ when $t = 2$, then what is m when $t = 6$?

83. If y varies inversely as v and $y = 8$ when $v = 6$, then what is y when $v = 24$?

84. If y varies inversely as r and $y = 9$ when $r = 3$, then what is y when $r = 9$?

85. Suppose y varies jointly as u and v, and $y = 72$ when $u = 3$ and $v = 4$. Find y when $u = 5$ and $v = 2$.

86. Suppose q varies jointly as s and t, and $q = 10$ when $s = 4$ and $t = 3$. Find q when $s = 25$ and $t = 6$.

87. ***Taxi fare.*** The cost of a taxi ride varies directly with the length of the ride. If a 12-minute ride costs $9.00, then what should be the cost of a 20-minute ride?

88. ***Installing shingles.*** The number of hours it takes to apply 296 bundles of shingles varies inversely with the number of roofers working on the job. If three roofers can complete the job in 40 hours, then how long would it take five roofers?

CHAPTER 6 TEST

For each point, name the quadrant in which it lies or the axis on which it lies.

1. $(-2, 7)$ **2.** $(-\pi, 0)$

3. $(3, -6)$ **4.** $(0, 1785)$

Find the slope of the line through each pair of points.

5. $(3, 3)$ and $(4, 4)$ **6.** $(-2, -3)$ and $(4, -8)$

Write the equation of each line. Give the answer in slope-intercept form.

7. The line through $(0, 3)$ with slope $-\frac{1}{2}$

8. The line through $(-1, -2)$ with slope $\frac{3}{7}$

Write the equation of each line. Give the answer in standard form using only integers as the coefficients.

9. The line through $(2, -3)$ that is perpendicular to the line $y = -3x + 12$

10. The line through $(3, 4)$ that is parallel to the line $5x + 3y = 9$

Sketch the graph of each equation.

11. $y = \frac{1}{2}x - 3$ **12.** $2x - 3y = 6$

13. $y = 4$ **14.** $x = -2$

Determine whether each set is a function.

15. $\{(1, 3), (2, 3), (5, 7)\}$ **16.** $\{(x, y) \mid x = y^4\}$

Determine the domain and range of each function.

17. $f(x) = |x| + 1$ **18.** $y = x^2$

Let $f(x) = 2x + 5$ and $g(x) = x^2 - 4$. Find the following.

19. $f(-2)$ **20.** $g(3)$

Solve each problem.

21. Determine whether the triangle whose vertices are $(-2, -3)$, $(4, 0)$, and $(3, 2)$ is a right triangle.

22. Julie's mail-order CD club charges a shipping and handling fee of $2.50 plus $0.75 per CD for each order shipped. Write the shipping and handling fee S as a function of the number n of CDs in the order.

23. The price P of a soft drink is a linear function of the volume v of the cup. A 10-ounce drink sells for 50 cents, and a 16-ounce drink sells for 68 cents. Write P as a linear function of v. What should the price be for a 20-ounce drink?

24. The price of a watermelon varies directly with its weight. If the price of a 30-pound watermelon is $4.20, then what is the price of a 20-pound watermelon?

25. The amount of time that Jason spends studying for an algebra test is inversely proportional to his score on the previous test. If Jason studied three hours when his previous test score was 60, then how many hours would he study when his previous test score was 90?

Tying It All Together CHAPTERS 1–6

Simplify each expression.

1. $3^2 - 2^3$

2. $3^2 \cdot 2^3$

3. $10^4 \cdot 10^9$

4. $2^{12} \div 2^{10}$

5. $(34 \cdot 258)^0$

6. $(8^0 - 3^2)^3$

7. $\left(\frac{1}{2}\right)^3 + \left(\frac{2}{3}\right)^2$

8. $\left(-\frac{3}{2}\right)^3 - \left(-\frac{3}{4}\right)^2$

9. $\left(\frac{3}{5}\right)^3 \cdot \left(\frac{5}{6}\right)^2$

10. $\left(-\frac{3}{5}\right)^3 \div \left(-\frac{6}{5}\right)^4$

11. $\dfrac{\frac{1}{4} - \frac{1}{8}}{\frac{3}{4} + \frac{1}{2}}$

12. $\dfrac{\frac{1}{3} - \frac{1}{5}}{\frac{3}{10} + \frac{1}{20}}$

Perform the indicated operations.

13. $-3(2x - 7)$

14. $x - 3(2x - 7)$

15. $(x - 3)(2x - 7)$

16. $(2x - 1)^2$

17. $(z + 5)^2$

18. $(w - 7)(w + 7)$

19. $3x^2y^3 \cdot 12xy^4$

20. $(2x^2y)^3 \cdot 5x^2y^6$

Sketch a graph of each equation.

21. $y = \frac{1}{3}x$

22. $y = 3x$

23. $y = -3x$

24. $y = -\frac{1}{3}x$

25. $y = 3x + 1$

26. $y = 3x - 2$

27. $y = 3$

28. $x = 3$

Solve each equation for y.

29. $3\pi y + 2 = t$

30. $x = \dfrac{y - b}{m}$

31. $3x - 3y - 12 = 0$

32. $2y - 3 = 9$

33. $y^2 - 3y - 40 = 0$

34. $\frac{y}{2} - \frac{y}{4} = \frac{1}{5}$

Solve each equation.

35. $5 = 4x - 7$

36. $5 = 4x^2 - 11$

37. $(3x - 4)(x + 9) = 0$

38. $\frac{2}{3} - \frac{x}{6} = \frac{1}{2} + \frac{x}{4}$

39. $2x^2 - 7x = 0$

40. $\frac{3}{x} = \frac{x - 1}{2}$

Solve.

41. ***Financial planning.*** Financial advisors at Fidelity Investments use the information in the accompanying graph as a guide for retirement investing.

a) What is the slope of the line segment for ages 35 through 50?

b) What is the slope of the line segment for ages 50 through 65?

c) If a 38-year-old man is making $40,000 per year, then what percent of his income should he be saving?

d) If a 58-year-old woman has an annual salary of $60,000, then how much should she have saved and how much should she be saving per year?

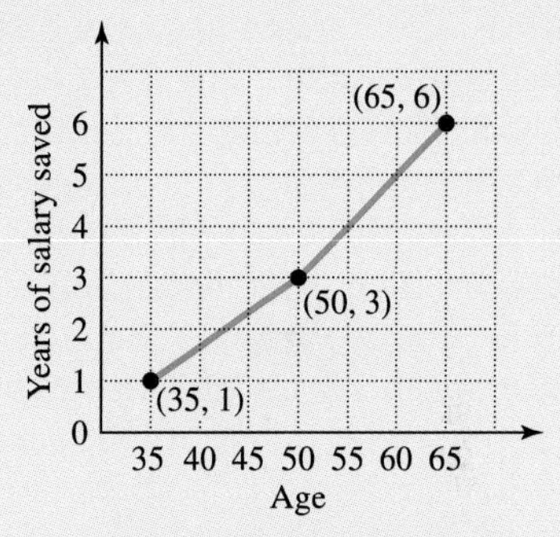

Figure for Exercise 41

CHAPTER 7

Systems of Equations and Inequalities

What determines the prices of the products that you buy? Why do prices of some products go down while the prices of others go up? Economists theorize that prices result from the collective decisions of consumers and producers. Ideally, the demand or quantity purchased by consumers depends only on the price, and price is a function of the supply. Theoretically, if the demand is greater than the supply, then prices rise and manufacturers produce more to meet the demand. As the supply of goods increases, the price comes down. The price at which the supply is equal to the demand is called the equilibrium price.

However, what happens in the real world does not always match the theory. Manufacturers cannot always control the supply, and factors other than price can affect a consumer's decision to buy. For example, droughts in Brazil decreased the supply of coffee and drove coffee prices up. Floods in California did the same to the prices of produce. With one of the most abundant wheat crops ever in 1994, cattle gained weight more quickly, increasing the supply of cattle ready for market. With supply going up, prices went down. Decreased demand for beef in Japan and Mexico drove the price of beef down further. With lower prices, consumers should be buying more beef, but increased competition from chicken and pork products, as well as health concerns, have kept consumer demand low.

The two functions that govern supply and demand form a system of equations. In this chapter you will learn how to solve systems of equations. In Exercise 31 of Section 7.2 you will see an example of supply and demand equations for ground beef.

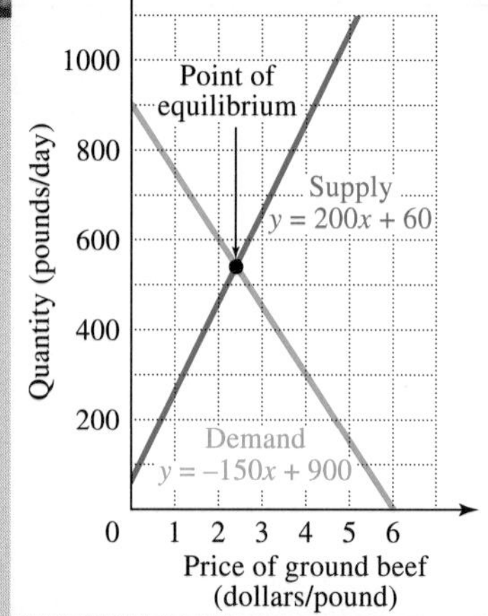

7.1 The Graphing Method

In this section:

- Solving a System of Linear Equations by Graphing
- Independent, Inconsistent, and Dependent Equations

You studied linear equations in two variables in Chapter 6. In this section you will learn to solve systems of linear equations in two variables and use systems to solve problems.

Solving a System of Linear Equations by Graphing

Consider the linear equation $y = 2x - 1$. The graph of this equation is a straight line, and every point on the line is a solution to the equation. Now consider a second linear equation, $x + y = 2$. The graph of this equation is also a straight line, and every point on the line is a solution to this equation. The pair of equations

$$y = 2x - 1$$
$$x + y = 2$$

is called a **system of equations.** A point that satisfies both equations is called a **solution to the system.**

EXAMPLE 1 **A solution to a system**

Determine whether the point $(-1, 3)$ is a solution to each system of equations.

a) $3x - y = -6$
$x + 2y = 5$

b) $y = 2x - 1$
$x + y = 2$

Solution

a) If we let $x = -1$ and $y = 3$ in both equations of the system, we get the following equations:

$$3(-1) - 3 = -6 \quad \text{Correct}$$
$$-1 + 2(3) = 5 \quad \text{Correct}$$

Because both of these equations are correct, $(-1, 3)$ is a solution to the system.

b) If we let $x = -1$ and $y = 3$ in both equations of the system, we get the following equations:

$$3 = 2(-1) - 1 \quad \text{Incorrect}$$
$$-1 + 3 = 2 \quad \text{Correct}$$

Because the first equation is not satisfied by $(-1, 3)$, the point $(-1, 3)$ is not a solution to the system. ■

If we graph each equation of a system on the same coordinate plane, then we may be able to see the points that they have in common. Any point that is on both graphs is a solution to the system.

EXAMPLE 2 **Using graphing to solve a system**

Solve the system by graphing:

$$y = 2x - 1$$
$$x + y = 2$$

Solution

We first write each equation in slope-intercept form:

$$y = 2x - 1$$
$$y = -x + 2$$

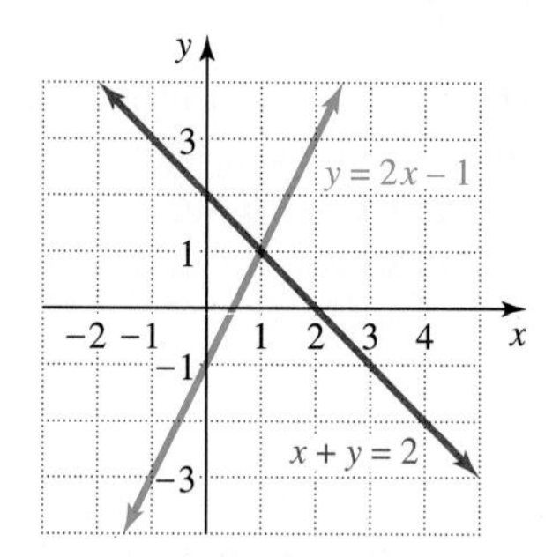

Figure 7.1

Use the y-intercept and the slope to draw the graphs as in Fig. 7.1. From the graph, it appears that these lines intersect at $(1, 1)$. To be certain, we check that $(1, 1)$ satisfies both equations. Let $x = 1$ and $y = 1$ in the original equations:

$$y = 2x - 1 \qquad x + y = 2$$
$$1 = 2(1) - 1 \qquad 1 + 1 = 2$$

Because these equations are both true, $(1, 1)$ is the solution to the system. ■

EXAMPLE 3 **A system of parallel lines**

Solve the system by graphing:

$$3y = 2x - 6$$
$$2x - 3y = 3$$

Solution

Write each equation in slope-intercept form to get the following system:

$$y = \frac{2}{3}x - 2$$
$$y = \frac{2}{3}x - 1$$

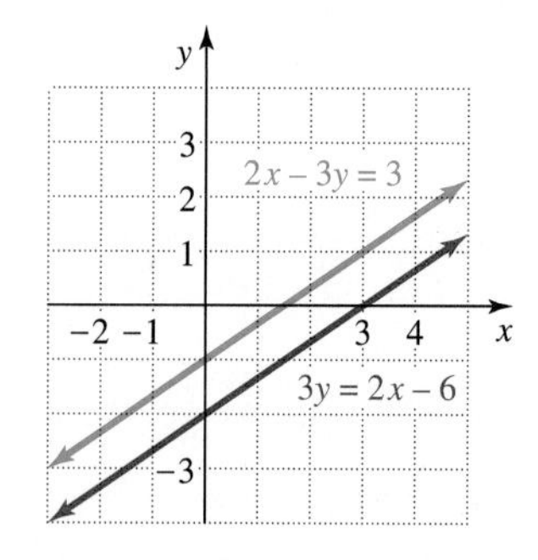

Figure 7.2

Each line has slope $\frac{2}{3}$, but they have different y-intercepts. Their graphs are shown in Fig. 7.2. Because these two lines have the same slope, they are parallel and there is no point on both lines. The system has no solution. ■

Independent, Inconsistent, and Dependent Equations

When two lines are positioned in a coordinate plane, they might intersect at a single point, they might have no intersection, or they might coincide. There are no other possibilities. If they intersect at a single point, then there is exactly one solution to the corresponding system of linear equations. If the lines are parallel, then there is no solution to the system. If the lines are coincident, then any point on the line satisfies both equations of the system.

Independent, Inconsistent, and Dependent

The equations of a system of two linear equations are

1. **independent** if the lines intersect in exactly one point,
2. **inconsistent** if the lines are parallel, and
3. **dependent** if the lines coincide.

Figure 7.3 shows each of the cases.

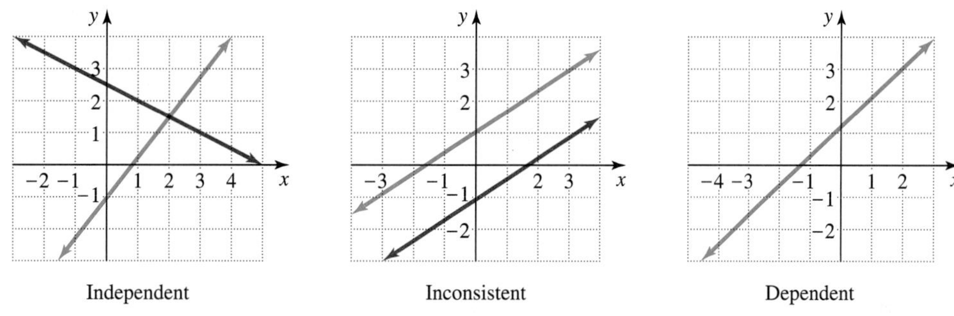

Figure 7.3

Because the system of Example 2 had a single point for the solution, the equations are independent (or the system is independent). Because the graphs of the equations in Example 3 were parallel, the equations are inconsistent and there is no solution to that system. The next example illustrates a system of dependent equations.

EXAMPLE 4 **A dependent system of equations**

Solve the system by graphing:

$$4x - 2y = 6$$
$$y - 2x = -3$$

Solution

Rewrite both equations in slope-intercept form for easy graphing:

$$\begin{aligned} 4x - 2y &= 6 \\ -2y &= -4x + 6 \\ y &= 2x - 3 \end{aligned} \qquad \begin{aligned} y - 2x &= -3 \\ y &= 2x - 3 \end{aligned}$$

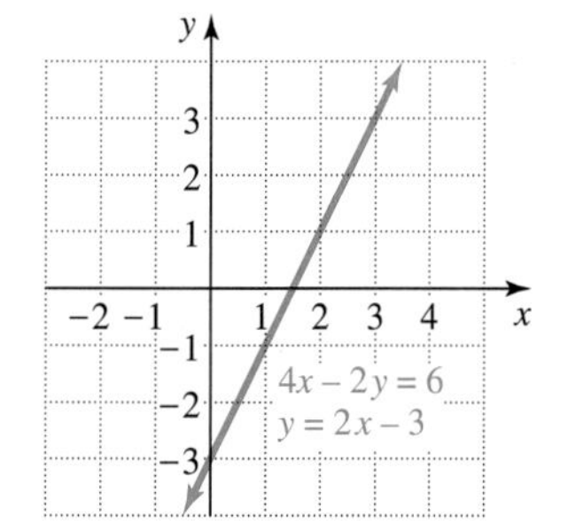

Figure 7.4

By writing the equations in slope-intercept form, we discover that they are identical. The graphs of the system are shown in Fig. 7.4. Because the graphs of the two equations are identical, any point on the line satisfies both equations. The set of points on that line is written as

$$\{(x, y) \mid y = 2x - 3\}.$$

The system is dependent. ■

Calculator Close-up

A graphing calculator can show the graphs of both equations of a system in a single viewing window. The **trace** feature can then be used to estimate the solution to an independent system. You can also use the **zoom** feature to "blow up" the intersection and get more accuracy. Consult your graphing calculator manual for details about these features on your calculator.

Figure 7.5 shows the graphs of $y = 1.2x - 8.6$ and $y = -3.4x + 2.1$. Graph these lines on your calculator, and verify that the intersection of the lines is approximately $(2.3, -5.8)$.

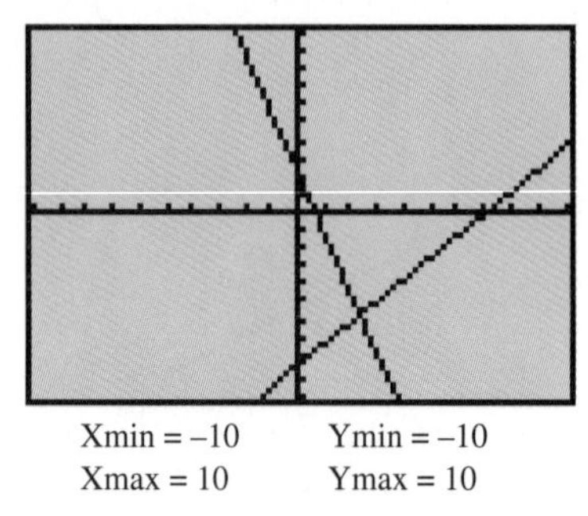

Xmin = −10 Ymin = −10
Xmax = 10 Ymax = 10

Figure 7.5

Warm-ups

True or false? Explain your answer.
The statements refer to the following systems:

a) $y = 2x - 5$ **b)** $y = 3x - 4$ **c)** $x + y = 9$
$y = -2x - 5$ $y = 3x + 5$ $y = 9 - x$

1. The ordered pair $(1, -3)$ satisfies the equation $y = 2x - 5$.
2. The ordered pair $(1, -3)$ satisfies the equation $y = -2x - 5$.
3. The ordered pair $(1, -3)$ is a solution to system (a).
4. System (a) is inconsistent.
5. System (b) has no solution.
6. The equations of system (b) are inconsistent.
7. System (c) is dependent.
8. The set of ordered pairs that satisfy system (c) is $\{(x, y) \mid y = 9 - x\}$.
9. Two distinct straight lines in the coordinate plane either are parallel or intersect each other in exactly one point.
10. Any system of two linear equations can be solved by graphing.

7.1 EXERCISES

Which of the given points is a solution to the given system? See Example 1.

1. $2x + y = 4$ $\quad (6, 1), (3, -2), (2, 4)$
$x - y = 5$

2. $2x - 3y = 5$ $\quad (-1, -1), (3, 4), (2, 3)$
$y = x + 1$

3. $6x - 2y = 4$ $\quad (0, -2), (2, 4), (3, 7)$
$y = 3x - 2$

4. $y = -2x + 5$ $\quad (9, -13), (-1, 7), (0, 5)$
$4x + 2y = 10$

5. $2x - y = 3$ $\quad (3, 3), (5, 7), (7, 11)$
$2x - y = 2$

6. $y = x + 5$ $\quad (1, -2), (3, 0), (6, 3)$
$y = x - 3$

Use the given graph to find an ordered pair that satisfies each system of equations. Check that your answer satisfies both equations of each system.

7. $y = 3x + 9$
$2x + 3y = 5$

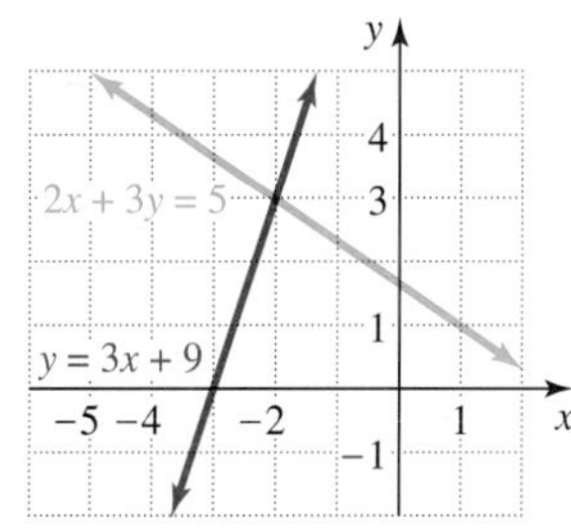

8. $x - 2y = 5$
$y = -\frac{2}{3}x + 1$

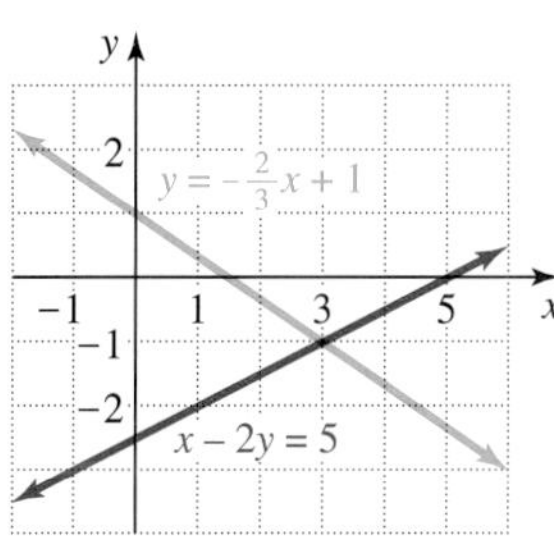

Solve each system by graphing. See Example 2.

9. $y = 2x$
$y = -x + 6$

10. $y = 3x$
$y = -x + 4$

11. $3x - y = 1$
$2y - 3x = 1$

12. $2x + y = 3$
$x + y = 1$

13. $x - y = 5$
$x + y = -5$

14. $y + 4x = 10$
$2x - y = 2$

15. $2y + x = 4$
$2x - y = -7$

16. $2x + y = -1$
$x + y = -2$

17. $y = x$
$x + y = 0$

18. $x = 2y$
$0 = 9x - y$

19. $y = 2x - 1$
$x - 2y = -4$

20. $y = x - 1$
$2x - y = 0$

Solve each system by graphing and indicate whether the system is independent, inconsistent, or dependent. See Examples 3 and 4.

21. $x - y = 3$
$3x = 3y + 12$

22. $x - y = 3$
$3x = 3y + 9$

23. $x - y = 3$
$3x = y + 5$

24. $3x + 2y = 6$
$2x - y = 4$

25. $2x + y = 3$
$6x - 9 = -3y$

26. $4y - 2x = -16$
$x - 2y = 8$

27. $x - y = 0$
$5x = 5y$

28. $y = -3x + 1$
$2 - 2y = 6x$

29. $x - y = -1$
$y = \frac{1}{2}x - 1$

30. $y = \frac{1}{3}x + 2$
$y = -\frac{1}{3}x$

31. $y - 4x = 4$
$y + 4x = -4$

32. $2y = -3x + 6$
$2y = -3x - 2$

Use the given graphs to find two ordered pairs that satisfy each nonlinear system of equations. Check that your answers satisfy both equations of each system.

33. $y = x^2$
$y = 4x - x^2$

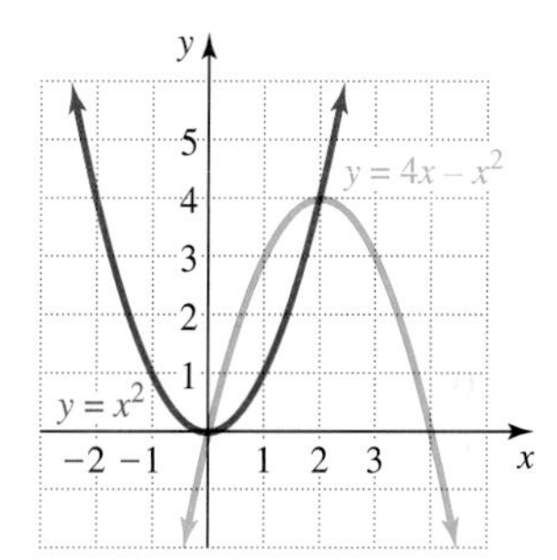

34. $y = x^2 - 2x$
$y = -x^2 - 2x + 2$

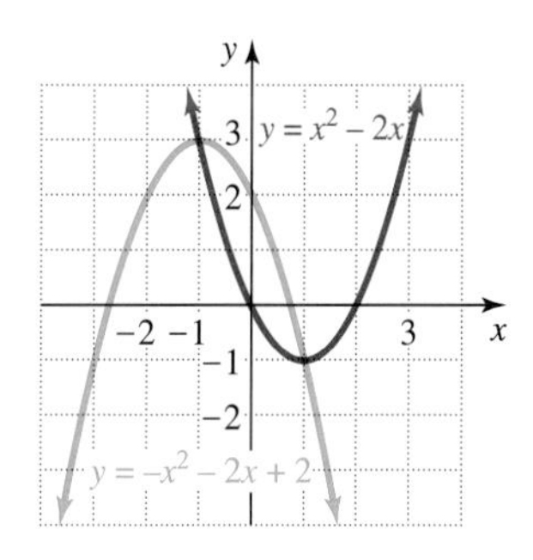

Solve each problem by using the graphing method.

35. ***Entertainment spending.*** The accompanying graph shows the amount spent on home video sales and rentals and the amount spent on movies (at the box office) since 1980 (*Fortune*, June 27, 1994). In what year did the amount spent on home video equal the amount spent on movies? What amount was spent that year on each source of entertainment? In what year did consumers spend twice as much on home video as they did at the movies?

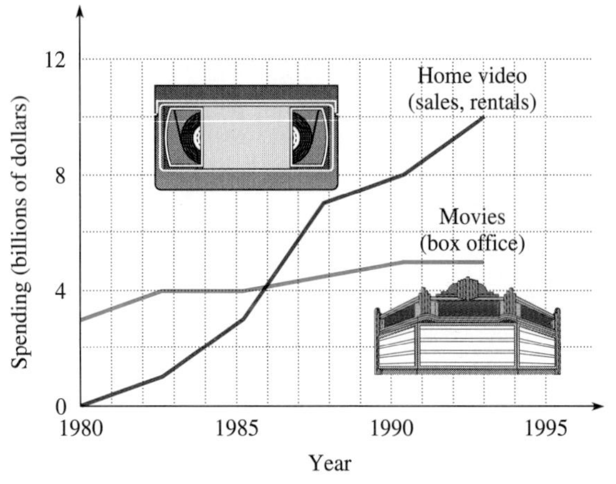

Figure for Exercise 35

36. ***Equilibrium price.*** A manufacturer plans to supply y units of its model 1020P CD player per month when the retail price is p dollars per player, where $y = 6p + 100$. Consumer studies show that consumer demand for the model 1020P is y units per month, where $y = -3p + 910$. Graph both linear equations on the same coordinate system. What is the price at which the supply is equal to the demand, the *equilibrium price*?

37. ***Cost of two copiers.*** An office manager figures the total cost in dollars for a certain used Xerox copier is given by $C = 800 + 0.05x$, where x is the number of copies made. She is also considering a used Panasonic copier for which the total cost is $C = 500 + 0.07x$. Graph both equations on the same coordinate system. For what number of copies is the total cost the same for either copier. If she plans to buy another copier before 10,000 copies are made, then which copier is cheaper?

38. ***Flat tax proposals.*** Representative Schneider has proposed a flat income tax of 15% on earnings in excess of \$10,000. Under his proposal the tax T for a person earning E dollars is given by $T = 0.15(E - 10{,}000)$. Representative Humphries has proposed that the income tax should be 20% on earnings in excess of \$20,000, or $T = 0.20(E - 20{,}000)$. Graph both linear equations on the same coordinate system. For what earnings would you pay the same amount of income tax under either plan? Under which plan does a rich person pay less income tax?

Getting More Involved

39. ***Discussion.*** If both $(-1, 3)$ and $(2, 7)$ satisfy a system of two linear equations, then what can you say about the system?

40. ***Cooperative learning.*** Working in groups, write an independent system of two linear equations whose solution is $(3, 5)$. Each group should then give its system to another group to solve.

41. ***Cooperative learning.*** Working in groups, write an inconsistent system of linear equations such that $(-2, 3)$ satisfies one equation and $(1, 4)$ satisfies the other. Each group should then give its system to another group to solve.

Graphing Calculator Exercise

42. Solve each system by graphing each pair of equations on a graphing calculator and using the trace feature to estimate the point of intersection. Find the coordinates of the intersection to the nearest tenth.

a) $y = 2.5x - 6.2$
$y = -1.3x + 8.1$

b) $2.2x - 3.1y = 3.4$
$5.4x + 6.2y = 7.3$

7.2 The Substitution Method

In this section:

- Solving a System of Linear Equations by Substitution
- Inconsistent and Dependent Systems
- Applications

Solving a system by graphing is certainly limited by the accuracy of the graph. If the lines intersect at a point whose coordinates are not integers, then it is difficult to identify the solution from a graph. In this section we introduce a method for solving systems of linear equations in two variables that does not depend on a graph and is totally accurate.

Solving a System of Linear Equations by Substitution

The next example shows how to solve a system without graphing. The method is called **substitution.**

EXAMPLE 1 **Solving a system by substitution**

Solve:

$$\begin{aligned} 2x - 3y &= 9 \\ y - 4x &= -8 \end{aligned}$$

Solution

First solve the second equation for y:

$$\begin{aligned} y - 4x &= -8 \\ y &= 4x - 8 \end{aligned}$$

Now substitute $4x - 8$ for y in the first equation:

$$\begin{aligned} 2x - 3y &= 9 \\ 2x - 3(4x - 8) &= 9 && \text{Substitute } 4x - 8 \text{ for } y. \\ 2x - 12x + 24 &= 9 && \text{Simplify.} \\ -10x + 24 &= 9 \\ -10x &= -15 \\ x &= \frac{-15}{-10} \\ &= \frac{3}{2} \end{aligned}$$

Use the value $x = \frac{3}{2}$ in $y = 4x - 8$ to find y:

$$\begin{aligned} y &= 4 \cdot \frac{3}{2} - 8 \\ &= -2 \end{aligned}$$

Check that $\left(\frac{3}{2}, -2\right)$ satisfies both of the original equations. The solution to the system is $\left(\frac{3}{2}, -2\right)$.

EXAMPLE 2 Solving a system by substitution

Solve:

$$\begin{aligned} 3x + 4y &= 5 \\ x &= y - 1 \end{aligned}$$

Solution

Because the second equation is already solved for x in terms of y, we can substitute $y - 1$ for x in the first equation:

$$\begin{aligned} 3x + 4y &= 5 \\ 3(y - 1) + 4y &= 5 && \text{Replace } x \text{ with } y - 1. \\ 3y - 3 + 4y &= 5 && \text{Simplify.} \\ 7y - 3 &= 5 \\ 7y &= 8 \\ y &= \frac{8}{7} \end{aligned}$$

Now use the value $y = \frac{8}{7}$ in one of the original equations to find x. The simplest one to use is $x = y - 1$:

$$\begin{aligned} x &= \frac{8}{7} - 1 \\ x &= \frac{1}{7} \end{aligned}$$

Check that $\left(\frac{1}{7}, \frac{8}{7}\right)$ satisfies both equations. The solution to the system is $\left(\frac{1}{7}, \frac{8}{7}\right)$. ■

The strategy for solving by substitution can be summarized as follows.

Strategy for Solving a System by Substitution

1. Solve one of the equations for one variable in terms of the other.
2. Substitute this value into the other equation to eliminate one of the variables.
3. Solve for the remaining variable.
4. Insert this value into one of the original equations to find the value of the other variable.
5. Check your solution in both equations.

Inconsistent and Dependent Systems

The following examples illustrate how the inconsistent and dependent cases appear when we use substitution to solve the system.

EXAMPLE 3 **An inconsistent system**

Solve by substitution:

$$3x - 6y = 9$$
$$x = 2y + 5$$

Solution

Use $x = 2y + 5$ to replace x in the first equation:

$$3x - 6y = 9$$
$$3(2y + 5) - 6y = 9 \quad \text{Replace } x \text{ by } 2y + 5.$$
$$6y + 15 - 6y = 9 \quad \text{Simplify.}$$
$$15 = 9$$

No values for x and y will make 15 equal to 9. So there is no ordered pair that satisfies both equations. This system is inconsistent. It has no solution. The equations are the equations of parallel lines. ■

EXAMPLE 4 **A dependent system**

Solve:

$$2(y - x) = x + y - 1$$
$$y = 3x - 1$$

Solution

Because the second equation is solved for y, we will eliminate the variable y in the substitution. Substitute $y = 3x - 1$ into the first equation:

$$2(3x - 1 - x) = x + (3x - 1) - 1$$
$$2(2x - 1) = 4x - 2$$
$$4x - 2 = 4x - 2$$

Any value for x makes the last equation true because both sides are identical. So any value for x can be used as a solution to the original system as long as we choose $y = 3x - 1$. The system is dependent. The two equations are equations for the same straight line. The solution to the system is the set of all points on that line,

$$\{(x, y) \mid y = 3x - 1\}.$$ ■

When solving a system by substitution, we can recognize an inconsistent system or dependent system as follows:

Inconsistent and Dependent Systems

An inconsistent system leads to a false statement.
A dependent system leads to a statement that is always true.

Applications

Many of the problems that we solved in previous chapters had two unknown quantities, but we wrote only one equation to solve the problem. For problems with two unknown quantities we can use two variables and a system of equations.

EXAMPLE 5 **Two investments**

Mrs. Robinson invested a total of \$25,000 in two investments, one paying 6% and the other paying 8%. If her total income from these investments was \$1790, then how much money did she invest in each?

Solution

Let x represent the amount invested at 6%, and let y represent the amount invested at 8%. The following table organizes the given information.

	Interest rate	Amount invested	Amount of interest
First investment	6%	x	$0.06x$
Second investment	8%	y	$0.08y$

Write one equation describing the total of the investments, and the other equation describing the total interest:

$$x + y = 25{,}000 \quad \text{Total investments}$$
$$0.06x + 0.08y = 1790 \quad \text{Total interest}$$

To solve the system, we solve the first equation for y:

$$y = 25{,}000 - x$$

Substitute $25{,}000 - x$ for y in the second equation:

$$\begin{aligned} 0.06x + 0.08(25{,}000 - x) &= 1790 \\ 0.06x + 2000 - 0.08x &= 1790 \\ -0.02x + 2000 &= 1790 \\ -0.02x &= -210 \\ x &= \frac{-210}{-0.02} \\ &= 10{,}500 \end{aligned}$$

Let $x = 10{,}500$ in the equation $y = 25{,}000 - x$ to find y:

$$\begin{aligned} y &= 25{,}000 - 10{,}500 \\ &= 14{,}500 \end{aligned}$$

Check these values for x and y in the original problem. Mrs. Robinson invested \$10,500 at 6% and \$14,500 at 8%. ■

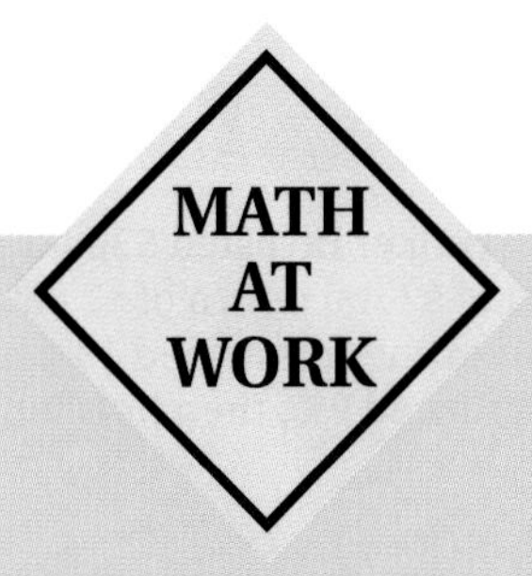

Race Car Driver

First the special handshake for luck, then climbing in the coupe, strapping the seat belt on as tightly as possible, eyes locked forward, thinking who is the person to beat, where are the bumps and curves . . . concentration! These are some of the thoughts and rituals Ossie Babson performs as the driver-partner of the Babson Brothers Racing Team. The special handshake is with his brother Dave Babson, who is the crew chief of the team. The car is a $\frac{5}{8}$-scale model of a 1940 Ford Coupe, specially built for the Legends of Nascar Racing Series.

There are strict rules on the weight distribution, frame height, tire size, and engine size for these cars, but for best results the car should be set up to lean to the left and front. The challenge is to meet all these criteria for a successful race. Before the race, Ossie drives on the track and makes observations and recommendations on how the car handles going into the turns and what adjustments to make for better performance under specific track conditions. Dave then supervises the changes to the car. This process continues until both driver and crew chief are satisfied. Ultimately, a combination of art, how the car feels, and science makes the car perform to its ultimate capabilities. As a result of their teamwork last year, the Babsons finished in one of the top ten positions in nine races including an impressive second place finish.

In Exercises 29 and 30 of this section you will see how the Babsons use a system of equations to determine the proper weight distribution for their car.

Warm-ups

True or false? Explain your answer.
For Exercises 1–7, use the following systems:

a) $y = x - 7$
$2x + 3y = 4$

b) $x + 2y = 1$
$2x - 4y = 0$

1. If we substitute $x - 7$ for y in system (a), we get $2x + 3(x - 7) = 4$.
2. The x-coordinate of the solution to system (a) is 5.
3. The solution to system (a) is $(5, -2)$.
4. The point $\left(\frac{1}{2}, \frac{1}{4}\right)$ satisfies system (b).
5. It would be difficult to solve system (b) by graphing.
6. Either x or y could be eliminated by substitution in system (b).
7. System (b) is a dependent system.
8. Solving an inconsistent system by substitution will result in a false statement.
9. Solving a dependent system by substitution results in an equation that is always true.
10. Any system of two linear equations can be solved by substitution.

7.2 EXERCISES

Solve each system by the substitution method. See Examples 1 and 2.

1. $y = x + 3$
$2x - 3y = -11$

2. $y = x - 5$
$x + 2y = 8$

3. $x = 2y - 4$
$2x + y = 7$

4. $x = y - 2$
$-2x + y = -1$

5. $2x + y = 5$
$5x + 2y = 8$

6. $5y - x = 0$
$6x - y = 2$

7. $x + y = 0$
$3x + 2y = -5$

8. $x - y = 6$
$3x + 4y = -3$

9. $x + y = 1$
$4x - 8y = -4$

10. $x - y = 2$
$3x - 6y = 8$

11. $2x + 3y = 2$
$4x - 9y = -1$

12. $x - 2y = 1$
$3x + 10y = -1$

Solve each system by substitution, and identify each system as independent, dependent, or inconsistent. See Examples 3 and 4.

13. $x - 2y = -2$
$x + 2y = 8$

14. $y = -3x + 1$
$y = 2x + 4$

15. $x = 4 - 2y$
$4y + 2x = -8$

16. $21x - 35 = 7y$
$3x - y = 5$

17. $y - 3 = 2(x - 1)$
$y = 2x + 3$

18. $y + 1 = 5(x + 1)$
$y = 5x - 1$

19. $3x - 2y = 7$
$3x + 2y = 7$

20. $2x + 5y = 5$
$3x - 5y = 6$

21. $x + 5y = 4$
$x + 5y = 4y$

22. $2x + y = 3x$
$3x - y = 2y$

Write a system of two equations in two unknowns for each problem. Solve each system by substitution. See Example 5.

23. ***Investing in the future.*** Mrs. Miller invested \$20,000 and received a total of \$1,600 in interest. If she invested part of the money at 10% and the remainder at 5%, then how much did she invest at each rate?

24. ***Stocks and bonds.*** Mr. Walker invested \$30,000 in stocks and bonds and had a total return of \$2,880 in one year. If his stock investment returned 10% and his bond investment returned 9%, then how much did he invest in each?

25. ***Unknown earnings.*** In 1993, director Steven Spielberg earned \$26 million more than 13-year-old actor Macaulay Culkin (*Parade*, June 26, 1994). If the total earnings of these two entertainers was \$58 million, then how much did each one earn?

26. ***Tennis court dimensions.*** The singles court in tennis is four yards longer than it is wide. If its perimeter is 44 yards, then what are the length and width?

27. ***Mowing and shoveling.*** When Mr. Wilson came back from his spring vacation, he paid Frank \$50 for mowing his lawn three times and shoveling his sidewalk two times. During Mr. Wilson's spring vacation last year, Frank earned \$45 for mowing the lawn two times and shoveling the sidewalk three times. How much does Frank make for mowing the lawn once? How much does Frank make for shoveling the sidewalk once?

28. ***Burgers and fries.*** Donna ordered four burgers and one order of fries at the Hamburger Palace. However, the waiter put three burgers and two orders of fries in the bag and charged Donna the correct price for three burgers and two orders of fries, \$3.15. When Donna discovered the mistake, she went back to complain. She found out that the price for four burgers and one order of fries is \$3.45 and decided to keep what she had. What is the price of one burger, and what is the price of one order of fries?

29. ***Racing rules.*** According to Nascar rules, no more than 52% of a car's total weight can be on any pair of tires. For optimal performance a driver of a 1150-pound car wants to have 50% of its weight on the left rear and left front tires and 48% of its weight on the left rear and right front tires. If the right front weight is determined to be 264 pounds, then what amount of weight should be on the left rear and left front? Are the Nascar rules satisfied with this weight distribution?

30. ***Weight distribution.*** A driver of a 1200-pound car wants to have 50% of the car's weight on the left front and left rear tires, 48% on the left rear and right front tires, and 51% on the left rear and right rear tires. How much weight should be on each of these tires?

31. ***Price of hamburger.*** A grocer will supply y pounds of ground beef per day when the retail price is x dollars per pound, where $y = 200x + 60$. Consumer studies show that consumer demand for ground beef is y pounds per day, where $y = -150x + 900$. What is the price at which the supply is equal to the demand, the equilibrium price? See the figure below.

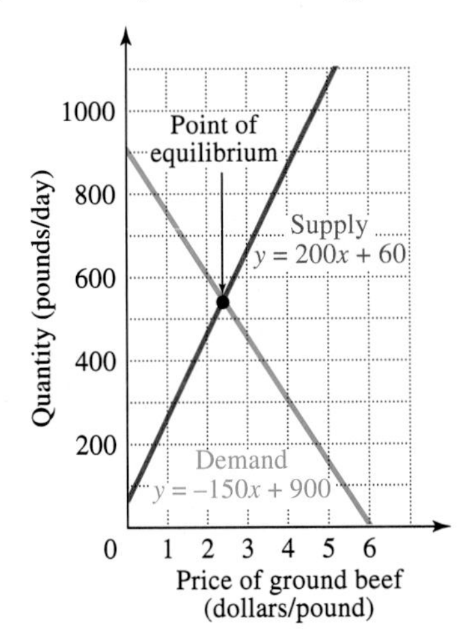

Figure for Exercise 31

32. ***Tax returns.*** In 1994 114 million individual tax returns were filed (*U.S.A. Today,* March 10, 1995). For every 17 returns filed on paper, 2 returns were filed electronically. How many returns were filed electronically in 1994?

Graphing Calculator Exercise

33. ***Life expectancy.*** Since 1950 the life expectancy y of a U.S. male born in year x is modeled by the formula

$$y = 0.159x - 244.3,$$

and the life expectancy of a U.S. female born in year x is modeled by

$$y = 0.203x - 325.7.$$

("Monitoring Your Health," *Readers Digest,* 1991).

a) Find the life expectancy of a U.S. male born in 1975 and a U.S. female born in 1975.
b) Graph both equations on your graphing calculator for $1950 < x < 2050$.
c) Will U.S. males ever catch up with U.S. females in life expectancy?
d) Assuming that these equations were valid before 1950, solve the system to find the year of birth for which U. S. males and females had the same life expectancy.

7.3 The Addition Method

In this section:

- Solving a System of Linear Equations by Addition
- Inconsistent and Dependent Systems
- Applications

In Section 7.2 we solved systems of equations by using substitution. We substituted one equation into the other to eliminate a variable. The addition method of this section is another method for eliminating a variable to solve a system of equations.

Solving a System of Linear Equations by Addition

In the substitution method we solve for one variable in terms of the other variable. When doing this, we may get an expression involving fractions, which must be substituted into the other equation. The addition method avoids fractions and is easier to use on certain systems.

EXAMPLE 1 **Solving a system by addition**

Solve: $3x - y = 5$

$2x + y = 10$

Solution

The addition property of equality allows us to add the same number to each side of an equation. We can also use the addition property of equality to add the two left-hand sides and add the two right-hand sides:

$$\begin{aligned} 3x - y &= 5 \\ 2x + y &= 10 \quad \text{Add.} \\ \hline 5x \quad &= 15 \quad -y + y = 0 \\ x &= 3 \end{aligned}$$

Note that the y-term was eliminated when we added the equations because the coefficients of y in the two equations were opposites. Now use $x = 3$ in either one of the original equations to find y:

$$\begin{aligned} 2x + y &= 10 \\ 2(3) + y &= 10 \quad \text{Let } x = 3. \\ y &= 4 \end{aligned}$$

Check that $(3, 4)$ satisfies both equations. The solution to the system is $(3, 4)$. ■

The addition method is based on the addition property of equality. We are adding equal quantities to each side of an equation. The form of the equations does not matter as long as the equal signs and the like terms are in line.

In Example 1, y was eliminated by the addition because the coefficients of y in the two equations were opposites. If no variable will be eliminated by addition, we can use the multiplication property of equality to change the coefficients of the variables. In the next example the coefficient of x in one equation is a multiple of the coefficient of x in the other equation. We use the multiplication property of equality to get opposite coefficients for x.

EXAMPLE 2 **Solving a system by addition**

Solve: $-x + 4y = -14$

$2x - 3y = 18$

Solution

If we add these equations as they are written, we will not eliminate any variables. However, if we multiply each side of the first equation by 2, then we will be adding $-2x$ and $2x$, and x will be eliminated:

$$\begin{aligned} 2(-x + 4y) &= 2(-14) \quad \text{Multiply each side by 2.} \\ 2x - 3y &= 18 \end{aligned}$$

$$\begin{aligned} -2x + 8y &= -28 \\ 2x - 3y &= 18 \quad \text{Add.} \\ \hline 5y &= -10 \\ y &= -2 \end{aligned}$$

Now replace y by -2 in one of the original equations:

$$-x + 4(-2) = -14$$
$$-x - 8 = -14$$
$$-x = -6$$
$$x = 6$$

Check $x = 6$ and $y = -2$ in the original equations.

$$-6 + 4(-2) = -14$$
$$2(6) - 3(-2) = 18$$

The solution to the system is $(6, -2)$. ■

In the next example we need to use a multiple of each equation to eliminate a variable by addition.

EXAMPLE 3 **Solving a system by addition**

Solve: $2x + 3y = 7$
$3x + 4y = 10$

Solution

To eliminate x by addition, the coefficients of x in the two equations must be opposites. So we multiply the first equation by -3 and the second by 2:

$$-3(2x + 3y) = -3(7)$$
$$2(3x + 4y) = 2(10)$$

$$\begin{aligned} -6x - 9y &= -21 \\ 6x + 8y &= 20 \quad \text{Add.} \\ \hline -y &= -1 \\ y &= 1 \end{aligned}$$

Replace y with 1 in one of the original equations:

$$2x + 3y = 7$$
$$2x + 3(1) = 7$$
$$2x + 3 = 7$$
$$2x = 4$$
$$x = 2$$

Check $x = 2$ and $y = 1$ in the original equations.

$$2(2) + 3(1) = 7$$
$$3(2) + 4(1) = 10$$

The solution to the system is $(2, 1)$. ■

If the equations have fractions, you can multiply each equation by the LCD to eliminate the fractions. Once the fractions are cleared, it is easier to see how to eliminate a variable by addition.

EXAMPLE 4 **A system involving fractions**

Solve: $\frac{1}{2}x - \frac{2}{3}y = 2$

$\frac{1}{4}x + \frac{1}{2}y = 6$

Solution

Multiply the first equation by 6 and the second by 4 to eliminate the fractions:

$$6\left(\frac{1}{2}x - \frac{2}{3}y\right) = 6 \cdot 2$$
$$4\left(\frac{1}{4}x + \frac{1}{2}y\right) = 4 \cdot 6$$
$$3x - 4y = 12$$
$$x + 2y = 24$$

Now multiply $x + 2y = 24$ by 2 to get $2x + 4y = 48$, and then add:

$$\begin{aligned} 3x - 4y &= 12 \\ 2x + 4y &= 48 \\ \hline 5x \quad\quad &= 60 \\ x &= 12 \end{aligned}$$

Let $x = 12$ in $x + 2y = 24$:

$$12 + 2y = 24$$
$$2y = 12$$
$$y = 6$$

Check $x = 12$ and $y = 6$ in the original equations. The solution is $(12, 6)$. ■

Use the following strategy to solve a system by addition.

Strategy for Solving a System by Addition

1. Write both equations in standard form.
2. If a variable will be eliminated by adding, then add the equations.
3. If necessary, obtain multiples of one or both equations so that a variable will be eliminated by adding the equations.
4. After one variable is eliminated, solve for the remaining variable.
5. Use the value of the remaining variable to find the value of the eliminated variable.
6. Check the solution in the original system.

Inconsistent and Dependent Systems

When the addition method is used, an inconsistent system will be indicated by a false statement. A dependent system will be indicated by an equation that is always true.

EXAMPLE 5 **Inconsistent and dependent systems**

Use the addition method to solve each system.

a) $-2x + 3y = 9$

$2x - 3y = 18$

b) $2x - y = 1$

$4x - 2y = 2$

Solution

a) Add the equations:

$$\begin{aligned} -2x + 3y &= 9 \\ 2x - 3y &= 18 \\ \hline 0 &= 27 \quad \text{False.} \end{aligned}$$

There is no solution to the system. The system is inconsistent.

b) Multiply the first equation by -2, and then add the equations:

$$\begin{aligned} -2(2x - y) &= -2(1) \\ 4x - 2y &= 2 \end{aligned}$$

$$\begin{aligned} -4x + 2y &= -2 \\ 4x - 2y &= 2 \\ \hline 0 &= 0 \quad \text{True.} \end{aligned}$$

Because the equation $0 = 0$ is correct for any value of x, the system is dependent. The set of points satisfying the system is $\{(x, y) \mid 2x - y = 1\}$. ■

Applications

In the next example we solve a problem using a system of equations and the addition method.

EXAMPLE 6 **Milk and bread**

Lea purchased two gallons of milk and three loaves of bread for $8.25. Yesterday she purchased five gallons of milk and two loaves of bread for $13.75. What is the price of a single gallon of milk? What is the price of a single loaf of bread?

Solution

Let x represent the price of one gallon of milk. Let y represent the price of one loaf of bread. We can write two equations about the milk and bread:

$$\begin{aligned} 2x + 3y &= 8.25 \quad \text{Today's purchase} \\ 5x + 2y &= 13.75 \quad \text{Yesterday's purchase} \end{aligned}$$

To eliminate x, multiply the first equation by -5 and the second by 2:

$$\begin{aligned} -5(2x + 3y) &= -5(8.25) \\ 2(5x + 2y) &= 2(13.75) \end{aligned}$$

$$\begin{aligned} -10x - 15y &= -41.25 \\ 10x + 4y &= 27.50 \quad \text{Add.} \\ \hline -11y &= -13.75 \\ y &= 1.25 \end{aligned}$$

Replace y by 1.25 in the first equation:

$$\begin{aligned} 2x + 3(1.25) &= 8.25 \\ 2x + 3.75 &= 8.25 \\ 2x &= 4.50 \\ x &= 2.25 \end{aligned}$$

A gallon of milk costs $2.25, and a loaf of bread costs $1.25. ■

Warm-ups

True or false? Explain your answer.

Use the following systems for these exercises:

a) $3x + 2y = 7$
$4x - 5y = -6$

b) $y = -3x + 2$
$2y + 6x - 4 = 0$

c) $y = x - 5$
$x = y + 6$

1. To eliminate x by addition in system (a), we multiply the first equation by 4 and the second equation by 3.
2. Either variable in system (a) can be eliminated by the addition method.
3. The ordered pair $(1, 2)$ is a solution to system (a).
4. The addition method can be used to eliminate a variable in system (b).
5. Both $(0, 2)$ and $(1, -1)$ satisfy system (b).
6. The solution to system (c) is $\{(x, y) \mid y = x - 5\}$.
7. System (c) is independent.
8. System (b) is inconsistent.
9. System (a) is dependent.
10. The graphs of the equations in system (c) are parallel lines.

7.3 EXERCISES

Solve each system by the addition method. See Examples 1–4.

1. $2x + y = 5$
$3x - y = 10$

2. $3x - y = 3$
$4x + y = 11$

3. $x + 2y = 7$
$-x + 3y = 18$

4. $x + 2y = 7$
$-x + 4y = 5$

5. $x + 2y = 2$
$-4x + 3y = 25$

6. $2x - 3y = -7$
$5x + y = -9$

7. $x + 3y = 4$
$2x - y = -1$

8. $x - y = 0$
$x - 2y = 0$

9. $y = 4x - 1$
$y = 3x + 7$

10. $x = 3y + 45$
$x = 2y + 40$

11. $4x = 3y + 1$
$2x = y - 1$

12. $2x = y - 9$
$x = -1 - 3y$

13. $2x - 5y = -22$
$-6x + 3y = 18$

14. $4x - 3y = 7$
$5x + 6y = -1$

15. $2x + 3y = 4$
$-3x + 5y = 13$

16. $-5x + 3y = 1$
$2x - 7y = 17$

17. $2x - 5y = 11$
$3x - 2y = 11$

18. $4x - 3y = 17$
$3x - 5y = 21$

19. $5x + 4y = 13$
$2x + 3y = 8$

20. $4x + 3y = 8$
$6x + 5y = 14$

Use either the addition method or substitution to solve each system. State whether the system is independent, inconsistent, or dependent. See Example 5.

21. $x + y = 5$
$x + y = 6$

22. $x + y = 5$
$x + 2y = 6$

23. $x + y = 5$
$2x + 2y = 10$

24. $2x + 3y = 4$
$2x - 3y = 4$

25. $2x = y + 3$
$2y = 4x - 6$

26. $y = 2x - 1$
$2x - y + 5 = 0$

27. $x + 3y = 3$
$5x = 15 - 15y$

28. $y - 3x = 2$
$5y = -15x + 10$

29. $6x - 2y = -2$
$\frac{1}{3}y = x + \frac{4}{3}$

30. $x + y = 8$
$\frac{1}{3}x - \frac{1}{2}y = 1$

31. $\frac{1}{2}x - \frac{2}{3}y = -6$
$-\frac{3}{4}x - \frac{1}{2}y = -18$

32. $\frac{1}{2}x - y = 3$
$\frac{1}{5}x + 2y = 6$

33. $0.04x + 0.09y = 7$
$x + y = 100$

34. $0.08x - 0.05y = 0.2$
$2x + y = 140$

35. $0.1x - 0.2y = -0.01$
$0.3x + 0.5y = 0.08$

36. $0.5y = 0.2x - 0.25$
$0.1y = 0.8x - 1.57$

Use a calculator to assist you in finding the exact solution to each system.

37. $2.33x - 4.58y = 16.319$
$4.98x + 3.44y = -2.162$

38. $234x - 499y = 1337$
$282x + 312y = 51{,}846$

Use two variables and a system of equations to solve each problem. See Example 6.

39. ***Cars and trucks.*** An automobile dealer had 250 vehicles on his lot during the month of June. He must pay a monthly inventory tax of \$3 per car and \$4 per truck. If his tax bill for June was \$850, then how many cars and how many trucks did he have on his lot during June?

40. ***Dimes and nickels.*** Kimberly opened a parking meter and removed 30 coins consisting of dimes and nickels. If the value of these coins is \$2.30, then how many of each type does she have?

41. ***Adults and children.*** The Audubon Zoo charges \$5.50 for each adult admission and \$2.75 for each child. The total bill for the 30 people on the Spring Creek Elementary School kindergarten field trip was \$99. How many adults and how many children went on the field trip?

42. ***Coffee and doughnuts.*** Jorge has worked at Dandy Doughnuts so long that he has memorized the amounts for many of the common orders. For example, six doughnuts and five coffees cost \$4.35, while four doughnuts and three coffees cost \$2.75. What are the prices of one cup of coffee and one doughnut?

43. ***Marketing research.*** The Independent Marketing Research Corporation found 130 smokers among 300 adults surveyed. If one-half of the men and one-third of the women were smokers, then how many men and how many women were in the survey?

44. ***Time and a half.*** In one month, Shelly earned \$1800 for 210 hours of work. If she earns \$8 per hour for regular time and \$12 per hour for overtime, then how many hours of each type did she work?

Getting More Involved

45. ***Discussion.*** Compare and contrast the three methods for solving systems of linear equations in two variables that were presented in this chapter. What are the advantages and disadvantages of each method? How do you choose which method to use?

46. ***Exploration.*** Consider the following system:

$$a_1x + b_1y = c_1$$
$$a_2x + b_2y = c_2$$

a) Multiply the first equation by a_2 and the second equation by $-a_1$. Add the resulting equations and solve for y to get a formula for y in terms of the a's, b's, and c's.

b) Multiply the first equation by b_2 and the second by $-b_1$. Add the resulting equations and solve for x to get a formula for x in terms of the a's, b's, and c's.

c) Use the formulas that you found in (a) and (b) to find the solution to the following system:

$$2x + 3y = 7$$
$$5x + 4y = 14$$

7.4 Linear Inequalities in Two Variables

In this section:

- Definition
- Graph of a Linear Inequality
- Using a Test Point to Graph an Inequality
- Applications

You studied linear equations and inequalities in one variable in Chapter 2. In this section we extend the ideas of linear equations in two variables to study linear inequalities in two variables.

Definition

Linear inequalities in two variables have the same form as linear equations in two variables. An inequality symbol is used in place of the equal sign.

Linear Inequality in Two Variables

If A, B, and C are real numbers with A and B not both zero, then

$$Ax + By < C$$

is called a **linear inequality in two variables.** In place of $<$, we can also use $\leq$, $>$, or $\geq$.

The inequalities

$$3x - 4y \leq 8, \qquad y > 2x - 3, \qquad \text{and} \qquad x - y + 9 < 0$$

are linear inequalities. Not all of these are in the form of the definition, but they could all be rewritten in that form.

An ordered pair is a solution to an inequality in two variables if the ordered pair satisfies the inequality.

EXAMPLE 1 **Satisfying a linear inequality**

Determine whether each point satisfies the inequality $2x - 3y \geq 6$.

a) $(4, 1)$

b) $(3, 0)$

c) $(3, -2)$

Solution

a) To determine whether $(4, 1)$ is a solution to the inequality, we replace x by 4 and y by 1 in the inequality $2x - 3y \geq 6$:

$$2(4) - 3(1) \geq 6$$
$$8 - 3 \geq 6$$
$$5 \geq 6 \quad \text{Incorrect}$$

So $(4, 1)$ does not satisfy the inequality $2x - 3y \geq 6$.

b) Replace x by 3 and y by 0:

$$2(3) - 3(0) \geq 6$$
$$6 \geq 6 \quad \text{Correct}$$

So the point $(3, 0)$ satisfies the inequality $2x - 3y \geq 6$.

c) Replace x by 3 and y by -2:

$$2(3) - 3(-2) \geq 6$$
$$6 + 6 \geq 6$$
$$12 \geq 6 \quad \text{Correct}$$

So the point $(3, -2)$ satisfies the inequality $2x - 3y \geq 6$. ■

Graph of a Linear Inequality

The graph of a linear inequality in two variables consists of all points in the rectangular coordinate system that satisfy the inequality. For example, the graph of the inequality

$$y > x + 2$$

consists of all points where the y-coordinate is larger than the x-coordinate plus 2. Consider the point $(3, 5)$ on the line

$$y = x + 2.$$

The y-coordinate of $(3, 5)$ is equal to the x-coordinate plus 2. If we choose a point with a larger y-coordinate, such as $(3, 6)$, it satisfies the inequality and it is above the line $y = x + 2$. In fact, any point above the line $y = x + 2$ satisfies $y > x + 2$. Likewise, all points below the line $y = x + 2$ satisfy the inequality $y < x + 2$. See Fig. 7.6.

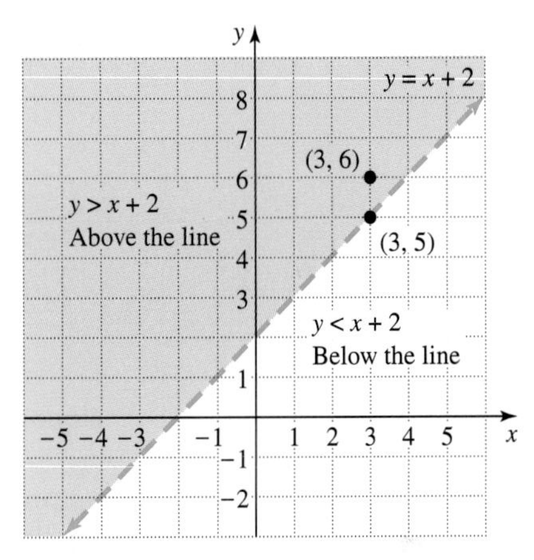

Figure 7.6

To graph the inequality, we shade all points above the line $y = x + 2$. To indicate that the line is not included in the graph of $y > x + 2$, we use a dashed line.

The procedure for graphing linear inequalities is summarized as follows.

Strategy for Graphing a Linear Inequality in Two Variables

1. Solve the inequality for y, then graph $y = mx + b$.

   $y > mx + b$ is the region above the line.
   $y = mx + b$ is the line itself.
   $y < mx + b$ is the region below the line.

2. If the inequality involves only x, then graph the vertical line $x = k$.

   $x > k$ is the region to the right of the line.
   $x = k$ is the line itself.
   $x < k$ is the region to the left of the line.

EXAMPLE 2 **Graphing a linear inequality**

Graph each inequality.

a) $y < \frac{1}{3}x + 1$ **b)** $y \geq -2x + 3$ **c)** $2x - 3y < 6$

Solution

a) The set of points satisfying this inequality is the region below the line

$$y = \frac{1}{3}x + 1.$$

To show this region, we first graph the boundary line. The slope of the line is $\frac{1}{3}$, and the y-intercept is $(0, 1)$. We draw the line dashed because it is not part of the graph of $y < \frac{1}{3}x + 1$. In Fig. 7.7 the graph is the shaded region.

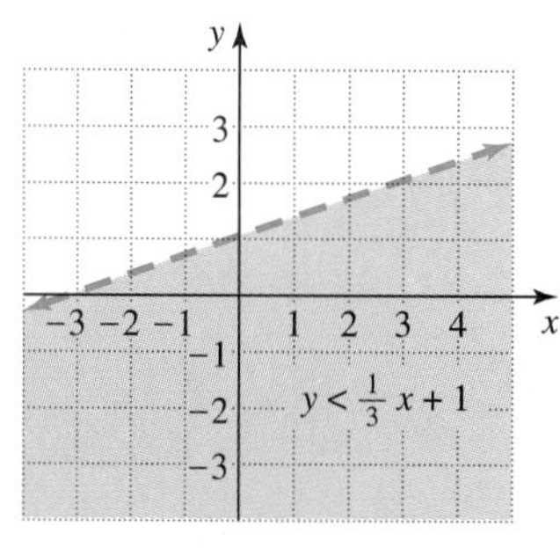

Figure 7.7

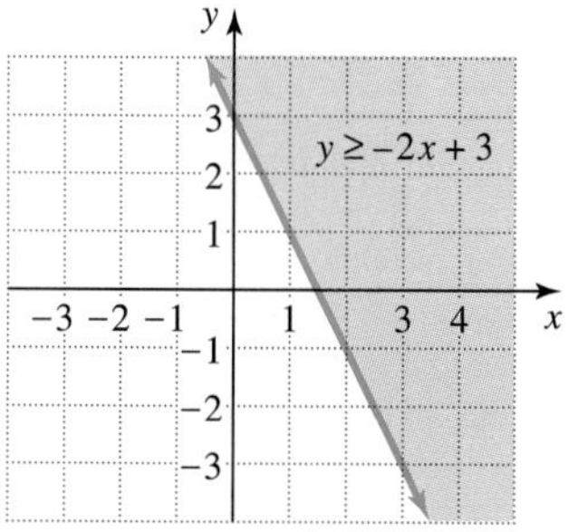

Figure 7.8

b) Because the inequality symbol is $\geq$, every point on or above the line satisfies this inequality. We use the fact that the slope of this line is -2 and the y-intercept is $(0, 3)$ to draw the graph of the line. To show that the line $y = -2x + 3$ is included in the graph, we make it a solid line and shade the region above. See Fig. 7.8.

c) First solve for y:

$$2x - 3y < 6$$
$$-3y < -2x + 6$$
$$y > \frac{2}{3}x - 2 \quad \text{Divide by } -3 \text{ and reverse the inequality.}$$

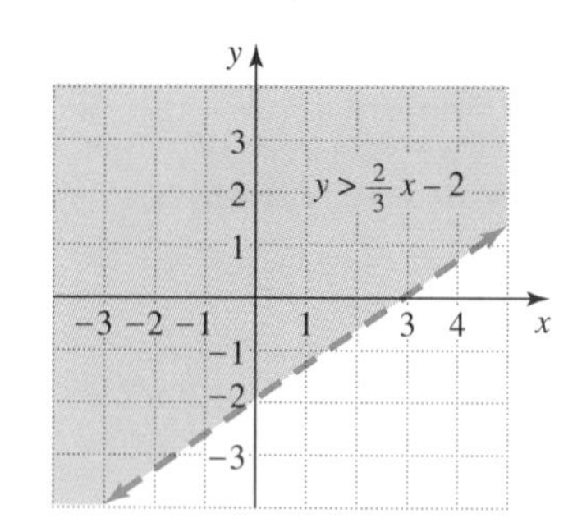

Figure 7.9

To graph this inequality, we first graph the line with slope $\frac{2}{3}$ and y-intercept $(0, -2)$. We use a dashed line for the boundary because it is not included, and we shade the region above the line. Remember, "less than" means below the line and "greater than" means above the line only when the inequality is solved for y. See Fig. 7.9 for the graph. ■

EXAMPLE 3 **Horizontal and vertical boundary lines**

Graph each inequality.

a) $y \leq 4$ **b)** $x > 3$

Solution

a) The line $y = 4$ is the horizontal line with y-intercept $(0, 4)$. We draw a solid horizontal line and shade below it as in Fig. 7.10.

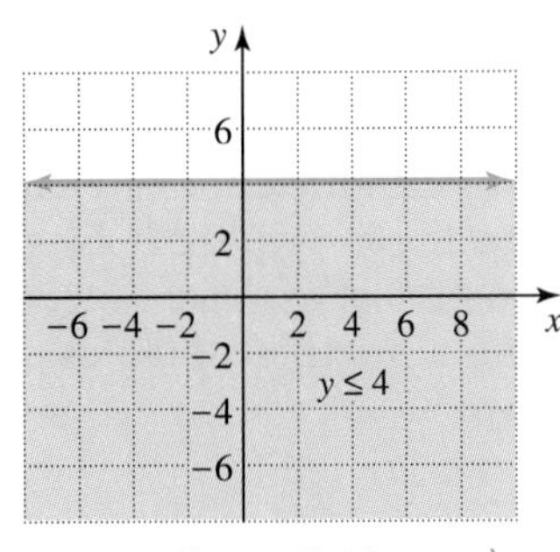

Figure 7.10

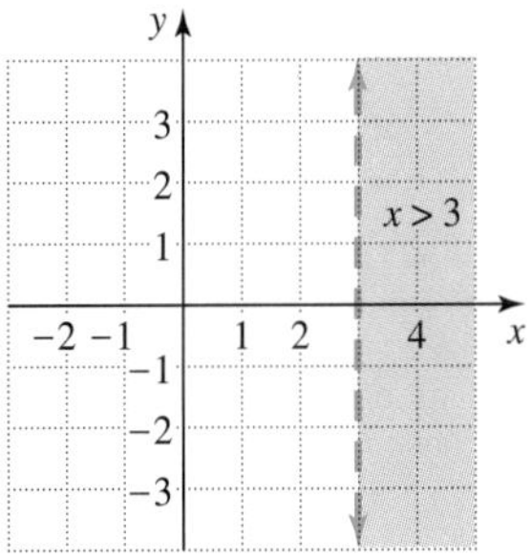

Figure 7.11

b) The line $x = 3$ is a vertical line through $(3, 0)$. Any point to the right of this line has an x-coordinate larger than 3. The graph is shown in Fig. 7.11.

Using a Test Point to Graph an Inequality

The graph of a linear equation such as $2x - 3y = 6$ separates the coordinate plane into two regions. One region satisfies the inequality $2x - 3y > 6$, and the other region satisfies the inequality $2x - 3y < 6$. We can tell which region satisfies which inequality by testing a point in one region. With this method it is not necessary to solve the inequality for y.

EXAMPLE 4 **Using a test point**

Graph the inequality $2x - 3y > 6$.

Solution

First graph the equation $2x - 3y = 6$ using the x-intercept $(3, 0)$ and the y-intercept $(0, -2)$ as shown in Fig. 7.12. Select a point on one side of the line, say $(0, 1)$, to test in the inequality. Because

$$2(0) - 3(1) > 6$$

is false, the region on the other side of the line satisfies the inequality. The graph of $2x - 3y > 6$ is shown in Fig. 7.13.

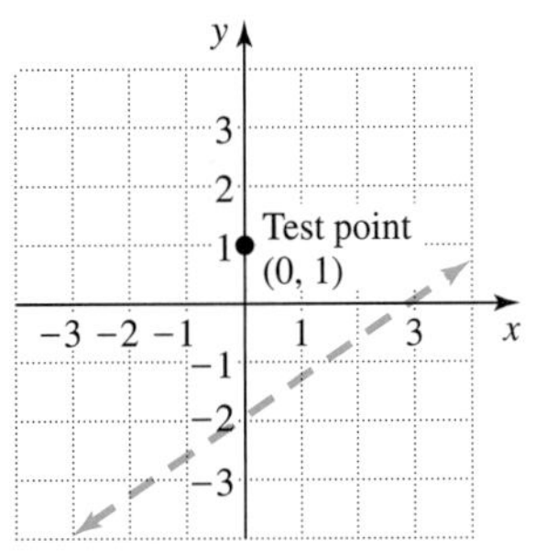

Figure 7.12

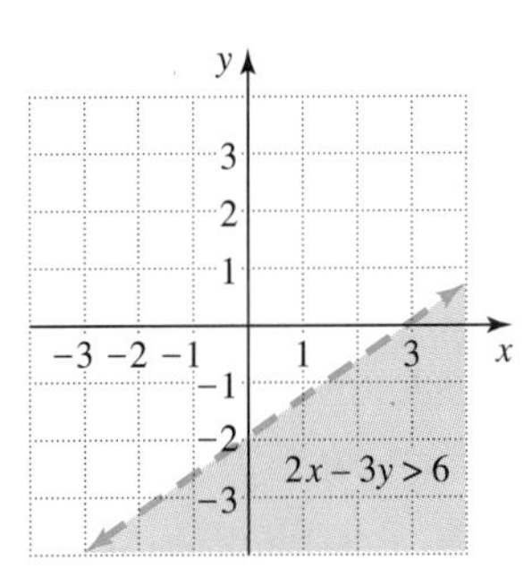

Figure 7.13

Applications

The values of variables used in applications are often restricted to nonnegative numbers. So solutions to inequalities in these applications are graphed in the first quadrant only.

EXAMPLE 5 **Manufacturing tables**

The Ozark Furniture Company can obtain at most 8,000 board feet of oak lumber for making two types of tables. It takes 50 board feet to make a round table and 80 board feet to make a rectangular table. Write an inequality that limits the possible number of tables of each type that can be made. Graph the inequality.

Solution

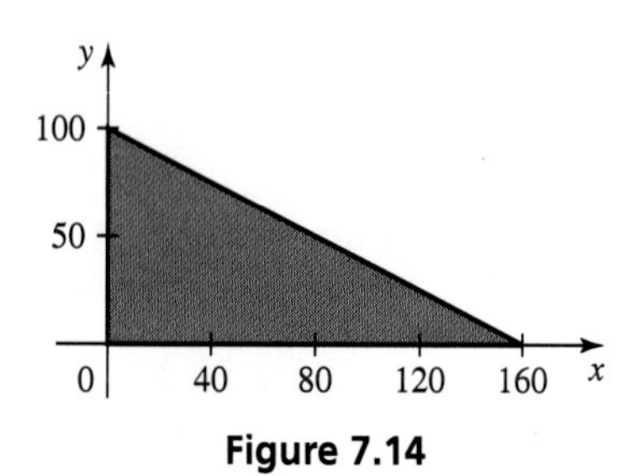

Figure 7.14

If x is the number of round tables and y is the number of rectangular tables, then x and y must satisfy the inequality $50x + 80y \le 8000$. The line $50x + 80y = 8000$ has intercepts $(0, 100)$ and $(160, 0)$. Because $(0, 0)$ satisfies $50x + 80y \le 8000$, the number of tables of each type must be in the region below the line and in the first quadrant as shown in Fig. 7.14. Assuming that Ozark is not interested in making a fraction of a table, only points in this region with whole-number coordinates are possible solutions to the problem. ■

Warm-ups

True or false? Explain your answer.

1. The point $(-1, 4)$ satisfies the inequality $y > 3x + 1$.
2. The point $(2, -3)$ satisfies the inequality $3x - 2y \ge 12$.
3. The graph of the inequality $y > x + 9$ is the region above the line $y = x + 9$.
4. The graph of the inequality $x < y + 2$ is the region below the line $x = y + 2$.
5. The graph of $x = 3$ is a single point on the x-axis.
6. The graph of $y \le 5$ is the region below the horizontal line $y = 5$.
7. The graph of $x < 3$ is the region to the left of the vertical line $x = 3$.
8. In graphing the inequality $y \ge x$ we use a dashed boundary line.
9. The point $(0, 0)$ is on the graph of the inequality $y \ge x$.
10. The point $(0, 0)$ lies above the line $y = 2x + 1$.

7.4 EXERCISES

Determine which of the points following each inequality satisfy that inequality. See Example 1.

1. $x - y > 5$ $(2, 3)$, $(-3, -9)$, $(8, 3)$
2. $2x + y < 3$ $(-2, 6)$, $(0, 3)$, $(3, 0)$
3. $y \ge -2x + 5$ $(3, 0)$, $(1, 3)$, $(-2, 5)$
4. $y \le -x + 6$ $(2, 0)$, $(-3, 9)$, $(-4, 12)$
5. $x > -3y + 4$ $(2, 3)$, $(7, -1)$, $(0, 5)$
6. $x < -y - 3$ $(1, 2)$, $(-3, -4)$, $(0, -3)$

Graph each inequality. See Examples 2 and 3.

7. $y < x + 4$

8. $y < 2x + 2$

9. $y > -x + 3$

10. $y < -2x + 1$

11. $y > \frac{2}{3}x - 3$

12. $y < \frac{1}{2}x + 1$

13. $y \leq -\frac{2}{5}x + 2$

14. $y \geq -\frac{1}{2}x + 3$

15. $y - x \geq 0$

16. $x - 2y \leq 0$

17. $x > y - 5$

18. $2x < 3y + 6$

19. $x - 2y + 4 \leq 0$

20. $2x - y + 3 \geq 0$

21. $y \geq 2$

22. $y < 7$

23. $x > 9$

24. $x \leq 1$

25. $x + y \leq 60$

26. $x - y \leq 90$

27. $x \leq 100y$

28. $y \geq 600x$

29. $3x - 4y \leq 8$

30. $2x + 5y \geq 10$

Graph each inequality. Use the test point method of Example 4.

31. $2x - 3y < 6$

32. $x - 4y > 4$

33. $x - 4y \leq 8$

34. $3y - 5x \geq 15$

35. $y - \frac{7}{2}x \leq 7$

36. $\frac{2}{3}x + 3y \leq 12$

37. $x - y < 5$

38. $y - x > -3$

39. $3x - 4y < -12$

40. $4x + 3y > 24$

41. $x < 5y - 100$

42. $-x > 70 - y$

Solve each problem. See Example 5.

43. ***Storing the tables.*** Ozark Furniture Company must store its oak tables before shipping. A round table is packaged in a carton with a volume of 25 cubic feet (ft^3), and a rectangular table is packaged in a carton with a volume of 35 ft^3. The warehouse has at most 3850 ft^3 of space available for these tables. Write an inequality that limits the possible number of tables of each type that can be stored, and graph the inequality in the first quadrant.

44. ***Maple rockers.*** Ozark Furniture Company can obtain at most 3000 board feet of maple lumber for making its classic and modern maple rocking chairs. A classic maple rocker requires 15 board feet of maple, and a modern rocker requires 12 board feet of maple. Write an inequality that limits the possible number of maple rockers of each type that can be made, and graph the inequality in the first quadrant.

45. ***Enzyme concentration.*** A food chemist tests enzymes for their ability to break down pectin in fruit juices (Dennis Callas, *Snapshots of Applications in Mathematics*). Excess pectin makes juice cloudy. In one test, the chemist measures the concentration of the enzyme, c, in milligrams per milliliter and the fraction of light absorbed by the liquid, a. If $a > 0.07c + 0.02$, then the enzyme is working as it should. Graph the inequality for $0 < c < 5$.

Getting More Involved

46. ***Discussion.*** When asked to graph the inequality $x + 2y < 12$, a student found that $(0, 5)$ and $(8, 0)$ both satisfied $x + 2y < 12$. The student then drew a dashed line through these two points and shaded the region below the line. What is wrong with this method? Do all of the points graphed by this student satisfy the inequality?

47. ***Writing.*** Compare and contrast the two methods presented in this section for graphing linear inequalities. What are the advantages and disadvantages of each method? How do you choose which method to use?

7.5 Systems of Linear Inequalities

In this section:

- The Solution to a System of Inequalities
- Graphing a System of Inequalities

In Section 7.4 you learned how to solve a linear inequality. In this section you will solve systems of linear inequalities.

The Solution to a System of Inequalities

A point is a solution to a system of equations if it satisfies both equations. Similarly, a point is a solution to a system of inequalities if it satisfies both inequalities.

EXAMPLE 1 **Satisfying a system of inequalities**

Determine whether each point is a solution to the system of inequalities:

$$2x + 3y < 6$$
$$y > 2x - 1$$

a) $(-3, 2)$ **b)** $(4, -3)$ **c)** $(5, 1)$

Solution

a) The point $(-3, 2)$ is a solution to the system if it satisfies both inequalities. Let $x = -3$ and $y = 2$ in each inequality:

$$\begin{aligned} 2x + 3y &< 6 & y &> 2x - 1 \\ 2(-3) + 3(2) &< 6 & 2 &> 2(-3) - 1 \\ 0 &< 6 & 2 &> -7 \end{aligned}$$

Because both inequalities are satisfied, the point $(-3, 2)$ is a solution to the system.

b) Let $x = 4$ and $y = -3$ in each inequality:

$$\begin{aligned} 2x + 3y &< 6 & y &> 2x - 1 \\ 2(4) + 3(-3) &< 6 & -3 &> 2(4) - 1 \\ -1 &< 6 & -3 &> 7 \end{aligned}$$

Because only one inequality is satisfied, the point $(4, -3)$ is not a solution to the system.

c) Let $x = 5$ and $y = 1$ in each inequality:

$$\begin{aligned} 2x + 3y &< 6 & y &> 2x - 1 \\ 2(5) + 3(1) &< 6 & 1 &> 2(5) - 1 \\ 13 &< 6 & 1 &> 9 \end{aligned}$$

Because neither inequality is satisfied, the point (5, 1) is not a solution to the system. ■

Graphing a System of Inequalities

There are infinitely many points that satisfy a typical system of inequalities. The best way to describe the solution to a system of inequalities is with a graph showing all points that satisfy the system. When we graph the points that satisfy a system, we say that we are graphing the system.

EXAMPLE 2 **Graphing a system of inequalities**

Graph all ordered pairs that satisfy the following system of inequalities:

$$y > x - 2$$
$$y < -2x + 3$$

Solution

We want a graph showing all points that satisfy both inequalities. The lines $y = x - 2$ and $y = -2x + 3$ divide the coordinate plane into four regions as shown in Fig. 7.15. To determine which of the four regions contains points that satisfy the system, we check one point in each region to see whether it satisfies both inequalities. The points are shown in Fig. 7.15.

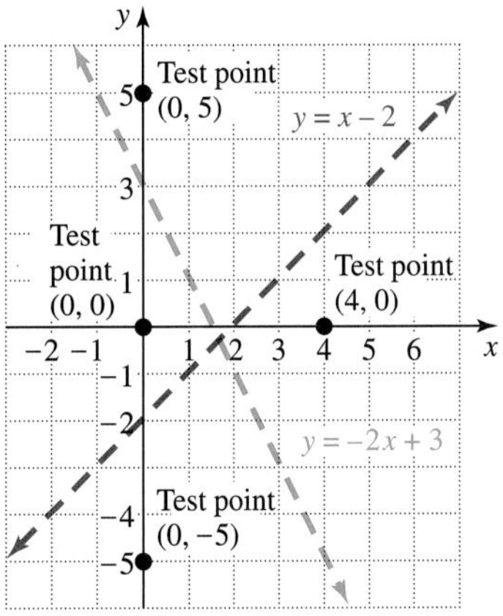

Figure 7.15

Check $(0, 0)$:		Check $(0, 5)$:	
$0 > 0 - 2$	Correct	$5 > 0 - 2$	Correct
$0 < -2(0) + 3$	Correct	$5 < -2(0) + 3$	Incorrect

Check $(0, -5)$:		Check $(4, 0)$:	
$-5 > 0 - 2$	Incorrect	$0 > 4 - 2$	Incorrect
$-5 < -2(0) + 3$	Correct	$0 < -2(4) + 3$	Incorrect

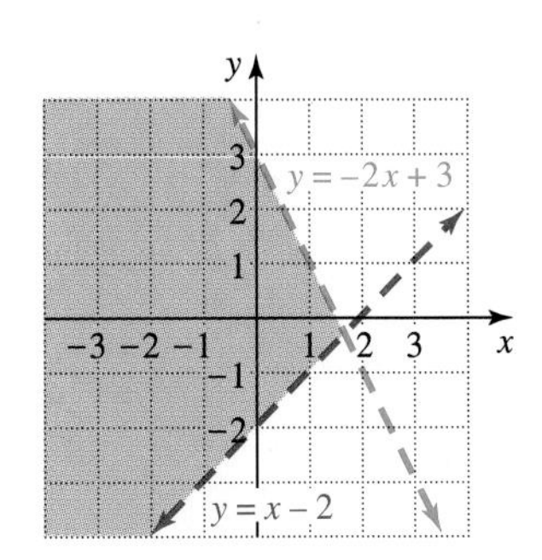

Figure 7.16

The only point that satisfies both inequalities of the system is $(0, 0)$. So every point in the region containing $(0, 0)$ also satisfies both inequalities. The points that satisfy the system are graphed in Fig. 7.16. ■

EXAMPLE 3 **Graphing a system of inequalities**

Graph all ordered pairs that satisfy the following system of inequalities:

$$y > -3x + 4$$
$$2y - x > 2$$

Solution

First graph the equations $y = -3x + 4$ and $2y - x = 2$. Now we select the points $(0, 0)$, $(0, 2)$, $(0, 6)$, and $(5, 0)$. We leave it to you to check each point in the system of inequalities. You will find that only $(0, 6)$ satisfies the system. So only the region containing $(0, 6)$ is shaded in Fig. 7.17.

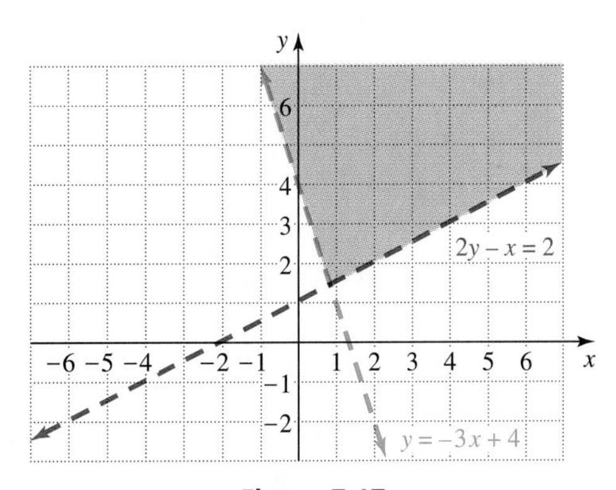

Figure 7.17 ■

EXAMPLE 4 **Horizontal and vertical boundary lines**

Graph the system of inequalities: $x > 4$

$y < 3$

Solution

We first graph the vertical line $x = 4$ and the horizontal line $y = 3$. The points that satisfy both inequalities are those points that lie to the right of the vertical line $x = 4$ and below the horizontal line $y = 3$. See Fig. 7.18 for the graph of the system. ■

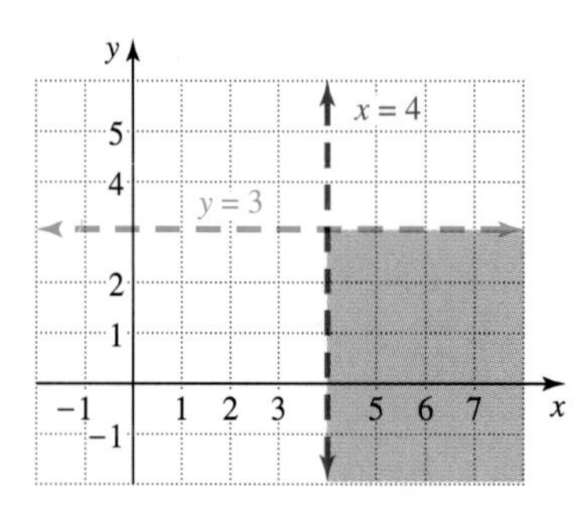

Figure 7.18

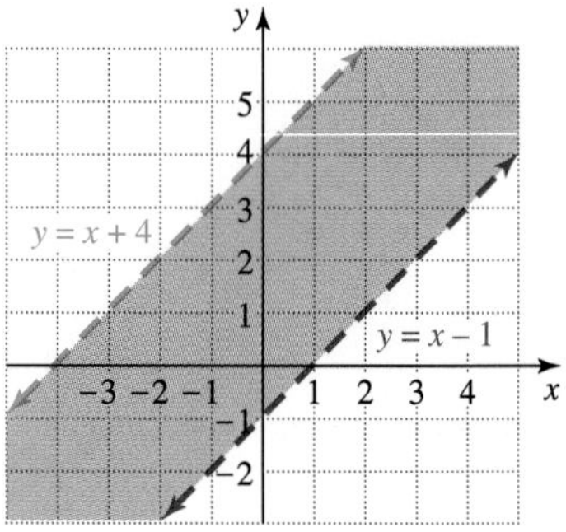

Figure 7.19

EXAMPLE 5 **Between parallel lines**

Graph the system of inequalities: $y < x + 4$

$y > x - 1$

Solution

First graph the parallel lines $y = x + 4$ and $y = x - 1$. These lines divide the plane into three regions. Check $(0, 0)$, $(0, 6)$, and $(0, -4)$ in the system. Only $(0, 0)$ satisfies the system. So the solution to the system consists of all points in between the parallel lines, as shown in Fig. 7.19. ■

Warm-ups

True or false? Explain your answer.

Use the following systems for Exercises 1–7.

a) $y > -3x + 5$, $y < 2x - 3$ **b)** $y > 2x - 3$, $y < 2x + 3$ **c)** $x + y > 4$, $x - y < 0$

1. The point $(2, -3)$ is a solution to system (a).
2. The point $(5, 0)$ is a solution to system (a).
3. The point $(0, 0)$ is a solution to system (b).
4. The graph of system (b) is the region between two parallel lines.
5. You can use $(0, 0)$ as a test point for system (c).
6. The point $(2, 2)$ satisfies system (c).
7. The point $(4, 5)$ satisfies system (c).
8. The inequality $x + y > 4$ is equivalent to the inequality $y < -x + 4$.
9. The graph of $y < 2x + 3$ is the region below the line $y = 2x + 3$.
10. There is no ordered pair that satisfies $y < 2x - 3$ and $y > 2x + 3$.

7.5 EXERCISES

Determine which of the points following each system is a solution to the system. See Example 1.

1. $x - y < 5$ $(4, 3), (8, 2), (-3, 0)$
$2x + y > 3$

2. $x + y < 4$ $(2, -3), (1, 1), (0, -1)$
$2x - y < 3$

3. $y > -2x + 1$ $(-3, 2), (-1, 5), (3, 6)$
$y < 3x + 5$

4. $y < -x + 7$ $(-3, 8), (0, 8), (-5, 15)$
$y < -x + 9$

5. $x > 3$ $(-5, 4), (9, -5), (6, 0)$
$y < -2$

6. $y < -5$ $(-2, 4), (0, -7), (6, -9)$
$x < 1$

Graph each system of inequalities. See Examples 2–5.

7. $y > -x - 1$
$y > x + 1$

8. $y < x + 3$
$y < -2x + 4$

9. $y < 2x - 3$
$y > -x + 2$

10. $y > 2x - 1$
$y < -x - 4$

11. $x + y > 5$
$x - y < 3$

12. $2x + y < 3$
$x - 2y > 2$

13. $2x - 3y < 6$
$x - y > 3$

14. $3x - 2y > 6$
$x + y < 4$

15. $x > 5$
$y > 5$

16. $x < 3$
$y > 2$

17. $y < -1$
$x > -3$

18. $y > -2$
$x < 1$

19. $y > 2x - 4$
$y < 2x + 1$

20. $y < -2x + 3$
$y > -2x$

21. $y > x$
$x > 3$

22. $y < x$
$y < 1$

23. $y > -x$
$x < -1$

24. $y < -x$
$y > -3$

25. $x > 1$
$y - 2x < 3$

26. $y < 2$
$2x + 3y < 6$

27. $2x - 5y < 5$
$x + 2y > 4$

28. $3x + 2y < 2$
$-x - 2y > 4$

29. $x + y > 3$
$x + y > 1$

30. $x - y < 5$
$x - y < 3$

31. $y > 3x + 2$
$y < 3x + 3$

32. $y > x$
$y < -x$

33. $x + y < 5$
$x - y > -1$

34. $2x - y > 4$
$x - 5y < 5$

35. $2x - 3y < 6$
$3x + 4y < 12$

36. $x - 3y > 3$
$x + 2y < 4$

37. $3x - 5y < 15$
$3x + 2y < 12$

38. $x - 4y < 0$
$x + y > 0$

Solve each problem.

39. ***Target heart rate.*** For beneficial exercise, experts recommend that your target heart rate h should be between 65% and 75% of the maximum heart rate for your age a. That is,

$$h > 0.65(220 - a) \text{ and } h < 0.75(220 - a).$$

Graph this system of inequalities for $20 < a < 70$.

40. ***Making and storing the tables.*** The Ozark Furniture Company can obtain at most 8,000 board feet of oak lumber for making round and rectangular tables. The tables must be stored in a warehouse that has at most 3850 ft^3 of space available for the tables. A round table requires 50 board feet of lumber and 25 ft^3 of warehouse space. A rectangular table requires 80 board feet of lumber and 35 ft^3 of warehouse space. Write a system of inequalities that limits the possible number of tables of each type that can be made and stored. Graph the system.

41. ***Allocating resources.*** Wausaukee Enterprises makes yard barns in two sizes. One small barn requires \$250 in materials and 20 hours of labor, and one large barn requires \$400 in materials and 30 hours of labor. Wausaukee has at most \$4000 to spend on materials and at most 300 hours of labor available. Write a system of inequalities that limits the possible number of barns of each type that can be built. Graph the system.

COLLABORATIVE ACTIVITIES

Which Cider?

Grouping: 3 to 4 students
Topic: Systems of equations

For the Fall Harvest Festival your math club decides to sell apple cider. You can get a good deal on bulk apple cider by buying 30 gallons for $115. At the club meeting, one member suggests buying the apple cider from her uncle, who has a nearby organic apple orchard. He will lower his price to the club to $5 a gallon. You can get paper cups at $2 per 100. To decide which cider to buy, you analyze the two options using the profit equation Profit = Sales − Cost.

1. If you want to sell all 30 gallons in 8-ounce cups, how many cups will you need to buy?
2. What will your total costs be for paper cups and bulk cider? For paper cups and local cider?
3. The club wants to sell the cider for $0.50 a cup. Write a profit equation for each type of cider in terms of number of cups sold. Let $c =$ number of cups sold and $P =$ profit.
4. Graph both equations using a graphing calculator or on graph paper. Have the vertical y-axis be profit and the horizontal x-axis be the number of cups sold. Let each mark on the axes represent 20 or 40 units.
5. How many cups do you need to sell to make a profit for local cider? For bulk cider?
6. Is there a point at which the number of cups and the profit are the same for both types?

The member who wants the local cider points out that your club could sell it for more, since it will be fresher and of higher quality. She suggests selling it for $0.75 a cup.

7. Write an equation for local cider at 75 cents a cup. Graph this equation with the one you had for bulk cider.
8. When are the profits greater for the local cider? Estimate this from your graph. Find this answer algebraically, using elimination or substitution.
9. Decide which cider your club should sell and at which price.

Wrap-up CHAPTER 7

SUMMARY

Systems of Linear Equations in Two Variables		Examples
Graphing method	Sketch each graph and identify the points they have in common.	
Substitution method	Solve one equation for one variable in terms of the other, then substitute into the other equation.	$y = x - 4$ $x + y = 9$ $x + (x - 4) = 9$

Addition method	Multiply each equation as necessary to eliminate a variable upon addition of the equations.	$5x - 3y = 4$ $x + 3y = 1$ $6x \quad = 5$
Independent	Only one point satisfies both equations. The graphs cross at one point.	$y = x - 4$ $y = 2x + 5$
Inconsistent	No solution The graphs are parallel lines.	$y = 5x - 3$ $y = 5x + 1$
Dependent	Infinitely many solutions One equation is a multiple of the other. The graphs coincide.	$5x + 3y = 2$ $10x + 6y = 4$

Linear Inequalities in Two Variables		Examples
Graphing the solution to an inequality in two variables	1. Solve the inequality for y, then graph $y = mx + b$. $y > mx + b$ is the region above the line. $y = mx + b$ is the line itself. $y < mx + b$ is the region below the line.	 $y > x + 3$ $y = x + 3$ $y < x + 3$
	Remember that "less than" means below the line and "greater than" means above the line only when the inequality is solved for y.	
	2. If the inequality involves only x, then graph the vertical line $x = k$. $x > k$ is the region to the right of the line. $x = k$ is the line itself. $x < k$ is the region to the left of the line.	$x > 5$ Region to right of vertical line $x = 5$
Test points	A linear inequality may also be graphed by graphing the equation and then testing a point to determine which region satisfies the inequality.	$x + y > 4$ (0, 6) satisfies the inequality.
Graphing a system of inequalities	Graph the equations and use test points to see which regions satisfy both inequalities.	$x + y > 4$ $x - y < 1$ (0, 6) satisfies the system.

REVIEW EXERCISES

7.1 *Solve each system by graphing.*

1. $y = 2x + 1$
$x + y = 4$

2. $y = -x + 1$
$y = -x + 3$

3. $y = 2x + 3$
$y = -2x - 1$

4. $x + y = 6$
$x - y = -10$

7.2 *Solve each system by the substitution method.*

5. $y = 3x$
$2x + 3y = 22$

6. $x + y = 3$
$3x - 2y = -11$

7. $x = y - 5$
$2x - 3y = -7$

8. $2x + y = 5$
$6x - 9 = 3y$

7.3 *In Exercises 9–20, solve each system by the addition method. Indicate whether each system is independent, inconsistent, or dependent.*

9. $x - y = 4$
$2x + y = 5$

10. $x + 2y = -5$
$x - 3y = 10$

11. $2x - 4y = 8$
$x - 2y = 4$

12. $x + 3y = 7$
$2x + 6y = 5$

13. $y = 3x - 5$
$2y = -x - 3$

14. $3x + 4y = 6$
$4x + 3y = 1$

15. $2x + 7y = 0$
$7x + 2y = 0$

16. $3x - 5y = 1$
$10y = 6x - 1$

17. $x - y = 6$
$2x - 12 = 2y$

18. $y = 4x$
$y = 3x$

19. $y = 4x$
$y = 4x + 3$

20. $3x - 5y = 21$
$4x + 7y = -13$

7.4 *Graph each inequality.*

21. $y > \frac{1}{3}x - 5$

22. $y < \frac{1}{2}x + 2$

23. $y \leq -2x + 7$

24. $y \geq x - 6$

25. $y \leq 8$

26. $x \geq -6$

27. $2x + 3y \leq -12$

28. $x - 3y < 9$

7.5 *Graph each system of inequalities.*

29. $x < 5$
$y < 4$

30. $y > -2$
$x < 1$

31. $x + y < 2$
$y > 2x - 3$

32. $x - y > 4$
$2y > x - 4$

33. $y > 5x - 7$
$y < 5x + 1$

34. $y > x - 6$
$y < x - 5$

35. $y < 3x + 5$
$y < 3x$

36. $y > -2x$
$y > -3x$

Miscellaneous

Use a system of equations in two variables to solve each problem. Solve the system by the method of your choice.

37. ***Apples and oranges.*** Two apples and three oranges cost \$1.95, and three apples and two oranges cost \$2.05. What are the costs of one apple and one orange?

38. ***Small or medium.*** Three small drinks and one medium drink cost \$2.30, and two small drinks and four medium drinks cost \$3.70. What is the cost of one small drink? What is the cost of one medium drink?

39. ***Gambling fever.*** After a long day at the casinos in Biloxi, Louis returned home and told his wife Lois that he had won \$430 in \$5 bills and \$10 bills. On counting them again, he realized that he had mixed up the number of bills of each denomination, and he had really won only \$380. How many bills of each denomination does Louis have?

40. ***Diversifying investments.*** Diane invested her \$10,000 bonus in a municipal bond fund and an emerging market fund. In one year the amount invested in the bond fund earned 8%, and the amount invested in the emerging market fund earned 10%. If the total income from these two investments for one year was \$880, then how much did she invest in each fund?

41. ***Protein and carbohydrates.*** One serving of green beans contains 1 gram of protein and 4 grams of carbohydrates. One serving of chicken soup contains 3 grams of protein and 9 grams of carbohydrates. The Westdale Diet recommends a lunch of 13 grams of protein and 43 grams of carbohydrates. How many servings of each are necessary to obtain the recommended amounts?

42. ***Advertising revenue.*** A television station aired four 30-second commercials and three 60-second commercials during the first hour of the midnight movie. During the second hour, it aired six 30-second commercials and five 60-second commercials. The advertising revenue for the first hour was \$7,700, and that for the second hour was \$12,300. What is the cost of each type of commercial?

CHAPTER 7 TEST

Solve the system by graphing.

1. $x + y = 2$
$y = 2x + 5$

Solve each system by substitution.

2. $y = 2x - 3$
$2x + 3y = 7$

3. $x - y = 4$
$3x - 2y = 11$

Solve each system by the addition method.

4. $2x + 5y = 19$
$4x - 3y = -1$

5. $3x - 2y = 10$
$2x + 5y = 13$

Determine whether each system is independent, inconsistent, or dependent.

6. $y = 4x - 9$
$y = 4x + 8$

7. $3x - 3y = 12$
$y = x - 4$

8. $y = 2x$
$y = 5x$

Graph each inequality.

9. $y > 3x - 5$

10. $x - y < 3$

11. $x - 2y \geq 4$

Graph each system of inequalities.

12. $x < 6$
$y > -1$

13. $2x + 3y > 6$
$3x - y < 3$

14. $y > 3x - 4$
$3x - y > 3$

For each problem, write a system of equations in two variables. Use the method of your choice to solve each system.

15. Kathy and Chris studied a total of 54 hours for the CPA exam. If Chris studied only one-half as many hours as Kathy, then how many hours did each of them study?

16. The Rest-Is-Easy Motel just outside Amarillo rented five singles and three doubles on Monday night for a total of $188. On Tuesday night it rented three singles and four doubles for a total of $170. On Wednesday night it rented only one single and one double. How much rent did the motel receive on Wednesday night?

Tying It All Together CHAPTERS 1–7

Solve each equation.

1. $2(x - 5) + 3x = 25$

2. $3x - 5 = 0$

3. $\frac{x}{3} - \frac{2}{5} = \frac{x}{2} - \frac{12}{5}$

4. $x^2 + 2x = 24$

5. $\frac{x + 5}{x - 1} = \frac{2}{3}$

6. $2x^2 - 7x + 3 = 0$

Solve each inequality in one variable, and sketch the graph on the number line.

7. $3(2 - x) < -6$

8. $-3 \leq 2x - 4 \leq 6$

9. $x > 1$

Sketch the graph of each equation.

10. $y = 3x - 7$

11. $y = 5 - x$

12. $y = x - 1$

13. $y = x + 1$

14. $y = -2x + 4$

15. $y = -4x - 1$

Graph each inequality in two variables.

16. $y \geq 3x - 7$

17. $x - 2y < 6$

18. $x > 1$

Write the equation of the line going through each pair of points.

19. (0, 36) and (8, 84)

20. (1, 88) and (12, 11)

Solve the problem.

21. ***Decreasing CD-ROM prices.*** The industry average price per title for a CD-ROM is projected to go from \$45 in 1994 to \$25 in 1998 as shown in the accompanying graph (*Fortune*, October 3, 1994).

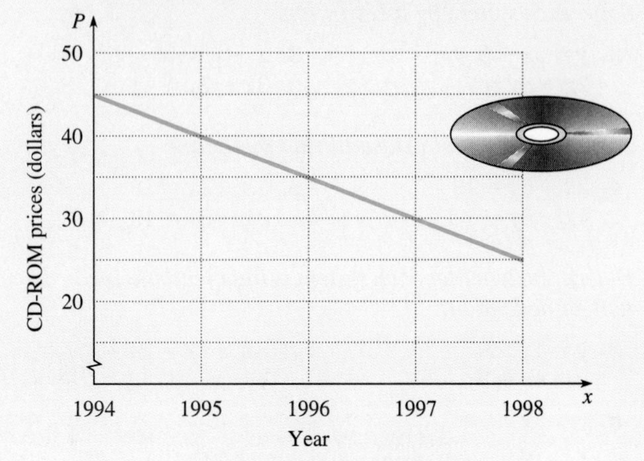

Figure for Exercise 21

a) Write the average price p as a linear function of x, where x is the number of years after 1994.

b) A new company, CDs Inc., plans to increase its average price from \$30 in 1994 to \$40 in 1998 by increasing its quality. Write the average price p for CDs Inc. as a linear function of x, where x is the number of years after 1994.

c) Solve the system of equations that you found in parts (a) and (b) to find the year in which the average price for CDs Inc. will equal the industry average.

CHAPTER 8

Powers and Roots

Recycling—the very word conjures up pictures of separating trash, composting gardens, and returning aluminum cans. But many industries are interested in recycling for both economic and environmental reasons.

Metals account for a large fraction of the mass of materials in industrial products. Although metals are relatively easy to reprocess and reuse, they are often combined with other substances or waste materials. When a manufacturer wants to reuse or reprocess a material containing a particular metal such as aluminum, bronze, or gold, it must separate the metal from other waste material. In each case there are expenditures involved in separating a specific metal from other components. Expenses are also involved in collecting and transporting used-up products, scrap, and waste. But if the metal is not separated and used, other costs are involved in not recycling but simply disposing of the waste. In addition, there are hidden costs in terms of environmental damage. When the hidden costs are ignored, it is often cheaper to extract certain metals from virgin ores than to recycle them.

In studying the costs of recycling and extracting metals from ores, Thomas Sherwood of MIT discovered some interesting relationships between the price of a metal and its concentration in the commercial ore. In Exercises 99 and 100 of Section 8.1 you will see how negative exponents are used in Sherwood's data and see how to determine which types of metallic waste are underexploited.

8.1 Negative Exponents and Scientific Notation

In this section:

- Negative Integral Exponents
- Rules for Integral Exponents
- Converting from Scientific Notation
- Converting to Scientific Notation
- Computations with Scientific Notation

We defined exponential expressions with positive integral exponents in Chapter 1 and learned the rules for positive integral exponents in Chapter 3. In this section you will first study negative exponents and then see how positive and negative integral exponents are used in scientific notation.

Negative Integral Exponents

If x is nonzero, the reciprocal of x is written as $\frac{1}{x}$. For example, the reciprocal of 2^3 is written as $\frac{1}{2^3}$. To write the reciprocal of an exponential expression in a simpler way, we use a negative exponent. For example, the reciprocal of 2^3 can be written as 2^{-3}. Since $2^3 = 8$, we have $2^{-3} = \frac{1}{8}$. In general we have the following definition.

Negative Integral Exponents

If a is a nonzero real number and n is a positive integer, then

$$a^{-n} = \frac{1}{a^n}. \quad \text{(If } n \text{ is positive, } -n \text{ is negative.)}$$

Later in this section you will see how the rules for positive integral exponents work with any integral exponents, and then you will understand why negative exponents are used to represent reciprocals.

EXAMPLE 1 **Simplifying expressions with negative exponents**

Simplify.

a) 2^{-5} **b)** $(-2)^{-5}$ **c)** $\frac{2^{-3}}{3^{-2}}$

Solution

a) $2^{-5} = \frac{1}{2^5} = \frac{1}{32}$

b) $(-2)^{-5} = \frac{1}{(-2)^5}$ Definition of negative exponent

$= \frac{1}{-32} = -\frac{1}{32}$

c) $\frac{2^{-3}}{3^{-2}} = 2^{-3} \div 3^{-2}$

$= \frac{1}{2^3} \div \frac{1}{3^2}$

$= \frac{1}{8} \div \frac{1}{9} = \frac{1}{8} \cdot \frac{9}{1} = \frac{9}{8}$ ■

CAUTION In simplifying -5^{-2}, the negative sign preceding the 5 is used after 5 is squared and the reciprocal is found. So $-5^{-2} = -(5^{-2}) = -\frac{1}{25}$. ⊘

To evaluate a^{-n}, you can first find the nth power of a and then find the reciprocal. However, the result is the same if you first find the reciprocal of a and then find the nth power of the reciprocal. For example,

$$3^{-2} = \frac{1}{3^2} = \frac{1}{9} \quad \text{or} \quad 3^{-2} = \left(\frac{1}{3}\right)^2 = \frac{1}{3} \cdot \frac{1}{3} = \frac{1}{9}.$$

So the power and the reciprocal can be found in either order. If the exponent is -1, we simply find the reciprocal. For example,

$$5^{-1} = \frac{1}{5}, \quad \left(\frac{1}{4}\right)^{-1} = 4, \quad \text{and} \quad \left(-\frac{3}{5}\right)^{-1} = -\frac{5}{3}.$$

Because $3^{-2} \cdot 3^2 = 1$, the reciprocal of 3^{-2} is 3^2, and we have

$$\frac{1}{3^{-2}} = 3^2.$$

These examples illustrate the following rules.

Rules for Negative Exponents

If a is a nonzero real number and n is a positive integer, then

$$a^{-n} = \left(\frac{1}{a}\right)^n, \quad a^{-1} = \frac{1}{a}, \quad \text{and} \quad \frac{1}{a^{-n}} = a^n.$$

EXAMPLE 2 **Using the rules for negative exponents**

Simplify.

a) $\left(\frac{3}{4}\right)^{-3}$ **b)** $10^{-1} + 10^{-1}$ **c)** $\frac{2}{10^{-3}}$

Solution

a) $\left(\frac{3}{4}\right)^{-3} = \left(\frac{4}{3}\right)^3 = \frac{64}{27}$

b) $10^{-1} + 10^{-1} = \frac{1}{10} + \frac{1}{10} = \frac{2}{10} = \frac{1}{5}$

c) $\frac{2}{10^{-3}} = 2 \cdot \frac{1}{10^{-3}} = 2 \cdot 10^3 = 2 \cdot 1000 = 2000$ ■

Rules for Integral Exponents

Negative exponents are used to make expressions involving reciprocals simpler looking and easier to write. Negative exponents have the added benefit of working in conjunction with all of the rules of exponents that you learned in Chapter 3. For example, we can use the product rule to get

$$x^{-2} \cdot x^{-3} = x^{-2+(-3)} = x^{-5}$$

and the quotient rule to get

$$\frac{y^3}{y^5} = y^{3-5} = y^{-2}.$$

With negative exponents there is no need to state the quotient rule in two parts as we did in Chapter 3. It can be stated simply as

$$\frac{a^m}{a^n} = a^{m-n}$$

for any integers m and n. We list the rules of exponents here for easy reference.

Rules for Integral Exponents

The following rules hold for nonzero real numbers a and b and any integers m and n.

1. $a^0 = 1$	Definition of zero exponent
2. $a^m \cdot a^n = a^{m+n}$	Product rule
3. $\frac{a^m}{a^n} = a^{m-n}$	Quotient rule
4. $(a^m)^n = a^{mn}$	Power rule
5. $(ab)^n = a^n \cdot b^n$	Power of a product rule
6. $\left(\frac{a}{b}\right)^n = \frac{a^n}{b^n}$	Power of a quotient rule

EXAMPLE 3 **The product and quotient rules for integral exponents**

Simplify. Write your answers without negative exponents. Assume that the variables represent nonzero real numbers.

a) $b^{-3}b^5$ **b)** $-3x^{-3} \cdot 5x^2$ **c)** $\frac{m^{-6}}{m^{-2}}$ **d)** $\frac{4y^5}{-12y^{-3}}$

Solution

a) $b^{-3}b^5 = b^{-3+5}$ Product rule

$= b^2$ Simplify.

b) $-3x^{-3} \cdot 5x^2 = -15x^{-1}$ Product rule

$= -\frac{15}{x}$ Definition of negative exponent

c) $\frac{m^{-6}}{m^{-2}} = m^{-6-(-2)}$ Quotient rule

$= m^{-4}$ Simplify.

$= \frac{1}{m^4}$ Definition of negative exponent

d) $\frac{4y^5}{-12y^{-3}} = \frac{y^{5-(-3)}}{-3} = \frac{-y^8}{3}$ ■

In the next example we use the power rules with negative exponents.

EXAMPLE 4 **The power rules for integral exponents**

Simplify each expression. Write your answers with positive exponents only. Assume that all variables represent nonzero real numbers.

a) $(a^{-3})^2$ b) $(10x^{-3})^{-2}$ c) $\left(\dfrac{4x^{-5}}{y^2}\right)^{-2}$

Solution

a)
$$\begin{aligned}(a^{-3})^2 &= a^{-3\cdot 2} && \text{Power rule}\\ &= a^{-6}\\ &= \frac{1}{a^6} && \text{Definition of negative exponent}\end{aligned}$$

b)
$$\begin{aligned}(10x^{-3})^{-2} &= 10^{-2}(x^{-3})^{-2} && \text{Power of a product rule}\\ &= 10^{-2}x^{(-3)(-2)} && \text{Power rule}\\ &= \frac{x^6}{10^2} && \text{Definition of negative exponent}\\ &= \frac{x^6}{100}\end{aligned}$$

c)
$$\begin{aligned}\left(\frac{4x^{-5}}{y^2}\right)^{-2} &= \frac{(4x^{-5})^{-2}}{(y^2)^{-2}} && \text{Power of a quotient rule}\\ &= \frac{4^{-2}x^{10}}{y^{-4}} && \text{Power of a product rule and power rule}\\ &= \frac{x^{10}y^4}{4^2} && \text{Definition of negative exponent}\\ &= \frac{x^{10}y^4}{16} && \text{Simplify.}\end{aligned}$$

Converting from Scientific Notation

Many of the numbers occurring in science are either very large or very small. The speed of light is 983,569,000 feet per second. One millimeter is equal to 0.000001 kilometer. In scientific notation, numbers larger than 10 or smaller than 1 are written by using positive or negative exponents.

Scientific notation is based on multiplication by integral powers of 10. Multiplying a number by a positive power of 10 moves the decimal point to the right:

$$\begin{aligned}10(5.32) &= 53.2\\ 10^2(5.32) &= 100(5.32) = 532\\ 10^3(5.32) &= 1000(5.32) = 5320\end{aligned}$$

Multiplying by a negative power of 10 moves the decimal point to the left:

$$10^{-1}(5.32) = \frac{1}{10}(5.32) = 0.532$$

$$10^{-2}(5.32) = \frac{1}{100}(5.32) = 0.0532$$

$$10^{-3}(5.32) = \frac{1}{1000}(5.32) = 0.00532$$

So if n is a positive integer, multiplying by 10^n moves the decimal point n places to the right and multiplying by 10^{-n} moves it n places to the left.

A number in scientific notation is written as a product of a number between 1 and 10 and a power of 10. The times symbol $\times$ indicates multiplication. For example, 3.27×10^9 and 2.5×10^{-4} are numbers in scientific notation. In scientific notation there is one digit to the left of the decimal point.

To convert 3.27×10^9 to standard notation, move the decimal point nine places to the right:

$$3.27 \times 10^9 = 3{,}\underbrace{270{,}000{,}000}_{\text{9 places to the right}}$$

Of course, it is not necessary to put the decimal point in when writing a whole number.

To convert 2.5×10^{-4} to standard notation, the decimal point is moved four places to the left:

$$2.5 \times 10^{-4} = 0.\underbrace{0002}_{\text{4 places to the left}}5$$

In general, we use the following strategy to convert from scientific notation to standard notation.

Strategy for Converting from Scientific Notation to Standard Notation

1. Determine the number of places to move the decimal point by examining the exponent on the 10.
2. Move to the right for a positive exponent and to the left for a negative exponent.

EXAMPLE 5 **Converting scientific notation to standard notation**

Write in standard notation.

a) 7.02×10^6 **b)** 8.13×10^{-5}

Solution

a) Because the exponent is positive, move the decimal point six places to the right:

$$7.02 \times 10^6 = 7020000. = 7{,}020{,}000$$

b) Because the exponent is negative, move the decimal point five places to the left.

$$8.13 \times 10^{-5} = 0.0000813$$

Converting to Scientific Notation

To convert a positive number to scientific notation, we just reverse the strategy for converting from scientific notation.

Strategy for Converting to Scientific Notation

1. Count the number of places (n) that the decimal must be moved so that it will follow the first nonzero digit of the number.
2. If the original number was larger than 10, use 10^n.
3. If the original number was smaller than 1, use 10^{-n}.

Remember that the scientific notation for a number larger than 10 will have a positive power of 10 and the scientific notation for a number between 0 and 1 will have a negative power of 10.

EXAMPLE 6 **Converting numbers to scientific notation**

Write in scientific notation.

a) 7,346,200 **b)** 0.0000348 **c)** 135×10^{-12}

Solution

a) Because 7,346,200 is larger than 10, the exponent on the 10 will be positive:

$$7{,}346{,}200 = 7.3462 \times 10^6$$

b) Because 0.0000348 is smaller than 1, the exponent on the 10 will be negative:

$$0.0000348 = 3.48 \times 10^{-5}$$

c) There should be only one nonzero digit to the left of the decimal point:

$$135 \times 10^{-12} = 1.35 \times 10^2 \times 10^{-12} \quad \text{Convert 135 to scientific notation.}$$
$$= 1.35 \times 10^{-10} \quad \text{Product rule}$$

Computations with Scientific Notation

An important feature of scientific notation is its use in computations. Numbers in scientific notation are nothing more than exponential expressions, and you have already studied operations with exponential expressions in this section. We use the same rules of exponents on numbers in scientific notation that we use on any other exponential expressions.

EXAMPLE 7 **Using the rules of exponents with scientific notation**

Perform the indicated computations. Write the answers in scientific notation.

a) $(3 \times 10^6)(2 \times 10^8)$ **b)** $\dfrac{4 \times 10^5}{8 \times 10^{-2}}$ **c)** $(5 \times 10^{-7})^3$

Solution

a) $(3 \times 10^6)(2 \times 10^8) = 3 \cdot 2 \cdot 10^6 \cdot 10^8 = 6 \times 10^{14}$

b) $\dfrac{4 \times 10^5}{8 \times 10^{-2}} = \dfrac{4}{8} \cdot \dfrac{10^5}{10^{-2}} = \dfrac{1}{2} \cdot 10^{5-(-2)}$ Quotient rule

$= (0.5)10^7$ $\frac{1}{2} = 0.5$

$= 5 \times 10^{-1} \cdot 10^7$ Write 0.5 in scientific notation.

$= 5 \times 10^6$ Product rule

c) $(5 \times 10^{-7})^3 = 5^3(10^{-7})^3$ Power of a product rule

$= 125 \cdot 10^{-21}$ Power rule

$= 1.25 \times 10^2 \times 10^{-21}$ $125 = 1.25 \times 10^2$

$= 1.25 \times 10^{-19}$ Product rule

EXAMPLE 8 **Converting to scientific notation for computations**

Perform these computations by first converting each number into scientific notation. Give your answer in scientific notation.

a) $(3{,}000{,}000)(0.0002)$ **b)** $(20{,}000{,}000)^3(0.0000003)$

Solution

a) $(3{,}000{,}000)(0.0002) = 3 \times 10^6 \cdot 2 \times 10^{-4}$ Scientific notation

$= 6 \times 10^2$ Product rule

b) $(20{,}000{,}000)^3(0.0000003) = (2 \times 10^7)^3(3 \times 10^{-7})$ Scientific notation

$= 8 \times 10^{21} \cdot 3 \times 10^{-7}$ Power of a product rule

$= 24 \times 10^{14}$

$= 2.4 \times 10^1 \times 10^{14}$ $24 = 2.4 \times 10^1$

$= 2.4 \times 10^{15}$ Product rule

Calculator Close-up

A scientific calculator automatically converts into scientific notation any result that will not fit onto its display screen. Most calculators are limited to a power of 10 from −99 to 99. To enter a number in scientific notation into a calculator, we usually use a key labeled [EXP] or [EE] (enter exponent). For example, the product $(6.3 \times 10^{20})(9.5 \times 10^{-6})$ is found on a scientific calculator as follows:

6.3 [EE] 20 [×] 9.5 [EE] 6 [+/−] [=]

The display should read

5.985 E15

This result indicates that the answer is 5.985×10^{15}. The exponent 15 may appear in another manner on your calculator. On a graphing calculator your display should look like the following:

6.3E20*9.5E⁻6 [ENTER]

5.985 E15

So the answer is again 5.985×10^{15}.

Warm-ups

True or false? Explain your answer.

1. $10^{-2} = \frac{1}{100}$

2. $\left(-\frac{1}{5}\right)^{-1} = 5$

3. $3^{-2} \cdot 2^{-1} = 6^{-3}$

4. $\frac{3^{-2}}{3^{-1}} = \frac{1}{3}$

5. $23.7 = 2.37 \times 10^{-1}$

6. $0.000036 = 3.6 \times 10^{-5}$

7. $25 \cdot 10^{7} = 2.5 \times 10^{8}$

8. $0.442 \times 10^{-3} = 4.42 \times 10^{-4}$

9. $(3 \times 10^{-9})^2 = 9 \times 10^{-18}$

10. $(2 \times 10^{-5})(4 \times 10^{4}) = 8 \times 10^{-20}$

8.1 EXERCISES

Variables in all exercises represent positive real numbers.

Evaluate each expression. See Example 1.

1. 3^{-1} **2.** 3^{-3} **3.** $(-2)^{-4}$

4. $(-3)^{-4}$ **5.** -4^{-2} **6.** -2^{-4}

7. $\frac{5^{-2}}{10^{-2}}$ **8.** $\frac{3^{-4}}{6^{-2}}$

Simplify. See Example 2.

9. $\left(\frac{5}{2}\right)^{-3}$ **10.** $\left(\frac{4}{3}\right)^{-2}$ **11.** $6^{-1} + 6^{-1}$

12. $2^{-1} + 4^{-1}$ **13.** $\frac{10}{5^{-3}}$ **14.** $\frac{1}{25 \cdot 10^{-4}}$

15. $\frac{1}{4^{-3}} + \frac{3^2}{2^{-1}}$ **16.** $\frac{2^3}{10^{-2}} - \frac{2}{7^{-2}}$

Simplify. Write answers without negative exponents. See Example 3.

17. $x^{-1}x^2$ **18.** $y^{-3}y^5$

19. $-2x^2 \cdot 8x^{-6}$ **20.** $5y^5(-6y^{-7})$

21. $-3a^{-2}(-2a^{-3})$ **22.** $(-b^{-3})(-b^{-5})$

23. $\frac{u^{-5}}{u^3}$ **24.** $\frac{w^{-4}}{w^6}$

25. $\frac{8t^{-3}}{-2t^{-5}}$ **26.** $\frac{-22w^{-4}}{-11w^{-3}}$

27. $\frac{-6x^5}{-3x^{-6}}$ **28.** $\frac{-51y^6}{17y^{-9}}$

Simplify each expression. Write answers without negative exponents. See Example 4.

29. $(x^2)^{-5}$ **30.** $(y^{-2})^4$ **31.** $(a^{-3})^{-3}$

32. $(b^{-5})^{-2}$ **33.** $(2x^{-3})^{-4}$ **34.** $(3y^{-1})^{-2}$

35. $(4x^2y^{-3})^{-2}$ **36.** $(6s^{-2}t^4)^{-1}$ **37.** $\left(\frac{2x^{-1}}{y^{-3}}\right)^{-2}$

38. $\left(\frac{a^{-2}}{3b^3}\right)^{-3}$ **39.** $\left(\frac{2a^{-3}}{ac^{-2}}\right)^{-4}$ **40.** $\left(\frac{3w^2}{w^4x^3}\right)^{-2}$

Simplify. Write answers without negative exponents.

41. $2^{-1} \cdot 3^{-1}$ **42.** $2^{-1} + 3^{-1}$

43. $(2 \cdot 3^{-1})^{-1}$ **44.** $(2^{-1} + 3)^{-1}$

45. $(x^{-2})^{-3} + 3x^7(-5x^{-1})$ **46.** $(ab^{-1})^2 - ab(-ab^{-3})$

47. $\frac{a^3b^{-2}}{a^{-1}} + \left(\frac{b^6a^{-2}}{b^5}\right)^{-2}$ **48.** $\left(\frac{x^{-3}y^{-1}}{2x}\right)^{-3} + \frac{6x^9y^3}{-3x^{-3}}$

Write each number in standard notation. See Example 5.

49. 9.86×10^9 **50.** 4.007×10^4

51. 1.37×10^{-3} **52.** 9.3×10^{-5}

53. 1×10^{-6} **54.** 3×10^{-1}

55. 6×10^5 **56.** 8×10^6

Write each number in scientific notation. See Example 6.

57. 9000 **58.** 5,298,000

59. 0.00078 **60.** 0.000214

61. 0.0000085 **62.** 5,670,000,000

63. 525×10^9 **64.** 0.0034×10^{-8}

Perform the computations. Write answers in scientific notation. See Example 7.

65. $(3 \times 10^5)(2 \times 10^{-15})$ **66.** $(2 \times 10^{-9})(4 \times 10^{23})$

67. $\frac{4 \times 10^{-8}}{2 \times 10^{30}}$ **68.** $\frac{9 \times 10^{-4}}{3 \times 10^{-6}}$

69. $\frac{3 \times 10^{20}}{6 \times 10^{-8}}$ **70.** $\frac{1 \times 10^{-8}}{4 \times 10^{7}}$

71. $(3 \times 10^{12})^2$ **72.** $(2 \times 10^{-5})^3$

73. $(5 \times 10^4)^3$ **74.** $(5 \times 10^{14})^{-1}$

75. $(4 \times 10^{32})^{-1}$ **76.** $(6 \times 10^{11})^2$

Perform the following computations by first converting each number into scientific notation. Write answers in scientific notation. See Example 8.

77. $(4300)(2{,}000{,}000)$

78. $(40{,}000)(4{,}000{,}000{,}000)$

79. $(4{,}200{,}000)(0.00005)$

80. $(0.00075)(4{,}000{,}000)$

81. $(300)^3(0.000001)^5$

82. $(200)^4(0.0005)^3$

83. $\dfrac{(4000)(90{,}000)}{0.00000012}$

84. $\dfrac{(30{,}000)(80{,}000)}{(0.000006)(0.002)}$

Perform the following computations with the aid of a scientific calculator. Write answers in scientific notation. Round to three decimal places.

85. $(6.3 \times 10^6)(1.45 \times 10^{-4})$

86. $(8.35 \times 10^9)(4.5 \times 10^3)$

87. $(5.36 \times 10^{-4}) + (3.55 \times 10^{-5})$

88. $(8.79 \times 10^8) + (6.48 \times 10^9)$

89. $\dfrac{(3.5 \times 10^5)(4.3 \times 10^{-6})}{3.4 \times 10^{-8}}$

90. $\dfrac{(3.5 \times 10^{-8})(4.4 \times 10^{-4})}{2.43 \times 10^{45}}$

91. $(3.56 \times 10^{85})(4.43 \times 10^{96})$

92. $(8 \times 10^{99}) + (3 \times 10^{99})$

Solve each problem.

93. ***Distance to the sun.*** The distance from the earth to the sun is 93 million miles. Express this distance in feet. (1 mile = 5,280 feet.)

94. ***Speed of light.*** The speed of light is 9.83569×10^8 feet per second. How long does it take light to travel from the sun to the earth? See Exercise 93.

95. ***Warp drive, Scotty.*** How long does it take a spacecraft traveling at 2×10^{35} miles per hour (warp factor 4) to travel 93 million miles.

96. ***Area of a dot.*** If the radius of a very small circle is 2.35×10^{-8} centimeters, then what is the circle's area?

97. ***Circumference of a circle.*** If the circumference of a circle is 5.68×10^9 feet, then what is its radius?

98. ***Diameter of a circle.*** If the diameter of a circle is 1.3×10^{-12} meters, then what is its radius?

99. ***Extracting metals from ore.*** Thomas Sherwood studied the relationship between the concentration of a metal in commercial ore and the price of the metal (*Physics Today,* November 1994). The accompanying graph shows the Sherwood plot with the locations of several metals marked. Even though the scales on this graph are not typical, the graph can be read in the same manner as other graphs. Note also that a concentration of 100 is 100%.

a) Use the figure to estimate the price of copper (Cu) and its concentration in commercial ore.

b) Use the figure to estimate the price of a metal that has a concentration of 10^{-6} percent in commercial ore.

c) Would the four points shown in the graph lie along a straight line if they were plotted in our usual coordinate system?

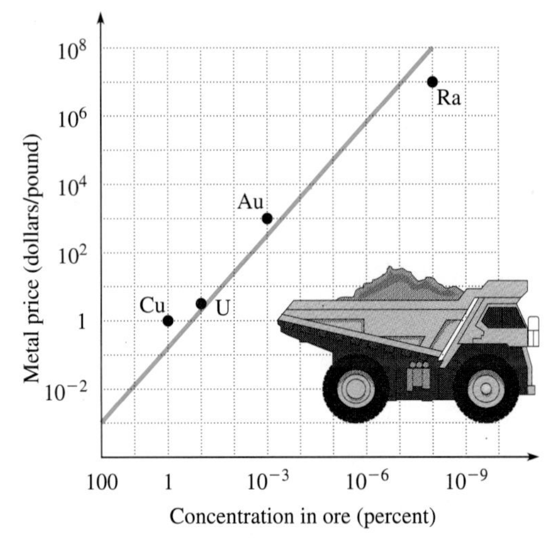

Figure for Exercise 99

100. ***Recycling metals.*** The accompanying graph shows the prices of various metals that are being recycled and the minimum concentration in waste required for recycling (*Physics Today,* November 1994). The straight line is the line from the figure for Exercise 99. Points above the line correspond to metals for which it is economically feasible to increase recycling efforts.

a) Use the figure to estimate the price of mercury (Hg) and the minimum concentration in waste required for recycling mercury.

b) Use the figure to estimate the price of silver (Ag) and the minimum concentration in waste required for recycling silver.

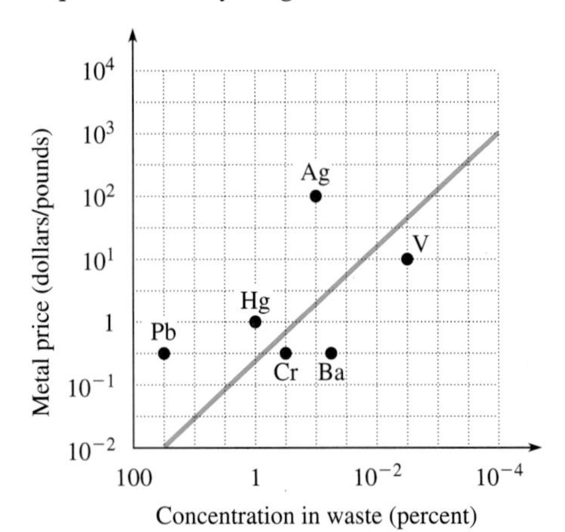

Figure for Exercise 100

101. ***Present value.*** The present value P that will amount to A dollars in n years with interest compounded annually at annual interest rate r, is given by

$$P = A(1 + r)^{-n}.$$

Find the present value that will amount to \$50,000 in 20 years at 8% compounded annually.

102. ***Investing in stocks.*** According to Fidelity Investments, Boston, U.S. common stocks have returned an average of 10% annually for the last 70 years. Use the present value formula from Exercise 101 to find the amount invested today in common stocks that would be worth \$1 million in 70 years, assuming that common stocks continued to return 10% annually for the next 70 years.

Getting More Involved

103. ***Exploration.*** **a)** If $w^{-3} < 0$, then what can you say about w? **b)** If $(-5)^m < 0$, then what can you say about m? **c)** What restriction must be placed on w and m so that $w^m < 0$?

104. ***Discussion.*** Which of the following expressions is not equal to -1? Explain your answer.

a) -1^{-1} **b)** -1^{-2}
c) $(-1^{-1})^{-1}$ **d)** $(-1)^{-1}$
e) $(-1)^{-2}$

8.2 Roots

In this section:

- Fundamentals
- Roots and Variables
- Product Rule for Radicals
- Quotient Rule for Radicals

In Section 8.1 you learned the basic facts about powers. In this section you will study roots and see how powers and roots are related.

Fundamentals

We use the idea of roots to reverse powers. Because $2^4 = 16$, we say that 2 is a fourth root of 16.

*n*th Roots

If $a = b^n$ for a positive integer n, then b is an ***n*th root of *a*.** If $a = b^2$, then b is a **square root** of a. If $a = b^3$, then b is the **cube root** of a.

Because $3^2 = 9$ and $(-3)^2 = 9$, there are two square roots of 9, namely 3 and -3. Because $2^4 = 16$ and $(-2)^4 = 16$, both 2 and -2 are fourth roots of 16. If n is a positive even integer and a is positive, then there are two real nth roots of a. We call these roots **even roots.** The positive even root of a positive number is called the **principal root.** The principal square root of 9 is 3, and the principal fourth root of 16 is 2.

Because $2^3 = 8$ and $(-2)^3 = -8$, the cube root of 8 is 2, and the cube root of -8 is -2. If n is a positive odd integer and a is any real number, there is only one real nth root of a. We call that root an **odd root.**

We use the **radical symbol** $\sqrt{}$ to signify roots.

$\sqrt[n]{a}$

If n is a positive *even* integer and a is positive, then $\sqrt[n]{a}$ denotes the *principal nth root* of a.

If n is a positive *odd* integer, then $\sqrt[n]{a}$ denotes the nth root of a.

If n is any positive integer, then $\sqrt[n]{0} = 0$.

We read $\sqrt[n]{a}$ as "the nth root of a." In the notation $\sqrt[n]{a}$, n is the **index of the radical** and a is the **radicand.** For square roots the index is omitted, and we simply write $\sqrt{a}$.

EXAMPLE 1 **Evaluating radical expressions**

Find the following roots:

a) $\sqrt{25}$ **b)** $\sqrt[3]{-27}$ **c)** $\sqrt[6]{64}$ **d)** $-\sqrt{4}$

Solution

a) Because $5^2 = 25$, $\sqrt{25} = 5$.

b) Because $(-3)^3 = -27$, $\sqrt[3]{-27} = -3$.

c) Because $2^6 = 64$, $\sqrt[6]{64} = 2$.

d) Because $\sqrt{4} = 2$, $-\sqrt{4} = -(\sqrt{4}) = -2$. ■

CAUTION In radical notation, $\sqrt{4}$ represents the *principal square root of 4*, so $\sqrt{4} = 2$. Note that -2 is also a square root of 4, but $\sqrt{4} \neq -2$.

Note that even roots of negative numbers are omitted from the definition of nth roots because even powers of real numbers are never negative. So no real number can be an even root of a negative number. Expressions such as

$$\sqrt{-9}, \quad \sqrt[4]{-81}, \quad \text{and} \quad \sqrt[6]{-64}$$

are not real numbers. Square roots of negative numbers will be discussed in Section 9.5 when we discuss the imaginary numbers.

Calculator Close-up

The expression $\sqrt{3}$ represents the unique positive real number whose square is 3. The number $\sqrt{3}$ is an irrational number. To find an approximation for $\sqrt{3}$ on a scientific calculator, try the following:

3 [√] [=]

The display (rounded to three decimal places) should read

1.732

On a graphing calculator your display should look like the following:

√ 3 [ENTER]

1.732

Note that 1.732 is a rational number that approximates $\sqrt{3}$. Because $\sqrt{3}$ is irrational, the simplest representation for the exact value of the square root of 3 is $\sqrt{3}$.

Roots and Variables

The expression $\sqrt{x}$ is not a real number if x is negative. So we will assume in this book that *the values of any variables in the radicand are nonnegative.*

If x is a nonnegative number, then $\sqrt{x^2}$ represents a nonnegative number whose square is x^2. So

$$\sqrt{x^2} = x.$$

Because $(x^3)^2 = x^6$, we have

$$\sqrt{x^6} = x^3$$

for any nonnegative number x. In fact, any even power of a variable is a perfect square.

> **Perfect Squares**
>
> The following expressions are perfect squares:
>
> $$x^2, \quad x^4, \quad x^6, \quad x^8, \quad x^{10}, \quad x^{12}, \quad \ldots$$

To find the square roots of any of these powers of x, we just use the same variable with one-half of the exponent. For example,

$$\sqrt{x^{48}} = x^{24}.$$

We have a similar situation for cube roots. Any power of a variable in which the exponent is divisible by 3 is a perfect cube.

> **Perfect Cubes**
>
> The following expressions are perfect cubes:
>
> $$x^3, \quad x^6, \quad x^9, \quad x^{12}, \quad x^{15}, \quad \ldots$$

To find the cube root of any of these powers of x, we use the same variable with one-third of the exponent. For example, because $(x^5)^3 = x^{15}$, we have

$$\sqrt[3]{x^{15}} = x^5.$$

If the exponent is divisible by 4, we have a perfect fourth power, and so on.

EXAMPLE 2 **Roots of exponential expressions**

Find each root. Assume that all variables represent nonnegative real numbers.

a) $\sqrt{x^{22}}$ **b)** $\sqrt[3]{t^{18}}$ **c)** $\sqrt[5]{s^{30}}$

Solution

a) $\sqrt{x^{22}} = x^{11}$ because $(x^{11})^2 = x^{22}$.

b) $\sqrt[3]{t^{18}} = t^6$ because $(t^6)^3 = t^{18}$.

c) $\sqrt[5]{s^{30}} = s^6$ because one-fifth of 30 is 6. ■

Product Rule for Radicals

Consider the expression $\sqrt{2} \cdot \sqrt{3}$. If we square this product, we get

$$
\begin{aligned}
(\sqrt{2} \cdot \sqrt{3})^2 &= (\sqrt{2})^2(\sqrt{3})^2 && \text{Power of a product rule} \\
&= 2 \cdot 3 && (\sqrt{2})^2 = 2 \text{ and } (\sqrt{3})^2 = 3 \\
&= 6.
\end{aligned}
$$

The number $\sqrt{6}$ is the unique positive number whose square is 6. Because we squared $\sqrt{2} \cdot \sqrt{3}$ and obtained 6, we must have $\sqrt{6} = \sqrt{2} \cdot \sqrt{3}$. This example illustrates the product rule for radicals.

> **Product Rule for Radicals**
>
> The nth root of a product is equal to the product of the nth roots. In symbols,
>
> $$\sqrt[n]{ab} = \sqrt[n]{a} \cdot \sqrt[n]{b},$$
>
> provided all of these roots are real numbers.

EXAMPLE 3 **Using the product rule for radicals**

Simplify each radical. Assume that all variables represent positive real numbers.

a) $\sqrt{4y}$ **b)** $\sqrt{3y^8}$

Solution

a)
$$
\begin{aligned}
\sqrt{4y} &= \sqrt{4} \cdot \sqrt{y} && \text{Product rule for radicals} \\
&= 2\sqrt{y} && \text{Simplify.}
\end{aligned}
$$

b)
$$
\begin{aligned}
\sqrt{3y^8} &= \sqrt{3} \cdot \sqrt{y^8} && \text{Product rule for radicals} \\
&= \sqrt{3} \cdot y^4 && \sqrt{y^8} = y^4 \\
&= y^4\sqrt{3} && \text{A radical is usually written last in a product.}
\end{aligned}
$$

Quotient Rule for Radicals

Because $\sqrt{2} \cdot \sqrt{3} = \sqrt{6}$, we have $\sqrt{6} \div \sqrt{3} = \sqrt{2}$, or

$$\sqrt{2} = \sqrt{\frac{6}{3}} = \frac{\sqrt{6}}{\sqrt{3}}.$$

This example illustrates the quotient rule for radicals.

> **Quotient Rule for Radicals**
>
> The nth root of a quotient is equal to the quotient of the nth roots. In symbols,
>
> $$\sqrt[n]{\frac{a}{b}} = \frac{\sqrt[n]{a}}{\sqrt[n]{b}},$$
>
> provided that all of these roots are real numbers and $b \neq 0$.

In the next example we use the quotient rule to simplify radical expressions.

EXAMPLE 4 **Using the quotient rule for radicals**

Simplify each radical. Assume that all variables represent positive real numbers.

a) $\sqrt{\frac{t}{9}}$ **b)** $\sqrt[3]{\frac{x^{21}}{y^6}}$

Solution

a) $\sqrt{\frac{t}{9}} = \frac{\sqrt{t}}{\sqrt{9}}$ Quotient rule for radicals

$= \frac{\sqrt{t}}{3}$

b) $\sqrt[3]{\frac{x^{21}}{y^6}} = \frac{\sqrt[3]{x^{21}}}{\sqrt[3]{y^6}}$ Quotient rule for radicals

$= \frac{x^7}{y^2}$

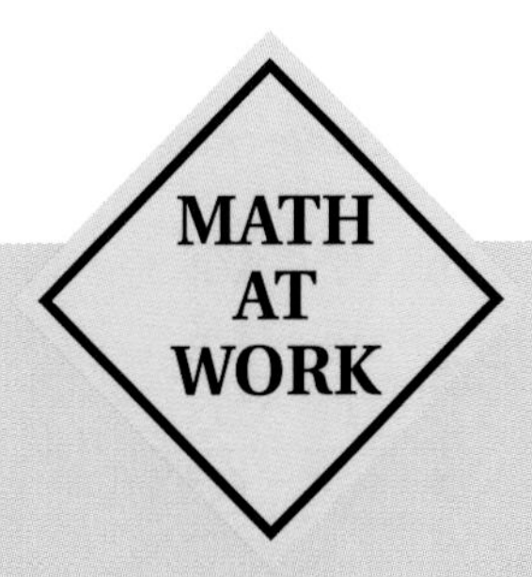

Computer Systems Designer

3–2–1–Blast off! Joseph Bursavich watches each Space Shuttle mission with particular attention. He is a Senior Computer Systems Designer for Martin Marietta, writing programs that support the building of the external tank that is the structural backbone of the Space Shuttle for NASA. Each tank, made up of three major parts, is 157 feet long and takes almost two years to build. To date, 74 tanks have been completed, and each has done its job in carrying cargo into space.

Currently, Mr. Bursavich supports a team whose objective is to develop a superlightweight tank made of a mixture of aluminum and other metals. Reducing the weight of the tank is vital to the space station program because a lighter tank means that each mission can carry a greater payload into space. Because of economic considerations, as many as ten external tanks might be produced at one time. Mr. Bursavich writes programs to determine the economic order quantity (EOQ) for components of the external tank. The EOQ depends on setup costs, labor costs, the quantity of the component to be used in one year, the cost of holding stock for one year, and maintenance costs.

In Exercise 77 of this section you will use the formula that Mr. Bursavich uses to determine the EOQ for component parts of the tanks.

Warm-ups

True or false? Explain your answer.

1. $\sqrt{2} \cdot \sqrt{2} = 2$ **2.** $\sqrt[3]{2} \cdot \sqrt[3]{2} = 2$ **3.** $\sqrt[3]{-27} = -3$
4. $\sqrt{-25} = -5$ **5.** $\sqrt[4]{16} = 2$ **6.** $\sqrt{9} = 3$
7. $\sqrt{2^9} = 2^3$ **8.** $\sqrt{17} \cdot \sqrt{17} = 289$
9. If $w \geq 0$, then $\sqrt{w^2} = w$. **10.** If $t \geq 0$, then $\sqrt[4]{t^{12}} = t^3$.

8.2 EXERCISES

For all of the exercises in this section assume that all variables represent positive real numbers.

Find each root. See Example 1.

1. $\sqrt{36}$ **2.** $\sqrt{49}$ **3.** $\sqrt[5]{32}$
4. $\sqrt[4]{81}$ **5.** $\sqrt[3]{1000}$ **6.** $\sqrt[4]{16}$
7. $\sqrt[4]{-16}$ **8.** $\sqrt{-1}$ **9.** $\sqrt{0}$
10. $\sqrt{1}$ **11.** $\sqrt[3]{-1}$ **12.** $\sqrt[3]{0}$
13. $\sqrt[3]{1}$ **14.** $\sqrt[4]{81}$ **15.** $\sqrt[4]{-81}$
16. $\sqrt[6]{-64}$ **17.** $\sqrt[6]{64}$ **18.** $\sqrt[7]{128}$
19. $\sqrt[5]{-32}$ **20.** $\sqrt[3]{-125}$ **21.** $-\sqrt{100}$
22. $-\sqrt{36}$ **23.** $\sqrt[4]{-50}$ **24.** $-\sqrt{-144}$

Find each root. See Example 2.

25. $\sqrt{m^2}$ **26.** $\sqrt{m^6}$ **27.** $\sqrt[5]{y^{15}}$
28. $\sqrt[4]{m^8}$ **29.** $\sqrt[3]{y^{15}}$ **30.** $\sqrt{m^8}$
31. $\sqrt[3]{m^3}$ **32.** $\sqrt[4]{x^4}$ **33.** $\sqrt{3^6}$
34. $\sqrt{4^2}$ **35.** $\sqrt{2^{10}}$ **36.** $\sqrt[3]{2^{99}}$
37. $\sqrt[3]{5^9}$ **38.** $\sqrt[3]{10^{18}}$ **39.** $\sqrt{10^{20}}$
40. $\sqrt{10^{18}}$

Use the product rule for radicals to simplify each expression. See Example 3.

41. $\sqrt{9y}$ **42.** $\sqrt{16n}$ **43.** $\sqrt{4a^2}$
44. $\sqrt{36n^2}$ **45.** $\sqrt{x^4y^2}$ **46.** $\sqrt{w^6t^2}$
47. $\sqrt{5m^{12}}$ **48.** $\sqrt{7z^{16}}$ **49.** $\sqrt[3]{8y}$
50. $\sqrt[3]{27z^2}$ **51.** $\sqrt[3]{-27w^3}$ **52.** $\sqrt[3]{-125m^6}$
53. $\sqrt[4]{16s}$ **54.** $\sqrt[4]{81w}$
55. $\sqrt[3]{-125a^9y^6}$ **56.** $\sqrt[3]{-27z^3w^{15}}$

Simplify each radical. See Example 4.

57. $\sqrt{\dfrac{t}{4}}$ **58.** $\sqrt{\dfrac{w}{36}}$ **59.** $\sqrt{\dfrac{625}{16}}$
60. $\sqrt{\dfrac{9}{144}}$ **61.** $\sqrt[3]{\dfrac{t}{8}}$ **62.** $\sqrt[3]{\dfrac{a}{27}}$
63. $\sqrt[3]{\dfrac{-8x^6}{y^3}}$ **64.** $\sqrt[3]{\dfrac{-27y^{36}}{1000}}$ **65.** $\sqrt{\dfrac{4a^6}{9}}$
66. $\sqrt{\dfrac{9a^2}{49b^4}}$ **67.** $\sqrt[4]{\dfrac{y}{16}}$ **68.** $\sqrt[4]{\dfrac{5w}{81}}$

Use a calculator to find the approximate value of each expression to three decimal places.

69. $\sqrt{3} + \sqrt{5}$ **70.** $\sqrt{7} - \sqrt{3}$
71. $\dfrac{\sqrt{5} + \sqrt{2}}{\sqrt{3} - 4}$ **72.** $\dfrac{\sqrt{2} - \sqrt{3}}{1 - \sqrt{5}}$
73. $\sqrt{7.1^2 - 4(1.2)(3)}$ **74.** $\sqrt{3^2 - 4(-2)(0.2)}$
75. $\dfrac{-3 + \sqrt{3^2 - 4(1)(-2.9)}}{2}$
76. $\dfrac{8 + \sqrt{(-8)^2 - 4(1.3)(-6.2)}}{2(1.3)}$

Solve each problem.

77. ***Economic order quantity.*** When a part is needed for a space shuttle external fuel tank, Joseph Bursavich at Martin Marietta determines the most economic order quantity E by using the formula $E = \sqrt{\dfrac{2AS}{I}}$, where A is the quantity that the plant will use in one year, S is the cost of setup for making the part, and I is the cost of holding one unit in stock for one year. Find the most economic order quantity if $S = \$5,290$, $A = 20$, and $I = \$100$.

78. ***Diagonal of a box.*** The length of the diagonal D of the box shown in the figure can be found from the formula

$$D = \sqrt{L^2 + W^2 + H^2},$$

where L, W, and H represent the length, width, and height of the box, respectively. If the box has length 6 inches, width 4 inches, and height 3 inches, then what is the length of the diagonal to the nearest tenth of an inch?

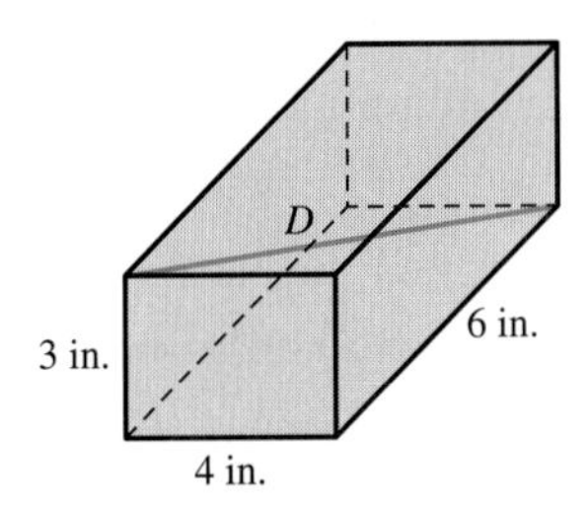

Figure for Exercise 78

79. ***Buena vista.*** The formula $V = 1.22\sqrt{A}$ gives the view in miles from horizon to horizon at an altitude of A feet (Delta Airlines brochure). Find the view to the nearest mile from an altitude of 25,000 feet.

80. ***Sailing speed.*** To find the maximum possible speed in knots (nautical miles per hour) for a sailboat, sailors use the formula $M = 1.3\sqrt{w}$, where w is the length of the boat's waterline in feet. If the waterline for the sloop John B. is 35 feet, then what is the maximum speed for the John B.?

Getting More Involved

81. ***Discussion.*** Determine whether each equation is correct.

a) $\sqrt{(-5)^2} = -5$ **b)** $\sqrt[3]{(-2)^3} = -2$

c) $\sqrt[4]{(-3)^4} = -3$ **d)** $\sqrt[5]{(-7)^5} = -7$

82. ***Writing.*** If x is a negative number and $\sqrt[n]{x^n} = x$, then what can you say about n? Explain your answer.

8.3 Simplifying Square Roots

In this section:

- Using the Product Rule
- Rationalizing the Denominator
- Simplified Form of a Square Root

In Section 8.2 you learned to simplify some radical expressions using the product rule. In this section you will learn three basic rules to follow for writing expressions involving square roots in simplest form. These rules can be extended to radicals with index greater than 2, but we will not do that in this text.

Using the Product Rule

We can use the product rule to simplify square roots of certain numbers. For example,

$$\begin{aligned} \sqrt{45} &= \sqrt{9 \cdot 5} && \text{Factor 45 as } 9 \cdot 5. \\ &= \sqrt{9} \cdot \sqrt{5} && \text{Product rule for radicals} \\ &= 3\sqrt{5} && \sqrt{9} = 3 \end{aligned}$$

Because 45 is not a perfect square, we cannot write $\sqrt{45}$ without the radical symbol. However, $3\sqrt{5}$ is considered a simpler expression that represents the exact value of $\sqrt{45}$. When simplifying square roots, we can factor the perfect squares out of the radical and replace them with their square roots. Look for the factors

$$4, \quad 9, \quad 16, \quad 25, \quad 36, \quad 49, \quad \text{and so on.}$$

EXAMPLE 1 **Simplifying radicals using the product rule**

Simplify.

a) $\sqrt{12}$ **b)** $\sqrt{50}$ **c)** $\sqrt{72}$

Solution

a) Because $12 = 4 \cdot 3$, we can use the product rule to write

$$\sqrt{12} = \sqrt{4} \cdot \sqrt{3} = 2\sqrt{3}.$$

b) $\sqrt{50} = \sqrt{25} \cdot \sqrt{2} = 5\sqrt{2}$

c) Note that 4, 9, and 36 are perfect squares and are factors of 72. In factoring out a perfect square, it is most efficient to use the largest perfect square:

$$\sqrt{72} = \sqrt{36} \cdot \sqrt{2} = 6\sqrt{2}$$

If we had factored out 9, we could still get the correct answer as follows:

$$\sqrt{72} = \sqrt{9} \cdot \sqrt{8} = 3 \cdot \sqrt{8} = 3 \cdot \sqrt{4} \cdot \sqrt{2} = 3 \cdot 2\sqrt{2} = 6\sqrt{2}$$

Rationalizing the Denominator

Radicals such as $\sqrt{2}$, $\sqrt{3}$, and $\sqrt{5}$ are irrational numbers. So a fraction such as $\frac{3}{\sqrt{5}}$ has an irrational denominator. Because fractions with rational denominators are considered simpler than fractions with irrational denominators, we usually convert fractions with irrational denominators to equivalent ones with rational denominators. That is, we **rationalize the denominator.**

EXAMPLE 2 **Rationalizing denominators**

Simplify each expression by rationalizing its denominator.

a) $\frac{3}{\sqrt{5}}$ **b)** $\frac{\sqrt{3}}{\sqrt{7}}$

Solution

a) Because $\sqrt{5} \cdot \sqrt{5} = 5$, we multiply numerator and denominator by $\sqrt{5}$:

$$\frac{3}{\sqrt{5}} = \frac{3 \cdot \sqrt{5}}{\sqrt{5} \cdot \sqrt{5}} \quad \text{Multiply numerator and denominator by } \sqrt{5}.$$

$$= \frac{3\sqrt{5}}{5} \quad \sqrt{5} \cdot \sqrt{5} = 5$$

b) Because $\sqrt{7} \cdot \sqrt{7} = 7$, multiply the numerator and denominator by $\sqrt{7}$:

$$\frac{\sqrt{3}}{\sqrt{7}} = \frac{\sqrt{3} \cdot \sqrt{7}}{\sqrt{7} \cdot \sqrt{7}} \quad \text{Multiply numerator and denominator by } \sqrt{7}.$$

$$= \frac{\sqrt{21}}{7} \quad \text{Product rule for radicals}$$

Simplified Form of a Square Root

When we simplify any expression, we try to write a "simpler" expression that is equivalent to the original. However, one person's idea of simpler is sometimes different from another person's. For a square root the expression must satisfy three conditions to be in simplified form. These three conditions provide specific rules to follow for simplifying square roots.

Simplified Form for Square Roots

An expression involving a square root is in **simplified form** if it has

1. *no* perfect-square factors inside the radical,
2. *no* fractions inside the radical, and
3. *no* radicals in the denominator.

EXAMPLE 3 **Simplified form for square roots**

Write each radical expression in simplified form.

a) $\sqrt{300}$ **b)** $\sqrt{\frac{2}{5}}$ **c)** $\frac{\sqrt{10}}{\sqrt{6}}$

Solution

a) We must remove the perfect square factor of 100 from inside the radical:

$$\sqrt{300} = \sqrt{100 \cdot 3} = \sqrt{100} \cdot \sqrt{3} = 10\sqrt{3}$$

b) We first use the quotient rule to remove the fraction $\frac{2}{5}$ from inside the radical:

$$\sqrt{\frac{2}{5}} = \frac{\sqrt{2}}{\sqrt{5}} \quad \text{Quotient rule for radicals}$$

$$= \frac{\sqrt{2} \cdot \sqrt{5}}{\sqrt{5} \cdot \sqrt{5}} \quad \text{Rationalize the denominator.}$$

$$= \frac{\sqrt{10}}{5} \quad \text{Product rule for radicals}$$

c) The numerator and denominator have a common factor of $\sqrt{2}$:

$$\frac{\sqrt{10}}{\sqrt{6}} = \frac{\sqrt{2} \cdot \sqrt{5}}{\sqrt{2} \cdot \sqrt{3}} \quad \text{Product rule for radicals}$$

$$= \frac{\sqrt{5}}{\sqrt{3}} \quad \text{Reduce.}$$

$$= \frac{\sqrt{5} \cdot \sqrt{3}}{\sqrt{3} \cdot \sqrt{3}} \quad \text{Rationalize the denominator.}$$

$$= \frac{\sqrt{15}}{3} \quad \text{Product rule for radicals}$$

Note that we could have simplified $\frac{\sqrt{10}}{\sqrt{6}}$ by first using the quotient rule to get $\frac{\sqrt{10}}{\sqrt{6}} = \sqrt{\frac{10}{6}}$ and then reducing $\frac{10}{6}$. Another way to simplify $\frac{\sqrt{10}}{\sqrt{6}}$ is to first multiply the numerator and denominator by $\sqrt{6}$. You should try these alternatives. Of course, the simplified form is $\frac{\sqrt{15}}{3}$ by any method. ■

In the next example we simplify some expressions involving variables. Remember that *any exponential expression with an even exponent is a perfect square.*

EXAMPLE 4 Radicals containing variables

Simplify each expression. All variables represent nonnegative real numbers.

a) $\sqrt{x^3}$

b) $\sqrt{8a^9}$

c) $\sqrt{18a^4b^7}$

Solution

a) $\sqrt{x^3} = \sqrt{x^2 \cdot x}$ The largest perfect square factor of x^3 is x^2.

$= \sqrt{x^2} \cdot \sqrt{x}$ Product rule for radicals

$= x\sqrt{x}$ For any nonnegative x, $\sqrt{x^2} = x$.

b) $\sqrt{8a^9} = \sqrt{4a^8} \cdot \sqrt{2a}$ The largest perfect square factor of $8a^9$ is $4a^8$.

$= 2a^4\sqrt{2a}$ $\sqrt{4a^8} = 2a^4$

c) $\sqrt{18a^4b^7} = \sqrt{9a^4b^6} \cdot \sqrt{2b}$ Factor out the perfect squares.

$= 3a^2b^3\sqrt{2b}$ $\sqrt{9a^4b^6} = 3a^2b^3$ ■

If square roots of variables appear in the denominator, then we rationalize the denominator.

EXAMPLE 5 Radicals containing variables

Simplify each expression. All variables represent positive real numbers.

a) $\dfrac{5}{\sqrt{a}}$

b) $\sqrt{\dfrac{a}{b}}$

c) $\dfrac{\sqrt{2}}{\sqrt{6a}}$

Solution

a) $\dfrac{5}{\sqrt{a}} = \dfrac{5 \cdot \sqrt{a}}{\sqrt{a} \cdot \sqrt{a}}$ Multiply numerator and denominator by $\sqrt{a}$.

$= \dfrac{5\sqrt{a}}{a}$ $\sqrt{a} \cdot \sqrt{a} = a$

b) $\sqrt{\dfrac{a}{b}} = \dfrac{\sqrt{a}}{\sqrt{b}}$ Quotient rule for radicals

$= \dfrac{\sqrt{a} \cdot \sqrt{b}}{\sqrt{b} \cdot \sqrt{b}}$ Rationalize the denominator.

$= \dfrac{\sqrt{ab}}{b}$ Product rule for radicals

c) $\dfrac{\sqrt{2}}{\sqrt{6a}} = \dfrac{\sqrt{2} \cdot \sqrt{6a}}{\sqrt{6a} \cdot \sqrt{6a}}$ Rationalize the denominator.

$= \dfrac{\sqrt{12a}}{6a}$ Product rule for radicals

$= \dfrac{\sqrt{4} \cdot \sqrt{3a}}{6a}$ Factor out the perfect square.

$= \dfrac{2\sqrt{3a}}{6a}$ $\sqrt{4} = 2$

$= \dfrac{2\sqrt{3a}}{2 \cdot 3a}$ Factor the denominator.

$= \dfrac{\sqrt{3a}}{3a}$ Divide out the common factor 2. ■

CAUTION Do not attempt to reduce an expression like the one in Example 5(c):

$$\frac{\sqrt{3a}}{3a}$$

You cannot divide out common factors when one is inside a radical. ⊘

Warm-ups

True or false? Explain your answer.

1. $\sqrt{20} = 2\sqrt{5}$
2. $\sqrt{18} = 9\sqrt{2}$
3. $\dfrac{1}{\sqrt{3}} = \dfrac{\sqrt{3}}{3}$
4. $\dfrac{9}{4} = \dfrac{3}{2}$
5. $\sqrt{a^3} = a\sqrt{a}$ for any positive value of a.
6. $\sqrt{a^9} = a^3$ for any positive value of a.
7. $\sqrt{y^{17}} = y^8\sqrt{y}$ for any positive value of y.
8. $\dfrac{\sqrt{6}}{2} = \sqrt{3}$
9. $\sqrt{4} = \sqrt{2}$
10. $\sqrt{283} = 17$

8.3 EXERCISES

Assume that all variables in the exercises represent positive real numbers.

Simplify each radical. See Example 1.

1. $\sqrt{8}$ **2.** $\sqrt{20}$ **3.** $\sqrt{24}$

4. $\sqrt{75}$ **5.** $\sqrt{28}$ **6.** $\sqrt{40}$

7. $\sqrt{90}$ **8.** $\sqrt{200}$ **9.** $\sqrt{500}$

10. $\sqrt{98}$ **11.** $\sqrt{150}$ **12.** $\sqrt{120}$

In Exercises 13–24, simplify each expression by rationalizing the denominator. See Example 2.

13. $\dfrac{1}{\sqrt{5}}$ **14.** $\dfrac{1}{\sqrt{6}}$ **15.** $\dfrac{3}{\sqrt{2}}$

16. $\dfrac{4}{\sqrt{3}}$ **17.** $\dfrac{\sqrt{3}}{\sqrt{2}}$ **18.** $\dfrac{\sqrt{7}}{\sqrt{6}}$

19. $\dfrac{-3}{\sqrt{10}}$ **20.** $\dfrac{-4}{\sqrt{5}}$ **21.** $\dfrac{-10}{\sqrt{17}}$

22. $\frac{-3}{\sqrt{19}}$ **23.** $\frac{\sqrt{11}}{\sqrt{7}}$ **24.** $\frac{\sqrt{10}}{\sqrt{3}}$

Write each radical expression in simplified form. See Example 3.

25. $\sqrt{63}$ **26.** $\sqrt{48}$ **27.** $\sqrt{\frac{3}{2}}$

28. $\sqrt{\frac{3}{5}}$ **29.** $\sqrt{\frac{5}{8}}$ **30.** $\sqrt{\frac{5}{18}}$

31. $\frac{\sqrt{6}}{\sqrt{10}}$ **32.** $\frac{\sqrt{12}}{\sqrt{20}}$ **33.** $\frac{\sqrt{75}}{\sqrt{3}}$

34. $\frac{\sqrt{45}}{\sqrt{5}}$ **35.** $\frac{\sqrt{15}}{\sqrt{10}}$ **36.** $\frac{\sqrt{30}}{\sqrt{21}}$

Simplify each expression. See Example 4.

37. $\sqrt{a^8}$ **38.** $\sqrt{y^{10}}$ **39.** $\sqrt{a^9}$

40. $\sqrt{t^{11}}$ **41.** $\sqrt{8a^6}$ **42.** $\sqrt{18w^9}$

43. $\sqrt{20a^4b^9}$ **44.** $\sqrt{12x^2y^3}$ **45.** $\sqrt{27x^3y^3}$

46. $\sqrt{45x^5y^3}$ **47.** $\sqrt{27a^3b^8c^2}$ **48.** $\sqrt{125x^3y^9z^4}$

Simplify each expression. See Example 5.

49. $\frac{1}{\sqrt{x}}$ **50.** $\frac{1}{\sqrt{2x}}$ **51.** $\frac{\sqrt{2}}{\sqrt{3a}}$

52. $\frac{\sqrt{5}}{\sqrt{2b}}$ **53.** $\frac{\sqrt{3}}{\sqrt{15y}}$ **54.** $\frac{\sqrt{5}}{\sqrt{10x}}$

55. $\sqrt{\frac{3x}{2y}}$ **56.** $\sqrt{\frac{6}{5w}}$ **57.** $\sqrt{\frac{10y}{15x}}$

58. $\sqrt{\frac{6x}{4y}}$ **59.** $\sqrt{\frac{8x^3}{y}}$ **60.** $\sqrt{\frac{8s^5}{t}}$

Simplify each expression.

61. $\sqrt{80x^3}$ **62.** $\sqrt{90y^{80}}$ **63.** $\sqrt{9y^9x^{15}}$

64. $\sqrt{48x^2y^7}$ **65.** $\frac{20x^6}{\sqrt{5x^5}}$ **66.** $\frac{7x^7y}{\sqrt{7x^9}}$

67. $\frac{-22p^2}{p\sqrt{6pq}}$ **68.** $\frac{-30t^5}{t^2\sqrt{3t}}$

69. $\frac{a^3b^7\sqrt{a^2b^3c^4}}{\sqrt{abc}}$ **70.** $\frac{3n^4b^5\sqrt{n^2b^2c^7}}{\sqrt{nbc}}$

71. $\frac{\sqrt{4xy^2}}{x^9y^3\sqrt{6xy^3}}$ **72.** $\frac{\sqrt{8m^3n^2}}{m^3n^2\sqrt{6mn^3}}$

Use a calculator to evaluate each expression.

73. $\frac{1}{\sqrt{2}} - \frac{\sqrt{2}}{2}$ **74.** $\frac{\sqrt{2}}{\sqrt{3}} - \frac{\sqrt{6}}{3}$

75. $\frac{\sqrt{6}}{\sqrt{2}} - \sqrt{3}$ **76.** $2 - \frac{\sqrt{20}}{\sqrt{5}}$

Solve each problem.

77. ***Economic order quantity.*** The formula for economic order quantity

$$E = \sqrt{\frac{2AS}{I}}$$

was used in Exercise 77 of Section 8.2. Express the right-hand side in simplified form.

78. ***Landing speed.*** Aircraft design engineers determine the proper landing speed V (in ft/sec) by using the formula

$$V = \sqrt{\frac{841L}{CS}},$$

where L is the gross weight of the aircraft in pounds, C is the coefficient of lift, and S is the wing surface area in square feet. Express the right-hand side in simplified form.

8.4 Operations with Radicals

In this section:

- Adding and Subtracting Radicals
- Multiplying Radicals
- Dividing Radicals

In this section you will learn how to perform the basic operations of arithmetic with radical expressions.

Adding and Subtracting Radicals

Consider the sum $2\sqrt{3} + 5\sqrt{3}$. When you studied like terms, you learned that

$$2x + 5x = 7x$$

is true for any value of x. If $x = \sqrt{3}$, then we get

$$2\sqrt{3} + 5\sqrt{3} = 7\sqrt{3}.$$

The expressions $2\sqrt{3}$ and $5\sqrt{3}$ are called **like radicals** because they can be combined just as we combine like terms. We can also add or subtract other radicals as long as they have *the same index and the same radicand.* For example,

$$8\sqrt[3]{w} - 6\sqrt[3]{w} = 2\sqrt[3]{w}.$$

EXAMPLE 1 **Combining like radicals**

Simplify the following expressions by combining like radicals. Assume that the variables represent nonnegative numbers.

a) $2\sqrt{5} + 7\sqrt{5}$ **b)** $3\sqrt[3]{2} - 9\sqrt[3]{2}$ **c)** $\sqrt{2} - 5\sqrt{a} + 4\sqrt{2} - 3\sqrt{a}$

Solution

a) $2\sqrt{5} + 7\sqrt{5} = 9\sqrt{5}$

b) $3\sqrt[3]{2} - 9\sqrt[3]{2} = -6\sqrt[3]{2}$

c) $\sqrt{2} - 5\sqrt{a} + 4\sqrt{2} - 3\sqrt{a} = \sqrt{2} + 4\sqrt{2} - 5\sqrt{a} - 3\sqrt{a}$

$= 5\sqrt{2} - 8\sqrt{a}$ Combine like radicals only. ■

CAUTION We cannot combine the terms in the expressions

$$\sqrt{2} + \sqrt{5}, \quad \sqrt[3]{x} - \sqrt{x}, \quad \text{or} \quad 3\sqrt{2} + \sqrt[4]{6}$$

because the radicands are unequal in the first expression, the indices are unequal in the second, and both the radicands and the indices are unequal in the third. ⊘

In the next example we simplify the radicals before adding or subtracting.

EXAMPLE 2 **Simplifying radicals before combining like terms**

Simplify. Assume that all variables represent nonnegative real numbers.

a) $\sqrt{12} + \sqrt{75}$ **b)** $\sqrt{8x^3} + x\sqrt{18x}$ **c)** $\frac{4}{\sqrt{2}} - \frac{\sqrt{3}}{\sqrt{6}}$

Solution

a) $\sqrt{12} + \sqrt{75} = \sqrt{4} \cdot \sqrt{3} + \sqrt{25} \cdot \sqrt{3}$ Product rule for radicals

$= 2\sqrt{3} + 5\sqrt{3}$ Simplify.

$= 7\sqrt{3}$ Combine like radicals.

b) $\sqrt{8x^3} + x\sqrt{18x} = \sqrt{4x^2} \cdot \sqrt{2x} + x\sqrt{9} \cdot \sqrt{2x}$ Product rule for radicals

$= 2x\sqrt{2x} + 3x\sqrt{2x}$ Simplify.

$= 5x\sqrt{2x}$ Combine like radicals.

c) $\frac{4}{\sqrt{2}} - \frac{\sqrt{3}}{\sqrt{6}} = \frac{4 \cdot \sqrt{2}}{\sqrt{2} \cdot \sqrt{2}} - \frac{\sqrt{3} \cdot \sqrt{6}}{\sqrt{6} \cdot \sqrt{6}}$ Rationalize the denominators.

$= \frac{4\sqrt{2}}{2} - \frac{\sqrt{18}}{6}$ Simplify.

$= \frac{4\sqrt{2}}{2} - \frac{3\sqrt{2}}{6}$ $\sqrt{18} = \sqrt{9} \cdot \sqrt{2} = 3\sqrt{2}$

$= \frac{4\sqrt{2}}{2} - \frac{\sqrt{2}}{2} = \frac{3\sqrt{2}}{2}$ $4\sqrt{2} - \sqrt{2} = 3\sqrt{2}$ ■

Multiplying Radicals

We have been using the product rule for radicals

$$\sqrt[n]{ab} = \sqrt[n]{a} \cdot \sqrt[n]{b}$$

to express a root of a product as a product of the roots of the factors. When we rationalized denominators in Section 8.3, we used the product rule to multiply radicals. We will now study multiplication of radicals in more detail.

EXAMPLE 3 **Multiplying radical expressions**

Multiply and simplify. Assume that variables represent positive numbers.

a) $\sqrt{2} \cdot \sqrt{5}$ **b)** $2\sqrt{5} \cdot 3\sqrt{6}$ **c)** $\sqrt{2a^2} \cdot \sqrt{6a}$ **d)** $\sqrt[3]{4} \cdot \sqrt[3]{2}$

Solution

a) $\sqrt{2} \cdot \sqrt{5} = \sqrt{2 \cdot 5}$ Product rule for radicals

$= \sqrt{10}$

b) $2\sqrt{5} \cdot 3\sqrt{6} = 2 \cdot 3 \cdot \sqrt{5} \cdot \sqrt{6}$

$= 6\sqrt{30}$ Product rule for radicals

c) $\sqrt{2a^2} \cdot \sqrt{6a} = \sqrt{12a^3}$ Product rule for radicals

$= \sqrt{4a^2} \cdot \sqrt{3a}$ Factor out the perfect square.

$= 2a\sqrt{3a}$ Simplify.

d) $\sqrt[3]{4} \cdot \sqrt[3]{2} = \sqrt[3]{8}$ Product rule for radicals

$= 2$ ■

A sum such as $\sqrt{6} + \sqrt{2}$ is in its simplest form, and so it is treated like a binomial when it occurs in a product.

EXAMPLE 4 **Using the distributive property with radicals**

Find the product: $3\sqrt{3}(\sqrt{6} + \sqrt{2})$

Solution

$3\sqrt{3}(\sqrt{6} + \sqrt{2}) = 3\sqrt{3} \cdot \sqrt{6} + 3\sqrt{3} \cdot \sqrt{2}$ Distributive property

$= 3\sqrt{18} + 3\sqrt{6}$ Product rule for radicals

$= 3 \cdot 3\sqrt{2} + 3\sqrt{6}$ $\sqrt{18} = \sqrt{9} \cdot \sqrt{2} = 3\sqrt{2}$

$= 9\sqrt{2} + 3\sqrt{6}$ ■

In the next example we use the FOIL method to find products of expressions involving radicals.

EXAMPLE 5 **Using FOIL to multiply radicals**

Multiply and simplify.

a) $(\sqrt{3} + 5)(\sqrt{3} - 2)$ **b)** $(\sqrt{5} - 2)(\sqrt{5} + 2)$

c) $(2\sqrt{3} + \sqrt{5})(\sqrt{3} - \sqrt{5})$

Solution

a) $(\sqrt{3} + 5)(\sqrt{3} - 2) = \overset{F}{3} - \overset{O}{2\sqrt{3}} + \overset{I}{5\sqrt{3}} - \overset{L}{10}$

$= 3\sqrt{3} - 7$ Add the like terms.

b) The product $(\sqrt{5} - 2)(\sqrt{5} + 2)$ is the product of a sum and a difference. Recall that $(a - b)(a + b) = a^2 - b^2$.

$$\begin{aligned}(\sqrt{5} - 2)(\sqrt{5} + 2) &= (\sqrt{5})^2 - 2^2 \\ &= 5 - 4 \\ &= 1\end{aligned}$$

c)
$$\begin{aligned}(2\sqrt{3} + \sqrt{5})(\sqrt{3} - \sqrt{5}) &= \overset{\text{F}}{2\sqrt{3}\sqrt{3}} - \overset{\text{O}}{2\sqrt{3}\sqrt{5}} + \overset{\text{I}}{\sqrt{5}\sqrt{3}} - \overset{\text{L}}{\sqrt{5}\sqrt{5}} \\ &= 6 - 2\sqrt{15} + \sqrt{15} - 5 \\ &= 1 - \sqrt{15}\end{aligned}$$

■

Dividing Radicals

In Section 8.2 we used the quotient rule for radicals to write a square root of a quotient as a quotient of square roots. We can also use the quotient rule for radicals to divide radicals of the same index. For example,

$$\frac{\sqrt{10}}{\sqrt{2}} = \sqrt{\frac{10}{2}} = \sqrt{5}.$$

Division of radicals is simplest when the quotient of the radicands is a whole number, as it was in the example $\sqrt{10} \div \sqrt{2} = \sqrt{5}$. If the quotient of the radicands is not a whole number, then we divide by rationalizing the denominator, as shown in the next example.

EXAMPLE 6 **Dividing radicals**

Divide and simplify.

a) $\sqrt{30} \div \sqrt{3}$ **b)** $(5\sqrt{2}) \div (2\sqrt{5})$ **c)** $(15\sqrt{6}) \div (3\sqrt{2})$

Solution

a) $\sqrt{30} \div \sqrt{3} = \dfrac{\sqrt{30}}{\sqrt{3}} = \sqrt{10}$

b)
$$\begin{aligned}(5\sqrt{2}) \div (2\sqrt{5}) &= \frac{5\sqrt{2}}{2\sqrt{5}} = \frac{5\sqrt{2} \cdot \sqrt{5}}{2\sqrt{5} \cdot \sqrt{5}} && \text{Rationalize the denominator.} \\ &= \frac{5\sqrt{10}}{2 \cdot 5} && \text{Product rule for radicals} \\ &= \frac{\sqrt{10}}{2} && \text{Reduce.}\end{aligned}$$

Note that $\sqrt{10} \div 2 \neq \sqrt{5}$.

c) $(15\sqrt{6}) \div (3\sqrt{2}) = \dfrac{15\sqrt{6}}{3\sqrt{2}} = 5\sqrt{3}$ $\quad \sqrt{6} \div \sqrt{2} = \sqrt{3}$ ■

CAUTION You can use the quotient rule to divide roots of the same index only. For example,

$$\frac{\sqrt{14}}{\sqrt{2}} = \sqrt{7} \quad \text{but} \quad \frac{\sqrt{14}}{2} \neq \sqrt{7}.$$

In the next example we simplify expressions with radicals in the numerator and whole numbers in the denominator.

EXAMPLE 7 **Simplifying radical expressions**

Simplify.

a) $\dfrac{4 - \sqrt{20}}{4}$ b) $\dfrac{-6 + \sqrt{27}}{3}$

Solution

a) $\dfrac{4 - \sqrt{20}}{4} = \dfrac{4 - 2\sqrt{5}}{4}$ $\sqrt{20} = \sqrt{4} \cdot \sqrt{5} = 2\sqrt{5}$

$= \dfrac{\cancel{2}(2 - \sqrt{5})}{\cancel{2} \cdot 2}$ Factor out the GCF, 2.

$= \dfrac{2 - \sqrt{5}}{2}$ Reduce.

b) $\dfrac{-6 + \sqrt{27}}{3} = \dfrac{-6 + 3\sqrt{3}}{3} = \dfrac{\cancel{3}(-2 + \sqrt{3})}{\cancel{3}} = -2 + \sqrt{3}$ ■

CAUTION In the expression $\frac{2 - \sqrt{5}}{2}$ you cannot divide out the remaining 2's because 2 is not a *factor* of the numerator.

In Example 5(b) we used the rule for the product of a sum and a difference to get $(\sqrt{5} - 2)(\sqrt{5} + 2) = 1$. If we apply the same rule to other products of this type, we also get a rational number as the result. For example,

$$(\sqrt{7} + \sqrt{2})(\sqrt{7} - \sqrt{2}) = 7 - 2 = 5.$$

Expressions such as $\sqrt{5} + 2$ and $\sqrt{5} - 2$ are called **conjugates** of each other. The conjugate of $\sqrt{7} + \sqrt{2}$ is $\sqrt{7} - \sqrt{2}$. We can use conjugates to simplify a radical expression that has a sum or a difference in its denominator.

EXAMPLE 8 **Rationalizing the denominator using conjugates**

Simplify each expression.

a) $\dfrac{\sqrt{3}}{\sqrt{7} - \sqrt{2}}$ b) $\dfrac{4}{6 + \sqrt{2}}$

Solution

a) $\dfrac{\sqrt{3}}{\sqrt{7} - \sqrt{2}} = \dfrac{\sqrt{3}(\sqrt{7} + \sqrt{2})}{(\sqrt{7} - \sqrt{2})(\sqrt{7} + \sqrt{2})}$ Multiply by $\sqrt{7} + \sqrt{2}$, the conjugate of $\sqrt{7} - \sqrt{2}$.

$= \dfrac{\sqrt{21} + \sqrt{6}}{7 - 2}$

$= \dfrac{\sqrt{21} + \sqrt{6}}{5}$

b) $\dfrac{4}{6 + \sqrt{2}} = \dfrac{4(6 - \sqrt{2})}{(6 + \sqrt{2})(6 - \sqrt{2})}$ Multiply by $6 - \sqrt{2}$, the conjugate of $6 + \sqrt{2}$.

$= \dfrac{24 - 4\sqrt{2}}{36 - 2}$

$= \dfrac{24 - 4\sqrt{2}}{34}$

$= \dfrac{\cancel{2}(12 - 2\sqrt{2})}{\cancel{2} \cdot 17} = \dfrac{12 - 2\sqrt{2}}{17}$ ■

Warm-ups

True or false? Explain your answer.

1. $\sqrt{9} + \sqrt{16} = \sqrt{25}$

2. $\frac{5}{\sqrt{5}} = \sqrt{5}$

3. $\sqrt{10} \div 2 = \sqrt{5}$

4. $3\sqrt{2} \cdot 3\sqrt{2} = 9\sqrt{2}$

5. $3\sqrt{5} \cdot 3\sqrt{2} = 9\sqrt{10}$

6. $\sqrt{5} + 3\sqrt{5} = 4\sqrt{10}$

7. $\frac{\sqrt{15}}{3} = \sqrt{5}$

8. $\sqrt{2} \div \sqrt{6} = \sqrt{3}$

9. $\frac{\sqrt{27}}{\sqrt{3}} = 3$

10. $(\sqrt{3} - 1)(\sqrt{3} + 1) = 2$

8.4 EXERCISES

Assume that all variables in these exercises represent only positive real numbers.

Simplify each expression by combining like radicals. See Example 1.

1. $4\sqrt{5} + 3\sqrt{5}$

2. $\sqrt{2} + \sqrt{2}$

3. $\sqrt[3]{2} + \sqrt[3]{2}$

4. $4\sqrt[3]{6} - 7\sqrt[3]{6}$

5. $3u\sqrt{11} + 5u\sqrt{11}$

6. $9m\sqrt{5} - 12m\sqrt{5}$

7. $\sqrt{2} + \sqrt{3} - 5\sqrt{2} + 3\sqrt{3}$

8. $8\sqrt{6} - \sqrt{2} - 3\sqrt{6} + 5\sqrt{2}$

9. $3\sqrt{y} - \sqrt{x} - 4\sqrt{y} - 3\sqrt{x}$

10. $5\sqrt{7} - \sqrt{a} + 3\sqrt{7} - 5\sqrt{a}$

11. $3x\sqrt{y} - \sqrt{a} + 2x\sqrt{y} + 3\sqrt{a}$

12. $a\sqrt{b} + 5a\sqrt{b} - 2\sqrt{a} + 3\sqrt{a}$

Simplify each expression. See Example 2.

13. $\sqrt{24} + \sqrt{54}$

14. $\sqrt{12} + \sqrt{27}$

15. $2\sqrt{27} - 4\sqrt{75}$

16. $\sqrt{2} - \sqrt{18}$

17. $\sqrt{3a} - \sqrt{12a}$

18. $\sqrt{5w} - \sqrt{45w}$

19. $\sqrt{x^3} + x\sqrt{4x}$

20. $\sqrt{27x^3} + 5x\sqrt{12x}$

21. $\frac{1}{\sqrt{3}} + \frac{\sqrt{2}}{\sqrt{6}}$

22. $\frac{3}{\sqrt{5}} + \frac{\sqrt{2}}{\sqrt{10}}$

23. $\frac{1}{\sqrt{3}} + \sqrt{12}$

24. $\frac{1}{\sqrt{2}} + 3\sqrt{8}$

Multiply and simplify. See Example 3.

25. $\sqrt{7} \cdot \sqrt{11}$

26. $\sqrt{3} \cdot \sqrt{13}$

27. $2\sqrt{6} \cdot 3\sqrt{6}$

28. $4\sqrt{2} \cdot 3\sqrt{2}$

29. $-3\sqrt{5} \cdot 4\sqrt{2}$

30. $-8\sqrt{3} \cdot 3\sqrt{2}$

31. $\sqrt{2a^3} \cdot \sqrt{6a^5}$

32. $\sqrt{3w^7} \cdot \sqrt{w^9}$

33. $\sqrt[3]{9} \cdot \sqrt[3]{3}$

34. $\sqrt[3]{-25} \cdot \sqrt[3]{5}$

35. $\sqrt[3]{-4m^2} \cdot \sqrt[3]{2m}$

36. $\sqrt[3]{100m^4} \cdot \sqrt[3]{10m^2}$

Multiply and simplify. See Example 4.

37. $\sqrt{2}(\sqrt{2} + \sqrt{3})$

38. $\sqrt{3}(\sqrt{3} - \sqrt{2})$

39. $3\sqrt{2}(2\sqrt{6} + \sqrt{10})$

40. $2\sqrt{3}(\sqrt{6} + 2\sqrt{15})$

41. $2\sqrt{5}(\sqrt{5} - 3\sqrt{10})$

42. $\sqrt{6}(\sqrt{24} - 6)$

Multiply and simplify. See Example 5.

43. $(\sqrt{5} - 4)(\sqrt{5} + 3)$

44. $(\sqrt{6} - 2)(\sqrt{6} - 3)$

45. $(\sqrt{3} - 1)(\sqrt{3} + 1)$

46. $(\sqrt{6} + 2)(\sqrt{6} - 2)$

47. $(\sqrt{5} - \sqrt{2})(\sqrt{5} + \sqrt{2})$

48. $(\sqrt{3} - \sqrt{6})(\sqrt{3} + \sqrt{6})$

49. $(2\sqrt{5} + 1)(3\sqrt{5} - 2)$

50. $(2\sqrt{2} + 3)(4\sqrt{2} + 4)$

51. $(2\sqrt{3} - 3\sqrt{5})(3\sqrt{3} + 4\sqrt{5})$

52. $(4\sqrt{3} + 3\sqrt{7})(2\sqrt{3} + 4\sqrt{7})$

53. $(2\sqrt{3} + 5)^2$

54. $(3\sqrt{2} + \sqrt{6})^2$

Divide and simplify. See Example 6.

55. $\sqrt{10} \div \sqrt{5}$

56. $\sqrt{14} \div \sqrt{2}$

57. $\sqrt{5} \div \sqrt{3}$

58. $\sqrt{3} \div \sqrt{2}$

59. $(4\sqrt{5}) \div (3\sqrt{6})$

60. $(3\sqrt{7}) \div (4\sqrt{3})$

61. $(5\sqrt{14}) \div (3\sqrt{2})$

62. $(4\sqrt{15}) \div (5\sqrt{2})$

Simplify each expression. See Example 7.

63. $\frac{2 + \sqrt{8}}{2}$

64. $\frac{3 + \sqrt{18}}{3}$

65. $\frac{-4 + \sqrt{20}}{2}$

66. $\frac{-6 + \sqrt{45}}{3}$

67. $\frac{4 - \sqrt{20}}{6}$

68. $\frac{-6 - \sqrt{27}}{6}$

69. $\frac{-4 - \sqrt{24}}{-6}$

70. $\frac{-3 - \sqrt{27}}{-3}$

Simplify each expression. See Example 8.

71. $\frac{5}{\sqrt{3} - \sqrt{2}}$

72. $\frac{3}{\sqrt{6} + \sqrt{2}}$

73. $\frac{\sqrt{3}}{\sqrt{5} - \sqrt{3}}$

74. $\frac{\sqrt{2}}{\sqrt{2} + \sqrt{5}}$

75. $\frac{2 + \sqrt{3}}{5 - \sqrt{3}}$

76. $\frac{\sqrt{2} - \sqrt{3}}{\sqrt{3} - 1}$

77. $\frac{\sqrt{7} - 5}{2\sqrt{7} + 1}$

78. $\frac{\sqrt{5} + 4}{3\sqrt{2} - \sqrt{5}}$

Simplify.

79. $\sqrt{5a} + \sqrt{20a}$

80. $a\sqrt{6} \cdot a\sqrt{12}$

81. $\sqrt{75} \div \sqrt{6}$

82. $\sqrt{24} - \sqrt{150}$

83. $(5 + 3\sqrt{5})^2$

84. $(\sqrt{6} - \sqrt{5})(\sqrt{6} + \sqrt{5})$

85. $\sqrt{5} + \frac{\sqrt{20}}{3}$

86. $\frac{5}{\sqrt{8} - \sqrt{3}}$

Use a calculator to find the approximate value of each expression to three decimal places.

87. $\frac{2 + \sqrt{3}}{2}$

88. $\frac{-2 + \sqrt{3}}{-6}$

89. $\frac{-4 - \sqrt{6}}{5 - \sqrt{3}}$

90. $\frac{-5 - \sqrt{2}}{\sqrt{3} + \sqrt{7}}$

Solve each problem.

91. Find the exact area and perimeter of the given rectangle.

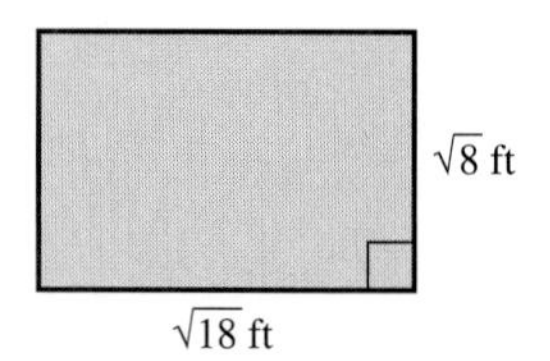

Figure for Exercise 91

92. Find the exact area and perimeter of the given triangle.

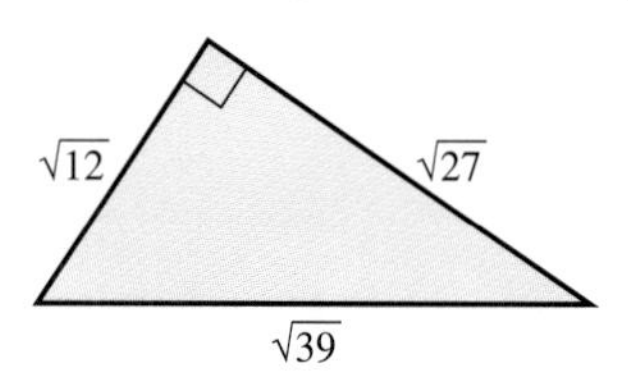

Figure for Exercise 92

93. Find the exact volume in cubic meters of the given rectangular box.

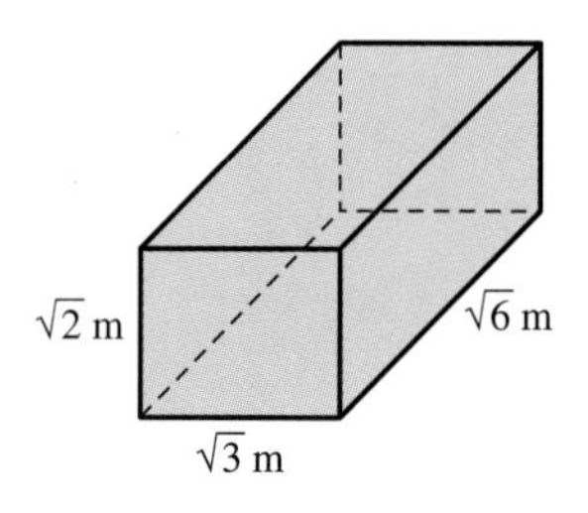

Figure for Exercise 93

94. Find the exact surface area of the box in Exercise 93.

8.5 Solving Equations with Radicals and Exponents

In this section:

- The Square Root Property
- Obtaining Equivalent Equations
- Squaring Each Side of an Equation
- Solving for the Indicated Variable
- Applications

Equations involving radicals and exponents occur in many applications. In this section you will learn to solve equations of this type, and you will see how these equations occur in applications.

The Square Root Property

An equation of the form $x^2 = k$ can have two solutions, one solution, or no solutions, depending on the value of k. For example,

$$x^2 = 4$$

has two solutions because $(2)^2 = 4$ and $(-2)^2 = 4$. So $x^2 = 4$ is equivalent to the compound equation

$$x = 2 \quad \text{or} \quad x = -2,$$

which is also written as $x = \pm 2$ and is read as "x equals positive or negative 2." The only solution to $x^2 = 0$ is 0. The equation

$$x^2 = -4$$

has no real solution because the square of every real number is nonnegative.

These examples illustrate the **square root property.**

Square Root Property (Solving $x^2 = k$)

For $k > 0$ the equation $x^2 = k$ is equivalent to the compound equation

$$x = \sqrt{k} \text{ or } x = -\sqrt{k}. \qquad \text{(Also written } x = \pm\sqrt{k}.\text{)}$$

For $k = 0$ the equation $x^2 = k$ is equivalent to $x = 0$.
For $k < 0$ the equation $x^2 = k$ has no real solution.

CAUTION The expression $\sqrt{9}$ has a value of 3 only, but the equation $x^2 = 9$ has two solutions: 3 and -3.

EXAMPLE 1 **Using the square root property**

Solve each equation.

a) $x^2 = 12$

b) $2(x + 1)^2 - 18 = 0$

c) $x^2 = -9$

d) $(x - 16)^2 = 0$

Solution

a) $x^2 = 12$

$x = \pm\sqrt{12}$ Square root property

$x = \pm 2\sqrt{3}$

Check: $(2\sqrt{3})^2 = 4 \cdot 3 = 12$, and $(-2\sqrt{3})^2 = 4 \cdot 3 = 12$.

b) $2(x + 1)^2 - 18 = 0$

$2(x + 1)^2 = 18$ Add 18 to each side.

$(x + 1)^2 = 9$ Divide each side by 2.

$x + 1 = \pm\sqrt{9}$ Square root property

$x + 1 = 3$ or $x + 1 = -3$

$x = 2$ or $x = -4$

Check 2 and -4 in the original equation. Both -4 and 2 are solutions to the equation.

c) The equation $x^2 = -9$ has no real solution because no real number has a square that is negative.

d) $(x - 16)^2 = 0$

$x - 16 = 0$ Square root property

$x = 16$

Check: $(16 - 16)^2 = 0$. The equation has only one solution, 16.

Obtaining Equivalent Equations

When solving equations, we use a sequence of equivalent equations with each one simpler than the last. To get an equivalent equation, we can

1. add the same number to each side,
2. subtract the same number from each side,
3. multiply each side by the same nonzero number, or
4. divide each side by the same nonzero number.

However, "doing the same thing to each side" is not the only way to obtain an equivalent equation. In Chapter 4 we used the zero factor property to obtain equivalent equations. For example, by the zero factor property the equation

$$(x - 3)(x + 2) = 0$$

is equivalent to the compound equation

$$x - 3 = 0 \quad \text{or} \quad x + 2 = 0.$$

In this section you just learned how to obtain equivalent equations by the square root property. This property tells us how to write an equation that is equivalent to the equation $x^2 = k$. Note that the square root property does not tell us to "take the square root of each side." To become proficient at solving equations, we must understand these methods. One of our main goals in algebra is to keep expanding our skills for solving equations.

Squaring Each Side of an Equation

Some equations involving radicals can be solved by squaring each side:

$$\sqrt{x} = 5$$
$$(\sqrt{x})^2 = 5^2 \quad \text{Square each side.}$$
$$x = 25$$

All three of these equations are equivalent. Because $\sqrt{25} = 5$ is correct, 25 satisfies the original equation.

However, squaring each side does not necessarily produce an equivalent equation. For example, consider the equation

$$x = 3.$$

Squaring each side, we get

$$x^2 = 9.$$

Both 3 and -3 satisfy $x^2 = 9$, but only 3 satisfies the original equation $x = 3$. So $x^2 = 9$ is not equivalent to $x = 3$. The extra solution to $x^2 = 9$ is called an **extraneous solution.**

These two examples illustrate the **squaring property of equality.**

Squaring Property of Equality

When we square each side of an equation, the solutions to the new equation include all of the solutions to the original equation. However, the new equation might have extraneous solutions.

This property means that *we may square each side of an equation, but we must check all of our answers to eliminate extraneous solutions.*

EXAMPLE 2 **Using the squaring property of equality**

Solve each equation.

a) $\sqrt{x^2 - 16} = 3$ **b)** $x = \sqrt{2x + 3}$ **c)** $\sqrt{x^2 - 4x} = \sqrt{2 - 3x}$

Solution

a)
$$\sqrt{x^2 - 16} = 3$$
$$(\sqrt{x^2 - 16})^2 = 3^2 \quad \text{Square each side.}$$
$$x^2 - 16 = 9$$
$$x^2 = 25$$
$$x = \pm 5 \quad \text{Square root property}$$

Check each solution:

$$\sqrt{5^2 - 16} = \sqrt{25 - 16} = \sqrt{9} = 3$$
$$\sqrt{(-5)^2 - 16} = \sqrt{25 - 16} = \sqrt{9} = 3$$

So both 5 and -5 are solutions to the equation.

b)
$$x = \sqrt{2x + 3}$$
$$x^2 = (\sqrt{2x + 3})^2 \quad \text{Square each side.}$$
$$x^2 = 2x + 3$$
$$x^2 - 2x - 3 = 0 \quad \text{Solve by factoring.}$$
$$(x - 3)(x + 1) = 0 \quad \text{Factor.}$$
$$x - 3 = 0 \quad \text{or} \quad x + 1 = 0 \quad \text{Zero factor property}$$
$$x = 3 \quad \text{or} \quad x = -1$$

Check in the original equation:

Check $x = 3$:	Check $x = -1$:
$3 = \sqrt{2 \cdot 3 + 3}$	$-1 = \sqrt{2(-1) + 3}$
$3 = \sqrt{9}$ Correct	$-1 = \sqrt{1}$ Incorrect

Because -1 does not satisfy the original equation, -1 is an extraneous solution. The only solution is 3.

c)
$$\sqrt{x^2 - 4x} = \sqrt{2 - 3x}$$
$$x^2 - 4x = 2 - 3x \quad \text{Square each side.}$$
$$x^2 - x - 2 = 0$$
$$(x - 2)(x + 1) = 0$$
$$x - 2 = 0 \quad \text{or} \quad x + 1 = 0 \quad \text{Zero factor property}$$
$$x = 2 \quad \text{or} \quad x = -1$$

Check each solution in the original equation:

Check $x = 2$:	Check $x = -1$:
$\sqrt{2^2 - 4 \cdot 2} = \sqrt{2 - 3 \cdot 2}$	$\sqrt{(-1)^2 - 4(-1)} = \sqrt{2 - 3(-1)}$
$\sqrt{-4} = \sqrt{-4}$	$\sqrt{5} = \sqrt{5}$

Because $\sqrt{-4}$ is not a real number, 2 is an extraneous solution. The only solution to the original equation is -1. ■

In the next example, one of the sides of the equation is a binomial. When we square each side, we must be sure to square the binomial properly.

EXAMPLE 3 **Squaring each side of an equation**

Solve the equation $x + 2 = \sqrt{-2 - 3x}$.

Solution

$$x + 2 = \sqrt{-2 - 3x}$$

$$(x + 2)^2 = (\sqrt{-2 - 3x})^2 \quad \text{Square each side.}$$

$$x^2 + 4x + 4 = -2 - 3x \quad \text{Square the binomial on the left side.}$$

$$x^2 + 7x + 6 = 0$$

$$(x + 6)(x + 1) = 0 \quad \text{Factor.}$$

$$x + 6 = 0 \quad \text{or} \quad x + 1 = 0$$

$$x = -6 \quad \text{or} \quad x = -1$$

Check these solutions in the original equation:

Check $x = -6$:

$$-6 + 2 = \sqrt{-2 - 3(-6)}$$

$$-4 = \sqrt{16} \quad \text{Incorrect}$$

Check $x = -1$:

$$-1 + 2 = \sqrt{-2 - 3(-1)}$$

$$1 = \sqrt{1} \quad \text{Correct}$$

The solution -6 does not check. The only solution to the original equation is -1. ■

Solving for the Indicated Variable

In the next example we use the square root property to solve a formula for an indicated variable.

EXAMPLE 4 **Solving for a variable**

Solve the formula $A = \pi r^2$ for r.

Solution

$$A = \pi r^2$$

$$\frac{A}{\pi} = r^2 \quad \text{Divide each side by } \pi.$$

$$\pm\sqrt{\frac{A}{\pi}} = r \quad \text{Square root property}$$

The formula solved for r is

$$r = \pm\sqrt{\frac{A}{\pi}}.$$

If r is the radius of a circle with area A, then r is positive and

$$r = \sqrt{\frac{A}{\pi}}.$$

■

Applications

Equations involving exponents occur in many applications. If the exact answer to a problem is an irrational number in radical notation, it is usually helpful to find a decimal approximation for the answer.

EXAMPLE 5 **Finding the side of a square with a given diagonal**

If the diagonal of a square window is 10 feet long, then what are the exact and approximate lengths of a side? Round the approximate answer to two decimal places.

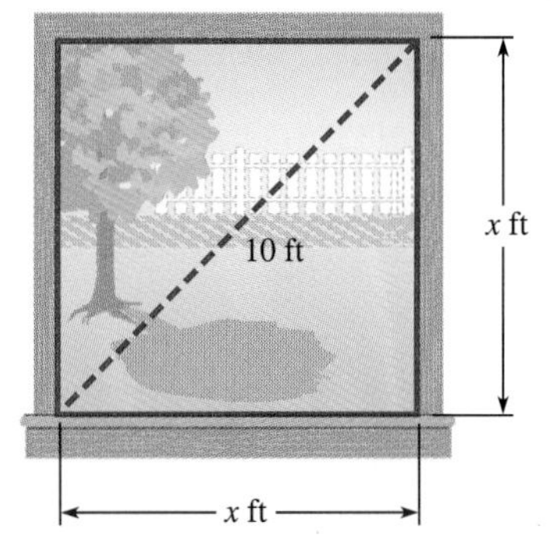

Figure 8.1

Solution

First make a sketch as in Fig. 8.1. Let x be the length of a side. The Pythagorean theorem tells us that the sum of the squares of the sides is equal to the diagonal squared:

$$
\begin{aligned}
x^2 + x^2 &= 10^2 \\
2x^2 &= 100 \\
x^2 &= 50 \\
x &= \pm\sqrt{50} \\
&= \pm 5\sqrt{2}
\end{aligned}
$$

Because the length of a side must be positive, we disregard the negative solution. The exact length of a side is $5\sqrt{2}$ feet. Use a calculator to get $5\sqrt{2} \approx 7.07$. The symbol $\approx$ means "is approximately equal to." The approximate length of a side is 7.07 feet.

Warm-ups

True or false? Explain your answer.

1. The equation $x^2 = 9$ is equivalent to the equation $x = 3$.
2. The equation $x^2 = -16$ has no real solution.
3. The equation $a^2 = 0$ has no solution.
4. Both $-\sqrt{5}$ and $\sqrt{5}$ are solutions to $x^2 + 5 = 0$.
5. The equation $-x^2 = 9$ has no solution.
6. To solve $\sqrt{x+4} = \sqrt{2x-9}$, first take the square root of each side.
7. All extraneous solutions give us a denominator of zero.
8. Squaring both sides of $\sqrt{x} = -1$ will produce an extraneous solution.
9. The equation $x^2 - 3 = 0$ is equivalent to $x = \pm\sqrt{3}$.
10. The equation $-2 = \sqrt{6x^2 - x - 8}$ has no solution.

8.5 EXERCISES

Solve each equation. See Example 1.

1. $x^2 = 16$
2. $x^2 = 49$
3. $x^2 - 40 = 0$
4. $x^2 - 24 = 0$
5. $3x^2 = 2$
6. $2x^2 = 3$
7. $9x^2 = -4$
8. $25x^2 + 1 = 0$
9. $(x - 1)^2 = 4$
10. $(x + 3)^2 = 9$
11. $2(x - 5)^2 + 1 = 7$
12. $3(x - 6)^2 - 4 = 11$
13. $(x + 19)^2 = 0$
14. $5x^2 + 5 = 5$

Solve each equation. See Example 2.

15. $\sqrt{x - 9} = 9$
16. $\sqrt{x + 3} = 4$
17. $\sqrt{2x - 3} = -4$
18. $\sqrt{3x - 5} = -9$
19. $4 = \sqrt{x^2 - 9}$
20. $1 = \sqrt{x^2 - 1}$
21. $x = \sqrt{18 - 3x}$
22. $x = \sqrt{6x + 27}$
23. $x = \sqrt{x}$
24. $x = \sqrt{2x}$
25. $\sqrt{x + 1} = \sqrt{2x - 5}$
26. $\sqrt{1 - 3x} = \sqrt{x + 5}$
27. $3\sqrt{2x - 1} + 3 = 5$
28. $4\sqrt{x + 5} - 3 = 9$

Solve each equation. See Example 3.

29. $x - 3 = \sqrt{2x - 6}$
30. $x - 1 = \sqrt{3x - 5}$
31. $\sqrt{x + 13} = x + 1$
32. $x + 1 = \sqrt{22 - 2x}$
33. $\sqrt{10x - 44} = x - 2$
34. $\sqrt{8x - 7} = x + 1$

Solve each formula for the indicated variable. See Example 4.

35. $V = \pi r^2 h$ for r
36. $V = \frac{4}{3}\pi r^2 h$ for r
37. $a^2 + b^2 = c^2$ for b
38. $y = ax^2 + c$ for x
39. $b^2 - 4ac = 0$ for b
40. $s = \frac{1}{2}gt^2 + v$ for t
41. $v = \sqrt{2pt}$ for t
42. $y = \sqrt{2x}$ for x

Solve each equation.

43. $3x^2 - 6 = 0$
44. $5x^2 + 3 = 0$
45. $\sqrt{2x - 3} = \sqrt{3x + 1}$
46. $\sqrt{2x - 4} = \sqrt{x - 9}$
47. $(2x - 1)^2 = 8$
48. $(3x - 2)^2 = 18$
49. $\sqrt{2x - 9} = 0$
50. $\sqrt{5 - 3x} = 0$
51. $x + 1 = \sqrt{2x + 10}$
52. $x - 3 = \sqrt{2x + 18}$
53. $3(x + 1)^2 - 27 = 0$
54. $2(x - 3)^2 - 50 = 0$
55. $(2x - 5)^2 = 0$
56. $(3x - 1)^2 = 0$

Use a calculator to find approximate solutions to each equation. Round your answers to three decimal places.

57. $x^2 = 3.25$
58. $(x + 1)^2 = 20.3$
59. $\sqrt{x + 2} = 1.73$
60. $\sqrt{2.3x - 1.4} = 3.3$
61. $1.3(x - 2.4)^2 = 5.4$
62. $-2.4x^2 = -9.55$

Find the exact answer to each problem. If the answer is irrational, then find an approximation to three decimal places. See Example 5.

63. ***Side of a square.*** Find the length of the side of a square whose area is 18 square feet.

64. ***Side of a field.*** Find the length of the side of a square wheat field whose area is 75 square miles.

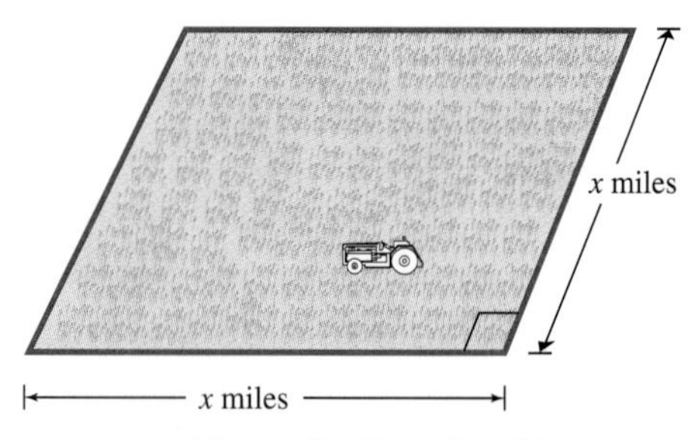

Figure for Exercise 64

65. ***Side of a table.*** Find the length of the side of a square coffee table whose diagonal is 6 feet.

66. ***Side of a square.*** Find the length of the side of a square whose diagonal measures 1 yard.

67. ***Diagonal of a tile.*** Find the length of the diagonal of a square floor tile whose sides measure 1 foot each.

68. ***Diagonal of a sandbox.*** The sandbox at Totland is shaped like a square with an area of 20 square meters. Find the length of the diagonal of the square.

69. ***Diagonal of a tub.*** Find the length of the diagonal of a rectangular bathtub with sides of 3 feet and 4 feet.

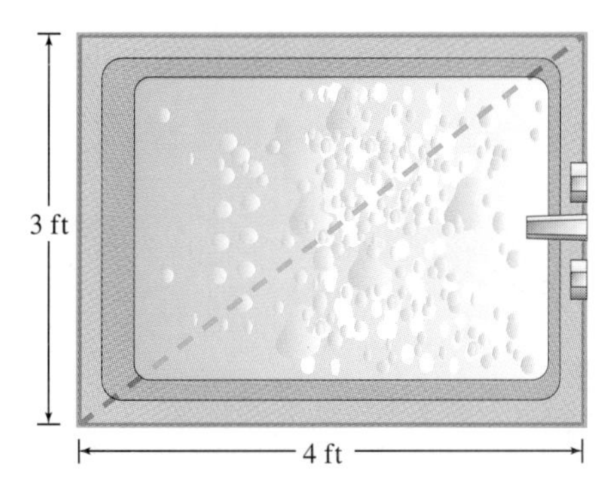

Figure for Exercise 69

70. ***Diagonal of a rectangle.*** What is the length of the diagonal of a rectangular office whose sides are 6 feet and 8 feet?

71. ***Falling bodies.*** If we neglect air resistance, then the number of feet that a body falls from rest during t seconds is given by $s = 16t^2$. How long does it take a pine cone to fall from the top of a 100-foot pine tree?

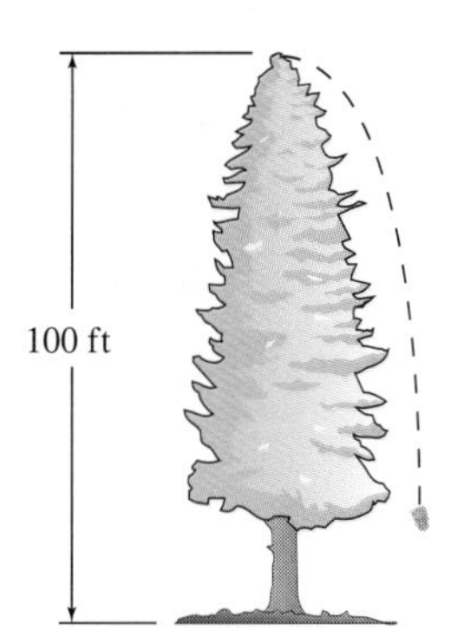

Figure for Exercise 71

72. ***America's favorite pastime.*** A baseball diamond is actually a square, 90 feet on each side. How far is it from home plate to second base?

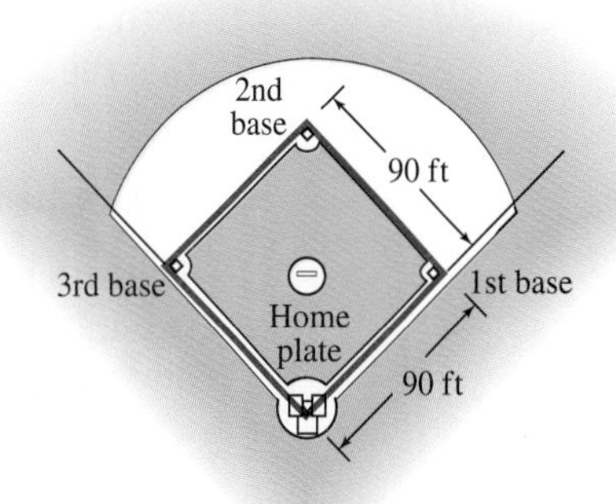

Figure for Exercise 72

73. ***Length of a guy wire.*** A guy wire from the top of a 200-foot tower is to be attached to a point on the ground whose distance from the base of the tower is $\frac{2}{3}$ of the height of the tower. Find the length of the guy wire?

74. ***America's favorite pastime.*** The size of a rectangular television screen is commonly given by the manufacturer as the length of the diagonal of the rectangle. If a television screen measures 10 inches wide and 8 inches high, then what is the exact length of the diagonal of the screen? What is the approximate size of this television screen to the nearest inch?

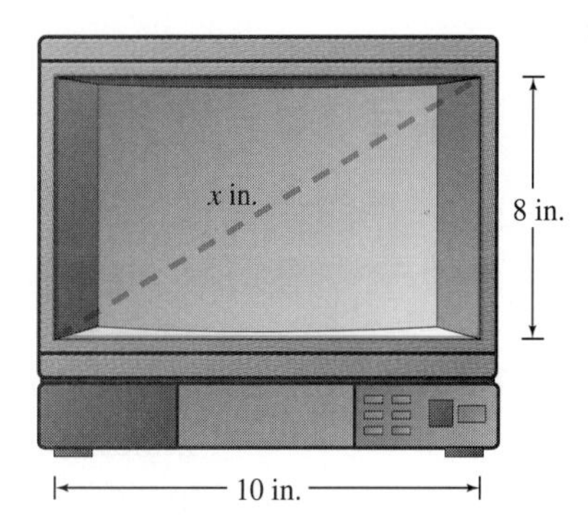

Figure for Exercise 74

8.6 Rational Exponents

In this section:

- Rational Exponents
- Using the Rules

You have learned how to use exponents to express powers of numbers and radicals to express roots. In this section you will see that roots can be expressed with exponents also. The advantage of using exponents to express roots is that the rules of exponents can be applied to the expressions.

Rational Exponents

The nth root of a number can be expressed by using radical notation or the exponent $1/n$. For example, $8^{1/3}$ and $\sqrt[3]{8}$ both represent the cube root of 8, and we have

$$8^{1/3} = \sqrt[3]{8} = 2.$$

Definition of $a^{1/n}$

If n is any positive integer, then

$$a^{1/n} = \sqrt[n]{a},$$

provided that $\sqrt[n]{a}$ is a real number.

Later in this section we will see that using exponent $1/n$ for nth root is compatible with the rules for integral exponents that we already know.

EXAMPLE 1 **Radicals or exponents**

Write each radical expression using exponent notation and each exponential expression using radical notation.

a) $\sqrt[3]{35}$ **b)** $\sqrt[4]{xy}$ **c)** $5^{1/2}$ **d)** $a^{1/5}$

Solution

a) $\sqrt[3]{35} = 35^{1/3}$

b) $\sqrt[4]{xy} = (xy)^{1/4}$

c) $5^{1/2} = \sqrt{5}$

d) $a^{1/5} = \sqrt[5]{a}$ ■

In the next example we evaluate some exponential expressions.

EXAMPLE 2 **Finding roots**

Evaluate each expression.

a) $4^{1/2}$ **b)** $(-8)^{1/3}$ **c)** $81^{1/4}$ **d)** $(-9)^{1/2}$

Solution

a) $4^{1/2} = \sqrt{4} = 2$

b) $(-8)^{1/3} = \sqrt[3]{-8} = -2$

c) $81^{1/4} = \sqrt[4]{81} = 3$

d) Because $(-9)^{1/2}$ is an even root of a negative number, it is not a real number. ■

We now extend the definition of exponent $1/n$ to include any rational number as an exponent. The numerator of the rational number indicates the power, and the denominator indicates the root. For example, the expression

$$8^{2/3}$$

(2 — Power; 3 — Root)

represents the square of the cube root of 8. So we have

$$8^{2/3} = (8^{1/3})^2 = (2)^2 = 4.$$

Definition of $a^{m/n}$

If m and n are positive integers, then

$$a^{m/n} = (a^{1/n})^m,$$

provided that $a^{1/n}$ is a real number.

We define negative rational exponents just like negative integral exponents.

Definition of $a^{-m/n}$

If m and n are positive integers and $a \neq 0$, then

$$a^{-m/n} = \frac{1}{a^{m/n}},$$

provided that $a^{1/n}$ is a real number.

EXAMPLE 3 **Radicals or exponents**

Write each radical expression using exponent notation and each exponential expression using radical notation.

a) $\sqrt[3]{x^2}$ **b)** $\dfrac{1}{\sqrt[4]{m^3}}$ **c)** $5^{2/3}$ **d)** $a^{-2/5}$

Solution

a) $\sqrt[3]{x^2} = x^{2/3}$

b) $\dfrac{1}{\sqrt[4]{m^3}} = \dfrac{1}{m^{3/4}} = m^{-3/4}$

c) $5^{2/3} = \sqrt[3]{5^2}$

d) $a^{-2/5} = \dfrac{1}{\sqrt[5]{a^2}}$ ■

To evaluate an expression with a negative rational exponent, remember that the denominator indicates root, the numerator indicates power, and the negative sign indicates reciprocal:

$$a^{-m/n}$$

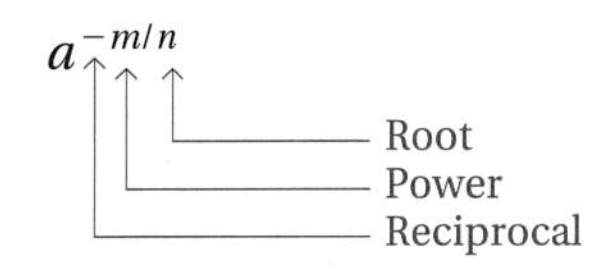

For example, to evaluate $8^{-2/3}$, we find the cube root of 8 (which is 2), square 2 to get 4, then take the reciprocal of 4 to get $\frac{1}{4}$. So

$$8^{-2/3} = \frac{1}{4}.$$

Note that we get the same value regardless of the order in which we perform the three steps, but to evaluate an expression mentally, it is usually easiest to do them in the order indicated.

EXAMPLE 4 **Rational exponents**

Evaluate each expression.

a) $27^{2/3}$ **b)** $4^{-3/2}$ **c)** $81^{-3/4}$ **d)** $(-8)^{-5/3}$

Solution

a) Because the exponent is 2/3, we find the cube root of 27 and then square it:

$$27^{2/3} = (27^{1/3})^2 = 3^2 = 9$$

b) Because the exponent is $-3/2$, we find the square root of 4, cube it, and find the reciprocal:

$$4^{-3/2} = \frac{1}{(4^{1/2})^3} = \frac{1}{2^3} = \frac{1}{8}$$

c) Because the exponent is $-3/4$, we find the fourth root of 81, cube it, and find the reciprocal:

$$81^{-3/4} = \frac{1}{(81^{1/4})^3} = \frac{1}{3^3} = \frac{1}{27} \quad \text{Definition of negative exponent}$$

d) $(-8)^{-5/3} = \frac{1}{((-8)^{1/3})^5} = \frac{1}{(-2)^5} = \frac{1}{-32} = -\frac{1}{32}$ ■

CAUTION An expression with a negative base and a negative exponent can have a positive or a negative value. For example,

$$(-8)^{-5/3} = -\frac{1}{32} \quad \text{and} \quad (-8)^{-2/3} = \frac{1}{4}.$$

Using the Rules

As we mentioned earlier, the advantage of using exponents to express roots is that the rules for integral exponents can also be used for rational exponents. We will use those rules in the next example.

EXAMPLE 5 **Using the rules of exponents**

Simplify each expression. Write answers with positive exponents. Assume that all variables represent positive real numbers.

a) $2^{1/2} \cdot 2^{3/2}$ **b)** $\dfrac{x}{x^{2/3}}$ **c)** $(b^{1/2})^{1/3}$

d) $(x^4y^{-6})^{1/2}$ **e)** $\left(\dfrac{x^6}{y^3}\right)^{-2/3}$

Solution

a) $2^{1/2} \cdot 2^{3/2} = 2^{1/2 + 3/2}$ Product rule

$= 2^2$ $\frac{1}{2} + \frac{3}{2} = \frac{4}{2} = 2$

$= 4$

b) $\frac{x}{x^{2/3}} = x^{1 - 2/3}$ Quotient rule

$= x^{1/3}$ $1 - \frac{2}{3} = \frac{3}{3} - \frac{2}{3} = \frac{1}{3}$

c) $(b^{1/2})^{1/3} = b^{(1/2) \cdot (1/3)}$ Power rule

$= b^{1/6}$ $\frac{1}{2} \cdot \frac{1}{3} = \frac{1}{6}$

d) $(x^4y^{-6})^{1/2} = (x^4)^{1/2}(y^{-6})^{1/2}$ Power of a product rule

$= x^2y^{-3}$ Power rule

$= \frac{x^2}{y^3}$ Definition of negative exponent

e) $\left(\frac{x^6}{y^3}\right)^{-2/3} = \left(\frac{y^3}{x^6}\right)^{2/3}$ Negative exponent rule

$= \frac{(y^3)^{2/3}}{(x^6)^{2/3}}$ Power of a quotient rule

$= \frac{y^2}{x^4}$ Power rule

Calculator Close-up

Expressions with rational exponents are evaluated on a calculator in the same way as expressions with integral exponents. For example, to evaluate $5^{2/3}$ on a scientific calculator, try the following keystrokes:

5 [x^y] [(] 2 [÷] 3 [)] [=]

The display (rounded to three decimal places) should read

2.924

On a graphing calculator your display should look like the following:

5 ^ (2/3) [ENTER]

2.924

Warm-ups

True or false? Explain your answer.

1. $9^{1/3} = \sqrt[3]{9}$

2. $8^{5/3} = \sqrt[5]{8^3}$

3. $(-16)^{1/2} = -16^{1/2}$

4. $9^{-3/2} = \frac{1}{27}$

5. $6^{-1/2} = \frac{\sqrt{6}}{6}$

6. $\frac{2}{2^{1/2}} = 2^{1/2}$

7. $2^{1/2} \cdot 2^{1/2} = 4^{1/2}$

8. $16^{-1/4} = -2$

9. $6^{1/6} \cdot 6^{1/6} = 6^{1/3}$

10. $(2^8)^{3/4} = 2^6$

8.6 EXERCISES

Write each radical expression using exponent notation and each exponential expression using radical notation. See Example 1.

1. $\sqrt[4]{7}$
2. $\sqrt[3]{cbs}$
3. $9^{1/5}$
4. $3^{1/2}$
5. $\sqrt{5x}$
6. $\sqrt{3y}$
7. $a^{1/2}$
8. $(-b)^{1/5}$

Evaluate each expression. See Example 2.

9. $25^{1/2}$
10. $16^{1/2}$
11. $(-125)^{1/3}$
12. $(-32)^{1/5}$
13. $16^{1/4}$
14. $8^{1/3}$
15. $(-4)^{1/2}$
16. $(-16)^{1/4}$

Write each radical expression using exponent notation and each exponential expression using radical notation. See Example 3.

17. $\sqrt[3]{w^7}$
18. $\sqrt{a^5}$
19. $\dfrac{1}{\sqrt[3]{2^{10}}}$
20. $\sqrt[3]{\dfrac{1}{a^2}}$
21. $w^{-3/4}$
22. $6^{-5/3}$
23. $(ab)^{3/2}$
24. $(3m)^{-1/5}$

Evaluate each expression. See Example 4.

25. $125^{2/3}$
26. $1000^{2/3}$
27. $25^{3/2}$
28. $16^{3/2}$
29. $27^{-4/3}$
30. $16^{-3/4}$
31. $4^{-3/2}$
32. $25^{-3/2}$
33. $(-27)^{-1/3}$
34. $(-8)^{-4/3}$
35. $(-16)^{-1/4}$
36. $(-100)^{-3/2}$

Simplify each expression. Write answers with positive exponents only. See Example 5.

37. $x^{1/4}x^{1/4}$
38. $y^{1/3}y^{2/3}$
39. $n^{1/2}n^{-1/3}$
40. $w^{-1/4}w^{3/5}$
41. $\dfrac{x^2}{x^{1/2}}$
42. $\dfrac{a^{1/2}}{a^{1/3}}$
43. $\dfrac{8t^{1/2}}{4t^{1/4}}$
44. $\dfrac{6w^{1/4}}{3w^{1/3}}$
45. $(x^6)^{1/3}$
46. $(y^{-4})^{1/2}$
47. $(5^{-1/4})^{-1/2}$
48. $(7^{-3/4})^6$
49. $(x^2y^6)^{1/2}$
50. $(t^3w^6)^{1/3}$
51. $(9x^{-2}y^8)^{-1/2}$
52. $(4w^{-2}t^{-4})^{-1/2}$

Evaluate each expression.

53. $16^{-1/2} + 2^{-1}$
54. $4^{-1/2} - 8^{-2/3}$
55. $27^{-1/6} \cdot 27^{-1/2}$
56. $32^{-1/10} \cdot 32^{-1/10}$
57. $\dfrac{81^{5/6}}{81^{1/12}}$
58. $\dfrac{25^{-3/4}}{25^{3/4}}$
59. $(3^{-4} \cdot 6^8)^{-1/4}$
60. $(-2^{-9} \cdot 3^6)^{-1/3}$

Solve each problem.

61. ***Yacht dimensions.*** Since 1988 a yacht competing for the America's Cup must satisfy the inequality

$$L + 1.25S^{1/2} - 9.8D^{1/3} \le 16.296,$$

where L is the boat's length in meters, S is the sail area in square meters, and D is the displacement in cubic meters (*Scientific American,* May 1992). Does a boat with a displacement of 21.8 m^3, a sail area of 305.4 m^2, and a length of 21.5 m satisfy the inequality? If the length and displacement are not changed, then what is the maximum number of square meters of sail that could be added and still have the boat satisfy the inequality?

62. ***Surface area.*** If A is the surface area of a cube and V is its volume, then

$$A = 6V^{2/3}.$$

Find the surface area of a cube that has a volume of 27 cubic centimeters. What is the surface area of a cube whose sides measure 5 centimeters each?

63. ***Average annual return.*** The average annual return on an investment, r, is given by the formula

$$r = \left(\frac{S}{P}\right)^{1/n} - 1,$$

where P is the original amount invested and S is the value of the investment after n years. An investment of \$10,000 in 1983 in Fidelity's Magellan Fund was worth \$52,796 in 1993 (Fidelity Magellan Annual Report). What was the average annual return on this investment for this period?

64. ***Population growth.*** The U.S. population grew from 203.3 million in 1970 to 248.7 million in 1990 (U.S. Census Bureau). Use the formula given in Exercise 63 to find the average annual rate of growth for the U.S. population for that period.

65. ***Piano tuning.*** The note middle C on a piano is tuned so that the string vibrates at 262 cycles per second, or 262 Hz (Hertz). The C note that is one octave higher is tuned to 524 Hz. Tuning for the 11 notes in between using the method of *equal temperament* is $262 \cdot 2^{n/12}$, where n takes the values 1 through 11. Find the tuning rounded to the nearest whole Hertz for those 11 notes.

Getting More Involved

66. ***Discussion.*** If $a^{-m/n} < 0$, then what can you conclude about the values of a, m, and n?

67. ***Discussion.*** If $a^{-m/n}$ is not a real number, then what can you conclude about the values of a, m, and n?

COLLABORATIVE ACTIVITIES

What's μ?

Grouping: 3 to 4 students
Topic: Radical equations

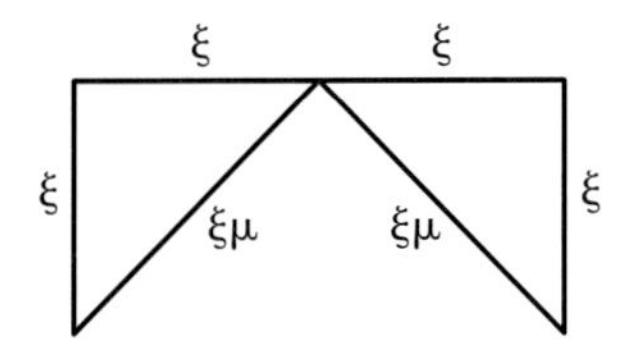

While on an archeological excavation, Arizona Barnes found this diagram. He attributed it to aliens from beyond the Milky Way. His more rational assistant claimed that she could use it as evidence that the civilization they were studying had knowledge of irrational numbers. By carefully measuring all the angles, she confirmed her suspicion that the three largest angles were right angles. She was then able to write two equations using the symbols from the diagram. By solving these, she was able to verify her hypothesis.

1. Write the two equations that the assistant found.
2. Solve the equations to find a value for μ.
3. Can you find the value of ξ from either equation?
4. Try substituting one or two values for ξ into the diagram and your equations.

State your conclusions.

Wrap-up CHAPTER 8

SUMMARY

Powers and Roots		Examples
Negative integral exponents	If n is a positive integer and a is a nonzero real number, then $a^{-n} = \frac{1}{a^n}$	$3^{-2} = \frac{1}{3^2}$, $x^{-5} = \frac{1}{x^5}$
Rules for negative exponents	If a is a nonzero real number and n is a positive integer, then $a^{-n} = \left(\frac{1}{a}\right)^n$, $a^{-1} = \frac{1}{a}$, and $\frac{1}{a^{-n}} = a^n$.	$\left(\frac{2}{3}\right)^{-3} = \left(\frac{3}{2}\right)^3$, $5^{-1} = \frac{1}{5}$ $\frac{1}{w^{-8}} = w^8$
nth roots	If $a = b^n$ for a positive integer n, then b is an nth root of a.	2 and -2 are fourth roots of 16.
Principal root	The positive even root of a positive number	The principal fourth root of 16 is 2.
Radical notation	If n is a positive even integer and a is positive, then the symbol $\sqrt[n]{a}$ denotes the principal nth root of a.	$\sqrt[4]{16} = 2$ $\sqrt[4]{16} \neq -2$
	If n is a positive odd integer, then the symbol $\sqrt[n]{a}$ denotes the nth root of a.	$\sqrt[3]{-8} = -2$, $\sqrt[3]{8} = 2$
	If n is any positive integer, then $\sqrt[n]{0} = 0$.	$\sqrt[5]{0} = 0$, $\sqrt[6]{0} = 0$
Definition of $a^{1/n}$	If n is any positive integer, then $a^{1/n} = \sqrt[n]{a}$ provided that $\sqrt[n]{a}$ is a real number.	$8^{1/3} = \sqrt[3]{8} = 2$ $(-4)^{1/2}$ is not real.

Definition of $a^{m/n}$	If m and n are positive integers, then $a^{m/n} = (a^{1/n})^m$, provided that $a^{1/n}$ is a real number.	$8^{2/3} = (8^{1/3})^2 = 2^2 = 4$ $(-16)^{3/4}$ is not real.
Definition of $a^{-m/n}$	If m and n are positive integers and $a \neq 0$, then $a^{-m/n} = \frac{1}{a^{m/n}}$, provided that $a^{1/n}$ is a real number.	$8^{-2/3} = \frac{1}{8^{2/3}} = \frac{1}{4}$

Rules of Exponents — Examples

The following rules hold for any rational numbers m and n and nonzero real numbers a and b, provided that all expressions represent real numbers.

Zero exponent	$a^0 = 1$	$8^0 = 1, (3x - y)^0 = 1$
Product rule	$a^m a^n = a^{m+n}$	$3^3 \cdot 3^4 = 3^7,\ x^5 x^{-2} = x^3$
Quotient rule	$\frac{a^m}{a^n} = a^{m-n}$	$\frac{3^5}{3^7} = 3^{-2},\ \frac{x}{x^{1/4}} = x^{3/4}$
Power rule	$(a^m)^n = a^{mn}$	$(2^2)^3 = 2^6$ $(w^{3/4})^4 = w^3$
Power of a product rule	$(ab)^n = a^n b^n$	$(2t)^{1/2} = 2^{1/2} t^{1/2}$
Power of a quotient rule	$\left(\frac{a}{b}\right)^n = \frac{a^n}{b^n}$	$\left(\frac{x}{3}\right)^{-3} = \frac{x^{-3}}{3^{-3}}$

Rules for Radicals — Examples

The following rules hold, provided that all roots are real numbers and n is a positive integer.

Product rule for radicals	$\sqrt[n]{ab} = \sqrt[n]{a} \cdot \sqrt[n]{b}$	$\sqrt{2} \cdot \sqrt{3} = \sqrt{6}$ $\sqrt{9y} = 3\sqrt{y}$
Quotient rule for radicals	$\sqrt[n]{\frac{a}{b}} = \frac{\sqrt[n]{a}}{\sqrt[n]{b}}$	$\sqrt{\frac{5}{4}} = \frac{\sqrt{5}}{2}$, $\frac{\sqrt{6}}{\sqrt{2}} = \sqrt{\frac{6}{2}} = \sqrt{3}$
Simplified form for square roots	A square root expression is in simplified form if it has	
	1. *no* perfect square factors inside the radical,	$\sqrt{12} = \sqrt{4 \cdot 3} = 2\sqrt{3}$
	2. *no* fractions inside the radical, and	$\sqrt{\frac{5}{2}} = \frac{\sqrt{5}}{\sqrt{2}}$
	3. *no* radicals in the denominator.	$\frac{\sqrt{5}}{\sqrt{2}} = \frac{\sqrt{5} \cdot \sqrt{2}}{\sqrt{2} \cdot \sqrt{2}} = \frac{\sqrt{10}}{2}$

Solving Equations Involving Squares and Square Roots — Examples

Square root property (solving $x^2 = k$)	If $k > 0$, the equation $x^2 = k$ is equivalent to $x = \sqrt{k}$ or $x = -\sqrt{k}$ (also written $x = \pm\sqrt{k}$).	$x^2 = 6$ $x = \pm\sqrt{6}$
	If $k = 0$, the equation $x^2 = k$ is equivalent to $x = 0$.	$t^2 = 0$ $t = 0$
	If $k < 0$, the equation $x^2 = k$ has no real solution.	$x^2 = -8$, no solution

Squaring property of equality	Squaring each side of an equation may introduce extraneous solutions. We must check all of our answers.	$\sqrt{x} = -3$ $(\sqrt{x})^2 = (-3)^2$ $x = 9$ Extraneous solution

Scientific Notation		**Examples**
Converting from scientific notation	1. Find the number of places to move the decimal point by examining the exponent on the 10.	$5.6 \times 10^3 = 5600$
	2. Move to the right for a positive exponent and to the left for a negative exponent.	$9 \times 10^{-4} = 0.0009$
Converting into scientific notation (positive numbers)	1. Count the number of places (n) that the decimal point must be moved so that it will follow the first nonzero digit of the number.	
	2. If the original number was larger than 10, use 10^n.	$304.6 = 3.046 \times 10^2$
	3. If the original number was smaller than 1, use 10^{-n}.	$0.0035 = 3.5 \times 10^{-3}$

REVIEW EXERCISES

For the following exercises, assume that all of the variables represent positive real numbers.

8.1 *Simplify each expression. Use only positive exponents in answers.*

1. 2^{-5}
2. -2^{-4}
3. 10^{-3}
4. $5^{-1} \cdot 5^0$
5. x^5x^{-8}
6. $a^{-3}a^{-9}$
7. $\dfrac{a^{-8}}{a^{-12}}$
8. $\dfrac{a^{10}}{a^{-4}}$
9. $\dfrac{a^3}{a^{-7}}$
10. $\dfrac{b^{-2}}{b^{-6}}$
11. $(x^{-3})^4$
12. $(x^5)^{-10}$
13. $(2x^{-3})^{-3}$
14. $(3y^{-5})^2$
15. $\left(\dfrac{a}{3b^{-3}}\right)^{-2}$
16. $\left(\dfrac{a^{-2}}{5b}\right)^{-3}$

Convert each number in scientific notation to a number in standard notation, and convert each number in standard notation to a number in scientific notation.

17. 5000
18. 0.00009
19. 3.4×10^5
20. 5.7×10^{-8}
21. 0.0000461
22. 44,000
23. 5.69×10^{-6}
24. 5.5×10^9

Perform each computation without using a calculator. Write answers in scientific notation.

25. $(3.5 \times 10^8)(2.0 \times 10^{-12})$
26. $(9 \times 10^{12})(2 \times 10^{17})$
27. $(2 \times 10^{-4})^4$
28. $(-3 \times 10^5)^3$
29. (0.00000004)(2,000,000,000)
30. (3,000,000,000) ÷ (0.000002)
31. $(0.0000002)^5$
32. $(50{,}000{,}000{,}000)^3$

8.2 *Find each root.*

33. $\sqrt[5]{32}$
34. $\sqrt[3]{-27}$
35. $\sqrt[3]{1000}$
36. $\sqrt{100}$
37. $\sqrt{x^{12}}$
38. $\sqrt{a^{10}}$
39. $\sqrt[3]{x^6}$
40. $\sqrt[3]{a^9}$
41. $\sqrt{4x^2}$
42. $\sqrt{9y^4}$
43. $\sqrt[3]{125x^6}$
44. $\sqrt[3]{8y^{12}}$
45. $\sqrt{\dfrac{4x^{16}}{y^{14}}}$
46. $\sqrt{\dfrac{9y^8}{t^{10}}}$
47. $\sqrt{\dfrac{w^2}{16}}$
48. $\sqrt{\dfrac{a^4}{25}}$

8.3 *Write each expression in simplified form.*

49. $\sqrt{72}$ **50.** $\sqrt{48}$ **51.** $\frac{1}{\sqrt{3}}$

52. $\frac{2}{\sqrt{5}}$ **53.** $\sqrt{\frac{3}{5}}$ **54.** $\sqrt{\frac{5}{6}}$

55. $\frac{\sqrt{33}}{\sqrt{3}}$ **56.** $\frac{\sqrt{50}}{\sqrt{5}}$ **57.** $\frac{\sqrt{3}}{\sqrt{8}}$

58. $\frac{\sqrt{2}}{\sqrt{18}}$ **59.** $\sqrt{y^6}$ **60.** $\sqrt{z^{10}}$

61. $\sqrt{24t^9}$ **62.** $\sqrt{8p^7}$ **63.** $\sqrt{12m^5t^3}$

64. $\sqrt{18p^3q^7}$ **65.** $\frac{\sqrt{2}}{\sqrt{x}}$ **66.** $\frac{\sqrt{5}}{\sqrt{y}}$

67. $\sqrt{\frac{3a^5}{2s}}$ **68.** $\sqrt{\frac{5x^7}{3w}}$

8.4 *Perform each computation and simplify.*

69. $2\sqrt{7} + 8\sqrt{7}$ **70.** $3\sqrt{6} - 5\sqrt{6}$

71. $\sqrt{12} - \sqrt{27}$ **72.** $\sqrt{18} + \sqrt{50}$

73. $2\sqrt{3} \cdot 5\sqrt{3}$ **74.** $-3\sqrt{6} \cdot 2\sqrt{6}$

75. $-3\sqrt{6} \cdot 5\sqrt{3}$ **76.** $4\sqrt{12} \cdot 6\sqrt{8}$

77. $-3\sqrt{3}(5 + \sqrt{3})$ **78.** $4\sqrt{2}(6 + \sqrt{8})$

79. $-\sqrt{3}(\sqrt{6} - \sqrt{15})$ **80.** $-\sqrt{2}(\sqrt{6} - \sqrt{2})$

81. $(\sqrt{3} - 5)(\sqrt{3} + 5)$

82. $(\sqrt{2} + \sqrt{7})(\sqrt{2} - \sqrt{7})$

83. $(2\sqrt{5} - \sqrt{6})^2$

84. $(3\sqrt{2} + \sqrt{6})^2$

85. $(4 - 3\sqrt{6})(5 - \sqrt{6})$

86. $(\sqrt{3} - 2\sqrt{5})(\sqrt{3} + 4\sqrt{5})$

87. $3\sqrt{5} \div (6\sqrt{2})$

88. $6\sqrt{5} \div (4\sqrt{3})$

89. $\frac{2 - \sqrt{20}}{10}$ **90.** $\frac{6 - \sqrt{12}}{-2}$

91. $\frac{3}{1 - \sqrt{5}}$ **92.** $\frac{\sqrt{2}}{\sqrt{6} + \sqrt{3}}$

8.5 *Solve each equation.*

93. $x^2 = 400$ **94.** $x^2 = 121$

95. $7x^2 = 3$ **96.** $3x^2 - 7 = 0$

97. $(x - 4)^2 - 18 = 0$ **98.** $2(x + 1)^2 - 40 = 0$

99. $\sqrt{x} = 9$ **100.** $\sqrt{x} - 20 = 0$

101. $x = \sqrt{36 - 5x}$ **102.** $x = \sqrt{2 - x}$

103. $x + 2 = \sqrt{52 + 2x}$ **104.** $x - 4 = \sqrt{x - 4}$

Solve each formula for t.

105. $t^2 - 8sw = 0$ **106.** $(t + b)^2 = b^2 - 4ac$

107. $3a = \sqrt{bt}$ **108.** $a - \sqrt{t} = w$

8.6 *Simplify each expression. Answers with exponents should have positive exponents only.*

109. $25^{-3/2}$ **110.** $9^{-5/2}$

111. $25^{1/2}$ **112.** $9^{3/2}$

113. $64^{-1/2}$ **114.** $125^{-2/3}$

115. $x^{-3/5}x^{-2/5}$ **116.** $t^{-1/3}t^{1/2}$

117. $(-8x^{-6})^{-1/3}$ **118.** $(-27x^{-9})^{-2/3}$

119. $w^{-3/2} \div w^{-7/2}$ **120.** $m^{1/3} \div m^{-1/4}$

121. $\left(\frac{9t^{-6}}{s^{-4}}\right)^{-1/2}$ **122.** $\left(\frac{8y^{-3}}{x^6}\right)^{-2/3}$

123. $\left(\frac{8x^{-12}}{y^{30}}\right)^{2/3}$ **124.** $\left(\frac{16y^{-3/4}}{t^{1/2}}\right)^{-2}$

Solve each problem.

125. ***National debt.*** If the debt of the United States government is 1.3×10^{12} dollars and the U.S. population is 2.6×10^8 people, then what is the debt in terms of dollars per person?

126. ***Energy consumption.*** According to the *1993 World Almanac,* total world energy consumption for 1990 was 3.38×10^{17} Btu. If the world population in 1990 was 5×10^9 people, then what was the energy consumption per person?

127. ***Depreciation rate.*** If the cost of a piece of equipment was C dollars and it is sold for S dollars after n years, then the annual depreciation rate r is given by $r = 1 - \left(\frac{S}{C}\right)^{1/n}$. A 1985 Mercedes-Benz 500 SEC sold for $56,800 in 1985 and for $21,000 in 1994 *(Edmund's 1994 Used Car Prices).* Find the annual depreciation rate for this car to the nearest tenth of a percent.

128. ***Depreciation rate.*** A 1985 Ford Escort that sold new for $6956 sells for $875 in 1994 *(Edmund's 1994 Used Car Prices).* Use the formula given in Exercise 127 to find the annual depreciation rate for this car to the nearest tenth of a percent.

129. ***Radius of a drop.*** The amount of water in a large raindrop is 0.25 cm^3. Use the formula

$$r = \left(\frac{3V}{4\pi}\right)^{1/3}$$

to find the radius of the spherical drop to the nearest tenth of a centimeter.

130. ***Radius of a circle.*** Solve the formula $A = \pi r^2$ for r.

131. ***Waffle cones.*** A large waffle cone has a height of 6 in. and a radius of 2 in. as shown in the accompanying figure. Find the exact amount of waffle in a cone this size. The formula $A = \pi r\sqrt{r^2 + h^2}$ gives the lateral surface area of a right circular cone with radius r and height h. Be sure to simplify the radical. Use a calculator to find the answer to the nearest square inch.

Figure for Exercise 131

132. ***Salting the roads.*** A city manager wants to find the amount of canvas required to cover a conical salt pile that is stored for the winter. The height of the pile is 10 yards, and the diameter of the base is 24 yards. Use the formula in Exercise 131 to find the exact number of square yards of canvas needed. Simplify the radical. Use a calculator to find the answer to the nearest square yard.

133. ***World's largest picture tube.*** The Mitsubishi CS-40503 color television has a screen with diagonal measure of 40 inches. Assuming that the screen is square, find the length and width of the screen to the nearest inch.

134. ***Projection television.*** The screen on the Mitsubishi VS-6051 Big Screen TV measures 37 in. by 47 in. Find the diagonal measure of the screen to the nearest inch.

CHAPTER 8 TEST

Simplify each expression.

1. 2^{-5} **2.** $\sqrt{144}$

3. $\sqrt[3]{-27}$ **4.** 9^{-2}

5. $16^{1/4}$ **6.** $\sqrt{24}$

7. $\sqrt{\frac{3}{8}}$ **8.** $(-4)^{3/2}$

9. $\sqrt{8} + \sqrt{2}$ **10.** $(2 + \sqrt{3})^2$

11. $(3\sqrt{2} - \sqrt{7})(3\sqrt{2} + \sqrt{7})$

12. $\sqrt{21} \div \sqrt{3}$ **13.** $\sqrt{20} \div \sqrt{3}$

14. $\frac{2 + \sqrt{8}}{2}$ **15.** $27^{4/3}$

16. $\sqrt{3}(\sqrt{6} - \sqrt{3})$

Simplify. Assume that all variables represent positive real numbers, and write answers with positive exponents only.

17. $3x^{-2} \cdot 5x^7$ **18.** $(2x^{-6})^3$

19. $(-3x^{-5}y^2)^{-3}$ **20.** $\frac{2y^{-5}}{8y^9}$

21. $\frac{t^{-7}}{t^{-3}}$ **22.** $(x^3y^9)^{1/3}$

23. $(-2s^{-3}t^2)^{-2}$ **24.** $\left(\frac{125w^3}{u^{-12}}\right)^{-1/3}$

25. $\sqrt{\frac{3}{t}}$ **26.** $\sqrt{4y^6}$

27. $\sqrt[3]{8y^{12}}$ **28.** $\sqrt{18t^7}$

Solve each equation.

29. $(x + 3)^2 = 36$ **30.** $\sqrt{x + 7} = 5$

31. $5x^2 = 2$ **32.** $(3x - 4)^2 = 0$

33. $x - 3 = \sqrt{5x + 9}$

Solve the equation for the specified variable.

34. $S = \pi r^2 h$ for r

Convert to scientific notation.

35. 5,433,000 **36.** 0.0000065

Perform each computation by converting to scientific notation. Give answers in scientific notation.

37. $(80{,}000)(0.000006)$ **38.** $(0.0000003)^4$

Show a complete solution to each problem.

39. Find the exact length of the side of a square whose diagonal is 5 meters.

40. To utilize a center-pivot irrigation system, a farmer planted his crop in a circular field of 100,000 square meters. Find the radius of the circular field to the nearest tenth of a meter.

Tying It All Together CHAPTERS 1–8

Solve each equation or inequality. For the inequalities, also sketch the graph of the inequality.

1. $2x + 3 = 0$

2. $2x = 3$

3. $2x + 3 > 0$

4. $-2x + 3 > 0$

5. $2(x + 3) = 0$

6. $2x^2 = 3$

7. $\frac{x}{3} = \frac{2}{x}$

8. $\frac{x - 1}{x} = \frac{x}{x - 2}$

9. $(2x + 3)^2 = 0$

10. $(2x + 3)(x - 3) = 0$

11. $2x^2 + 3 = 0$

12. $(2x + 3)^2 = 1$

13. $(2x + 3)^2 = -1$

14. $\sqrt{2x^2 - 14} = x - 1$

Let $a = 2$, $b = -3$, and $c = -9$. Find the value of each algebraic expression.

15. b^2

16. $-4ac$

17. $b^2 - 4ac$

18. $\sqrt{b^2 - 4ac}$

19. $-b + \sqrt{b^2 - 4ac}$

20. $-b - \sqrt{b^2 - 4ac}$

21. $\frac{-b + \sqrt{b^2 - 4ac}}{2a}$

22. $\frac{-b - \sqrt{b^2 - 4ac}}{2a}$

Factor each trinomial completely.

23. $x^2 - 6x + 9$

24. $x^2 + 10x + 25$

25. $x^2 + 12x + 36$

26. $x^2 - 20x + 100$

27. $2x^2 - 8x + 8$

28. $3x^2 + 6x + 3$

Perform the indicated operation.

29. $(3 + 2x) - (6 - 5x)$

30. $(5 + 3t)(4 - 5t)$

31. $(8 - 6j)(3 + 4j)$

32. $(1 - u) + (5 + 7u)$

33. $(3 - 4v) - (2 - 5v)$

34. $(2 + t)^2$

35. $(t - 7)(t + 7)$

36. $(3 - 2n)(3 + 2n)$

37. $(1 - m)^2$

38. $(-4 - 6t) - (-3 - 8t)$

39. $(1 + r)(3 - 4r)$

40. $(2 - 6y)(1 + 3y)$

41. $(1 - 2j) + (-6 + 5j)$

42. $(-2 - j) + (4 - 5j)$

43. $\frac{4 - 6x}{2}$

44. $\frac{-3 - 9p}{3}$

45. $\frac{8 - 12q}{-4}$

46. $\frac{20 - 5z}{-5}$

Solve the problem.

47. ***Oxygen uptake.*** In studying the oxygen uptake rate for marathon runners, Costill and Fox calculate the power expended P in kilocalories per minute using the formula $P = M(av - b)$, where M is the mass of the runner in kilograms and v is the speed in meters per minute (*Medicine and Science in Sports,* Vol. 1). The constants a and b have values $a = 1.02 \times 10^{-3}$ and $b = 2.62 \times 10^{-2}$.

a) Find P for a 60-kg runner who is running at 300 m/min.

b) Find the velocity of a 55-kg runner who is expending 14 kcal/min.

C H A P T E R 9

Quadratic Equations and Functions

Throughout time, humans have been building bridges over waterways. Primitive people threw logs across streams or attached ropes to branches to cross the waters. Later, the Romans built stone structures to span rivers and chasms. Throughout the centuries, bridges have been made of wood and stone and later from cast iron, concrete, and steel. Today's bridges are among the most beautiful and complex creations of modern engineering. Whether the bridge spans a small creek or a four-mile-wide stretch of water, mathematics is a part of its very foundation.

The function of a bridge, the length it must span, and the load it must carry often determine the type of bridge that is built. Some common types designed by civil engineers are cantilevered, arch, cable-stayed, and suspension bridges. The military is known for building trestle bridges and floating or pontoon bridges.

New technology has enabled engineers to build bridges that are stronger, lighter, and less expensive than in the past, as well as being esthetically pleasing. Currently, some engineers are working on making bridges earthquake resistant. Another idea that is being explored is incorporating carbon fibers in cement to warn of small cracks through electronic signals.

In Exercise 39 of Section 9.6 you will see how quadratic equations and functions are used in designing suspension bridges.

9.1 Familiar Quadratic Equations

In this section:

- Definition
- Using the Square Root Property
- Solving Equations by Factoring

We solved some quadratic equations in Chapters 4 and 5 by factoring. In Chapter 8 we solved some quadratic equations using the square root property. In this section we will review the types that you have already learned to solve. In the next section you will learn a method by which you can solve any quadratic equation.

Definition

We saw the definition of a quadratic equation in Chapter 4, but we will repeat it here.

> **Quadratic Equation**
>
> A **quadratic equation** is an equation of the form
>
> $$ax^2 + bx + c = 0,$$
>
> where a, b, and c are real numbers with $a \neq 0$.

Equations that can be written in the form of the definition may also be called quadratic equations. In Chapters 4, 5, and 8 we solved quadratic equations such as

$$x^2 = 10, \qquad 5(x - 2)^2 = 20, \qquad \text{and} \qquad x^2 - 5x = -6.$$

Using the Square Root Property

If $b = 0$ in $ax^2 + bx + c = 0$, then the quadratic equation can be solved by using the square root property.

EXAMPLE 1 **Using the square root property**

Solve the equations.

a) $x^2 - 9 = 0$

b) $2x^2 - 3 = 0$

c) $-3(x + 1)^2 = -6$

Solution

a) Solve the equation for x^2, and then use the square root property:

$$x^2 - 9 = 0$$

$$x^2 = 9 \qquad \text{Add 9 to each side.}$$

$$x = \pm 3 \qquad \text{Square root property}$$

Check 3 and -3 in the original equation. Both 3 and -3 are solutions to $x^2 - 9 = 0$.

b) $2x^2 - 3 = 0$

$$2x^2 = 3$$

$$x^2 = \frac{3}{2}$$

$$x = \pm\sqrt{\frac{3}{2}} \quad \text{Square root property}$$

$$x = \pm\frac{\sqrt{3} \cdot \sqrt{2}}{\sqrt{2} \cdot \sqrt{2}} \quad \text{Rationalize the denominator.}$$

$$x = \pm\frac{\sqrt{6}}{2}$$

Check. The solutions to $2x^2 - 3 = 0$ are $\frac{\sqrt{6}}{2}$ and $-\frac{\sqrt{6}}{2}$.

c) This equation is a bit different from the previous two equations. If we actually squared the quantity $(x + 1)$, then we would get a term involving x. Then b would not be equal to zero as it is in the other equations. However, we can solve this equation like the others if we do not square $x + 1$.

$$-3(x + 1)^2 = -6$$

$$(x + 1)^2 = 2 \quad \text{Divide each side by } -3.$$

$$x + 1 = \pm\sqrt{2} \quad \text{Square root property}$$

$$x = -1 \pm \sqrt{2} \quad \text{Subtract 1 from each side.}$$

Check $x = -1 \pm \sqrt{2}$ in the original equation:

$$-3(-1 \pm \sqrt{2} + 1)^2 = -3(\pm\sqrt{2})^2 = -3(2) = -6$$

The solutions are $-1 + \sqrt{2}$ and $-1 - \sqrt{2}$. ■

EXAMPLE 2 **A quadratic equation with no real solution**

Solve $x^2 + 12 = 0$.

Solution

The equation $x^2 + 12 = 0$ is equivalent to $x^2 = -12$. Because the square of any real number is nonnegative, this equation has no solution. ■

Solving Equations by Factoring

In Chapter 4 you learned to factor trinomials and to use factoring to solve some quadratic equations. Recall that quadratic equations are solved by factoring as follows.

Strategy for Solving Quadratic Equations by Factoring

1. Write the equation with 0 on one side.
2. Factor the other side.
3. Use the zero factor property. (Set each factor equal to 0.)
4. Solve the two linear equations.
5. Check the answers in the original quadratic equation.

EXAMPLE 3 **Solving a quadratic equation by factoring**

Solve by factoring.

a) $x^2 + 2x = 8$

b) $3x^2 + 13x - 10 = 0$

c) $\frac{1}{6}x^2 - \frac{1}{2}x = 3$

Solution

a)

$$x^2 + 2x = 8$$

$$x^2 + 2x - 8 = 0 \quad \text{Get 0 on the right-hand side.}$$

$$(x + 4)(x - 2) = 0 \quad \text{Factor.}$$

$$x + 4 = 0 \quad \text{or} \quad x - 2 = 0 \quad \text{Zero factor property}$$

$$x = -4 \quad \text{or} \quad x = 2 \quad \text{Solve the linear equations.}$$

Check in the original equation:

$$(-4)^2 + 2(-4) = 16 - 8 = 8$$

$$2^2 + 2 \cdot 2 = 4 + 4 = 8$$

Both -4 and 2 are solutions to the equation.

b)

$$3x^2 + 13x - 10 = 0$$

$$(3x - 2)(x + 5) = 0 \quad \text{Factor.}$$

$$3x - 2 = 0 \quad \text{or} \quad x + 5 = 0 \quad \text{Zero factor property}$$

$$3x = 2 \quad \text{or} \quad x = -5$$

$$x = \frac{2}{3} \quad \text{or} \quad x = -5$$

Check in the original equation. Both -5 and $\frac{2}{3}$ are solutions to the equation.

c)

$$\frac{1}{6}x^2 - \frac{1}{2}x = 3$$

$$x^2 - 3x = 18 \quad \text{Multiply each side by 6.}$$

$$x^2 - 3x - 18 = 0 \quad \text{Get 0 on the right-hand side.}$$

$$(x - 6)(x + 3) = 0 \quad \text{Factor.}$$

$$x - 6 = 0 \quad \text{or} \quad x + 3 = 0 \quad \text{Zero factor property}$$

$$x = 6 \quad \text{or} \quad x = -3$$

Check in the original equation. The solutions are -3 and 6. ■

CAUTION You can set each factor equal to zero only when the product of the factors is zero. Note that $x^2 - 3x = 18$ is equivalent to $x(x - 3) = 18$, but you can make no conclusion about two factors that have a product of 18. ⊘

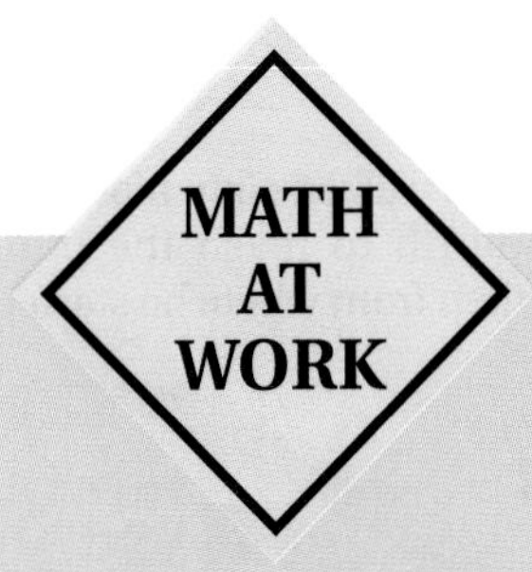

Even as a child, Gregory Brown was building tunnels and bridges in the sand. Now as a structural engineer for Stone and Webster, he is designing and analyzing real-life bridges and buildings. Many factors must be considered in analyzing a new or existing structure. For example, in the northern part of the United States the effects of the weight of snowfall on a roof must be considered, while in the southern part the strength of hurricane winds must be taken into account. Of course, earthquakes pose yet another consideration.

At the present time, Mr. Brown is working on bridge ratings. To rate a bridge, he first studies the design and construction of the bridge. Then he examines the structure for any kind of deterioration such as rust, cracks, or holes. In addition to the traffic load a bridge carries, the weight and size of trucks using the bridge are evaluated. When all this information is collected, collated, and analyzed, a bridge rating report is submitted. This report notes the extent of the deterioration, provides recommendations regarding vehicle weight limitations, and provides suggested repairs to strengthen the structure.

In Exercises 51 and 52 of this section you will see how an engineer can use a quadratic equation to find the length of a diagonal brace on a bridge.

Warm-ups

True or false? Explain your answer.

1. Both -4 and 4 satisfy the equation $x^2 - 16 = 0$.
2. The equation $(x - 3)^2 = 8$ is equivalent to $x - 3 = 2\sqrt{2}$.
3. Every quadratic equation can be solved by factoring.
4. Both -5 and 4 are solutions to $(x - 4)(x + 5) = 0$.
5. The quadratic equation $x^2 = -3$ has no real solutions.
6. The equation $x^2 = 0$ has no real solutions.
7. The equation $(2x + 3)(4x - 5) = 0$ is equivalent to $x = \frac{3}{2}$ or $x = \frac{5}{4}$.
8. The only solution to the equation $(x + 2)^2 = 0$ is -2.
9. $(x - 3)(x - 5) = 4$ is equivalent to $x - 3 = 2$ or $x - 5 = 2$.
10. All quadratic equations have two distinct solutions.

9.1 EXERCISES

Solve each equation. See Examples 1 and 2.

1. $x^2 - 36 = 0$

2. $x^2 - 81 = 0$

3. $x^2 + 10 = 0$

4. $x^2 + 4 = 0$

5. $5x^2 = 50$

6. $7x^2 = 14$

7. $3t^2 - 5 = 0$

8. $5y^2 - 7 = 0$

9. $-3y^2 + 8 = 0$

10. $-5w^2 + 12 = 0$

11. $(x - 3)^2 = 4$

12. $(x + 5)^2 = 9$

13. $(y - 2)^2 = 18$

14. $(m - 5)^2 = 20$

15. $2(x + 1)^2 = \frac{1}{2}$

16. $-3(x - 1)^2 = -\frac{3}{4}$

17. $(x - 1)^2 = \frac{1}{2}$

18. $(y + 2)^2 = \frac{1}{2}$

19. $\left(x + \frac{1}{2}\right)^2 = \frac{1}{2}$

20. $\left(x - \frac{1}{2}\right)^2 = \frac{3}{2}$

21. $(x - 11)^2 = 0$

22. $(x + 45)^2 = 0$

Solve each equation by factoring. See Example 3.

23. $x^2 - 2x - 15 = 0$

24. $x^2 - x - 12 = 0$

25. $x^2 + 6x + 9 = 0$

26. $x^2 + 10x + 25 = 0$

27. $4x^2 - 4x = 8$

28. $3x^2 + 3x = 90$

29. $3x^2 - 6x = 0$

30. $-5x^2 + 10x = 0$

31. $-4t^2 + 6t = 0$

32. $-6w^2 + 15w = 0$

33. $2x^2 + 11x - 21 = 0$

34. $2x^2 - 5x + 2 = 0$

35. $x^2 - 10x + 25 = 0$

36. $x^2 - 4x + 4 = 0$

37. $x^2 - \frac{7}{2}x = 15$

38. $3x^2 - \frac{2}{5}x = \frac{1}{5}$

39. $\frac{1}{10}a^2 - a + \frac{12}{5} = 0$

40. $\frac{2}{9}w^2 + \frac{5}{3}w - 3 = 0$

Solve each equation.

41. $x^2 - 2x = 2(3 - x)$

42. $x^2 + 2x = \dfrac{1 + 4x}{2}$

43. $x = \dfrac{27}{12 - x}$

44. $x = \dfrac{6}{x + 1}$

45. $\sqrt{3x - 8} = x - 2$

46. $\sqrt{3x - 14} = x - 4$

Solve each problem.

47. ***Side of a square.*** If the diagonal of a square is 5 meters, then what is the length of a side?

48. ***Diagonal of a square.*** If the side of a square is 5 meters, then what is the length of the diagonal?

49. ***Howard's journey.*** Howard walked eight blocks east and then four blocks north to reach the public library. How far was he then from where he started?

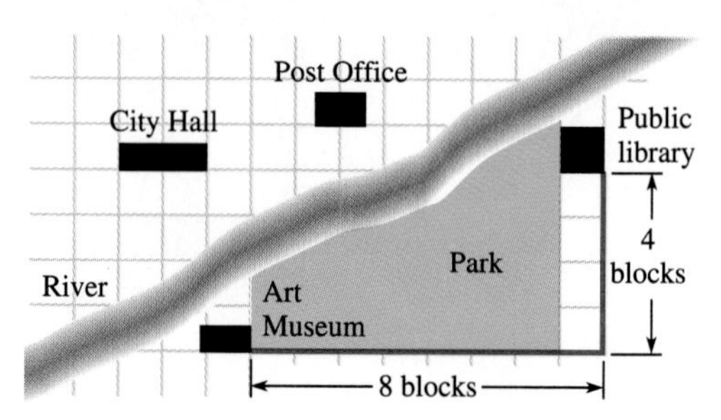

Figure for Exercise 49

50. ***Side and diagonal.*** Each side of a square has length s, and its diagonal has length d. Write a formula for s in terms of d.

51. ***Designing a bridge.*** Find the length d of the diagonal brace shown in the accompanying diagram.

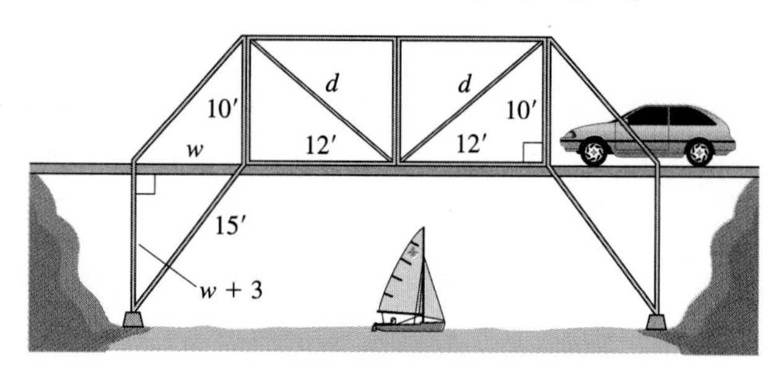

Figure for Exercises 51 and 52

52. ***Designing a bridge.*** Find the length labeled w in the accompanying diagram.

53. ***Two years of interest.*** Tasha deposited \$500 into an account that paid interest compounded annually. At the end of two years she had \$565. Solve the equation $565 = 500(1 + r)^2$ to find the annually rate r.

54. ***Rate of increase.*** The price of a new Ford Taurus went from \$15,251 in 1991 to \$16,441 in 1993 (*Edmond's Used Car Prices, 1994*). Find the average annual rate of increase in the price by solving the equation $16{,}441 = 15{,}251(1 + r)^2$ for r.

55. ***Projectile motion.*** If an object is projected upward with initial velocity v_0 ft/sec from an initial height of s_0 feet, then its height s (in feet) t seconds after it is projected is given by the formula $s = -16t^2 + v_0t + s_0$. If a stone is projected upward at 80 ft/sec from a height of 6 feet, then for what values of t is the stone 102 feet above the ground?

56. ***Diving time.*** A springboard diver can perform complicated maneuvers in a short period of time. If a diver springs into the air at 24 ft/sec from a board that is 16 feet above the water, then in how many seconds will she hit the water? Use the formula from Exercise 55.

57. ***Sum of integers.*** The formula $S = \frac{n^2 + n}{2}$ gives the sum of the first n positive integers. For what value of n is this sum equal to 45?

58. ***Serious reading.*** Kristy's New Year's resolution is to read one page of *Training Your Boa to Squeeze* on January 1, two pages on January 2, three pages on January 3, and so on. On what date will she finish the 136-page book? See Exercise 57.

Getting More Involved

59. ***Writing.*** One of the following equations has no real solutions. Find it by inspecting all of the equations (without solving). Explain your answer.

a) $x^2 - 99 = 0$ **b)** $2(v + 77)^2 = 0$
c) $3(y - 22)^2 + 11 = 0$ **d)** $5(w - 8)^2 - 9 = 0$

60. ***Cooperative learning.*** For each of three soccer teams A, B, and C to play the other two teams once, it takes three games (AB, AC, and BC). Work in groups to answer the following questions.

a) How many games are required for each team of a four-team league to play every other team once?
b) How many games are required in a five-team soccer league?
c) Find an expression of the form $an^2 + bn + c$ that gives the number of games required in a soccer league of n teams.
d) The Urban Soccer League has fields available for a 120-game season. If the organizers want each team to play every other team once, then how many teams should be in the league?

9.2 Solving Any Quadratic Equation

In this section:

- Perfect Square Trinomials
- Solving a Quadratic Equation by Completing the Square
- Applications

The quadratic equations in Section 9.1 were solved by factoring or the square root property, but some quadratic equations cannot be solved by either of those methods. In this section you will learn a method that works on *any* quadratic equation.

Perfect Square Trinomials

The new method for solving any quadratic equation depends on perfect square trinomials. Recall that a perfect square trinomial is the square of a binomial. Just as we recognize the numbers

$$1, \quad 4, \quad 9, \quad 16, \quad 25, \quad 36, \quad \ldots$$

as being the squares of the positive integers, we can recognize a perfect square trinomial. The following is a list of some perfect square trinomials with a leading coefficient of 1:

$$x^2 + 2x + 1 = (x + 1)^2 \qquad x^2 - 2x + 1 = (x - 1)^2$$
$$x^2 + 4x + 4 = (x + 2)^2 \qquad x^2 - 4x + 4 = (x - 2)^2$$
$$x^2 + 6x + 9 = (x + 3)^2 \qquad x^2 - 6x + 9 = (x - 3)^2$$
$$x^2 + 8x + 16 = (x + 4)^2 \qquad x^2 - 8x + 16 = (x - 4)^2$$

To solve quadratic equations using perfect square trinomials, we must be able to determine the last term of a perfect square trinomial when given the first two terms. This process is called **completing the square.** For example, the perfect square trinomial whose first two terms are $x^2 + 6x$ is $x^2 + 6x + 9$.

If the coefficient of x^2 is 1, there is a simple rule for finding the last term in a perfect square trinomial.

Finding the Last Term

The last term of a perfect square trinomial is the square of one-half of the coefficient of the middle term. In symbols, the perfect square trinomial whose first two terms are $x^2 + bx$ is $x^2 + bx + \left(\frac{b}{2}\right)^2$.

EXAMPLE 1 **Completing the square**

Find the perfect square trinomial whose first two terms are given, and factor the trinomial.

a) $x^2 + 10x$ **b)** $x^2 - 20x$ **c)** $x^2 + 3x$ **d)** $x^2 - x$

Solution

a) One-half of 10 is 5, and 5 squared is 25. So the perfect square trinomial is $x^2 + 10x + 25$. Factor as follows:

$$x^2 + 10x + 25 = (x + 5)^2$$

b) One-half of -20 is -10, and -10 squared is 100. So the perfect square trinomial is $x^2 - 20x + 100$. Factor as follows:

$$x^2 - 20x + 100 = (x - 10)^2$$

c) One-half of 3 is $\frac{3}{2}$, and $\frac{3}{2}$ squared is $\frac{9}{4}$. So the perfect square trinomial is $x^2 + 3x + \frac{9}{4}$. Factor as follows:

$$x^2 + 3x + \frac{9}{4} = \left(x + \frac{3}{2}\right)^2$$

d) One-half of -1 is $-\frac{1}{2}$, and $\left(-\frac{1}{2}\right)^2 = \frac{1}{4}$. So the perfect square is $x^2 - x + \frac{1}{4}$. Factor as follows:

$$x^2 - x + \frac{1}{4} = \left(x - \frac{1}{2}\right)^2$$

Solving a Quadratic Equation by Completing the Square

To complete the squares in Example 1, we simply found the missing last terms. In the next three examples we use that process along with the square root property to solve equations of the form $ax^2 + bx + c = 0$. When we use completing the square to solve an equation, we add the appropriate last term to both sides of the equation to obtain an equivalent equation. We first consider an equation in which the coefficient of x^2 is 1.

EXAMPLE 2 **Solving by completing the square ($a = 1$)**

Solve $x^2 + 6x - 7 = 0$ by completing the square.

Solution

Add 7 to each side of the equation to isolate $x^2 + 6x$:

$$x^2 + 6x = 7$$

Now complete the square for $x^2 + 6x$. One-half of 6 is 3, and $3^2 = 9$.

$$x^2 + 6x + 9 = 7 + 9 \quad \text{Add 9 to each side.}$$

$$(x + 3)^2 = 16 \quad \text{Factor the left side, and simplify the right side.}$$

$$x + 3 = \pm 4 \quad \text{Square root property}$$

$$x = -3 \pm 4$$

$$x = -3 + 4 \quad \text{or} \quad x = -3 - 4$$

$$x = 1 \quad \text{or} \quad x = -7$$

Check these answers in the original equation. The solutions are -7 and 1. ■

All of the perfect square trinomials in Examples 1 and 2 have 1 as the leading coefficient. If the leading coefficient is not 1, then we must divide each side of the equation by the leading coefficient to get an equation with a leading coefficient of 1. The steps to follow in solving a quadratic equation by completing the square are summarized as follows.

Strategy for Solving a Quadratic Equation by Completing the Square

1. The coefficient of x^2 must be 1.
2. Write the equation with only the x^2-terms and the x-terms on the left-hand side.
3. Complete the square on the left-hand side by adding the square of $\frac{1}{2}$ the coefficient of x to both sides of the equation.
4. Factor the perfect square trinomial as the square of a binomial.
5. Apply the square root property.
6. Solve for x and simplify the answer.
7. Check in the original equation.

In the next example we solve a quadratic equation in which the coefficient of x^2 is not 1.

EXAMPLE 3 Solving by completing the square ($a \neq 1$)

Solve $2x^2 - 5x - 3 = 0$ by completing the square.

Solution

Our perfect square trinomial must begin with x^2 and not $2x^2$:

$$\frac{2x^2 - 5x - 3}{2} = \frac{0}{2}$$ Divide each side by 2 to get 1 for the coefficient of x^2.

$$x^2 - \frac{5}{2}x - \frac{3}{2} = 0$$ Simplify.

$$x^2 - \frac{5}{2}x = \frac{3}{2}$$ Write only the x^2- and x-terms on the left-hand side.

$$x^2 - \frac{5}{2}x + \frac{25}{16} = \frac{3}{2} + \frac{25}{16}$$ Complete the square: $\frac{1}{2}\left(-\frac{5}{2}\right) = -\frac{5}{4}, \left(-\frac{5}{4}\right)^2 = \frac{25}{16}$

$$\left(x - \frac{5}{4}\right)^2 = \frac{49}{16}$$ Factor the left-hand side.

$$x - \frac{5}{4} = \pm\frac{7}{4}$$ Square root property

$$x = \frac{5}{4} \pm \frac{7}{4}$$

$$x = \frac{5}{4} + \frac{7}{4} \quad \text{or} \quad x = \frac{5}{4} - \frac{7}{4}$$

$$x = \frac{12}{4} \quad \text{or} \quad x = -\frac{2}{4}$$

$$x = 3 \quad \text{or} \quad x = -\frac{1}{2}$$

Check these answers in the original equation. The solutions to the equation are $-\frac{1}{2}$ and 3. ■

The equations in Examples 2 and 3 could have been solved by factoring. The quadratic equation in the next example cannot be solved by factoring, but it can be solved by completing the square. In fact, every quadratic equation can be solved by completing the square.

EXAMPLE 4 A quadratic equation with irrational solutions

Solve $x^2 + 4x - 3 = 0$ by completing the square.

Solution

$$x^2 + 4x - 3 = 0$$ Original equation

$$x^2 + 4x = 3$$ Add 3 to each side to isolate the x^2- and x-terms.

$$x^2 + 4x + 4 = 3 + 4$$ Complete the square by adding 4 to both sides.

$$(x + 2)^2 = 7$$ Factor the left-hand side.

$$x + 2 = \pm\sqrt{7}$$ Square root property

$$x = -2 \pm \sqrt{7}$$

$$x = -2 + \sqrt{7} \quad \text{or} \quad x = -2 - \sqrt{7}$$

Checking answers involving radicals can be done by using the operations with radicals that you learned in Chapter 8. Replace x with $-2 + \sqrt{7}$ in $x^2 + 4x - 3$:

$$(-2 + \sqrt{7})^2 + 4(-2 + \sqrt{7}) - 3 = 4 - 4\sqrt{7} + 7 - 8 + 4\sqrt{7} - 3$$
$$= 0$$

You should check $-2 - \sqrt{7}$. Both $-2 + \sqrt{7}$ and $-2 - \sqrt{7}$ satisfy the equation. ■

Applications

In the next example we use completing the square to solve a geometric problem.

EXAMPLE 5 **A geometric problem**

The sum of the lengths of the two legs of a right triangle is 8 feet. If the area of the right triangle is 5 square feet, then what are the lengths of the legs?

Solution

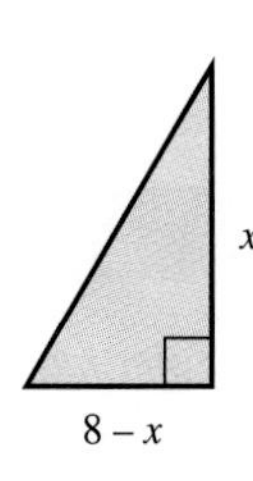

Figure 9.1

If x represents the length of one leg, then $8 - x$ represents the length of the other. See Fig. 9.1. The area of a triangle is given by the formula $A = \frac{1}{2}bh$. Let $A = 5$, $b = x$, and $h = 8 - x$ in this formula:

$$5 = \frac{1}{2}x(8 - x)$$

$$2 \cdot 5 = 2 \cdot \frac{1}{2}x(8 - x) \quad \text{Multiply each side by 2.}$$

$$10 = 8x - x^2$$

$$x^2 - 8x + 10 = 0$$

$$x^2 - 8x = -10 \quad \text{Subtract 10 from each side to isolate the } x^2\text{- and } x\text{-terms.}$$

$$x^2 - 8x + 16 = -10 + 16 \quad \text{Complete the square: } \tfrac{1}{2}(-8) = -4,\ (-4)^2 = 16$$

$$(x - 4)^2 = 6 \quad \text{Factor.}$$

$$x - 4 = \pm\sqrt{6}$$

$$x - 4 = \sqrt{6} \quad \text{or} \quad x - 4 = -\sqrt{6}$$

$$x = 4 + \sqrt{6} \quad \text{or} \quad x = 4 - \sqrt{6}$$

If $x = 4 + \sqrt{6}$, then

$$8 - x = 8 - (4 + \sqrt{6}) = 4 - \sqrt{6}.$$

If $x = 4 - \sqrt{6}$, then

$$8 - x = 8 - (4 - \sqrt{6}) = 4 + \sqrt{6}.$$

So there is only one pair of possible lengths for the legs: $4 + \sqrt{6}$ and $4 - \sqrt{6}$. Check that the area is 5 square feet:

$$A = \frac{1}{2}(4 + \sqrt{6})(4 - \sqrt{6}) = \frac{1}{2}(16 - 6) = \frac{1}{2}(10) = 5$$ ■

Warm-ups

True or false? Explain your answer.

1. Completing the square is used for finding the area of a square.
2. The polynomial $x^2 + \frac{2}{3}x + \frac{4}{9}$ is a perfect square trinomial.
3. Every quadratic equation can be solved by factoring.
4. The polynomial $x^2 - x + 1$ is a perfect square trinomial.
5. Every quadratic equation can be solved by completing the square.
6. The solutions to the equation $x - 2 = \pm\sqrt{3}$ are $2 + \sqrt{3}$ and $2 - \sqrt{3}$.
7. There are no real numbers that satisfy $(x + 7)^2 = -5$.
8. To solve $x^2 - 5x = 4$ by completing the square, we can add $\frac{25}{4}$ to each side.
9. One-half of four-fifths is two-fifths.
10. One-half of three-fourths is three-eighths.

9.2 EXERCISES

Find the perfect square trinomial whose first two terms are given, then factor the trinomial. See Example 1.

1. $x^2 + 6x$ **2.** $x^2 - 4x$

3. $x^2 + 14x$ **4.** $x^2 + 16x$

5. $x^2 - 16x$ **6.** $x^2 - 14x$

7. $t^2 - 18t$ **8.** $w^2 + 18w$

9. $m^2 + 3m$ **10.** $n^2 - 5n$

11. $z^2 + z$ **12.** $v^2 - v$

13. $x^2 - \frac{1}{2}x$ **14.** $y^2 + \frac{1}{3}y$

15. $y^2 + \frac{1}{4}y$ **16.** $z^2 - \frac{4}{3}z$

Factor each perfect square trinomial as the square of a binomial.

17. $x^2 + 10x + 25$ **18.** $x^2 - 6x + 9$

19. $m^2 - 2m + 1$ **20.** $n^2 + 4n + 4$

21. $x^2 + x + \frac{1}{4}$ **22.** $y^2 - y + \frac{1}{4}$

23. $t^2 + \frac{1}{3}t + \frac{1}{36}$ **24.** $v^2 - \frac{2}{3}v + \frac{1}{9}$

25. $x^2 + \frac{2}{5}x + \frac{1}{25}$ **26.** $y^2 - \frac{1}{4}y + \frac{1}{64}$

Solve each quadratic equation by completing the square. See Examples 2 and 3.

27. $x^2 + 2x - 15 = 0$ **28.** $x^2 + 2x - 24 = 0$

29. $x^2 - 4x - 21 = 0$ **30.** $x^2 - 4x - 12 = 0$

31. $x^2 + 6x + 9 = 0$ **32.** $x^2 - 10x + 25 = 0$

33. $2t^2 - 3t + 1 = 0$ **34.** $2t^2 - 3t - 2 = 0$

35. $2w^2 - 7w + 6 = 0$ **36.** $4t^2 + 5t - 6 = 0$

37. $3x^2 + 2x - 1 = 0$ **38.** $3x^2 - 8x - 3 = 0$

Solve each quadratic equation by completing the square. See Example 4.

39. $x^2 + 2x - 6 = 0$ **40.** $x^2 + 4x - 4 = 0$

41. $x^2 + 6x + 1 = 0$ **42.** $x^2 - 6x - 3 = 0$

43. $y^2 - y - 3 = 0$ **44.** $t^2 + t - 1 = 0$

45. $v^2 + 3v - 3 = 0$ **46.** $u^2 - 3u + 1 = 0$

47. $2m^2 - m - 4 = 0$ **48.** $4q^2 + 2q - 1 = 0$

Solve each equation by whichever method is appropriate.

49. $(x - 5)^2 = 7$ **50.** $x^2 + x = 12$

51. $3n^2 - 5 = 0$ **52.** $2m^2 + 16 = 0$

53. $3x^2 + 1 = 0$ **54.** $x^2 + 6x + 7 = 0$

55. $x^2 + 5 = 8x - 3$ **56.** $2x^2 + 3x = 42 - 2x$

57. $(2x - 7)^2 = 0$

58. $x^2 - 7 = 0$

59. $y^2 + 6y = 11$

60. $y^2 + 6y = 0$

61. $\frac{1}{4}w^2 + \frac{1}{2} = w$

62. $\frac{1}{2}z^2 + \frac{1}{2} = 2z$

63. $t^2 + 0.2t = 0.24$

64. $p^2 - 0.9p + 0.18 = 0$

Use a quadratic equation and completing the square to solve each problem. See Example 5.

65. ***Area of a triangle.*** The sum of the measures of the base and height of a triangle is 10 inches. If the area of the triangle is 11 square inches, then what are the measures of the base and height?

66. ***Dimensions of a rectangle.*** A rectangle has a perimeter of 12 inches and an area of 6 square inches. What are the length and width of the rectangle?

67. ***Missing numbers.*** The sum of two numbers is 12, and their product is 34. What are the numbers?

68. ***More missing numbers.*** The sum of two numbers is 8, and their product is 11. What are the numbers?

69. ***Saving candles.*** Joan has saved the candles from her birthday cake for every year of her life. If Joan has 78 candles, then how old is Joan? (See Exercise 57 of Section 9.1.)

70. ***Lottery tickets.*** A charitable organization is selling chances to win a used Corvette. If the tickets are x dollars each, then the members will sell $5000 - 200x$ tickets. So the total revenue for the tickets is given by $R = x(5000 - 200x)$. What is the revenue if the tickets are sold at \$8 each? For what ticket price is the revenue \$31,250?

Getting More Involved

71. ***Exploration.***

a) Find the product $[x - (5 + \sqrt{3})][x - (5 - \sqrt{3})]$.

b) Use completing the square to solve the quadratic equation formed by setting the answer to part (a) equal to zero.

c) Write a quadratic equation (in the form $ax^2 + bx + c = 0$) that has solutions $\frac{3 + \sqrt{2}}{2}$ and $\frac{3 - \sqrt{2}}{2}$.

d) Explain how to find a quadratic equation in the form $ax^2 + bx + c = 0$ for any two given solutions.

9.3 The Quadratic Formula

In this section:

- Developing the Quadratic Formula
- The Discriminant
- Which Method to Use

In Section 9.2 you learned that every quadratic equation can be solved by completing the square. In this section we use completing the square to get a formula, the quadratic formula, for solving any quadratic equation.

Developing the Quadratic Formula

To develop a formula for solving any quadratic equation, we start with the general quadratic equation

$$ax^2 + bx + c = 0$$

and solve it by completing the square. Assume that a is positive for now, and divide each side by a:

$$\frac{ax^2 + bx + c}{a} = \frac{0}{a} \qquad \text{Divide by } a \text{ to get 1 for the coefficient of } x^2.$$

$$x^2 + \frac{b}{a}x + \frac{c}{a} = 0 \qquad \text{Simplify.}$$

$$x^2 + \frac{b}{a}x = -\frac{c}{a} \qquad \text{Isolate the } x^2\text{- and } x\text{-terms.}$$

Now complete the square on the left. One-half of $\frac{b}{a}$ is $\frac{b}{2a}$, and $\left(\frac{b}{2a}\right)^2 = \frac{b^2}{4a^2}$.

$$x^2 + \frac{b}{a}x + \frac{b^2}{4a^2} = \frac{b^2}{4a^2} - \frac{c}{a}$$ Add $\frac{b^2}{4a^2}$ to each side.

$$\left(x + \frac{b}{2a}\right)^2 = \frac{b^2}{4a^2} - \frac{4ac}{4a^2}$$ Factor on the left-hand side, and get a common denominator on the right-hand side.

$$\left(x + \frac{b}{2a}\right)^2 = \frac{b^2 - 4ac}{4a^2}$$

$$x + \frac{b}{2a} = \pm\sqrt{\frac{b^2 - 4ac}{4a^2}}$$ Square root property

$$x = -\frac{b}{2a} \pm \frac{\sqrt{b^2 - 4ac}}{2a}$$ Because $a > 0$, $\sqrt{4a^2} = 2a$.

$$x = \frac{-b \pm \sqrt{b^2 - 4ac}}{2a}$$ Combine the two expressions.

We assumed that a was positive so that $\sqrt{4a^2} = 2a$ would be correct. If a is negative, then $\sqrt{4a^2} = -2a$. Either way, the result is the same. It is called the **quadratic formula.** The formula gives x in terms of the coefficients a, b, and c. The quadratic formula is generally used instead of completing the square to solve a quadratic equation that cannot be factored.

The Quadratic Formula

The solutions to $ax^2 + bx + c = 0$, where $a \neq 0$, are given by

$$x = \frac{-b \pm \sqrt{b^2 - 4ac}}{2a}.$$

EXAMPLE 1 **Equations with rational solutions**

Use the quadratic formula to solve each equation.

a) $x^2 + 2x - 3 = 0$

b) $4x^2 = -9 + 12x$

Solution

a) To use the formula, we first identify a, b, and c. For the equation

$$\underset{\substack{\uparrow \\ a}}{1}x^2 + \underset{\substack{\uparrow \\ b}}{2}x - \underset{\substack{\uparrow \\ c}}{3} = 0,$$

$a = 1$, $b = 2$, and $c = -3$. Now use these values in the quadratic formula:

$$x = \frac{-b \pm \sqrt{b^2 - 4ac}}{2a}$$

$$x = \frac{-2 \pm \sqrt{2^2 - 4(1)(-3)}}{2(1)} \qquad 2^2 - 4(1)(-3) = 4 + 12 = 16$$

$$x = \frac{-2 \pm \sqrt{16}}{2}$$

$$x = \frac{-2 \pm 4}{2}$$

$$x = \frac{-2 + 4}{2} \quad \text{or} \quad x = \frac{-2 - 4}{2}$$

$$x = 1 \quad \text{or} \quad x = -3$$

Check these answers in the original equation. The solutions are -3 and 1.

b) Write the equation in the form $ax^2 + bx + c = 0$ to identify a, b, and c:

$$4x^2 = -9 + 12x$$

$$4x^2 - 12x + 9 = 0$$

Now $a = 4$, $b = -12$, and $c = 9$. Use these values in the formula:

$$x = \frac{12 \pm \sqrt{(-12)^2 - 4(4)(9)}}{2(4)} \qquad b = -12, -b = 12$$

$$x = \frac{12 \pm \sqrt{0}}{8} = \frac{12}{8} = \frac{3}{2}$$

Check. The only solution to the equation is $\frac{3}{2}$. ■

The equations in Example 1 could have been solved by factoring. (Try it.) The quadratic equation in the next example has an irrational solution and cannot be solved by factoring.

EXAMPLE 2 **An equation with an irrational solution**

Solve $3x^2 - 6x + 1 = 0$.

Solution

For this equation, $a = 3$, $b = -6$, and $c = 1$:

$$x = \frac{6 \pm \sqrt{(-6)^2 - 4(3)(1)}}{2(3)} \qquad \text{Since } b = -6, -b = 6.$$

$$= \frac{6 \pm \sqrt{24}}{6}$$

$$= \frac{6 \pm 2\sqrt{6}}{6} \qquad \sqrt{24} = \sqrt{4}\sqrt{6} = 2\sqrt{6}$$

$$= \frac{2(3 \pm \sqrt{6})}{2(3)} \qquad \text{Numerator and denominator have 2 as a common factor.}$$

$$= \frac{3 \pm \sqrt{6}}{3}$$

The two solutions are the irrational numbers $\frac{3 + \sqrt{6}}{3}$ and $\frac{3 - \sqrt{6}}{3}$. ■

Calculator Close-up

Irrational solutions like those found in Example 2 can be checked in the original equation, but checking can be tedious. An easier way to check is to check an approximate value in the original equation:

$$\frac{3 + \sqrt{6}}{3} \approx 1.8165$$

Now evaluate $3x^2 - 6x + 1$ using this value for x:

$$3(1.8165)^2 - 6(1.8165) + 1 \approx 0.00001675$$

Of course, the answer is not exactly 0 because we did not check the exact answer. If you use more digits in the approximate answer, then the value that you get in the check will be closer to 0. If you use all of the digits available on your calculator, you should get 0 in the check. To use all of the accuracy available on your calculator, you can store the calculator value for x in memory and then use it from the memory in your check.

We have seen quadratic equations such as $x^2 = -9$ that do not have any real number solutions. In general, you can conclude that a quadratic equation has no real number solutions if you get a square root of a negative number in the quadratic formula.

EXAMPLE 3 **A quadratic equation with no real number solutions**

Solve $5x^2 - x + 1 = 0$.

Solution

For this equation we have $a = 5$, $b = -1$, and $c = 1$:

$$x = \frac{1 \pm \sqrt{(-1)^2 - 4(5)(1)}}{2(5)} \qquad b = -1, -b = 1$$

$$x = \frac{1 \pm \sqrt{-19}}{10}$$

The equation has no real solutions because $\sqrt{-19}$ is not real. ■

The Discriminant

A quadratic equation can have two real solutions, one real solution, or no real number solutions, depending on the value of $b^2 - 4ac$. If $b^2 - 4ac$ is positive, as in Example 1(a) and Example 2, we get two solutions. If $b^2 - 4ac$ is 0, we get only one solution, as in Example 1(b). If $b^2 - 4ac$ is negative, there are no real number solutions, as in Example 3. Table 9.1 summarizes these facts.

Value of $b^2 - 4ac$	Number of real solutions to $ax^2 + bx + c = 0$
Positive	2
Zero	1
Negative	0

Table 9.1

The quantity $b^2 - 4ac$ is called the **discriminant** because its value determines the number of real solutions to the quadratic equation.

EXAMPLE 4 **The number of real solutions**

Find the value of the discriminant, and determine the number of real solutions to each equation.

a) $3x^2 - 5x + 1 = 0$ **b)** $x^2 + 6x + 9 = 0$ **c)** $2x^2 + 1 = x$

Solution

a) For the equation $3x^2 - 5x + 1 = 0$ we have $a = 3$, $b = -5$, and $c = 1$. Now find the value of the discriminant:

$$b^2 - 4ac = (-5)^2 - 4(3)(1) = 25 - 12 = 13$$

Because the discriminant is positive, there are two real solutions to this quadratic equation.

b) For the equation $x^2 + 6x + 9 = 0$, we have $a = 1$, $b = 6$, and $c = 9$:

$$b^2 - 4ac = (6)^2 - 4(1)(9) = 36 - 36 = 0$$

Since the discriminant is zero, there is only one real solution to the equation.

c) We must first rewrite the equation:

$$2x^2 + 1 = x$$

$$2x^2 - x + 1 = 0 \quad \text{Subtract } x \text{ from each side.}$$

Now $a = 2$, $b = -1$, and $c = 1$.

$$b^2 - 4ac = (-1)^2 - 4(2)(1) = 1 - 8 = -7$$

Because the discriminant is negative, the equation has no real number solutions. ■

Which Method to Use

If the quadratic equation is simple enough, we can solve it by factoring or by the square root property. These methods should be considered first. *All quadratic equations can be solved by the quadratic formula.* Remember that the quadratic formula is just a shortcut to completing the square and is usually easier to use. However, you should learn completing the square because it is used elsewhere in algebra. The available methods are summarized as follows.

Solving the Quadratic Equation $ax^2 + bx + c = 0$

Method	Comments	Examples
Square root property	Use when $b = 0$.	If $x^2 = 7$, then $x = \pm\sqrt{7}$. If $(x - 2)^2 = 9$, then $x - 2 = \pm 3$
Factoring	Use when the polynomial can be factored.	$x^2 + 5x + 6 = 0$ $(x + 2)(x + 3) = 0$
Quadratic formula	Use when the first two methods do not apply.	$x^2 + 2x - 6 = 0$ $x = \frac{-2 \pm \sqrt{2^2 - 4 \cdot 1 \cdot (-6)}}{2 \cdot 1}$
Completing the square	Use only when this method is specified.	$x^2 + 4x - 9 = 0$ $x^2 + 4x + 4 = 9 + 4$ $(x + 2)^2 = 13$

Warm-ups

True or false? Explain your answer.

1. Completing the square is used to develop the quadratic formula.

2. For the equation $x^2 - x + 1 = 0$, we have $a = 1$, $b = -x$, and $c = 1$.

3. For the equation $x^2 - 3 = 5x$, we have $a = 1$, $b = -3$, and $c = 5$.

4. The quadratic formula can be expressed as $x = -b \pm \frac{\sqrt{b^2 - 4ac}}{2a}$.

5. The quadratic equation $2x^2 - 6x = 0$ has two real solutions.

6. All quadratic equations have two distinct real solutions.

7. Some quadratic equations cannot be solved by the quadratic formula.

8. We could solve $2x^2 - 6x = 0$ by factoring, completing the square, or the quadratic formula.

9. The equation $x^2 = x$ is equivalent to $\left(x - \frac{1}{2}\right)^2 = \frac{1}{4}$.

10. The only solution to $x^2 + 6x + 9 = 0$ is -3.

9.3 EXERCISES

Solve by the quadratic formula. See Examples 1–3.

1. $x^2 + 2x - 15 = 0$

2. $x^2 - 3x - 18 = 0$

3. $x^2 + 10x + 25 = 0$

4. $x^2 - 12x + 36 = 0$

5. $2x^2 + x - 6 = 0$

6. $2x^2 + x - 15 = 0$

7. $4x^2 + 4x - 3 = 0$

8. $4x^2 + 8x + 3 = 0$

9. $2y^2 - 6y + 3 = 0$

10. $3y^2 + 6y + 2 = 0$

11. $2t^2 + 4t = -1$

12. $w^2 + 2 = 4w$

13. $2x^2 - 2x + 3 = 0$

14. $-2x^2 + 3x - 9 = 0$

15. $8x^2 = 4x$

16. $9y^2 + 3y = -6y$

17. $5w^2 - 3 = 0$

18. $4 - 7z^2 = 0$

19. $\frac{1}{2}h^2 + 7h + \frac{1}{2} = 0$

20. $\frac{1}{4}z^2 - 6z + 3 = 0$

Find the value of the discriminant, and state how many real solutions there are to each quadratic equation. See Example 4.

21. $4x^2 - 4x + 1 = 0$

22. $9x^2 + 6x + 1 = 0$

23. $6x^2 - 7x + 4 = 0$

24. $-3x^2 + 5x - 7 = 0$

25. $-5t^2 - t + 9 = 0$

26. $-2w^2 - 6w + 5 = 0$

27. $4x^2 - 12x + 9 = 0$

28. $9x^2 + 12x + 4 = 0$

29. $x^2 + x + 4 = 0$

30. $y^2 - y + 2 = 0$

31. $x - 5 = 3x^2$

32. $4 - 3x = x^2$

Use the method of your choice to solve each equation.

33. $x^2 + \frac{3}{2}x = 1$

34. $x^2 - \frac{7}{2}x = 2$

35. $(x - 1)^2 + (x - 2)^2 = 5$

36. $x^2 + (x - 3)^2 = 29$

37. $\frac{1}{x} + \frac{1}{x + 2} = \frac{5}{12}$

38. $\frac{1}{x} + \frac{1}{x + 1} = \frac{5}{6}$

39. $x^2 + 6x + 8 = 0$

40. $2x^2 - 5x - 3 = 0$

41. $x^2 - 9x = 0$

42. $x^2 - 9 = 0$

43. $(x + 5)^2 = 9$

44. $(3x - 1)^2 = 0$

45. $x(x - 3) = 2 - 3(x + 4)$

46. $(x - 1)(x + 4) = (2x - 4)^2$

47. $\frac{x}{3} = \frac{x + 2}{x}$

48. $\frac{x - 2}{x} = \frac{5}{x + 2}$

49. $2x^2 - 3x = 0$

50. $x^2 = 5$

Use a calculator to find the approximate solutions to each quadratic equation. Round answers to two decimal places.

51. $x^2 - 3x - 3 = 0$

52. $x^2 - 2x - 2 = 0$

53. $x^2 - x - 3.2 = 0$

54. $x^2 - 4.3x + 3 = 0$

55. $5.29x^2 - 3.22x + 0.49 = 0$

56. $2.6x^2 + 3.1x - 5 = 0$

Use a calculator to solve each problem.

57. ***Women in the workplace.*** The percentage of females in the 20–34 age group in the labor force was increasing in the 1970s, but peaked in the 1980s, and is now on the decline (*Fortune*, June 27, 1994). The percentage of young women in the labor force, p, in the year $1970 + x$ can be modeled by the formula $p = -0.000625x^2 + 0.025x + 0.50$. In what year was $p = 70\%$, and in what year will p again be 70% according to this model?

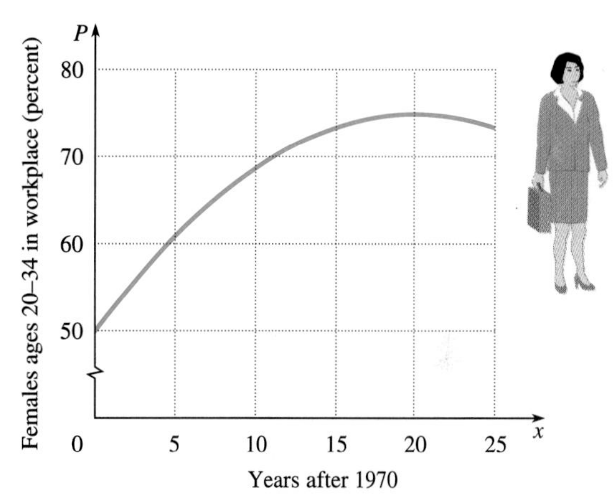

Figure for Exercise 57

58. ***Lottery tickets.*** The formula $R = -200x^2 + 5000x$ was used in Exercise 70 of Section 9.2 to predict the revenue when lottery tickets are sold for x dollars each. For what ticket price is the revenue $25,000?

9.4 Applications

In this section:

- Geometric Applications
- Work Problems
- Vertical Motion

In this section we will solve problems that involve quadratic equations.

Geometric Applications

Quadratic equations can be used to solve problems involving area.

EXAMPLE 1 **Dimensions of a rectangle**

The length of a rectangular flower bed is 2 feet longer than the width. If the area is 6 square feet, then what are the exact length and width. Also find the approximate dimensions of the rectangle to the nearest tenth of a foot.

Solution

Let x represent the width, and $x + 2$ represent the length as shown in Fig. 9.2. Write an equation using the formula for the area of a rectangle, $A = LW$:

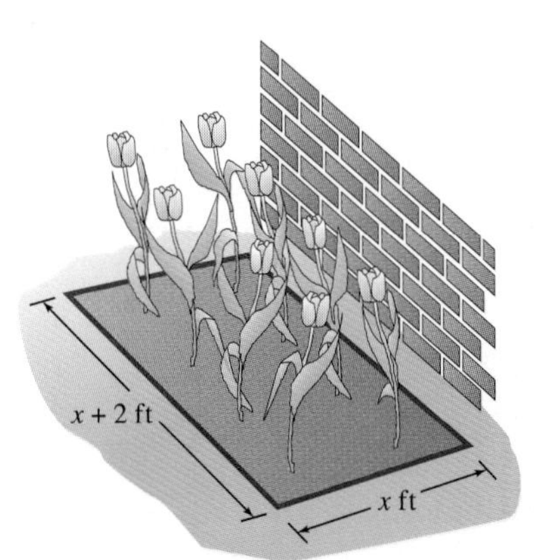

Figure 9.2

$$x(x + 2) = 6 \quad \text{The area is 6 square feet.}$$
$$x^2 + 2x - 6 = 0$$

We use the quadratic formula to solve the equation:

$$x = \frac{-2 \pm \sqrt{2^2 - 4(1)(-6)}}{2(1)} = \frac{-2 \pm \sqrt{28}}{2}$$
$$= \frac{-2 \pm 2\sqrt{7}}{2} = \frac{2(-1 \pm \sqrt{7})}{2} = -1 \pm \sqrt{7}$$

Because $-1 - \sqrt{7}$ is negative, it cannot be the width of a rectangle. If

$$x = -1 + \sqrt{7},$$

then

$$x + 2 = -1 + \sqrt{7} + 2 = 1 + \sqrt{7}.$$

So the exact width is $-1 + \sqrt{7}$ feet, and the exact length is $1 + \sqrt{7}$ feet. We can check that these dimensions give an area of 6 square feet as follows:

$$LW = (1 + \sqrt{7})(-1 + \sqrt{7}) = -1 - \sqrt{7} + \sqrt{7} + 7 = 6$$

Use a calculator to find the approximate dimensions of 1.6 and 3.6 feet.

Work Problems

The work problems in this section are similar to the work problems that you solved in Chapter 4. However, you will need the quadratic formula to solve the work problems presented in this section.

EXAMPLE 2 **Working together**

Amy can mow the lawn by herself in 2 hours less time than Bob takes to mow the lawn by himself. When they work together, it takes them only 6 hours to mow the lawn. How long would it take each of them to mow the lawn working alone? Find the exact and approximate answers.

Solution

If x is the number of hours it takes Amy by herself to mow the lawn, then Amy mows at the rate of $\frac{1}{x}$ of the lawn per hour. If $x + 2$ is the number of hours it takes Bob to mow the lawn by himself, then Bob mows at the rate of $\frac{1}{x+2}$ of the lawn per hour. Make a table using the fact that the product of the rate and the time gives the amount of work completed (or the fraction of the lawn mowed).

	Rate	Time	Amount of work
Amy	$\frac{1}{x}\frac{\text{lawn}}{\text{hr}}$	6 hr	$\frac{6}{x}$ lawn
Bob	$\frac{1}{x+2}\frac{\text{lawn}}{\text{hr}}$	6 hr	$\frac{6}{x+2}$ lawn

Because the *total* amount of work done is 1 lawn, we can write the following equation:

$$\frac{6}{x} + \frac{6}{x+2} = 1$$

$$x(x+2)\frac{6}{x} + x(x+2)\frac{6}{x+2} = x(x+2)1 \quad \text{Multiply by the LCD.}$$

$$6x + 12 + 6x = x^2 + 2x$$

$$12x + 12 = x^2 + 2x$$

$$-x^2 + 10x + 12 = 0$$

$$x^2 - 10x - 12 = 0 \quad \text{Multiply each side by } -1.$$

Use the quadratic formula with $a = 1$, $b = -10$, and $c = -12$:

$$x = \frac{10 \pm \sqrt{(-10)^2 - 4(1)(-12)}}{2(1)}$$

$$x = \frac{10 \pm \sqrt{148}}{2} = \frac{10 \pm 2\sqrt{37}}{2} = 5 \pm \sqrt{37}$$

Use a calculator to find that

$$x = 5 - \sqrt{37} \approx -1.08 \quad \text{and} \quad x = 5 + \sqrt{37} \approx 11.08.$$

Because x must be positive, Amy's time alone is $5 + \sqrt{37}$, or approximately 11.1 hours. Because Bob's time alone is 2 hours more than Amy's, Bob's time is $7 + \sqrt{37}$ or approximately 13.1 hours. ■

Vertical Motion

If an object is projected upward or downward with an initial velocity of v_0 feet per second from an altitude of s_0 feet, then its altitude s in feet after t seconds is given by the formula

$$s = -16t^2 + v_0 t + s_0.$$

We use this formula in the next example.

EXAMPLE 3 **Vertical motion**

A soccer ball bounces straight up into the air off of the head of a soccer player from an altitude of 6 feet with an initial velocity of 40 feet per second. How long does it take the ball to reach the earth? Find the exact answer and an approximate answer.

Solution

The time that it takes the ball to reach the earth is the value of t for which s has a value of 0 in the formula $s = -16t^2 + v_0 t + s_0$. To find t, we use $s = 0$, $v_0 = 40$, and $s_0 = 6$:

$$0 = -16t^2 + 40t + 6$$

$$16t^2 - 40t - 6 = 0$$

$$8t^2 - 20t - 3 = 0 \quad \text{Divide each side by 2.}$$

$$t = \frac{20 \pm \sqrt{(-20)^2 - 4(8)(-3)}}{2(8)}$$

$$= \frac{20 \pm \sqrt{496}}{16} = \frac{20 \pm 4\sqrt{31}}{16}$$

$$= \frac{5 \pm \sqrt{31}}{4}$$

Because the time must be positive, we have

$$t = \frac{5 + \sqrt{31}}{4} \approx 2.64 \text{ seconds.}$$

It takes the ball $\frac{5 + \sqrt{31}}{4}$ or 2.64 seconds to reach the earth. ■

Warm-ups

True or false? Explain your answer.

1. Two numbers that have a sum of 10 are represented by x and $x + 10$.
2. The area of a right triangle is one-half the product of the lengths of the legs.
3. If the speed of a boat in still water is x mph and the current is 5 mph, then the speed of the boat with the current is $5x$ mph.
4. If Boudreaux eats a 50-pound bag of crawfish in x hours, then his eating rate is $\frac{50}{x}$ bag/hr.
5. If the Concorde flew 1800 miles in $x + 2$ hours, then its average speed was $\frac{1800}{x + 2}$ mph.
6. The quantity $\frac{7 - \sqrt{50}}{2}$ is negative.
7. The quantity $(-5 + \sqrt{27})$ is positive.
8. If the length of one side of a square is $x + 9$ meters, then the area of the square is $x^2 + 81$ square meters.
9. If Julia mows an entire lawn in x hours, then her mowing rate is $\frac{1}{x}$ lawn/hr.
10. If John's boat goes 20 miles per hour in still water, then against a 5-mph current it will go 15 miles per hour.

9.4 EXERCISES

Find the exact solution to each problem. See Example 1.

1. ***Length and width.*** The length of a rectangle is 2 meters longer than the width. If the area is 10 square meters, then what are the length and width?

2. ***Unequal legs.*** One leg of a right triangle is 4 centimeters longer than the other leg. If the area of the triangle is 8 square centimeters, then what are the lengths of the legs?

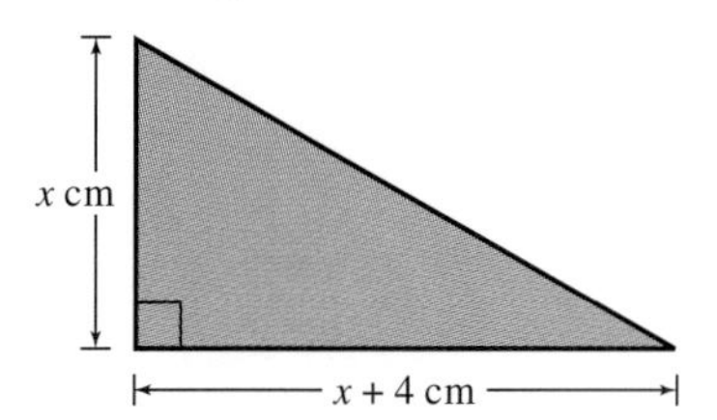

Figure for Exercise 2

3. ***Bracing a gate.*** If the diagonal brace of the square gate shown in the figure is 8 feet long, then what is the length of a side of the square gate?

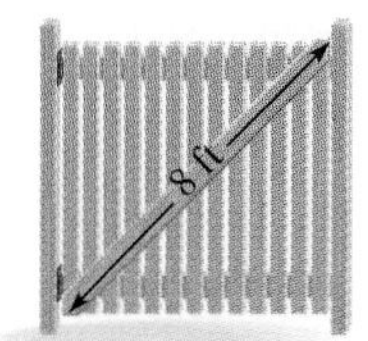

Figure for Exercise 3

4. ***Dimensions of a rectangle.*** If one side of a rectangle is 2 meters shorter than the other side and the diagonal is 10 meters long, then what are the dimensions of the rectangle?

5. ***Area of a parallelogram.*** The base of a parallelogram is 6 inches longer than its height. If the area of the parallelogram is 10 square inches, then what are the base and height?

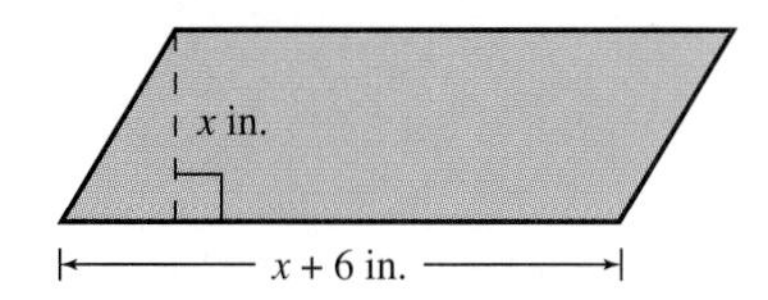

Figure for Exercise 5

6. ***Positive numbers.*** Find two positive real numbers that have a sum of 8 and a product of 4.

In Exercises 7–18, solve each problem. Give the exact answer and an approximate answer rounded to two decimal places. See Examples 2 and 3.

7. ***In the berries.*** On Monday, Alberta picked the strawberry patch, and Ernie sold the berries. On Tuesday, Ernie picked and Alberta sold, but it took him 2 hours longer to get the berries picked than it took Alberta. On Wednesday they worked together and got all of the berries picked in 2 hours. How long did it take Ernie to pick the berries by himself?

8. ***Meter readers.*** Claude and Melvin read the water meters for the city of Ponchatoula. When Claude reads all of the meters by himself, it takes him a full day longer than it takes Melvin to read all of the meters by himself. If they can get the job done working together in 2 days, then how long does it take Claude by himself?

9. ***Hanging wallpaper.*** Working alone, Tasha can hang all of the paper in the McLendons' new house in 8 hours less time than it takes Tena working alone. Working together, they completed the job in 20 hours. How long would it take Tasha working alone?

10. ***Laying bricks.*** Chau's team of bricklayers can lay all of the bricks in the McLendons' new house in 3 working days less than Hong's team. To speed things up, the McLendons hire both teams and get the job done in 10 working days. How many working days do the McLendons save by using both teams rather than just the faster team?

11. ***Hang time.*** A punter kicks a football straight up from a height of 4 feet with an initial velocity of 60 feet per second. How long will it take the ball to reach the earth?

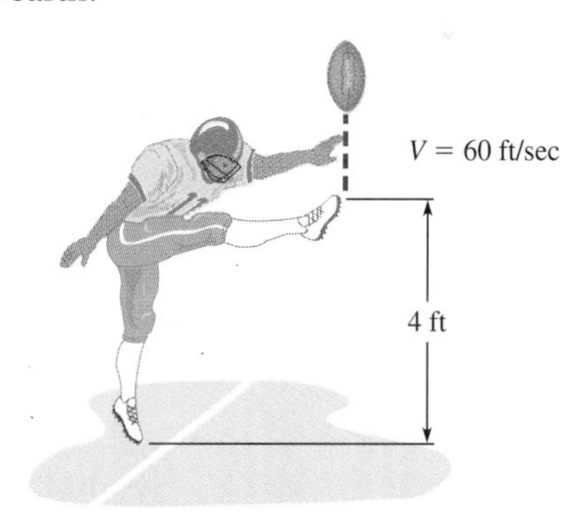

Figure for Exercise 11

12. ***Hunting accident.*** Dwight accidentally fired his rifle straight into the air while sitting in his deer stand 30 feet off the ground. If the bullet left the barrel with a velocity of 200 feet per second, then how long did it take the bullet to fall to the earth?

13. ***Going up.*** A ball is tossed into the air at 20 feet per second from a height of 5 feet. How long (to the nearest tenth of a second) will it take the ball to reach the ground?

14. ***Going down.*** A comedian throws a watermelon downward at 30 feet per second from a height of 200 feet. How long (to the nearest tenth of a second) will it take the watermelon to reach the ground? (The initial velocity of the watermelon is negative.)

15. ***Gone fishing.*** Nancy traveled 6 miles upstream to do some fly fishing. It took her 20 minutes longer to get there than to return. If the current in the river is 2 miles per hour, then how fast will her boat go in still water?

16. ***Commuting to work.*** Gladys and Bonita commute to work daily. Bonita drives 40 miles and averages 9 miles per hour more than Gladys. Gladys drives 50 miles, and she is on the road one-half hour longer than Bonita. How fast does each of them drive?

17. ***Expanding garden.*** Olin's garden is 5 feet wide and 8 feet long. He bought enough okra seed to plant 100 square feet in okra. If he wants to increase the width and the length by the same amount to plant all of his okra, then what should the increase be?

18. ***Spring flowers.*** Lillian has a 5-foot-square bed of tulips. She plans to surround this bed with a crocus bed of uniform width. If she has enough crocus bulbs to plant 100 square feet of crocus, then how wide should the crocus bed be?

9.5 Complex Numbers

In this section:

- Definition
- Operations with Complex Numbers
- Square Roots of Negative Numbers
- Complex Solutions to Quadratic Equations

In this chapter we have seen quadratic equations that have no solution in the set of real numbers. In this section you will learn that the set of real numbers is contained in the set of complex numbers. *Quadratic equations that have no real number solutions have solutions that are complex numbers.*

Definition

If we try to solve the equation $x^2 + 1 = 0$ by the square root property, we get

$$x^2 = -1 \text{ or } x = \pm\sqrt{-1}.$$

Because $\sqrt{-1}$ has no meaning in the real number system, the equation has no real solution. However, there is an extension of the real number system, called the *complex numbers,* in which $x^2 = -1$ has two solutions.

The complex numbers are formed by adding the **imaginary unit** i to the real number system. We make the definitions that

$$i = \sqrt{-1} \qquad \text{and} \qquad i^2 = -1.$$

In the complex number system, $x^2 = -1$ has two solutions: i and $-i$.

The set of complex numbers is defined as follows.

Complex Numbers

The set of complex numbers is the set of all numbers of the form

$$a + bi,$$

where a and b are real numbers, $i = \sqrt{-1}$, and $i^2 = -1$.

In the complex number $a + bi$, a is called the **real part** and b is called the **imaginary part.** If $b \neq 0$, the number $a + bi$ is called an **imaginary number.**

In dealing with complex numbers, we treat $a + bi$ as if it were a binomial with variable i. Thus we would write $2 + (-3)i$ as $2 - 3i$. We agree that $2 + i3$,

$3i + 2$, and $i3 + 2$ are just different ways of writing $2 + 3i$. Some examples of complex numbers are

$$2 + 3i, \quad -2 - 5i, \quad 0 + 4i, \quad 9 + 0i, \quad \text{and} \quad 0 + 0i.$$

For simplicity we will write only $4i$ for $0 + 4i$. The complex number $9 + 0i$ is the real number 9, and $0 + 0i$ is the real number 0. Any complex number with $b = 0$, is a real number. The diagram in Fig. 9.3 shows the relationships between the complex numbers, the real numbers, and the imaginary numbers.

Complex numbers

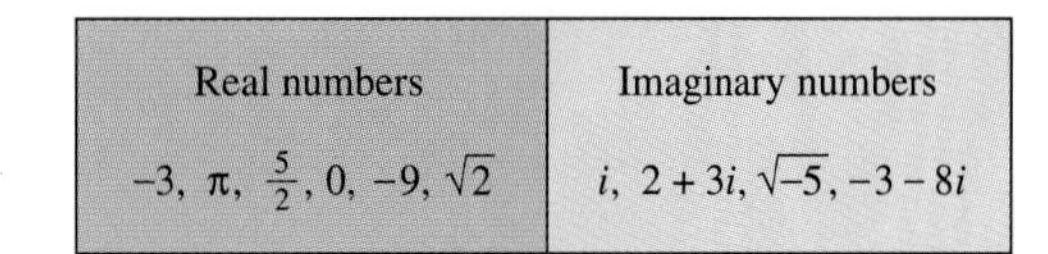

Real numbers
−3, π, 5/2, 0, −9, √2
Imaginary numbers
i, 2 + 3i, √−5, −3 − 8i

Figure 9.3

Operations with Complex Numbers

We define addition and subtraction of complex numbers as follows.

Addition and Subtraction of Complex Numbers

$$(a + bi) + (c + di) = (a + c) + (b + d)i$$
$$(a + bi) - (c + di) = (a - c) + (b - d)i$$

According to the definition, we perform these operations as if the complex numbers were binomials with i a variable.

EXAMPLE 1 **Adding and subtracting complex numbers**

Perform the indicated operations.

a) $(2 + 3i) + (4 + 5i)$
b) $(2 - 3i) + (-1 - i)$
c) $(3 + 4i) - (1 + 7i)$
d) $(2 - 3i) - (-2 - 5i)$

Solution

a) $(2 + 3i) + (4 + 5i) = 6 + 8i$

b) $(2 - 3i) + (-1 - i) = 1 - 4i$

c) $(3 + 4i) - (1 + 7i) = 3 + 4i - 1 - 7i$
$= 2 - 3i$

d) $(2 - 3i) - (-2 - 5i) = 2 - 3i + 2 + 5i$
$= 4 + 2i$

We define multiplication of complex numbers as follows.

Multiplication of Complex Numbers

$$(a + bi)(c + di) = (ac - bd) + (ad + bc)i$$

There is no need to memorize the definition of multiplication of complex numbers. We multiply complex numbers in the same way we multiply polynomials: We use the distributive property. We can also use the FOIL method and the fact that $i^2 = -1$.

EXAMPLE 2 **Multiplying complex numbers**

Perform the indicated operations.

a) $2i(1 - 3i)$ **b)** $(-2 - 5i)(6 - 7i)$ **c)** $(5i)^2$

d) $(-5i)^2$ **e)** $(3 - 2i)(3 + 2i)$

Solution

a) $2i(1 - 3i) = 2i - 6i^2$ Distributive property

$= 2i - 6(-1)$ $i^2 = -1$

$= 6 + 2i$

b) Use FOIL to multiply these complex numbers:

$(-2 - 5i)(6 - 7i) = -12 + 14i - 30i + 35i^2$

$= -12 - 16i + 35(-1)$ $i^2 = -1$

$= -12 - 16i - 35$

$= -47 - 16i$

c) $(5i)^2 = 25i^2 = 25(-1) = -25$

d) $(-5i)^2 = (-5)^2 i^2 = 25(-1) = -25$

e) $(3 - 2i)(3 + 2i) = 9 - 4i^2$

$= 9 - 4(-1)$

$= 9 + 4$

$= 13$

Notice that the product of the imaginary numbers $3 - 2i$ and $3 + 2i$ in Example 2(e) is a real number. We call $3 - 2i$ and $3 + 2i$ *complex conjugates* of each other.

Complex Conjugates

The complex numbers $a + bi$ and $a - bi$ are called **complex conjugates** of each other. Their product is the real number $a^2 + b^2$.

EXAMPLE 3 **Complex conjugates**

Find the product of the given complex number and its conjugate.

a) $4 - 3i$ **b)** $-2 + 5i$ **c)** $-i$

Solution

a) The complex conjugate of $4 - 3i$ is $4 + 3i$:

$$(4 - 3i)(4 + 3i) = 16 - 9i^2 = 16 + 9 = 25$$

b) The conjugate of $-2 + 5i$ is $-2 - 5i$:

$$(-2 + 5i)(-2 - 5i) = 4 - 25i^2 = 4 + 25 = 29$$

c) The conjugate of $-i$ is i:

$$(-i)(i) = -i^2 = -(-1) = 1$$

To divide a complex number by a real number, we divide each part by the real number. For example,

$$\frac{4 - 6i}{2} = 2 - 3i.$$

We use the idea of complex conjugates to divide by a complex number. The process is similar to rationalizing the denominator.

Dividing by a Complex Number

To divide by a complex number, multiply the numerator and denominator of the quotient by the complex conjugate of the denominator.

EXAMPLE 4 **Dividing complex numbers**

Perform the indicated operations.

a) $\dfrac{2}{3 - 4i}$ **b)** $\dfrac{6}{2 + i}$ **c)** $\dfrac{3 - 2i}{i}$

Solution

a) Multiply the numerator and denominator by $3 + 4i$, the conjugate of $3 - 4i$:

$$\begin{aligned}\frac{2}{3 - 4i} &= \frac{2(3 + 4i)}{(3 - 4i)(3 + 4i)}\\ &= \frac{6 + 8i}{9 - 16i^2}\\ &= \frac{6 + 8i}{25} \qquad 9 - 16i^2 = 9 - 16(-1) = 25\\ &= \frac{6}{25} + \frac{8}{25}i\end{aligned}$$

b) Multiply the numerator and denominator by $2 - i$, the conjugate of $2 + i$:

$$\begin{aligned}\frac{6}{2 + i} &= \frac{6(2 - i)}{(2 + i)(2 - i)}\\ &= \frac{12 - 6i}{4 - i^2}\\ &= \frac{12 - 6i}{5} \qquad 4 - i^2 = 4 - (-1) = 5\\ &= \frac{12}{5} - \frac{6}{5}i\end{aligned}$$

c) Multiply the numerator and denominator by $-i$, the conjugate of i:

$$\frac{3 - 2i}{i} = \frac{(3 - 2i)(-i)}{i(-i)} = \frac{-3i + 2i^2}{-i^2} = \frac{-3i - 2}{1} = -2 - 3i$$

Square Roots of Negative Numbers

In Example 2 we saw that both

$$(5i)^2 = -25 \qquad \text{and} \qquad (-5i)^2 = -25.$$

Because the square of each of these complex numbers is -25, both $5i$ and $-5i$ are square roots of -25. When we use the radical notation, we write

$$\sqrt{-25} = 5i.$$

The square root of a negative number is not a real number, it is a complex number.

Square Root of a Negative Number

For any positive number b,

$$\sqrt{-b} = i\sqrt{b}.$$

For example, $\sqrt{-9} = i\sqrt{9} = 3i$ and $\sqrt{-7} = i\sqrt{7}$. Note that the expression $\sqrt{7}\,i$ could easily be mistaken for the expression $\sqrt{7i}$, where i is under the radical. For this reason, when the coefficient of i contains a radical, we write i preceding the radical.

EXAMPLE 5 **Square roots of negative numbers**

Write each expression in the form $a + bi$, where a and b are real numbers.

a) $2 + \sqrt{-4}$ **b)** $\dfrac{2 + \sqrt{-12}}{2}$ **c)** $\dfrac{-2 - \sqrt{-18}}{3}$

Solution

a) $2 + \sqrt{-4} = 2 + i\sqrt{4}$

$= 2 + 2i$

b) $\dfrac{2 + \sqrt{-12}}{2} = \dfrac{2 + i\sqrt{12}}{2}$

$= \dfrac{2 + 2i\sqrt{3}}{2}$ $\quad \sqrt{12} = \sqrt{4} \cdot \sqrt{3} = 2\sqrt{3}$

$= 1 + i\sqrt{3}$

c) $\dfrac{-2 - \sqrt{-18}}{3} = \dfrac{-2 - i\sqrt{18}}{3}$

$= \dfrac{-2 - 3i\sqrt{2}}{3}$ $\quad \sqrt{18} = \sqrt{9}\sqrt{2} = 3\sqrt{2}$

$= -\dfrac{2}{3} - i\sqrt{2}$ ■

Complex Solutions to Quadratic Equations

The equation $x^2 = -4$ has no real number solutions, but it has two complex solutions, which can be found as follows:

$$x^2 = -4$$
$$x = \pm\sqrt{-4} = \pm i\sqrt{4} = \pm 2i$$

Check:

$$(2i)^2 = 4i^2 = 4(-1) = -4$$
$$(-2i)^2 = 4i^2 = -4$$

Both $2i$ and $-2i$ are solutions to the equation.

Consider the general quadratic equation

$$ax^2 + bx + c = 0,$$

where a, b, and c are real numbers. If the discriminant $b^2 - 4ac$ is positive, then the quadratic equation has two real solutions. If the discriminant is 0, then the equation has one real solution. If the discriminant is negative, then the equation has two complex solutions. *In the complex number system, all quadratic equations have solutions.* We can use the quadratic formula to find them.

EXAMPLE 6 **Quadratics with imaginary solutions**

Find the complex solutions to the quadratic equations.

a) $x^2 - 2x + 5 = 0$ **b)** $2x^2 + 3x + 5 = 0$

Solution

a) To solve $x^2 - 2x + 5 = 0$, use $a = 1$, $b = -2$, and $c = 5$ in the quadratic formula:

$$x = \frac{2 \pm \sqrt{(-2)^2 - 4(1)(5)}}{2(1)}$$
$$= \frac{2 \pm \sqrt{-16}}{2} = \frac{2 \pm 4i}{2} = 1 \pm 2i$$

We can use the operations with complex numbers to check these solutions:

$$(1 + 2i)^2 - 2(1 + 2i) + 5 = 1 + 4i + 4i^2 - 2 - 4i + 5$$
$$= 1 + 4i - 4 - 2 - 4i + 5 = 0$$

You should verify that $1 - 2i$ also satisfies the equation. The solutions are $1 - 2i$ and $1 + 2i$.

b) To solve $2x^2 + 3x + 5 = 0$, use $a = 2$, $b = 3$, and $c = 5$ in the quadratic formula:

$$x = \frac{-3 \pm \sqrt{3^2 - 4(2)(5)}}{2(2)}$$
$$= \frac{-3 \pm \sqrt{-31}}{4} = \frac{-3 \pm i\sqrt{31}}{4}$$

Check these answers. The solutions are $\frac{-3 + i\sqrt{31}}{4}$ and $\frac{-3 - i\sqrt{31}}{4}$. ■

The following box summarizes the basic facts about complex numbers.

Complex Numbers

1. $i = \sqrt{-1}$ and $i^2 = -1$.
2. A complex number has the form $a + bi$, where a and b are real numbers.
3. The complex number $a + 0i$ is the real number a.
4. If b is a positive real number, then $\sqrt{-b} = i\sqrt{b}$.
5. The complex conjugate of $a + bi$ is $a - bi$.
6. Add, subtract, and multiply complex numbers as if they were binomials with variable i.
7. Divide complex numbers by multiplying the numerator and denominator by the conjugate of the denominator.
8. In the complex number system, all quadratic equations have solutions.

Warm-ups

True or false? Explain your answer.

1. $(3 + i) + (2 - 4i) = 5 - 3i$

2. $(4 - 2i)(3 - 5i) = 2 - 26i$

3. $(4 - i)(4 + i) = 17$

4. $i^4 = 1$

5. $\sqrt{-5} = 5i$

6. $\sqrt{-36} = \pm 6i$

7. The complex conjugate of $-2 + 3i$ is $2 - 3i$.

8. Zero is the only real number that is also a complex number.

9. Both $2i$ and $-2i$ are solutions to the equation $x^2 = 4$.

10. Every quadratic equation has at least one complex solution.

9.5 EXERCISES

Perform the indicated operations. See Example 1.

1. $(3 + 5i) + (2 + 4i)$

2. $(8 + 3i) + (1 + 2i)$

3. $(-1 + i) + (2 - i)$

4. $(-2 - i) + (-3 + 5i)$

5. $(4 - 5i) - (2 + 3i)$

6. $(3 - 2i) - (7 + 6i)$

7. $(-3 - 5i) - (-2 - i)$

8. $(-4 - 8i) - (-2 - 3i)$

9. $(8 - 3i) - (9 - 3i)$

10. $(5 + 6i) - (-3 + 6i)$

11. $\left(\frac{1}{2} + i\right) + \left(\frac{1}{4} - \frac{1}{2}i\right)$

12. $\left(\frac{2}{3} - i\right) - \left(\frac{1}{4} - \frac{1}{2}i\right)$

Perform the indicated operations. See Example 2.

13. $3(2 - 3i)$

14. $-4(3 - 2i)$

15. $(6i)^2$

16. $(3i)^2$

17. $(-6i)^2$

18. $(-3i)^2$

19. $(2 + 3i)(3 - 5i)$

20. $(4 - i)(3 - 6i)$

21. $(5 - 2i)^2$

22. $(3 + 4i)^2$

23. $(4 - 3i)(4 + 3i)$

24. $(-3 + 5i)(-3 - 5i)$

25. $(1 - i)(1 + i)$

26. $(3 - i)(3 + i)$

Find the product of the given complex number and its conjugate. See Example 3.

27. $2 + 5i$

28. $3 + 4i$

29. $4 - 6i$

30. $2 - 7i$

31. $-3 + 2i$

32. $-4 - i$

33. i

34. $-2i$

Perform the indicated operations. See Example 4.

35. $(2 - 6i) \div 2$

36. $(-3 + 6i) \div (-3)$

37. $\dfrac{-2 + 8i}{2}$

38. $\dfrac{6 - 9i}{-3}$

39. $\dfrac{4 + 6i}{-2i}$

40. $\dfrac{3 - 8i}{i}$

41. $\dfrac{4i}{3 + 2i}$

42. $\dfrac{5}{4 - 5i}$

43. $\dfrac{2 + i}{2 - i}$

44. $\dfrac{i - 5}{5 - i}$

45. $\dfrac{4 - 12i}{3 + i}$

46. $\dfrac{-4 + 10i}{5 - i}$

Write each expression in the form $a + bi$, where a and b are real numbers. See Example 5.

47. $5 + \sqrt{-9}$

48. $6 + \sqrt{-16}$

49. $-3 - \sqrt{-7}$

50. $2 - \sqrt{-3}$

51. $\dfrac{-2 + \sqrt{-12}}{2}$

52. $\dfrac{-6 + \sqrt{-18}}{3}$

53. $\dfrac{-8 - \sqrt{-20}}{-4}$

54. $\dfrac{6 + \sqrt{-24}}{-2}$

55. $\dfrac{-4 + \sqrt{-28}}{6}$

56. $\dfrac{6 - \sqrt{-45}}{6}$

57. $\dfrac{-2 + \sqrt{-100}}{-10}$

58. $\dfrac{-3 + \sqrt{-81}}{-9}$

Find the complex solutions to each quadratic equation. See Example 6.

59. $x^2 + 81 = 0$

60. $x^2 + 100 = 0$

61. $x^2 + 5 = 0$

62. $x^2 + 6 = 0$

63. $3y^2 + 2 = 0$

64. $5y^2 + 3 = 0$

65. $x^2 - 4x + 5 = 0$

66. $x^2 - 6x + 10 = 0$

67. $y^2 + 13 = 6y$

68. $y^2 + 29 = 4y$

69. $x^2 - 4x + 7 = 0$

70. $x^2 - 10x + 27 = 0$

71. $9y^2 - 12y + 5 = 0$

72. $2y^2 - 2y + 1 = 0$

73. $x^2 - x + 1 = 0$

74. $4x^2 - 20x + 27 = 0$

75. $-4x^2 + 8x - 9 = 0$

76. $-9x^2 + 12x - 10 = 0$

Solve each problem.

77. Evaluate $(2 - 3i)^2 + 4(2 - 3i) - 9$.

78. Evaluate $(3 + 5i)^2 - 2(3 + 5i) + 5$.

79. What is the value of $x^2 - 8x + 17$ if $x = 4 - i$?

80. What is the value of $x^2 - 6x + 34$ if $x = 3 + 5i$?

81. Find the product $[x - (6 - i)][x - (6 + i)]$.

82. Find the product $[x - (3 + 7i)][x - (3 - 7i)]$.

83. Write a quadratic equation that has $3i$ and $-3i$ as its solutions.

84. Write a quadratic equation that has $2 + 5i$ and $2 - 5i$ as its solutions.

Getting More Involved

85. ***Discussion.*** Determine whether each given number is in each of the following sets: the natural numbers, the integers, the rational numbers, the irrational numbers, the real numbers, the imaginary numbers, and the complex numbers.

a) 54 **b)** $-\dfrac{3}{8}$ **c)** $3\sqrt{5}$

d) $6i$ **e)** $\pi + i\sqrt{5}$

86. ***Discussion.*** Which of the following equations have real solutions? Imaginary solutions? Complex solutions?

a) $3x^2 - 2x + 9 = 0$ **b)** $5x^2 - 2x - 10 = 0$

c) $\dfrac{1}{2}x^2 - x + 3 = 0$ **d)** $7w^2 + 12 = 0$

9.6 Quadratic Functions and Their Graphs

In this section:

- Definition
- Graphing Quadratic Functions
- The Vertex and Intercepts
- Applications

We have seen *quadratic functions* on several occasions in this text, but we have not yet defined the term. In this section we study quadratic functions and their graphs.

Definition

In Section 9.1 you used the formula $s = -16t^2 + v_0t + s_0$ to find the height s (in feet) of an object that is projected upward with initial velocity v_0 ft/sec from an initial height of s_0 feet. This equation expresses s as a quadratic function of t. In general, if y is determined from x by a formula involving a quadratic polynomial, then we say that y is a *quadratic function of x.*

Quadratic Function

A **quadratic function** is a function of the form

$$y = ax^2 + bx + c,$$

where a, b, and c are real numbers and $a \neq 0$.

Without the term ax^2, this function would be a linear function. That is why we specify that $a \neq 0$.

EXAMPLE 1 **Finding ordered pairs of a quadratic function**

Complete each ordered pair so that it satisfies the given equation.

a) $y = x^2 - x - 6$; $(2,\quad)$, $(\quad, 0)$

b) $s = -16t^2 + 48t + 84$; $(0,\quad)$, $(\quad, 20)$

Solution

a) If $x = 2$, then $y = 2^2 - 2 - 6 = -4$. So the ordered pair is $(2, -4)$. To find x when $y = 0$, replace y by 0 and solve the resulting quadratic equation:

$$x^2 - x - 6 = 0$$
$$(x - 3)(x + 2) = 0$$
$$x - 3 = 0 \quad \text{or} \quad x + 2 = 0$$
$$x = 3 \quad \text{or} \quad x = -2$$

The ordered pairs are $(-2, 0)$ and $(3, 0)$.

b) If $t = 0$, then $s = -16 \cdot 0^2 + 48 \cdot 0 + 84 = 84$. The ordered pair is $(0, 84)$. To find t when $s = 20$, replace s by 20 and solve the equation for t:

$$-16t^2 + 48t + 84 = 20$$
$$-16t^2 + 48t + 64 = 0 \quad \text{Subtract 20 from each side.}$$
$$t^2 - 3t - 4 = 0 \quad \text{Divide each side by } -16.$$
$$(t - 4)(t + 1) = 0 \quad \text{Factor.}$$
$$t - 4 = 0 \quad \text{or} \quad t + 1 = 0 \quad \text{Zero factor property}$$
$$t = 4 \quad \text{or} \quad t = -1$$

The ordered pairs are $(-1, 20)$ and $(4, 20)$. ■

CAUTION When variables other than x and y are used, the independent variable is the first coordinate of an ordered pair, and the dependent variable is the second coordinate. In Example 1(b), t is the independent variable and first coordinate because s depends on t by the formula $s = -16t^2 + 48t + 84$.

Graphing Quadratic Functions

Any real number may be used for x in the formula $y = ax^2 + bx + c$. So the domain (the set of x-coordinates) for any quadratic function is the set of all real numbers, R. The range (the set of y-coordinates) can be determined from the graph. All quadratic functions have graphs that are similar in shape. The graph of any quadratic function is called a **parabola.**

EXAMPLE 2 **Graphing the simplest quadratic function**

Graph the function $y = x^2$, and state the domain and range.

Solution

Make a table of values for x and y:

x	-2	-1	0	1	2
$y = x^2$	4	1	0	1	4

See Fig. 9.4 for the graph. The domain is the set of all real numbers, R, because we can use any real number for x. From the graph we see that the smallest y-coordinate of the function is 0. So the range is the set of real numbers that are greater than or equal to 0, $\{y \mid y \geq 0\}$.

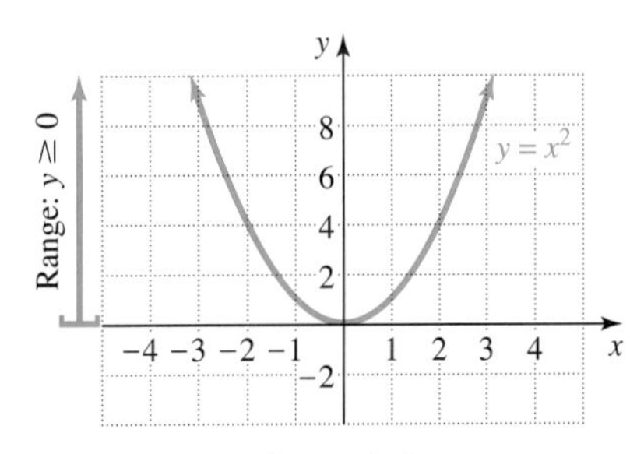

Figure 9.4

The parabola in Fig. 9.4 is said to **open upward.** In the next example we see a parabola that **opens downward.** If $a > 0$ in the equation $y = ax^2 + bx + c$, then the parabola opens upward. If $a < 0$, then the parabola opens downward.

EXAMPLE 3 **A quadratic function**

Graph the function $y = 4 - x^2$, and state the domain and range.

Solution

We plot enough points to get the correct shape of the graph:

x	-2	-1	0	1	2
$y = 4 - x^2$	0	3	4	3	0

See Fig. 9.5 for the graph. The domain is the set of all real numbers, R. From the graph we see that the largest y-coordinate is 4. So the range is $\{y \mid y \leq 4\}$.

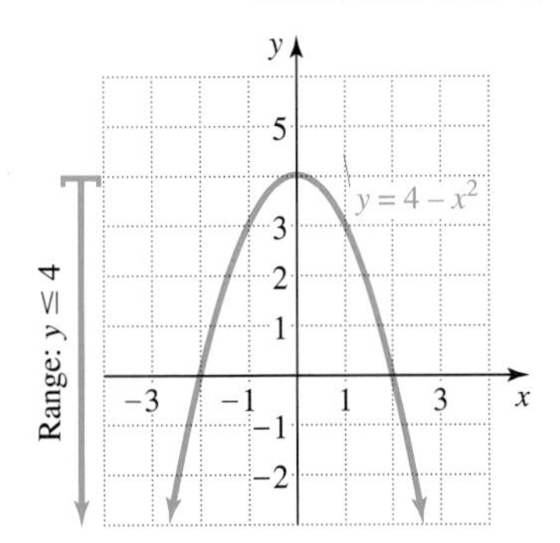

Figure 9.5

The Vertex and Intercepts

The lowest point on a parabola that opens upward or the highest point on a parabola that opens downward is called the **vertex.** The y-coordinate of the vertex is the **minimum value** of the function if the parabola opens upward, and it is the **maximum value** of the function if the parabola opens downward. For $y = x^2$ the vertex is $(0, 0)$, and 0 is the minimum value of the function. For $g(x) = 4 - x^2$ the vertex is $(0, 4)$, and 4 is the maximum value of the function.

Because the vertex is either the highest or lowest point on a parabola, it is an important point to find before drawing the graph. The vertex can be found by using the following fact.

Vertex of a Parabola

The x-coordinate of the vertex of $y = ax^2 + bx + c$ is $\frac{-b}{2a}$, provided that $a \neq 0$.

You can remember $\frac{-b}{2a}$ by observing that it is part of the quadratic formula

$$x = \frac{-b \pm \sqrt{b^2 - 4ac}}{2a}.$$

When you graph a parabola, you should always locate the vertex because it is the point at which the graph "turns around." With the vertex and several nearby points you can see the correct shape of the parabola.

EXAMPLE 4 **Using the vertex in graphing a quadratic function**

Graph $y = -x^2 - x + 2$, and state the domain and range.

Solution

First find the x-coordinate of the vertex:

$$x = \frac{-b}{2a} = \frac{-(-1)}{2(-1)} = \frac{1}{-2} = -\frac{1}{2}$$

Now find y for $x = -\frac{1}{2}$:

$$y = -\left(-\frac{1}{2}\right)^2 - \left(-\frac{1}{2}\right) + 2 = -\frac{1}{4} + \frac{1}{2} + 2 = \frac{9}{4}$$

The vertex is $\left(-\frac{1}{2}, \frac{9}{4}\right)$. Now find a few points on either side of the vertex:

x	-2	-1	$-\frac{1}{2}$	0	1
$y = -x^2 - x + 2$	0	2	$\frac{9}{4}$	2	0

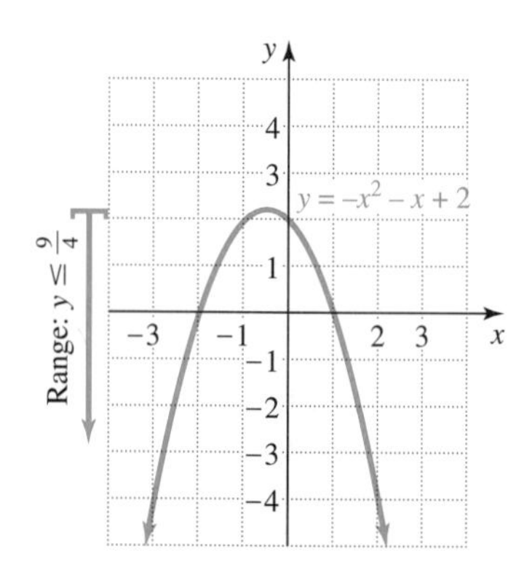

Figure 9.6

Sketch a parabola through these points as in Fig. 9.6. The domain is R. Because the graph goes no higher than $\frac{9}{4}$, the range is $\left\{y \mid y \leq \frac{9}{4}\right\}$. ■

The y-intercept of a parabola is the point that has 0 as the first coordinate. The x-intercepts are the points that have 0 as their second coordinates.

EXAMPLE 5 **Using the intercepts in graphing a quadratic function**

Find the vertex and intercepts, and sketch the graph of each function.

a) $y = x^2 - 2x - 8$ **b)** $s = -16t^2 + 64t$

Solution

a) Use $x = \frac{-b}{2a}$ to get $x = 1$ as the x-coordinate of the vertex. If $x = 1$, then

$$\begin{aligned} y &= 1^2 - 2 \cdot 1 - 8 \\ &= -9. \end{aligned}$$

So the vertex is $(1, -9)$. If $x = 0$, then

$$\begin{aligned} y &= 0^2 - 2 \cdot 0 - 8 \\ &= -8. \end{aligned}$$

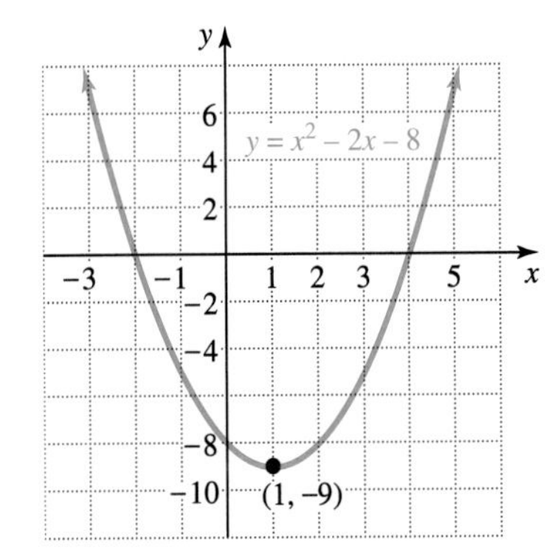

Figure 9.7

The y-intercept is $(0, -8)$. To find the x-intercepts, replace y by 0:

$$\begin{aligned} x^2 - 2x - 8 &= 0 \\ (x - 4)(x + 2) &= 0 \end{aligned}$$

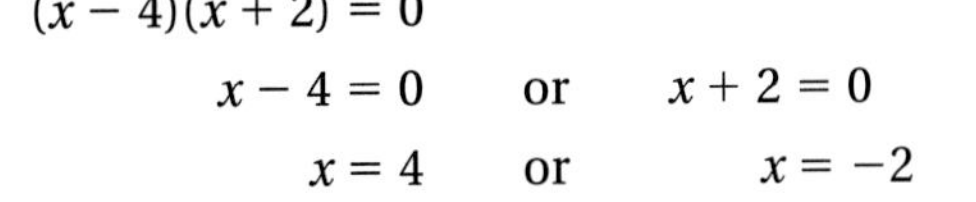

$$x - 4 = 0 \quad \text{or} \quad x + 2 = 0$$
$$x = 4 \quad \text{or} \quad x = -2$$

The x-intercepts are $(-2, 0)$ and $(4, 0)$. The graph is shown in Fig. 9.7.

b) Because s is expressed as a function of t, the first coordinate is t. Use $t = \frac{-b}{2a}$ to get

$$t = \frac{-64}{2(-16)} = 2.$$

If $t = 2$, then

$$\begin{aligned} s &= -16 \cdot 2^2 + 64 \cdot 2 \\ &= 64. \end{aligned}$$

So the vertex is $(2, 64)$. If $t = 0$, then

$$\begin{aligned} s &= -16 \cdot 0^2 + 64 \cdot 0 \\ &= 0. \end{aligned}$$

So the s-intercept is $(0, 0)$. To find the t-intercepts, replace s by 0:

$$\begin{aligned} -16t^2 + 64t &= 0 \\ -16t(t - 4) &= 0 \end{aligned}$$

$$-16t = 0 \quad \text{or} \quad t - 4 = 0$$
$$t = 0 \quad \text{or} \quad t = 4$$

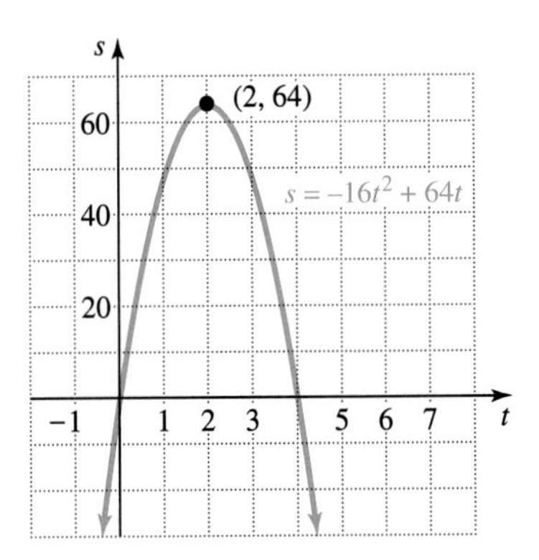

Figure 9.8

The t-intercepts are $(0, 0)$ and $(4, 0)$. The graph is shown in Fig. 9.8. ■

Applications

In applications we are often interested in finding the maximum or minimum value of a variable. If the graph of a quadratic function opens downward, then the maximum value of the second coordinate is the second coordinate of the vertex. If the parabola opens upward, then the minimum value of the second coordinate is the second coordinate of the vertex.

EXAMPLE 6 **Finding the maximum height**

A ball is tossed upward with a velocity of 64 feet per second from a height of 5 feet. What is the maximum height reached by the ball?

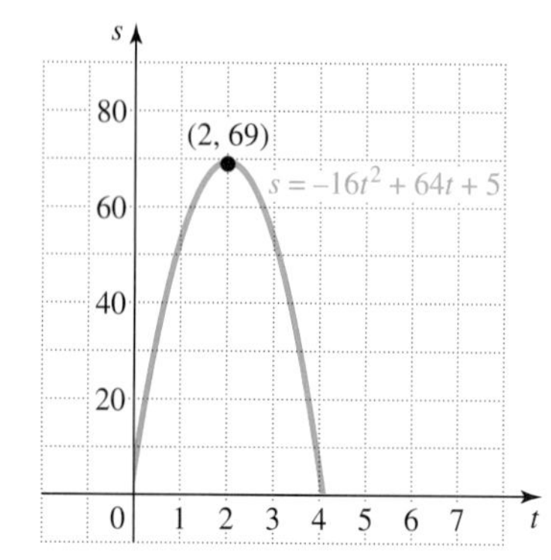

Figure 9.9

Solution

The height s of the ball for any time t is given by $s = -16t^2 + 64t + 5$. Because the maximum height occurs at the vertex of the parabola, we use $t = \frac{-b}{2a}$ to find the vertex:

$$t = \frac{-64}{2(-16)} = 2$$

Now use $t = 2$ to find the second coordinate of the vertex:

$$s = -16t^2 + 64t + 5 = -16(2)^2 + 64(2) + 5 = 69$$

The maximum height reached by the ball is 69 feet. See Fig. 9.9.

Warm-ups

True or false? Explain your answer.

1. The ordered pair $(-2, -1)$ satisfies $y = x^2 - 5$.
2. The y-intercept for $y = x^2 - 3x + 9$ is $(9, 0)$.
3. The x-intercepts for $y = x^2 - 5$ are $(\sqrt{5}, 0)$ and $(-\sqrt{5}, 0)$.
4. The graph of $y = x^2 - 12$ opens upward.
5. The graph of $y = 4 + x^2$ opens downward.
6. The vertex of $y = x^2 + 2x$ is $(-1, -1)$.
7. The parabola $y = x^2 + 1$ has no x-intercepts.
8. The y-intercept for $y = ax^2 + bx + c$ is $(0, c)$.
9. If $w = -2v^2 + 9$, then the maximum value of w is 9.
10. If $y = 3x^2 - 7x + 9$, then the maximum value of y occurs when $x = \frac{7}{6}$.

9.6 EXERCISES

Complete each ordered pair so that it satisfies the given equation. See Example 1.

1. $y = x^2 - x - 12$ $(3, \quad), (\quad, 0)$
2. $y = -\frac{1}{2}x^2 - x + 1$ $(0, \quad), (\quad, -3)$
3. $s = -16t^2 + 32t$ $(4, \quad), (\quad, 0)$
4. $a = b^2 + 4b + 5$ $(-2, \quad), (\quad, 2)$

Graph each quadratic function, and state its domain and range. See Examples 2 and 3.

5. $y = x^2 + 2$

6. $y = x^2 - 4$

7. $y = \frac{1}{2}x^2 - 4$

8. $y = \frac{1}{3}x^2 - 6$

9. $y = -2x^2 + 5$

10. $y = -x^2 - 1$

11. $y = -\frac{1}{3}x^2 + 5$

12. $y = -\frac{1}{2}x^2 + 3$

13. $y = (x - 2)^2$

14. $y = (x + 3)^2$

Find the vertex and intercepts for each quadratic function. Sketch the graph, and state the domain and range. See Examples 4 and 5.

15. $y = x^2 - x - 2$

16. $y = x^2 + 2x - 3$

17. $y = x^2 + 2x - 8$

18. $y = x^2 + x - 6$

19. $y = -x^2 - 4x - 3$

20. $y = -x^2 - 5x - 4$

21. $y = -x^2 + 3x + 4$

22. $y = -x^2 - 2x + 8$

23. $a = b^2 - 6b - 16$

24. $v = -u^2 - 8u + 9$

Find the maximum or minimum value of y for each function.

25. $y = x^2 - 8$

26. $y = 33 - x^2$

27. $y = -3x^2 + 14$

28. $y = 6 + 5x^2$

29. $y = x^2 + 2x + 3$

30. $y = x^2 - 2x + 5$

31. $y = -2x^2 - 4x$

32. $y = -3x^2 + 24x$

Solve each problem. See Example 6.

33. ***Maximum height.*** If a baseball is projected upward from ground level with an initial velocity of 64 feet per second, then its height is a function of time, given by $s = -16t^2 + 64t$. Graph this function for $0 \le t \le 4$. What is the maximum height reached by the ball?

34. ***Maximum height.*** If a soccer ball is kicked straight up with an initial velocity of 32 feet per second, then its height above the earth is a function of time given by $s = -16t^2 + 32t$. Graph this function for $0 \le t \le 2$. What is the maximum height reached by this ball?

35. ***Maximum area.*** Jason plans to fence a rectangular area with 100 meters of fencing. He has written the formula $A = w(50 - w)$ to express the area in terms of the width w. What is the maximum possible area that he can enclose with his fencing?

36. ***Minimizing cost.*** A company uses the function $C = 0.02x^2 - 3.4x + 150$ to model the unit cost in dollars for producing x stabilizer bars. For what number of bars is the unit cost at its minimum? What is the unit cost at that level of production?

37. ***Air pollution.*** The amount of nitrogen dioxide A in parts per million (ppm) that was present in the air in the city of Homer on a certain day in June is modeled by the function

$$A = -2t^2 + 32t + 12,$$

where t is the number of hours after 6:00 A.M. Use this function to find the time at which the nitrogen dioxide level was at its maximum.

38. ***Replacing secretaries.*** The decline in the number of secretarial jobs has been attributed to the dramatic increases in computing power in recent years (*Fortune*, June 27, 1994). The number of U.S. secretarial jobs s in the year $1983 + x$ can be modeled by the function $s = -0.02x^2 + 0.16x + 3.95$, where s is in millions.

a) Use the graph to estimate the year in which the number of secretarial jobs was at its maximum.

b) Use the function to find the year in which the number of secretarial jobs was at its maximum.

c) How many secretarial jobs were there in the year of part (b)?

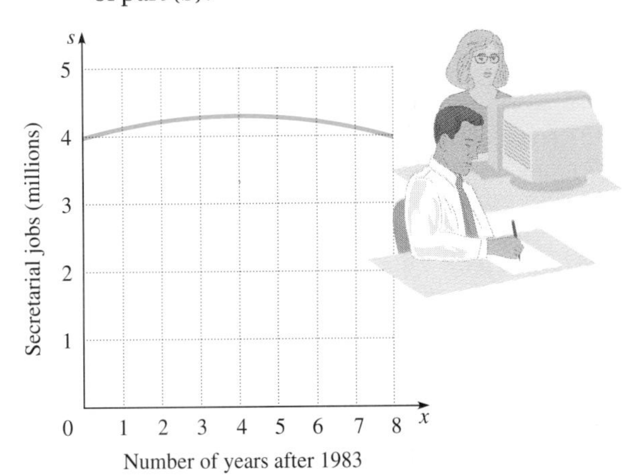

Figure for Exercise 38

39. ***Suspension bridge.*** The cable of the suspension bridge shown in the accompanying figure hangs in the shape of a parabola with equation $y = 0.0375x^2$, where x and y are in meters. What is the height of each tower above the roadway? What is the length z for the cable bracing the tower?

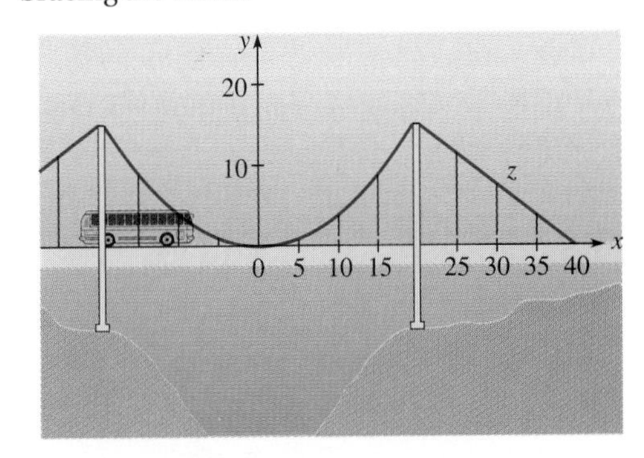

Figure for Exercise 39

Getting More Involved

40. ***Exploration.***

a) Write the function $y = 3(x - 2)^2 + 6$ in the form $y = ax^2 + bx + c$, and find the vertex of the parabola using the formula $x = \frac{-b}{2a}$.

b) Repeat part (a) with the functions $y = -4(x - 5)^2 - 9$ and $y = 3(x + 2)^2 - 6$.

c) What is the vertex for a parabola that is written in the form $y = a(x - h)^2 + k$? Explain your answer.

Graphing Calculator Exercises

41. Graph $y = x^2$, $y = \frac{1}{2}x^2$, and $y = 2x^2$ on the same coordinate system. What can you say about the graph of $y = kx^2$?

42. Graph $y = x^2$, $y = (x - 3)^2$, and $y = (x + 3)^2$ on the same coordinate system. How does the graph of $y = (x - k)^2$ compare to the graph of $y = x^2$?

43. The equation $x = y^2$ is equivalent to $y = \pm\sqrt{x}$. Graph both $y = \sqrt{x}$ and $y = -\sqrt{x}$ on a graphing calculator. How does the graph of $x = y^2$ compare to the graph of $y = x^2$?

44. Graph each of the following equations by solving for y.

a) $x = y^2 - 1$ **b)** $x = -y^2$

c) $x^2 + y^2 = 4$

COLLABORATIVE ACTIVITIES

Designing a River Park

Grouping: 4 students
Topic: Solving quadratic equations

At the last election, voters approved the creation of a park along the river on the edge of the city. You are student representatives on the citizen's committee working with the park designer. The city is allocating 300 acres for the park. The park designer says that the committee needs to decide the dimensions of the park. In terms of the land available, the park dimensions can vary between a length that is twice the width to a length that is 10 times the width. The committee decides to consider the following three cases:

A. The length is twice the width.

B. The length is 10 times the width.

C. The length is 5 times the width.

Because you have all studied algebra, you offer to find the dimensions and draw sketches of the three cases.

The following suggestions will help you in solving this problem.

1. Determine what units of measurement you wish to use (feet, yards, miles, meters or kilometers).
2. A calculator with a conversion feature will make your work easier.
3. Assign roles. You might use the roles Moderator, Recorder, Calculator user, and Sketcher.

In your report to the committee, do the following:

1. For each case, include the equations used, the width and length you found (rounded to the nearest whole number), and your work showing how you found the length and width.
2. Include a small sketch of each case showing the given proportions.
3. Choose one of the three cases that your group feels gives the best dimensions for the park. Considering what features (flowerbeds, playground, soccer field, walking or bike paths, etc.) you would like to have in the park will help you decide which dimensions you would choose. Make a larger-scale drawing of this case and include the features you would want in a park.
4. Write a short paragraph explaining why you chose the case you did.

Wrap-up CHAPTER 9

SUMMARY

Quadratic Equations		Examples
Quadratic equation	An equation of the form $ax^2 + bx + c = 0,$ where a, b, and c are real numbers with $a \neq 0$	$x^2 = 10$ $(x + 3)^2 = 8$ $x^2 + 5x - 7 = 0$
Methods for solving quadratic equations	Factoring	$x^2 + 5x + 6 = 0$ $(x + 3)(x + 2) = 0$
	Square root property	$(x - 3)^2 = 6$ $x - 3 = \pm\sqrt{6}$
	Completing the square (works on any quadratic): Take one-half of middle term, square it, then add it to each side.	$x^2 + 6x = 7$ $x^2 + 6x + 9 = 7 + 9$ $(x + 3)^2 = 16$
	Quadratic formula (works on any quadratic): $x = \dfrac{-b \pm \sqrt{b^2 - 4ac}}{2a}$	$2x^2 - 3x - 6 = 0$ $x = \dfrac{3 \pm \sqrt{9 - 4(2)(-6)}}{2(2)}$
Number and types of solutions	Determined by the discriminant $b^2 - 4ac$ $b^2 - 4ac > 0$: two real solutions	$x^2 + 5x - 9 = 0$ has two real solutions because $5^2 - 4(1)(-9) > 0$.
	$b^2 - 4ac = 0$: one real solution	$x^2 + 6x + 9 = 0$ has one real solution because $6^2 - 4(1)(9) = 0$.
	$b^2 - 4ac < 0$: no real solutions (two complex solutions)	$x^2 + 3x + 10 = 0$ has no real solutions because $3^2 - 4(1)(10) < 0$.

Complex Numbers		Examples
Complex numbers	Numbers of the form $a + bi$, where a and b are real $i = \sqrt{-1}$ and $i^2 = -1$	$12, -3i, 5 + 4i, \sqrt{2} - i\sqrt{3}$
Imaginary numbers	Numbers of the form $a + bi$, where $b \neq 0$	$5i, 13 + i\sqrt{6}$
Square root of a negative number	If b is a positive real number, then $\sqrt{-b} = i\sqrt{b}$.	$\sqrt{-3} = i\sqrt{3}$ $\sqrt{-4} = i\sqrt{4} = 2i$
Complex conjugates	The complex numbers $a + bi$ and $a - bi$ are called complex conjugates of each other. Their product is a real number.	$(1 + 2i)(1 - 2i) = 1 + 4 = 5$

Complex number operations	Add, subtract, and multiply as if the complex numbers were binomials with variable i. Use the distributive property for multiplication. Remember that $i^2 = -1$.	$(2 + 3i) + (3 - 5i) = 5 - 2i$ $(2 - 5i) - (4 - 2i) = -2 - 3i$ $(3 - 4i)(2 + 5i) = 26 + 7i$
	Divide complex numbers by multiplying the numerator and denominator by the conjugate of the denominator, then simplify.	$\frac{4 - 6i}{5 + 2i} = \frac{(4 - 6i)(5 - 2i)}{(5 + 2i)(5 - 2i)}$

Quadratic Functions		Examples
Quadratic function	A function of the form $y = ax^2 + bx + c$, where a, b, and c are real numbers and $a \neq 0$	$y = 3x^2 - 8x + 9$ $p = -3q^2 - 8q + 1$
Graphing a quadratic function	The graph is a parabola opening upward if $a > 0$ and downward if $a < 0$. The first coordinate of the vertex is $\frac{-b}{2a}$. The second coordinate of the vertex is either the minimum value of the function if $a > 0$ or the maximum value of the function if $a < 0$.	$y = x^2 - 2x + 5$ Opens upward Vertex: (1, 4) Minimum value of y is 4.

REVIEW EXERCISES

9.1 *Solve each equation.*

1. $x^2 - 9 = 0$ **2.** $x^2 - 1 = 0$

3. $x^2 - 9x = 0$ **4.** $x^2 - x = 0$

5. $x^2 - x = 2$ **6.** $x^2 - 9x = 10$

7. $(x - 9)^2 = 10$ **8.** $(x + 5)^2 = 14$

9. $4x^2 - 12x + 9 = 0$ **10.** $9x^2 + 6x + 1 = 0$

11. $t^2 - 9t + 20 = 0$ **12.** $s^2 - 4s + 3 = 0$

13. $\frac{x}{2} = \frac{7}{x + 5}$ **14.** $\sqrt{x + 4} = \frac{2x - 1}{3}$

15. $\frac{1}{2}x^2 + \frac{7}{4}x = 1$ **16.** $\frac{2}{3}x^2 - 1 = -\frac{1}{3}x$

9.2 *Solve each equation by completing the square.*

17. $x^2 + 4x - 7 = 0$ **18.** $x^2 + 6x - 3 = 0$

19. $x^2 + 3x - 28 = 0$ **20.** $x^2 - x - 6 = 0$

21. $x^2 + 3x - 5 = 0$ **22.** $x^2 + \frac{4}{3}x - \frac{1}{3} = 0$

23. $2x^2 + 9x - 5 = 0$ **24.** $2x^2 + 6x - 5 = 0$

9.3 *Find the value of the discriminant, and tell how many real solutions each equation has.*

25. $25t^2 - 10t + 1 = 0$ **26.** $3x^2 + 2 = 0$

27. $-3w^2 + 4w - 5 = 0$ **28.** $5x^2 - 7x = 0$

29. $-3v^2 + 4v = -5$ **30.** $49u^2 + 42u + 9 = 0$

Use the quadratic formula to solve each equation.

31. $6x^2 + x - 2 = 0$ **32.** $-6x^2 + 11x + 10 = 0$

33. $x^2 - x = 4$ **34.** $y^2 - 2y = 4$

35. $5x^2 - 6x - 1 = 0$ **36.** $t^2 - 6t + 4 = 0$

37. $3x^2 - 5x = 0$ **38.** $2w^2 - w = 15$

9.4 *For each problem, find the exact and approximate answers. Round the decimal answers to three decimal places.*

39. ***Bird watching.*** Chuck is standing 12 meters from a tree, watching a bird's nest that is 5 meters above eye level. Find the distance from Chuck's eyes to the nest.

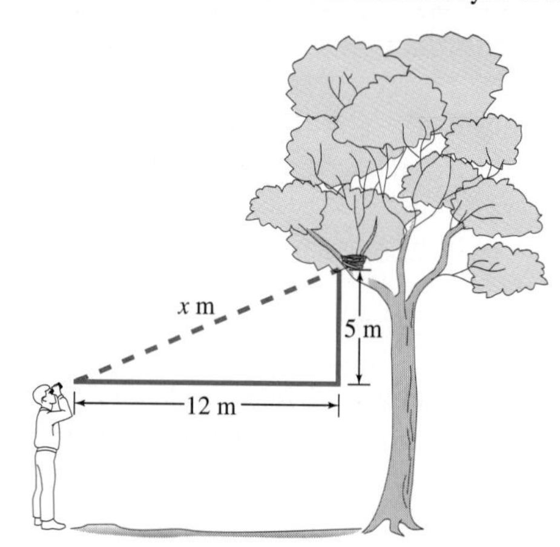

Figure for Exercise 39

40. ***Diagonal of a square.*** Find the diagonal of a square if the length of each side is 20 yards.

41. ***Lengthy legs.*** The hypotenuse of a right triangle measures 5 meters, and one leg is 2 meters longer than the other. Find the lengths of the legs.

42. ***Width and height.*** The width of a rectangular bookcase is 3 feet shorter than the height. If the diagonal is 7 feet, then what are the dimensions of the bookcase?

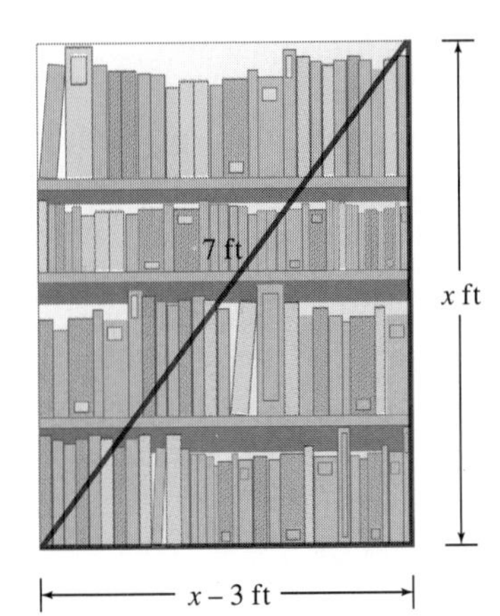

Figure for Exercise 42

43. ***Base and height.*** The base of a triangle is 4 inches longer than the height. If the area of the triangle is 20 square inches, then what are the lengths of the base and height?

44. ***Dimensions of a parallelogram.*** The base of a parallelogram is 1 meter longer than the height. If the area of the parallelogram is 8 square meters, then what are the lengths of the base and height?

45. ***Unknown numbers.*** Find two positive real numbers whose sum is 6 and whose product is 7.

46. ***Dimensions of a rectangle.*** The perimeter of a rectangle is 16 feet, and its area is 13 square feet. What are the dimensions of the rectangle?

47. ***Printing time.*** The old printer took 2 hours longer than the new printer to print 100,000 mailing labels. With both printers working on the job, the 100,000 labels can be printed in 8 hours. How long would it take each printer working alone to do the job?

48. ***Tilling the garden.*** When Blake uses his old tiller, it takes him 3 hours longer to till the garden than it takes Cassie using her new tiller. If Cassie will not let Blake use her new tiller and they can till the garden together in 6 hours then how long would it take each one working alone?

9.5 *Perform the indicated operations. Write answers in the form $a + bi$.*

49. $(2 + 3i) + (5 - 6i)$

50. $(2 - 5i) + (-9 - 4i)$

51. $(-5 + 4i) - (-2 - 3i)$

52. $(1 - i) - (1 + i)$

53. $(2 - 9i)(3 + i)$

54. $2i - 3(6 - 2i)$

55. $(3 + 8i)^2$

56. $(-5 - 2i)(-5 + 2i)$

57. $\dfrac{-2 - \sqrt{-8}}{2}$

58. $\dfrac{-6 + \sqrt{-54}}{-3}$

59. $\dfrac{1 + 3i}{6 - i}$

60. $\dfrac{3i}{8 + 3i}$

61. $\dfrac{5 + i}{4 - i}$

62. $\dfrac{3 + 2i}{i}$

Find the complex solutions to the quadratic equations.

63. $x^2 + 121 = 0$

64. $x^2 + 120 = 0$

65. $x^2 - 16x + 65 = 0$

66. $x^2 - 10x + 28 = 0$

67. $2x^2 - 3x + 9 = 0$

68. $3x^2 - 6x + 4 = 0$

9.6 *Find the vertex and intercepts for each quadratic function, and sketch its graph.*

69. $y = x^2 - 6x$

70. $y = x^2 + 4x$

71. $y = x^2 - 4x - 12$

72. $y = x^2 + 2x - 24$

73. $y = -2x^2 + 8x$

74. $y = -3x^2 + 6x$

75. $y = -x^2 + 2x + 3$

76. $y = -x^2 - 3x - 2$

Find the domain and range of each quadratic function.

77. $y = x^2 + 4x + 1$

78. $y = x^2 - 6x + 2$

79. $y = -2x^2 - x + 4$

80. $y = -3x^2 + 2x + 7$

Solve each problem.

81. ***Minimizing cost.*** The unit cost in dollars for manufacturing n starters is given by $C = 0.004n^2 - 3.2n + 660$. What is unit cost when 390 starters are manufactured? For what number of starters is the unit cost at a minimum?

82. ***Maximizing profit.*** The total profit (in dollars) for sales of n rowing machines is given by $P = -0.2x^2 + 300x - 200$. What is the profit if 500 are sold? For what value of n will the profit be at a maximum?

CHAPTER 9 TEST

Calculate the value of $b^2 - 4ac$ *and state how many real solutions each equation has.*

1. $9x^2 - 12x + 4 = 0$

2. $-2x^2 + 3x - 5 = 0$

3. $-2x^2 + 5x - 1 = 0$

Solve by using the quadratic formula.

4. $5x^2 + 2x - 3 = 0$

5. $2x^2 - 4x - 3 = 0$

Solve by completing the square.

6. $x^2 + 4x - 21 = 0$

7. $x^2 + 3x - 5 = 0$

Solve by any method.

8. $x(x + 1) = 20$

9. $x^2 - 28x + 75 = 0$

10. $\dfrac{x - 1}{3} = \dfrac{x + 1}{2x}$

Perform the indicated operations. Write answers in the form $a + bi$.

11. $(2 - 3i) + (8 + 6i)$

12. $(-2 - 5i) - (4 - 12i)$

13. $(-6i)^2$

14. $(3 - 5i)(4 + 6i)$

15. $(8 - 2i)(8 + 2i)$

16. $(4 - 6i) \div 2$

17. $\dfrac{-2 + \sqrt{-12}}{2}$

18. $\dfrac{6 - \sqrt{-18}}{-3}$

19. $\dfrac{5i}{4 + 3i}$

Find the complex solutions to the quadratic equations.

20. $x^2 + 6x + 12 = 0$

21. $-5x^2 + 6x - 5 = 0$

Graph each quadratic function. State the domain and range.

22. $y = 16 - x^2$

23. $y = x^2 - 3x$

Solve each problem.

24. Find the x-intercepts for the parabola $y = x^2 - 6x + 5$.

25. The height in feet for a ball thrown upward at 48 feet per second is give by $s = -16t^2 + 48t$, where t is the time in seconds after the ball is tossed. What is the maximum height that the ball will reach?

26. Find two positive numbers that have a sum of 10 and a product of 23. Give exact answers.

Tying It All Together CHAPTERS 1–9

Solve each equation.

1. $2x - 1 = 0$
2. $2(x - 1) = 0$
3. $2x^2 - 1 = 0$
4. $(2x - 1)^2 = 8$
5. $2x^2 - 4x - 1 = 0$
6. $2x^2 - 4x = 0$
7. $2x^2 + x = 1$
8. $x - 2 = \sqrt{2x - 1}$
9. $\frac{1}{x} = \frac{x}{2x - 15}$
10. $\frac{1}{x} - \frac{1}{x - 1} = -\frac{1}{2}$

Solve each equation for y.

11. $5x - 4y = 8$
12. $3x - y = 9$
13. $\frac{y - 4}{x + 2} = \frac{2}{3}$
14. $ay + b = 0$
15. $ay^2 + by + c = 0$
16. $y - 1 = -\frac{2}{3}(x - 9)$
17. $\frac{2}{3}x + \frac{1}{2}y = \frac{1}{9}$
18. $x^2 + y^2 = a^2$

Suppose that each side of a square has length s, the diagonal has length d, the area of the square is A, and its perimeter is P.

19. Write P in terms of s.
20. Write A in terms of s.
21. Write P in terms of d.
22. Write d in terms of A.

Solve each system of equations.

23. $3x - 2y = 12$
    $2x + 5y = -11$
24. $y = 3x + 1$
    $3x - 0.6y = 3$

Graph each function.

25. $y = x - 3$
26. $y = 2 - x$
27. $y = x^2 - 3$
28. $y = 2 - x^2$
29. $y = \frac{2}{3}x - 4$
30. $y = -\frac{4}{3}x + 5$

Solve the problem.

31. ***Maximizing revenue.*** For the last three years the Lakeland Air Show has raised the price of its tickets and has sold fewer and fewer tickets, as shown in the table.

Ticket price	\$10	\$12	\$16
Tickets sold	8000	7500	6500

a) Use this information to write the number of tickets sold s as a linear function of the ticket price p.
b) Has the revenue from ticket sales increased or decreased as the ticket price was raised?
c) Write the revenue R as a function of the ticket price p.
d) What ticket price would produce the maximum revenue?

APPENDIX A

Geometry Review

GEOMETRIC FIGURES AND FORMULAS

Triangle: A three-sided figure

Area: $A = \frac{1}{2}bh$, Perimeter: $P = a + b + c$

Sum of the measures of the angles is 180°.

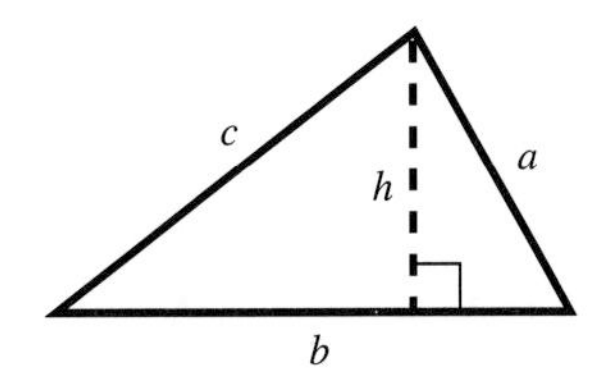

Right Triangle: A triangle with a 90° angle

Area $= \frac{1}{2}ab$, Perimeter: $P = a + b + c$

Pythagorean Theorem: $c^2 = a^2 + b^2$

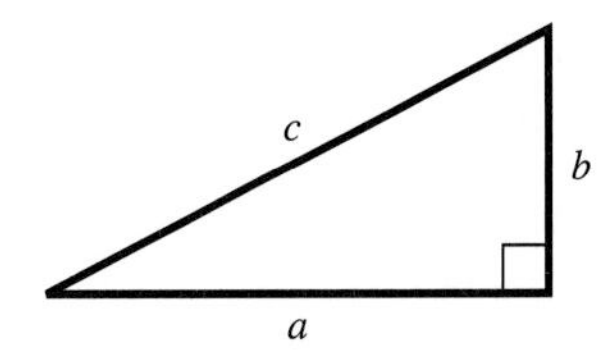

30-60-90 Right Triangle

The side opposite 30° is one-half the length of the hypotenuse.

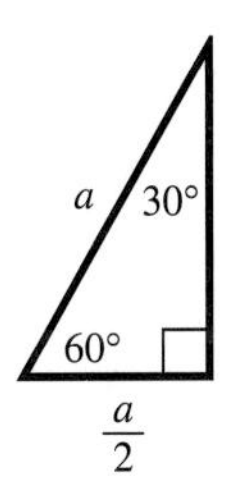

Parallelogram: A four-sided figure with opposite sides parallel

Area: $A = bh$

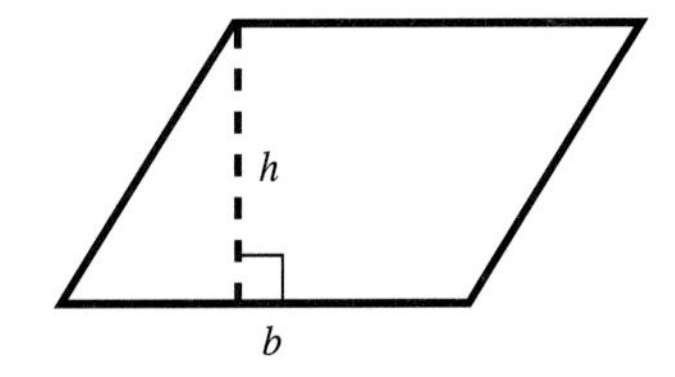

Trapezoid: A four-sided figure with one pair of parallel sides

Area: $A = \frac{1}{2}h(b_1 + b_2)$

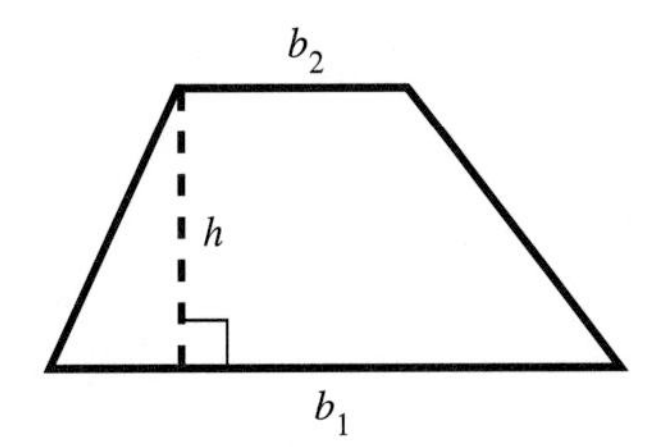

Rhombus: A four-sided figure with four equal sides

Perimeter: $P = 4a$

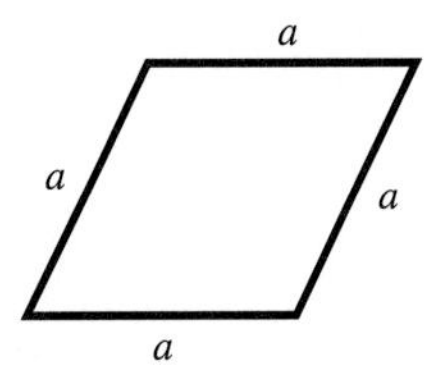

Rectangle: A four-sided figure with four right angles
Area: $A = LW$
Perimeter: $P = 2L + 2W$

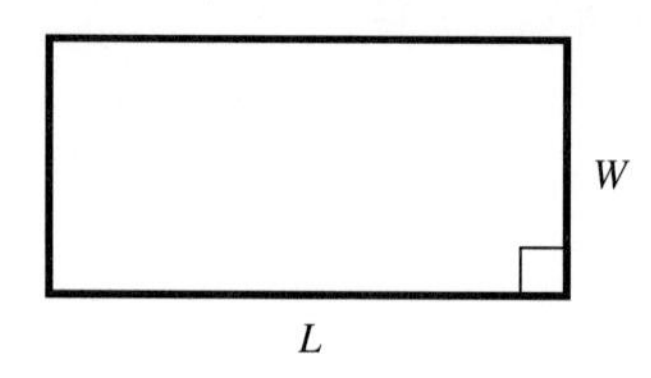

Square: A four-sided figure with four equal sides and four right angles
Area: $A = s^2$
Perimeter: $P = 4s$

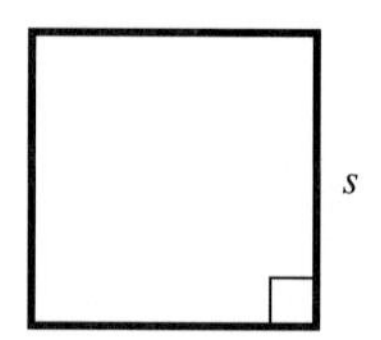

Circle
Area: $A = \pi r^2$
Circumference: $C = 2\pi r$
Diameter: $d = 2r$

Sphere
Volume: $V = \frac{4}{3}\pi r^3$
Surface Area: $S = 4\pi r^2$

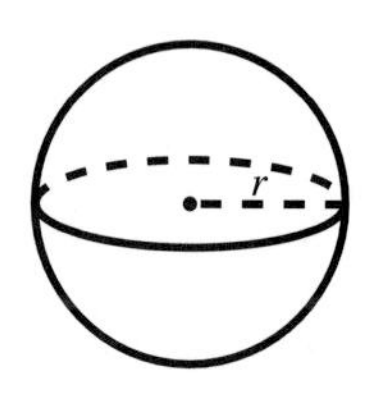

Right Circular Cone
Volume: $V = \frac{1}{3}\pi r^2 h$
Lateral Surface Area: $S = \pi r\sqrt{r^2 + h^2}$

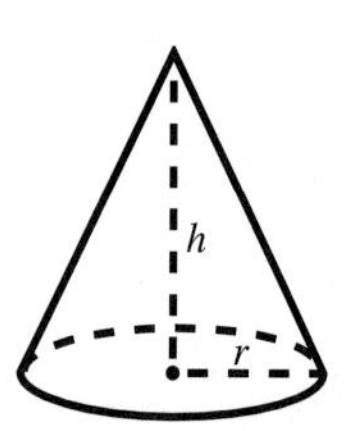

Right Circular Cylinder
Volume: $V = \pi r^2 h$
Lateral Surface Area: $S = 2\pi rh$

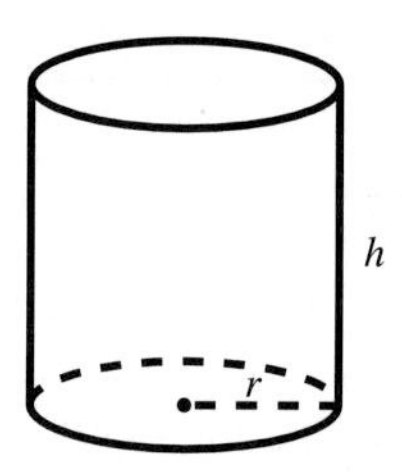

Rectangular Solid
Volume: $V = LWH$
Surface Area: $A = 2LW + 2WH + 2LH$

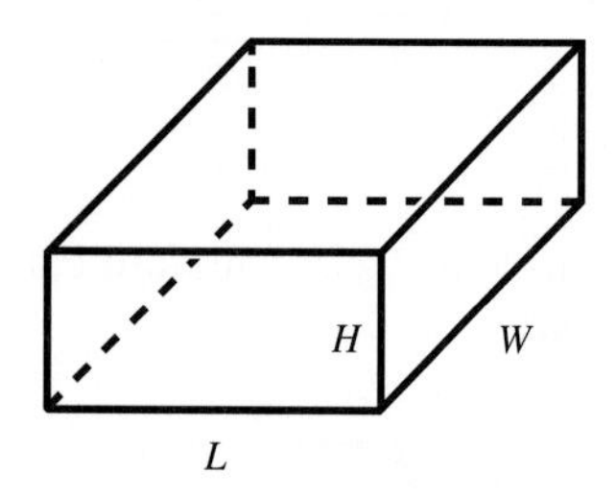

GEOMETRIC TERMS

An **angle** is a union of two rays with a common endpoint.

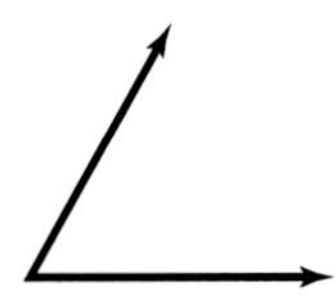

An **acute angle** is an angle with a measure between 0° and 90°.

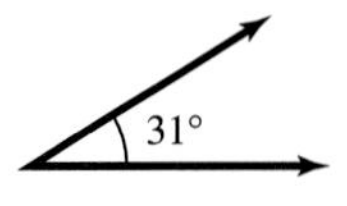

A **right angle** is an angle with a measure of 90°.

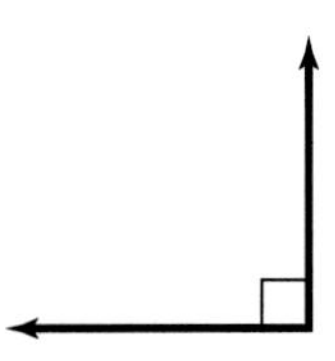

An **obtuse angle** is an angle with a measure between 90° and 180°.

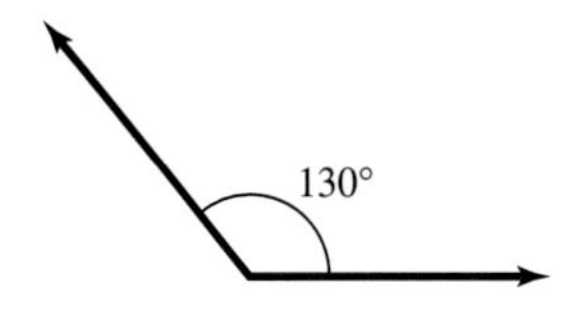

Two angles are **complementary** if the sum of their measures is 90°.

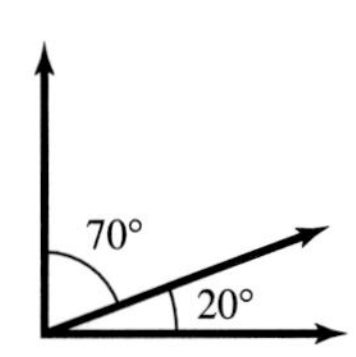

Two angles are **supplementary** if the sum of their measures is 180°.

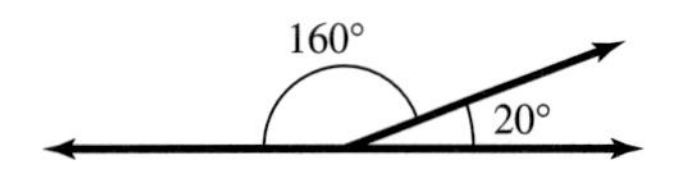

An **isosceles triangle** is a triangle that has two equal sides.

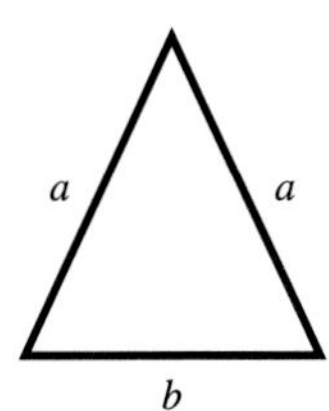

An **equilateral triangle** is a triangle that has three equal sides.

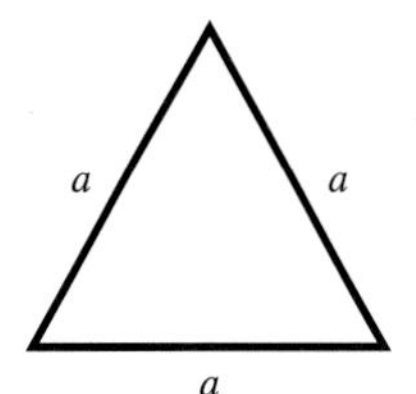

Similar triangles are triangles that have the same shape. Their corresponding angles are equal and corresponding sides are proportional:

$$\frac{a}{d} = \frac{b}{e} = \frac{c}{f}$$

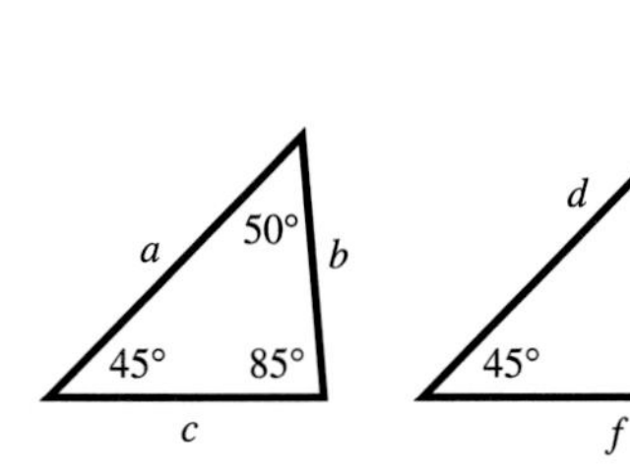

APPENDIX B

Table of Squares and Square Roots

n	n^2	$\sqrt{n}$	n	n^2	$\sqrt{n}$	n	n^2	$\sqrt{n}$
1	1	1.0000	41	1681	6.4031	81	6561	9.0000
2	4	1.4142	42	1764	6.4807	82	6724	9.0554
3	9	1.7321	43	1849	6.5574	83	6889	9.1104
4	16	2.0000	44	1936	6.6332	84	7056	9.1652
5	25	2.2361	45	2025	6.7082	85	7225	9.2195
6	36	2.4495	46	2116	6.7823	86	7396	9.2736
7	49	2.6458	47	2209	6.8557	87	7569	9.3274
8	64	2.8284	48	2304	6.9282	88	7744	9.3808
9	81	3.0000	49	2401	7.0000	89	7921	9.4340
10	100	3.1623	50	2500	7.0711	90	8100	9.4868
11	121	3.3166	51	2601	7.1414	91	8281	9.5394
12	144	3.4641	52	2704	7.2111	92	8464	9.5917
13	169	3.6056	53	2809	7.2801	93	8649	9.6437
14	196	3.7417	54	2916	7.3485	94	8836	9.6954
15	225	3.8730	55	3025	7.4162	95	9025	9.7468
16	256	4.0000	56	3136	7.4833	96	9216	9.7980
17	289	4.1231	57	3249	7.5498	97	9409	9.8489
18	324	4.2426	58	3364	7.6158	98	9604	9.8995
19	361	4.3589	59	3481	7.6811	99	9801	9.9499
20	400	4.4721	60	3600	7.7460	100	10000	10.0000
21	441	4.5826	61	3721	7.8102	101	10201	10.0499
22	484	4.6904	62	3844	7.8740	102	10404	10.0995
23	529	4.7958	63	3969	7.9373	103	10609	10.1489
24	576	4.8990	64	4096	8.0000	104	10816	10.1980
25	625	5.0000	65	4225	8.0623	105	11025	10.2470
26	676	5.0990	66	4356	8.1240	106	11236	10.2956
27	729	5.1962	67	4489	8.1854	107	11449	10.3441
28	784	5.2915	68	4624	8.2462	108	11664	10.3923
29	841	5.3852	69	4761	8.3066	109	11881	10.4403
30	900	5.4772	70	4900	8.3666	110	12100	10.4881
31	961	5.5678	71	5041	8.4261	111	12321	10.5357
32	1024	5.6569	72	5184	8.4853	112	12544	10.5830
33	1089	5.7446	73	5329	8.5440	113	12769	10.6301
34	1156	5.8310	74	5476	8.6023	114	12996	10.6771
35	1225	5.9161	75	5625	8.6603	115	13225	10.7238
36	1296	6.0000	76	5776	8.7178	116	13456	10.7703
37	1369	6.0828	77	5929	8.7750	117	13689	10.8167
38	1444	6.1644	78	6084	8.8318	118	13924	10.8628
39	1521	6.2450	79	6241	8.8882	119	14161	10.9087
40	1600	6.3246	80	6400	8.9443	120	14400	10.9545

Answers to Selected Exercises

CHAPTER 1

Section 1.1 Warm-ups T T F T T T T T F T

1. $\frac{6}{8}$ **3.** $\frac{32}{12}$ **5.** $\frac{10}{2}$ **7.** $\frac{75}{100}$ **9.** $\frac{30}{100}$ **11.** $\frac{70}{42}$
13. $\frac{1}{2}$ **15.** $\frac{2}{3}$ **17.** 3 **19.** $\frac{1}{2}$ **21.** 2 **23.** $\frac{3}{8}$
25. $\frac{13}{21}$ **27.** $\frac{12}{13}$ **29.** $\frac{10}{27}$ **31.** 5 **33.** $\frac{7}{10}$ **35.** $\frac{7}{13}$
37. $\frac{21}{5}$ **39.** 15 **41.** 3 **43.** $\frac{1}{15}$ **45.** 4 **47.** $\frac{4}{5}$
49. $\frac{7}{8}$ **51.** $\frac{1}{2}$ **53.** $\frac{1}{3}$ **55.** $\frac{1}{4}$ **57.** $\frac{7}{12}$ **59.** $\frac{1}{12}$
61. $\frac{19}{24}$ **63.** $\frac{11}{72}$ **65.** $\frac{199}{48}$ **67.** 60%, 0.6
69. $\frac{9}{100}$, 0.09 **71.** 8%, $\frac{2}{25}$ **73.** 0.75, 75%
75. $\frac{1}{50}$, 0.02 **77.** $\frac{1}{100}$, 1% **79.** 3 **81.** 1 **83.** $\frac{71}{96}$
85. $\frac{17}{120}$ **87.** $\frac{65}{16}$ **89.** $\frac{69}{4}$ **91.** $\frac{13}{12}$ **93.** $\frac{1}{8}$
95. 2625 in.^{2}, 140.7 ft^3 **97.** $54\frac{7}{8}$, 0.2%

Section 1.2 Warm-ups T T F F T F T T F F

1. 6 **3.** 0 **5.** −2 **7.** −12
9. 1, 2, 3, 4, 5 

11. 0, 1, 2, 3, 4 

13. 0, 1, 2, 3, 4 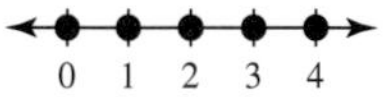

15. 1, 2, 3, 4, 5, . . . 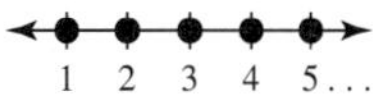

17. 1, 2, 3, 4, 5, . . . 

19. True **21.** False **23.** True **25.** True **27.** True
29. False **31.** 6 **33.** 0 **35.** 7 **37.** 9 **39.** 45
41. $\frac{3}{4}$ **43.** 5.09 **45.** −16 **47.** $-\frac{5}{2}$ **49.** 2 **51.** 3
53. −9 **55.** 16 **57.** True **59.** True **61.** True

Section 1.3 Warm-ups T T T F F F T F T F

1. 13 **3.** −13 **5.** −1.15 **7.** $-\frac{1}{2}$ **9.** 0 **11.** 0
13. 2 **15.** −6 **17.** 5.6 **19.** −2.9 **21.** $-\frac{1}{4}$
23. 8 + (−2) **25.** 4 + (−12) **27.** −3 + 8
29. 8.3 + 1.5 **31.** −4 **33.** −10 **35.** 11 **37.** −11
39. $-\frac{1}{4}$ **41.** $\frac{3}{4}$ **43.** 7 **45.** 0.93 **47.** 9.3
49. −5.03 **51.** 3 **53.** −9 **55.** −120 **57.** 78
59. −27 **61.** −7 **63.** −201 **65.** −322 **67.** −15.97
69. −2.92 **71.** −3.73 **73.** 3.7 **75.** $\frac{3}{20}$ **77.** $\frac{7}{24}$
79. −3.49 **81.** −0.3422 **83.** −48.84 **85.** −8.85
87. −\$8.85 **89.** −6°F

Section 1.4 Warm-ups T F T F T T T F T F

1. −27 **3.** 132 **5.** $-\frac{1}{3}$ **7.** −0.3 **9.** 144 **11.** 0
13. −1 **15.** 3 **17.** $-\frac{2}{3}$ **19.** $\frac{5}{6}$ **21.** Undefined
23. 0 **25.** −80 **27.** 0.25 **29.** −100 **31.** 27
33. −3 **35.** −4 **37.** −30 **39.** 19 **41.** −0.18

43. 0.3 **45.** -6 **47.** 1.5 **49.** 22 **51.** $-\frac{1}{3}$
53. -164.25 **55.** 1529.41 **57.** 16 **59.** -8 **61.** 0
63. 0 **65.** -3.9 **67.** -40 **69.** 0.4 **71.** 0.4
73. -0.2 **75.** -7.5 **77.** $-\frac{1}{30}$ **79.** $-\frac{1}{10}$
81. 7.562 **83.** 19.35 **85.** 0 **87.** Undefined

Section 1.5 Warm-ups F F T F F F F T F T

1. -4 **3.** 1 **5.** -8 **7.** -7 **9.** -16 **11.** -4
13. 4^4 **15.** $(-5)^4$ **17.** $(-y)^3$ **19.** $\left(\frac{3}{7}\right)^5$ **21.** $5 \cdot 5 \cdot 5$
23. $b \cdot b$ **25.** $\left(-\frac{1}{2}\right)\left(-\frac{1}{2}\right)\left(-\frac{1}{2}\right)\left(-\frac{1}{2}\right)\left(-\frac{1}{2}\right)$
27. $(0.22)(0.22)(0.22)(0.22)$ **29.** 81 **31.** 0 **33.** 625
35. -216 **37.** 100,000 **39.** -0.001 **41.** $\frac{1}{8}$
43. $\frac{1}{4}$ **45.** -64 **47.** -4096 **49.** 27 **51.** -13
53. 36 **55.** 18 **57.** -19 **59.** -17 **61.** -44
63. 18 **65.** -78 **67.** 0 **69.** 27 **71.** 1 **73.** 8
75. 7 **77.** 11 **79.** 111 **81.** 21 **83.** -1 **85.** -11
87. 9 **89.** 16 **91.** 28 **93.** 121 **95.** -73
97. 25 **99.** 0 **101.** -2 **103.** 12 **105.** 82
107. -54 **109.** -79 **111.** -24 **113.** 41.92
115. 184.643547 **117.** 8.0548 **119.** 270.6 million

Section 1.6 Warm-ups T F T F T F F F T F

1. Difference **3.** Cube **5.** Sum **7.** Difference
9. Product **11.** Square **13.** The difference of x^2 and a^2
15. The square of $x - a$
17. The quotient of $x - 4$ and 2
19. The difference of $\frac{x}{2}$ and 4 **21.** The cube of ab
23. $2x + 3y$ **25.** $8 - 7x$ **27.** $(a + b)^2$
29. $(x + 9)(x + 12)$ **31.** $\frac{x-7}{7-x}$ **33.** 3 **35.** 3
37. 16 **39.** -9 **41.** -3 **43.** -8 **45.** $-\frac{2}{3}$ **47.** 4
49. -1 **51.** 1 **53.** -4 **55.** 0 **57.** Yes **59.** No
61. Yes **63.** Yes **65.** Yes **67.** Yes **69.** No
71. No **73.** $5x + 3x = 8x$ **75.** $3(x + 2) = 12$
77. $\frac{x}{3} = 5x$ **79.** $(a + b)^2 = 9$ **81.** 14.65 **83.** 37.12
85. 169.3 cm, 41 cm **87.** 4, 5.5, 10.5 **89.** 920 feet

Section 1.7 Warm-ups F F T F T T T T T T

1. $r + 9$ **3.** $3(x + 2)$ **5.** $-5x + 4$ **7.** $6x$
9. $-2(x - 4)$ **11.** $4 - 8y$ **13.** $4w^2$ **15.** $3a^2b$
17. $9x^3z$ **19.** -3 **21.** -10 **23.** -21 **25.** 0.6
27. -22.4 **29.** $3x - 15$ **31.** $2(m + 6)$ **33.** $2a + at$
35. $-3w + 18$ **37.** $-20 + 4y$ **39.** $4(x - 1)$
41. $-a + 7$ **43.** $-t - 4$ **45.** $4(y - 4)$ **47.** $4(a + 2)$
49. 2 **51.** $-\frac{1}{5}$ **53.** $\frac{1}{7}$ **55.** 1 **57.** -4 **59.** $\frac{2}{5}$
61. Commutative property of multiplication
63. Distributive property **65.** Associative property
67. Inverse property
69. Commutative property of multiplication
71. Identity property **73.** Distributive property
75. Inverse property **77.** Multiplication property of 0
79. Distributive property **81.** $y + a$ **83.** $(5a)w$
85. $\frac{1}{2}(x + 1)$ **87.** $3(2x + 5)$ **89.** 1 **91.** 0
93. $\frac{100}{33}$ **95.** 45 bricks/hour
97. 2.9 people/second; 1,753,920 people/week

Section 1.8 Warm-ups T F T T F F F F F T

1. 7000 **3.** 1 **5.** 356 **7.** 350 **9.** 36 **11.** 36,000
13. 0 **15.** 98 **17.** $11w$ **19.** $3x$ **21.** $5x$ **23.** $-a$
25. $-2a$ **27.** $10 - 6t$ **29.** $8x^2$ **31.** $-4x + 2x^2$
33. $-7mw^2$ **35.** $12h$ **37.** $-18b$ **39.** $-9m^2$
41. $12d^2$ **43.** y^2 **45.** $-15ab$ **47.** $-6a - 3ab$
49. $-k + k^2$ **51.** y **53.** $-3y$ **55.** y **57.** $2y^2$
59. $2a - 1$ **61.** $3x - 2$ **63.** $-2x + 1$ **65.** $8 - y$
67. $m - 6$ **69.** $w - 5$ **71.** $8x + 15$ **73.** $5x - 1$
75. $-2a - 1$ **77.** $5a - 2$ **79.** $3m - 18$ **81.** $-3x - 7$
83. $0.95x - 0.5$ **85.** $-3.2x + 12.71$ **87.** $4x - 4$
89. $2y + 4$ **91.** $2y + m - 1$ **93.** 3 **95.** $0.15x - 0.4$
97. $-14k + 23$ **99.** 45
101. $0.28x - 4940$, \$9620, \$40,000
103. $4x + 80$, 200 feet

Chapter 1 Review

1. $\frac{17}{24}$ **3.** 6 **5.** $\frac{3}{7}$ **7.** $\frac{14}{3}$ **9.** $\frac{13}{12}$ **11.** 0, 1, 2, 10
13. $-2, 0, 1, 2, 10$ **15.** $-\sqrt{5}$, π **17.** True **19.** False
21. False **23.** True **25.** 2 **27.** -13 **29.** -7
31. -7 **33.** 11.95 **35.** -0.05 **37.** $-\frac{1}{6}$ **39.** $-\frac{11}{15}$
41. -15 **43.** 4 **45.** 5 **47.** $\frac{1}{6}$ **49.** -0.3
51. -0.24 **53.** 1 **55.** 66 **57.** 49 **59.** 41
61. 1 **63.** 50 **65.** -135 **67.** -2 **69.** -16
71. 16 **73.** 5 **75.** 9 **77.** 7 **79.** $-\frac{1}{3}$ **81.** 1
83. -9 **85.** Yes **87.** No **89.** Yes **91.** No
93. Distributive property **95.** Inverse property
97. Identity property
99. Associative property of addition
101. Commutative property of multiplication
103. Inverse property **105.** Identity property
107. $-a + 12$ **109.** $6a^2 - 6a$ **111.** $-12t + 39$
113. $-0.9a - 0.57$ **115.** $-0.05x - 4$
117. $27x^2 + 6x + 5$ **119.** $-2a$ **121.** $x^2 + 4x - 3$
123. 0 **125.** 8 **127.** -21 **129.** $\frac{1}{2}$ **131.** -0.5
133. -1 **135.** $x + 2$ **137.** $4 + 2x$ **139.** $2x$

141. $-4x + 8$ **143.** $6x$ **145.** x **147.** $8x$
149. 18 memberships per hour

Chapter 1 Test

1. 0, 8 **2.** -3, 0, 8 **3.** -3, $-\frac{1}{4}$, 0, 8
4. $-\sqrt{3}$, $\sqrt{5}$, π **5.** -21 **6.** -4 **7.** 9 **8.** -7
9. -0.95 **10.** -56 **11.** 978 **12.** 13 **13.** -1
14. 0 **15.** 9740 **16.** $-\frac{7}{24}$ **17.** -20 **18.** $-\frac{1}{6}$
19. -39 **20.** Distributive property
21. Commutative property of multiplication
22. Associative property of addition
23. Inverse property **24.** Identity property
25. Multiplication property of 0 **26.** $3(x + 10)$
27. $7(w - 1)$ **28.** $6x + 6$ **29.** $4x - 2$ **30.** $7x - 3$
31. $0.9x + 7.5$ **32.** $14a^2 + 5a$ **33.** $x + 2$ **34.** $4t$
35. $54x^2y^2$ **36.** 41 **37.** 5 **38.** -12 **39.** No
40. Yes **41.** Yes **42.** 9 deliveries per hour
43. $3.66R - 0.06A + 82.205$, 168.905 cm

CHAPTER 2

Section 2.1 Warm-ups T T F F F F T T T T

1. 1 **3.** -17 **5.** 1 **7.** 0 **9.** -2 **11.** 5 **13.** 1
15. -6 **17.** -12 **19.** -4 **21.** -13 **23.** 1.7
25. 0.1 **27.** 4.6 **29.** -8 **31.** 1.8 **33.** -5 **35.** 5
37. $\frac{2}{3}$ **39.** $\frac{1}{2}$ **41.** 20 **43.** $-\frac{2}{3}$ **45.** 12
47. -10 **49.** -14 **51.** 4 **53.** $\frac{1}{4}$ **55.** $-\frac{2}{5}$
57. $\frac{3}{2}$ **59.** -1 **61.** -6 **63.** 8 **65.** 34 **67.** 6
69. 0 **71.** -6 **73.** 60 **75.** -10 **77.** 18 **79.** -20
81. -3 **83.** $\frac{1}{2}$ **85.** -4.3 **87.** 20°C **89.** 9 feet

Section 2.2 Warm-ups T T F F F T T F T T

1. R, identity **3.** No solution, inconsistent
5. 0, conditional **7.** No solution, inconsistent
9. No solution, inconsistent **11.** 1, conditional
13. No solution, inconsistent **15.** R, identity
17. All nonzero real numbers, identity **19.** R, identity
21. 7 **23.** 24 **25.** 16 **27.** -12 **29.** 60 **31.** 24
33. 90 **35.** 6 **37.** -2 **39.** 80 **41.** 60 **43.** 200
45. 800 **47.** $\frac{9}{2}$ **49.** 3 **51.** 25 **53.** -2 **55.** -3
57. 5 **59.** -10 **61.** 2 **63.** -4 **65.** R **67.** R
69. 100 **71.** $-\frac{3}{2}$ **73.** 30 **75.** 6 **77.** 0.5
79. 19,608 **81.** \$88,000

Section 2.3 Warm-ups F F F F F F T T T F T

1. $R = \frac{D}{T}$ **3.** $\pi = \frac{C}{D}$ **5.** $P = \frac{I}{rt}$ **7.** $C = \frac{5}{9}(F - 32)$
9. $h = \frac{2A}{b}$ **11.** $L = \frac{P - 2W}{2}$ **13.** $a = 2A - b$
15. $r = \frac{S - P}{Pt}$ **17.** $a = \frac{2A - hb}{h}$ **19.** $x = \frac{b - a}{2}$
21. $x = -7a$ **23.** $x = 12 - a$ **25.** $x = 7ab$
27. $y = -x - 9$ **29.** $y = -x + 6$ **31.** $y = 2x - 2$
33. $y = 3x + 4$ **35.** $y = -\frac{1}{2}x + 2$ **37.** $y = x - \frac{1}{2}$
39. $y = 3x - 14$ **41.** $y = \frac{1}{2}x$ **43.** $y = \frac{3}{2}x + 6$ **45.** 2
47. 7 **49.** $-\frac{9}{5}$ **51.** 1 **53.** 1.33 **55.** 4%
57. 4 years **59.** 7 yards **61.** 225 feet **63.** \$300
65. 20% **67.** 160 feet **69.** 24 cubic feet
71. 4 inches **73.** 8 feet **75.** 12 inches
77. 640 milligrams, age 13 **79.** 3.75 milliliters

Section 2.4 Warm-ups T T T F T F F F T F

1. $x + 3$ **3.** $x - 3$ **5.** $5x$ **7.** $0.1x$ **9.** $\frac{x}{3}$
11. $\frac{1}{3}x$ **13.** x and $x + 15$ **15.** x and $6 - x$
17. x and $-4 - x$ **19.** x and $x + 3$ **21.** x and $0.05x$
23. x and $1.30x$ **25.** x and $90 - x$ **27.** x and $120 - x$
29. x and $x + 2$ **31.** x and $x + 1$ **33.** m and $m - 2$
35. y, $y + 2$, $y + 4$ **37.** $3x$ miles **39.** $0.25q$ dollars
41. $\frac{x}{20}$ hour **43.** $\frac{x - 100}{12}$ meters per second
45. $5x$ square meters **47.** $2w + 2(w + 3)$ inches
49. $150 - x$ feet **51.** $2x + 1$ feet
53. $x(x + 5)$ square meters **55.** $0.18(x + 1000)$
57. $\frac{16.50}{x}$ dollars per pound **59.** $90 - x$ degrees
61. $x(x + 5) = 8$ **63.** $x - 0.07x = 84{,}532$
65. $500x = 100$ **67.** $0.05x + 0.10(x + 2) = 3.80$
69. $x + 5 = 13$ **71.** $x + (x + 1) + (x + 2) = 42$
73. $x(x + 1) = 182$ **75.** $0.12x = 3000$ **77.** $0.05x = 13$
79. $x(x + 5) = 126$ **81.** $5n + 10(n - 1) = 95$
83. $x + x - 38 = 180$ **85.** $x(x + 3) = 24$
87. $w(w - 4) = 24$

Section 2.5 Warm-ups F T T F F T T T F T

1. 46, 47, 48 **3.** 75, 77 **5.** 47, 48, 49, 50
7. Length 50 meters, width 25 meters
9. Width 42 inches, length 46 inches **11.** 13 inches
13. 35° **15.** 65 miles per hour **17.** 55 miles per hour
19. 4 hours, 2048 miles **21.** Raiders 32, Vikings 14
23. 3 hours, 106 miles
25. Crawford 1906, Wayne 1907, Stewart 1908

Section 2.6 Warm-ups T T F T F T

1. \$320 **3.** \$400 **5.** \$80,000 **7.** \$30.24
9. 100 Fund \$10,000, 101 Fund \$13,000
11. Fidelity \$14,000, Price \$11,000 **13.** 30 gallons

15. 20 liters of 5% alcohol, 10 liters of 20% alcohol
17. 55,700 **19.** \$15,000 **21.** 75% **23.** 600
25. 42 private rooms, 30 semiprivate rooms
27. 12 pounds **29.** 4 nickels, 6 dimes
31. 800 gallons

Section 2.7 Warm-ups T T F T F T F F T F

1. True **3.** True **5.** False **7.** True **9.** True

11. True **13.** True **15.** −1 0 1 2 3 4 5

17. −4 −3 −2 −1 0 1 2

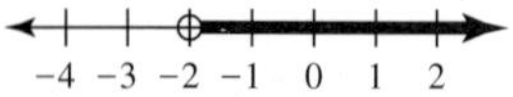

19. −5 −4 −3 −2 −1 0 1

21. −4 −3 −2 −1 0 1 2

23. $\frac{1}{2}$ −1 0 1 2 3 **25.** 5.3 1 2 3 4 5 6 7

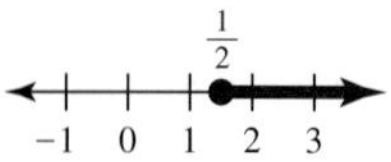

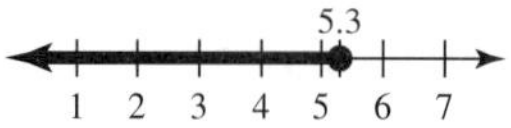

27. −4 −3 −2 −1 0 1 2

29. 2 3 4 5 6 7 8

31. −5 −4 −3 −2 −1 0 1

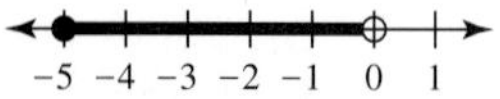

33. 20 40 60 80 100 120 **35.** $x > 3$ **37.** $x \le 2$

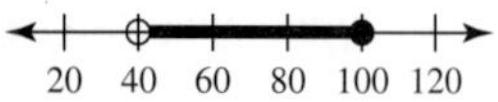

39. $0 < x < 2$ **41.** $-5 < x \le 7$ **43.** $x > -4$ **45.** Yes
47. No **49.** No **51.** Yes **53.** Yes **55.** Yes
57. No **59.** Yes **61.** No **63.** 0, 5.1 **65.** 5.1
67. 5.1 **69.** −5.1, 0, 5.1 **71.** $0.08x > 1500$
73. $\frac{44 + 72 + x}{3} \ge 60$ **75.** $396 < 8x < 453$
77. $60 < 90 - x < 70$

Section 2.8 Warm-ups T F F T F T F T T F

1. $x > -7$ **3.** $w \ge 3$ **5.** $k > 1$ **7.** $y \le -8$

9. $x > -3$ −5 −4 −3 −2 −1 0 1

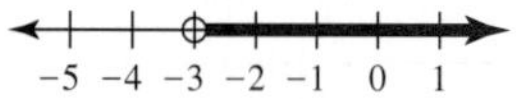

11. $w > -2$ −2 −1 0 1 2 3

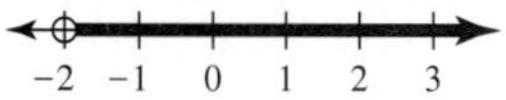

13. $b < 4$ −1 0 1 2 3 4 5

15. $z \ge -2$ −4 −3 −2 −1 0 1 2

17. $y < 3$ −1 0 1 2 3 4 5

19. $z \ge -2$ −2 −1 0 1 2 3

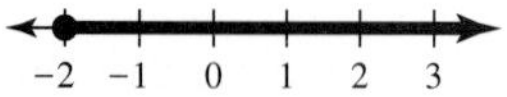

21. $r > -3$ −4 −3 −2 −1 0 1

23. $p < 13$ 9 10 11 12 13 14 15

25. $q \le 24$ 20 22 24 26

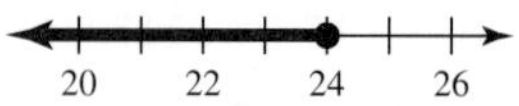

27. $t \le 12$ 8 9 10 11 12 13 14

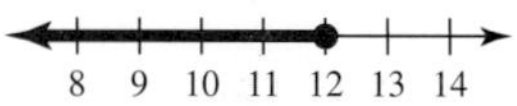

29. $x < -11$ −15 −14 −13 −12 −11 −10 −9

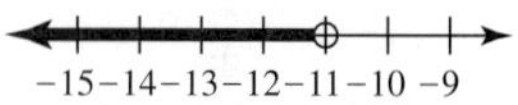

31. $x > -10$ −12 −11 −10 −9 −8 −7 −6

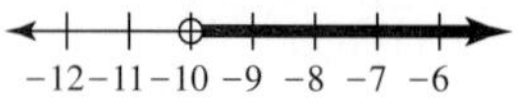

33. $x < 614.3$ 614.3

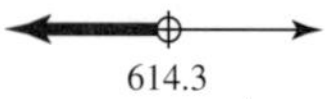

35. $8 < x < 10$ 6 7 8 9 10 11 12

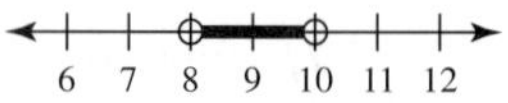

37. $1 < v < \frac{9}{2}$ $\frac{9}{2}$ 0 1 2 3 4 5 6

39. $-2 \le k \le 9$ 9 −2 0 2 4 6 8 10

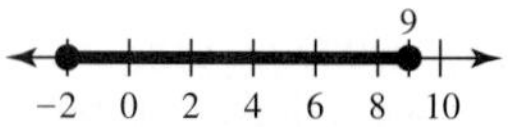

41. $-5 \le y < 3$ −5 −4 −3 −2 −1 0 1 2 3

43. $12 < u < 30$ 12 18 24 30

45. $-5 < m \le 5$ −5 −3 −1 0 1 3 5

47. $102.1 < x < 108.3$ 102.1 108.3

49. $x \le 6$ 1 2 3 4 5 6 7

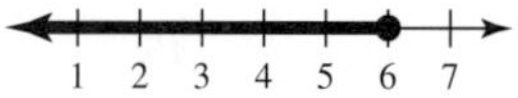

51. $x < 0$ −2 −1 0 1 2 3

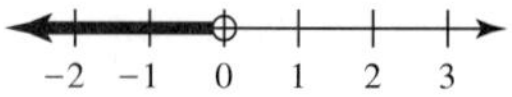

53. $2 < x < 3$ 1 2 3 4

55. At least 28 meters **57.** Less than \$9358
59. At most \$550 **61.** At least 64
63. Between 81 and 94.5 inclusive
65. Between 49.5 and 56.625 miles per hour
67. Between 55° and 85°

Chapter 2 Review

1. 7 **3.** $\frac{7}{3}$ **5.** −2 **7.** 0 **9.** −8 **11.** −30

13. $\frac{3}{7}$ **15.** 400 **17.** 100

19. No solution, inconsistent **21.** R, identity
23. All nonzero real numbers, identity
25. 24, conditional **27.** 80, conditional
29. 1000, conditional **31.** $\frac{1}{4}$ **33.** $\frac{21}{8}$ **35.** $-\frac{4}{5}$

37. 4 **39.** 24 **41.** −100 **43.** $x = -\frac{b}{a}$

45. $x = \frac{b+2}{a}$ **47.** $x = \frac{V}{LW}$ **49.** $x = -\frac{b}{3}$

51. $y = -\frac{5}{2}x + 3$ **53.** $y = -\frac{1}{2}x + 4$

55. $y = -2x + 16$ **57.** -13 **59.** $-\frac{2}{5}$ **61.** 17

63. $x + 9$ **65.** x and $x + 8$ **67.** $0.65x$

69. $x(x + 5) = 98$ **71.** $2(x + 10) = 3x$

73. $x + x + 2 + x + 4 = 88$ **75.** $t + 2t + t - 10 = 180$

77. 77, 79, 81 **79.** Betty 45 mph, Lawanda 60 mph

81. Wanda \$36,000, husband \$30,000 **83.** No

85. No **87.** $x > 1$ **89.** $x \geq 2$ **91.** $-3 \leq x < 3$

93. $x < -1$ **95.** $x > -1$

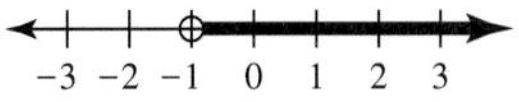

97. $x < 3$

99. $x \leq -4$

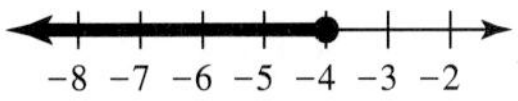

101. $x > -4$

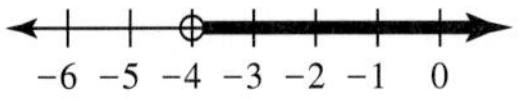

103. $-1 < x < 5$

105. $-2 < x \leq \frac{1}{2}$

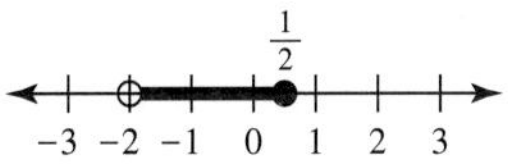

107. $0 \leq x \leq 3$

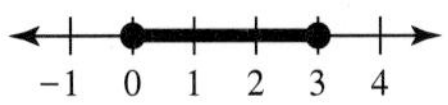

109. $0 < x < 1$

111. \$700, \$60 **113.** 400 **115.** 31°

117. Less than 6 feet

Chapter 2 Test

1. -7 **2.** 2 **3.** -9 **4.** 700 **5.** $y = \frac{2}{3}x - 3$

6. $a = \frac{m + w}{P}$ **7.** $-3 < x \leq 2$ **8.** $x > 1$

9. $w > 19$

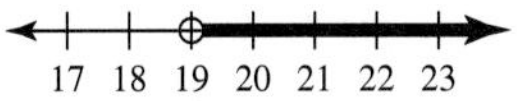

10. $-7 < x < -1$

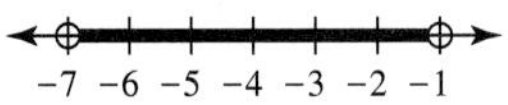

11. $1 < x < 3$

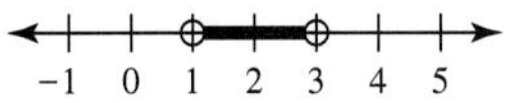

12. $y > -6$

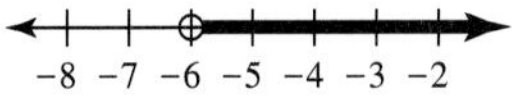

13. No solution **14.** R **15.** 1 **16.** $\frac{7}{6}$

17. Width 14 meters **18.** Height 9 inches

19. 150 liters **20.** At most \$2000 **21.** 30°, 60°, 90°

Tying It All Together Chapters 1–2

1. $8x$ **2.** $15x^2$ **3.** $2x + 1$ **4.** $4x - 7$ **5.** $-2x + 13$

6. 60 **7.** 72 **8.** -10 **9.** $-2x^3$ **10.** -1 **11.** 1

12. R **13.** 0 **14.** 1 **15.** 2 **16.** 2 **17.** $\frac{13}{2}$

18. 200 **19.** \$13,600, \$10,000

CHAPTER 3

Section 3.1 Warm-ups F F T T F T F T T F

1. -3, 7 **3.** 0, 6 **5.** $\frac{1}{3}, \frac{7}{2}$ **7.** Monomial, 0

9. Monomial, 3 **11.** Binomial, 1 **13.** Trinomial, 10

15. Binomial, 6 **17.** Trinomial, 3 **19.** 6, 10

21. -5, -85 **23.** -3.5975, 22.7296

25. -4.97665, -8.0251072 **27.** $4x - 8$ **29.** $2q$

31. $x^2 + 3x - 2$ **33.** $x^3 + 9x - 7$ **35.** $3a^2 - 7a - 4$

37. $-3w^2 - 8w + 5$ **39.** $9.66x^2 - 1.93x - 1.49$

41. $-4x + 6$ **43.** -5 **45.** $-z^2 + 2z$

47. $w^5 + w^4 - w^3 - w^2$

49. $2t + 13$ **51.** $-8y + 7$ **53.** $-22.85x - 423.2$

55. $4a + 2$ **57.** $-2x + 4$ **59.** $2a$ **61.** $-5m + 7$

63. $4x^2 + 1$ **65.** $a^3 - 9a^2 + 2a + 7$ **67.** $-3x + 9$

69. $2y^3 + 7y^2 - 4y - 14$ **71.** $5m - 5$ **73.** $2y + 1$

75. $2x^2 + 7x - 5$ **77.** $-3m + 3$ **79.** $-11y - 3$

81. $2x^2 - 6x + 12$ **83.** $-5z^4 - 8z^3 + 3z^2 + 7$

85. $100x + 500$ dollars, \$5,500

87. $6x + 3$ meters, 27 meters

89. $5x + 40$ miles, 140 miles **91.** 800 feet, 800 feet

93. $0.17x + 74.47$ dollars, \$244.47

95. 1321.39 calories

Section 3.2 Warm-ups F F T F T T T T T F

1. $27x^5$ **3.** $14a^{11}$ **5.** $-30x^4$ **7.** $27x^{17}$ **9.** $-54s^2t^2$

11. $24t^7w^8$ **13.** $25y^2$ **15.** $4x^6$ **17.** $4y^7 - 8y^3$

19. $-18y^2 + 12y$ **21.** $-3y^3 + 15y^2 - 18y$

23. $-xy^2 + x^3$ **25.** $15a^4b^3 - 5a^5b^2 - 10a^6b$

27. $-2t^5v^3 + 3t^3v^2 + 2t^2v^2$ **29.** $x^2 + 3x + 2$

31. $x^2 + 2x - 15$ **33.** $t^2 - 13t + 36$

35. $x^3 + 3x^2 + 4x + 2$ **37.** $6y^3 + y^2 + 7y + 6$

39. $2y^8 - 3y^6z - 5y^4z^2 + 3y^2z^3$ **41.** $2a^2 + 7a - 15$

43. $14x^2 + 95x + 150$ **45.** $20x^2 - 7x - 6$

47. $2am - 6an + mb - 3nb$ **49.** $x^3 + 9x^2 + 16x - 12$

51. $-4a^4 + 9a^2 - 8a - 12$ **53.** $x^2 - y^2$ **55.** $x^3 + y^3$

57. $u - 3t$ **59.** $-3x - y$ **61.** $3a^2 + a - 6$

63. $-3v^2 - v + 6$ **65.** $-6x^2 + 27x$

67. $-6x^2 + 27x + 2$ **69.** $-x - 7$ **71.** $36x^{12}$

73. $-6a^3b^{10}$ **75.** $25x^2 + 60x + 36$ **77.** $25x^2 - 36$

79. $6x^7 - 8x^4$ **81.** $m^3 - 1$

83. $3x^3 - 5x^2 - 25x + 18$

85. $x^2 + 4x$ square feet, 140 square feet

87. $x^2 + \frac{1}{2}x$ square feet, 27.5 square feet **89.** $x^2 + 5x$

91. $8.05x^2 + 15.93x + 6.12$ square meters

93. 30,000, \$300,000, $40{,}000p - 1000p^2$

95. $10x^5 + 10x^4 + 10x^3 + 10x^2 + 10x$, \$67.16

Section 3.3 Warm-ups F T T T T F T F F F

1. $x^2 + 6x + 8$ **3.** $a^2 - a - 6$ **5.** $2x^2 - 5x + 2$
7. $2a^2 - a - 3$ **9.** $w^2 - 60w + 500$
11. $y^2 - ay + 5y - 5a$ **13.** $5w - w^2 + 5m - mw$
15. $10m^2 - 9mt - 9t^2$ **17.** $45a^2 + 53ab + 14b^2$
19. $x^4 - 3x^2 - 10$ **21.** $h^6 + 10h^3 + 25$
23. $3b^6 + 14b^3 + 8$ **25.** $y^3 - 2y^2 - 3y + 6$
27. $6m^6 + 7m^3n^2 - 3n^4$ **29.** $12u^4v^2 + 10u^2v - 12$
31. $b^2 + 9b + 20$ **33.** $x^2 + 6x - 27$
35. $a^2 + 10a + 25$ **37.** $4x^2 - 4x + 1$ **39.** $z^2 - 100$
41. $a^2 + 2ab + b^2$ **43.** $a^2 - 3a + 2$
45. $2x^2 + 5x - 3$ **47.** $5t^2 - 7t + 2$
49. $h^2 - 16h + 63$ **51.** $h^2 + 14hw + 49w^2$
53. $4h^4 - 4h^2 + 1$ **55.** $8a^2 + a - \frac{1}{4}$
57. $\frac{1}{8}x^2 + \frac{1}{6}x - \frac{1}{6}$ **59.** $-12x^6 - 26x^5 + 10x^4$
61. $x^3 + 3x^2 - x - 3$ **63.** $9x^3 + 45x^2 - 4x - 20$
65. $2x + 10$ **67.** $2x^2 + 5x - 3$ square feet
69. $5.2555x^2 + 0.41095x - 1.995$ square meters

Section 3.4 Warm-ups F T T T F T T T F F

1. $x^2 + 2x + 1$ **3.** $y^2 + 8y + 16$ **5.** $9x^2 + 48x + 64$
7. $s^2 + 2st + t^2$ **9.** $4x^2 + 4xy + y^2$
11. $4t^2 + 12ht + 9h^2$ **13.** $a^2 - 6a + 9$
15. $t^2 - 2t + 1$ **17.** $9t^2 - 12t + 4$ **19.** $s^2 - 2st + t^2$
21. $9a^2 - 6ab + b^2$ **23.** $9z^2 - 30yz + 25y^2$
25. $a^2 - 25$ **27.** $y^2 - 1$ **29.** $9x^2 - 64$ **31.** $r^2 - s^2$
33. $64y^2 - 9a^2$ **35.** $25x^4 - 4$ **37.** $x^3 + 3x^2 + 3x + 1$
39. $8a^3 - 36a^2 + 54a - 27$
41. $a^4 - 12a^3 + 54a^2 - 108a + 81$
43. $a^4 + 4a^3b + 6a^2b^2 + 4ab^3 + b^4$ **45.** $a^2 - 400$
47. $x^2 + 15x + 56$ **49.** $16x^2 - 1$ **51.** $81y^2 - 18y + 1$
53. $6t^2 - 7t - 20$ **55.** $4t^2 - 20t + 25$ **57.** $4t^2 - 25$
59. $x^4 - 1$ **61.** $4y^6 - 36y^3 + 81$
63. $4x^6 + 12x^3y^2 + 9y^4$ **65.** $\frac{1}{4}x^2 + \frac{1}{3}x + \frac{1}{9}$
67. $0.04x^2 - 0.04x + 0.01$ **69.** $a^3 + 3a^2b + 3ab^2 + b^3$
71. $2.25x^2 + 11.4x + 14.44$ **73.** $12.25t^2 - 6.25$
75. $x^2 - 25$ square feet, 25 square feet smaller
77. $3.14b^2 + 6.28b + 3.14$ square meters
79. $v = k(R^2 - r^2)$ **81.** $P + 2Pr + Pr^2$, \$242
83. \$14,380.95

Section 3.5 Warm-ups F F T F T F T T T T

1. 1 **3.** 1 **5.** 1 **7.** 1 **9.** x^6 **11.** $\frac{3}{a^5}$ **13.** $\frac{-4}{x^4}$
15. $-y$ **17.** $-3x$ **19.** $\frac{-3}{x^3}$ **21.** $x - 2$
23. $x^3 + 3x^2 - x$ **25.** $4xy - 2x + y$ **27.** $y^2 - 3xy$
29. Quotient $x + 2$, remainder 7
31. Quotient 2, remainder -10
33. Quotient $a^2 + 2a + 8$, remainder 13
35. Quotient $x - 4$, remainder 4
37. Quotient $h^2 + 3h + 9$, remainder 0
39. Quotient $2x - 3$, remainder 1
41. Quotient $x^2 + 1$, remainder -1
43. $3 + \frac{15}{x - 5}$ **45.** $-1 + \frac{3}{x + 3}$ **47.** $1 - \frac{1}{x}$
49. $3 + \frac{1}{x}$ **51.** $x - 1 + \frac{1}{x + 1}$ **53.** $x - 2 + \frac{8}{x + 2}$
55. $x^2 + 2x + 4 + \frac{8}{x - 2}$ **57.** $x^2 + \frac{3}{x}$ **59.** $-3a$
61. $\frac{4t^4}{w^5}$ **63.** $-a + 4$ **65.** $x - 3$ **67.** $-6x^2 + 2x - 3$
69. $t + 4$ **71.** $2w + 1$ **73.** $4x^2 - 6x + 9$
75. $t^2 - t + 3$ **77.** $v^2 - 2v + 1$ **79.** $x - 5$ meters

Section 3.6 Warm-ups F F F F F F T F F T

1. 200 **3.** 2^{15} **5.** $6u^{10}v^8$ **7.** a^4b^{10} **9.** $\frac{-1}{2a^4}$
11. $\frac{2a^6}{5}$ **13.** x^6 **15.** $2x^{12}$ **17.** $\frac{1}{t^2}$ **19.** $\frac{1}{2}$
21. x^3y^6 **23.** $-8t^{15}$ **25.** $-8x^6y^{15}$ **27.** $a^9b^2c^{14}$
29. $\frac{x^{12}}{64}$ **31.** $\frac{16a^8}{b^{12}}$ **33.** $-\frac{x^6}{8y^3}$ **35.** $\frac{4z^{12}}{x^8}$
37. 35 **39.** 125 **41.** $\frac{8}{27}$ **43.** 200 **45.** $15x^{11}$
47. $-125x^{12}$ **49.** $-27y^6z^{19}$ **51.** $\frac{3v}{u}$ **53.** $-16x^9t^6$
55. $\frac{8}{x^6}$ **57.** $\frac{-32a^{15}b^{20}}{c^{25}}$ **59.** $\frac{y^5}{32x^5}$ **61.** $P(1 + r)^{15}$

Chapter 3 Review

1. $5w - 2$ **3.** $-6x + 4$ **5.** $2w^2 - 7w - 4$
7. $-2m^2 + 3m - 1$ **9.** $-50x^{11}$ **11.** $121a^{14}$
13. $-4x + 15$ **15.** $3x^2 - 10x + 12$
17. $15m^5 - 3m^3 + 6m^2$ **19.** $x^3 - 7x^2 + 20x - 50$
21. $3x^3 - 8x^2 + 16x - 8$ **23.** $q^2 + 2q - 48$
25. $2t^2 - 21t + 27$ **27.** $20y^2 - 7y - 6$
29. $6x^4 + 13x^2 + 5$ **31.** $z^2 - 49$ **33.** $y^2 + 14y + 49$
35. $w^2 - 6w + 9$ **37.** $x^4 - 9$ **39.** $9a^2 + 6a + 1$
41. $y^2 - 8y + 16$ **43.** $-5x^2$ **45.** $\frac{-2a^2}{b^2}$
47. $-x + 3$ **49.** $-3x^2 + 2x - 1$ **51.** -1
53. $m^3 + 2m^2 + 4m + 8$
55. Quotient $m^2 - 3m + 6$, remainder 0
57. Quotient $b - 5$, remainder 15
59. Quotient $2x - 1$, remainder -8
61. Quotient $x^2 + 2x - 9$, remainder 1
63. $2 + \frac{6}{x - 3}$ **65.** $-2 + \frac{2}{1 - x}$ **67.** $x - 1 - \frac{2}{x + 1}$
69. $x - 1 + \frac{1}{x + 1}$ **71.** $6y^{30}$ **73.** $\frac{-5}{c^6}$ **75.** b^{30}

77. $-8x^9y^6$ **79.** $\frac{8a^3}{b^3}$ **81.** $\frac{8x^6y^{15}}{z^{18}}$
83. $x^2 + 10x + 21$ **85.** $t^2 - 7ty + 12y^2$ **87.** 2
89. $-27h^3t^{18}$ **91.** $2w^2 - 9w - 18$ **93.** $9u^2 - 25v^2$
95. $9h^2 + 30h + 25$ **97.** $x^3 + 9x^2 + 27x + 27$
99. $14s^5t^6$ **101.** $\frac{k^8}{16}$ **103.** $x^2 - 9x - 5$
105. $5x^2 - x - 12$ **107.** $x^3 - x^2 - 19x + 4$
109. $x + 6$
111. $P = 4w + 88$, $A = w^2 + 44w$, $P = 288$ ft, $A = 4700$ ft^2
113. $R = -15p^2 + 600p$, \$5040, \$20

Chapter 3 Test

1. $7x^3 + 4x^2 + 2x - 11$ **2.** $-x^2 - 9x + 2$ **3.** $-35x^8$
4. $12x^5y^9$ **5.** $2ab^4$ **6.** -1 **7.** $-2y^2 + 3y$
8. $\frac{3a^4}{b^2}$ **9.** $\frac{-32a^5}{b^5}$ **10.** $\frac{16v^6}{u^{10}}$ **11.** $x^2 + x - 1$
12. $15x^5 - 21x^4 + 12x^3 - 3x^2$ **13.** $x^2 + 3x - 10$
14. $6a^2 + a - 35$ **15.** $a^2 - 14a + 49$
16. $16x^2 + 24xy + 9y^2$ **17.** $b^2 - 9$ **18.** $9t^4 - 49$
19. $4x^4 + 5x^2 - 6$ **20.** $x^3 - 3x^2 - 10x + 24$
21. $2 + \frac{6}{x - 3}$ **22.** $x - 5 + \frac{15}{x + 2}$
23. Quotient $x - 2$, remainder 3
24. $-2x^2 + x + 15$
25. $x^2 + 4x$ ft^2, $4x + 8$ ft, 32 ft^2, 24 ft
26. $R = -150q^2 + 3000q$, \$14,400

Tying It All Together Chapters 1–3

1. 8 **2.** -9 **3.** 41 **4.** 2^{25} **5.** 32 **6.** 992 **7.** 144
8. -1 **9.** 64 **10.** 34 **11.** 899 **12.** 961
13. $x^2 + 8x + 15$ **14.** $x + 3$ **15.** $4x + 15$
16. $x^3 + 13x^2 + 55x + 75$ **17.** $-15t^5v^7$ **18.** $5tv$
19. $3y - 4$ **20.** $4y^2 - 5y - 3$ **21.** $-\frac{1}{2}$ **22.** 7
23. $\frac{3}{2}$ **24.** 4 **25.** -3 **26.** $-\frac{2}{3}$
27. $\frac{2.25n + 100{,}000}{n}$, \$102.25, \$3.25, \$2.35

CHAPTER 4

Section 4.1 Warm-ups F F F T T T T F F T

1. $2 \cdot 3^2$ **3.** $2^2 \cdot 13$ **5.** $2 \cdot 7^2$ **7.** $2^2 \cdot 5 \cdot 23$
9. $2^2 \cdot 3 \cdot 7 \cdot 11$ **11.** 4 **13.** 12 **15.** 8 **17.** 4
19. 1 **21.** $2x$ **23.** $2x$ **25.** xy **27.** $12ab$
29. 1 **31.** $6ab$ **33.** $9(3x)$ **35.** $8t(3t)$ **37.** $4y^2(9y^3)$
39. $uv(u^3v^2)$ **41.** $2m^4(-7n^3)$ **43.** $-3x^3yz\,(11xy^2z)$
45. $x(x^2 - 6)$ **47.** $5a(x + y)$ **49.** $h^3(h^2 - 1)$
51. $2k^3m^4(-k^4 + 2m^2)$ **53.** $2x(x^2 - 3x + 4)$
55. $6x^2t(2x^2 + 5x - 4t)$ **57.** $(x - 3)(a + b)$
59. $(y + 1)^2(a + b)$ **61.** $9a^2b^4(4ab - 3 + 2b^5)$
63. $8(x - y)$, $-8(-x + y)$
65. $4x(-1 + 2x)$, $-4x(1 - 2x)$
67. $1(x - 5)$, $-1(-x + 5)$ **69.** $1(4 - 7a)$, $-1(-4 + 7a)$
71. $8a^2(-3a + 2)$, $-8a^2(3a - 2)$
73. $6x(-2x - 3)$, $-6x(2x + 3)$
75. $2x(-x^2 - 3x + 7)$, $-2x(x^2 + 3x - 7)$
77. $2ab(2a^2 - 3ab - 2b^2)$, $-2ab(-2a^2 + 3ab + 2b^2)$
79. $x + 2$ hours **81.** $S = 2\pi r(r + h)$

Section 4.2 Warm-ups F T F F T T F F T T

1. $(a - 2)(a + 2)$ **3.** $(x - 7)(x + 7)$
5. $(y + 3x)(y - 3x)$ **7.** $(5a + 7b)(5a - 7b)$
9. $(11m + 1)(11m - 1)$ **11.** $(3w - 5c)(3w + 5c)$
13. Perfect square trinomial **15.** Neither
17. Perfect square trinomial **19.** Neither
21. Difference of two squares
23. Perfect square trinomial
25. $(x + 6)^2$ **27.** $(a - 2)^2$ **29.** $(2w + 1)^2$
31. $(4x - 1)^2$ **33.** $(2t + 5)^2$ **35.** $(3w + 7)^2$
37. $(n + t)^2$ **39.** $5(x - 5)(x + 5)$
41. $-2(x - 3)(x + 3)$ **43.** $a(a - b)(a + b)$
45. $3(x + 1)^2$ **47.** $-5(y - 5)^2$ **49.** $x(x - y)^2$
51. $-3(x - y)(x + y)$ **53.** $2a(x - 7)(x + 7)$
55. $3a(b - 3)^2$ **57.** $-4m(m - 3n)^2$
59. $(b + c)(x + y)$ **61.** $(x - 2)(x + 2)(x + 1)$
63. $(3 - x)(a - b)$ **65.** $(a^2 + 1)(a + 3)$
67. $(a + 3)(x + y)$ **69.** $(c - 3)(ab + 1)$
71. $(a + b)(x - 1)(x + 1)$ **73.** $(y + b)(y + 1)$
75. $6ay(a + 2y)^2$ **77.** $6ay(2a - y)(2a + y)$
79. $2a^2y(ay - 3)$ **81.** $(b - 4w)(a + 2w)$
83. $h = -16(t - 20)(t + 20)$, 6336 feet
85. $y - 3$ inches

Section 4.3 Warm-ups T T F F T F T F F F

1. $(x + 3)(x + 1)$ **3.** $(x + 3)(x + 6)$ **5.** $(a - 3)(a - 4)$
7. $(b - 6)(b + 1)$ **9.** $(y + 2)(y + 5)$ **11.** $(a - 2)(a - 4)$
13. $(m - 8)(m - 2)$ **15.** $(w + 10)(w - 1)$
17. $(w - 4)(w + 2)$ **19.** Prime **21.** $(m + 16)(m - 1)$
23. Prime **25.** $(z - 5)(z + 5)$ **27.** Prime
29. $(m + 2)(m + 10)$ **31.** Prime
33. $(m - 18)(m + 1)$ **35.** Prime **37.** $(t + 8)(t - 3)$
39. $(t - 6)(t + 4)$ **41.** $(t - 20)(t + 10)$
43. $(x - 15)(x + 10)$ **45.** $(y + 3)(y + 10)$
47. $(x - 6y)(x + 2y)$ **49.** $(x - 12y)(x - y)$
51. $(x - 8s)(x + 3s)$ **53.** $w(w - 8)$
55. $2(w - 9)(w + 9)$ **57.** $x^2(w^2 + 9)$ **59.** $(w - 9)^2$
61. $6(w - 3)(w + 1)$ **63.** $2x^2(4 - x)(4 + x)$
65. $3(w + 3)(w + 6)$ **67.** $w(w^2 + 18w + 36)$
69. $8v(w + 2)^2$ **71.** $6xy(x + 3y)(x + 2y)$
73. $x + 4$ feet **75.** 3 feet and 5 feet

Section 4.4 Warm-ups T F T F T F F F F T

1. 2 and 10 **3.** -6 and 2 **5.** 3 and 4 **7.** -2 and -9
9. -3 and 4 **11.** $(2x + 1)(x + 1)$ **13.** $(2x + 1)(x + 4)$
15. $(3t + 1)(t + 2)$ **17.** $(2x - 1)(x + 3)$
19. $(3x - 1)(2x + 3)$ **21.** $(2x - 3)(x - 2)$
23. $(5b - 3)(b - 2)$ **25.** $(4y + 1)(y - 3)$ **27.** Prime

29. $(4x+1)(2x-1)$ **31.** $(3t-1)(3t-2)$
33. $(5x+1)(3x+2)$ **35.** $(5x-1)(3x-2)$
37. $(x+2)(3x+1)$ **39.** $(5x+1)(x+2)$
41. $(3a-1)(2a-5)$ **43.** $(5a+1)(a+2)$
45. $(2w+3)(2w+1)$ **47.** $(5x-2)(3x+1)$
49. $(4x-1)(2x-1)$ **51.** $(15x-1)(x-2)$
53. $2(x^2+9x-45)$ **55.** $(3x-5)(x+2)$
57. $(5x+y)(2x-y)$ **59.** $(6a-b)(7a-b)$
61. $w(9w-1)(9w+1)$ **63.** $2(2w-5)(w+3)$
65. $3(2x+3)^2$ **67.** $(3w+5)(2w-7)$
69. $3z(x-3)(x+2)$ **71.** $y^2(10x-9)(x+1)$
73. $(a+5b)(a-3b)$ **75.** $-t(3t+2)(2t-1)$
77. $2t^2(3t-2)(2t+1)$ **79.** $y(2x-y)(2x-3y)$
81. $-1(w-1)(4w-3)$ **83.** $-2a(2a-3b)(3a-b)$
85. $h=-8(2t+1)(t-3)$, 0 feet

Section 4.5 Warm-ups F F T T F T F T T F

1. $(x+4)(x-3)(x+2)$ **3.** $(x-1)(x+3)(x+2)$
5. $(x-2)(x^2+2x+4)$ **7.** $(x+5)(x^2-x+2)$
9. $(x+1)(x^2+x+1)$ **11.** $(m-1)(m^2+m+1)$
13. $(x+2)(x^2-2x+4)$ **15.** $(2w+1)(4w^2-2w+1)$
17. $(2t-3)(4t^2+6t+9)$ **19.** $(x-y)(x^2+xy+y^2)$
21. $(2t+y)(4t^2-2ty+y^2)$ **23.** $2(x-3)(x+3)$
25. $4(x+5)(x-3)$ **27.** $x(x+2)^2$ **29.** $5am(x^2+4)$
31. $(3x+1)^2$ **33.** $y(3x+2)(2x-1)$ **35.** Prime
37. $2(4m+1)(2m-1)$ **39.** $(3a+4)^2$
41. $2(3x-1)(4x-3)$ **43.** $3a(a-9)$
45. $2(2-x)(2+x)$ **47.** $x(6x^2-5x+12)$
49. $ab(a-2)(a+2)$ **51.** $(x-2)(x+2)^2$
53. $2w(w-2)(w^2+2w+4)$ **55.** $3w(a-3)^2$
57. $5(x-10)(x+10)$ **59.** $(2-w)(m+n)$
61. $3x(x+1)(x^2-x+1)$ **63.** $4(w^2+w-1)$
65. $a^2(a+10)(a-3)$ **67.** $aw(2w-3)^2$ **69.** $(t+3)^2$
71. Length $x+5$ centimeters, width $x+3$ centimeters

Section 4.6 Warm-ups F F T T T F T T T F

1. $-4, -5$ **3.** $-\frac{5}{2}, \frac{4}{3}$ **5.** 2, 7 **7.** 0, -7 **9.** -5, 4
11. $\frac{1}{2}, -3$ **13.** 0, -8 **15.** $-\frac{9}{2}, 2$ **17.** $\frac{2}{3}, -4$ **19.** 5
21. $\frac{3}{2}$ **23.** 0, -3, 3 **25.** $-4, -2, 2$ **27.** -1, 1, 3
29. 0, 4, 5 **31.** -4, 4 **33.** -3, 3 **35.** 0, -1, 1
37. $-3, -2$ **39.** $-\frac{3}{2}, -4$ **41.** -6, 4 **43.** -1, 3
45. -4, 2 **47.** $-5, -3, 5$
49. Length 12 feet, width 5 feet
51. Width 5 feet, length 12 feet
53. 2 and 3, or -3 and -2 **55.** 5 and 6
57. 25 seconds **59.** 6 seconds
61. Base 6 inches, height 13 inches
63. 20 feet by 20 feet **65.** 80 feet
67. 3 yards by 3 yards, 6 yards by 6 yards
69. 12 miles **71.** 25%

Chapter 4 Review

1. $2^4 \cdot 3^2$ **3.** $2 \cdot 29$ **5.** $2 \cdot 3 \cdot 5^2$ **7.** 18 **9.** $4x$
11. $3(x+2)$ **13.** $-2(-a+10)$ **15.** $a(2-a)$
17. $3x^2y(2y-3x^3)$ **19.** $3y(x^2-4x-3y)$
21. $(y-20)(y+20)$ **23.** $(w-4)^2$ **25.** $(2y+5)^2$
27. $(r-2)^2$ **29.** $2t(2t-3)^2$ **31.** $(x+6y)^2$
33. $(x-y)(x+5)$ **35.** $(b+8)(b-3)$
37. $(r-10)(r+6)$ **39.** $(y-11)(y+5)$
41. $(u+20)(u+6)$ **43.** $3t^2(t+4)$
45. $5w(w^2+5w+5)$ **47.** $ab(2a+b)(a+b)$
49. $x(3x-y)(3x+y)$ **51.** $(7t-3)(2t+1)$
53. $(3x+1)(2x-7)$ **55.** $(3p+4)(2p-1)$
57. $-2p(5p+2)(3p-2)$ **59.** $(6x+y)(x-5y)$
61. $2(4x+y)^2$ **63.** $5x(x^2+8)$ **65.** $(3x-1)(3x+2)$
67. $(x+2)(x-1)(x+1)$ **69.** $xy(x-16y)$
71. $(a+1)^2$ **73.** $(x^2+1)(x-1)$ **75.** $(a+2)(a+b)$
77. $-2(x-6)(x-2)$ **79.** $(m-10)(m^2+10m+100)$
81. $(x+2)(x^2-2x+5)$ **83.** $(x+4)(x+5)(x-3)$
85. 0, 5 **87.** 0, 5 **89.** $-\frac{1}{2}, 5$ **91.** $-4, -3, 3$
93. $-2, -1$ **95.** $-\frac{1}{2}, \frac{1}{4}$ **97.** 5, 11
99. 6 inches by 8 inches **101.** $v=k(R-r)(R+r)$
103. 6 feet

Chapter 4 Test

1. $2 \cdot 3 \cdot 11$ **2.** $2^4 \cdot 3 \cdot 7$ **3.** 16 **4.** 6 **5.** $3y^2$
6. $6ab$ **7.** $5x(x-2)$ **8.** $6y^2(x^2+2x+2)$
9. $3ab(a-b)(a+b)$ **10.** $(a+6)(a-4)$
11. $(2b-7)^2$ **12.** $3m(m^2+9)$ **13.** $(a+b)(x-y)$
14. $(a-5)(x-2)$ **15.** $(3b-5)(2b+1)$
16. $(m+2n)^2$ **17.** $(2a-3)(a-5)$
18. $z(z+3)(z+6)$ **19.** $(x-1)(x-2)(x-3)$
20. $\frac{3}{2}, -4$ **21.** 0, -2, 2 **22.** $-2, \frac{5}{6}$
23. Length 12 feet, width 9 feet **24.** -4 and 8

Tying It All Together Chapters 1–4

1. -1 **2.** 2 **3.** -3 **4.** 57 **5.** 16 **6.** 7 **7.** $2x^2$
8. $3x$ **9.** $3+x$ **10.** $6x$ **11.** $24yz$ **12.** $6y+8z$
13. $4z-1$ **14.** t^6 **15.** t^{10} **16.** $4t^6$
17. $x < -9$
−13 −12 −11 −10 −9 −8 −7
18. $x \geq 3$
1 2 3 4 5 6 7
19. $x > 12$
10 11 12 13 14 15 16
20. $x < 600$
0 200 400 600 800
21. $\frac{3}{2}$ **22.** $-\frac{1}{2}$ **23.** 3, -5 **24.** $\frac{3}{2}, -\frac{1}{2}$ **25.** 0, 3
26. 0, 1 **27.** R **28.** No solution **29.** 10
30. 40 **31.** -3, 3 **32.** $-5, \frac{3}{2}$
33. Length 21 feet, width 13.5 feet

CHAPTER 5

Section 5.1 Warm-ups F T T F F T T T F F T

1. -1 **3.** $\frac{5}{3}$ **5.** $4, -4$ **7.** Any number can be used.
9. $\frac{2}{9}$ **11.** $\frac{7}{15}$ **13.** $\frac{2a}{5}$ **15.** $\frac{13}{5w}$ **17.** $\frac{3x+1}{3}$
19. $\frac{2}{3}$ **21.** $w-7$ **23.** $\frac{a-1}{a+1}$ **25.** $\frac{x+1}{2x-2}$
27. $\frac{x+3}{7}$ **29.** x^3 **31.** $\frac{1}{z^5}$ **33.** $-2x^2$
35. $\frac{-3m^3n^2}{2}$ **37.** $\frac{-3}{4c^3}$ **39.** $\frac{5c}{3a^4b^{16}}$ **41.** $\frac{35}{44}$
43. $\frac{11}{8}$ **45.** $\frac{21}{10x^4}$ **47.** $\frac{33a^4}{16}$ **49.** -1 **51.** $-h-t$
53. $\frac{-2}{3h+g}$ **55.** $\frac{-x-2}{x+3}$ **57.** -1 **59.** $\frac{-2y}{3}$
61. $\frac{x+2}{2-x}$ **63.** $\frac{-6}{a+3}$ **65.** $\frac{x^4}{2}$ **67.** $\frac{x+2}{2x}$ **69.** -1
71. $\frac{-2}{c+2}$ **73.** $\frac{x+2}{x-2}$ **75.** $\frac{-2}{x+3}$ **77.** q^2
79. $\frac{u+2}{u-8}$ **81.** $\frac{a^2+2a+4}{2}$ **83.** $y+2$
85. $-0.6, 9, 401, -199$ **87.** $\frac{300}{x+10}$ hours
89. $\frac{4.50}{x+4}$ dollars per pound **91.** $\frac{1}{x}$ pool per hour
93. a) \$0.75 **b)** \$0.75, \$0.63, \$0.615 **c)** Approaches \$0.60

Section 5.2 Warm-ups T T T F T F F T T T

1. $\frac{7}{9}$ **3.** $\frac{18}{5}$ **5.** $\frac{42}{5}$ **7.** $\frac{a}{44}$ **9.** $\frac{-x^5}{a^3}$
11. $\frac{18t^8y^7}{w^4}$ **13.** $\frac{2a}{a-b}$ **15.** $3x-9$ **17.** $\frac{8a+8}{5a^2+5}$
19. 30 **21.** $\frac{2}{3}$ **23.** $\frac{10}{9}$ **25.** $\frac{7x}{2}$ **27.** $\frac{2m^2}{3n^6}$ **29.** -3
31. $\frac{2}{x+2}$ **33.** $\frac{1}{4t-20}$ **35.** x^2-1 **37.** $2x-4y$
39. $\frac{x+2}{2}$ **41.** $\frac{x^2+9}{15}$ **43.** $9x+9y$ **45.** -3
47. $\frac{a+b}{a}$ **49.** $\frac{2b}{a}$ **51.** $\frac{y}{x}$ **53.** $\frac{-a^6b^8}{2}$
55. $\frac{1}{9m^3n}$ **57.** $\frac{x^2+5x}{3x-1}$ **59.** $\frac{a^3+8}{2a-4}$ **61.** 1
63. $\frac{(m+3)^2}{(m-3)(m+k)}$ **65.** 5 m^2

Section 5.3 Warm-ups F F T T T F F F F T T

1. $\frac{9}{27}$ **3.** $\frac{14x}{2x}$ **5.** $\frac{15t}{3bt}$ **7.** $\frac{-36z^2}{8awz}$ **9.** $\frac{10a^2}{15a^3}$
11. $\frac{8xy^3}{10x^2y^5}$ **13.** $\frac{-20}{-8x-8}$ **15.** $\frac{-32ab}{20b^2-20b^3}$
17. $\frac{3x-6}{x^2-4}$ **19.** $\frac{3x^2+3x}{x^2+2x+1}$ **21.** $\frac{y^2-y-30}{y^2+y-20}$
23. 48 **25.** 180 **27.** $30a^2$ **29.** $12a^4b^6$
31. $(x-4)(x+4)^2$ **33.** $x(x+2)(x-2)$
35. $2x(x-4)(x+4)$ **37.** $\frac{4}{24}, \frac{9}{24}$ **39.** $\frac{9b}{252ab}, \frac{20a}{252ab}$
41. $\frac{2x^3}{6x^5}, \frac{9}{6x^5}$ **43.** $\frac{4x^4}{36x^3y^5z}, \frac{3y^6z}{36x^3y^5z}, \frac{6xy^4z}{36x^3y^5z}$
45. $\frac{2x^2+4x}{(x-3)(x+2)}, \frac{5x^2-15x}{(x-3)(x+2)}$ **47.** $\frac{4}{a-6}, \frac{-5}{a-6}$
49. $\frac{x^2-3x}{(x-3)^2(x+3)}, \frac{5x^2+15x}{(x-3)^2(x+3)}$
51. $\frac{w^2+3w+2}{(w-5)(w+3)(w+1)}, \frac{-2w^2-6w}{(w-5)(w+1)(w+3)}$
53. $\frac{-5x-10}{6(x-2)(x+2)}, \frac{6x}{6(x-2)(x+2)}, \frac{9x-18}{6(x-2)(x+2)}$
55. $\frac{2q+8}{(2q+1)(q-3)(q+4)}, \frac{3q-9}{(2q+1)(q-3)(q+4)}, \frac{8q+4}{(2q+1)(q-3)(q+4)}$

Section 5.4 Warm-ups F T T T T F T F T F

1. $\frac{1}{5}$ **3.** $\frac{3}{4}$ **5.** $-\frac{2}{3}$ **7.** $-\frac{3}{4}$ **9.** $\frac{5}{9}$ **11.** $\frac{103}{144}$
13. $-\frac{31}{40}$ **15.** $\frac{5}{24}$ **17.** $\frac{5}{w}$ **19.** 3 **21.** -2 **23.** $\frac{3}{h}$
25. $\frac{x-4}{x+2}$ **27.** $\frac{17}{10a}$ **29.** $\frac{w}{36}$ **31.** $\frac{b^2-4ac}{4a}$
33. $\frac{2w+3z}{w^2z^2}$ **35.** $\frac{-x-3}{x(x+1)}$ **37.** $\frac{3a+b}{(a-b)(a+b)}$
39. $\frac{15-4x}{5x(x+1)}$ **41.** $\frac{a^2+5a}{(a-3)(a+3)}$ **43.** 0
45. $\frac{7}{2a-2}$ **47.** $\frac{-2x+1}{(x-5)(x+2)(x-2)}$
49. $\frac{7x+17}{(x+2)(x-1)(x+3)}$ **51.** $\frac{2x^2-x-4}{x(x-1)(x+2)}$
53. $\frac{a+51}{6a(a-3)}$ **55.** $\frac{11}{x}$ feet
57. $\frac{315x+600}{x(x+5)}$ hours, 5 hours **59.** $\frac{4x+6}{x(x+3)}, \frac{5}{9}$

Section 5.5 Warm-ups F T F F F F F T T T

1. $-\frac{10}{3}$ **3.** $\frac{22}{7}$ **5.** $\frac{14}{17}$ **7.** $\frac{45}{23}$ **9.** $\frac{3a+b}{a-3b}$

11. $\frac{5a-3}{3a+1}$ **13.** $\frac{x^2-4x}{6x^2-2}$ **15.** $\frac{10b}{3b^2-4}$

17. $\frac{y-2}{3y+4}$ **19.** $\frac{x^2-2x+4}{x^2-3x-1}$ **21.** $\frac{5x-14}{2x-7}$

23. $\frac{a-6}{3a-1}$ **25.** $\frac{-3m+12}{4m-3}$ **27.** $\frac{-w+5}{9w+1}$

29. -1 **31.** $\frac{6x-27}{4x-6}$ **33.** $\frac{2x^2}{3y}$ **35.** $\frac{a^2+7a+6}{a+3}$

37. $1-x$ **39.** $\frac{32}{95}, \frac{11}{35}$

Section 5.6 Warm-ups F F F F F T T T T T

1. 12 **3.** 30 **5.** 5 **7.** 4 **9.** 4 **11.** 4 **13.** 3
15. 2 **17.** -5, 2 **19.** 2, 3 **21.** -3, 3 **23.** 2
25. No solution **27.** No solution **29.** 3 **31.** 10
33. 0 **35.** -5, 5 **37.** 3, 5 **39.** 1 **41.** 3 **43.** 0
45. 4 **47.** -20 **49.** 3 **51.** 3 **53.** $54\frac{6}{11}$ mm

Section 5.7 Warm-ups T F F T T T F F F T

1. $\frac{5}{7}$ **3.** $\frac{8}{15}$ **5.** $\frac{7}{2}$ **7.** $\frac{9}{14}$ **9.** $\frac{5}{2}$ **11.** $\frac{15}{1}$
13. 3 to 2 **15.** 9 to 16 **17.** 31 to 1 **19.** 2 to 3
21. 6 **23.** $-\frac{2}{5}$ **25.** $-\frac{27}{5}$ **27.** 5 **29.** $-\frac{3}{4}$ **31.** $\frac{5}{4}$
33. 108 **35.** 176,000 **37.** Lions 85, Tigers 51
39. 40 luxury cars, 60 sports cars **41.** 84 inches
43. 15 minutes **45.** 536.7 miles **47.** 3920 pounds
49. 6,000 **51.** 2 to 3, 1000 **53.** 4074

Section 5.8 Warm-ups T T F T T F F T F T

1. $y=2x-5$ **3.** $y=-\frac{1}{2}x-2$ **5.** $y=mx-mb-a$

7. $y=-\frac{1}{3}x-\frac{1}{3}$ **9.** $C=\frac{B}{A}$ **11.** $p=\frac{a}{1+am}$

13. $m_1=\frac{r^2F}{km_2}$ **15.** $a=\frac{bf}{b-f}$ **17.** $r=\frac{S-a}{S}$

19. $P_2=\frac{P_1V_1T_2}{T_1V_2}$ **21.** $h=\frac{3V}{4\pi r^2}$ **23.** $\frac{5}{12}$ **25.** $-\frac{6}{23}$

27. $\frac{128}{3}$ **29.** -6 **31.** $\frac{6}{5}$
33. Marcie 4 mph, Frank 3 mph
35. Bob 25 mph, Pat 20 mph **37.** 5 mph
39. 6 hours **41.** 40 minutes **43.** 1 hour 36 minutes
45. Bananas 8 pounds, apples 10 pounds
47. 80 gallons

Chapter 5 Review

1. $\frac{6}{7}$ **3.** $\frac{c^2}{4a^2}$ **5.** $\frac{2w-3}{3w-4}$ **7.** $-\frac{x+1}{3}$ **9.** $\frac{1}{2}k$

11. $\frac{2x}{3y}$ **13.** a^2-a-6 **15.** $\frac{1}{2}$ **17.** 108

19. $24a^7b^3$ **21.** $12x(x-1)$ **23.** $(x+1)(x-2)(x+2)$

25. $\frac{15}{36}$ **27.** $\frac{10x}{15x^2y}$ **29.** $\frac{-10}{12-2y}$ **31.** $\frac{x^2+x}{x^2-1}$

33. $\frac{29}{63}$ **35.** $\frac{3x-4}{x}$ **37.** $\frac{2a-b}{a^2b^2}$

39. $\frac{27a^2-8a-15}{(2a-3)(3a-2)}$ **41.** $\frac{3}{a-8}$

43. $\frac{3x+8}{2(x+2)(x-2)}$ **45.** $-\frac{3}{14}$ **47.** $\frac{6b+4a}{3a-18b}$

49. $\frac{-2x+9}{3x-1}$ **51.** $\frac{x^2+x-2}{-4x+13}$ **53.** $-\frac{15}{2}$ **55.** 9

57. -3 **59.** $\frac{21}{2}$ **61.** 5 **63.** 8

65. 56 cups water, 28 cups rice **67.** $y=mx+b$

69. $m=\frac{1}{F-v}$ **71.** $y=4x-13$ **73.** 200 hours

75. Bert 60 cars, Ernie 50 cars **77.** 150,000 metric tons

79. $\frac{10}{2x}$ **81.** $\frac{-2}{5-a}$ **83.** $\frac{3x}{x}$ **85.** $2m$ **87.** $\frac{1}{6}$

89. $\frac{1}{a+1}$ **91.** $\frac{5-a}{5a}$ **93.** $\frac{a-2}{2}$ **95.** $b-a$

97. $\frac{1}{10a}$ **99.** $\frac{3}{2x}$ **101.** $\frac{4+y}{6xy}$ **103.** $\frac{8}{a-5}$

105. -1, 2 **107.** $-\frac{5}{3}$ **109.** 6 **111.** $\frac{1}{2}$

113. $\frac{3x+7}{(x-5)(x+5)(x+1)}$

115. $\frac{-5a}{(a-3)(a+3)(a+2)}$ **117.** $\frac{2}{5}$

Chapter 5 Test

1. -1, 1 **2.** $\frac{2}{3}$ **3.** 0 **4.** $-\frac{14}{45}$ **5.** $\frac{1+3y}{y}$

6. $\frac{4}{a-2}$ **7.** $\frac{-x+4}{(x+2)(x-2)(x-1)}$

8. $\frac{2}{3}$ **9.** $\frac{-2}{a+b}$ **10.** $\frac{a^3}{18b^4}$ **11.** $-\frac{4}{3}$ **12.** $\frac{3x-4}{-2x+6}$

13. $\frac{15}{7}$ **14.** 2, 3 **15.** 12 **16.** $y=-\frac{1}{5}x+\frac{13}{5}$

17. $c=\frac{3M-bd}{b}$ **18.** 7.2 minutes

19. Brenda 15 mph and Randy 20 mph, or Brenda 10 mph and Randy 15 mph
20. \$72 billion

Tying It All Together Chapters 1–5

1. $\frac{7}{3}$ **2.** $-\frac{10}{3}$ **3.** -2 **4.** No solution **5.** 0
6. $-4, -2$ **7.** $-1, 0, 1$ **8.** $-\frac{15}{2}$ **9.** $-6, 6$
10. $-2, 4$ **11.** 5 **12.** 3 **13.** $y = \frac{c - 2x}{3}$
14. $y = \frac{1}{2}x + \frac{1}{2}$ **15.** $y = \frac{c}{2 - a}$ **16.** $y = \frac{AB}{C}$
17. $y = 3B - 3A$ **18.** $y = \frac{6A}{5}$ **19.** $y = \frac{8}{3 - 5a}$
20. $y = 0$ or $y = B$ **21.** $y = \frac{2A - hb}{h}$ **22.** $y = -\frac{b}{2}$
23. 64 **24.** 16 **25.** 49 **26.** 121 **27.** $-2x - 2$
28. $2a^2 - 11a + 15$ **29.** x^4 **30.** $\frac{2x + 1}{5}$ **31.** $\frac{1}{2x}$
32. $\frac{x + 2}{2x}$ **33.** $\frac{x}{2}$ **34.** $\frac{x - 2}{2x}$ **35.** $-\frac{7}{5}$ **36.** $\frac{3a}{4}$
37. $x^2 - 64$ **38.** $3x^3 - 21x$ **39.** $10a^{14}$ **40.** x^{10}
41. $k^2 - 12k + 36$ **42.** $j^2 + 10j + 25$ **43.** -1
44. $3x^2 - 4x$ **45.** $P = \frac{r + 2}{(1 + r)^2}$, \$1.81

CHAPTER 6

Section 6.1 Warm-ups F F F F T T T F F T

1. (0, 9), (5, 24), (2, 15) **3.** (0, −7), (−4, 5), (−2, −1)
5. (0, 5), (10, −115), and (−1, 17)
7. (3, 0), (0, −2), (12, 6) **9.** (5, −3), (5, 5), (5, 0)
11.–25. (odd)

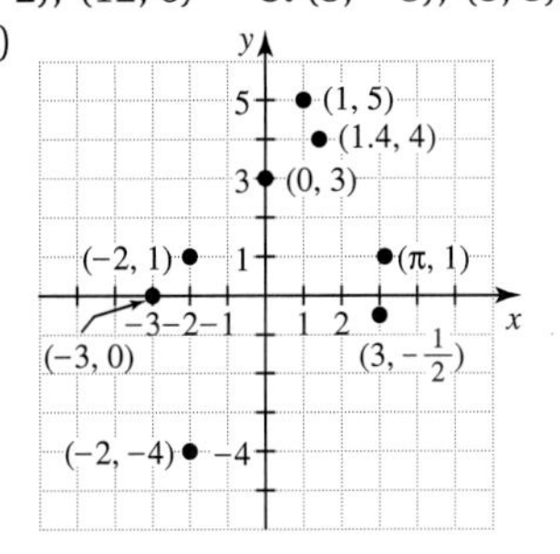

27.

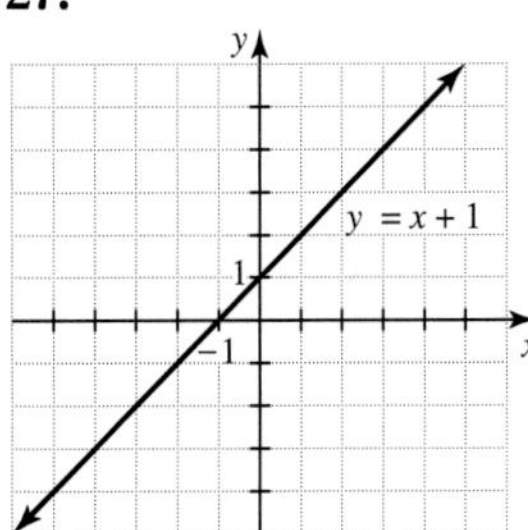

29.

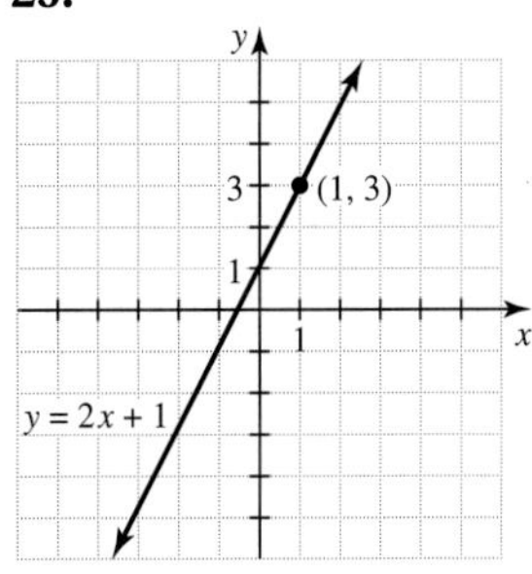

31.

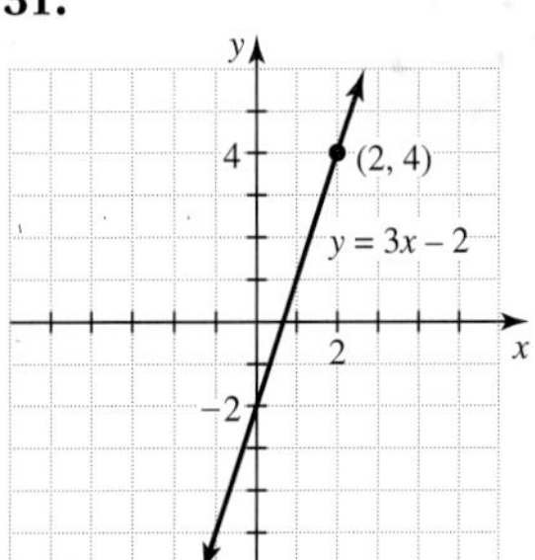

33.

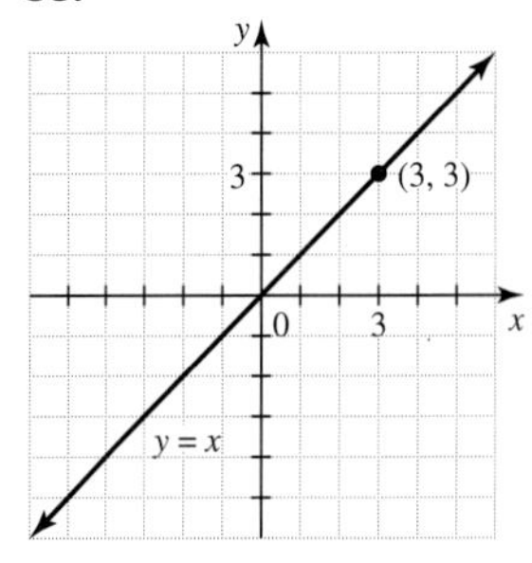

35.

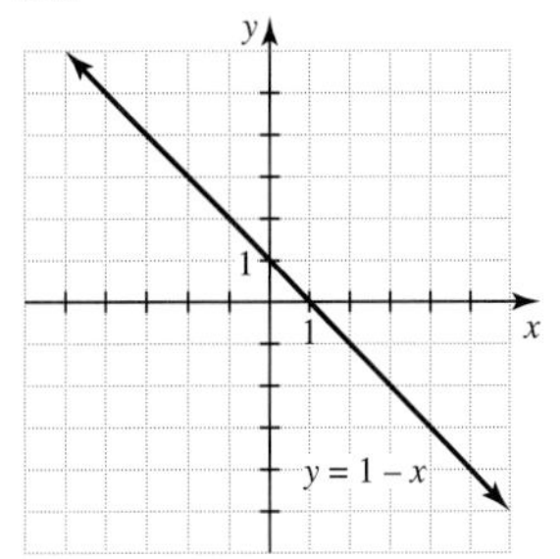

37.

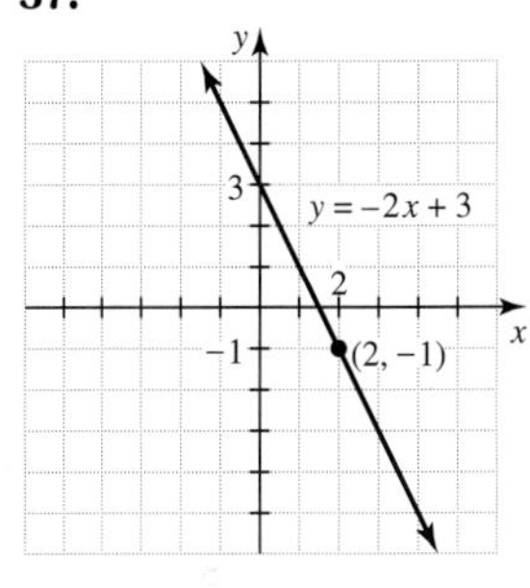

39.

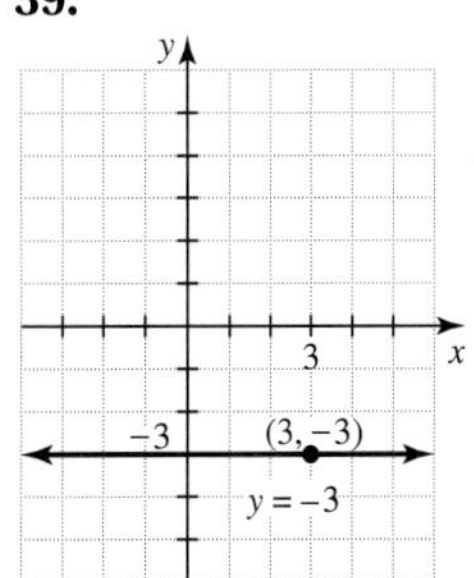

41.

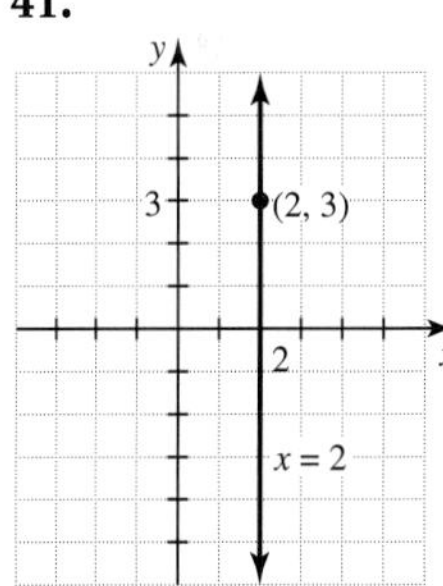

43.

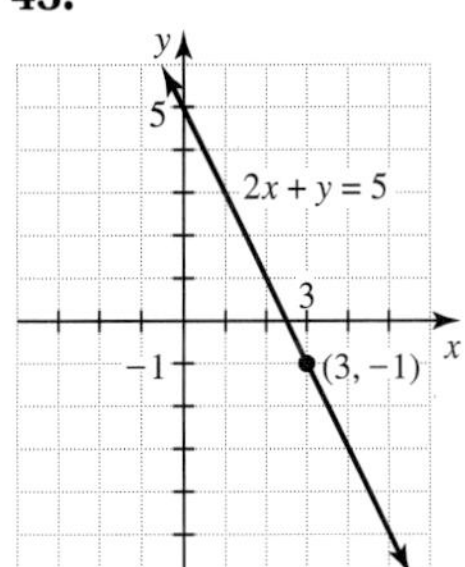

45.

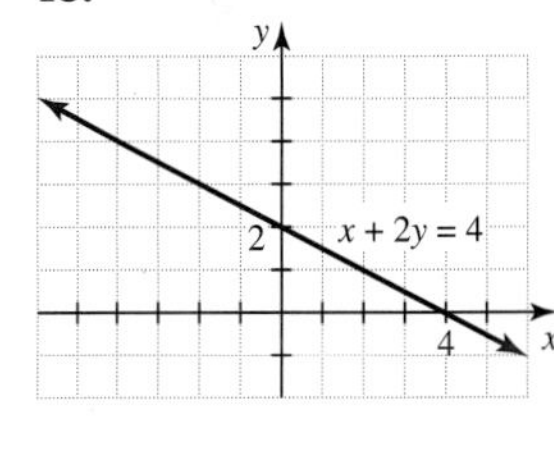

47.

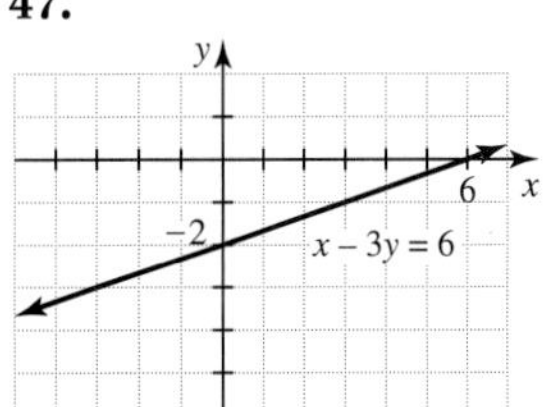

49.

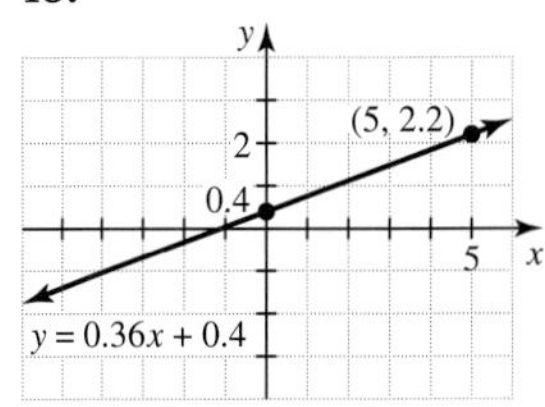

51. Quadrant II **53.** x-axis **55.** Quadrant III
57. Quadrant I **59.** Quadrant II **61.** y-axis

63.

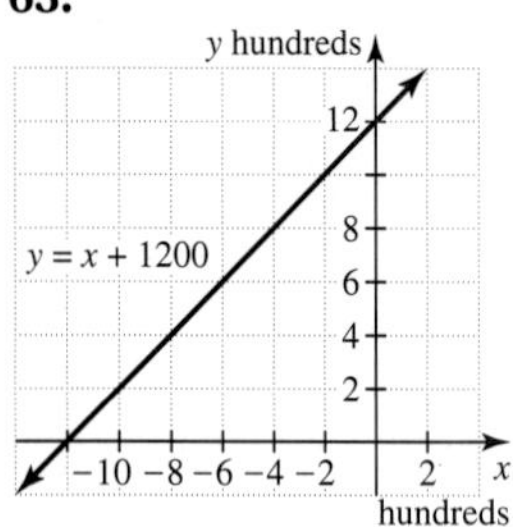

65.

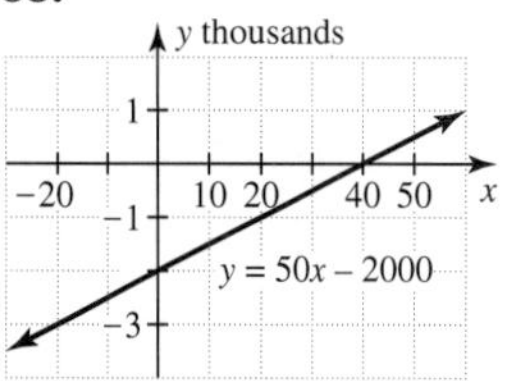

67.

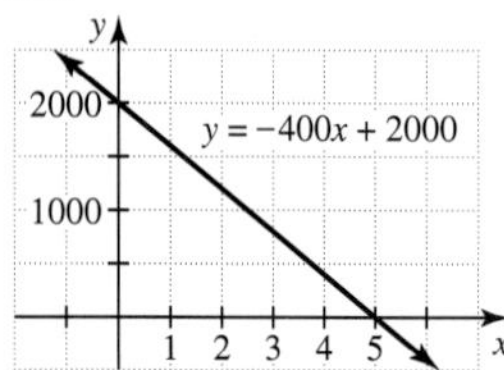

69.

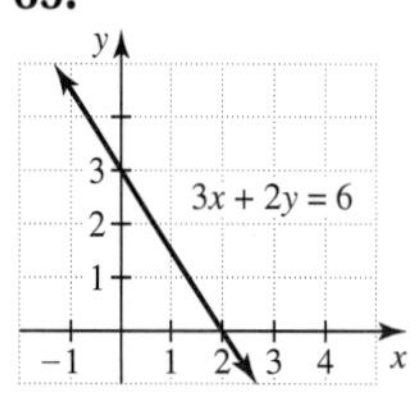

71.

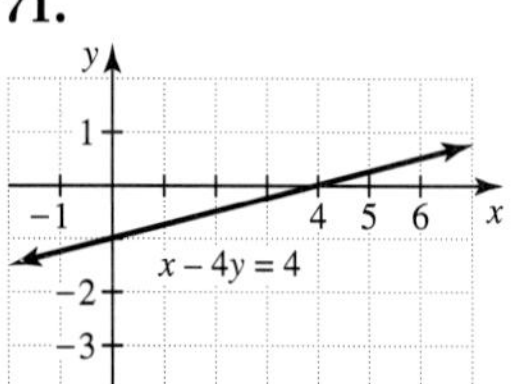

73.

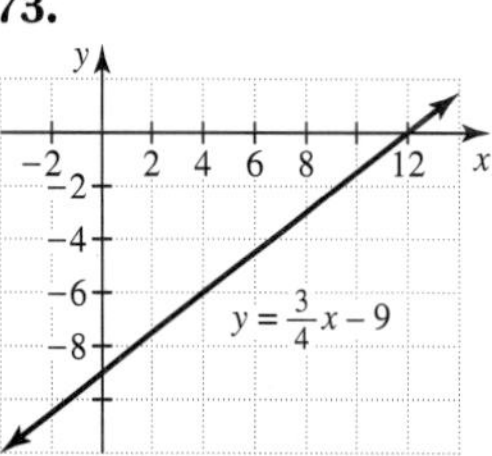

75.

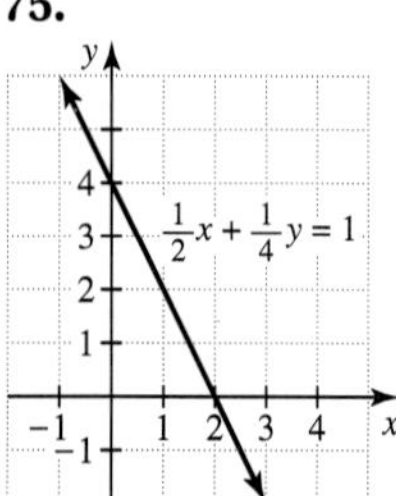

77. 21

79. 75%, 67, 68 and up

Section 6.2 Warm-ups T T F T F F F F T T

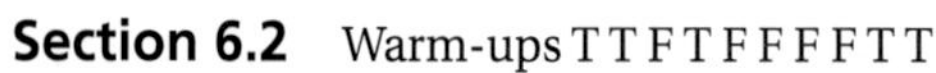

1. $-\frac{2}{3}$ **3.** $\frac{3}{2}$ **5.** 2 **7.** 0 **9.** $\frac{2}{5}$ **11.** $\frac{1}{5}$ **13.** 2

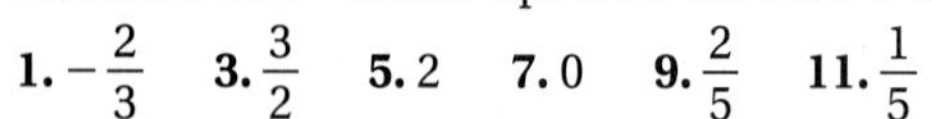

15. $-\frac{5}{3}$ **17.** $\frac{5}{7}$ **19.** $-\frac{4}{3}$ **21.** −1 **23.** 1

25. Undefined **27.** 0 **29.** 3

31.

33.

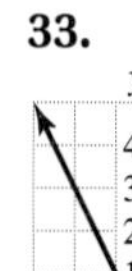

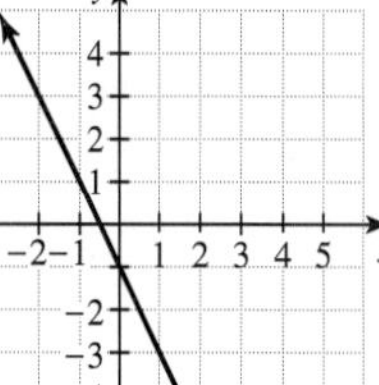

35.

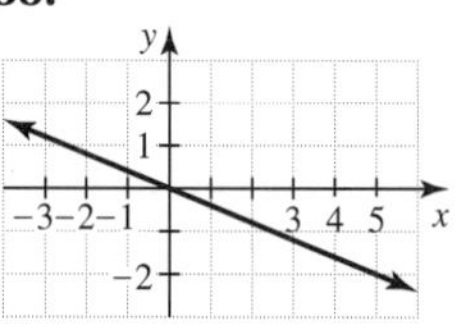

37.

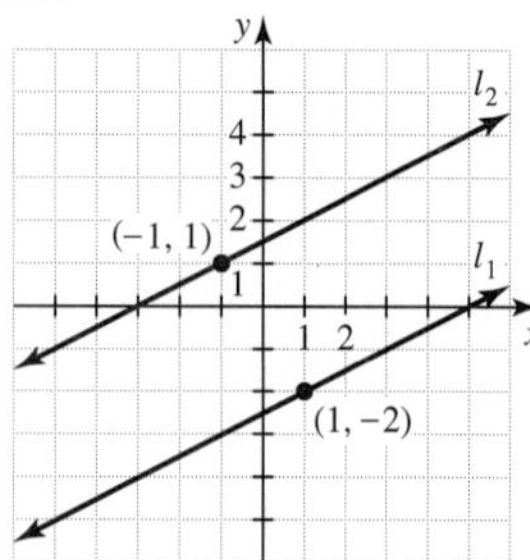

39.

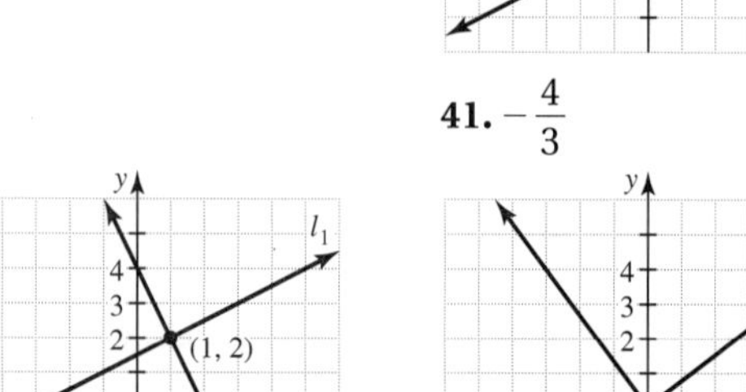

41. $-\frac{4}{3}$

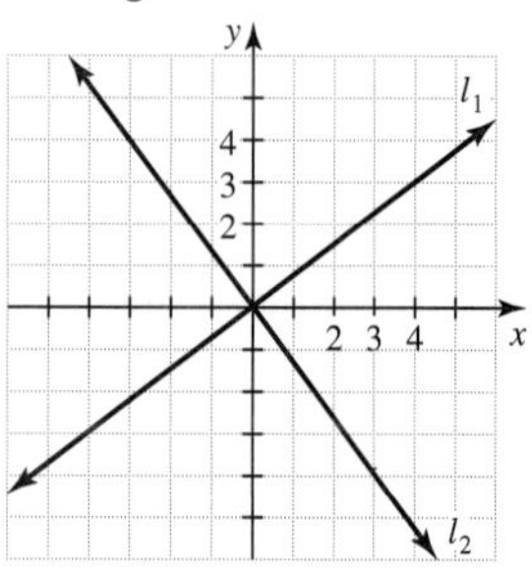

43. $\frac{1}{2}$

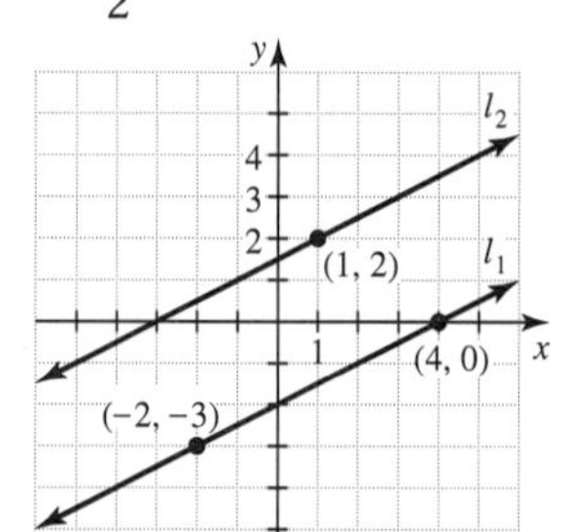

45. 1

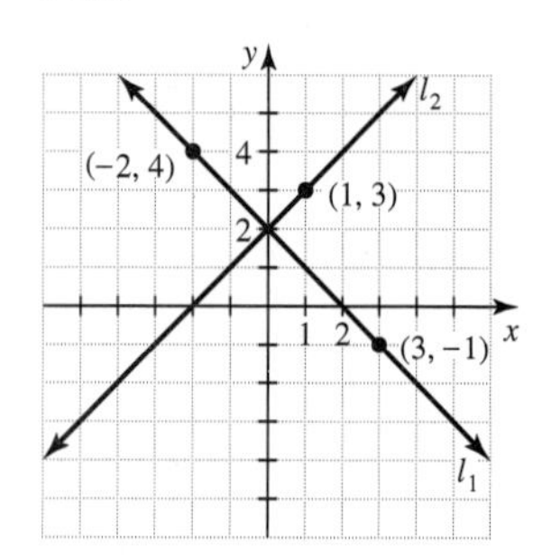

53. No

55. 60,000, average yearly increase, \$880,000, \$1,300,000

Section 6.3 Warm-ups T F T T T F F T T F

1. $y = \frac{3}{2}x + 1$ **3.** $y = -2x + 2$ **5.** $y = x - 2$

7. $y = -x$ **9.** $y = -1$ **11.** $x = -2$ **13.** 3, (0, −9)

15. 0, (0, 4) **17.** −3, (0, 0) **19.** −1, (0, 5)

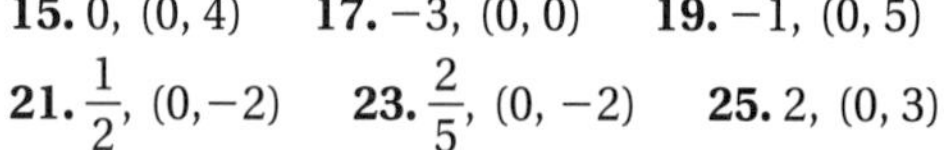

21. $\frac{1}{2}$, (0, −2) **23.** $\frac{2}{5}$, (0, −2) **25.** 2, (0, 3)

27. Undefined slope, no y-intercept **29.** $x + y = 2$

31. $x - 2y = -6$ **33.** $9x - 6y = 2$ **35.** $6x + 10y = 7$

37. $x = -10$ **39.** $3y = 10$ **41.** $5x - 6y = 0$

43. $x - 50y = -25$

45.

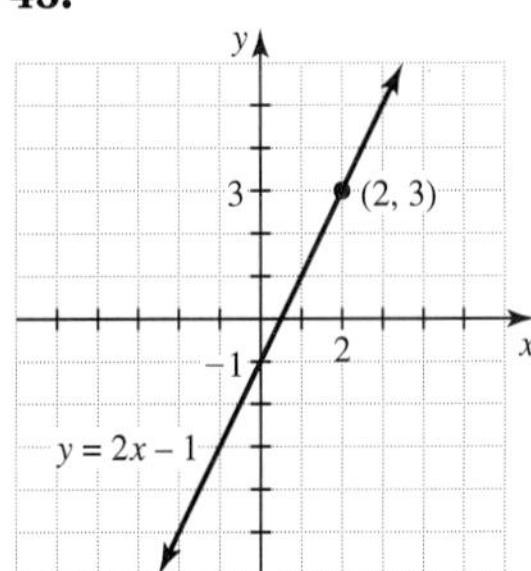

47.

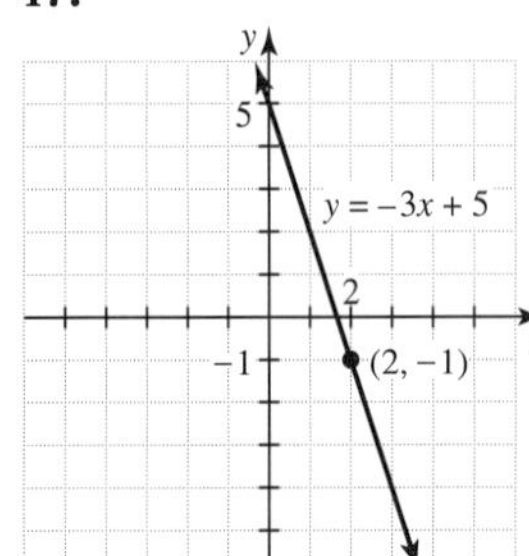

49.

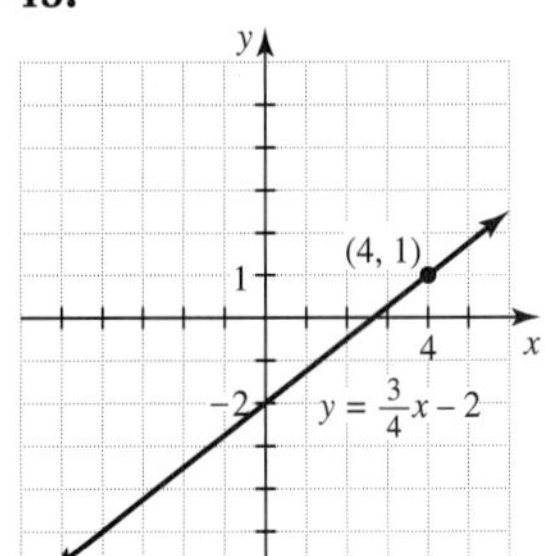

51.

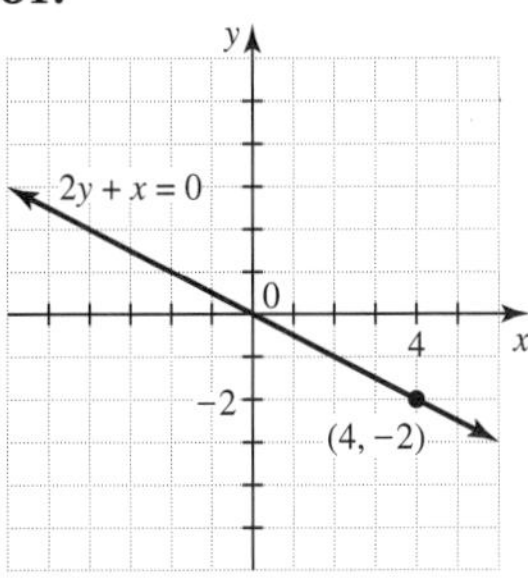

53.

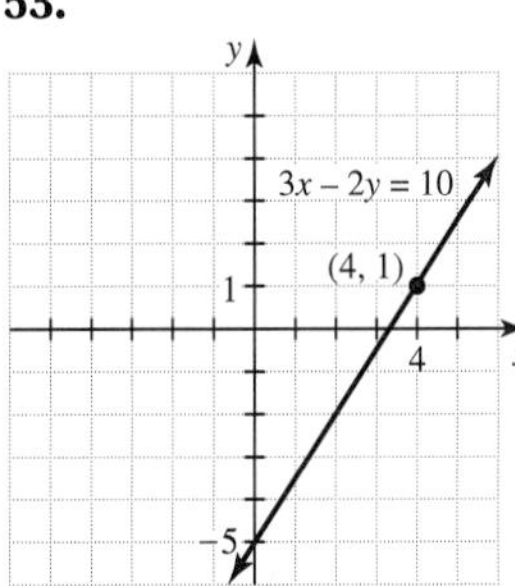

55.

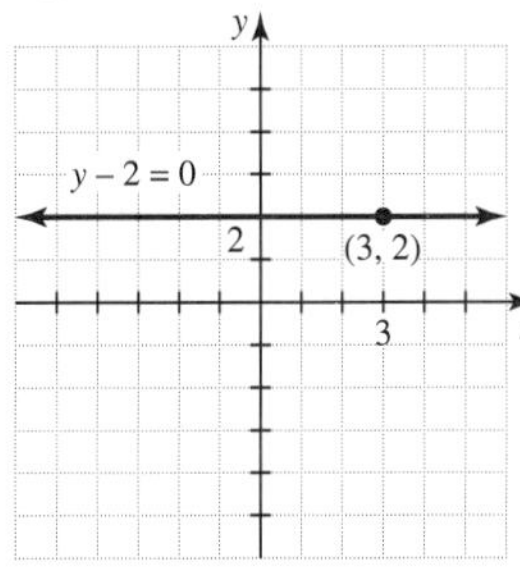

57. $y = -\frac{1}{3}x + 6$ **59.** $y = -2x + 3$ **61.** $y = 3$

63. $y = -\frac{4}{5}x + 4$ **65.** $1,150,000, $1,150,200, $200

Section 6.4 Warm-ups F F T T F T T T T T

1. $y = 5x + 11$ **3.** $y = \frac{3}{4}x - 20$ **5.** $y = \frac{2}{3}x + \frac{1}{3}$

7. $y = \frac{1}{3}x + \frac{7}{3}$ **9.** $y = -\frac{1}{2}x + 4$ **11.** $y = -6x - 13$

13. $2x - y = 7$ **15.** $x - 2y = 6$ **17.** $2x - 3y = 2$

19. $3x - 2y = -1$ **21.** $3x + 5y = -11$

23. $x - y = -2$ **25.** $y = -x + 4$ **27.** $y = \frac{5}{3}x - 1$

29. $y = -\frac{1}{3}x + 5$ **31.** $y = x + 3$ **33.** $y = -\frac{2}{3}x + \frac{5}{3}$

35. $y = -2x - 5$ **37.** $y = \frac{1}{3}x + \frac{7}{3}$ **39.** $y = 2x - 1$

41. $y = 0.625x - 50.5$, 12 billion

47. $-1 \le x \le 1$, $-1 \le y \le 1$

Section 6.5 Warm-ups T T F T T F T F T T

1. $F = 3Y$ **3.** $D = 65T$ **5.** $C = \pi D$ **7.** $P = 7.8H$

9.

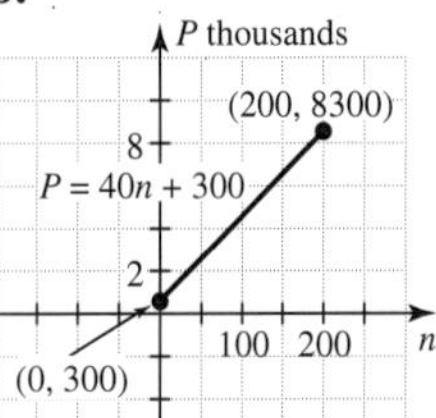

11.

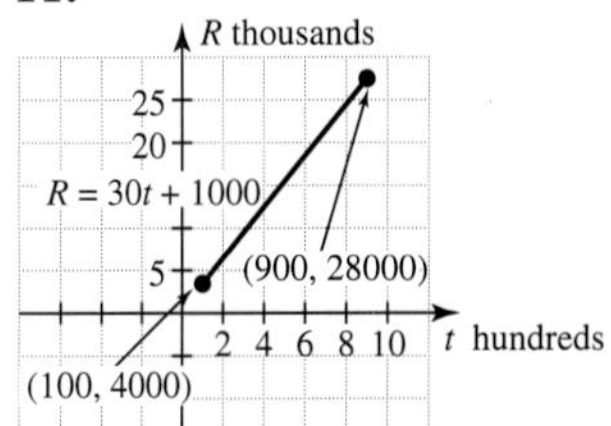

13.

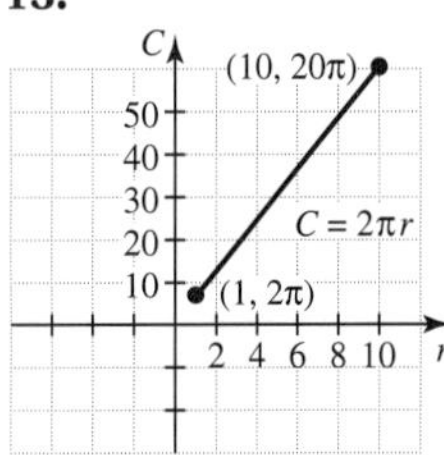

15.

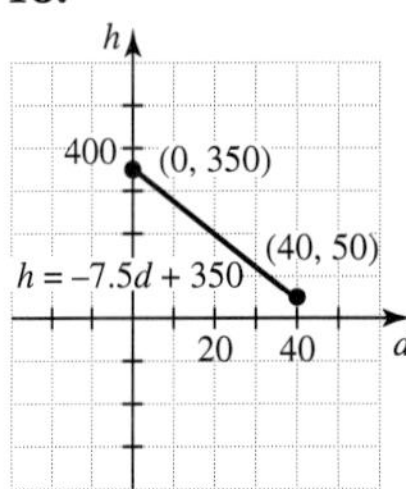

17. $7.05

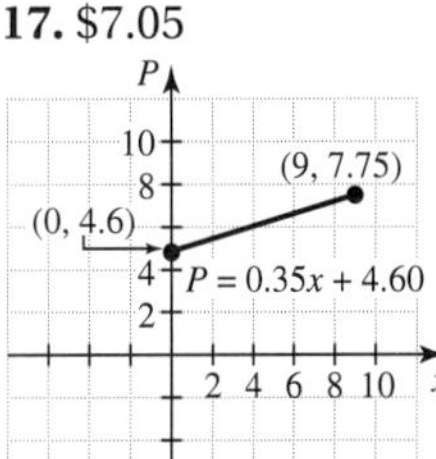

19. $C = 20n + 30$, $170

21. $S = 3L - \frac{41}{4}$, 8.5 **23.** $v = 32t + 10$, 122 ft/sec

25. $w = -\frac{1}{120}t + \frac{3}{2}$, $\frac{5}{6}$ inch **27.** $A = 0.6w$, 3.6 inches

29. $a = 0.08c$, 0.24, 6.25

31. a) Female **b)** Male **c)** 65 **d)** 235

Section 6.6 Warm-ups F T F T T F F T F T

1. $C = 0.50t + 5$ **3.** $T = 1.09S$ **5.** $C = 2\pi r$

7. $P = 4s$ **9.** $A = 5h$ **11.** Yes **13.** Yes **15.** No

17. Yes **19.** Yes **21.** Yes **23.** No **25.** Yes

27. Yes **29.** No **31.** Yes **33.** No **35.** Yes

37. Yes **39.** Yes **41.** No **43.** Yes **45.** No

47. No **49.** {1, 2, 3}, {3, 5, 7}

51. R, nonnegative real numbers **53.** R, R

55. R, nonnegative real numbers **57.** $\{s \mid s > 0\}$, $\{A \mid A > 0\}$

59. −1 **61.** 0 **63.** 13 **65.** −2.75 **67.** 2 **69.** 1

71. 150.988 **73.** 0.31 **75.** 183 pounds, 216 pounds

77. $C = \frac{r}{0.96}$

Section 6.7 Warm-ups T F T F F T T T F F

1. $T = kh$ **3.** $y = \frac{k}{r}$ **5.** $R = kts$ **7.** $i = kb$

9. $A = kym$ **11.** $y = \frac{5}{3}x$ **13.** $A = \frac{6}{B}$ **15.** $m = \frac{198}{p}$

17. $A = 2tu$ **19.** $T = \frac{9}{2}u$ **21.** 25 **23.** 1 **25.** 105
27. 100.3 pounds **29.** 50 minutes **31.** \$17.40
33. 80 mph

Chapter 6 Review

1. Quadrant II **3.** x-axis **5.** y-axis **7.** Quadrant IV
9. $(0, -5)$, $(-3, -14)$, $(4, 7)$
11. $\left(0, -\frac{8}{3}\right)$, $\left(3, -\frac{2}{3}\right)$, $\left(-6, -\frac{20}{3}\right)$

13.

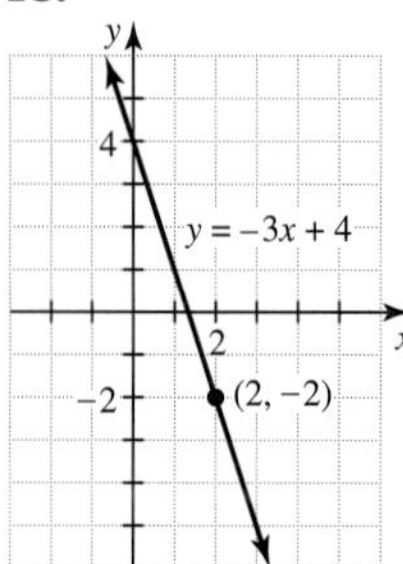

15.

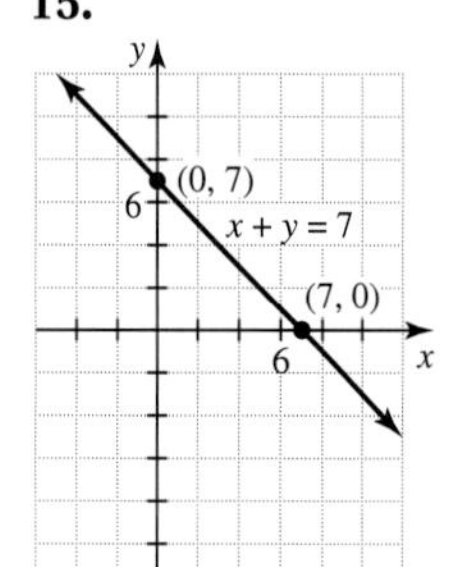

17. 1 **19.** $\frac{3}{2}$ **21.** $\frac{3}{7}$ **23.** Yes **27.** 3, $(0, -18)$
29. 2, $(0, -3)$ **31.** 2, $(0, -4)$

33.

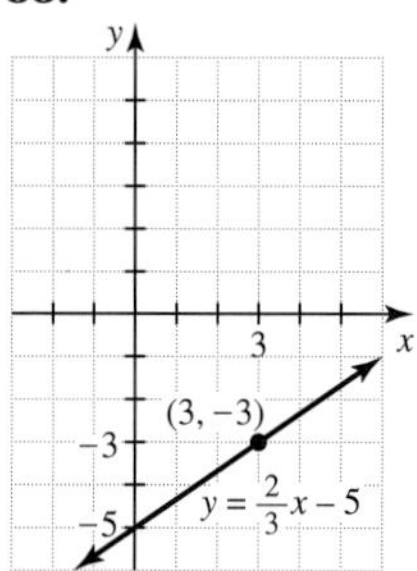

35.

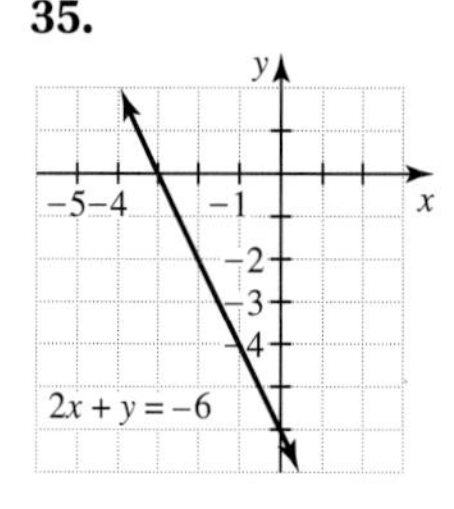

37.

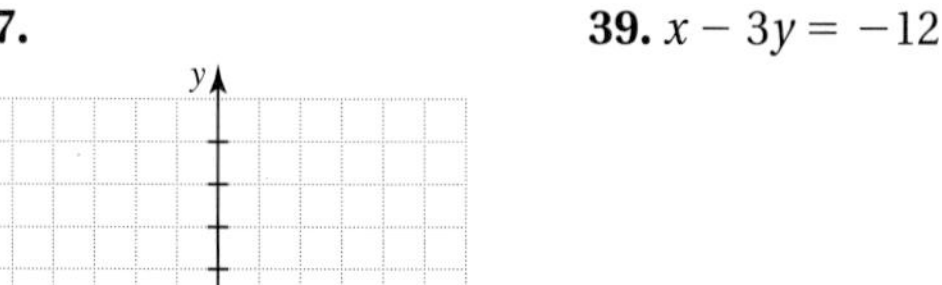
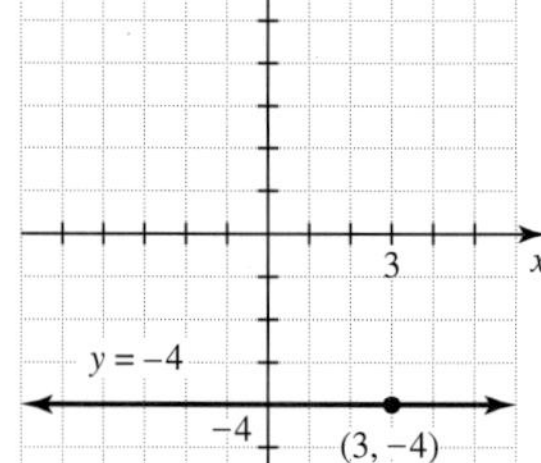

39. $x - 3y = -12$

41. $x + 2y = 0$ **43.** $y = 5$ **45.** $y = \frac{2}{3}x + 7$

47. $y = \frac{3}{7}x - 2$ **49.** $y = -\frac{3}{4}x + \frac{17}{4}$ **51.** $y = -2x - 1$

53. $y = \frac{6}{5}x + \frac{17}{5}$ **55.** $y = 3x - 14$

57.

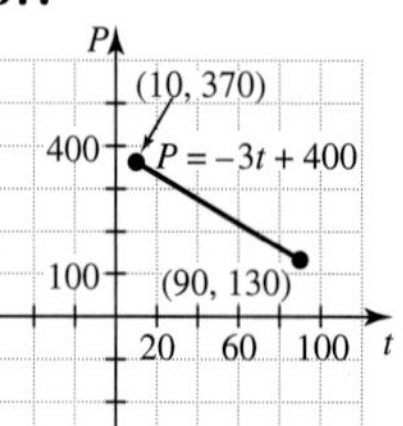

59.

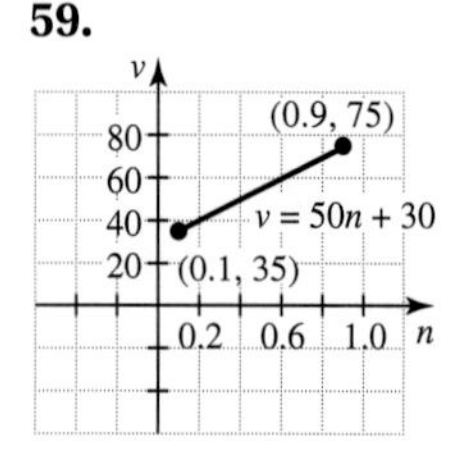

61. $C = 32n + 49$, \$177 **63.** $q = 1 - p$ **65.** Yes
67. No **69.** Yes **71.** Yes **73.** No **75.** R, R
77. $\{1, 2, 3\}$, $\{0, 2\}$ **79.** R, nonnegative real numbers
81. 132 **83.** 2 **85.** 60 **87.** \$15

Chapter 6 Test

1. Quadrant II **2.** x-axis **3.** Quadrant IV **4.** y-axis
5. 1 **6.** $-\frac{5}{6}$ **7.** $y = -\frac{1}{2}x + 3$ **8.** $y = \frac{3}{7}x - \frac{11}{7}$
9. $x - 3y = 11$ **10.** $5x + 3y = 27$

11.

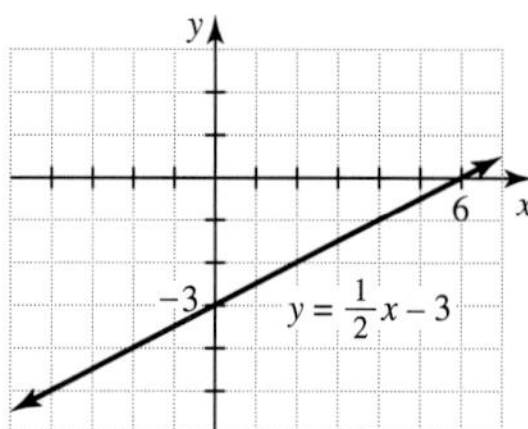

12.

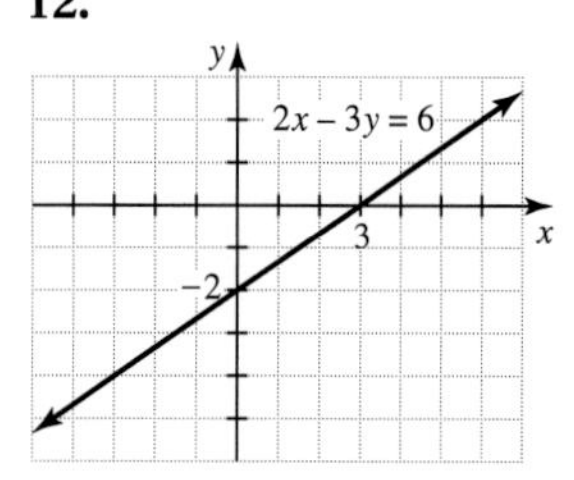

13.

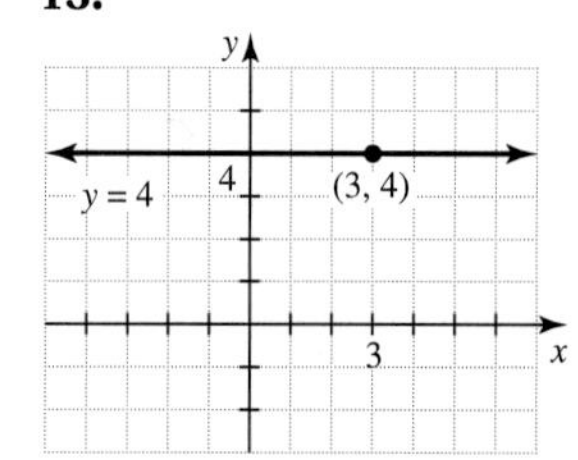

14.

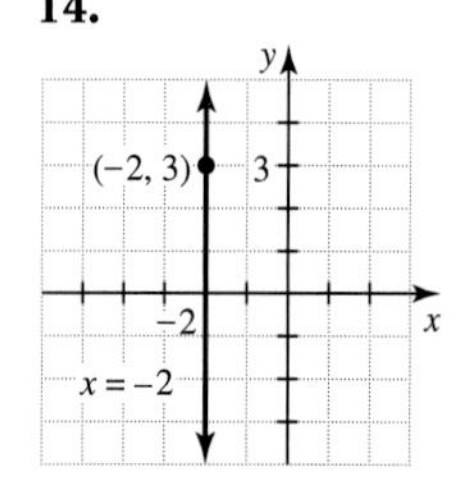

15. Yes **16.** No **17.** R, $\{y \mid y \geq 1\}$ **18.** R, $\{y \mid y \geq 0\}$
19. 1 **20.** 5 **21.** Yes **22.** $S = 0.75n + 2.50$
23. $P = 3v + 20$, 80 cents **24.** \$2.80 **25.** 2 hours

Tying It All Together Chapters 1–6

1. 1 **2.** 72 **3.** 10^{13} **4.** 4 **5.** 1 **6.** -512
7. $\frac{41}{72}$ **8.** $-\frac{63}{16}$ **9.** $\frac{3}{20}$ **10.** $-\frac{5}{48}$ **11.** $\frac{1}{10}$
12. $\frac{8}{21}$ **13.** $-6x + 21$ **14.** $-5x + 21$
15. $2x^2 - 13x + 21$ **16.** $4x^2 - 4x + 1$
17. $z^2 + 10z + 25$ **18.** $w^2 - 49$ **19.** $36x^3y^7$

20. $40x^8y^9$

21.

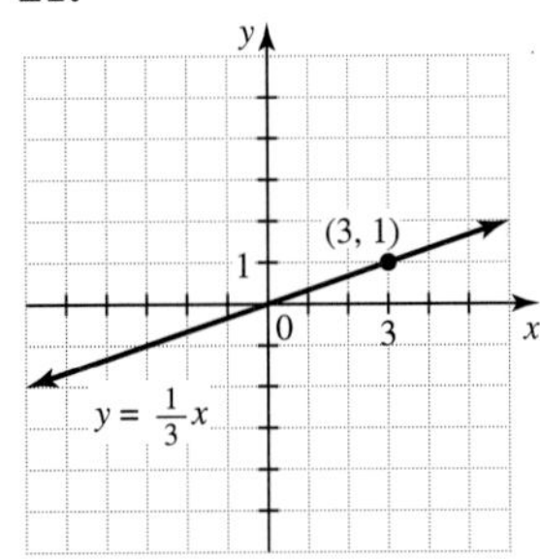

22.

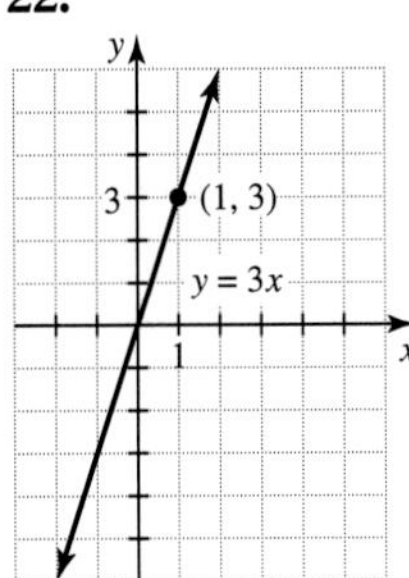

23.

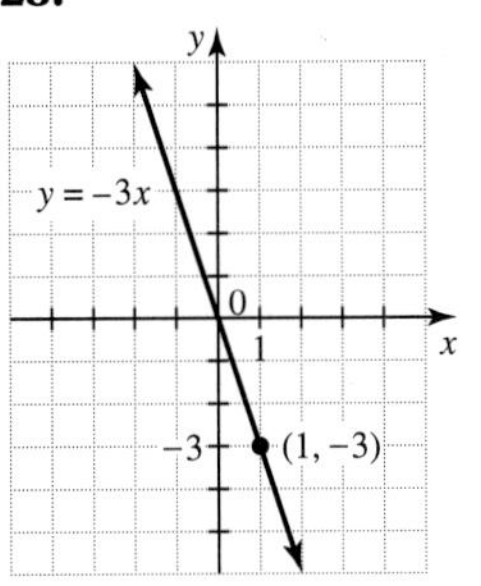

24.

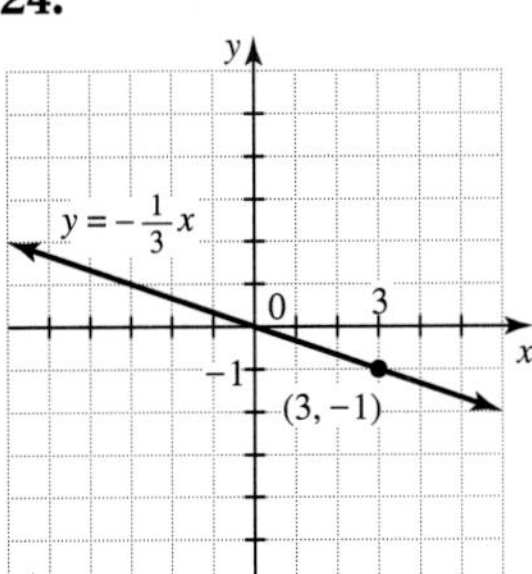

25.

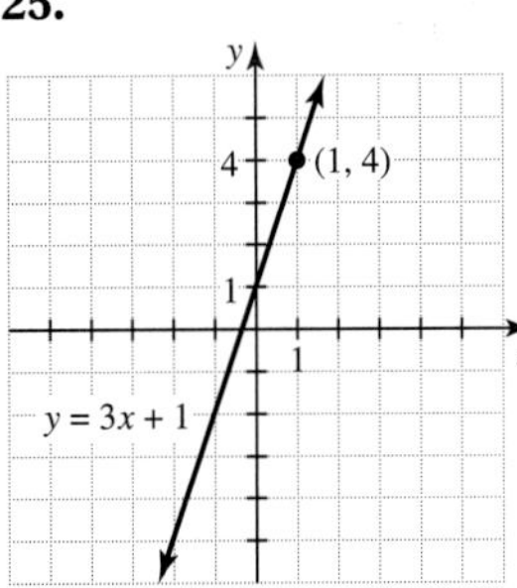

26.

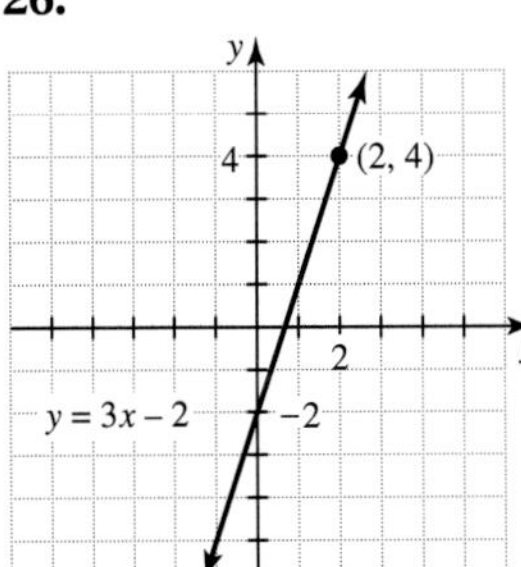

27.

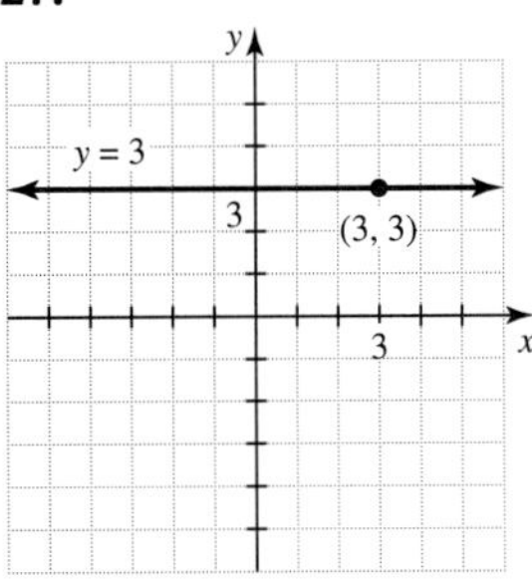

28.

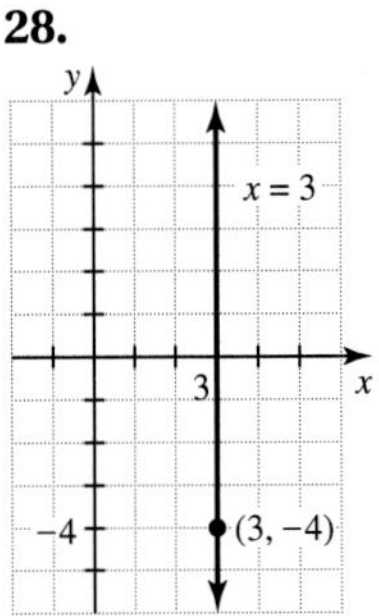

29. $y = \frac{t-2}{3\pi}$

30. $y = mx + b$ **31.** $y = x - 4$ **32.** $y = 6$
33. $y = 8$ or $y = -5$ **34.** $y = \frac{4}{5}$ **35.** 3 **36.** $-2, 2$
37. $-9, \frac{4}{3}$ **38.** $\frac{2}{5}$ **39.** $0, \frac{7}{2}$ **40.** $-2, 3$
41. a) $\frac{2}{15}$ **b)** $\frac{1}{5}$ **c)** About 13% per year
d) \$276,000 saved, \$12,000 per year

CHAPTER 7

Section 7.1 Warm-ups T F F F T T T T T F

1. (3, −2) **3.** All three **5.** None **7.** (−2, 3)
9. (2, 4) **11.** (1, 2) **13.** (0, −5) **15.** (−2, 3)
17. (0, 0) **19.** (2, 3) **21.** No solution, inconsistent
23. (1, −2), independent
25. $\{(x, y) \mid 2x + y = 3\}$, dependent
27. $\{(x, y) \mid y = x\}$, dependent
29. (−4, −3), independent **31.** (−1, 0), independent
33. (0, 0), (2, 4) **35.** 1986, \$4 billion, 1993
37. 15,000, Panasonic

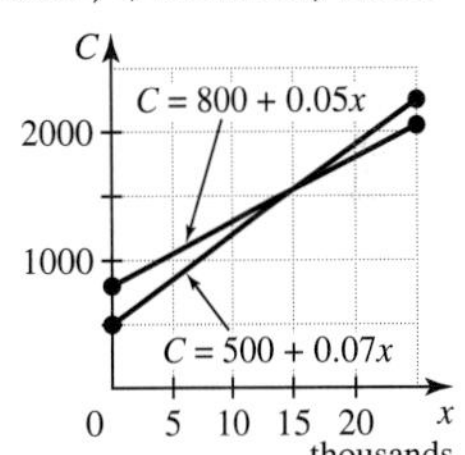

Section 7.2 Warm-ups T T T T T T F T T T

1. (2, 5) **3.** (2, 3) **5.** (−2, 9) **7.** (−5, 5)
9. $\left(\frac{1}{3}, \frac{2}{3}\right)$ **11.** $\left(\frac{1}{2}, \frac{1}{3}\right)$ **13.** $\left(3, \frac{5}{2}\right)$, independent
15. No solution, inconsistent
17. No solution, inconsistent
19. $\left(\frac{7}{3}, 0\right)$, independent **21.** (−1, 1), independent
23. \$12,000 at 10%, \$8,000 at 5%
25. Spielberg \$42 million, Culkin \$16 million
27. Lawn \$12, sidewalk \$7
29. Left rear 288 pounds, left front 287 pounds, no
31. \$2.40 per pound
33. a) 69.7 years, 75.2 years **c)** No **d)** 1850

Section 7.3 Warm-ups F T T T T F F F F T

1. (3, −1) **3.** (−3, 5) **5.** (−4, 3) **7.** $\left(\frac{1}{7}, \frac{9}{7}\right)$
9. (8, 31) **11.** (−2, −3) **13.** (−1, 4) **15.** (−1, 2)
17. (3, −1) **19.** (1, 2) **21.** No solution, inconsistent
23. $\{(x, y) \mid x + y = 5\}$, dependent
25. $\{(x, y) \mid 2x = y + 3\}$, dependent
27. $\{(x, y) \mid x + 3y = 3\}$, dependent
29. No solution, inconsistent
31. (12, 18), independent **33.** (40, 60), independent

35. (0.1, 0.1), independent **37.** (1.5, −2.8)
39. 150 cars, 100 trucks **41.** 6 adults, 24 children
43. 180 men, 120 women

Section 7.4

Warm-ups T T T F F F T F T F

1. (−3, −9) **3.** (3, 0), (1, 3) **5.** (2, 3), (0, 5)

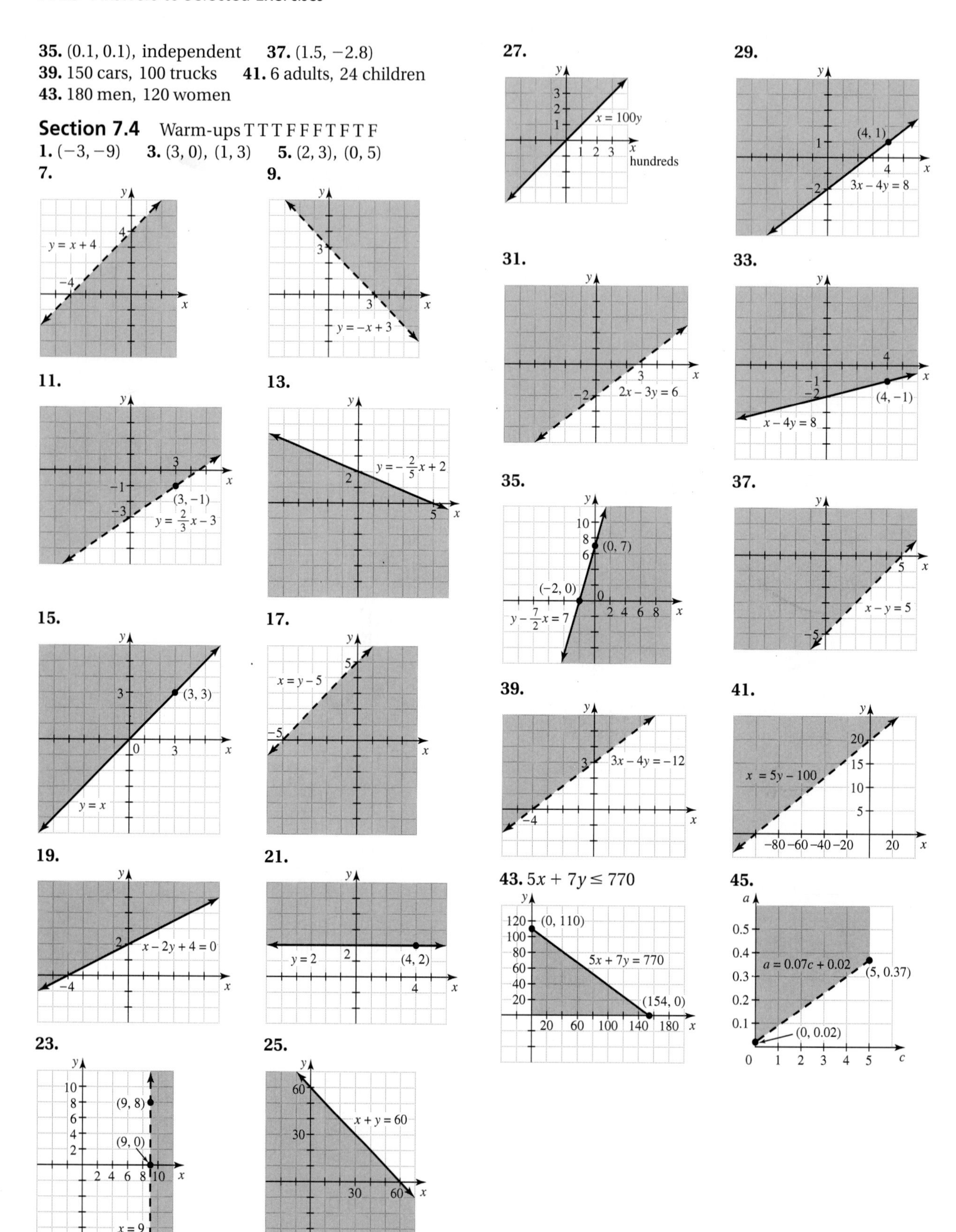

Section 7.5 Warm-ups F T T T F F T F T T

1. (4, 3) **3.** (3, 6) **5.** (9, −5)

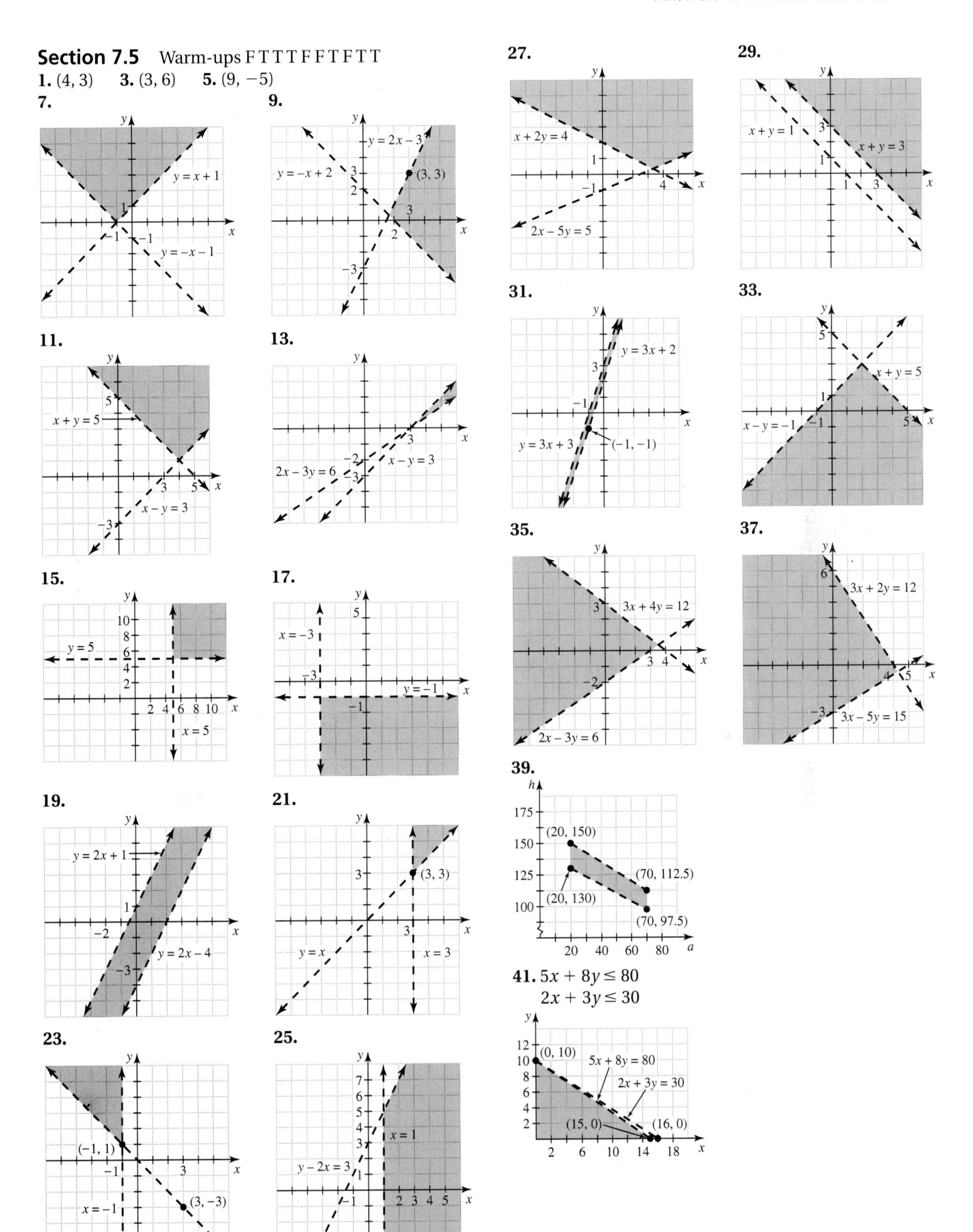

Chapter 7 Review

1. $(1, 3)$ **3.** $(-1, 1)$ **5.** $(2, 6)$ **7.** $(-8, -3)$
9. $(3, -1)$, independent
11. $\{(x, y) \mid x - 2y = 4\}$, dependent
13. $(1, -2)$, independent
15. $(0, 0)$, independent
17. $\{(x, y) \mid x - y = 6\}$, dependent
19. No solution, inconsistent

21.

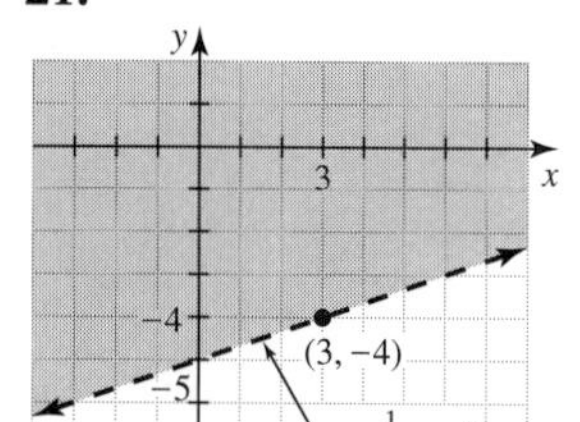

23.

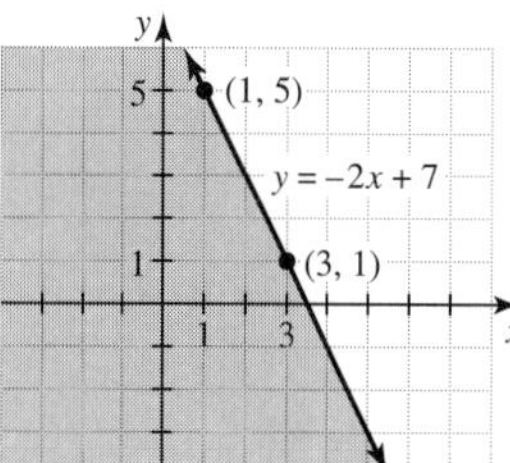

25.

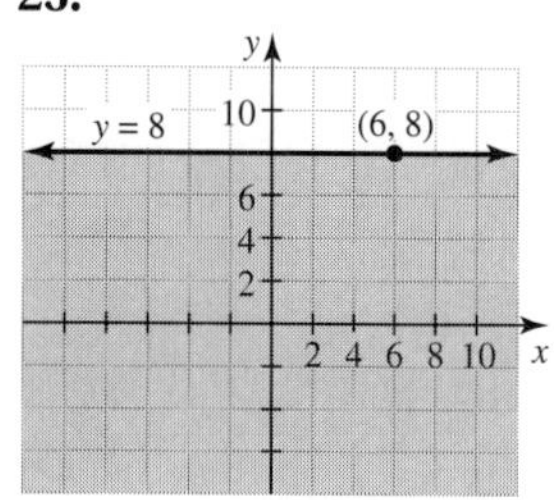

27.

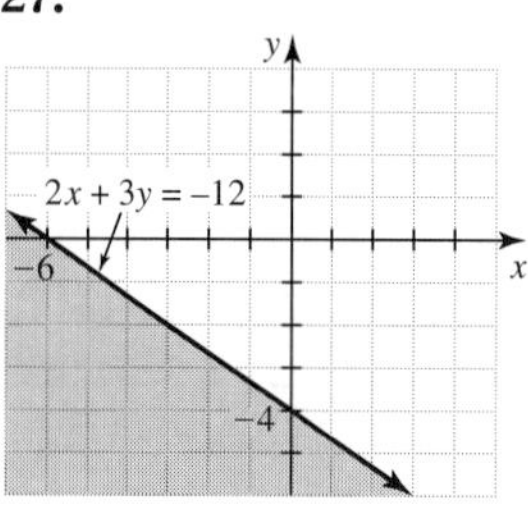

29.

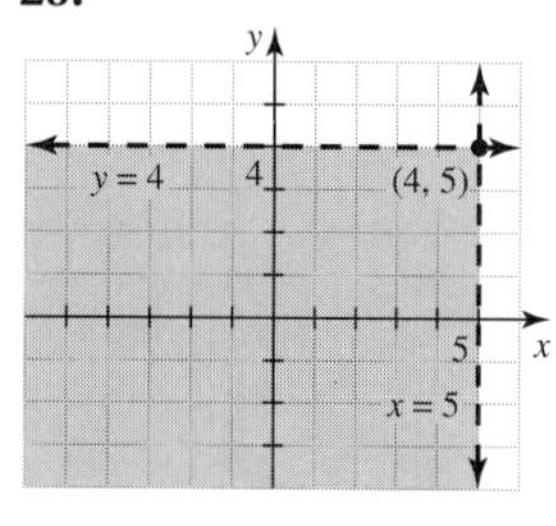

31.

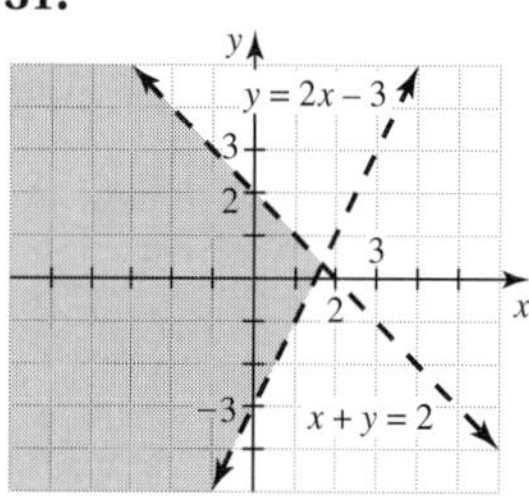

33.

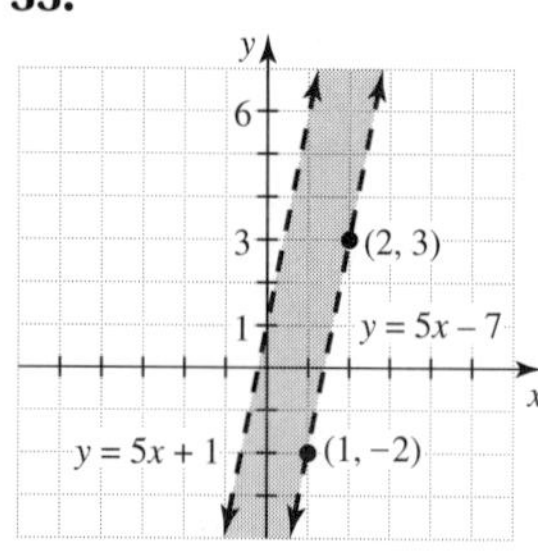

35.

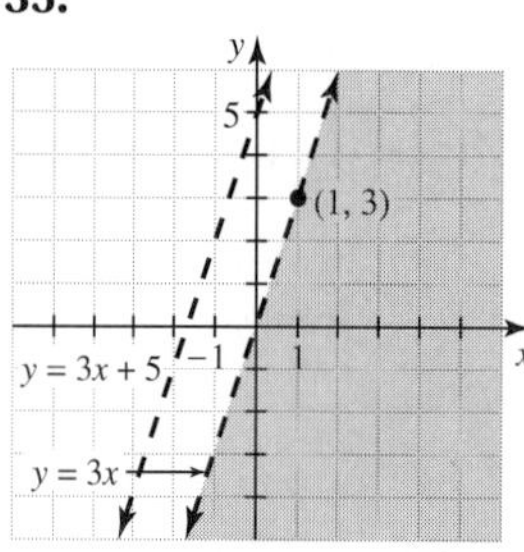

37. Apple \$0.45, orange \$0.35 **39.** 32 fives, 22 tens
41. 4 servings green beans, 3 servings chicken soup

Chapter 7 Test

1. $(-1, 3)$ **2.** $(2, 1)$ **3.** $(3, -1)$ **4.** $(2, 3)$ **5.** $(4, 1)$
6. Inconsistent **7.** Dependent **8.** Independent

9.

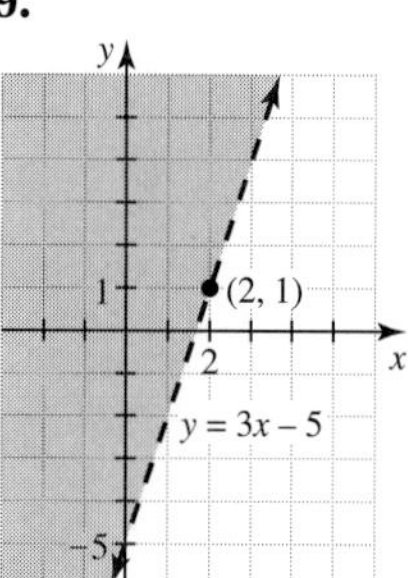

10.

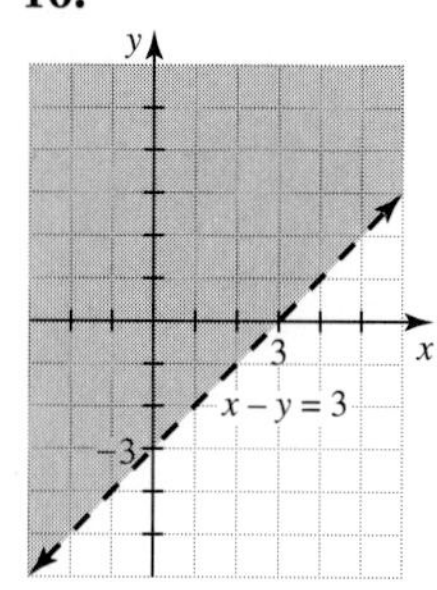

11.

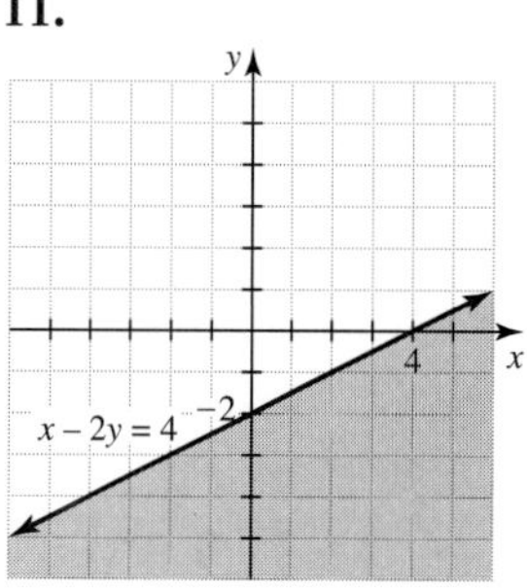

12.

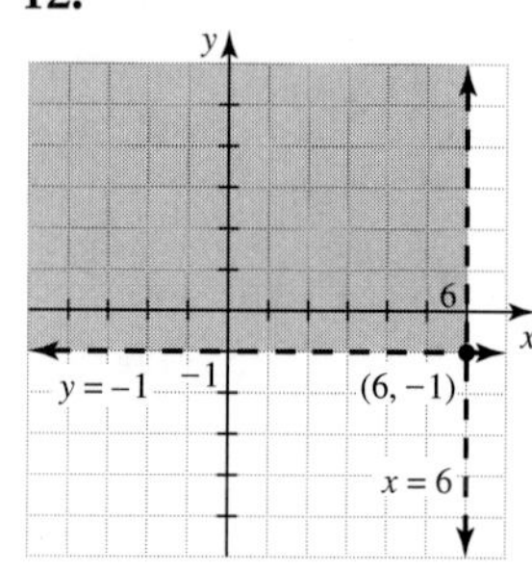

13.

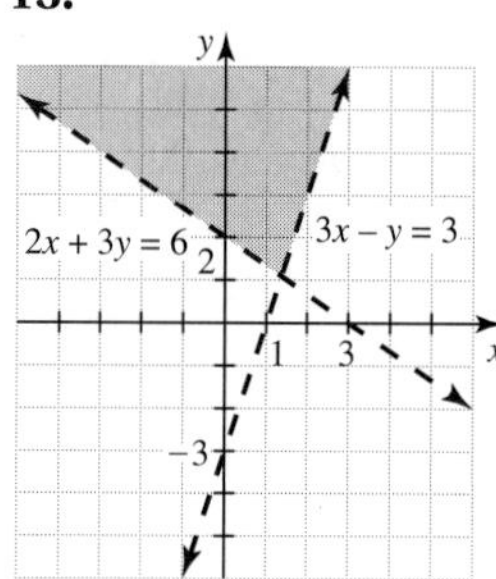

14.

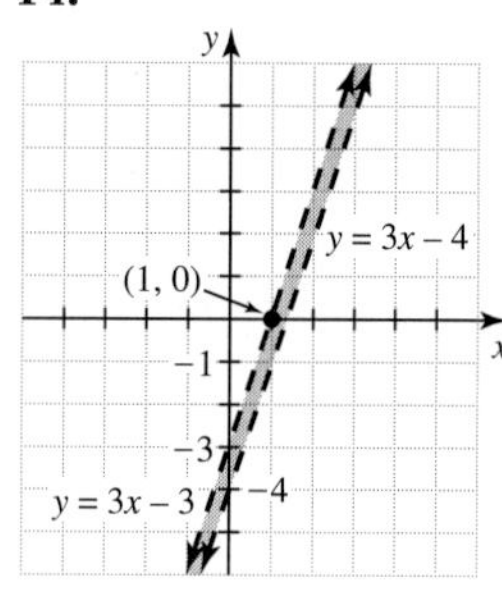

15. Kathy 36 hours, Chris 18 hours **16.** \$48

Tying It All Together Chapters 1–7

1. 7 **2.** $\frac{5}{3}$ **3.** 12 **4.** $-6, 4$ **5.** -17 **6.** $\frac{1}{2}$, 3

7. $x > 4$

2 3 4 5 6 7 8

8. $\frac{1}{2} \le x \le 5$

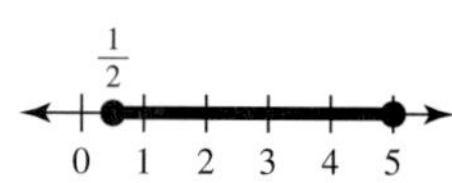

9. $x > 1$

−1 0 1 2 3 4 5

10.

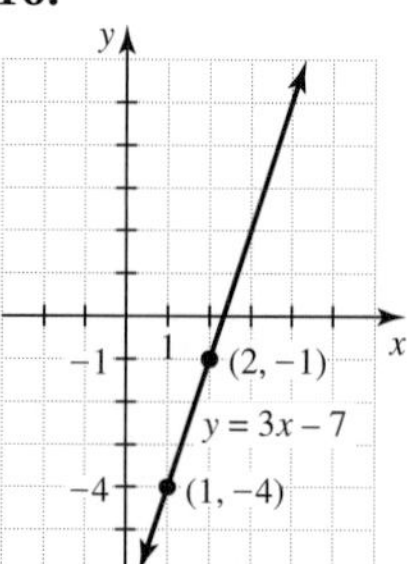

11.

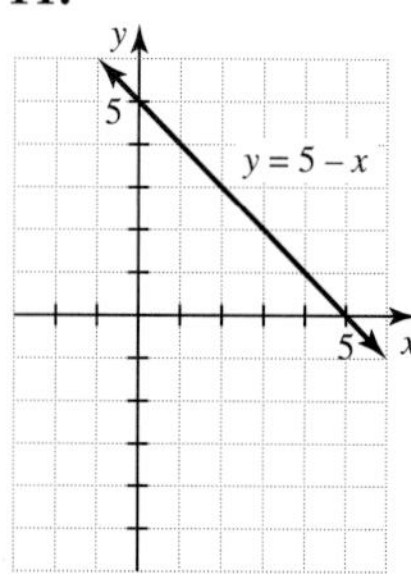

12.

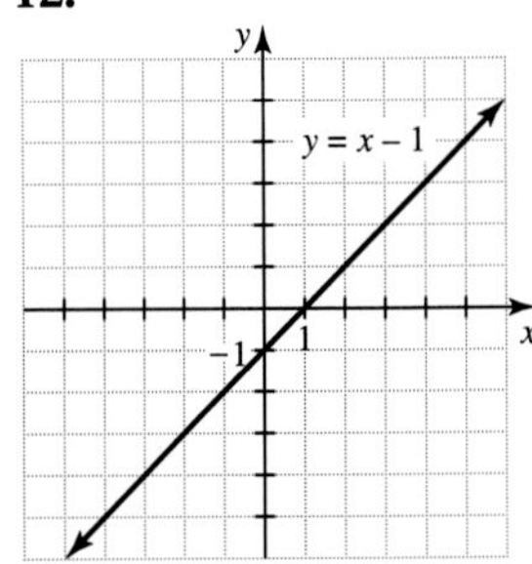

13.

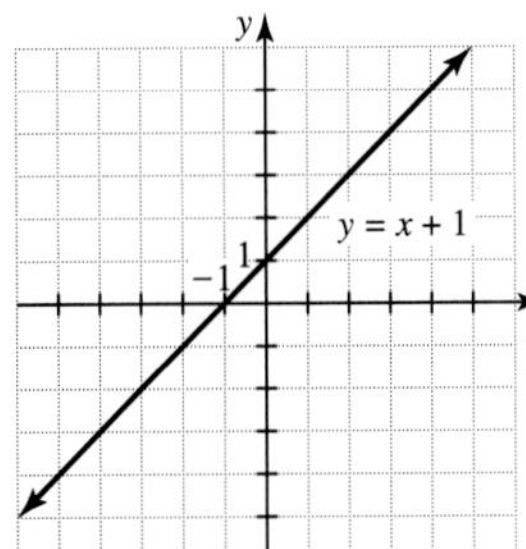

14.

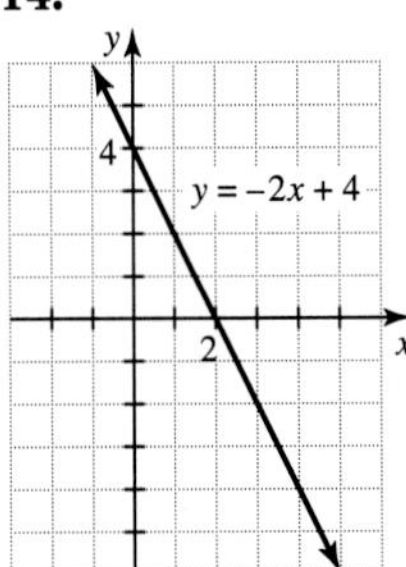

15.

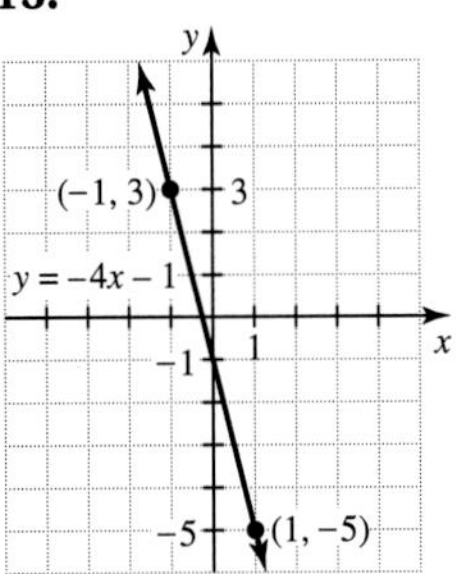

16.

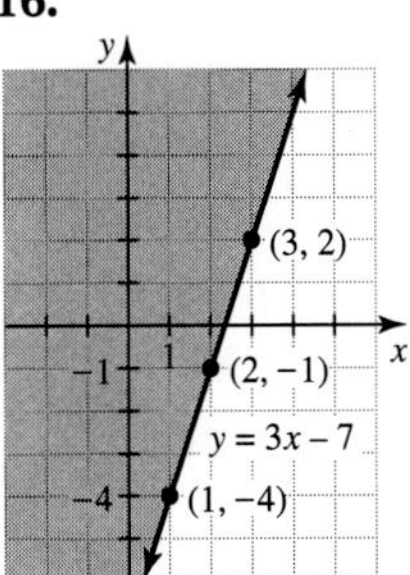

17.

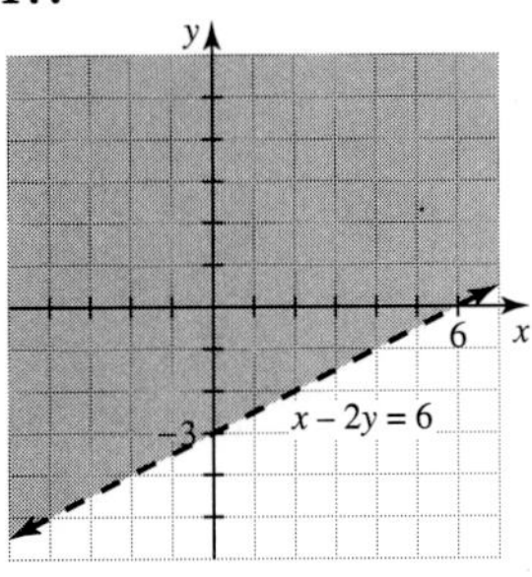

18.

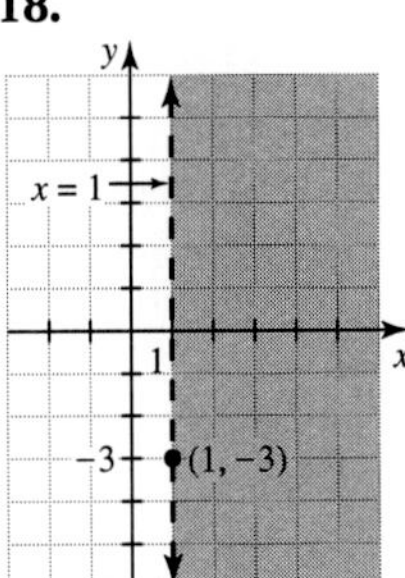

19. $y = 6x + 36$

20. $y = -7x + 95$ **21. a)** $p = -5x + 45$
b) $p = 2.5x + 30$ **c)** 1996

CHAPTER 8

Section 8.1 Warm-ups T F F T F T T T T F

1. $\frac{1}{3}$ **3.** $\frac{1}{16}$ **5.** $-\frac{1}{16}$ **7.** 4 **9.** $\frac{8}{125}$ **11.** $\frac{1}{3}$

13. 1250 **15.** 82 **17.** x **19.** $-\frac{16}{x^4}$ **21.** $\frac{6}{a^5}$

23. $\frac{1}{u^8}$ **25.** $-4t^2$ **27.** $2x^{11}$ **29.** $\frac{1}{x^{10}}$ **31.** a^9

33. $\frac{x^{12}}{16}$ **35.** $\frac{y^6}{16x^4}$ **37.** $\frac{x^2}{4y^6}$ **39.** $\frac{a^{16}}{16c^8}$ **41.** $\frac{1}{6}$

43. $\frac{3}{2}$ **45.** $-14x^6$ **47.** $\frac{2a^4}{b^2}$ **49.** 9,860,000,000

51. 0.00137 **53.** 0.000001 **55.** 600,000 **57.** 9×10^3
59. 7.8×10^{-4} **61.** 8.5×10^{-6} **63.** 5.25×10^{11}
65. 6×10^{-10} **67.** 2×10^{-38} **69.** 5×10^{27}
71. 9×10^{24} **73.** 1.25×10^{14} **75.** 2.5×10^{-33}
77. 8.6×10^9 **79.** 2.1×10^2 **81.** 2.7×10^{-23}
83. 3×10^{15} **85.** 9.135×10^2 **87.** 5.715×10^{-4}
89. 4.426×10^7 **91.** 1.577×10^{182}
93. 4.910×10^{11} feet **95.** 4.65×10^{-28} hours
97. 9.040×10^8 feet
99. a) \$1 per pound and 1%
b) \$1,000,000 dollars per pound **c)** No
101. \$10,727.41

Section 8.2 Warm-ups T F T F T T F F T T

1. 6 **3.** 2 **5.** 10 **7.** Not a real number
9. 0 **11.** −1 **13.** 1
15. Not a real number **17.** 2 **19.** −2 **21.** −10
23. Not a real number **25.** m **27.** y^3 **29.** y^5
31. m **33.** 27 **35.** 32 **37.** 125 **39.** 10^{10}
41. $3\sqrt{y}$ **43.** $2a$ **45.** x^2y **47.** $m^6\sqrt{5}$ **49.** $2\sqrt[3]{y}$

51. $-3w$ **53.** $2\sqrt[4]{s}$ **55.** $-5a^3y^2$ **57.** $\frac{\sqrt{t}}{2}$

59. $\frac{25}{4}$ **61.** $\frac{\sqrt[3]{t}}{2}$ **63.** $\frac{-2x^2}{y}$ **65.** $\frac{2a^3}{3}$ **67.** $\frac{\sqrt[4]{y}}{2}$

69. 3.968 **71.** −1.610 **73.** 6.001 **75.** 0.769
77. 46 **79.** 193 miles

Section 8.3 Warm-ups T F T F T F T F F F

1. $2\sqrt{2}$ **3.** $2\sqrt{6}$ **5.** $2\sqrt{7}$ **7.** $3\sqrt{10}$ **9.** $10\sqrt{5}$

11. $5\sqrt{6}$ **13.** $\frac{\sqrt{5}}{5}$ **15.** $\frac{3\sqrt{2}}{2}$ **17.** $\frac{\sqrt{6}}{2}$

19. $\frac{-3\sqrt{10}}{10}$ **21.** $\frac{-10\sqrt{17}}{17}$ **23.** $\frac{\sqrt{77}}{7}$ **25.** $3\sqrt{7}$

27. $\frac{\sqrt{6}}{2}$ **29.** $\frac{\sqrt{10}}{4}$ **31.** $\frac{\sqrt{15}}{5}$ **33.** 5 **35.** $\frac{\sqrt{6}}{2}$

37. a^4 **39.** $a^4\sqrt{a}$ **41.** $2a^3\sqrt{2}$ **43.** $2a^2b^4\sqrt{5b}$

45. $3xy\sqrt{3xy}$ **47.** $3ab^4c\sqrt{3a}$ **49.** $\frac{\sqrt{x}}{x}$

51. $\frac{\sqrt{6a}}{3a}$ **53.** $\frac{\sqrt{5y}}{5y}$ **55.** $\frac{\sqrt{6xy}}{2y}$ **57.** $\frac{\sqrt{6xy}}{3x}$

59. $\frac{2x\sqrt{2xy}}{y}$ **61.** $4x\sqrt{5x}$ **63.** $3y^4x^7\sqrt{yx}$

65. $4x^3\sqrt{5x}$ **67.** $\frac{-11\sqrt{6pq}}{3q}$ **69.** $a^3b^8c\sqrt{ac}$

71. $\frac{\sqrt{6y}}{3x^9y^4}$ **73.** 0 **75.** 0 **77.** $E = \frac{\sqrt{2AIS}}{I}$

Section 8.4 Warm-ups F T F F T F F F T T

1. $7\sqrt{5}$ **3.** $2\sqrt[3]{2}$ **5.** $8u\sqrt{11}$ **7.** $4\sqrt{3}-4\sqrt{2}$ **9.** $-4\sqrt{x}-\sqrt{y}$ **11.** $5x\sqrt{y}+2\sqrt{a}$ **13.** $5\sqrt{6}$ **15.** $-14\sqrt{3}$ **17.** $-\sqrt{3a}$ **19.** $3x\sqrt{x}$ **21.** $\frac{2\sqrt{3}}{3}$ **23.** $\frac{7\sqrt{3}}{3}$ **25.** $\sqrt{77}$ **27.** 36 **29.** $-12\sqrt{10}$ **31.** $2a^4\sqrt{3}$ **33.** 3 **35.** $-2m$ **37.** $2+\sqrt{6}$ **39.** $12\sqrt{3}+6\sqrt{5}$ **41.** $10-30\sqrt{2}$ **43.** $-7-\sqrt{5}$ **45.** 2 **47.** 3 **49.** $28-\sqrt{5}$ **51.** $-42-\sqrt{15}$ **53.** $37+20\sqrt{3}$ **55.** $\sqrt{2}$ **57.** $\frac{\sqrt{15}}{3}$ **59.** $\frac{2\sqrt{30}}{9}$ **61.** $\frac{5\sqrt{7}}{3}$ **63.** $1+\sqrt{2}$ **65.** $-2+\sqrt{5}$ **67.** $\frac{2-\sqrt{5}}{3}$ **69.** $\frac{2+\sqrt{6}}{3}$ **71.** $5\sqrt{3}+5\sqrt{2}$ **73.** $\frac{\sqrt{15}+3}{2}$ **75.** $\frac{13+7\sqrt{3}}{22}$ **77.** $\frac{19-11\sqrt{7}}{27}$ **79.** $3\sqrt{5a}$ **81.** $\frac{5\sqrt{2}}{2}$ **83.** $70+30\sqrt{5}$ **85.** $\frac{5\sqrt{5}}{3}$ **87.** 1.866 **89.** -1.974 **91.** 12 ft^2, $10\sqrt{2}$ ft **93.** 6 m^3

Section 8.5 Warm-ups F T F F T F F T T T

1. -4, 4 **3.** $-2\sqrt{10}$, $2\sqrt{10}$ **5.** $-\frac{\sqrt{6}}{3}$, $\frac{\sqrt{6}}{3}$ **7.** No solution **9.** -1, 3 **11.** $5-\sqrt{3}$, $5+\sqrt{3}$ **13.** -19 **15.** 90 **17.** No solution **19.** -5, 5 **21.** 3 **23.** 0, 1 **25.** 6 **27.** $\frac{13}{18}$ **29.** 3, 5 **31.** 3 **33.** 6, 8 **35.** $r=\pm\sqrt{\frac{V}{\pi h}}$ **37.** $b=\pm\sqrt{c^2-a^2}$ **39.** $b=\pm 2\sqrt{ac}$ **41.** $t=\frac{v^2}{2p}$ **43.** $-\sqrt{2}$, $\sqrt{2}$ **45.** No solution **47.** $\frac{1+2\sqrt{2}}{2}$, $\frac{1-2\sqrt{2}}{2}$ **49.** $\frac{9}{2}$ **51.** 3 **53.** -4, 2 **55.** $\frac{5}{2}$ **57.** -1.803, 1.803 **59.** 0.993 **61.** 0.362, 4.438 **63.** $3\sqrt{2}$ or 4.243 feet **65.** $3\sqrt{2}$ or 4.243 feet **67.** $\sqrt{2}$ or 1.414 feet **69.** 5 feet **71.** 2.5 seconds **73.** $\frac{200\sqrt{13}}{3}$ or 240.37 feet

Section 8.6 Warm-ups T F F T T T F F T T

1. $7^{1/4}$ **3.** $\sqrt[5]{9}$ **5.** $(5x)^{1/2}$ **7.** $\sqrt{a}$ **9.** 5 **11.** -5 **13.** 2 **15.** Not a real number **17.** $w^{7/3}$ **19.** $2^{-10/3}$ **21.** $\sqrt[4]{\frac{1}{w^3}}$ **23.** $\sqrt{(ab)^3}$ **25.** 25 **27.** 125 **29.** $\frac{1}{81}$ **31.** $\frac{1}{8}$ **33.** $-\frac{1}{3}$ **35.** Not a real number **37.** $x^{1/2}$ **39.** $n^{1/6}$ **41.** $x^{3/2}$ **43.** $2t^{1/4}$ **45.** x^2 **47.** $5^{1/8}$ **49.** xy^3 **51.** $\frac{x}{3y^4}$ **53.** $\frac{3}{4}$ **55.** $\frac{1}{9}$ **57.** 27 **59.** $\frac{1}{12}$ **61.** Yes, 9.2 m^2 **63.** 18.1% **65.** 278, 294, 312, 330, 350, 371, 393, 416, 441, 467, 495 Hz

Chapter 8 Review

1. $\frac{1}{32}$ **3.** $\frac{1}{1000}$ **5.** $\frac{1}{x^3}$ **7.** a^4 **9.** a^{10} **11.** $\frac{1}{x^{12}}$ **13.** $\frac{x^9}{8}$ **15.** $\frac{9}{a^2b^6}$ **17.** 5×10^3 **19.** 340,000 **21.** 4.61×10^{-5} **23.** 0.00000569 **25.** 7×10^{-4} **27.** 1.6×10^{-15} **29.** 8×10^1 **31.** 3.2×10^{-34} **33.** 2 **35.** 10 **37.** x^6 **39.** x^2 **41.** $2x$ **43.** $5x^2$ **45.** $\frac{2x^8}{y^7}$ **47.** $\frac{w}{4}$ **49.** $6\sqrt{2}$ **51.** $\frac{\sqrt{3}}{3}$ **53.** $\frac{\sqrt{15}}{5}$ **55.** $\sqrt{11}$ **57.** $\frac{\sqrt{6}}{4}$ **59.** y^3 **61.** $2t^4\sqrt{6t}$ **63.** $2m^2t\sqrt{3mt}$ **65.** $\frac{\sqrt{2x}}{x}$ **67.** $\frac{a^2\sqrt{6as}}{2s}$ **69.** $10\sqrt{7}$ **71.** $-\sqrt{3}$ **73.** 30 **75.** $-45\sqrt{2}$ **77.** $-15\sqrt{3}-9$ **79.** $-3\sqrt{2}+3\sqrt{5}$ **81.** -22 **83.** $26-4\sqrt{30}$ **85.** $38-19\sqrt{6}$ **87.** $\frac{\sqrt{10}}{4}$ **89.** $\frac{1-\sqrt{5}}{5}$ **91.** $-\frac{3+3\sqrt{5}}{4}$ **93.** -20, 20 **95.** $-\frac{\sqrt{21}}{7}$, $\frac{\sqrt{21}}{7}$ **97.** $4-3\sqrt{2}$, $4+3\sqrt{2}$ **99.** 81 **101.** 4 **103.** 6 **105.** $t=\pm 2\sqrt{2sw}$ **107.** $t=\frac{9a^2}{b}$ **109.** $\frac{1}{125}$ **111.** 5 **113.** $\frac{1}{8}$ **115.** $\frac{1}{x}$ **117.** $-\frac{1}{2}x^2$ **119.** w^2 **121.** $\frac{t^3}{3s^2}$ **123.** $\frac{4}{x^8y^{20}}$ **125.** \$5000 **127.** 10.5% **129.** 0.4 cm **131.** $4\pi\sqrt{10}$ or 40 in.^{2} **133.** 28 inches

Chapter 8 Test

1. $\frac{1}{32}$ **2.** 12 **3.** -3 **4.** $\frac{1}{81}$ **5.** 2 **6.** $2\sqrt{6}$ **7.** $\frac{\sqrt{6}}{4}$ **8.** Not a real number **9.** $3\sqrt{2}$ **10.** $7+4\sqrt{3}$ **11.** 11 **12.** $\sqrt{7}$ **13.** $\frac{2\sqrt{15}}{3}$ **14.** $1+\sqrt{2}$ **15.** 81 **16.** $3\sqrt{2}-3$ **17.** $15x^5$ **18.** $\frac{8}{x^{18}}$ **19.** $\frac{-x^{15}}{27y^6}$ **20.** $\frac{1}{4y^{14}}$ **21.** $\frac{1}{t^4}$ **22.** xy^3 **23.** $\frac{s^6}{4t^4}$ **24.** $\frac{1}{5u^4w}$ **25.** $\frac{\sqrt{3t}}{t}$ **26.** $2y^3$ **27.** $2y^4$ **28.** $3t^3\sqrt{2t}$ **29.** -9, 3 **30.** 18 **31.** $-\frac{\sqrt{10}}{5}$, $\frac{\sqrt{10}}{5}$ **32.** $\frac{4}{3}$ **33.** 11

34. $r = \pm\sqrt{\dfrac{S}{\pi h}}$ **35.** 5.433×10^6 **36.** 6.5×10^{-6}
37. 4.8×10^{-1} **38.** 8.1×10^{-27} **39.** $\dfrac{5\sqrt{2}}{2}$ meters
40. 178.4 meters

Tying It All Together Chapters 1–8

1. $-\dfrac{3}{2}$ **2.** $\dfrac{3}{2}$ **3.** $x > -\dfrac{3}{2}$

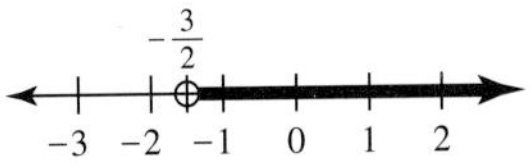

4. $x < \dfrac{3}{2}$

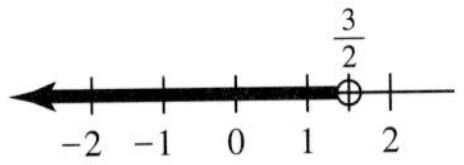

5. −3
6. $-\dfrac{\sqrt{6}}{2}, \dfrac{\sqrt{6}}{2}$ **7.** $-\sqrt{6}, \sqrt{6}$ **8.** $\dfrac{2}{3}$ **9.** $-\dfrac{3}{2}$
10. $-\dfrac{3}{2}, 3$ **11.** No solution **12.** −2, −1
13. No solution **14.** 3 **15.** 9 **16.** 72 **17.** 81
18. 9 **19.** 12 **20.** −6 **21.** 3 **22.** $-\dfrac{3}{2}$
23. $(x-3)^2$ **24.** $(x+5)^2$ **25.** $(x+6)^2$
26. $(x-10)^2$ **27.** $2(x-2)^2$ **28.** $3(x+1)^2$
29. $7x - 3$ **30.** $-15t^2 - 13t + 20$
31. $-24j^2 + 14j + 24$ **32.** $6u + 6$ **33.** $v + 1$
34. $t^2 + 4t + 4$ **35.** $t^2 - 49$ **36.** $-4n^2 + 9$
37. $m^2 - 2m + 1$ **38.** $2t - 1$ **39.** $-4r^2 - r + 3$
40. $-18y^2 + 2$ **41.** $3j - 5$ **42.** $-6j + 2$ **43.** $2 - 3x$
44. $-1 - 3p$ **45.** $-2 + 3q$ **46.** $-4 + z$
47. a) 16.788 kilocalories per minute **b)** 275.24 m/min

CHAPTER 9

Section 9.1 Warm-ups T F F T T F F T F F

1. −6, 6 **3.** No real solution **5.** $-\sqrt{10}, \sqrt{10}$
7. $-\dfrac{\sqrt{15}}{3}, \dfrac{\sqrt{15}}{3}$ **9.** $-\dfrac{2\sqrt{6}}{3}, \dfrac{2\sqrt{6}}{3}$ **11.** 1, 5
13. $2 - 3\sqrt{2}, 2 + 3\sqrt{2}$ **15.** $-\dfrac{3}{2}, -\dfrac{1}{2}$
17. $\dfrac{2-\sqrt{2}}{2}, \dfrac{2+\sqrt{2}}{2}$ **19.** $\dfrac{-1-\sqrt{2}}{2}, \dfrac{-1+\sqrt{2}}{2}$
21. 11 **23.** −3, 5 **25.** −3 **27.** −1, 2 **29.** 0, 2
31. $0, \dfrac{3}{2}$ **33.** $-7, \dfrac{3}{2}$ **35.** 5 **37.** $-\dfrac{5}{2}, 6$ **39.** 4, 6
41. $-\sqrt{6}, \sqrt{6}$ **43.** 3, 9 **45.** 3, 4 **47.** $\dfrac{5\sqrt{2}}{2}$ meters
49. $4\sqrt{5}$ blocks **51.** $2\sqrt{61}$ feet **53.** 6.3%
55. 2 seconds and 3 seconds **57.** 9

Section 9.2 Warm-ups F F F F T T T T T T

1. $x^2 + 6x + 9 = (x+3)^2$ **3.** $x^2 + 14x + 49 = (x+7)^2$
5. $x^2 - 16x + 64 = (x-8)^2$ **7.** $t^2 - 18t + 81 = (t-9)^2$
9. $m^2 + 3m + \dfrac{9}{4} = \left(m + \dfrac{3}{2}\right)^2$
11. $z^2 + z + \dfrac{1}{4} = \left(z + \dfrac{1}{2}\right)^2$
13. $x^2 - \dfrac{1}{2}x + \dfrac{1}{16} = \left(x - \dfrac{1}{4}\right)^2$
15. $y^2 + \dfrac{1}{4}y + \dfrac{1}{64} = \left(y + \dfrac{1}{8}\right)^2$ **17.** $(x+5)^2$
19. $(m-1)^2$ **21.** $\left(x + \dfrac{1}{2}\right)^2$ **23.** $\left(t + \dfrac{1}{6}\right)^2$
25. $\left(x + \dfrac{1}{5}\right)^2$ **27.** −5, 3 **29.** −3, 7 **31.** −3
33. $\dfrac{1}{2}, 1$ **35.** $\dfrac{3}{2}, 2$ **37.** $-1, \dfrac{1}{3}$
39. $-1 - \sqrt{7}, -1 + \sqrt{7}$ **41.** $-3 - 2\sqrt{2}, -3 + 2\sqrt{2}$
43. $\dfrac{1-\sqrt{13}}{2}, \dfrac{1+\sqrt{13}}{2}$
45. $\dfrac{-3-\sqrt{21}}{2}, \dfrac{-3+\sqrt{21}}{2}$
47. $\dfrac{1-\sqrt{33}}{4}, \dfrac{1+\sqrt{33}}{4}$ **49.** $5 - \sqrt{7}, 5 + \sqrt{7}$
51. $-\dfrac{\sqrt{15}}{3}, \dfrac{\sqrt{15}}{3}$ **53.** No real solution
55. $4 - 2\sqrt{2}, 4 + 2\sqrt{2}$ **57.** $\dfrac{7}{2}$
59. $-3 - 2\sqrt{5}, -3 + 2\sqrt{5}$ **61.** $2 \pm \sqrt{2}$
63. −0.6, 0.4 **65.** $5 - \sqrt{3}$ inches and $5 + \sqrt{3}$ inches
67. $6 - \sqrt{2}$ and $6 + \sqrt{2}$ **69.** 12 years old

Section 9.3 Warm-ups T F F F T F F T T T

1. −5, 3 **3.** −5 **5.** $-2, \dfrac{3}{2}$ **7.** $-\dfrac{3}{2}, \dfrac{1}{2}$
9. $\dfrac{3-\sqrt{3}}{2}, \dfrac{3+\sqrt{3}}{2}$ **11.** $\dfrac{-2-\sqrt{2}}{2}, \dfrac{-2+\sqrt{2}}{2}$
13. No real solution **15.** $0, \dfrac{1}{2}$ **17.** $-\dfrac{\sqrt{15}}{5}, \dfrac{\sqrt{15}}{5}$
19. $-7 + 4\sqrt{3}, -7 - 4\sqrt{3}$ **21.** 0, one
23. −47, none **25.** 181, two **27.** 0, one
29. −15, none **31.** −59, none **33.** $-2, \dfrac{1}{2}$ **35.** 0, 3
37. $-\dfrac{6}{5}, 4$ **39.** −4, −2 **41.** 0, 9 **43.** −8, −2
45. No real solution **47.** $\dfrac{3-\sqrt{33}}{2}, \dfrac{3+\sqrt{33}}{2}$
49. $0, \dfrac{3}{2}$ **51.** −0.79, 3.79 **53.** −1.36, 2.36
55. 0.30 **57.** 1981, 1999

Section 9.4 Warm-ups F T F F T T T F T T

1. Width $-1 + \sqrt{11}$ meters, length $1 + \sqrt{11}$ meters
3. $4\sqrt{2}$ feet
5. Height $-3 + \sqrt{19}$ inches, base $3 + \sqrt{19}$ inches
7. $3 + \sqrt{5}$ or 5.24 hours **9.** $16 + 4\sqrt{26}$ or 36.4 hours
11. $\dfrac{15 + \sqrt{241}}{8}$ or 3.82 seconds **13.** 1.5 seconds
15. $2\sqrt{19}$ or 8.72 mph **17.** $\dfrac{-13 + \sqrt{409}}{2}$ or 3.61 feet

Section 9.5 Warm-ups T T T T F F F F F T

1. $5 + 9i$ **3.** 1 **5.** $2 - 8i$ **7.** $-1 - 4i$ **9.** -1

11. $\frac{3}{4} + \frac{1}{2}i$ **13.** $6 - 9i$ **15.** -36 **17.** -36

19. $21 - i$ **21.** $21 - 20i$ **23.** 25 **25.** 2 **27.** 29

29. 52 **31.** 13 **33.** 1 **35.** $1 - 3i$ **37.** $-1 + 4i$

39. $-3 + 2i$ **41.** $\frac{8}{13} + \frac{12}{13}i$ **43.** $\frac{3}{5} + \frac{4}{5}i$ **45.** $-4i$

47. $5 + 3i$ **49.** $-3 - i\sqrt{7}$ **51.** $-1 + i\sqrt{3}$

53. $2 + \frac{1}{2}i\sqrt{5}$ **55.** $-\frac{2}{3} + \frac{1}{3}i\sqrt{7}$ **57.** $\frac{1}{5} - i$

59. $-9i, 9i$ **61.** $-i\sqrt{5}, i\sqrt{5}$ **63.** $-i\frac{\sqrt{6}}{3}, i\frac{\sqrt{6}}{3}$

65. $2 - i, 2 + i$ **67.** $3 - 2i, 3 + 2i$

69. $2 - i\sqrt{3}, 2 + i\sqrt{3}$ **71.** $\frac{2 - i}{3}, \frac{2 + i}{3}$

73. $\frac{1 - i\sqrt{3}}{2}, \frac{1 + i\sqrt{3}}{2}$ **75.** $\frac{2 - i\sqrt{5}}{2}, \frac{2 + i\sqrt{5}}{2}$

77. $-6 - 24i$ **79.** 0 **81.** $x^2 - 12x + 37$

83. $x^2 + 9 = 0$

Section 9.6 Warm-ups T F T T F T T T T F

1. $(3, -6), (4, 0), (-3, 0)$ **3.** $(4, -128), (0, 0), (2, 0)$

5. Domain R, range $\{y \mid y \geq 2\}$

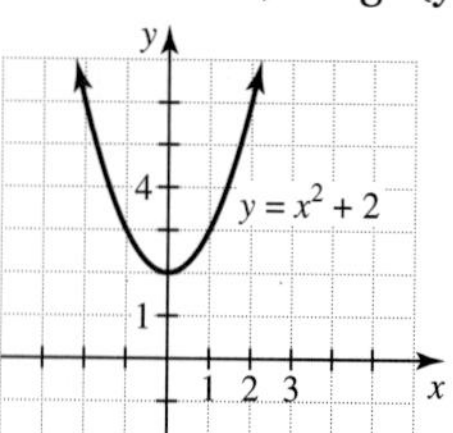

7. Domain R, range $\{y \mid y \geq -4\}$

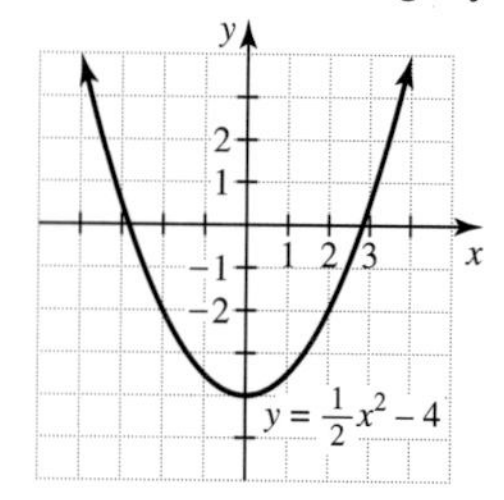

9. Domain R, range $\{y \mid y \leq 5\}$

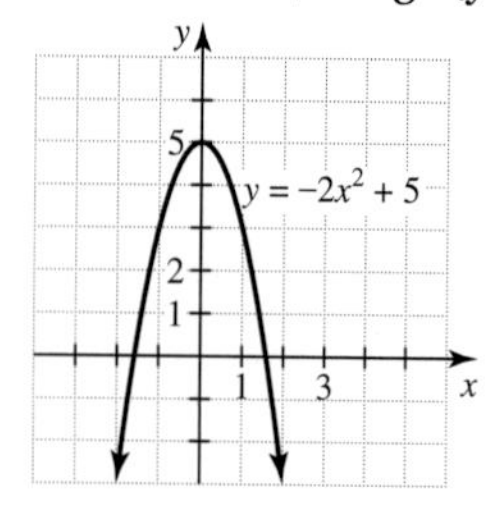

11. Domain R, range $\{y \mid y \leq 5\}$

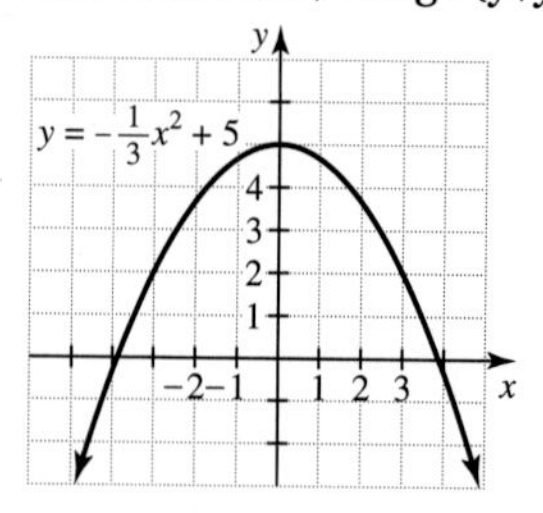

13. Domain R, range $\{y \mid y \geq 0\}$

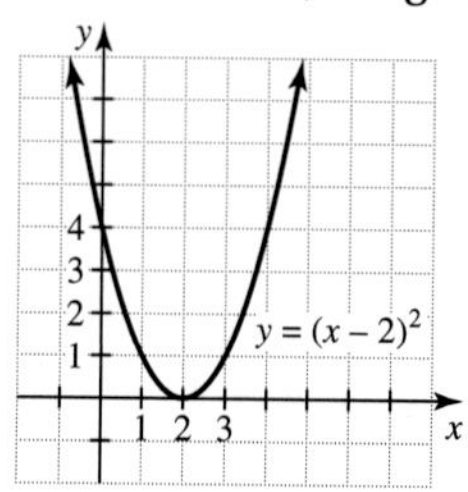

15. Vertex $\left(\frac{1}{2}, -\frac{9}{4}\right)$, intercepts $(0, -2), (-1, 0), (2, 0)$, domain R, range $\left\{y \mid y \geq -\frac{9}{4}\right\}$

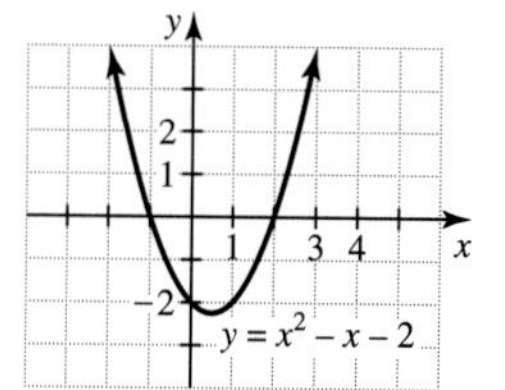

17. Vertex $(-1, -9)$, intercepts $(0, -8), (-4, 0), (2, 0)$, domain R, range $\{y \mid y \geq -9\}$

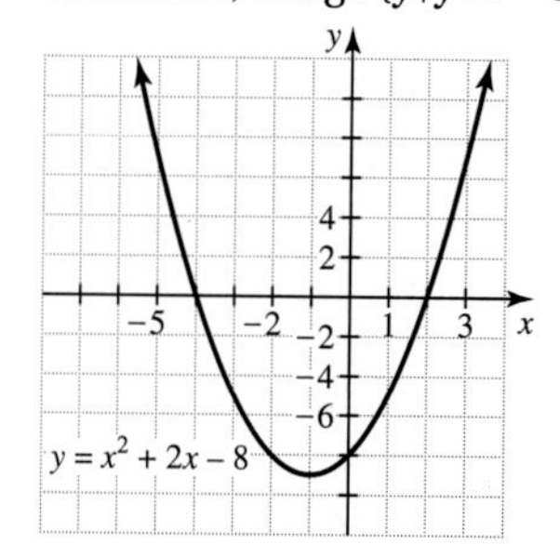

19. Vertex $(-2, 1)$, intercepts $(0, -3), (-1, 0), (-3, 0)$, domain R, range $\{y \mid y \leq 1\}$

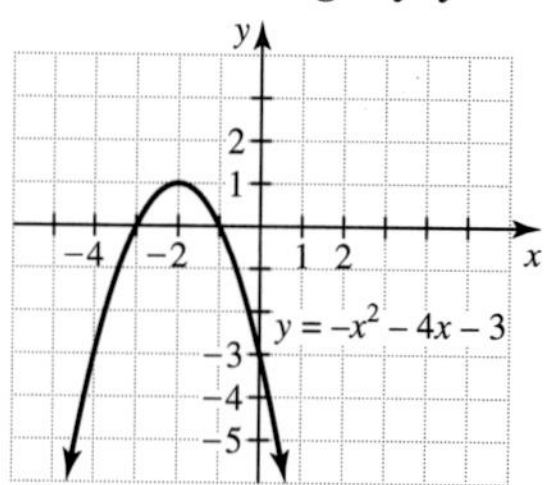

21. Vertex $\left(\frac{3}{2}, \frac{25}{4}\right)$, intercepts (0, 4), (4, 0), (−1, 0), domain R, range $\left\{y \mid y \le \frac{25}{4}\right\}$

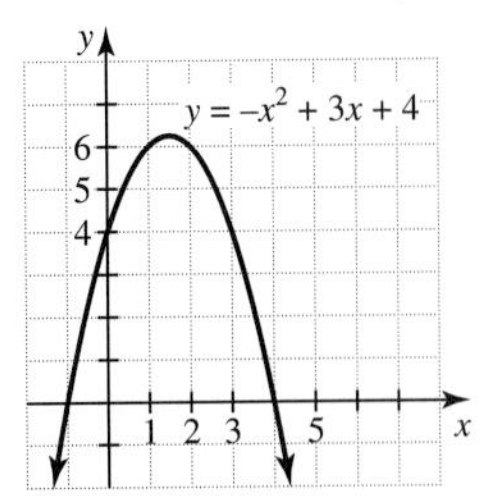

23. Vertex (3, −25), intercepts (0, −16), (8, 0), (−2, 0), domain R, range $\{a \mid a \ge -25\}$

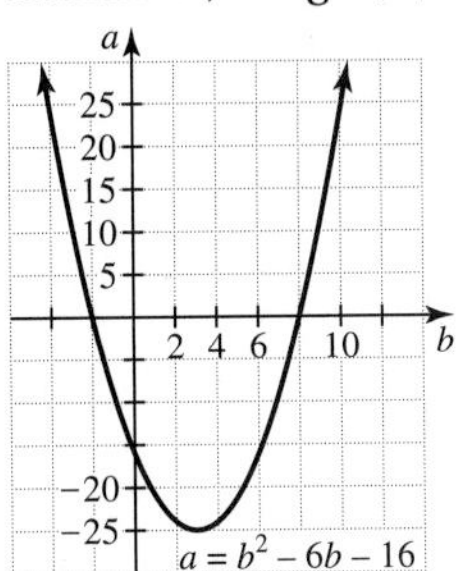

25. Minimum −8 **27.** Maximum 14
29. Minimum 2 **31.** Maximum 2
33. Maximum 64 feet

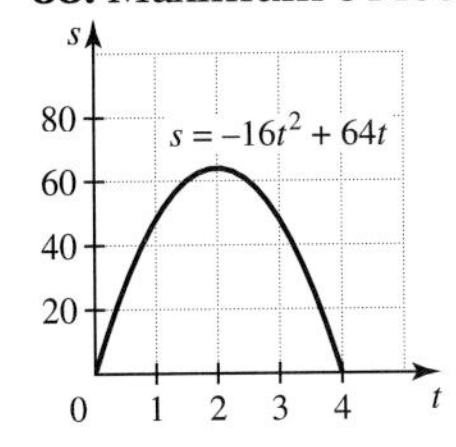

35. 625 square meters **37.** 2:00 P.M.
39. 15 meters, 25 meters

43. Height $-2 + 2\sqrt{11}$ or 4.633 inches, base $2 + 2\sqrt{11}$ or 8.633 inches
45. $3 + \sqrt{2}$ or 4.414 and $3 - \sqrt{2}$ or 1.586
47. New printer $7 + \sqrt{65}$ or 15.062 hours, old printer $9 + \sqrt{65}$ or 17.062 hours
49. $7 - 3i$ **51.** $-3 + 7i$ **53.** $15 - 25i$
55. $-55 + 48i$ **57.** $-1 - i\sqrt{2}$ **59.** $\frac{3}{37} + \frac{19}{37}i$
61. $\frac{19}{17} + \frac{9}{17}i$ **63.** $-11i$, $11i$ **65.** $8 - i$, $8 + i$
67. $\frac{3 - 3i\sqrt{7}}{4}, \frac{3 + 3i\sqrt{7}}{4}$
69. Vertex (3, −9), intercepts (0, 0), (6, 0)

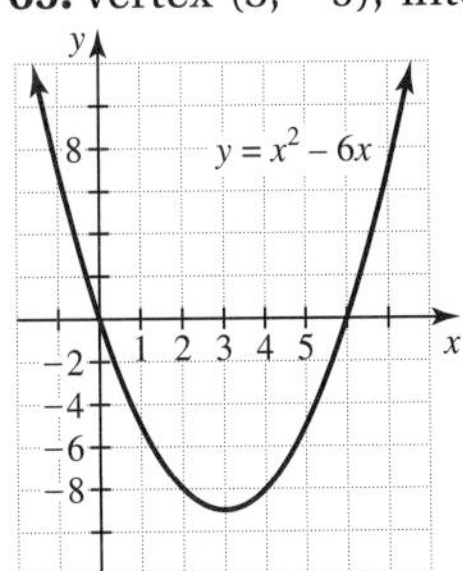

71. Vertex (2, −16), intercepts (0, −12), (−2, 0), and (6, 0)

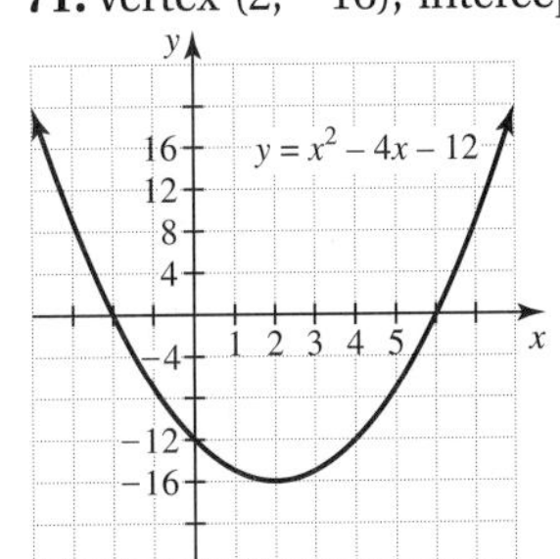

73. Vertex (2, 8), intercepts (0, 0), (4, 0)

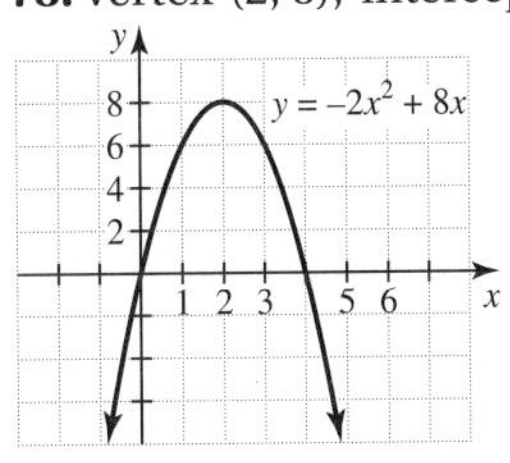

75. Vertex (1, 4), intercepts (0, 3), (−1, 0), (3, 0)

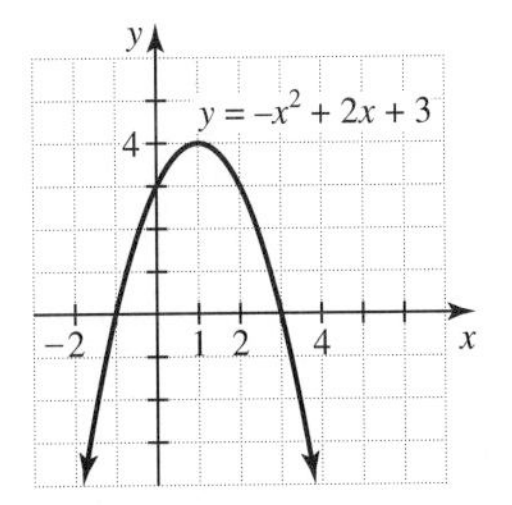

77. Domain R, range $\{y \mid y \ge -3\}$
79. Domain R, range $\{y \mid y \le 4.125\}$ **81.** \$20.40, 400

Chapter 9 Review

1. −3, 3 **3.** 0, 9 **5.** −1, 2 **7.** $9 - \sqrt{10}$, $9 + \sqrt{10}$
9. $\frac{3}{2}$ **11.** 4, 5 **13.** −7, 2 **15.** $-4, \frac{1}{2}$
17. $-2 - \sqrt{11}$, $-2 + \sqrt{11}$ **19.** −7, 4
21. $\frac{-3 - \sqrt{29}}{2}, \frac{-3 + \sqrt{29}}{2}$ **23.** $-5, \frac{1}{2}$
25. 0, one **27.** −44, none **29.** 76, two
31. $-\frac{2}{3}, \frac{1}{2}$ **33.** $\frac{1 - \sqrt{17}}{2}, \frac{1 + \sqrt{17}}{2}$
35. $\frac{3 - \sqrt{14}}{5}, \frac{3 + \sqrt{14}}{5}$ **37.** $0, \frac{5}{3}$ **39.** 13 meters
41. $\frac{-2 + \sqrt{46}}{2}$ or 2.391 meters and $\frac{2 + \sqrt{46}}{2}$ or 4.391 meters

Chapter 9 Test

1. 0, one **2.** −31, none **3.** 17, two **4.** $-1, \frac{3}{5}$

5. $\frac{2-\sqrt{10}}{2}, \frac{2+\sqrt{10}}{2}$ **6.** −7, 3

7. $\frac{-3-\sqrt{29}}{2}, \frac{-3+\sqrt{29}}{2}$ **8.** −5, 4 **9.** 3, 25

10. $-\frac{1}{2}, 3$ **11.** $10 + 3i$ **12.** $-6 + 7i$ **13.** −36

14. $42 - 2i$ **15.** 68 **16.** $2 - 3i$ **17.** $-1 + i\sqrt{3}$

18. $-2 + i\sqrt{2}$ **19.** $\frac{3}{5} + \frac{4}{5}i$

20. $-3 - i\sqrt{3}, -3 + i\sqrt{3}$ **21.** $\frac{3}{5} - \frac{4}{5}i, \frac{3}{5} + \frac{4}{5}i$

22. Domain R, range $\{y \mid y \le 16\}$

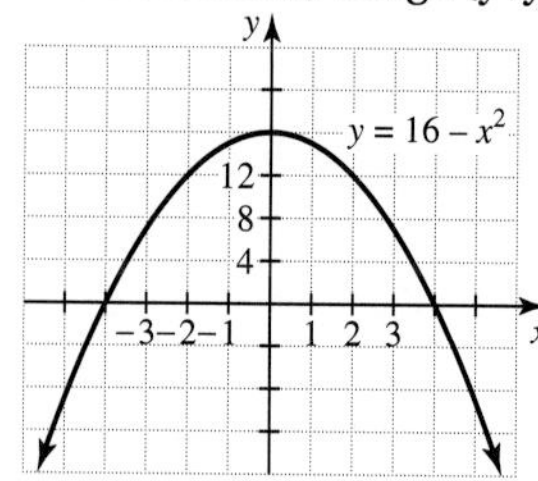

23. Domain R, range $\left\{y \mid y \ge -\frac{9}{4}\right\}$

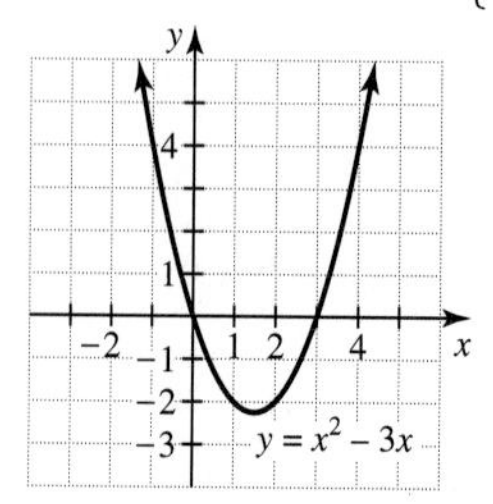

24. (1, 0), (5, 0) **25.** 36 feet **26.** $5 - \sqrt{2}$ and $5 + \sqrt{2}$

25.

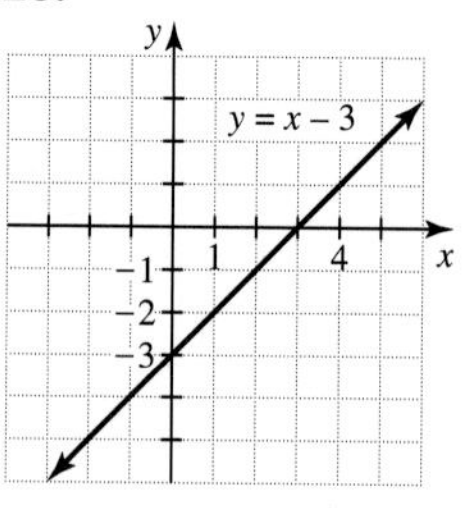

26.

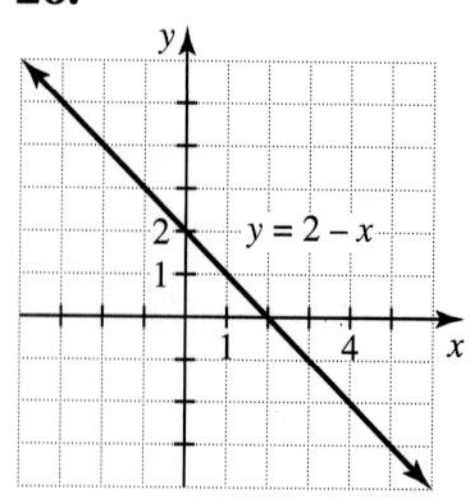

27.

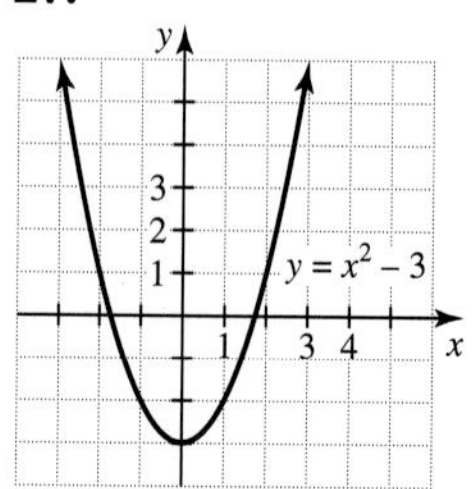

28.

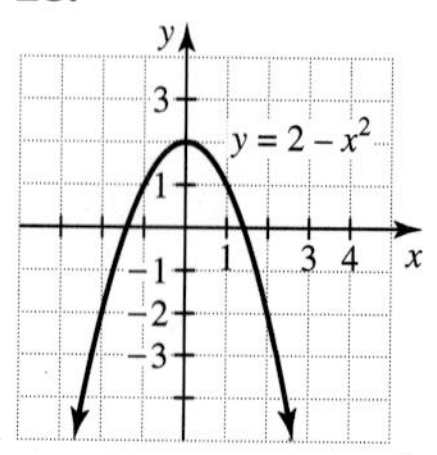

29.

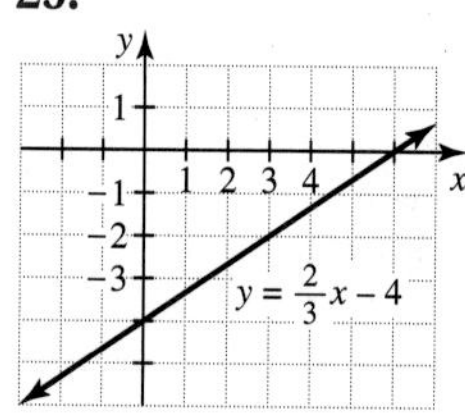

30.

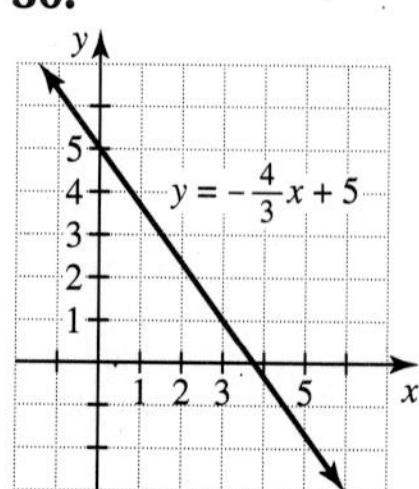

31. a) $s = -250p + 10{,}500$ **b)** Increased
c) $R = -250p^2 + 10{,}500p$ **d)** \$21

Tying It All Together Chapters 1–9

1. $\frac{1}{2}$ **2.** 1 **3.** $-\frac{\sqrt{2}}{2}, \frac{\sqrt{2}}{2}$ **4.** $\frac{1-2\sqrt{2}}{2}, \frac{1+2\sqrt{2}}{2}$

5. $\frac{2-\sqrt{6}}{2}, \frac{2+\sqrt{6}}{2}$ **6.** 0, 2 **7.** $-1, \frac{1}{2}$ **8.** 5

9. $1 - i\sqrt{14}, 1 + i\sqrt{14}$ **10.** −1, 2 **11.** $y = \frac{5}{4}x - 2$

12. $y = 3x - 9$ **13.** $y = \frac{2}{3}x + \frac{16}{3}$ **14.** $y = -\frac{b}{a}$

15. $y = \frac{-b \pm \sqrt{b^2 - 4ac}}{2a}$ **16.** $y = -\frac{2}{3}x + 7$

17. $y = -\frac{4}{3}x + \frac{2}{9}$ **18.** $y = \pm\sqrt{a^2 - x^2}$ **19.** $P = 4s$

20. $A = s^2$ **21.** $P = 2d\sqrt{2}$ **22.** $d = \sqrt{2A}$

23. (2, −3) **24.** (3, 10)

Index

Q

R

S

DEFINITIONS, RULES, AND FORMULAS

Subsets of the Real Numbers

Natural Numbers = $\{1, 2, 3, \ldots\}$

Whole Numbers = $\{0, 1, 2, 3, \ldots\}$

Integers = $\{\ldots -3, -2, -1, 0, 1, 2, 3, \ldots\}$

Rational = $\left\{\frac{a}{b} \middle| a \text{ and } b \text{ are integers with } b \neq 0\right\}$

Irrational = $\{x | x \text{ is not rational}\}$

Properties of the Real Numbers

For all real numbers a, b, and c

$a + b = b + a$; $a \cdot b = b \cdot a$ Commutative

$(a + b) + c = a + (b + c)$; $(ab)c = a(bc)$ Associative

$a(b + c) = ab + ac$; $a(b - c) = ab - ac$ Distributive

$a + 0 = a$; $1 \cdot a = a$ Identity

$a + (-a) = 0$; $a \cdot \frac{1}{a} = 1\ (a \neq 0)$ Inverse

$a \cdot 0 = 0$ Multiplication property of 0

Absolute Value

$$|a| = \begin{cases} a & \text{for } a \geq 0 \\ -a & \text{for } a < 0 \end{cases}$$

Order of Operations

No parentheses or absolute value present:

1. Exponential expressions
2. Multiplication and division
3. Addition and subtraction

With parentheses of absolute value:

First evaluate within each set of parentheses or absolute value, using the order of operations.

Exponents

$a^0 = 1$ $\qquad a^{-1} = \frac{1}{a}$

$a^{-r} = \frac{1}{a^r} = \left(\frac{1}{a}\right)^r$ $\qquad \frac{1}{a^{-r}} = a^r$

$a^r a^s = a^{r+s}$ $\qquad \frac{a^r}{a^s} = a^{r-s}$

$(a^r)^s = a^{rs}$ $\qquad (ab)^r = a^r b^r$

$\left(\frac{a}{b}\right)^r = \frac{a^r}{b^r}$ $\qquad \left(\frac{a}{b}\right)^{-r} = \left(\frac{b}{a}\right)^r$

Roots and Radicals

$a^{1/n} = \sqrt[n]{a}$ $\qquad a^{m/n} = \left(\sqrt[n]{a}\right)^m = \sqrt[n]{a^m}$

$\sqrt[n]{ab} = \sqrt[n]{a} \cdot \sqrt[n]{b}$ $\qquad \sqrt[n]{\frac{a}{b}} = \frac{\sqrt[n]{a}}{\sqrt[n]{b}}$

Factoring

$a^2 + 2ab + b^2 = (a + b)^2$

$a^2 - 2ab + b^2 = (a - b)^2$

$a^2 - b^2 = (a + b)(a - b)$

$a^3 - b^3 = (a - b)(a^2 + ab + b^2)$

$a^3 + b^3 = (a + b)(a^2 - ab + b^2)$

Rational Expressions

$\frac{a}{b} + \frac{c}{b} = \frac{a + c}{b}$ $\qquad \frac{a}{b} - \frac{c}{b} = \frac{a - c}{b}$

$\frac{ac}{bc} = \frac{a}{b}$ $\qquad \frac{a}{b} + \frac{c}{d} = \frac{ad + bc}{bd}$

$\frac{a}{b} \cdot \frac{c}{d} = \frac{ac}{bd}$ $\qquad \frac{a}{b} \div \frac{c}{d} = \frac{a}{b} \cdot \frac{d}{c}$

If $\frac{a}{b} = \frac{c}{d}$, then $ad = bc$.